AF567055

Sequenced Reactive Barriers *for* Groundwater Remediation

Sequenced Reactive Barriers *for* Groundwater Remediation

Edited by

Stephanie Fiorenza
Rice University, Houston, TX

Carroll L. Oubre
Rice University, Houston, TX

C. Herb Ward
Rice University, Houston, TX

Authors

James F. Barker
Barbara J. Butler
Evan Cox
John F. Devlin
Robert Focht
Susan M. Froud
Dennis J. Katic
Michaye McMaster
Mary Morkin
John Vogan

LEWIS PUBLISHERS
Boca Raton London New York Washington, D.C.

Library of Congress Cataloging-in-Publication Data

Sequenced reactive barriers for groundwater remediation / edited by
Stephanie Fiorenza, Carroll L. Oubre, C. Herb Ward.
p. cm. — (AATDF monographs)
Includes bibliographical references and index.
ISBN 1-56670-446-4 (alk. paper)
1. Groundwater—Purification. 2. Membranes (Technology)
I. Fiorenza, Stephanie. II. Oubre, Carroll L. III. Ward, C. H.
(Calvin Herbert), 1933- . IV. Series.
TD426.S47 1999
628.1′.68—dc21 99-38934
CIP

Although the information described herein has been funded wholly or in part by the United States Department of Defense (DOD) under Grant No. DACA39-93-1-0001 to Rice University for the Advanced Applied Technology Demonstration Facility for Environmental Technology Program (AATDF), it may necessarily reflect the views of the DOD or Rice University, and no official endorsement should be inferred.

International Standard Book Number 1-56670-446-4
Library of Congress Card Number 99-38934
Printed in the United States of America 1 2 3 4 5 6 7 8 9 0
Printed on acid-free paper

Foreword and Acknowledgments

The field of permeable, reactive barriers for groundwater remediation has grown rapidly since the University of Waterloo proposed this technology demonstration to the Advanced Applied Technology Demonstration Facility (AATDF) Program. The AATDF was interested in permeable reactive barriers because they constitute an *in situ* technique with the potential to minimize long-term operation and maintenance costs while limiting the migration of a contaminated groundwater plume. Most field tests of reactive barriers at the time that the University of Waterloo proposed this work were investigations of only granular iron for remediation of dissolved chlorinated hydrocarbons.

The University of Waterloo's project advanced the demonstration of reactive barriers by testing combinations of several technologies in sequence. Many groundwater plumes contain contaminants that are best treated with different remediation methods. The field work focused on plumes containing a mix of chlorinated and aromatic hydrocarbons and tested abiotic and anaerobic and aerobic bioremediation methods. The *in situ* sequencing of reactive barriers is a natural extension of the "treatment train" concept.

The University of Waterloo conducted tests at two field sites, Canadian Forces Base Borden in Ontario and the former Naval Air Station in Alameda, California. All this work was possible because this project was a team effort. Several engineering consulting firms aided with installation of the sequenced barrier systems. Canadian Forces Base Borden, the U.S. Navy, and the State of California and U.S. regulatory officials streamlined the Work Plan review and approval process.

This monograph records the results of the field demonstrations. The reader will find it to be complete and quantitative. The monograph also includes a section on the potential costs of a full-scale implementation of the demonstration system at the Alameda site. This design and evaluation was conducted by BEAK International, Inc. and EnviroMetal Technologies, Inc. (ETI).

The authors worked closely with the AATDF staff and project advisors throughout the demonstration. The advisors who reviewed and commented on this monograph were Dr. James Mercer, HSI GeoTrans and Dr. Charles J. Newell, Groundwater Services, Inc.

The authors would like to thank the following individuals from the University of Waterloo for valuable technical assistance, advice, and support of the expansive fieldwork undertaken for this project. Mike Brown, Sam Vales, Paul Johnson, Bob Ingleton, Jesse Ingleton, and Rick Gibson were all involved in the construction and setup of the CFB Borden experiment. The management of day-to-day field activities would not have been possible without the support of the co-op students: Karl Bilyj, Dave Bertrand, Matt Bogaart, Don Forbes, Sebastian Kirstine, Ryan Lyle, and Sam Piperakis. During the winter, Al Corbeil provided peace of mind by making daily visits to the CFB Borden site to ensure that the heat was always working and that the CFB Borden Experiment was operational. In order to complete large sampling rounds in short time periods, the following students provided additional field support: Kevin Allin, Eric Angevine, Jean Birks, Fred Blaine, Leslie Major, Cory Repta, and Sharon Wadley. Carey Austrins and John Molson provided excellent groundwater modeling. Jamey Rosen and Ryan Lyle managed the databases needed to handle the large amount of data generated from both experiments.

Shirley Chatten, Kim Hamilton, and Marianne van der Griendt of the Organic Geochemistry Lab at the University of Waterloo provided most of the analytical services for this project. Their ability to turn around hundreds of samples in short time periods, and arrange their vacations around field schedules, is gratefully acknowledged.

The construction team from Remedial Solutions Inc. made the Alameda field installation proceed smoothly. Rob Langdon and Murray Einarson and the rest of the staff at Einarson, Fowler, and Watson were invaluable in providing UW with local site support in Alameda.

Stephanie O'Hannesin, formerly of UW and now at ETI, provided substantial review and technical contributions to the Alameda experiment. Greg Friday and Wayne Noble of Dr. Gillham's laboratory are acknowledged for their assistance at Borden and with numerous lab experiments.

Dave Major and Janice Tessman, from BEAK, provided support and technical review for the TER. The work by Jeff Croxall and Tom Krug, of BEAK, to provide pump-and-treat comparison systems and costs for the TER is greatly appreciated.

A post-graduate scholarship from the Natural Science and Engineering Research Council (NSERC) of Canada was received by Ms. S. Fround. Support by Regenesis Bioremediation Products in the form of ORC™ "socks" is gratefully acknowledged. Technical advice from Dr. W. A. Farone and Dr. S. Koenigsberg of Regenesis is also acknowledged.

The Technology Advisory Board and other AATDF advisory committees provided valuable assistance at the start and during the project in such areas as technology selection, directional changes in projects, technology transfer, and reporting reviews. Mr. Richard Conway played a pivotal role in facilitating the publication process.

The AATDF Program was funded by the U.S. Department of Defense under Grant No. DACA39-93-1-0001 to Rice University. The program was given oversight by the U.S. Army Engineer Waterways Experiment Station in Vicksburg, Mississippi.

AATDF Monographs

This monograph is one of a ten-volume series that records the results of the AATDF Program:

- Surfactants and Cosolvents for NAPL Remediation: A Technology Practices Manual
- Sequenced Reactive Barriers for Groundwater Remediation
- Modular Remediation Testing System
- Phytoremediation of Hydrocarbon-Contaminated Soil
- Steam and Electroheating Remediation of Tight Soils
- Soil Vapor Extraction Using Radio Frequency Heating: Resource Manual and Technology Demonstration
- Subsurface Contamination Monitoring Using Laser Fluorescence
- Remediation of Firing-Range Impact Berms
- Reuse of Surfactants and Cosolvents for NAPL Remediation
- NAPL Removal: Surfactants, Foams, and Microemulsions

Advanced Applied Technology Demonstration Facility
(AATDF)
Energy and Environmental Systems Institute MS-316
Rice University
6100 Main Street
Houston, TX 77005-1892

U.S. Army Corps of Engineers
Waterways Experiment Station
3909 Halls Ferry Road
Vicksburg, MS 39180-6199

Rice University
6100 Main Street
Houston, TX 77005-1892

Preface

Following a national competition, the U.S. Department of Defense (DOD) awarded a $19.3 million grant to a university consortium of environmental research centers led by Rice University and directed by Dr. C. Herb Ward, Foyt Family Chair of Engineering. The DOD Advanced Applied Technology Demonstration Facility (AATDF) Program for Environmental Remediation Technologies was established on May 1, 1993 to enhance the development of innovative remediation technologies for DOD by facilitating the process from academic research to full-scale utilization. The AATDF's focus is to select, test, and document performance of innovative environmental technologies for the remediation of DOD sites.

Participating universities include Stanford University, University of Texas at Austin, Rice University, Lamar University, University of Waterloo, and Louisiana State University. The directors of the environmental research centers at these universities serve as the Technology Advisory Board (TAB). The U.S. Army Engineer Waterways Experiment Station manages the AATDF Grant for DOD. Dr. John Keeley is the Technical Grant Officer. The DOD/AATDF is supported by five leading consulting engineering firms: Remediation Technologies, Inc., Battelle Memorial Institute, GeoTrans, Inc., Arcadis Geraghty and Miller, Inc., and Groundwater Services, Inc., along with advisory groups from the DOD, industry, and commercialization interests.

Starting with 170 preproposals that were submitted in response to a broadly disseminated announcement, 12 projects were chosen by a peer-review process for field demonstrations. The technologies chosen were targeted at DOD's most serious problems of soil and groundwater contamination. The primary objective was to provide more cost-effective solutions, preferably using *in situ* treatment. Eight projects were led by university researchers, two projects were managed by government agencies, and two others were conducted by engineering companies. Engineering partners were paired with the academic teams to provide field demonstration experience. Technology experts helped guide each project.

DOD sites were evaluated for their potential to support quantative technology demonstrations. More than 75 sites were evaluated in order to match test sites to technologies. Following the development of detailed work plans, carefully monitored field tests were conducted and the performance and economics of each technology were evaluated.

One AATDF project designed and developed two portable Experimental Controlled Release Systems (ECRS) for testing and field simulations of emerging remediation concepts and technologies. The ECRS is modular and portable and allows researchers, at their sites, to safely simulate contaminant spills and study remediation techniques without contaminant loss to the environment. The completely contained system allows for accurate material and energy balances.

The results of the DOD/AATDF Program provide DOD and others with detailed performance and cost data for a number of emerging, field-tested technologies. The program also provides information on the niches and limitations of the technologies to allow for more informed selection of remedial solutions for environmental cleanup.

The AATDF Program can be contacted at Energy and Environmental Systems Institute, MS-316, Rice University, 6100 Main, Houston, TX, 77005; phone 713-527-4700; fax 713-285-5948; e-mail <eesi@rice.edu>.

The DOD/AATDF Program staff includes:

Director:
Dr. C. Herb Ward

Program Manager:
Dr. Carroll L. Oubre

Assistant Program Manager:
Dr. Kathy Balshaw-Biddle

Assistant Program Manager:
Dr. Stephanie Fiorenza

Assistant Program Manager:
Dr. Donald F. Lowe

Financial/Data Manager:
Mr. Robert M. Dawson

Publications Coordinator/Graphic Designer:
Ms. Mary Cormier

Meeting Coordinator:
Ms. Susie Spicer

This volume, *Sequenced Reactive Barriers for Groundwater Remediation*, is one of a ten-monograph series that records the results of DOD/AATDF environmental technology demonstrations. Many have contributed to the success of the AATDF program and to the knowledge gained. We trust that our efforts to fully disclose and record our findings will contribute valuable lessons learned and help further innovative technology development for environmental cleanup.

Stephanie Fiorenza

Carroll L. Oubre

C. Herb Ward

About the Editors

Stephanie Fiorenza

Stephanie Fiorenza is an Assistant Program Manager with AATDF at Rice University, where she has managed five projects involving the field demonstration of innovative remediation technologies. Dr. Fiorenza has a Ph.D. in Environmental Science and Engineering from Rice University and a B.A. in Environmental Studies from Brown University. In her role as an assistant program manager for AATDF, Dr. Fiorenza provided the managerial guidance and technical expertise that were required for the successful field demonstration of each technology. She has also been an active participant in the preparation of all the projects' reports. Prior to joining AATDF, Dr. Fiorenza worked as an environmental specialist for Amoco Corporation in their Groundwater Management Section. Her areas of interest are the biodegradation of organic chemicals, microbial ecology of the subsurface, and development of remediation technologies.

Carroll L. Oubre

Carroll L. Oubre is the Program Manager for the DOD/AATDF Program. As Program Manager he is responsible for the day-to-day management of the $19.3 million DOD/AATDF Program. This includes guidance of the AATDF staff and overview of the 12 demonstration projects, assuring project milestones are met within budget and that complete reporting of the results is timely.

Dr. Oubre has a B.S. in chemical engineering from the University of Southwestern Louisiana, an M.S. in chemical engineering from Ohio State University, and a Ph.D. in chemical engineering from Rice University. He worked for Shell Oil Co. for 28 years, with his last job as Manager of Environmental Research and Development for Royal Dutch Shell in England. Prior to that, he was Director of Environmental Research and Development at Shell Development Co. in Houston, Texas.

C. H. (Herb) Ward

C. H. Ward is the Foyt Family Chair of Engineering in the George R. Brown School of Engineering at Rice University. He is also Professor of Environmental Science and Engineering and Ecology and Evolutionary Biology.

Dr. Ward has undergraduate (B.S.) and graduate (M.S. and Ph.D.) degrees from New Mexico State University and Cornell University, respectively. He also earned the M.P.H. in environmental health from the University of Texas.

Following 22 years as Chair of the Department of Environmental Science and Engineering at Rice University, Dr. Ward is now Director of the Energy and Environmental Systems Institute (EESI), a university-wide program designed to mobilize industry, government, and academia to focus on problems related to energy production and environmental protection.

Dr. Ward is also Director of the Department of Defense Advanced Applied Technology Demonstration Facility (AATDF) Program, a distinguished consortium of university-based environmental research centers supported by consulting environmental engineering firms to guide selection, development, demonstration, and commercialization of advanced applied environmental restoration technologies for the DOD. For the past 18 years he has directed the activities of the National Center for Ground Water Research (NCGWR), a consortium of universities charged with conducting long-range exploratory research to help anticipate and solve the nation's emerging groundwater problems. He is also Co-Director of the EPA-sponsored Hazardous Substances Research Center/South and Southwest (HSRC/S and SW), which focuses its research on contaminated sediments and dredged materials.

Dr. Ward has served as president of both the American Institute of Biological Sciences and the Society for Industrial Microbiology. He is the founding and current Editor-in-Chief of the international journal *Environmental Toxicology and Chemistry.*

About the Authors

James F. (Jim) Barker

Jim Barker is a Professor in the Earth Sciences Department at the University of Waterloo, Waterloo, Ontario, Canada. His research interests include the migration and fate of organic chemicals in groundwater and their remediation by natural attenuation and with *in situ* techniques. Field studies have emphasized the migration and fate of organic contaminants in landfill leachate plumes and in controlled release field experiments. Petroleum hydrocarbons and related compounds were involved.

Barbara J. Butler

Barbara J. Butler is a Lecturer, Research Associate, and Adjunct Assistant Professor in the Department of Biology, University of Waterloo, Waterloo, Ontario, Canada. She obtained her Ph.D. from Waterloo in 1985. Her current research interests include biotransformation of organic contaminants in soil and groundwater systems, and microbial ecology of subsurface environments.

Evan Cox

Evan Cox was a Senior Environmental Microbiologist at BEAK International, Inc. with 9 years professional experience in bioremediation of hazardous organic compounds in subsurface environments. He has an M.Sc. in microbiology and a B.Sc. in biology with a geology minor, both from the University of Waterloo, Waterloo, Ontario, Canada. He has managed numerous field investigations of intrinsic and enhanced bioremediation of chlorinated solvents and recalcitrant compounds, laboratory treatability studies for groundwater, soil and soil-gas remediation, pilot- and full-scale demonstrations of *in situ* bioremediation, and development of innovative bioremediation technologies for chlorinated solvents, energetics, and recalcitrant compounds. Mr. Cox has published more than 25 papers regarding the biodegradation of hazardous contaminants in subsurface environments. He recently joined GeoSyntec Consultants, Inc. in Guelph, Ontario, Canada.

John F. (Rick) Devlin

J. F. Devlin is a Research Assistant Professor at the University of Waterloo, Waterloo, Ontario, Canada, where he conducts research on the fate and transport of organic and inorganic contaminants in groundwater, development of semi-passive groundwater remediation systems, and the processes underlying reactive barrier design. He has a Ph.D. in earth science (hydrogeology) from the University of Waterloo, an M.Sc. in geology from Queen's University, Kingston, Ontario, Canada, and a B.Sc. (Hons.) in chemistry also from Queen's University. Prior to obtaining his doctorate, Dr. Devlin worked as a hydrogeological consultant for 4 years. His project experience ranges from the characterization of landfill leachate plumes to *in situ* bioremediation of chlorinated solvents in groundwater. Dr. Devlin's remediation experience includes projects involving soil vapor extraction, pump and treat, reactive barrier treatment of solvent-contaminated groundwater using granular iron, and aerobic and anaerobic bioremediation of hydrocarbons, solvents, and nitrate in groundwater. Dr. Devlin has presented papers and published articles in the areas of bioremediation, sampling protocols and analytical QA/QC, redox manipulation of aquifers, abiotic organic transformations in aquifers, abiotic reactions involving granular iron, and modeling reactive transport.

Robert (Rob) Focht

Robert Focht received both his B.A.Sc. in chemical engineering and M.Sc. in hydrogeology from the University of Waterloo, Waterloo, Ontario, Canada. Mr. Focht has been involved with the granular-iron reactive barrier technology since he started his M.Sc. in 1992. After completion of his M.Sc. he continued to be involved in research and development efforts with the University of Waterloo, focusing on enhancements to this technology. During this time, he became involved with EnviroMetal Technologies Inc., joining the company in 1995. Mr. Focht has been mainly involved in reviewing data from potential test sites, chemical and groundwater modeling, and development of field designs, and has overseen many of the *in situ* remediation installations.

Susan M. Froud

Susan Froud is an environmental consultant with AquaTerre Solutions, Inc. of Toronto, Ontario. Ms. Froud completed her M.Sc. at the University of Waterloo, where the focus of her thesis was the investigation of sequential anaerobic–aerobic processes, using zero-valent iron and ORC at CFB Borden (as described in this text). She completed her B.A.Sc. in civil engineering, with a minor in environmental engineering, at McGill University, Montreal, Quebec, Canada. Her consulting interests focus on the assessment and remediation of contaminated soil and groundwater at industrial sites.

Dennis J. Katic

Dennis Katic received his M.Sc. from the University of Waterloo. The focus of his thesis was the investigation of sequential anaerobic–aerobic processes using a nutrient injection wall and a biosparge gate at CFB Borden (as described in this book). His background is in chemistry, environmental science, and geography. He obtained his B.Sc. at the University of Toronto, Toronto, Ontario, Canada with majors in chemistry and environmental science, and a minor in geography. His interests include field implementation of innovative groundwater remediation techniques. After completion of his M.Sc., Mr. Katic joined Golder Associates in Mississauga, Ontario.

Michaye McMaster

Michaye McMaster is an Environmental Microbiologist with GeoSyntec Consultants, Inc., where she provides technical expertise and review for groundwater remediation projects. Ms. McMaster obtained her M.Sc. degree in earth sciences and her B.Sc. degree in biology, both from the University of Waterloo, Waterloo, Ontario, Canada. After receiving her B.Sc., she worked for the University of Waterloo in the area of innovative on-site wastewater treatment processes. After completion of her M.Sc., she joined Beak International, Inc. Prior to joining BEAK and during her employment at BEAK, she served as Project Manager for the University of Waterloo Rice Project.

Mary Morkin

Mary Morkin received an M.Sc. from the Department of Earth Sciences at the University of Waterloo. The focus of her thesis was the investigation of sequential anaerobic–aerobic processes using zero-valent iron and a biosparge gate at NAS Alameda (as described in this book). Her background is in geology, with a B.Sc. from the University of Western Ontario, London, Ontario, Canada. Her interests include innovative *in situ* remediation of groundwater. She is currently employed by GeoSyntec Consultants, Inc. in Walnut Creek, California.

John Vogan

John Vogan has been Manager of EnviroMetal Technologies, Inc. (ETI) since October 1994, after joining ETI in February 1993 as a hydrogeologist. Mr. Vogan was named President in October 1997. He provides technical and administrative oversight in all aspects of the technology application. His specific responsibilities include contract negotiation and administration, review of hydrogeologic data from potential sites, and design and reporting of field applications of the technology. He has provided technical input into all commercial applications of the EnviroMetal Process to date. Mr. Vogan is also involved in international sublicensing arrangements with a number of firms in Europe and the Pacific Rim. He has co-authored several technical publications concerning permeable barrier installations. He obtained his M.Sc and B.Sc. from the Department of Earth Sciences, University of Waterloo, Waterloo, Ontario, Canada. Before joining ETI, Mr. Vogan had worked for more than 4 years for Dames and Moore on a variety of groundwater contamination and groundwater supply investigations.

AATDF Advisors

UNIVERSITY ENVIRONMENTAL RESEARCH CENTERS

National Center for Ground Water Research
Dr. C. H. Ward
Rice University
Houston, TX

Hazardous Substances Research Center South and Southwest
Dr. Danny Reible and
Dr. Louis Thibodeaux
Louisiana State University
Baton Rouge, LA

Waterloo Centre for Groundwater Research
Dr. John Cherry and Mr. David Smyth
University of Waterloo
Ontario, Canada

Western Region Hazardous Substances Research Center
Dr. Perry McCarty
Stanford University
Stanford, CA

Gulf Coast Hazardous Substances Research Center
Dr. Jack Hopper and Dr. Alan Ford
Lamar University
Beaumont, TX

Environmental Solutions Program
Dr. Raymond C. Loehr
University of Texas
Austin, TX

DOD/ADVISORY COMMITTEE

Dr. John Keeley, Co-Chair
Assistant Director
Environmental Laboratory
U.S. Army Corps of Engineers
Waterways Experiment Station
Vicksburg, MS

Mr. James I. Arnold, Co-Chair
Acting Division Chief, Technical Support
U.S. Army Environmental Center
Aberdeen, MD

Dr. John M. Cullinane
Program Manager, Installation Restoration
U.S. Army Corps of Engineers
Waterways Experiment Station
Vicksburg, MS

Mr. Scott Markert and Dr. Shun Ling
Naval Facilities Engineering Center
Alexandria, VA

Dr. Jimmy Cornette, Dr. Michael Katona and Major Mark Smith
Environics Directorate
Armstrong Laboratory
Tyndall AFB, FL

COMMERCIALIZATION AND TECHNOLOGY TRANSFER ADVISORY COMMITTEE

Mr. Benjamin Bailar, Chair
Dean, Jones Graduate School of Administration
Rice University
Houston, TX

Dr. James H. Johnson, Jr., Associate Chair
Dean of Engineering
Howard University
Washington, DC

Dr. Corale L. Brierley
Consultant
VistaTech Partnership, Ltd.
Salt Lake City, UT

Dr. Walter Kovalick
Director, Technology Innovation Office
Office of Solid Wastes and
Emergency Response
U.S. EPA
Washington, DC

Mr. M. R. (Dick) Scalf
U.S. EPA Robert S. Kerr
Environmental Research Laboratory (retired)
Ada, OK

Mr. Terry A. Young
Executive Director
Technology Licensing Office
Texas A&M University
College Station, TX

Mr. Stephen J. Banks
President
BCM Technologies, Inc.
Houston, TX

CONSULTING ENGINEERING PARTNERS

Remediation Technologies, Inc.
Dr. Robert W. Dunlap, Chair
President and CEO
Concord, MA

Parsons Engineering
Dr. Robert E. Hinchee (Originally with Battelle Memorial Institute)
Research Leader
South Jordan, UT

GeoTrans, Inc.
Dr. James W. Mercer
President and Principal Scientist
Sterling, VA

Arcadis Geraghty and Miller, Inc.
Mr. Nicholas Valkenburg and
Mr. David Miller, Vice Presidents
Plainview, NY

Groundwater Services, Inc.
Dr. Charles J. Newell
Vice President
Houston, TX

INDUSTRIAL ADVISORY COMMITTEE

Mr. Richard A. Conway, Chair
Senior Corporate Fellow
Union Carbide
S. Charleston, WV

Dr. Ishwar Murarka, Associate Chair
Electric Power Research Institute
Currently with Ish, Inc.
Cupertino, CA

Dr. Philip H. Brodsky
Director, Research and
Environmental Technology
Monsanto Company
St. Louis, MO

Dr. David E. Ellis
Bioremediation Technology Manager
DuPont Chemicals
Wilmington, DE

Dr. Paul C. Johnson
Arizona State University
Department of Civil Engineering
Tempe, AZ

Dr. Bruce Krewinghaus
Shell Development Company
Houston, TX

Dr. Frederick G. Pohland
Department of Civil Engineering
University of Pittsburgh
Pittsburgh, PA

Dr. Edward F. Neuhauser, Consultant
Niagara Mohawk Power Corporation
Syracuse, NY

Dr. Arthur Otermat, Consultant
Shell Development Company
Houston, TX

Mr. Michael S. Parr, Consultant
DuPont Chemicals
Wilmington, DE

Mr. Walter Simons, Consultant
Atlantic Richfield Company
Los Angeles, CA

List of Acronyms and Abbreviations

AATDF	Advanced Applied Technology Demonstration Facility
asl	above sea level
BEAK	Beak International Incorporated
bgs	below ground surface
BOD	biochemical oxygen demand
Br^-	bromide ion
BTEX	benzene, toluene, ethylbenzene, and xylenes
Ca	calcium
CA	Califormia
CF	chloroform
CFB	Canadian Forces Base
CH_4	methane
Cl^-	chloride ion
CM	chloromethane
cm	eentimeter
CO_2	carbon dioxide
COD	chemical oxygen demand
conc.	concentration
CS_2	carbon disulfide
CT	carbon tetrachloride
CVOC	chlorinated volatile organic compounds
1,2 DCE	1,2-dichloroethane
1,1 DCE	1,1-dichloroethene
*c*DCE	*cis*1,2-dichloroethene
*t*DCE	*trans*1,2-dichloroethane
DCE	all isomers of DCE
DCM	dichloromethane (or methylene chloride)
DHG	dissolved hydrogen gas
DIMP	diisopropyl methylphosphonate
DMA	dimethylamine
DNAPL	dense nonaqueous phase liquid
DO	dissolved oxygen
DOD	Department of Defense (U.S.)
DOE	Department of Energy (U.S.)
Eh	redox potential
ETI	EnviroMetal Technologies, Inc.
Fe	iron
ft	feet
FTD	Final Technical Document
F&G	Funnel-and-Gate
GAC	granular activated carbon
GC	gas chromatography
GES	groundwater extraction system
gpm	gallons per minute
GPC	gas phase carbon
GTS	groundwater treatment system
H_2	hydrogen
HCl	hydrochloric acid
HCO_3^-	bicarbonate ion
hr	hour
ID	inside diameter
IR	intrinsic remediation
ISPFS	*In Situ* Permeable Flow Sensors

K	hydraulic conductivity
K^+	potassium ion
$K_{r(g)}$	residual gas phase volume as % of total void volume
L	liter
LDPE	low density polyethylene
LNAPL	light nonaqueous phase liquid
LOQ	Limit of Quantification
m	meter
MBH	modified Bushell Haas
MCL	Maximum Contaminant Level
MDL	Method Detection Limit
mg	milligram
Mg	magnesium
min	minute
Mn	manganese
μg	microgram
μS	microsiemens
mV	millivolt
N_2	nitrogen
Na	sodium
NAPL	nonaqueous phase liquid
NAS	Naval Air Station
NDMA	n-nitroso dimethylamine
NFESC	Naval Facilities Engineering Service Center
NIW	Nutirent Injection Wall
NO_3^-	nitrate ion
NPV	net present value
O_2	oxygen
OD	outside diameter
OGL	Organic Geochemistry Lab at UW
OP1	Operating Phase 1
OP2	Operating Phase 2
ORC™	Oxygen Release Compound™
OSHA	Occupational Safety and Health Administration
PAH	polynuclear aromatic hydrocarbons
PCE	perchloroethylene (or tetrachloroethene)
PO_4^{-3}	phosphate ion
PRB	permeable reactive barriers
psi	pounds per square inch
P&T	pump-and-treat
PVC	polyvinyl chloride
QA/QC	Quality Assurance/Quality Control
RO	Remedial Objectives
s	second
SD	standard deviation
SO_4^{-2}	sulfate ion
SPRB	sequenced permeable reactive barrier
ss	stainless steel
STP	standard temperature and pressure
SVOC	semivolatile organic compounds
SW	source well
t	metric ton
TCE	trichloroethene
TDS	total dissolved solids
TER	Technology Evaluation Report

TOL	toluene
TWA	time-weighted average
U.S.	United States
U.S.A.F.	United States Air Force
USG	U.S. gallons
UW	University of Waterloo
WQL	Water Quality Lab at UW
WTI	Water Technology International
VC	vinyl chloride
VOC	Volatile Organic Compound
2-D	two dimensional

Contents

List of Figures in the Text

List of Figures in the Appendices

List of Tables in the Text

List of Tables in the Appendices

PART I

Technology Demonstrations

James F. Barker
University of Waterloo

Barbara J. Butler
University of Waterloo

John F. Devlin
University of Waterloo

Susan M. Froud
University of Waterloo

Dennis J. Katic
University of Waterloo

Michaye McMaster
GeoSyntec Consultants, Inc.

Mary Morkin
GeoSyntec Consultants, Inc.

Executive Summary

Passive and semipassive *in situ* remediation technologies were assessed in anaerobic–aerobic sequences for the treatment of a mixed petroleum and chlorinated hydrocarbon plume in a controlled release experiment at the Canadian Forces Base (CFB) Borden site in Ontario, Canada. A parallel experiment, using similar technologies, was undertaken at Naval Air Station (NAS) Alameda, near Oakland, California.

The CFB Borden experiment was performed in a large Funnel-and-Gate structure (5.5 m wide by 25 m long) consisting of three side-by-side treatment gates. Altogether, five technologies were tested, including two anaerobic–aerobic treatment sequences, and the intrinsic remediation option. Granular iron in a removable cassette system and biodegradation stimulated from a nutrient injection wall (NIW) were selected as the anaerobic treatment methods to reductively dechlorinate the target organics carbon tetrachloride (CT) and tetrachloroethene (PCE). The two aerobic treatment technologies, both based on *in situ* biodegradation, were included for the degradation of toluene (TOL) (representing petroleum aromatics) and partially dechlorinated products of the anaerobic treatment. Either oxygen-releasing compound, ORC™, or a biosparge system supplied dissolved oxygen. The intrinsic remediation option was evaluated in Gate 2 and included no engineered enhancements.

The treatment technologies were installed within the gates sequentially in pairs, with the order of encounter being anaerobic treatment first, then aerobic treatment. This order of encounter was considered the most reasonable, since the highly oxidized chlorinated hydrocarbons PCE and CT are especially susceptible to treatment by reductive processes. If the reductions are incomplete, resulting in the production of less chlorinated compounds, there is a reasonable expectation that these products will be treatable in the aerobic portion of the gate, along with the petroleum hydrocarbons present. Also, in some cases, there may be aesthetic reasons for producing aerobic water at the final stage of treatment. Reversing the order of encounter runs the risk of producing chlorinated transformation products at the end of the treatment sequence. If this occurred, an additional treatment system would have to be added to the gate at increased expense.

The first gate contained two granular iron cassettes and an oxygen-releasing compound (ORC™) cassette. The cassettes are a novel method for the emplacement of reactive media; ultimately, a cassette could be removed and the reactive medium then rejuvenated or replaced. The second gate was kept free of any instrumentation other than monitoring wells and served the double purpose of acting as a control gate and a gate for the assessment of intrinsic remediation. The third gate was instrumented with two permeable barriers, the NIW and a biosparge wall. Benzoate was injected into the NIW at intervals of approximately 1 month and established a zone of sulfate reduction and methanogenesis within 4 to 5 m downgradient. The biosparge wall was sparged with pure oxygen gas, first semiweekly, and later on a daily basis, to maintain the desired oxygen flux from the residual gas bubbles to the groundwater.

The target contaminants were introduced to the gates from source wells containing diffusion emitters. Diffusion emitters are devices composed of tubes of a plastic polymer that is permeable to hydrophobic organics. Near-saturated solutions of the target compounds (handled as pure compounds, not mixtures) were circulated from source drums containing the solutions to the source wells through thin-walled polyethylene tubes. The contaminants were delivered to the groundwater in the wells via diffusion through the tubing. Source concentrations were designed to be in the range of 1 to 2 mg/L for the chlorinated compounds and 10 mg/L for toluene. The diffusion emitters transmitted more mass than expected, leading to concentrations in the source wells up to an order of magnitude above the targets. However, the high concentrations in the source wells were not transmitted to the aquifer. They were counterbalanced, to some extent, by nonuniform passive flushing of the source wells, due to aquifer heterogeneity and well inefficiencies. Thus, the diffusion emitters performed satisfactorily for the purposes of the experiment, although additional developmental work is required.

The granular iron in the first gate completely dechlorinated PCE to ethene and ethane (other nonchlorinated products may have been formed but were not identified) with a half-life of about 0.5 days. Carbon tetrachloride was almost entirely transformed to chloroform (CF) in the aquifer before reaching the treatment cassettes. This phenomenon was also observed over the same distances in the other two gates. Within the iron cassettes, the CF was transformed to dichloromethane (DCM), or methylene chloride, which did not appear to undergo any further transformations. The estimated half-lives for CT and CF were about 11 days each for the intrinsic processes, and CF exhibited a half-life of about 0.7 days in the granular iron. There was no evidence of DCM transformation in the granular iron, and no DCM was ever detected in either of the other two gates.

ORC™ was initially installed in a cassette immediately downgradient of the granular iron, but did not release oxygen in significant quantities at any time while in this configuration. Laboratory experiments suggested that the high pH water emanating from the granular iron cassettes might have been detrimental to the ORC™ oxygen-releasing process. However, attempts to reduce the pH of the groundwater in the ORC™ cassette failed to stimulate oxygen release. Ultimately, the ORC™ was moved to three 25-cm-diameter contingency wells located about 2 m downgradient of the cassette system. The aquifer between the two structures buffered the groundwater and maintained a constant pH between 7.5 to 8 flowing into the ORC™ wells. Oxygen release began almost immediately and, within a month, a limited plume of oxygenated water was established in the gate. With the establishment of aerobic conditions, the apparent degradation of toluene followed.

A leak in the seals between the cassette system and the sheet pile walls of the gate was found to conduct increasing proportions of flow as the experiment progressed. This led to the development of untreated PCE and CF plumes (recall that CT was intrinsically transformed in the aquifer) downgradient of the granular iron cassettes, and the occurrence of these compounds in the aerobic portion of the gate. No aerobic degradation of PCE could be discerned from the available data — nor was any anticipated. However, there were indications that CF transformed aerobically, possibly the result of cometabolism. Further work is needed to clarify this issue.

The NIW was used to introduce pulses of benzoate to the aquifer to develop and sustain anaerobic microbial communities in the third gate. The chloromethanes, CT, and CF (and DCM if it was produced — it was never detected) were completely attenuated within the anaerobic zone with half lives on the order of 11 days, as observed in Gate 2. In addition, virtually complete reductive dechlorination of PCE was also observed after an acclimation period of less than 175 days, with *cis*-1,2-dichloroethene (*c*DCE) being the only identifiable product. Toluene may also have undergone some anaerobic degradation, but further analysis is required before firm conclusions can be drawn.

The biosparge wall successfully maintained an aerobic condition in the downgradient end of Gate 3 with daily sparging. The possibility that this sparging frequency caused some volatilization of contaminants cannot be discounted. However, volatilization rates were considered minimal on the basis of calculated maximum air stripping rates as well as nondetectable levels of any contaminants in unsaturated zone gas samples collected immediately after sparging. The aerobic conditions established in the gate appeared to stimulate aerobic biodegradation of toluene, *c*DCE, benzoate, and methane. Sufficient oxygen was delivered to the groundwater to treat all of the contaminant mass as well as the organic loading from benzoate and methane. Groundwater leaving Gate 3 at the close of the experiment was considered remediated with average PCE and *c*DCE concentrations below LOQs (7 μg/L) and CT and CF levels below MDLs (5 and 9 μg/L, respectively). At the end of the experiment, the only target organic found leaving the gate in quantifiable concentrations was TOL (40 μg/L), but this was considered to be a temporary situation, since the groundwater had been rendered sufficiently aerobic to handle the oxygen demand imposed by TOL and the other sources of dissolved carbon (benzoate, acetate, methane).

The technologies tested at CFB Borden each exhibit strengths and weaknesses. The preferred technology for application at a real site would depend on the contaminants and conditions encountered there. For example, for anaerobic treatment of chlorinated solvents, the granular iron was shown to promote the fastest and most complete reductive dechlorination (see half-lives above). However, at sites where depth to source or start-up costs are limiting factors, bioremediation may prove to be more viable. The same considerations apply to the aerobic treatment approaches. Where start-up costs and depth to contamination are restrictive, ORC™ may prove more useful. However, where pH-related interferences are anticipated and/or where a long-term steady oxygen source is required, biosparging may be more desirable.

There is clear evidence that during the time of this experiment, the passive and semipassive treatment systems were more effective than intrinsic remediation at the CFB Borden test site. However, additional work is required to fully assess the long-term performance of coupled anaerobic and aerobic systems operating in sequence.

At the Alameda site, the treatment sequence was contained in a gate structure approximately 3 m wide (the cross-gradient direction) and 6 m deep, with longitudinal dimensions as follows: 0.6 m of a 5% by weight mixture of sand and granular iron, followed by 1.5 m of 100% granular iron, followed by 0.9 m of pea gravel. This porous media sequence was followed by the biosparge zone, which consisted of a 0.6-m open-water zone partially filled with plastic Yeager™ Tripak balls to serve as a microbial biomass support medium. This, in turn, was followed by a final 0.6 m of pea gravel, which served as the final zone for compliance monitoring. The downgradient side of the gate was sealed with sheet piling and a constant pumping rate was maintained to establish an average groundwater flow velocity of about 6 cm/day through the gate. Sheet piling funnels (total length 9 m or 30 ft) were installed at the upgradient end of the gate for the capture of an approximately 7.6-m (25-ft)-wide section of the contaminant plume.

The granular iron zone was designed to handle input concentrations of 20 to 30 mg/L of *c*DCE and vinyl chloride (VC), as well as 1 mg/L or less of TCE, at average linear velocities of 23 cm/day. However, most of the experiment was conducted with a velocity of only 6 to 9 cm/day, resulting in residence times in the iron of at least 17 days. Over the course of the experiment, the contaminant loading was dominated by *c*DCE, which entered the treatment gate at a maximum concentration in excess of 200 mg/L in very localized sections of the granular iron barrier. This occurred at a time when VC was entering the system at about 50 mg/L. This level of input exceeded that used for the design of the system, and between 1 and 5 mg/L of VC and *c*DCE, respectively, broke through the granular iron treatment zone and into the biosparge zone. Mass removal (66%) of both these contaminants was observed in the biosparge zone, but only part of this (35 to 70% of the mass removed for *c*DCE and VC, respectively) could be attributed to volatilization, suggesting that a biodegradation treatment zone was successfully in place.

The high pH water emanating from the granular iron treatment zone at the Borden site caused concern in the design of the Alameda treatment gate. In particular, it was feared that high pH water would inhibit microbial growth in the biosparge zone. To overcome this potential limitation, carbon dioxide gas was sparged into the water of the biosparge zone along with oxygen. As a result, the average pH of the biosparge zone water never exceeded a value of 8.5. Thus the only major known interference in coupling the anaerobic and aerobic technologies was overcome for the purposes of this experiment. However, as in the Borden work, further monitoring of the system over a longer term is needed to identify interferences that were not manifested over the relatively short duration of this experiment and to quantify long-term performance of the treatment zones.

At the time that *c*DCE and VC were entering the biosparge zone, more than 99% of all contaminants were being removed from the groundwater as it passed through the entire treatment gate. This level of performance improved still more, due mainly to the ability of the granular iron to reduce the chlorinated compounds, when the input concentrations of chlorinated hydrocarbons declined later in the project. By January 1998, the system was performing sufficiently well that the remedial objectives (ROs) for the Alameda phase of the project were achieved for all compounds. However, over the course of the experiment, the petroleum hydrocarbons benzene, toluene, and xylenes were not detected in sufficient concentrations downgradient of the granular iron to evaluate their degradation rates in the biosparge zone. These compounds were apparently sorbed to a significant degree to the granular iron, resulting in very low net velocities through the anaerobic treatment section. Thus, there was no opportunity to rigorously assess the removal rates of petroleum hydrocarbons from groundwater in the biosparge zone. Additional work is needed to address this performance issue, although the prognosis is favorable given the high aerobic degradability of these compounds.

This project has shown, at two field sites, that the treatment of mixed contaminant plumes by *in situ*, semipassive remedial technologies is a viable option. The sequential operation of anaerobic and aerobic treatment technologies was anticipated to be challenging from the standpoint of geochemical and microbiological interferences. In practice, these kinds of interferences proved to be less problematic than first thought. During the experiment, the only case of permeability loss due to biofouling or precipitate formation occurred in the cassette system and the severity of the problem was magnified by the failure of seals between the cassette assembly and the sheet piles. It remains uncertain if the problem would have been noticed at all if not for the faulty seals. Further work is needed to evaluate the longer-term behavior of this kind of treatment system, to see if the anticipated interferences become more important with time, and to evaluate the sensitivity of the system to differing native geochemical environments.

CHAPTER 1

Introduction and Technology Review

1.1 RATIONALE FOR THE CURRENT RESEARCH

The understanding of groundwater contamination by organic chemicals has improved considerably over the past decade. This is particularly evident in the evolution of conceptual models under which hydrogeologists have operated over that time. Prior to the 1980s, organic chemicals were largely disregarded in hydrogeological circles. A few specific compounds, such as DDT, drew some attention, but were not subject to much hydrogeological interest because of their relatively low mobilities in groundwater. When organics were considered at all, they were generally assumed to behave much like petroleum hydrocarbons. According to the conceptual model of the day, these chemicals constituted a routine shallow contamination problem (Figure 1.1a). Their importance was, in part, underestimated because until the late 1970s there was a lack of empirical evidence that any major problem existed; it was rare for organic chemicals to be observed in well water. In particular, deep wells were thought to be safe from this kind of contamination.

With the development of improved methods in gas chromatography (GC), it became apparent that organic contamination of groundwater was much more common than had previously been thought. Among the most frequently encountered contaminants were the chlorinated solvents, which include the chlorinated ethenes, tetrachloroethene (also known as perchloroethene (PCE), trichloroethene (TCE), the dichloroethene isomers (*cis*-1,2- (*c*DCE), *trans*-1,2- (*t*DCE), and 1,1-DCE), vinyl chloride (VC), and the chloromethanes including carbon tetrachloride (CT), chloroform (CF), dichloromethane (DCM) (also known as methylene chloride), and chloromethane (CM). It was soon realized that these chemicals possess physical and chemical properties that enable them to contaminate vast quantities of water and to exhibit low retardation in aquifers, and that make them difficult to remediate. Specifically, as nonaqueous phases the above solvents have densities greater than that of water, allowing them to penetrate far below the water table (Figure 1.1b). In addition, they have solubilities in water that are low (most are immiscible with water), but are nonetheless orders of magnitude above the recommended drinking water standards of most jurisdictions in North America. These characteristics have led to the phrase "dense nonaqueous phase liquids" (DNAPLs) to describe such chemicals. Another consideration is the resistance these compounds exhibit toward degradation in aquifers; if they do degrade, it is often along pathways that can lead to products of greater concern than the original contaminants (e.g., TCE to VC). These combined factors lead to long-term contamination problems when chlorinated solvents are spilled into aquifers.

The recognition and acceptance of the DNAPL paradigm (Figure 1.1b) in the early part of this decade put an end to the belief that a site could be quickly remediated by simply implementing programs of excavation and/or pump-and-treat, and led to the development of new strategies for the cleanup of contaminated sites. Many of these strategies focused on the removal of source zone

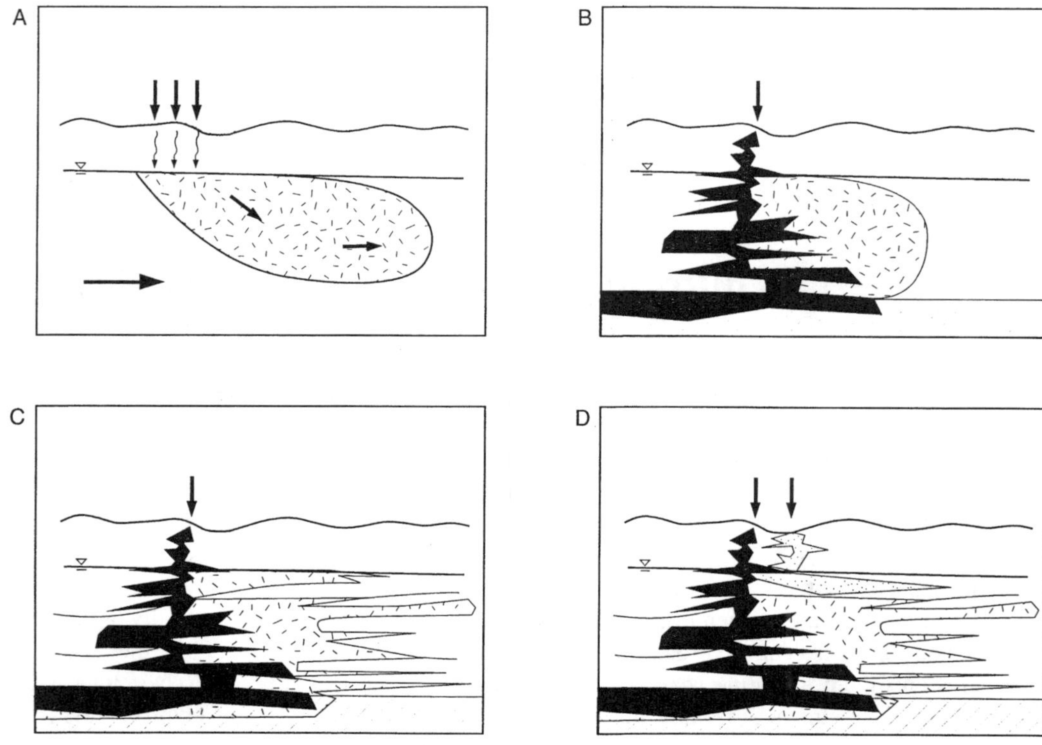

Figure 1.1 Development of conceptual models for groundwater contamination by organic chemicals.

mass by flushing techniques using mobile phases such as surfactant solutions, alcohols, or steam. These technologies continue to be researched and have been shown to be reasonably effective at recovering DNAPL mass from aquifers. However, they remain very expensive to implement, leave the problem of ultimate disposal unresolved, and, in most cases, fail to remove all of the DNAPL mass from the subsurface. Incomplete source removal is a serious limitation, because even small quantities of residual can continue to render water undrinkable for decades.

Operating under the conceptual model depicted in Figure 1.1c, in which aqueous and nonaqueous contamination penetrates relatively low permeability strata, the long-term problem is seen to be one of plume control, since complete source zone remediation is very likely unachievable. To minimize expenditures, a long-term plume control methodology should be implemented with the emphasis on low operation and maintenance costs. This goal may be attainable with passive or semipassive *in situ* treatment systems. Over the past 5 years, considerable experience has been gained with such technologies, at the lab scale and in the field. Examples of these are discussed in Section 1.3.

At the same time that chlorinated solvents were discovered in municipal water supplies, other organic chemicals were detected. Although none was found to be as ubiquitous as the chlorinated solvents, known and potential carcinogens that were detected with notable frequency included a mixture of petroleum derived aromatics, such as benzene, toluene, ethylbenzene, and the xylene isomers (ortho-, meta-, para-), collectively dubbed BTEX. These compounds are among the most soluble components of petroleum products such as gasoline, and hence the most mobile in groundwater.

The fact that two of the most common classes of organic contaminants in groundwater are chlorinated solvents and petroleum aromatics is unfortunate. The chlorinated solvents contain carbon in a relatively oxidized form; the oxidation state increases with the number of chlorine atoms on the molecule. In contrast, the petroleum organics contain carbon in a relatively reduced form. The geochemical conditions required for natural degradation of these two classes of chemicals in an aquifer are quite different. Dissolved BTEX compounds are quite rapidly biodegraded under aerobic conditions, while most chlorinated solvents are biodegraded, albeit more slowly than aerobic biodegradation of BTEX, under anaerobic conditions. In certain cases, abiotic reactions may also contribute to contaminant degradation (e.g., CT $\rightarrow$ CF) (Kriegman-King and Reinhard, 1994), but these reactions are constrained by the redox environment in the same way as the biotransformations. Hydrolysis reactions are redox-independent, but proceed at too slow a rate to be a useful attenuation mechanism for most chlorinated solvents or petroleum hydrocarbons (Vogel et al., 1987). A generalized conceptual model includes the complexities discussed above and the presence of two or more classes of contaminants requiring fundamentally different treatment systems (Figure 1.1d). The U.S. Navy has reported 420 sites and the U.S. Department of Defense (DOD) has 1750 sites that may have mixed contaminant plumes (J. Costanza, Naval Facilities Engineering Service Center, personal communication), so this problem is quite common. Where two kinds of contaminants such as BTEX and solvents are both present, the implementation of passive and semipassive remediation requires at least two quite different *in situ* technologies to function in sequence.

1.2 GOALS AND APPROACH

This project was divided between two field sites, the first at CFB Borden, located in Ontario, Canada (Figure 1.2), and the second at the Naval Air Station Alameda, located near Oakland, California (Figure 1.3). The sites are discussed individually in the following sections.

1.2.1 Borden Experiment

The specific goals of the work performed at CFB Borden were stated in the work plan (University of Waterloo, 1996) as follows:

1. To compare the performances of two anaerobic treatment options for removal of dissolved CT and PCE from groundwater, and two oxygen delivery methods for aerobic biodegradation of toluene in groundwater.
2. To determine the lowest solute concentrations that can be achieved by each treatment technology under the conditions of the test.
3. To quantify performance-related parameters for the experimental design at the U.S. facility.

This report focuses on the results pertaining to the first two objectives. Data related to the second objective were collected over the course of the experiment and are reported, within the limits of the analytical methods used, in the chapters discussing each technology and are summarized in Chapter 6. The third objective was partially fulfilled. The U.S. facility experiment was begun before the completion of the CFB Borden experiment. Nevertheless, sufficient data were available to identify the granular iron and biosparge technologies as the preferred ones to operate sequentially. The scientific basis for the decision is presented in Chapters 2 and 4.

The technologies selected for evaluation at CFB Borden included a permeable barrier containing granular iron for reductive dechlorination of solvents, a nutrient injection wall (NIW) for the stimulation of anaerobic microbial activity, and two methods of stimulating aerobic microbial activity: ORC™ (Oxygen Release Compound, supplied by Regenesis, Inc., San Juan Capistrano, CA) socks and biosparging. These technologies were installed within three sheet-pile alleys up to 25 m in length and 1 to 2 m wide, and 4 m deep, referred to hereafter as *gates*. The width of the gates was based on the contrasting needs for (1) sufficient space between sheet-pile walls to minimize edge effects on the system performance, and (2) a space small enough to accommodate detailed monitoring with minimal effort and expense. These considerations were also paramount

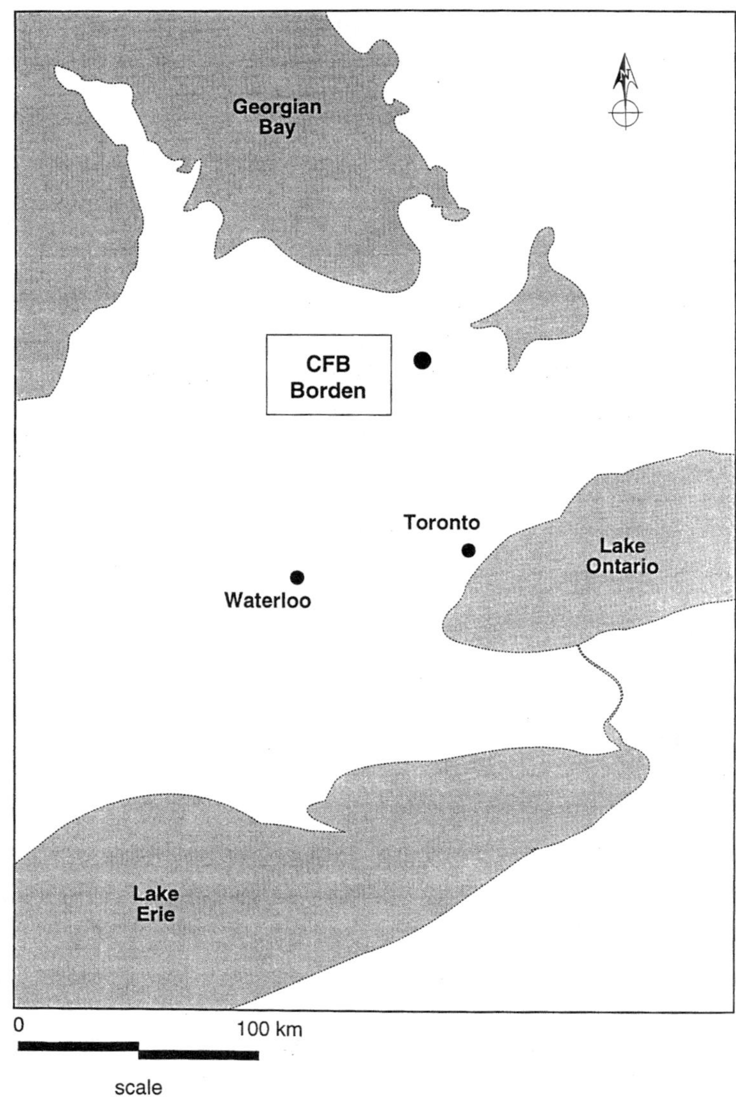

Figure 1.2 Location of CFB Borden.

in the selection of a site with limited vertical thickness for the experiment. The lengths of the gates were determined based on the anticipated distances required for the treatment processes to occur. The reactions in the granular iron were expected to be quite rapid so a short distance in Gate 1 was dedicated to this process. Aerobic biodegradation downgradient of the ORC™ in Gate 1 and following the biosparge zone at the end of Gate 3 was expected to require up to 4 m to occur. Approximately 10 m was allocated to the anaerobic biodegradation treatment system. The experiences that led to these design decisions are related in Section 1.3.

The gates afforded a passive means of control over the groundwater flow direction. The upgradient ends of the gates were open, while the downgradient ends were sealed. Water table gradients were maintained in the gates by means of pumping wells located at the furthest ends of the gates. In the first gate, a treatment module containing four cassettes was installed about 3 m from the opening (Figure 1.4). Briefly, the first cassette was filled with media containing a mixture of about 75% sand and 25% granular iron by weight. The second cassette was packed with 100%

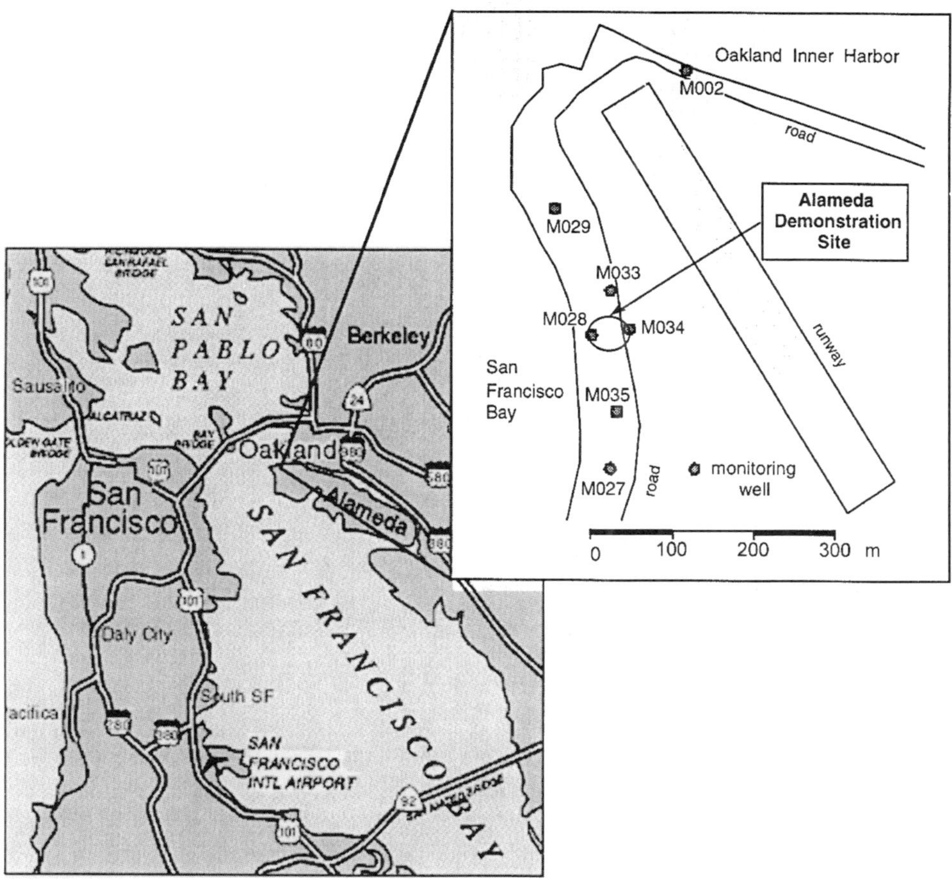

Figure 1.3 Location of the AATDF Demonstration Site, NAS Alameda, CA.

granular iron. The third cassette was filled with crushed limestone to serve as a substrate for precipitation of solids from the iron cassettes effluent, and ORC™ socks were suspended in open water within the final cassette. The use of a cassette system to support the reactive materials was considered to have the important advantage of permitting any single cassette to be exchanged for a fresh one should the media inside become clogged or nonreactive. Although cassette exchanges were not planned or performed as part of this work, the system was included to evaluate its performance in a straightforward application.

Downgradient of the cassettes, three contingency wells were installed to serve as backup oxygen source wells in the event that the ORC™ in the cassette system provided insufficient dissolved oxygen for complete contaminant biodegradation.

The second gate was instrumented with monitoring wells but not with engineered treatment technologies. The data from this gate provide an indication of the intrinsic degradation of the test contaminants (CT, PCE, and TOL) as well as serving as a control for the semipassive treatment gates.

The third gate was instrumented with a permeable wall for nutrient injections leading to the creation of an anaerobic, bioactive zone. The pulses, which were injected at approximately 1-month intervals, merged due to longitudinal dispersion within about 5 m of the wall. Further downgradient in the third gate, a second permeable wall was equipped with oxygen injectors, and is referred to as the *biosparge system*. In the biosparge section, periodic pulses of oxygen were bubbled through the coarse porous medium to establish and renew residual oxygen bubbles throughout the vertical and horizontal (width) extent of the gate. These were permitted to dissolve into passing groundwater

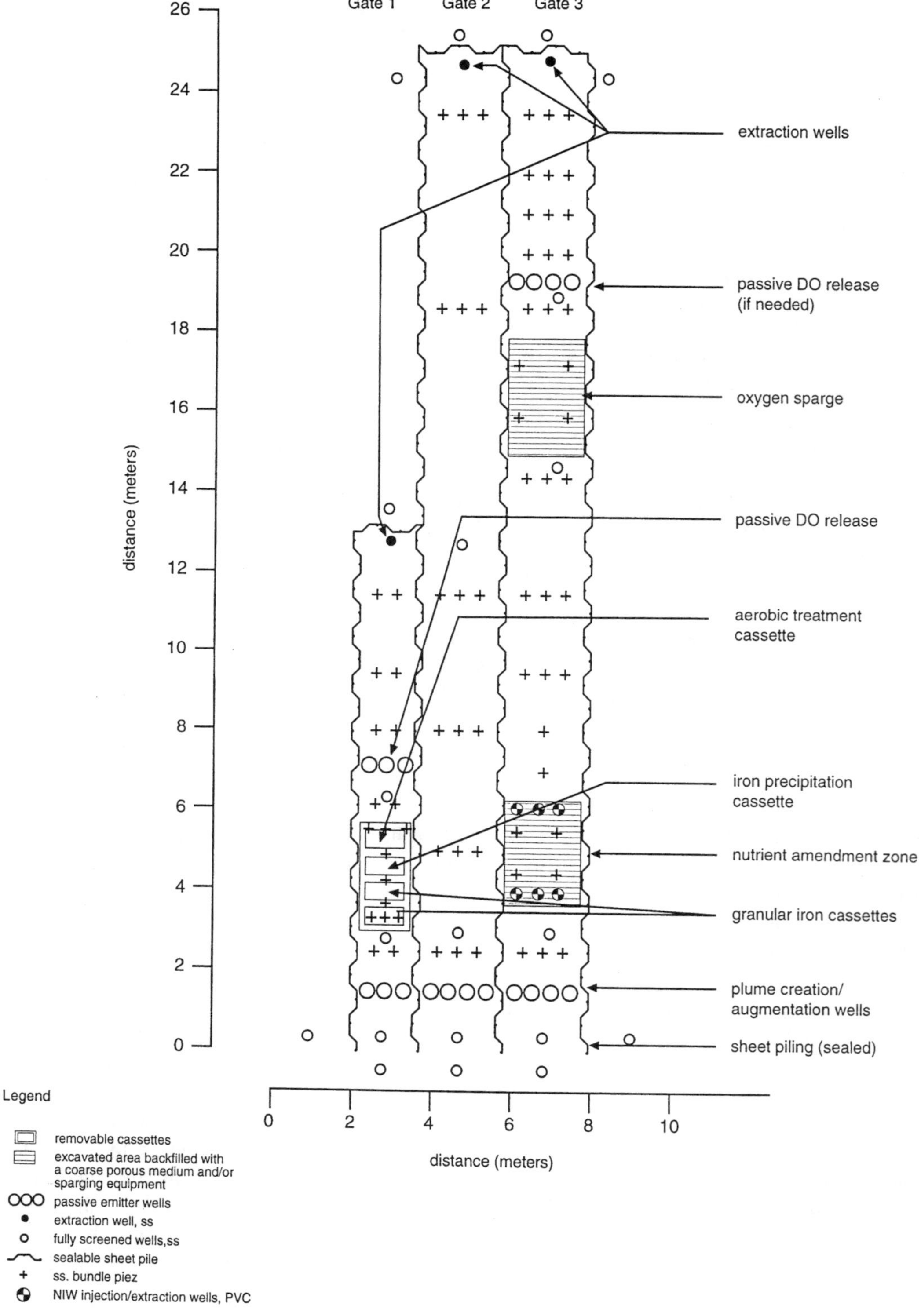

Figure 1.4 Experimental design for the Borden experiment.

prior to initiating subsequent pulses. The objective was to deliver oxygen to the groundwater efficiently, without purging volatile organics from the system.

At the inlet side of each gate, three to four source wells were installed for the introduction of the three test compounds, CT, PCE, and TOL. The source wells were instrumented with racks of coiled polyethylene tubing through which saturated solutions of the three test compounds were periodically pumped. The system was engineered to limit the flux of contaminants from the tubing so that a steady-state concentration of about 1 to 2 mg/L of each compound would develop in the aquifer. As groundwater passed through the source wells and into the gates, these contaminants would be picked up and transported through the treatment systems. The diffuser sources made it possible, in principle, to exercise a high degree of control over the input concentrations while simulating a NAPL source zone. In practice, the source concentrations were not maintained at the desired levels, due to minor leaks (these were corrected as soon as they were discovered) and variations in tubing properties. In some cases (PCE and TOL), the concentrations that were achieved in the source wells were higher than planned. In the worst case, this resulted in TOL being introduced to the aquifer at concentrations on the order of 10 mg/L rather than 1 mg/L. However, other factors, related to aquifer heterogeneity and connection between the source wells and the aquifer, offset the high PCE source well concentration and led to near target concentrations being achieved in the gates. Near target concentrations of CT were also achieved.

From the above description, it is clear that several independently developed technologies were combined to make the sequential treatment systems possible. The overall treatment strategy was to have the contaminants encounter the anaerobic treatment sections first, followed by the aerobic sections. This order of encounter was selected because the anaerobic treatments were known to produce transformation products that could be degraded aerobically (e.g., PCE → VC; CT → DCM). Also, aerobic water exiting the treatment system would be of a generally higher aesthetic quality than anaerobic water.

1.2.2 The Alameda Experiment

The specific goals of the work performed at NAS Alameda were stated in the revised work plan for Site 1, Alameda (University of Waterloo, 1996, Section 1.2) as follows:

1. To demonstrate the attainment of remedial objectives for target organics, iron (Fe), and the attainment of near-neutral pH (5 to 8.5).
2. To provide clear guidance and advice for applying this remedial technology at full scale here and at other sites.

This report focuses on the results pertaining to the first objective. The second objective is fulfilled in Part II of this volume.

The technologies selected for implementation at NAS Alameda were a permeable barrier containing granular iron for reductive dechlorination of solvents followed by biosparging for biodegradation of petroleum hydrocarbons. The granular iron proved reliable in Gate 1 at Borden, while the biosparge system in Gate 3 was initially judged to outperform the ORC™ cassette system in Gate 1, mainly due to the apparent inhibition of oxygen release from ORC™ in the presence of high pH water generated in the granular iron zone. These technologies were installed in a Funnel-and-Gate™ system, within an approximately 3.3 m wide by 4.6 m long treatment gate (Figure 1.5). Waterloo Barrier™ sealable sheet piling funnels were installed to a depth of about 10 m, 3.5 m below the top of the aquitard. The gate was excavated down to the aquitard at about 6.5 m below ground surface. On the basis of a treatability study conducted by Envirometal Technologies, Inc. (ETI), approximately 1.5 m of granular iron flow-through thickness was judged to be sufficient to treat the solvents (mainly *c*DCE, VC, and lesser concentrations of TCE) in the bulk of the contaminated groundwater, as characterized at the outset of the study. It was recognized that small volumes of more highly concentrated groundwater might not be treated to the remedial objective concentrations. The biosparge technology, designed to promote complete aerobic biodegradation of hydrocarbons, mainly TOL and benzene at the Alameda site, may also support aerobic biodegradation of the *c*DCE and VC sufficient to meet the remedial objectives.

It was proposed to control the groundwater flux through the treatment gate, at least for the initial period, in order to assess the treatment performance at a known flux. The downgradient end of the gate remained sealed by sheet piling and the groundwater flux was controlled by pumping at known rates from two wells located in the downgradient end of the gate. In this way, groundwater was drawn through the gate under controlled rates and the gate performance was assessed at these rates. While it was proposed to subsequently remove the downgradient barrier and operate the gate under natural flux conditions, this action was delayed until 1998, in a project to follow the AATDF project. Consequently, the results of the gate treatment are reported here under two rates of controlled groundwater flow.

1.3 DESCRIPTION OF THE STUDY SITES

1.3.1 Selection of the Site Location: CFB Borden Experiment

The experiment was conducted in a shallow sand aquifer located on the CFB Borden test site, in Ontario, Canada. This location was selected, in part, because the Borden aquifer is extremely well characterized and generally regarded as hydrogeologically simple. Thus, complexities in groundwater flow patterns were unlikely to obscure the performances of the treatment gates. Details of the site geology, hydrogeology, and aquifer characteristics can be found in the published literature (Egboka et al., 1983; Mackay et al., 1986; Ball et al., 1990; Sudicky, 1986).

Previous experiments at CFB Borden have been conducted in several locations across the base. The best known of these is the Stanford–Waterloo natural gradient experiment conducted in the early 1980s, in the abandoned sand quarry (Figure 1.6). The initial NIW experiment by Devlin and Barker (1996) (Section 1.4.2 of this chapter) was completed immediately next to the site of the Stanford–Waterloo experiment. The aquifer beneath and immediately downgradient of the sand quarry is on the order of 10 m thick, with flow predominantly in a westerly to northwesterly direction at a velocity of about 10 cm/day, and a water table within a meter of the surface. The sand quarry section of aquifer was considered as a location for the current experiment, but was eventually ruled out on the basis of its depth.

Experiments aimed at elucidating the processes of DNAPL source zone dissolution and multiphase transport in aquifers were initiated in the early 1990s. The emplaced source experiment (Rivett et al., 1995) was performed to the west and downgradient of the sand quarry, while a series of PCE release experiments were performed to the northeast of the sand quarry, in the sheet-pile area (Brewster et al., 1995) (Figure 1.6). The sheet-pile area is characterized by sandy material visually similar to that observed in the sand quarry. Groundwater flow is generally to the west at about 10 cm/day and the water table is shallow, although a nearby ephemeral stream exerts a seasonal influence on the shallow flow regime. The depth of the aquifer is only about 4 m in this area, making it possible, at a reasonable cost, to install the treatment gates keyed into the underlying clay aquitard. The proximity of the PCE release experiments to the current project location was of some concern. However, the prevailing ambient flow direction (to the west) was such that there was only a minimal threat of contamination reaching the AATDF project site from the PCE release site. This, coupled with activities at the PCE release site aimed at containing contaminant mass, was judged sufficient to limit the risk of interferences between the two experiments.

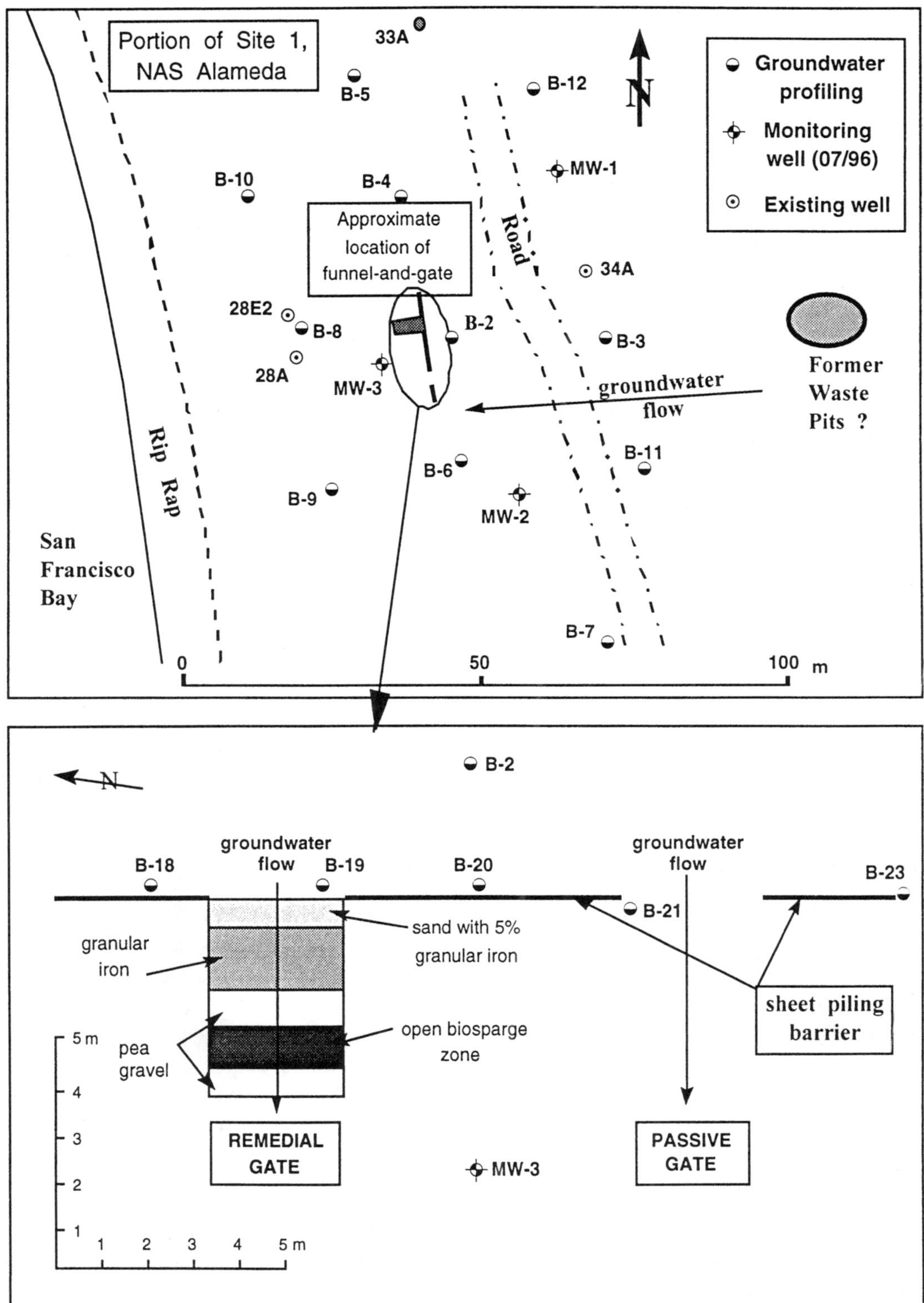

Figure 1.5 Experimental design for the Alameda demonstration.

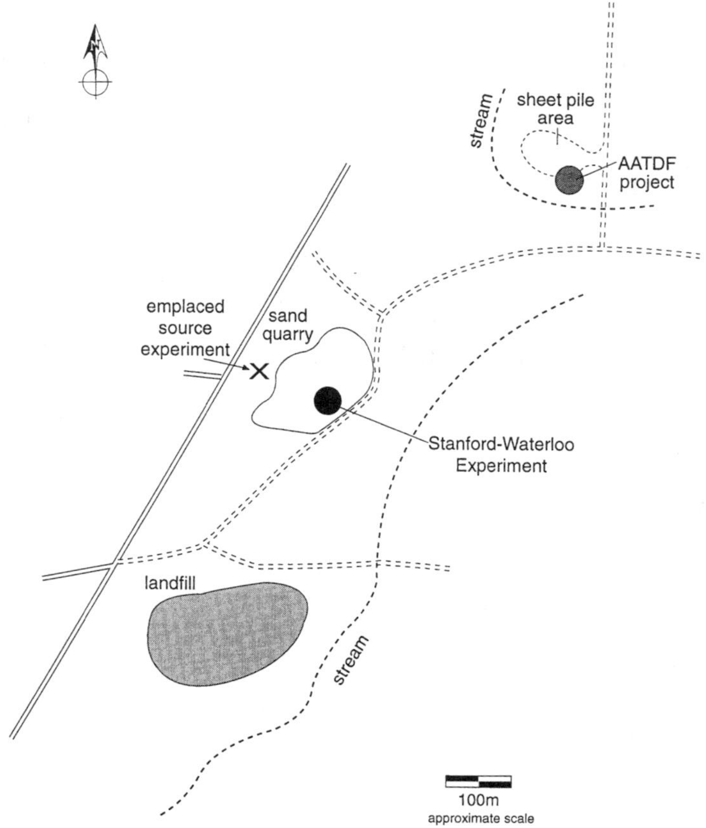

Figure 1.6 Location of the AATDF experimental site, CFB Borden in relation to other experiments conducted at CFB Borden.

1.3.2 Background Geochemistry: CFB Borden

Background samples collected from the site were tested for inorganic and organic constituents, and the data are presented in Appendix 6. The inorganic chemistry of the site groundwater was dominated by calcium (50 to 160 mg/L), sulfate (3 to 20 mg/L), and bicarbonate (150 to 770 mg/L). Although the bulk groundwater was found to be aerobic, containing low levels of dissolved oxygen (DO) (<2 mg/L), there were indications that portions of the aquifer were mildly reducing in character. For example, dissolved iron concentrations ranged from nondetectable (<0.05 mg/L) to about 0.5 mg/L and methane concentrations ranged from nondetectable to 71 μg/L. The highest concentrations of iron and methane corresponded to the shallowest sampling point on monitor G102L (Appendix 6), indicating that the anaerobic zones existed throughout the vertical thickness of the aquifer. Both nitrate and phosphate were absent (at detectable levels) in the samples tested, suggesting the possibility that inorganic nutrients could be limiting for microbial activity in the aquifer.

Background chloride concentrations ranged from 1 to 50 mg/L, with a median value of about 3 mg/L. Of 24 samples tested, only three exhibited chloride concentrations in excess of 20 mg/L, and all were located in Gate 3. Bromide was not detected in the aquifer. These generally low halogen ion concentrations in the background water made it possible to use chloride and bromide as tracers to investigate the groundwater flow within the gates.

1.3.3 Selection of the Site Location: Alameda Experiment

Sites at five DOD sites were assessed by AATDF and University of Waterloo staff, with site visits, data reviews, and additional data gathering at three sites. The suitability of the sites for demonstration of this technology was evaluated in terms of 12 criteria, considering site hydrogeology, contaminant concentrations, site access, and facility support. The activities and results are presented in a summary report (University of Waterloo,1997b). Site 1 at Naval Air Station Alameda, in Alameda, CA was selected for this demonstration.

As part of the site selection process, subsurface assessment was conducted at Site 1 in July 1996. Additional assessment was conducted in October 1996, after the site was selected to help locate the pilot-scale, Funnel-and-Gate system to be installed for the gate treatment demonstration (University of Waterloo, 1996).

At Site 1, NAS Alameda, an apparently narrow plume of solvents (*c*DCE, VC, minor TCE) and petroleum hydrocarbons (toluene and benzene especially) was selected for the demonstration. The plume occurs in a sandy-silt, unconfined aquifer, having a water table about 1.5 m below ground surface, underlain at about 6.3 m by a clay-rich aquitard. The groundwater flows laterally at 1–3 cm/day toward San Francisco Bay. A minor tidal influence has been noted at the demonstration site, about 30 m from the Bay.

1.3.4 Background Geochemistry: Alameda

At the demonstration site the groundwater is generally fresh, Na (Ca, Mg)–Cl or Na (Ca, Mg)–HCO_3 type, but somewhat brackish at depth (Cl up to 1200 mg/L), likely reflecting the influence of the adjacent Bay. Iron concentrations, locally up to 10 mg/L, and methane occurrence reflect the reducing character of the organic contaminant plume. A plume of organic contaminants is apparently migrating laterally from areas of waste disposal toward the Bay. In the area of the demonstration trial, groundwater contains up to 220 mg/L *c*DCE, 26 mg/L VC, <1 mg/L concentrations of other solvents such as TCE, 1,1-DCE, and trans-1,2dichloroethene (*t*DCE), and up to 8 mg/L toluene, 0.3 mg/L benzene, and <1 mg/L concentrations of ethylbenzene and xylenes.

1.4 TECHNOLOGY REVIEW

The passive and semipassive technologies included in this project were selected on the basis of their performances during development. Work continues, independent of this project, on some of the technologies. Following is an overview of the state of the art of these technologies.

1.4.1 Granular Iron for Reductive Dechlorination

The use of granular iron for removal of chlorinated compounds from water began to attract considerable attention following the suggestion by Gillham and O'Hannesin, in 1992, that granular iron could be used in establishing *in situ* reactive barriers for the interception and treatment of contaminant plumes, as well as in aboveground treatment systems. Preliminary field and laboratory studies of the performance of such systems were extremely encouraging but there was no fundamental understanding of the reaction(s) on which the systems were based. The mechanisms of the reduction reactions have yet to be completely elucidated, although progress is rapidly being made. There is compelling evidence to suggest that contact between the organic solute and the solid surface is required (Weber, 1996), and that sorption of chlorinated solvents (TCE and PCE) to iron does in fact occur (Burris et al., 1995). The reductions themselves may occur, at least partly, by two electron transfers (Totten and Roberts, 1995) resulting in the oxidative dissolution (corrosion) of the iron (Matheson and Tratnyek, 1994) and, in the case of the chlorinated ethenes, the possible bypassing of undesirable products such as VC (Roberts et al., 1996). Reaction rates have been found to be dependent on both iron surface area as well as solute concentration (Gillham and O'Hannesin, 1994; Johnson et al., 1996), although pseudo-first-order kinetics are usually observed. It has been suggested that the reaction kinetics are diffusion-controlled (Matheson and Tratnyek, 1994; Agrawal and Tratnyek, 1996), on the basis of batch tests involving acid-washed granular iron. However, recent electrochemical evidence suggests that, in the case of CT, transport times through the diffusion layer surrounding an oxide-free metal surface are minimal (Scherer et al., 1997). It has also been observed that hydrogen is evolved shortly after water (at neutral or slightly acid pH) contacts the Fe^0 (Reardon, 1995). Hydrogenolysis of PCE on palladium has been demonstrated in the laboratory and shown to be faster than PCE reductive dechlorination in the presence of iron (Schreier and Reinhard, 1995). Thus, with hydrogen present, the possibility that organic reductions occur by hydrogenolysis cannot be easily ruled out.

Several field trials of granular iron in permeable barriers have been undertaken recently (Focht et al., 1996). One, in Sunnyvale, CA, was emplaced across a plume of TCE and its dechlorinated products (DCE, VC). The reactive wall was about 1.3 m wide and received organic concentrations on the order of 200 to 500 μg/L. Over the first 2 years of operation, the effluent emerging from the granular iron wall was within maximum contaminant levels (MCLs in the low μg/L range). Phospholipid fatty acid (PLFA) testing showed no evidence of the development of microbial populations other than those observed in the background water. Similar successes were reported by Focht et al. (1996) for trials at sites in New York and Kansas, both incorporating Funnel-and-Gate structures constructed from sheet piling and slurry walls, respectively.

At the U.S. Coast Guard Support Center, near Elizabeth City, NC, a 46-m long granular iron wall was installed to cut off a plume of TCE and chromium at concentrations of 5600 and 6000 μg/L, respectively. The wall was 0.6 m wide and 7.3 m deep and took only 6 hours to install. Preliminary results indicate the wall reduces concentrations of TCE from influent levels to 5.3 μg/L and chromium to below detection limits (<10 μg/L) (Blowes et al., 1997).

1.4.2 Nutrient Injection Wall (NIW) for Anaerobic Biodegradation

The NIW operates on the principle that groundwater can be amended with solutes, designed to stimulate microbial activity, by flushing a permeable barrier at regular intervals with a spiked solution. The resulting pulses are permitted to migrate downgradient, into the aquifer, and merge as a result of longitudinal dispersion. Where the pulses merge, *in situ* microorganisms constantly receive a minimum supply of the required nutrients and are able to increase in numbers, thereby establishing a permanent bioactive zone where biodegradation can occur (Devlin, 1994). A field prototype of the system was tested by Devlin (1994) and reported by Devlin and Barker (1996). They found that sequential acetate injections into the Borden aquifer, at approximately 1 month spacing, were sufficient to establish and maintain a population of sulfate-reducing bacteria within 5 m of the injection wall. Pulses of TCE and CT were injected at concentrations of about 1 mg/L and monitored as they migrated through the geochemically modified aquifer. TCE underwent no detectable transformations, but CT was transformed to CF, DCM, and carbon disulfide (CS_2). There was also indirect evidence of CF and DCM transformations within 10 m of the injection wall. Concurrent laboratory studies indicated that some of these transformations were biologically driven, providing further evidence of the successful operation of the treatment system. The limitations were primarily microbiological in nature; within the time frame of the experiment, no adapted population of TCE degraders developed in the aquifer. It was unclear whether the development of such a population could have been accelerated with the addition of a different carbon substrate (other than acetate). Benzoate (Beeman et al., 1993) and methanol (Major et al., 1991) have been reported as substrates in field tests where reductive dechlorination of PCE was observed. Benzoate was selected for this project because it was considered less hazardous to handle than methanol. It was also unclear whether a more highly oxidized compound such as PCE might have shown signs of degrading sooner than TCE. Microcosm tests conducted concurrently with the field experiment demonstrated PCE degradation in Borden aquifer material, but only after a 3-year incubation time (Devlin, 1994). On this basis, it was decided that PCE should be the compound to represent the chlorinated ethenes in the current project.

1.4.3 ORC™ for Delivering Oxygen to Aquifers

The first field testing of ORC™ was reported by Bianchi-Mosquera et al. (1994) in a test conducted at CFB Borden. They mixed ORC™ powder with cement, and then broke the cement into pieces roughly 1 in. in diameter. The pieces were dropped into large-diameter wells and the groundwater up- and downgradient of these wells was monitored for DO. With groundwater velocities of 25 to 38 cm/day imposed between two short lengths of sheet piling (a crude gate design), clear increases in DO concentrations were observed in the downgradient wells (up to 14 mg/L from 2 mg/L). In addition, the DO-charged aquifer was shown to be capable of biodegrading benzene and toluene.

The use of ORC™ in cement as an oxygen source was also employed at a site near Leland, NC, where gasoline had leaked from an underground storage tank (Borden et al., 1997). A barrier comprising a series of 15-cm-diameter wells was installed across the plume and filled with ORC™-cement briquettes. Later on, briquettes containing sodium nitrate were also added to the barrier wells. Groundwater velocities across the treatment area were on the order of 30 cm/day. Over the first 242 days of the experiment, total BTEX concentrations were reduced from 17 to 3.4 mg/L, with a coinciding rise in DO concentrations from 0.4 to 1.8 mg/L. However, over this period, BTEX treatment efficiencies declined, probably due to declining rates of oxygen release from the ORC™. At the same time, the barrier wells became clogged with iron precipitates, also reducing the oxygen flux into the aquifer.

In another field trial of ORC™, Chapman et al. (1997) installed a fence of 20-cm-diameter wells across a BTEX plume emanating from a gas station in southern Ontario. They introduced the ORC™ in packets held together by rope, hereafter referred to as "socks." Groundwater velocities at the site were found to vary in direction and magnitude, with an average value of 17 cm/day. BTEX concentrations were also found to be variable, with maximum total concentrations on the order of 80 mg/L. Three dense monitoring fences were installed, consisting of multilevel wells with a grid spacing of 15 cm by 30 cm. All monitoring was done within 1.5 m of the water table. There was an apparent 70% contaminant mass reduction between the upgradient and downgradient fences (a distance of 2 to 2.5 m in which the treatment wells were located) over the first 51 days of the test. This efficiency declined over time due to increasing BTEX concentrations entering the system and reductions in the ORC™ oxygen delivery rate. It was found that complete BTEX removal was accomplished when input concentrations were at or below about 5 mg/L. Petroleum compounds other than BTEX in the plume were considered to exert a substantial oxygen demand and limited the system's ability to degrade the target (BTEX) compounds.

1.4.4 Biosparging for Oxygen Delivery to an Aquifer

The concept of delivering oxygen to groundwater by creating a curtain of residual bubbles in a Funnel-and-Gate structure was introduced by Barker et al. (1993). A pilot study was undertaken at a petroleum refinery in western Canada (Gorman, 1995; Austrins, 1997). Sheet piling was installed on three sides of the source area, which was contaminated with petroleum hydrocarbons; the upgradient side was left open. An opening, or gate, was also left in a portion of the downgradient wall. Using a large caisson, this opening was filled with pea gravel and instrumented with three horizontal sparging pipes each at a different depth. The system functioned by sparging the gate on a semiweekly basis to ensure a constant supply of oxygen bubbles. The concentrations of BTEX compounds in the source area were several tens to hundreds of milligrams/liter. However, the flux through the gate under natural gradient conditions was quite low, due to an extremely low hydraulic gradient across the site. For the purposes of the experiment, a pumping well was installed immediately downgradient of the gate to impose an artificially high gradient across the system (and flow through the gate). Preliminary results of the system were encouraging; oxygen concentrations in the gate increased by several milligrams per liter, and BTEX concentrations decreased. However, during tests to determine the optimum sparging pressure to apply, too high a pressure was applied, changing the structure of the gravel backfill. Preferred channels for gas migration developed and the system was no longer able to produce a uniform curtain of oxygen bubbles.

Subsequent testing on the above biosparge gate revealed a highly complex flow system within the gate. The act of pulsing gas through the system caused unpredictable movement of water throughout the gate, distributing the dissolved oxygen heterogeneously. Work is ongoing to better understand the mixing limitations within the gate.

At another site in Alberta, *in situ* treatment of BTEX-contaminated groundwater was conducted in open, 3-m-diameter caissons by the continuous addition of air bubbles (Granger, 1997). BTEX concentrations of a few tens of micrograms per liter were successfully treated in this way (Bowles et al., 1997). Higher influent BTEX levels (a few milligrams per liter of total BTEX) were introduced and these levels were also successfully treated to concentrations <1 μg/L by sparging with pure oxygen and by providing phosphorus in the form of solid apatite and solid sodium phosphate. Research continues into the correct distinction and apportioning of BTEX removal by biodegradation and volatilization, but initial conclusions are that >80% of the removal was from biodegradation (Granger, 1997).

1.5 ORGANIZATION OF THIS BOOK

The CFB Borden experiment consisted of the evaluation of two sequential remedial systems, employing a total of five technologies in three separate gates. Graduate students were assigned responsibility for Gates 1 and 3, the granular iron–ORC™ system and the NIW–biosparge system, respectively. Undergraduates monitored the center gate on work terms or as part of fourth-year theses. Each of the gates is discussed in a single chapter. Details concerning methodologies, supporting laboratory work, QA/QC analysis, and other common elements are provided in the appendices. Thus, the chapters are appropriately brief, focusing only on the designs and results of the experiments.

Chapter 1 is the introduction, presenting the rationale for the research, an overview of the experimental design, goals, and a review of the technologies used in the experiment. Chapter 2 is concerned with the performance of the granular iron and ORC™ in Gate 1, and Chapter 3 presents an assessment of intrinsic remediation at the site, as observed in Gate 2. Chapter 4 summarizes the results of the NIW and biosparge gate. Comparisons are made, in particular between Gates 2 and 3, to assess the benefits of passive or semipassive remediation.

Chapter 5 summarizes the pilot-scale *in situ*, sequential groundwater treatment system, operated at NAS Alameda to treat an existing, mixed (chlorinated solvents and petroleum hydrocarbons) groundwater. The system was installed in December 1996 and continues to be operated. Finally, in Chapter 6, an overview of the project is presented along with overall conclusions.

Reports concerning the workplan (University of Waterloo, 1996) and installation (University of Waterloo, 1997a and 1997c) were previously submitted to AATDF and are archived there. However, details concerning methodologies, data reduction, site assessments, supporting design and laboratory work, QA/QC analyzes, etc. are provided in appendices to this book.

REFERENCES

Agrawal, A., Tratnyek, P.G. 1996. Reduction of nitro- aromatic compounds in Fe^0-CO_2-H_2O systems: implications for groundwater remediation with iron metal. *Environmental Science and Technology*, 30(1): 153-160.

Austrins, C. 1997. Enhanced *In Situ* Bioremediation of Hydrocarbon Contaminated Groundwater Using Oxygen and Ammonia Gas Injection in a Funnel-and-Gate Scheme. M.Sc., University of Waterloo.

Ball, W.P., Buehler, C.H., Harmon, T.C., Mackay, D.M., Roberts, P.V. 1990. Characterization of a sandy aquifer material at the grain scale. *Journal of Contaminant Hydrology*, 5: 253-295.

Barker, J.F., Weber, J., Fyfe, J.S., Devlin, J.F., Mackay, D.M., Cherry, J.A. 1993. Efficient addition technology for *in situ* bioremediation. *Proceedings of the Third Annual GASReP Symposium on Groundwater and Soil Remediation*, Quebec City, September.

Beeman, R.E., Howell, J.E., Shoemaker, S.H., Salazar, E.A., Buttram, J.R. 1993. A field evaluation of *in situ* microbial reductive dehalogenation by the biotransformation of chlorinated ethenes. *Proceedings of the In-Situ and On Site Bioremediation Conference*, Battelle, San Diego.

Bianchi-Mosquera, G.C., Allen-King, R.M., Mackay, D.M. 1994. Enhanced degradation of dissolved benzene and toluene using a solid oxygen-releasing compound. *Ground Water Monitoring and Remediation*, 14(1): 120-128.

Blowes, D.W., Puls, R.W., Bennett, T.A., Gillham, R.W., Hanton-Fong, C.J., Ptacek, C.J. 1997. In-situ porous reactive wall for treatment of Cr(VI) and trichloroethylene in groundwater. Presented at the 1997 International Containment Technology Conference and Exhibition, St. Petersburg, FL, February 9-12.

Borden, R.C., Goin, R.T., Kao, C.M. 1997. Control of BTEX migration using a biologically enhanced permeable barrier. *Ground Water Monitoring and Remediation*, 17(1): 70-80.

Bowles, M., Bentley, L., Barker, J., Thomas, D., Granger, D., Jacobs, H. 1997. The East Garrington trench and gate program: it works. *Proceedings of the 6th Annual Symposium on Groundwater and Soil Remediation*. Montreal, March 18-21.

Brewster, M.L., Annan, A.P., Greenhouse, J.P., Kueper, B.H., Olhoeft, G.R., Redman, J.D., Sander, K.A. 1995. Observed migration of a controlled DNAPL release by geophysical methods. *Ground Water*, 33(6): 977-987.

Burris, D.R., Campbell, T.J., Manoranjan, V.S. 1995. Sorption of trichloroethylene and tetrachloroethylene in a batch reactive metallic iron-water system. *Environmental Science and Technology*, 29(11): 2850-2855.

Chapman, S.W., Byerley, B.T., Smyth, D.J.A., Mackay, D.M. 1997. A pilot test of passive oxygen release for enhancement of in situ bioremediation of BTEX-contaminated ground water. *Ground Water Monitoring and Remediation*, 17(2): 93-105.

Devlin, J.F., Barker, J.F. 1994. A semipassive nutrient injection scheme for enhanced *in situ* bioremediation. *Ground Water*, 32(3): 374-380.

Devlin, J.F. 1994. Enhanced *In Situ* Biodegradation of Carbon Tetrachloride and Trichloroethene Using a Permeable Wall Injection System. Ph.D. thesis, Department of Earth Sciences, University of Waterloo, Waterloo, Ontario, Canada.

Devlin, J.F., Barker, J.F. 1996. Field investigation of nutrient pulse mixing in an *in situ* biostimulation experiment. *Water Resources Research*, 32(9): 2869-2877.

Egboka, B.C.E., Cherry, J.A., Farvolden, R.N., Frind, E.O. 1983. Migration of contaminants in groundwater at a landfill: a case study. *Journal of Hydrology*, 63: 51-80.

ETI and BEAK, 1998. Technology Evaluation Report for *In Situ* Sequenced Permeable Treatment Systems for Groundwater Remediation. DRAFT Report March 1998. Submitted to AATDF March 1998.

Focht, R., Vogan, J., O'Hannesin, S. 1996. Field application of reactive iron walls for *in situ* degradation of volatile organic compounds in groundwater. *Remediation*, Summer, 1996.

Gillham, R.W., O'Hannesin, S.F. 1992. Metal-catalysed abiotic degradation of halogenated organic compounds. *Proceedings of the IAH Conference, Modern Trends in Hydrogeology*. Hamilton, Ontario, May 10-13, 94-103.

Gorman, J.K. 1995. Bioremediation of Hydrocarbon Contaminated Soil and Groundwater Using an *In Situ* Funnel-and-Gate Strategy. M.Sc. Project, University of Waterloo.

Granger, D.M. 1997. Nutrient amendment to Enhance Aerobic Biodegradation within a Funnel-and-Gate System. M.Sc. Project, University of Waterloo.

Johnson, T.L., Scherer, M.M., Tratnyek, P.G. 1996. Kinetics of halogenated organic compound degradation by iron metal. *Environmental Science and Technology*, 30(8): 2634-2640.

Kriegman-King, M.R., Reinhard, M. 1994. Transformation of carbon tetrachloride by pyrite in aqueous solution. *Environmental Science and Technology*, 28(4): 692-700.

Mackay, D.M., Freyberg, D.L., Roberts, P.V. 1986. A natural gradient experiment on solute transport in a sand aquifer. I. Approach and overview of plume movement. *Water Resources Research*, 22(13): 2017-2029.

Major, D.W., Hodgins, E.W., Butler, B.J. 1991. Field and laboratory evidence of in-situ biotransformation of tetrachloroethene to ethene and ethane at a chemical transfer facility in North Toronto. In *On Site Bioreclamation*, R.E. Hinchee and R.F. Olfenbuttel, Eds., Butterworth-Heinemann, Stoneham, MA, 147-171.

Matheson, L.J., Tratnyek, P.G. 1994. Reductive dehalogenation of chlorinated methanes by iron metal. *Environmental Science and Technology*, 28(12): 2045-2053.

Roberts, A.L., Totten, L.A., Arnold, W.A., Burris, D.R., Campbell, T.J. 1996. Reductive elimination of chlorinated ethylenes by zero-valent metals. *Environmental Science and Technology*, 30(8): 2654-2659.

Reardon, E.J. 1995. Anaerobic corrosion of granular iron: measurement and interpretation of hydrogen evolution rates. *Environmental Science and Technology*, 29(12): 2936-2945.

Rivett, M.O. 1995. Soil-gas signatures from volatile chlorinated solvents: Borden field experiment. *Ground Water*, 33(1): 84-98.

Scherer, M.M., Westall, J.C., Tratnyek, P.G. 1997. Kinetics of carbon tetrachloride reduction at an iron rotating disk electrode. *Proceedings of the 213th ACS National Meeting; Division of Environmental Chemistry.* San Francisco, CA, 37(1): 85-86.

Schreier, C.M., Reinhard, M. 1995. Catalytic dehydrohalogenation and hydrogenation using H_2 and supported palladium as a method for the removal of tetrachloroethylene from water. *Proceedings of the 209th ACS National Meeting*; Division of Environmental Chemistry. Anaheim, CA, April 2-7, 35(1): 749-751.

Sudicky, E.A. 1986. A natural gradient experiment on solute transport in a sand aquifer: spatial variability of hydraulic conductivity and its role in the dispersion process. *Water Resources Research*, 22(13): 2069-2082.

Totten, L.A., Roberts, A.L. 1995. Investigating electron transfer pathways during reductive dehalogenation reactions promoted by zero-valent metals. *Proceedings of the 209th ACS National Meeting*; Division of Environmental Chemistry. Anaheim, CA, April 2-7, 35(1): 706-709.

University of Waterloo, 1996. Revised Workplan for Semipassive Groundwater Remediation Demonstration Project at Site 1, Alameda Naval Air Station, California. Submitted to AATDF, December 1996.

University of Waterloo, 1997a. Report on the Installation of the CFB Borden Experiments for Passive and Semipassive Techniques for Groundwater Remediation. Submitted to AATDF, April 1997.

University of Waterloo, 1997b. Summary of Potential DOD Sites for Demonstration of the Passive and Semipassive, In-Situ Groundwater Remedial System. Submitted to AATDF, May 21, 1997.

University of Waterloo, 1997c. Report on the Installation of the Pilot-Scale, *In Situ* Groundwater Treatment System at NAS Alameda, Alameda, CA. Submitted to AATDF, November 1997.

Vogel, T.M., Criddle, C.S., McCarty, P.L. 1987. Transformations of halogenated aliphatic compounds. *Environmental Science and Technology*, 21(8): 722-736.

Weber, E.J. 1996. Iron-mediated reductive transformations: investigation of reaction mechanism. *Environmental Science and Technology*, 30(2): 716-719.

Chapter 2

Sequential *In Situ* Treatment Using Granular Iron and ORC™ — Gate 1 CFB Borden Experiment

2.1 INTRODUCTION

Remedial technologies selected for use in Gate 1 at the Borden site have received considerable scientific and public attention in the past few years. Granular iron and ORC™ have both shown considerable promise for use as practical field alternatives for treatment of chlorinated solvents and petroleum hydrocarbons, respectively. The application studied here involved the use of these technologies to sequentially treat a multicontaminant plume. Interactions, particularly those limiting the effectiveness of either technology, were of particular interest in this study.

A controlled field experiment, using a Funnel-and-Gate arrangement, was conducted in the shallow aquifer at CFB Borden in Ontario, Canada at the University of Waterloo (UW) experimental site. The site was originally uncontaminated, so a dissolved-phase groundwater plume consisting of PCE, CT, and toluene was created with solute concentrations of approximately 1 to 10 mg/L. Routine monitoring to assess the performance of the system was conducted from November 1996 until October 1997. Groundwater flow conditions were investigated prior to the injection of contaminants through the implementation of a conservative tracer test.

Additional investigations were completed to address discovered or suspected limitations associated with the use of these very different technologies (iron creates a highly reducing environment, while ORC™ provides oxygen to the system at concentrations approaching saturation) in close proximity. Laboratory and small field experiments were initiated to assess an apparent pH incompatibility between the ORC™ and the alkaline water emanating from the iron. Also, the formation of inorganic precipitates on the surfaces of the reactive materials, or in pore space between reactive grains, was a concern due to the high pH of the system and the complex geochemical environments created. The surface characteristics of the granular iron after exposure to natural Borden groundwater were investigated, and any modifications that occurred over the project duration were documented.

2.1.1 Technology Process Description

2.1.1.1 Granular Iron

Remediation of groundwater by granular iron has received considerable interest over the past few years. Generally, the process is accepted to be abiotic reductive dechlorination.

Under anaerobic conditions, iron is oxidized by water according to

$$Fe^0 + 2H_20 \rightarrow Fe^{2+} + H_2 + 2OH^- \tag{2.1}$$

Based on this reaction, the release of hydroxide ions causes an increase in pH and hydrogen gas is produced. Increases in dissolved iron concentrations are also expected (Gillham, 1995). In the presence of chlorinated organics (RX), the reaction involves the simultaneous oxidation of the iron metal and the reduction of the chlorinated compound:

$$Fe^0 + RX + H^+ \rightarrow Fe^{2+} + RH + X^- \tag{2.2}$$

This reaction can only proceed given a source of protons, such as water (Gillham, 1995), and will produce dissolved iron and an anion (chloride). Although the mechanism of the reaction is uncertain, it is believed to rely on direct contact of the metal and the organic contaminant (Matheson and Tratnyek, 1994; Weber, 1996).

The reaction is believed to be pseudo-first order with respect to the organic contaminant, over a broad concentration range (Orth and Gillham, 1996). Degradation rates were shown by Gillham and O'Hannesin (1994) to be directly related to the iron surface area-to-volume ratio and inversely proportional to the degree of chlorination (more highly oxidized compounds degrade at a faster rate, giving a shorter half-life).

Degradation of toluene was not expected to be enhanced by the presence of zero-valent metal since it, like other nonchlorinated aromatic compounds such as benzene and xylene, is a reduced compound (Gillham, 1995). Transport of toluene through the system might, however, be slowed due to sorption to iron surfaces.

Demonstrations of the granular iron technology in the laboratory and the field have led to its acceptance as a useful remedial alternative for chlorinated organics. Laboratory tests, including batch tests and column studies (Gillham and O'Hannesin, 1994; Agrawal and Tratnyek, 1994; and others), have shown rapid and complete reductive dechlorination of dissolved solvents including PCE and TCE in the presence of iron. *In situ* field demonstrations by O'Hannesin (1993), and Focht et al. (1996) have provided considerable evidence of the low-cost and low-maintenance benefits of the technology, as well as high dechlorination rates.

Uncertainty about the long-term performance of granular iron comes from concern over precipitate formation that could be favored in the geochemical environment created by the iron corrosion reaction (pH > 9 and increased dissolved iron concentrations). The effective lifetime of an iron barrier may be reduced due to the formation of precipitates on the iron surfaces themselves, causing reduced surface reactivity, or within pore spaces, clogging the flow system and decreasing system hydraulic conductivity (Schuhmacher, 1995).

2.1.1.2 Enhanced Aerobic Bioremediation

Intrinsic and enhanced bioremediation of petroleum-hydrocarbon-contaminated soil and groundwater are quickly gaining acceptance as treatment alternatives (NRC, 1993). While intrinsic bioremediation relies on the ability of naturally occurring microbial populations to degrade contaminants without intervention, various enhancements can provide improved degradation rates and more complete remediation. Petroleum hydrocarbons can be degraded under both aerobic and anaerobic conditions, although an aerobic environment typically provides quicker results (Borden et al., 1995).

Hydrocarbon biodegradation can be generalized by the reaction (CPPI, 1995)

$$\begin{gathered}\text{hydrocarbon + electron acceptors + microorganisms + nutrients} \rightarrow \\ \text{carbon dioxide + water + microorganisms + waste products}\end{gathered} \tag{2.3}$$

Bioremediation is often limited by a lack of electron acceptors (oxygen in the case of aerobic degradation), particularly in the core of a plume. The oxygen demand imposed by the aerobic degradation of the contaminant exceeds the oxygen supply, and thus an anaerobic environment is created. Only at the periphery of the plume can oxygen and contaminants mix sufficiently to maintain rapid degradation rates.

Various methods of enhancing bioremediation through the addition of oxygen or air to the subsurface environments have been used. A common approach is to pump the groundwater to the surface, add nutrients and oxygen, possibly as liquid peroxide, and then reinject using an infiltration gallery or wells. Recently, other alternatives have been used or proposed, including the use of solid metal peroxides (Borden et al., 1997), diffusive solute emitters (Wilson and Mackay, 1995), and biosparging (Austrins, 1997). An important limitation for the use of any technique is the ability to provide the oxygen within the core of the plume, not to the periphery. Through the use of reactive barriers that intersect the plume, or use in conjunction with a Funnel-and-Gate system, an adequate oxygen supply can be provided where it is most required. For petroleum hydrocarbons, a 3:1 stoichiometric ratio of oxygen to contaminant is typically required for completed aerobic mineralization (Norris et al., 1994).

Oxygen-releasing compound (ORC™, manufactured by Regenesis Bioremediation Products, San Juan Capistrano, CA) was selected for use at this site. ORC™ is a proprietary formulation of magnesium peroxide (MgO_2) in a solid form with some KH_2PO_4 or K_2HPO_4 added to control the kinetics of the oxygen-releasing reaction. The ORC™ releases oxygen upon contact with water according to the following reaction:

$$2MgO_2 + 2H_2O \rightarrow 2Mg(OH)_2 + O_2 \tag{2.4}$$

ORC™ was supplied in filter socks constructed of polyester material and enclosed in plastic mesh to limit swelling. Socks were approximately 10 cm in diameter and 30 cm in length and contained approximately 3.5 kg of a 50:50 mix of ORC™ and silica sand.

ORC™ has been implemented at both pilot- and full-scale in the field (Chapman et al., 1997; Smyth et al., 1995; Bianchi-Mosquera et al., 1994). Dissolved oxygen (DO) concentrations up to 10 mg/L were typically sustained over a period of approximately 6 months, with no maintenance to the installations required. Although not evident by the overall reaction given (Eq. 2.4), pH increases observed in previous work would imply the release of hydroxide ions as part of the reaction (likely due to the dissolution of $Mg(OH)_2$ to Mg^{2+} and $2OH^-$). Smyth et al. (1995) observed pH values greater than 10 during early stages of monitoring.

Performance of the ORC™ has been shown to be limited by the presence of elevated concentrations of metals (particularly iron) in target groundwater. Ferrous iron consumes oxygen released from ORC™ and precipitates on the surface of the socks as ferrous and ferric oxides and hydroxides. Active microbial populations accumulate as microbial slimes on the socks, and the formation of impermeable magnesium hydroxide ($Mg(OH)_2$) at the outside of the socks (Dr. William Farone, personal communication) also limits ORC™ performance. All of these factors result in the "lock-up" of reactive compounds within the sock that reduce the effective life of the ORC™.

2.1.2 Coupling of Technologies

The focus of this experiment was to evaluate the effectiveness of the selected remedial technologies for the degradation of the specific target contaminants. Use of these technologies in sequence to completely remediate a multicontaminant plume was the ultimate goal of the project. The technologies were installed in close proximity to minimize distance and consequently time for treatment to occur, but also to best investigate any limitations or interferences that might result from the coupling of the technologies. These limitations were of primary interest since they could ultimately eliminate the application of this treatment concept for full-scale sites.

At the outset of the project, potential limitations for the success of the system were thought to include conflicts between the highly reduced and highly oxidized environments. These conflicts created by the two selected technologies were as follows:

- Increased precipitate formation and reduction of hydraulic conductivity within the iron or fouling of the ORC™ surface;
- The creation of a highly alkaline environment, due to both iron and ORC™ reactions, which may be unfavorable for microbial populations and, consequently, aerobic biodegradation; and
- Problems directly related to selected field installation techniques.

During the early operation of the project, an unforeseen interaction that limited oxygen release from the ORC™, and was possibly related to the high pH of the water emanating from the iron, was discovered and was subsequently addressed.

2.2 FIELD DEMONSTRATION

2.2.1 Experimental Design

General details of the Borden experimental site including site location, hydrogeology, and geochemistry are provided in Appendix 1. A detailed description of construction activities and field installations was provided in the Phase 1 report submitted to AATDF in April 1997 (UW, 1997). A chronology of the field activities completed over the experimental phase of the project is provided in Appendix 2.

Figure 2.1 shows a plan view and cross section of Gate 1, including the locations of the source wells (SW-1 to 3), the groundwater monitoring network (consisting of multilevel piezometers and fully screened, stainless steel performance wells), the alternate oxygen addition wells (SW-12 to 14), and the extraction well (G113P). The cassette module containing four removable cassettes, which was installed and sealed within the sheet-piling walls, is also shown in the figure.

The removable cassette system (Figure 2.2) was used to implement the various treatment methods. The entire system was constructed of steel and consisted of an outer box and inner removable cassettes into which remedial materials could be placed. Conceptually, these cassettes can be used for remediation materials for this experiment (granular iron and ORC™) and reused in subsequent experiments to evaluate other combinations of materials. For a full-scale application, the ability to remove the cassettes would allow for changing or replenishing of remedial materials over the course of a project, thus allowing for reduced operating and maintenance costs.

The cassette structure was installed within the sheet-piling walls and its base was seated on the clay aquitard underlying the site. Bentonite seals were created between the sheet-piling walls at the four corners of the outer box. During installation of the outer cassette box, temporary sheet-piling walls isolated the construction area from the remainder of the gate. Figure 2.2 shows a detailed view of the cassette system and its location relative to sheet-piling walls. Items of particular interest include bentonite seals between the outer cassette box and the sheet piles, and the presence of pea gravel at both ends of the structure. The pea gravel was installed to provide a highly permeable entrance to the cassette structure to help to direct all groundwater flow through the remedial system.

During installation of the cassette system in December 1995, field observations indicated that a potential problem with the sealing of the outer cassette box within the sheet-piling walls existed. Mixing of the pea gravel placed at the ends of the cassette structure and the bentonite placed in the corners may have inadvertently occurred as the two materials were installed in their respective lifts. Although the importance of creating a good seal was recognized at the time, and every effort was made to achieve this, there remained potential for a leak. To assess this potential leak, a multilevel piezometer was installed on either side of the cassette box, between the box and the sheet-piling walls (labeled as G1CKL and G1CKR in Figure 2.1). The two piezometers were installed on June 25, 1996, and monitoring for tracer, volatile organic compounds (VOCs), and inorganics occurred regularly to document the status of any leakage that might exist. A complete discussion related to this work is provided in Appendix 14.

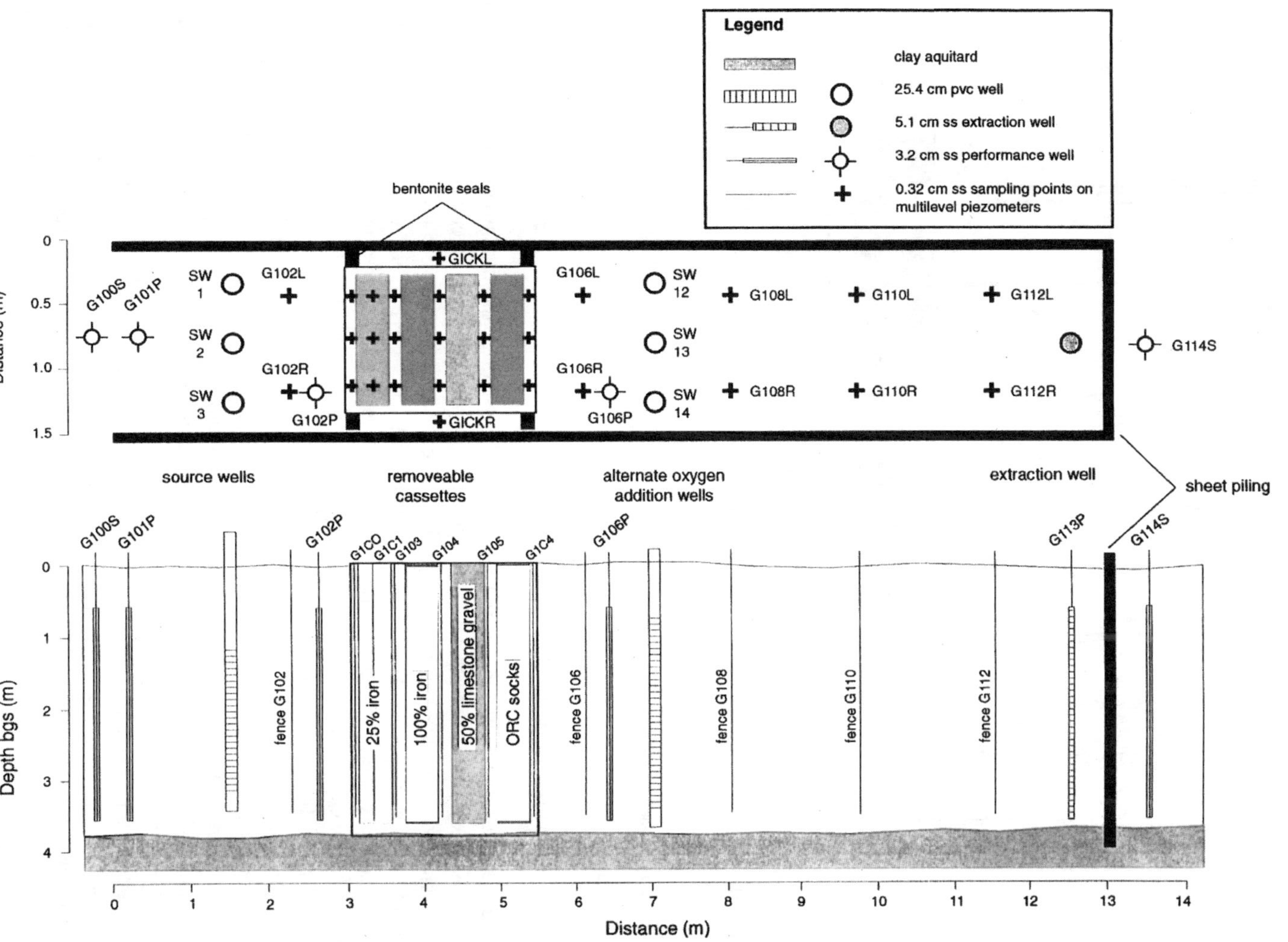

Figure 2.1 Plan view and cross section of Gate 1.

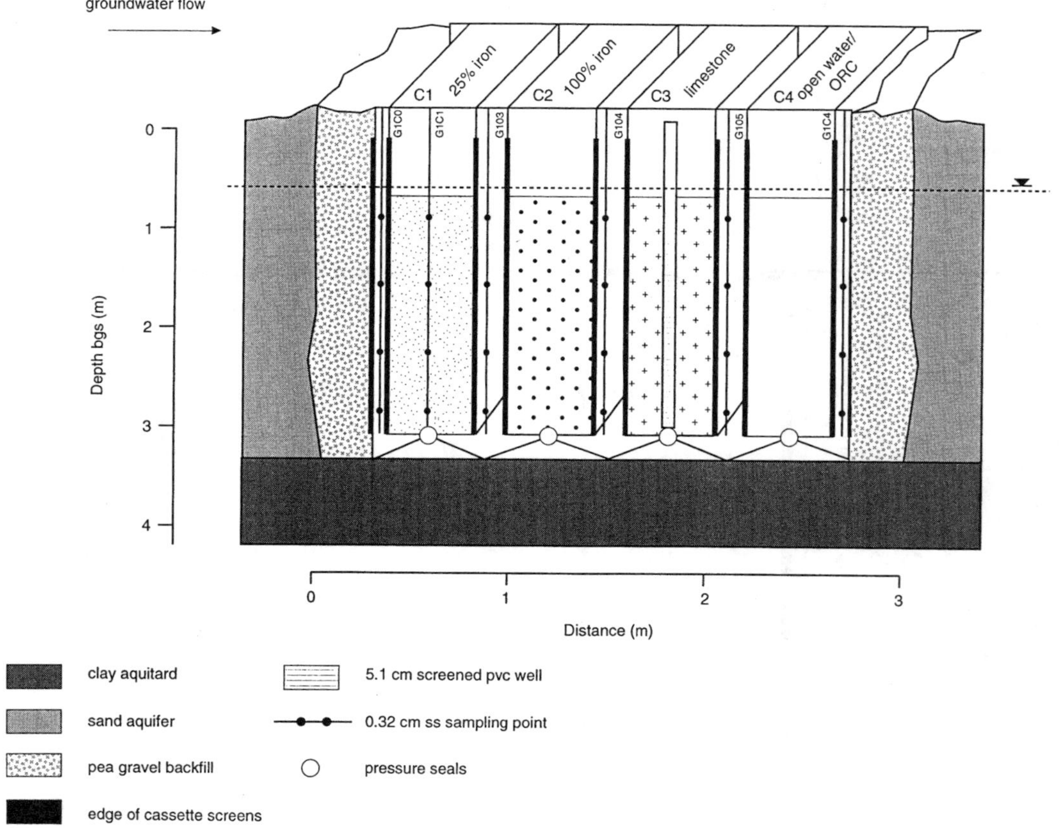

Figure 2.2 Detailed cross section through cassettes.

The cassettes themselves measured approximately 0.4 m in the direction of flow, 1 m wide (transverse to flow) and 4 m deep, and fit snugly into grooves within the outer cassette box. Pressure seals around each cassette consisted of flexible silicone tubing and were inflated using water to minimum pressures of 5 psi. These seals prevented water from bypassing remedial materials once inside the outer cassette structure. Open water zones measuring 20 cm in the direction of flow existed between each cassette. Monitoring devices were installed within these zones (fences G1C0, G103, G104, G105, and G1C4, as shown in Figure 2.2).

In the original system design (outlined in the Phase 1 Installation Report; UW, 1997), all reactive media were to be installed within the removable cassette system, providing a very compact arrangement for field implementation and close proximity of treatment zones to study possible interactions. A separation of 60 cm was achieved between the iron and ORC™ zones (40-cm flow-through cassette C3 and 20 cm of open water between cassettes).

Cassette 1 (C1) contained 25% granular iron and 75% silica sand by weight. The commercial-grade iron was obtained from Peerless Metal Powder and Abrasive (Detroit, MI), and the silica sand was typical of that used in the water well industry (No. 3 mesh). C1 was designed with a low iron content to allow observable breakthrough of parent compounds at sampling points located midway through the cassette, and to permit the identification of degradation products. Also, breakthrough was required to permit the calculation of transformation rates. Typical of the "real world," cassette 2 (C2) contained 100% granular iron. Based on the anticipated groundwater flow velocities and influent contaminant concentrations, this design was adopted to ensure complete degradation of target chlorinated compounds. Cassettes C1 and C2 were filled on April 23, 1996.

Cassette 3 (C3) contained a 50:50 mix, by volume, of silica sand and crushed limestone screenings. The limestone was washed over a 0.12-cm mesh sieve to remove fine particles, ensuring a hydraulic conductivity corresponding to that of a coarse sand. The presence of inorganic oxygen sinks, particularly reduced iron, could reduce the effectiveness of the ORC™. The limestone was provided as a solid substrate to scavenge excess dissolved iron, limiting its entry into cassette C4. Backfilling of C3 was completed on June 10, 1996.

Cassette 4 (C4) remained open for the suspension of ORC™ socks. The socks were strung together with rope and installed in C4 in three time intervals: October 8 to November 12, 1996 (56 socks); November 15, 1996 to April 15, 1997 (56 socks); and April 15 to May 6, 1997 (21 socks). The system operated in an entirely passive mode from October 8, 1996 until March 18, 1997, at which time acid addition experiments involving ORC™ were started in the field (Section 2.3.1). The various emplacements of ORC™ allowed evaluation of problems identified and permitted assessment using unexposed reactive materials. The quantity of socks used was determined based on two criteria:

1. To provide at least the minimum ORC™ required to satisfy the biochemical oxygen demand (BOD; based strictly on input concentrations of target organics) of the contaminant plume (21 socks should produce sufficient DO levels of approximately 5 mg/L within the aquifer, for a 6-month period); and
2. To provide adequate coverage of the flow cross section to ensure good contact of the water with the ORC™ (56 socks permitted the installation of 7 lines with 8 socks each, creating a wall of socks through the entire saturated depth).

Three 25-cm PVC wells, located downgradient of the cassette module (Figure 2.1), provided a backup location for supplying oxygen to the aquifer in the event that oxygen release in cassette C4 was insufficient. These wells were used, beginning on June 19, 1997, based on results discussed in Section 2.2.3.6. Ninety socks were installed — 27 each in SW-12 and SW-14, and 36 in SW-13. Three or four strings of socks (each string consisting of nine ORC™ socks) were fastened around 3-m-long sections of 5-cm-diameter PVC pipe, and were inserted into the wells (Figure 2.3 shows a three-string arrangement). Ideally, the ORC™ should fit snugly inside the well to ensure good contact between the flowing water and the socks. The PVC-pipe–ORC™ arrangement served to maximize ORC™–water contact, while minimizing the amount of ORC™ required.

2.2.2 Groundwater Sampling Program

The groundwater sampling program (as described in the Work Plan; UW, 1995) was divided into various monitoring programs completed over the duration of the project. A description of each program, the types of analyses completed, and the specific parameters analyzed are provided in Appendix 4, as are sampling and field analysis methods. Laboratory analytical methods are detailed in Appendix 3. Quality assurance/quality control (QA/QC) procedures and results are given in Appendix 5. Timing for all sampling events is included in the field activities chronology (Appendix 2).

2.2.3 Results

The following sections detail the results of the field experiment. Initial site characterization and preparation results related to geochemistry, flow conditions, hydraulics, and plume generation are provided in Sections 2.2.3.1 to 2.2.3.4. Sections 2.2.3.5 and 2.2.3.6 provide detailed results for the selected remedial technologies, granular iron, and enhanced aerobic bioremediation, respectively. The behavior of inorganic compounds through the gate is discussed in Section 2.2.3.7. A discussion of data handling procedures, particularly related to the interpretation of multilevel piezometer results, is provided in Appendix 24.

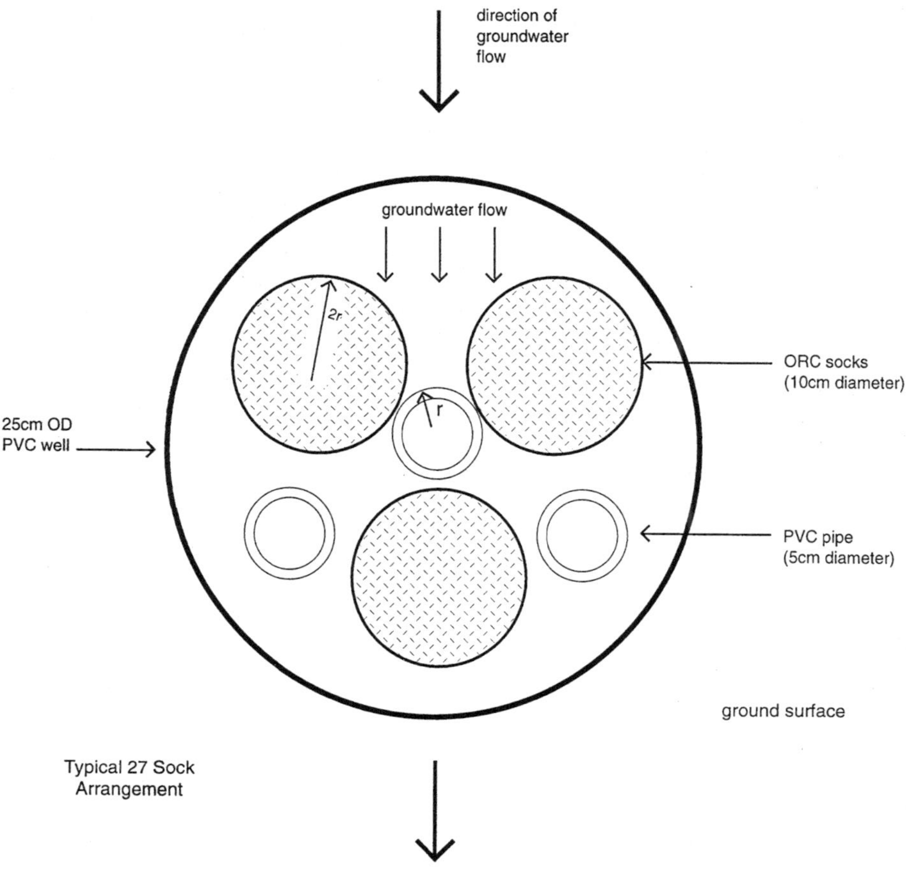

Figure 2.3 Arrangement of ORC™ socks in the well (schematic).

2.2.3.1 *Background Sampling*

Sampling to provide a comprehensive picture of background geochemical conditions at the experimental site was conducted approximately 1 month following flow system start-up. A complete description of background sampling activities and the results are provided in Appendix 6. A summary of Gate 1 results is provided in Table 2.1. The groundwater exhibited near-neutral pH, moderate to low electrical conductivity, DO concentrations between 1.5 and 3.8 mg/L, E_h measurements ranging from 70 to 180 mV, and an inorganic geochemistry typical of the Borden aquifer (Nicholson et al., 1983; Mackay et al., 1986). No volatile organic compounds (VOCs) were detected in sampling points in Gate 1. Dissolved hydrocarbon gas (DHG) sampling detected methane concentrations up to 71.5 μg/L, but essentially undetectable concentrations of other parameters (ethene, ethane, propene, and propane).

2.2.3.2 *Tracer Test Results*

Theoretical considerations, sample calculations, and a complete discussion of the tracer test are provided in Appendix 7. The following paragraphs summarize the important results for Gate 1.

Figure 2.4 shows breakthrough curves, based on fence-averaged bromide concentrations, for selected fences (G102, G104, G108, and G112). The peak of the curve at fence G102 corresponded to a bromide concentration of approximately 700 mg/L, while at fence G104 the maximum

Table 2.1 Results of Background Sampling for Gate 1 (May 1997)

Perimeter Analyzed	Average Concentrations (all depths included) G102L	G112L
Field Parameters (units noted)		
pH	7.46	7.59
E_h (mV)	116	108
Conductivity (μS)	517	398
DO (mg/L)	2.5	3.3
Alkalinity (mg/L $CaCO_3$)	284	201
Inorganics (mg/L)		
Iron	0.23	0.04
Calcium	93.3	66.8
Potassium	0.59	3.95
Magnesium	8.04	7.28
Manganese	nd	0.07
Sodium	13.8	2.76
Chloride	6.23	1.56
Phosphate	nd	nd
Sulfate	12.8	11.3
Nitrate	nd	nd
Bromide	nd	nd
Organics (μg/L)		
PCE	nd	nd
CT	nd	nd
Toluene	nd	nd
Ethene	nd	0.2
Ethane	nd	2.1
Methane	31.5	9.35

Notes: nd = not detected; concentrations below MDL (analysis specific).

concentration was approximately 400 mg/L, and further decreased at subsequent fences. The breakthrough curves became broader at fences further downgradient from the source. This response was expected due to dispersion.

The input to Gate 1 at fence G102 was highly variable, as shown in the breakthrough curves in Figure 2.5. Although tracer concentrations in the source wells were relatively uniform throughout the depth of the wells, the majority of the tracer was detected in points G102L-2, G102L-3, G102R-1, and G102R-3, while other points contributed minimally to the total mass crossing the fence. This evidence suggests that heterogeneities within the natural sand aquifer resulted in the formation of preferential flow pathways for the tracer solution. Inadequate well development efforts could also have contributed to this behavior. Organics emanating from these source wells later in the project were expected to be transported in a similar manner.

The next monitoring fence sampled during the tracer test was located in the middle of cassette C1 (fence G1C1) and showed a more even distribution of tracer at all sampling points (Figure 2.6). Very good lateral mixing was observed, while the vertical profile showed higher concentration with depth, raising the possibility of density driven flow. Although the tracer concentration was selected to avoid problems of density flow in aquifer materials, density flow in open water zones may have occurred. However, it was expected that the dissolved phase organic plume would not be affected. The observed lateral mixing was promising in terms of the expected behavior of the organics. The highly fingered plume observed in the upgradient aquifer was eliminated and a more uniform distribution resulted.

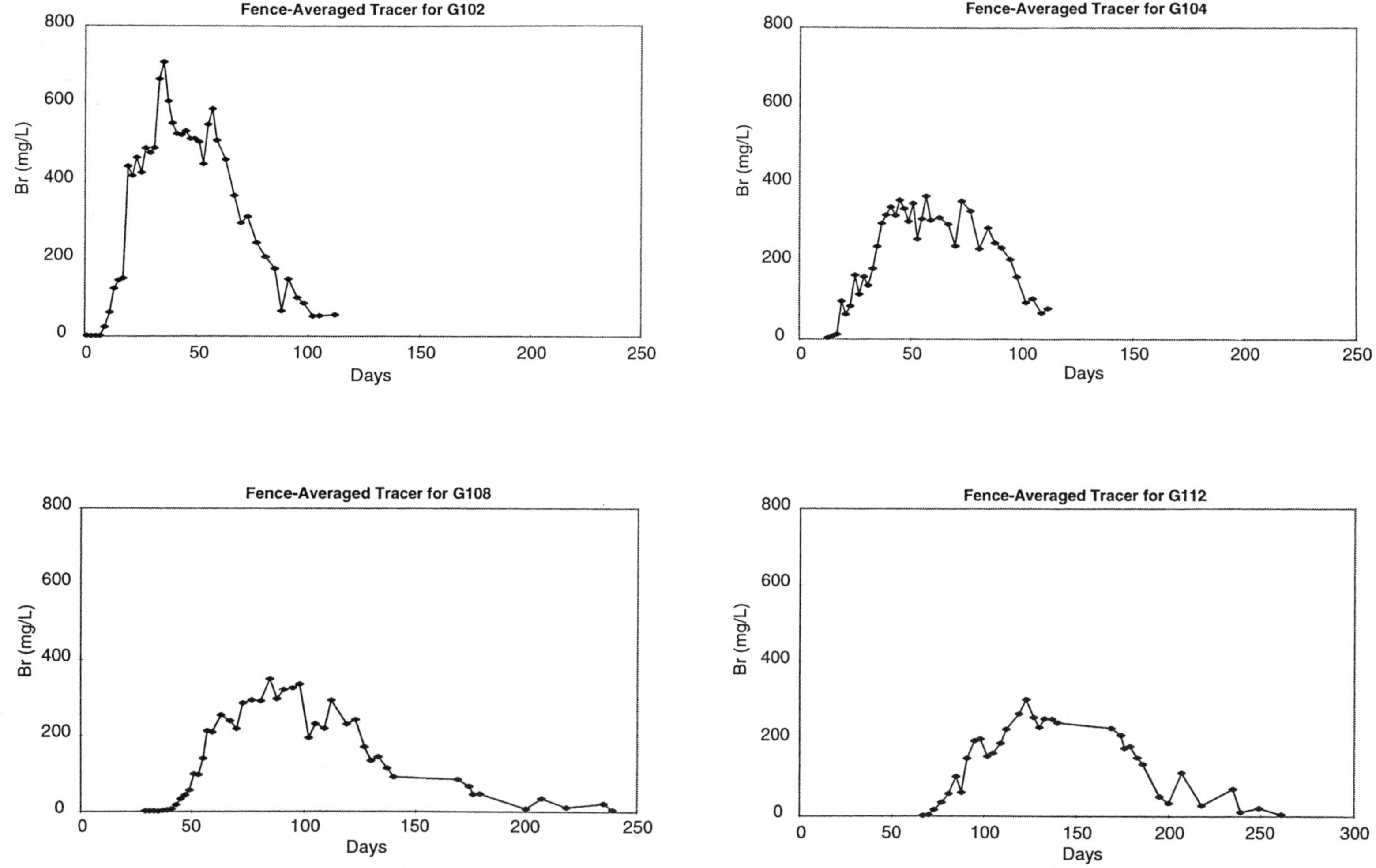

Figure 2.4 Tracer test results (fence-averaged breakthrough curves for selected fences).

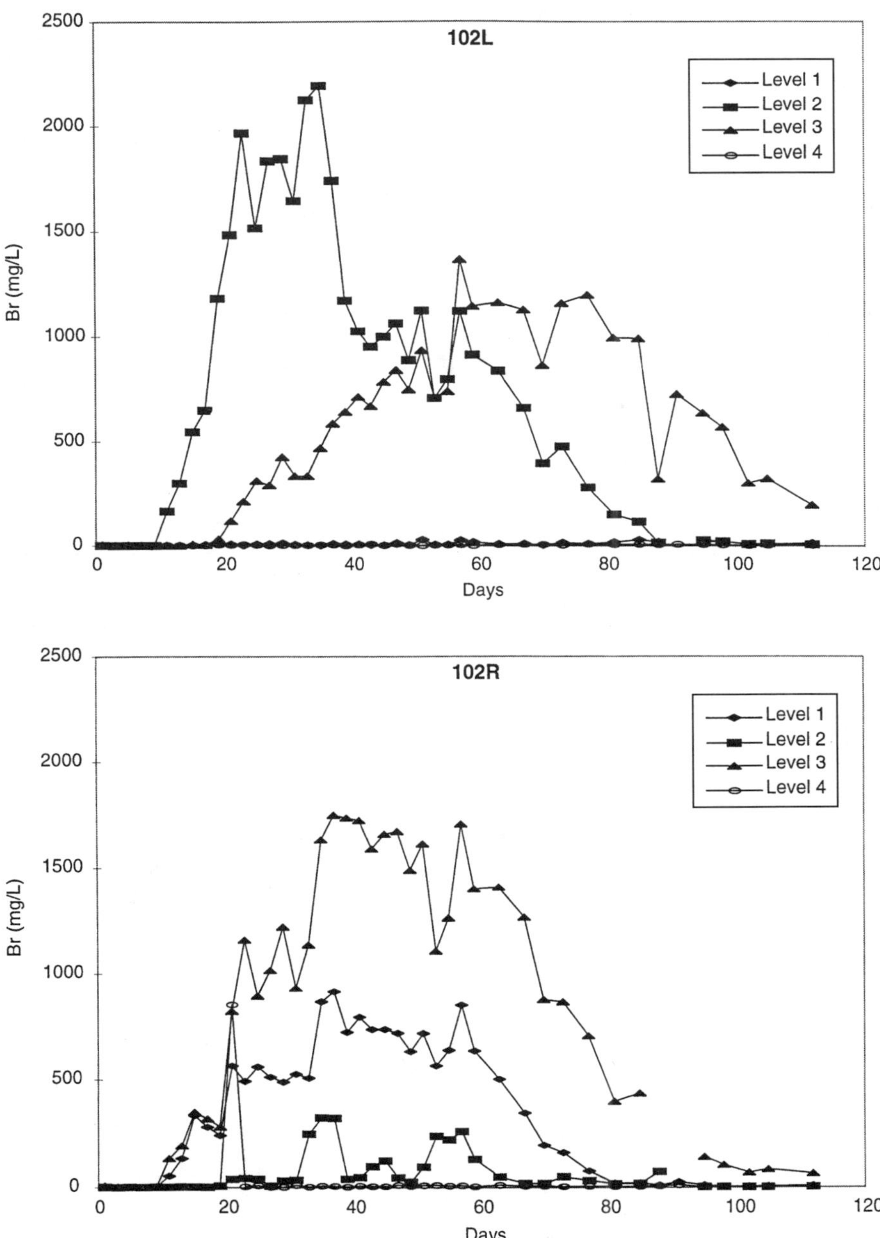

Figure 2.5 Tracer breakthrough curves at G102L and G102R.

Tracer behavior at fences G106 to G112 was dominated by transport at the base of the gate. These data were used to obtain a bulk estimate of groundwater velocity within the natural aquifer.

Inorganic tracer monitoring indicated the presence of a leak along the right side of Gate 1, between the sheet-piling wall and the cassette module (Figure 2.1). Bromide was detected in samples from G1CKR-3, while the remaining depths showed little evidence of the tracer. On the basis of the limited vertical extent of the bromide, the leak was thought to be of minor importance and unlikely to interfere with the fulfillment of the study objectives. The movement of tracer through the cassettes supported this view, and no effort was made to correct the problem. A full description

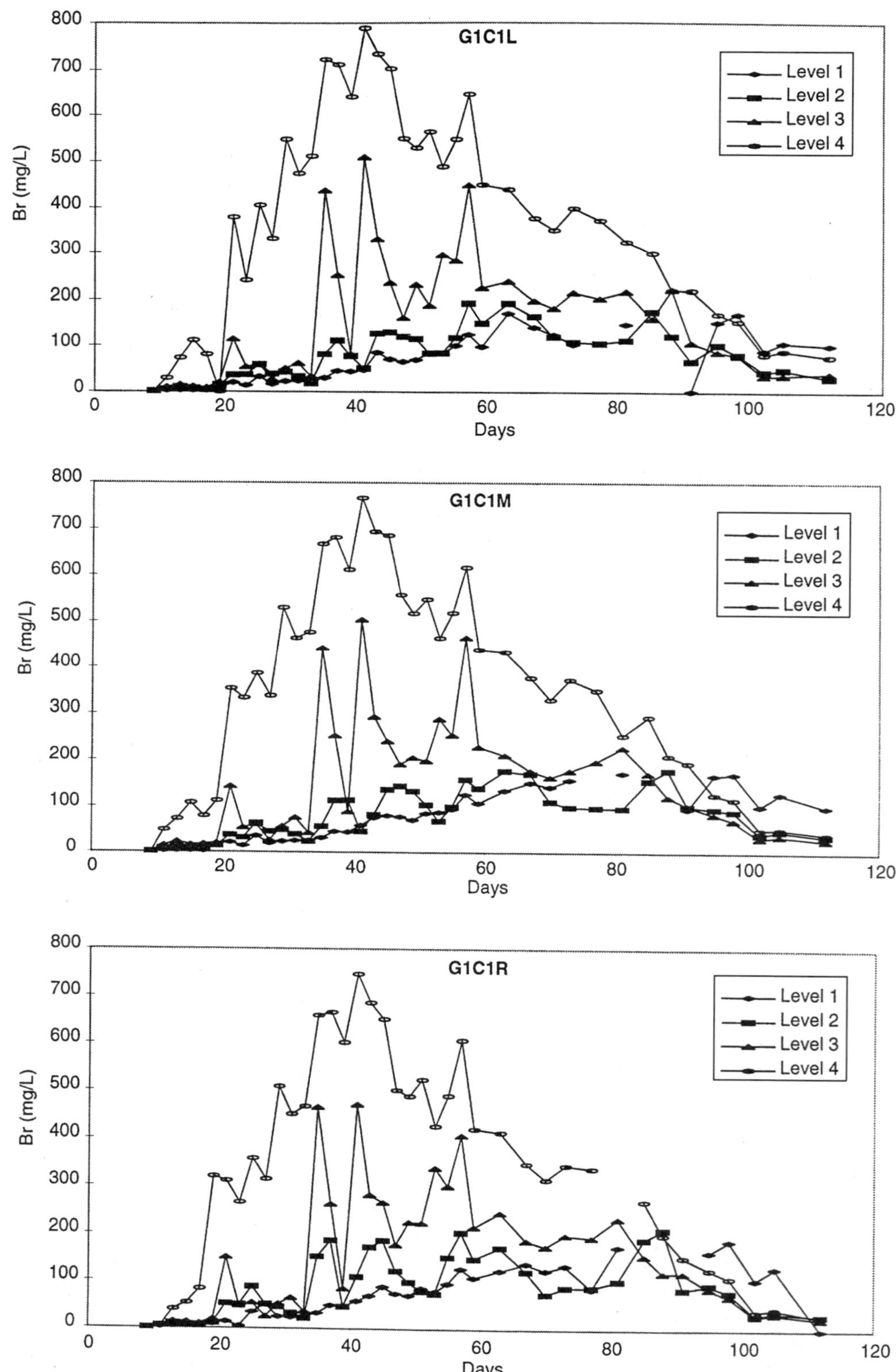

Figure 2.6 Tracer breakthrough curves at G1C1L, G1C1M, and G1C1R.

of the leak, including its detection and possible implications, is provided in Appendix 14. The seal on the left side of the cassette appeared to be secure, as no tracer was detected at G1CKL.

Table 2.2 provides a summary of the bulk velocity estimates, measured from fence G102 and calculated using fence-averaged tracer breakthrough curves. The results indicated that groundwater velocity in the natural aquifer (at fences G106 to G112) ranged from 10 to 12 cm/day and was considerably lower than in the cassettes (at fences G1C1 to G105), where the velocity was approximately 20 cm/day. This variation was mainly attributed to the lower cross-sectional area available to flow within the cassettes relative to that of the entire gate. Velocities determined in the tracer test compare favorably with values estimated using Darcy's equation, based on an average extraction rate of 100 mL/min (UW, 1997).

Table 2.2 Tracer-Determined Velocities for Gate 1 (from Fence G102)

Location	Calculated Velocity (cm/d)
G1C1	19
G103	21
G104	20
G105	18
G106	12
G108	11
G110	11
G112	10

An additional tracer test conducted only within the cassette system of Gate 1 was completed in September and October 1997. This test was implemented to address suspicions that flow through the cassettes had decreased over the duration of the project, and that groundwater flow through the leak between the sheet-piling wall and the outer cassette box had increased (Appendix 14). Results of this tracer test indicated that flow through the cassettes was minimal. Modeling of the flow system demonstrated that a two-order-of-magnitude difference between the hydraulic conductivities of the reactive media and the leak could have caused this shift in the flow pattern. A decrease in hydraulic conductivity of the reactive media in the cassettes could have resulted from the formation of precipitates on the influent screen of cassette C1, while the worsening of the leak over time could have led to an increase in hydraulic conductivity for this material. Nonetheless, at some time between the completion of the main tracer test (completed in the cassette portion of Gate 1 in December 1996) and the cassette tracer test (October 1997), a substantial change to the flow system occurred. Various lines of evidence combine to suggest that this change did not occur before mid-July 1997. These suggestions are provided in detail in Appendix 14.

2.2.3.3 *Water Level/Hydraulic Head Measurements*

Water level data are provided in Appendix 8. Figure 2.7 shows the potentiometric surface through the length of the gate on June 29 and November 13, 1996, and on October 24, 1997. Measurements were taken in all accessible piezometers. All plots demonstrate that the expected hydraulic gradient was achieved through the length of the gate and that groundwater flow was toward the extraction well. Plots at early time show a relatively uneven potentiometric surface with a region of high water table in the vicinity of the alternate oxygen addition wells in June 1996 and a region of low water table in the same region in November 1996. Pumping of the extraction well began in late May 1996. The surface variability in June 1996 could be due to slow development of a uniform flow regime. By the end of the project (October 1997), the potentiometric surface contours were very evenly distributed along the length of the gate, and the potentiometric high was

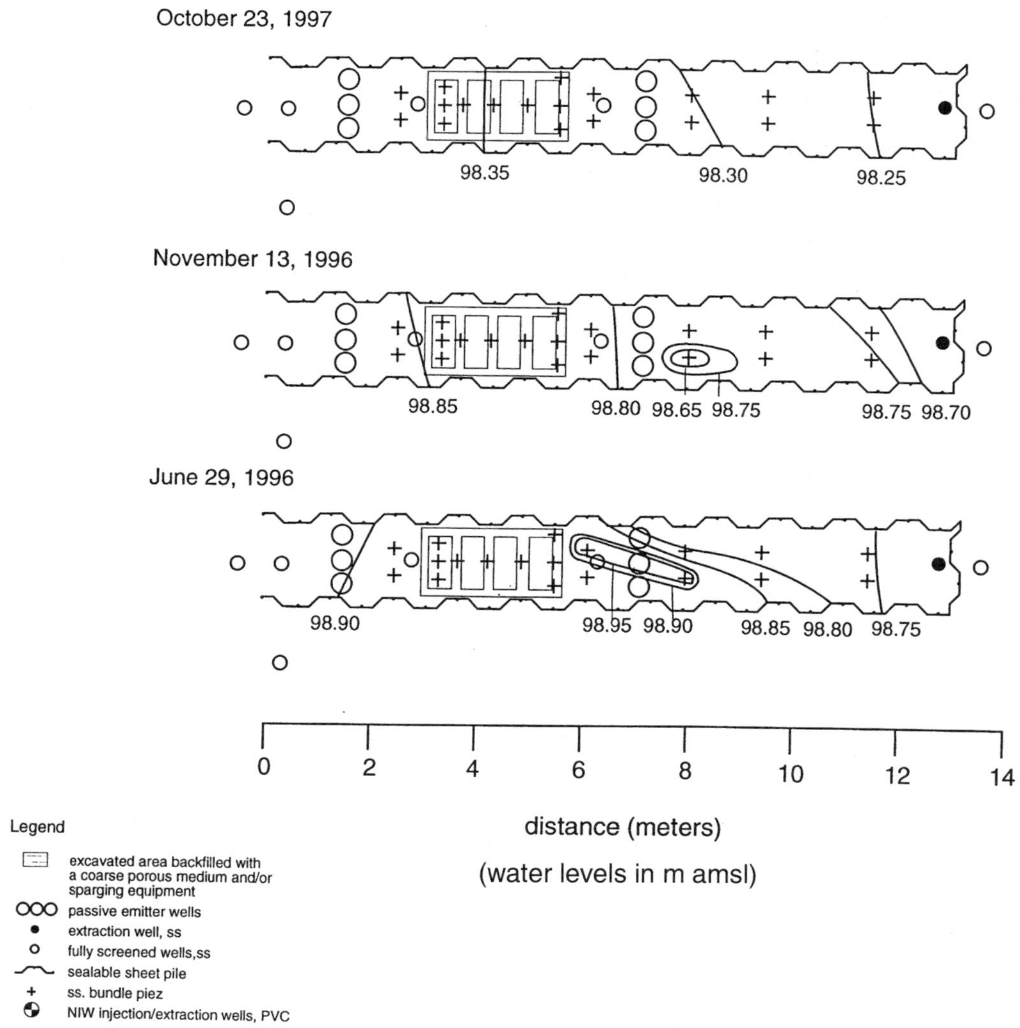

Figure 2.7 Potentiometric surface plots for Gate 1.

located at the upgradient end of the gate as would be expected. Water levels were considerably lower at all piezometers in October 1997 as compared to both of the other sampling dates, likely due to seasonal fluctuations.

Figure 2.8 shows water level fluctuations with time for wells G102L and G112L. An overall variation of approximately 0.8 m was observed for the duration of the entire experiment, with changes seemingly due to seasonal fluctuations. Lowest levels were observed during the summer months, increasing over the fall to the highest levels recorded. Data analyses related to contaminant degradation can be complicated by a fluctuating water table since contaminants can be trapped in the vadose zone due to a falling water table or can reenter the groundwater during times of rising water table. In this case, the influence of the changing water table can be disregarded. The performance of the granular iron can be conducted during times of high water table (March to July), while the performance of the biodegradation portion of the gate was only possible during times of low water table (July to October).

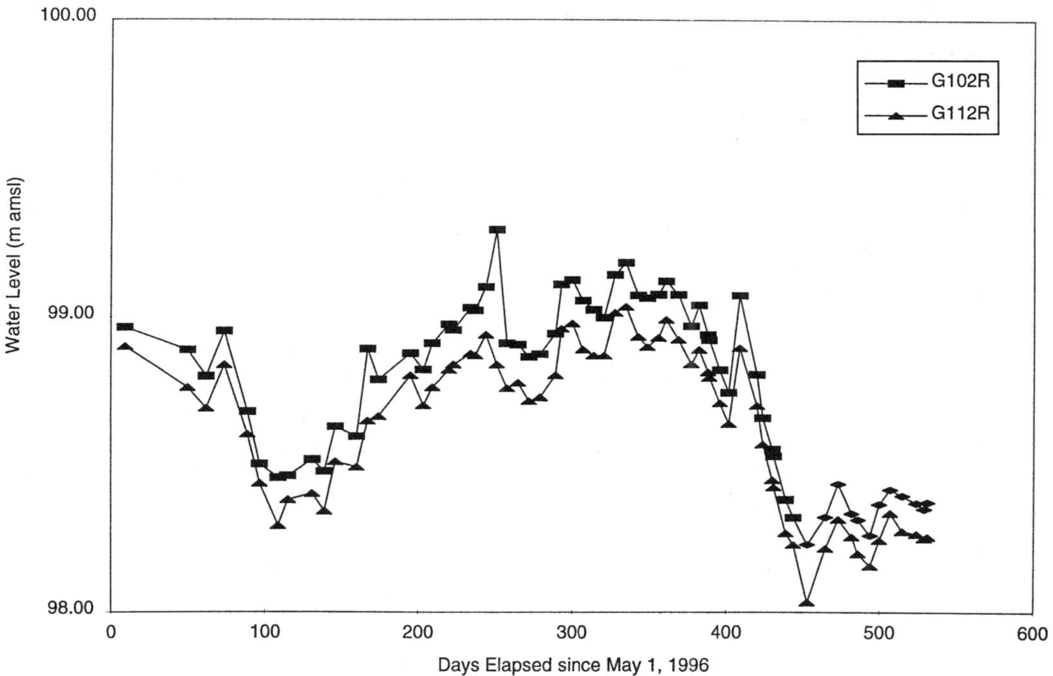

Figure 2.8 Water level fluctuations over time at G102R and G112R.

Although the water level itself was quite variable over time, the hydraulic gradient through the gate was relatively constant (i.e., the difference between levels at G102 and G112 was similar). The imposed hydraulic gradient through the length of the gate was approximately 0.013 m/m. In the sandpit experimental area, an average natural hydraulic gradient of 0.0043 m/m and an average groundwater velocity of 10 cm/day were determined (MacFarlane et al., 1983; Mackay et al., 1986). Interestingly, an apparently larger hydraulic gradient measured at the AATDF site resulted in a tracer-determined groundwater velocity within the desired range (10 to 12 cm/day in the natural aquifer; see Section 2.2.3.2). The difference between the two sites is possibly due to geological variations in grain size distribution, aquifer deposition, and hence layering.

2.2.3.4 Plume Generation

A mixed organic plume consisting of PCE, CT, and toluene was created in the shallow Borden aquifer from November 19, 1996 until October 28, 1997. The source was created using semipassive, diffusive release emitters that were designed based on work by Wilson and Mackay (1995) and installed in the source wells (SW-1 to 3) shown in Figure 2.1. Source well data and details of the plume generation are provided in Appendix 9.

Modeling, laboratory, and prototype field testing of the proposed source generation configuration were completed prior to full-scale implementation of these devices (UW, 1997). Delivery of contaminants to the aquifer was found to depend upon several factors, including tubing length, tubing wall thickness, the concentrations of the circulating contaminants within the diffusive tubing, and the degree of hydraulic connection between the source wells and the aquifer (i.e., the rate of flushing of the wells). All aspects of the design were considered, although ultimately, problems with the use of this method of source generation at this site remained and included the following:

- Contaminant concentrations within the source wells were quite variable (Appendix 9). Possible reasons for this behavior include differences in flushing rates within each well (depends on well development and/or natural heterogeneities at a particular well location) or variation in the mass delivered (depends on the contaminant delivery system — pumping rates, tubing thickness, etc.).
- Target ranges for the injected organics were 1 to 10 mg/L for toluene, and approximately 1 mg/L each for PCE and CT. The concentrations of contaminants within the source wells themselves far exceeded these goals. Maximum concentrations of 9.5, 38, and 150 mg/L were achieved for PCE, CT, and toluene, respectively. However, average concentrations delivered to the aquifer (measured at fence G102; Figure 2.9) were typically lower than desired for PCE and CT (maximums of approximately 400 and 700 μg/L, respectively), but were within the target range for toluene (between 1 and 6 mg/L). On one occasion, toluene concentrations were very high, particularly within the source wells themselves, and the system was disconnected temporarily to allow levels to decrease to within the desired range.
- As would be expected in a natural aquifer environment, delivery of organic contaminant concentrations to the aquifer (measured at fence G102, located approximately 1 m downgradient of the sources) was quite variable. At fence G102, only two of the eight sampling points (G102L-2 and G102R-3) showed significant mass fluxes. Trends in organic contaminant distribution, however, were similar to those observed for the inorganic tracer, confirming the presence of heterogeneities and preferential pathways (see Appendix 9).
- Steady-state profiles for the injected contaminants were not achieved (Figure 2.9). Concentrations of all components increased to a maximum approximately 50 days after the sources were started. Subsequently, toluene and PCE concentrations decreased, while CT concentrations increased significantly.

The overall variability of the input contaminant concentrations did not unduly restrict data interpretation. When degradation occurred it was obvious, irrespective of source generation complications.

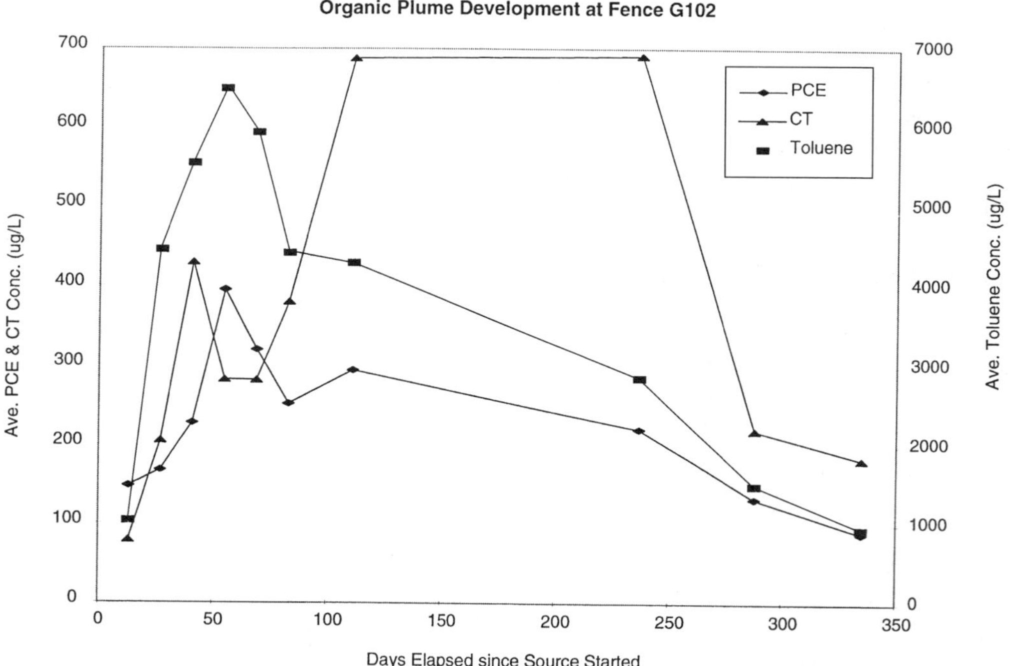

Figure 2.9 Plume development at fence G102.

2.2.3.5 Granular Iron

Snapshots of pH and E_h (fence-averaged) through cassettes C1 and C2 on December 16, 1996, March 10 and July 14, 1997 (238, 321, and 446 days following iron installation, respectively) show that the expected geochemical environment was established and maintained over the duration of the experiment (Figure 2.10). These results are typical of those obtained and show that iron-contacted water exhibited a pH of approximately 9.5 and a highly reducing character (E_h ranged from approximately –100 to –400 mV). Considerable variability in E_h within the iron cassettes was observed over the duration of the project; however, this behavior is common for this parameter and could be related to variable probe performance.

PCE and CT behavior through the iron is demonstrated by snapshots from sampling events on days 125, 195, and 238 (March 24, June 2, and July 14, 1997). September 4, 1997 data are not included due to the bias imposed by the lack of flow through the cassettes (Appendix 14). Figures 2.11 and 2.12 show fence-averaged concentrations of both input compounds and expected breakdown products: TCE, *c*DCE, VC, ethene, and ethane for PCE; CF, DCM, CM, and methane for CT. Of the aliphatic compounds expected, only ethene and ethane were detected at fence G104, indicating that at least some of the PCE underwent complete degradation to innocuous dissolved gases. In fact, no intermediate degradation products (TCE, *c*DCE, or VC) were detected. This behavior was similar to laboratory observations of TCE degradation made by Orth and Gillham (1996), where only 3 to 3.5% of the input TCE mass could be accounted for as chlorinated products, and approximately 80% of the mass was converted to ethene and ethane.

Dechlorination of CT to CF was observed in the natural aquifer upgradient of the cassettes. Based on results from Gate 2 (described in Chapter 3), this behavior was expected. Concentrations of CT entering the cassettes (fence G1C0) were quite low (44 µg/L to values below analytical limits of quantification); however, influent CF concentrations remained quite steady between sampling events (87 µg/L decreasing to 69 µg/L). Degradation of CF to DCM was observed in the iron environment, and CF was not detected in groundwater leaving the iron. DCM did not appear to undergo further degradation, as CM was not detected and methane concentrations did not appear to increase. This result is consistent with the stepwise degradation of CT to DCM observed in the laboratory by Matheson and Tratnyek (1994) and the persistence of DCM observed by Gillham and O'Hannesin (1994). The sequence of anaerobic followed by aerobic treatment zones was selected, in part, to study the behavior of the DCM, produced in the iron, in the aerobic environment.

Average DCM concentrations in water leaving the iron ranged from 18 to 25 µg/L, but these quickly dropped to concentrations below detection at monitoring points in the downgradient aquifer (fence G106). Dispersive effects may have caused this decrease, although biodegradation may also be involved. DCM has been shown to degrade under aerobic conditions (Brunner et al., 1980); however, aerobic microcosm studies completed as part of this study showed no response to DCM by the natural population (Appendix 11).

A complete description of the theory, assumptions, calculations, and results of the iron performance calculations is provided in Appendix 12. Estimates of first-order degradation rates were completed based on results of the tracer test ($v = 20$ cm/day) and fence-averaged VOC concentration data. Due to low concentrations of CT entering the cassette system, half-life calculations for this compound were not possible. The results are summarized in the following paragraphs.

Table 2.3 provides a summary of the half-lives calculated for the three sampling dates. Half-lives calculated for sampling events in March and June 1997 are very similar. Slightly lower half-life values observed for the July 14, 1997 data may be due to minor flow reductions within the cassettes. Half-lives for the first two sampling dates are expected to be more reliable than later estimates. To permit an effective comparison of these half-lives and those obtained in previous studies, half-life values were normalized to 1 m^2 of iron surface area per milliliter of solution. Normalized half-life values of 8 hours for PCE, and 10 and 14 hours for CF were estimated based on the results of this study (Table 2.3). The estimate for PCE falls within the range (2.1 to 10.8 hours) provided by Gillham (1995) for

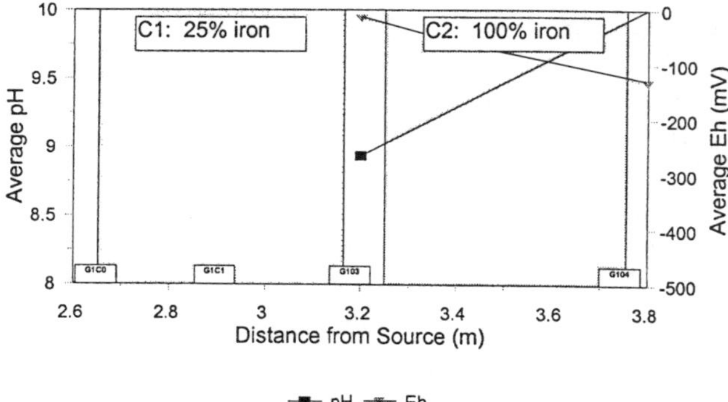

Reducing Conditions Achieved

Data from March 10, 1997

C1: 25% iron

C2: 100% iron

Average pH

Average Eh (mV)

Distance from Source (m)

pH Eh

Reducing Conditions Achieved

Data from July 14, 1997

C1: 25% iron

C2: 100% iron

Average pH

Average Eh (mV)

Distance from Source (m)

pH Eh

Figure 2.10 Reducing conditions achieved (pH and E_h in iron).

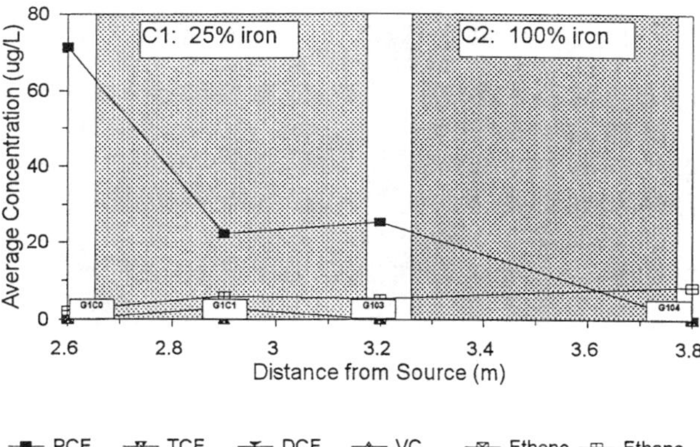

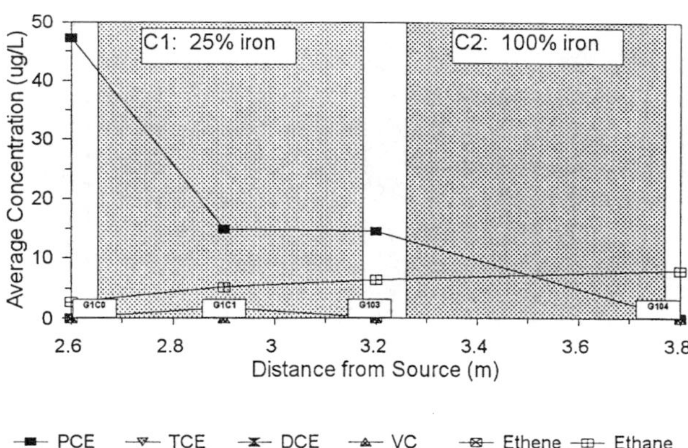

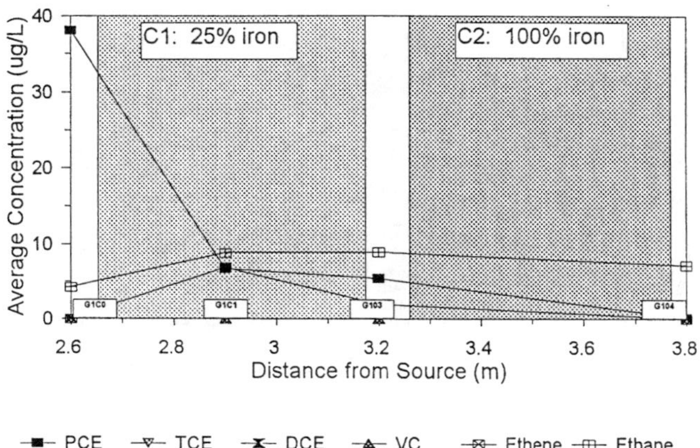

Figure 2.11 PCE degradation in iron (with breakdown products).

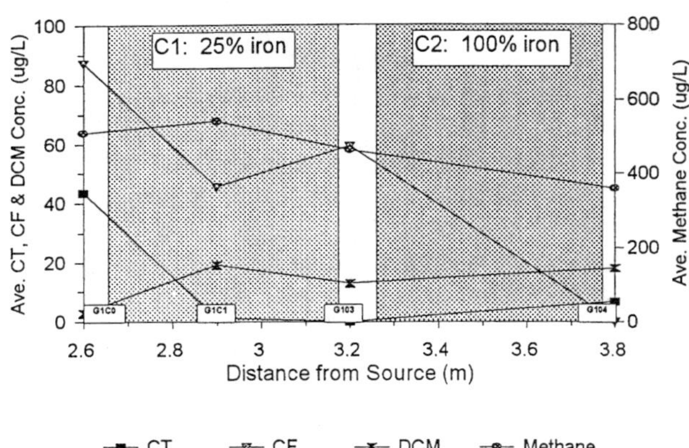

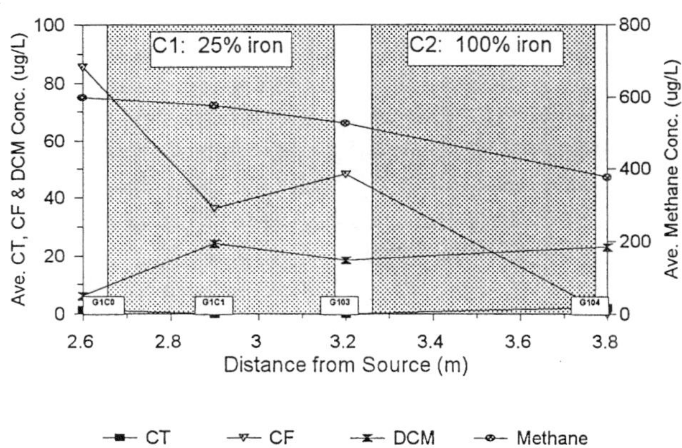

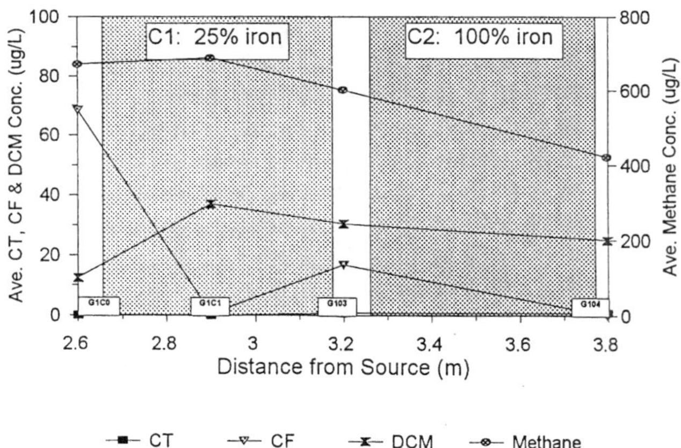

Figure 2.12 CT degradation in iron (with breakdown products).

Table 2.3 Summary of Calculated Half-Lives for Chlorinated Compounds in Granular Iron

		C1 (25% Iron) Half-Life (hr)	
Sampling Date	**Analyte**	**Calculated**	**Normalized**
March 24	PCE	14.2	9
(day 125; 16 PV)	CF	25.2	17
June 2	PCE	14.6	10
(day 195; 28 PV)	CF	19.1	13
July 14	PCE	9.8	6
(day 238; 35 PV)	CF	<8.2	<5

Notes: day x = number of days since contaminant injection started (November 18, 1996).

PV = number of pore volumes, calculated based on a 6-day retention time within the iron cassettes (accounting for plume development and time for transport to the cassettes).

<value = maximum half-life noted; calculation used an effluent concentration value that was uncertain (<MDL, taken as 9 μg/L for this calculation).

calculated half-life = half-life calculated as $t_{1/2} = \ln 2 / k$, using two concentration data points; half-life v = 20 cm/day; flow path = 20 cm.

normalized half-life = calculated half-life normalized to 1 m^2 of iron surface area per milliliter of solution half-life using normalization factor determined as 0.55.

laboratory experiments conducted using commercial-grade iron, while the half-lives estimated for CF compare poorly to the value of 4.8 hours given by Gillham (1995) for similar experiments. A field study using granular iron by O'Hannesin (1993) estimated a normalized half-life of 55 hours for PCE. No estimates based on field studies were obtained for CF. In general, slower degradation rates are expected for field studies due to temperature effects, the presence of inorganic compounds that might form precipitates on the iron surfaces and consequently reduce the reactivity of the media, and to nonuniform, influent contaminant distribution that could overload the system and cause contaminant breakthrough on the downgradient side of the wall. Toluene did not appear to degrade within the iron environment (Figure 2.13), but may have been slightly retarded due to sorption.

2.2.3.6 Enhanced Aerobic Biodegradation using ORC™

Provision of oxygen to stimulate the natural microbial population to biodegrade toluene, and any persistent breakdown products of reductive dechlorination, was the objective of the enhanced aerobic biodegradation system. ORC™ was selected to provide the required oxygen and is discussed in Section 2.2.3.6.1. Results related specifically to the aerobic degradation of the organics are provided in Section 2.2.3.6.2. Other microbial processes acting in the system are discussed in Section 2.2.3.6.3.

2.2.3.6.1 ORC™ Performance — The performance of the ORC™ was based on delivery of oxygen to the aquifer to promote the degradation of the target organics. This goal was assessed using measurements of DO and pH in the vicinity of the socks. The first monitoring program was designed to assess the performance when ORC™ was installed in C4 (October 8, 1996 to March 18, 1997). The second program monitored the dissolved oxygen and pH levels when the ORC™ socks were installed in the alternate oxygen addition wells (SW-12, 13, and 14) from June 19 to October 29, 1997. Details of the monitoring programs and the results are provided in Appendix 13.

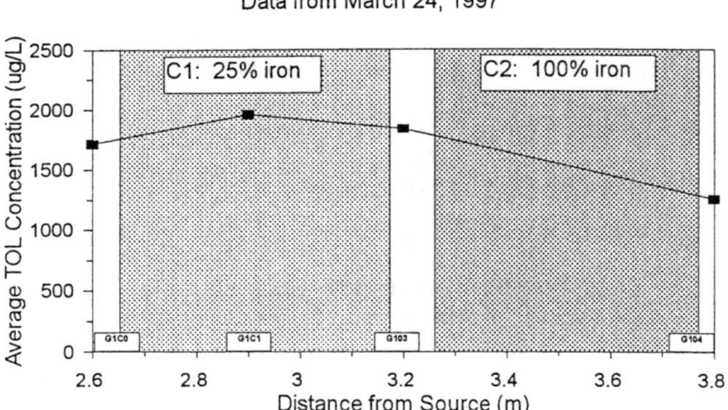

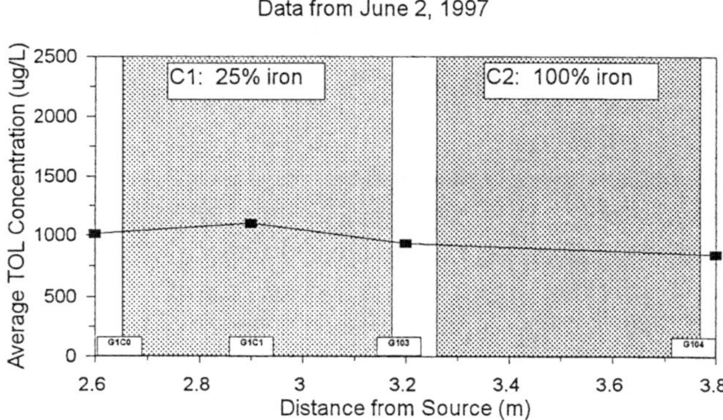

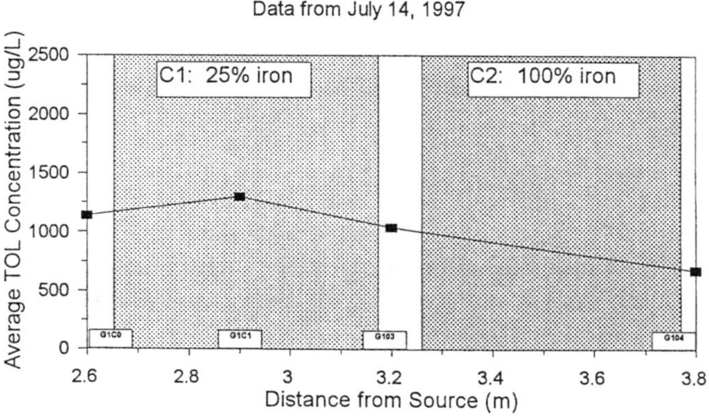

Figure 2.13 Toluene behavior in iron cassettes.

Figure 2.14 provides a summary of the monitoring results when the ORC™ socks were installed in C4. No significant increase in DO concentration was observed at fence G106, located in the aquifer downgradient of cassette C4 over that observed upgradient of cassette C4 (at fence G105). This lack of oxygen release was possibly due to inhibition of the ORC™ constituents in the presence of high pH water emanating from the iron cassettes (C1 and C2).

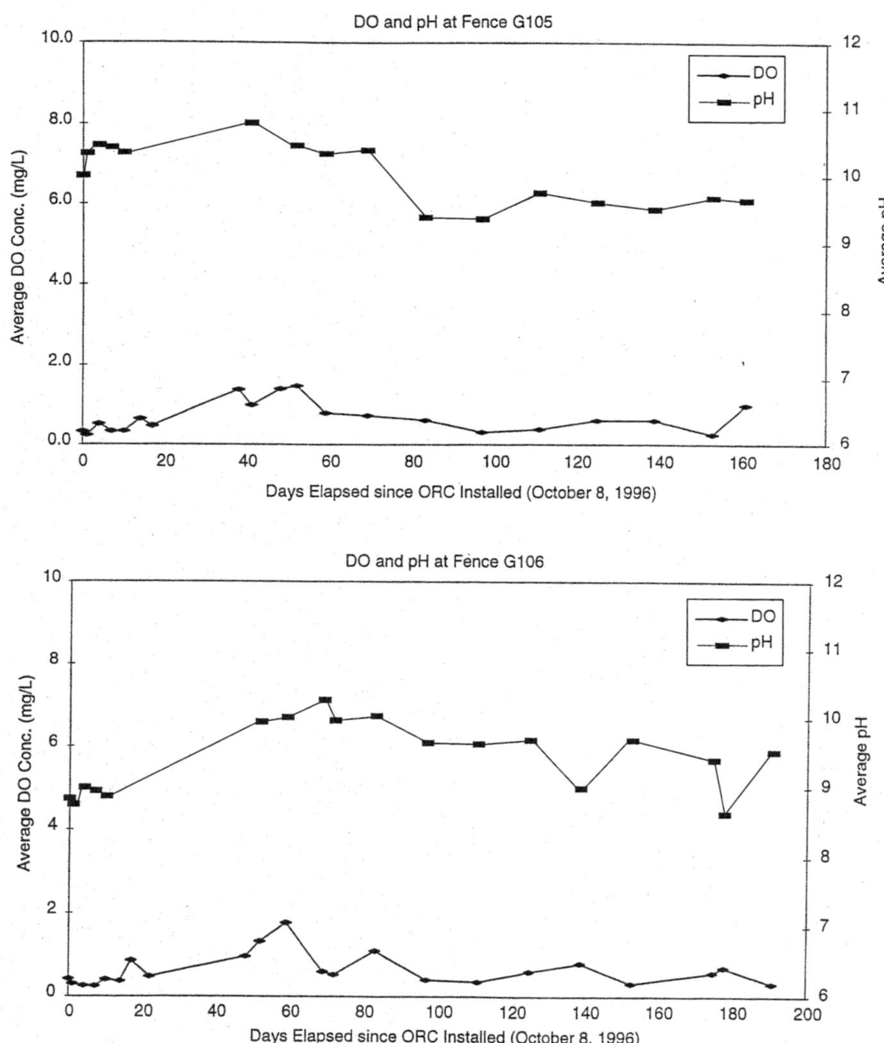

Figure 2.14 pH and DO profiles over time downgradient of ORC™ in C4.

Additional laboratory investigations involving the addition of ORC™ to water were conducted to evaluate this hypothesis. The intolerance of the ORC™ to high pH water was confirmed and possible solutions to this problem were investigated. Some attempts at implementing these remedies were also completed (details in Section 2.3.1 and Appendix 15).

Following several attempts to utilize the ORC™ in cassette C4 as originally designed, it was determined that design limitations, geochemical and possibly hydraulic, could not be reasonably overcome within the time frame of the experiment. It was thus determined that the use of the alternate oxygen addition wells (SW-12 to 14; refer to Figure 2.1 for location) would provide a means to assess, at least to some degree, the sequential use of granular iron and ORC™ for the treatment of a multicontaminant plume. It was expected that the presence of the natural aquifer materials between the two treatment regions should provide the necessary pH buffering to overcome the pH incompatibility problem (discussed in Appendix 15) and allow the release of oxygen from the ORC™. Prior to activating the wells, they were vigorously developed and tested.

Figure 2.15 shows the DO monitoring results for wells SW-12 to 14. Following the installation of the ORC™, significant increases in DO concentrations (values above 10 mg/L) were observed within the wells themselves, 10 days after ORC™ installation. DO concentrations continued to increase over time, but varied considerably with depth (typically ranging from 20 to 50 mg/L).

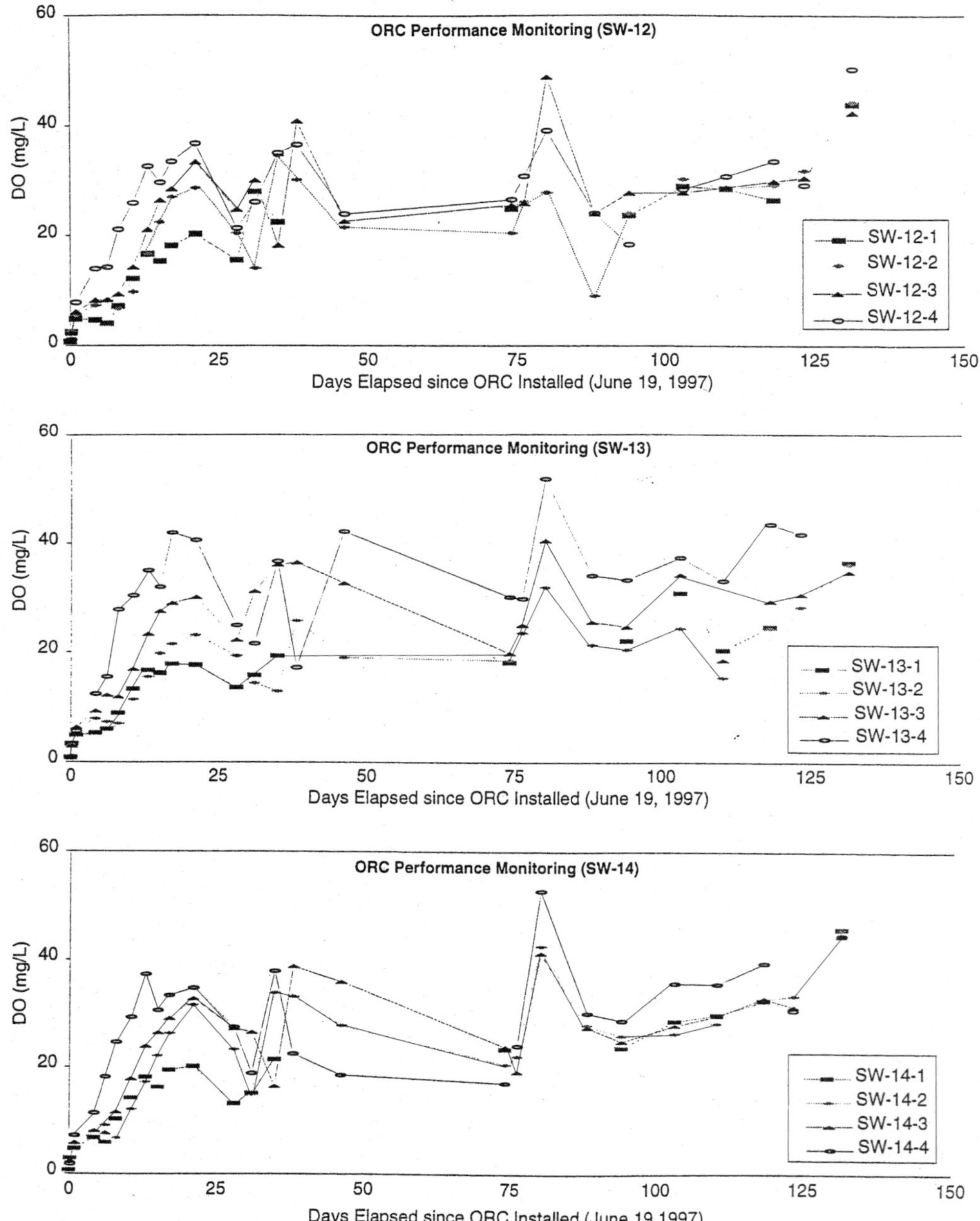

Figure 2.15 DO concentration in SWs 12 to 14 over time.

However, average DO concentrations were approximately 30 mg/L over the duration of the experiment (ORC™ installed for 124 days). Significant declines in DO concentration observed by Chapman et al. (1997; DO values decreased to less than 16 mg/L after 80 days) were not evident at this site. The pH within the alternate oxygen addition wells increased to approximately 11 almost immediately after the installation of the ORC™ socks, but the release of oxygen from the ORC™ did not appear to decrease (intolerance of ORC™ to high pH water is discussed in Appendix 15). Although no definite explanation for this inconsistency has been determined, it was suggested that the flow environment within the wells permitted contact of in-flowing, near-neutral water with the ORC™, and thus activated the release of oxygen.

Monitoring results for fence G108, located in the aquifer immediately downgradient of the wells, are provided in Figure 2.16. The results indicate that DO concentrations were quite variable,

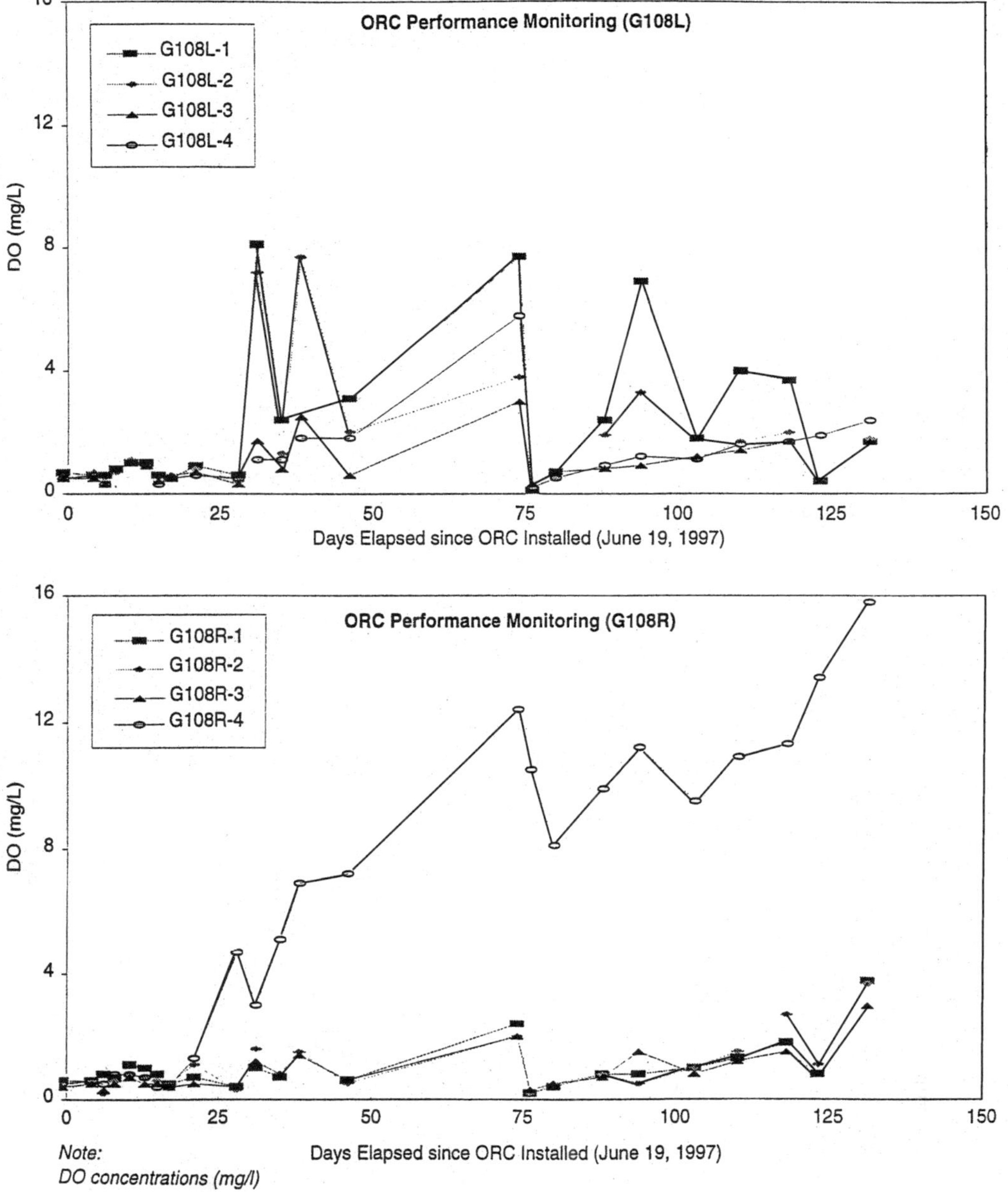

Figure 2.16 DO provided to aquifer at fence G108.

indicating the existence of preferential flow paths, as suggested by tracer test results. At sampling point G108R-4, DO concentrations up to 15 mg/L were observed in October 1997, while at other sampling points, concentrations rarely exceeded 4 mg/L. These DO values are considerably lower than those observed within the oxygen addition wells themselves (average between September 4 and October 29, 1997 of approximately 3 mg/L at fence G108 vs. 30 mg/L in SW-12 to 14). Also, tracer dilution tests, completed following well development, showed breakthrough of tracer compounds at fence G108 within approximately 200 hours, while increased oxygen levels were only observed approximately 35 days after the installation of the socks. The lower observed concentrations and the apparent time difference could indicate oxygen consumption through microbial uptake or other, abiotic, oxygen sinks (oxidation of iron or manganese, for example). These suggestions and their implications are discussed in detail in Appendix 13, while a summary related to the degradation of the target organics is provided in Section 2.2.3.6.2.

Occasional DO monitoring downgradient of fence G108 showed very little evidence of an increase in oxygen in the subsurface (DO concentrations typically <1 mg/L). The reasons suggested above for low DO levels also apply to this portion of the aquifer, but based on the degradation observed and discussed subsequently, some oxygen must make its way into the aquifer. It is perhaps not detected because it is rapidly consumed. Anaerobic zones may still persist within the aquifer and anaerobic microbes may also contribute to the degradation of the target organics. Nonetheless, for the purposes of the following discussion, the region in Gate 1 between fences G106 and G112, where marginally aerobic conditions were expected due to the use of the ORC™, will be referred to as the *aerobic zone*.

Increases in pH (as high as 10) were also observed at several monitoring points at fence G108. This increase was expected based on previous experience (Smyth et al., 1995); however, its implications with respect to the behavior of microbial populations and their degradation efficiencies are unknown. Natural pH buffering by aquifer materials was expected to reduce the pH to near neutral within a short distance from the ORC™, and thus prevent adverse impacts to the existing microbial population. This behavior was observed by Chapman et al. (1997).

2.2.3.6.2 Behavior of Organic Contaminants in an Aerobic Environment — Provision of oxygen for the biodegradation of toluene, and any persistent breakdown products of reductive dechlorination, was the objective of the ORC™. However, the performance problems associated with the ORC™ (discussed above) caused considerable delay in creating the desired aerobic environment. Ultimately, however, the ORC™ did provide at least an average of 3 mg/L of DO to the aquifer in September and October 1997 (measured at fence G108), and an assessment of the biodegradation that ensued can be provided.

Figure 2.17 shows the fence-averaged concentrations of toluene through the aerobic portion of the gate on several sampling dates. Relatively constant concentration profiles through the length of the gate were observed on June 2 and June 18, 1997, suggesting that a quasi-steady-state condition had been achieved. The profile for July 14, 1997 (25 days after ORC™ installation) also appears relatively constant, although a slight decline at fence G108 might exist. This profile was expected due to the delayed oxygen release from the ORC™ and the slow delivery of oxygen to the aquifer (increases in DO at fence G108 were not evident until 30 days after installation).

The first evidence of a concentration decline downgradient of the ORC™ was provided by sampling results from September 4, 1997 (77 days after ORC™ installation). In this snapshot, the fence-averaged concentration decreased from approximately 2000 to 500 μg/L in a distance of 5.8 m (between fences G106 and G112). Further evidence of toluene concentration decreases were observed for sampling events on September 29 and October 21, 1997 (102 and 124 days after ORC™ installation, respectively). Fence-averaged concentrations declined from approximately 1500 μg/L at fence G106 to less than 200 μg/L at fence G112, on both occasions.

Complete degradation of toluene was not observed, possibly due to the lack of oxygen within the subsurface. Measurable increases in DO concentration were observed only at a few points at fence G108, and DO levels at fences G108, G110, and G112 were typically less than 1 mg/L. The

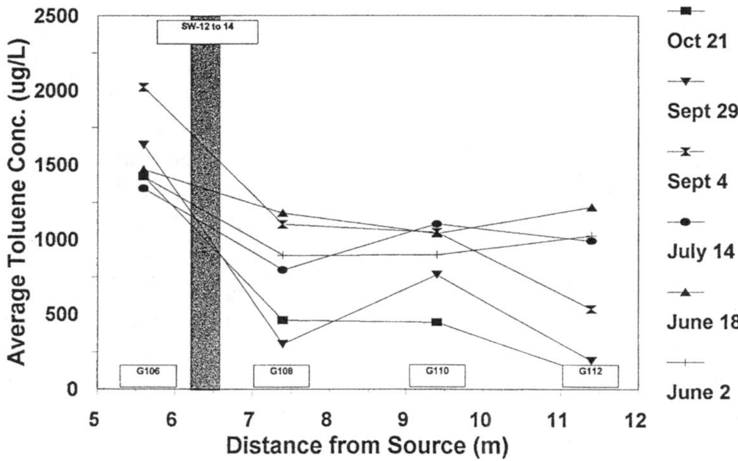

Figure 2.17 Toluene behavior in aerobic zone (sampling snapshots on days noted).

variable nature of the toluene and DO concentration data can be seen in the data from October 21, 1997 (Table 2.4). Also, a correlation between monitoring points with high DO and low toluene concentrations can be made. At points where high DO concentrations were measured (e.g., at point G108R-4, DO reached a maximum of 15 mg/L), toluene concentrations were below method detection limits. At G108L-4, where DO concentrations up to 2.5 mg/L were observed, a toluene concentration of approximately 400 μg/L was measured compared to other sampling points at this piezometer where toluene levels ranged from 760 to 1260 μg/L. In general, toluene concentrations measured in piezometers along the right side of the gate were considerably lower than those on the left side. Again, this suggests some preferential flow paths, and possibly a more thorough delivery of oxygen to the right portion of the gate as compared to the left.

Table 2.4 Toluene and DO Concentrations in the Aerobic Zone on October 21, 1997 (Gate 1)

		Left Piezometer		Right Piezometer	
Sampling Fence	**Depth**	**Toluene (μg/L)**	**DO (mg/L)**	**Toluene (μg/L)**	**DO (mg/L)**
G106	1	1130	0.4	1230	0.4
	2	1400	0.6	1520	0.4
	3	1610	0.5	1650	0.6
	4	1550	0.8	1310	0.5
G108	1	832	0.4	248	0.8
	2	1260	0.4	12.1	1.1
	3	761	0.4	149	0.8
	4	429	1.9	<MDL	13.4
G110	1	1150	0.4	1020	0.5
	2	182	0.8	195	0.4
	3	771	0.6	15.7	0.4
	4	28.3	0.3	208	0.9
G112	1	25.2	0.4	<MDL	0.5
	2	<MDL	0.4	<MDL	0.4
	3	504	0.5	<MDL	0.5
	4	139	0.8	N/A	0.4

Notes: LOQ refers to the limit of quantification; range for toluene (13.5–17.4) on this date; MDL refers to the method of detection limit; concentrations noted as <MDL were not detected; "N/A" indicates that a sample was not analyzed.

Concentrations of chlorinated compounds (PCE and CF) downgradient of the cassette system were above background levels due to a suspected leak between the sheet-piling walls and the outer cassette box (details provided in Appendix 14). Although aerobic degradation of highly chlorinated compounds is not expected, their inadvertent presence downgradient of the cassette system permitted an assessment of this potential. The persistence of DCM through the granular iron was expected based on previous work (Gillham and O'Hannesin, 1994). Aerobic biodegradation of this compound has been suggested (Brunner et al., 1980), and the sequential treatment system implemented here was designed to study this possibility. Due to low chlorinated methane (CT, CF, CM, and DCM) concentrations entering the iron, and consequently low concentrations of DCM exiting the cassettes (18 to 25 μg/L), an assessment of DCM degradation was not possible.

Figure 2.18 shows PCE and chloroform behavior through the length of the aerobic zone on several sampling occasions. In both plots, input concentrations (i.e., at fence G106) are relatively constant for the first three sampling events (June 2, June 18, and July 14, 1997), indicating a time when flow through the iron cassettes (and the leak) was believed to be relatively steady (Appendix 14). The constant contaminant input over the early part of the sampling interval allows for an evaluation of the degradation of these compounds, at least up to the time when the pore volume of known input exits the aerobic zone. Without accounting for retardation (thus, giving the most conservative estimate of travel time) a pore volume travels the length of the aerobic zone in approximately 60 days. Thus, groundwater arriving at fence G106 on July 14, 1997 should pass fence G112 on approximately September 14, 1997. All data within this time frame can be considered in the degradation assessment. However, for slow-moving compounds like PCE (retardation factor, R = 2.7 to 3.9; Roberts et al., 1986), this time frame was extended.

PCE concentrations through the aerobic zone appeared to decline with distance; however, this observation is more likely related to slow plume development rather than the result of degradation. Aerobic biodegradation of PCE is not typical (Mohn and Tiedje, 1992).

Chloroform behavior in the aerobic portion of the gate suggests some possibility for degradation. Slow plume development is unlikely in this case, since retardation of chloroform was minimal in other experiments at CFB Borden (R = 1.0 to 1.2; Devlin, 1994). Cometabolism of chloroform by toluene-degrading bacteria, as observed by McClay et al. (1996), or by methanotrophic activity (Little et al., 1988) are suggested as possible mechanisms for this behavior, based on chloroform, methane, and toluene concentrations over time (Table 2.5). By June 1997, all compounds showed relatively stable concentrations at fence G106, with chloroform, toluene, and methane averaging approximately 150, 1500, and 800 μg/L, respectively. At downgradient fences, slight decreases in all compounds were observed due to dispersive effects, sorption, and possibly some intrinsic anaerobic losses. By September 1997, notable chloroform losses were seen at fences G108, G110, and G112, particularly at sampling points where toluene concentrations also decreased, and toluene was assumed to be available. Methane data were not collected for this sampling event. Reduced concentrations of all compounds were evident in the October 1997 data. Beyond fence G106, chloroform was not detected, and in several instances both toluene and methane concentrations also decreased to less than 100 μg/L. However, these data are difficult to assess since this water reflects a changing input function (concentrations of chloroform at fence G106 decreased considerably between July 14 and September 4, 1997).

2.2.3.6.3 Other Microbial Processes Responsible for Declines in Contaminant Concentrations — Anaerobic intrinsic biodegradation could be at least partly responsible for the declining concentrations of both PCE and CF observed in Gate 1. Similar behavior was observed in Gate 2 (discussed in Chapter 3). The presence of the PCE breakdown product, *c*DCE, was likely due to intrinsic anaerobic biodegradation of PCE within the natural aquifer, since TCE and *c*DCE are typically observed as a result of this process. Also, the presence of elevated methane concentrations (Table 2.5) at fence G102 might suggest the development of an environment conducive to the biodegradation of chlorinated compounds, particularly CF. Intrinsic anaerobic degradation was

Table 2.5 CF, Methane, and Toluene Behavior

DATE:	10-Mar-97			02-Jun-97			04-Sep-97			21-Oct-97		
Well No.	CF (ug/L)	Methane (ug/L)	Toluene (ug/L)	CF (ug/L)	Methane (ug/L)	Toluene (ug/L)	CF (ug/L)	Methane (ug/L)	Toluene (ug/L)	CF (ug/L)	Methane (ug/L)	Toluene (ug/L)
LOQ	14.4-19.2	0.2	5.7-19.1	18.5-32.9	0.2	11.0-17.1	28.4-37.8	1.4	24.3-33.5	7.87-21.9	1.4	13.5-17.4
G102L-1	<MDL	4630	58.3	N/A	N/A	N/A	12.4	105	791	N/A	N/A	N/A
G102L-2	655	120	17840	N/A	N/A	N/A	<MDL	32.3	1660	<MDL	210	126
G102L-3	<MDL	6.5	35.6	N/A	N/A	N/A	<MDL	17.2	45.5	<MDL	8.8	<MDL
G102L-4	<MDL	10.4	9.6	N/A	N/A	N/A	<MDL	20.4	23.4	<MDL	13.4	<MDL
G102R-1	<MDL	1980	5.8	N/A	N/A	N/A	40.0	7670	291	32.3	535	490
G102R-2	<MDL	4.6	3.8	N/A	N/A	N/A	<MDL	10.3	21.2	<MDL	84	79.5
G102R-3	315	37.7	16330	N/A	N/A	N/A	361	22.0	9090	148	12.0	5890
G102R-4	<MDL	6.8	41.8	N/A	N/A	N/A	<MDL	10.0	8.9	<MDL	10.0	11.9
G106L-1	<MDL	243	489	100	370	1130	35.6	794	1140	<MDL	9790	1130
G106L-2	56.0	281	1570	183	944	1710	53.7	1450	2270	<MDL	1120	1400
G106L-3	21.8	258	803	167	756	1460	221	1250	2170	120	1060	1610
G106L-4	51.6	226	1440	196	787	1680	195	1260	2420	21.5	1420	1550
G106R-1	<MDL	263	610	110	658	1200	<MDL	851	1330	<MDL	1060	1230
G106R-2	121	311	2500	155	1000	1530	169	1450	2230	69.6	1120	1520
G106R-3	47.2	256	1040	164	712	1610	290	1130	2250	151	924	1650
G106R-4	41.1	408	841	136	800	1070	<MDL	1310	2380	<MDL	1910	1310
G108L-1	<MDL	254	351	47.6	398	750	<MDL	N/A	930	<MDL	607	832
G108L-2	28.9	259	638	104	466	933	<MDL	N/A	2410	<MDL	986	1260
G108L-3	63.8	302	1640	85.8	495	913	36.1	N/A	2020	<MDL	260	761
G108L-4	40.8	250	845	63.9	518	882	231	N/A	1120	<MDL	108	429
G108R-1	<MDL	249	308	84.8	489	860	<MDL	N/A	796	<MDL	143	248
G108R-2	43.7	263	892	131	557	977	<MDL	N/A	565	<MDL	22.3	12.1
G108R-3	26.3	164	167	114	519	973	<MDL	N/A	886	<MDL	680	149
G108R-4	36.0	289	624	99.6	524	856	18.9	N/A	101	<MDL	26.9	<MDL
G110L-1	<MDL	230	63.7	29.7	304	592	<MDL	N/A	682	<MDL	735	1150
G110L-2	32.4	222	347	83.4	433	957	<MDL	N/A	1150	<MDL	557	182
G110L-3	32.2	153	230	124	438	1040	<MDL	N/A	1360	<MDL	812	771
G110L-4	<MDL	236	13.2	120	557	953	<MDL	N/A	1210	<MDL	997	28.3
G110R-1	<MDL	296	90.7	29.2	574	469	<MDL	N/A	952	<MDL	862	1020
G110R-2	34.8	235	667	125	589	1030	<MDL	N/A	1080	<MDL	893	195
G110R-3	70.5	160	2060	167	767	1160	13.7	N/A	843	<MDL	661	15.7
G110R-4	43.4	249	506	113	513	980	38.4	N/A	1120	<MDL	103	208
G112L-1	N/A	N/A	N/A	<MDL	290	577	<MDL	N/A	187	<MDL	596	25.2
G112L-2	N/A	N/A	N/A	94.1	378	1420	<MDL	N/A	851	<MDL	756	<MDL
G112L-3	N/A	N/A	N/A	38.4	256	815	<MDL	N/A	1420	<MDL	1030	504
G112L-4	N/A	N/A	N/A	47.7	326	934	<MDL	N/A	934	<MDL	868	139
G112R-1	N/A	N/A	N/A	8.8	321	513	<MDL	N/A	125	<MDL	397	<MDL
G112R-2	58.2	174	<MDL	<MDL	387	1890	<MDL	N/A	3.2	<MDL	823	<MDL
G112R-3	25.6	217	<MDL	59.4	260	1210	<MDL	N/A	57.1	<MDL	139	<MDL
G112R-4	N/A	N/A	N/A	61.3	292	851	<MDL	N/A	726	N/A	495	N/A

NOTES:

- LOQ REFERS TO THE LIMIT OF QUANTIFICATION FOR THE LABORATORY ANALYSES; RANGE FOR ANALYTICAL EVENT GIVEN
- MDL REFERS TO THE METHOD DETECTION LIMIT; CONCENTRATIONS NOTED AS "<MDL" WERE NOT DETECTED
- "N/A" INDICATES THAT A SAMPLE WAS NOT ANALYZED

likely responsible for the reduction in chloroform concentrations observed at fence G106 for the October 1997 sampling (Figure 2.18). The granular iron was unlikely involved in the partial degradation of PCE discussed here, since concentrations of TCE and *c*DCE were below detection limits at fence G104, as they were throughout the entire project. Also, a suspected lack of flow through the cassettes (Appendix 14) would minimize persistence of any breakdown products in water reentering the aquifer, since time for complete degradation within the granular iron medium would surely exist.

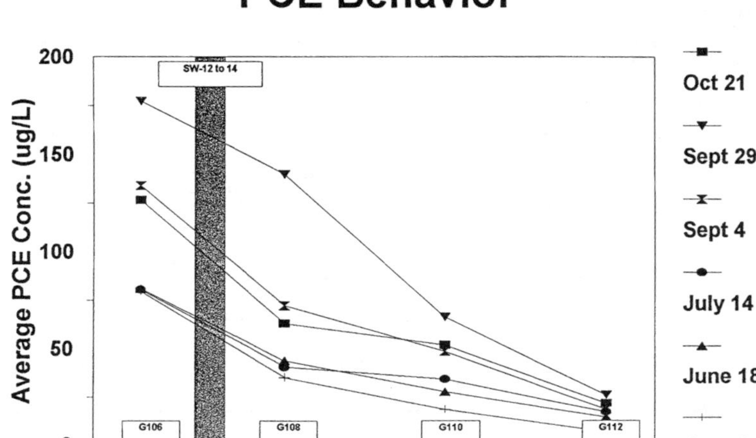

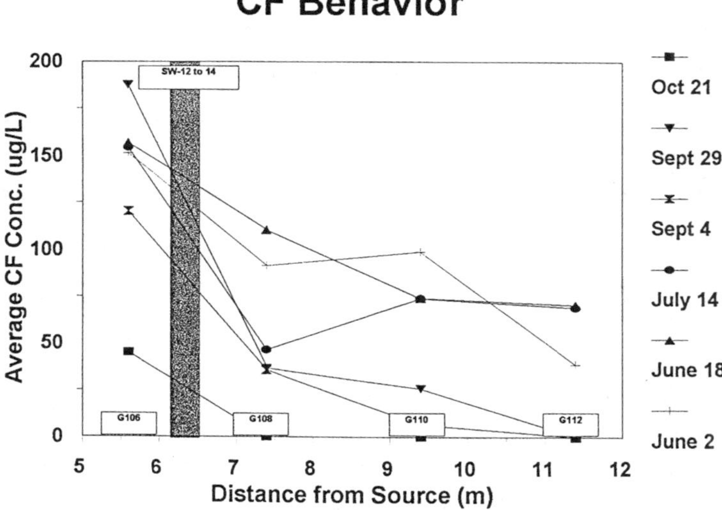

Figure 2.18 PCE and CF behavior in aerobic zone (sampling snapshots on days noted).

2.2.3.6.4 Influence of pH on Biodegradation — The significant pH increase observed at fence G108 following ORC™ installation was expected based on previous experience (Smyth et al., 1995); however, its implications with respect to the behavior of microbial populations and their degradation efficiencies are unknown. Natural pH buffering by aquifer materials was expected to reduce the pH to near neutrality within a short distance from the ORC™, and thus prevent adverse impacts to the existing microbial population. Figure 2.19 shows the pH profile downgradient of the ORC™ and demonstrates that, in fact, elevated pH levels did not persist. The desired pH range, close to that of the natural aquifer (pH 7 to 8.5), was achieved within 3 m of the oxygen source.

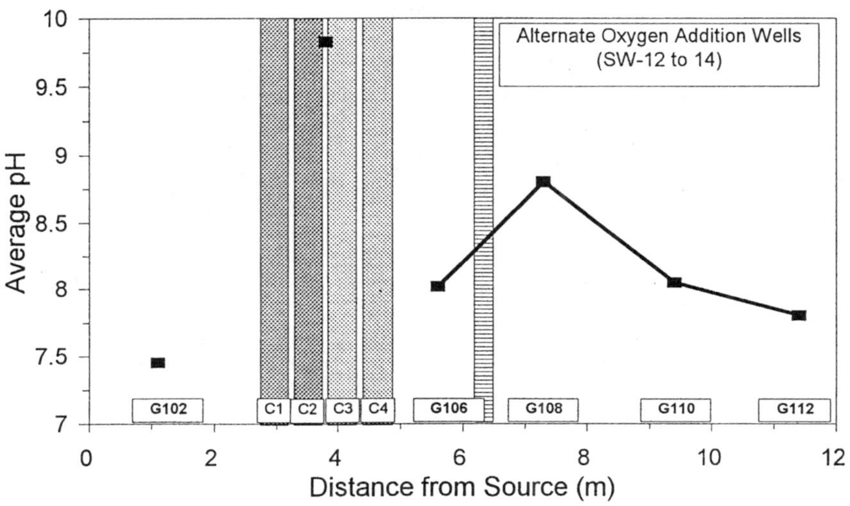

Figure 2.19 pH buffering downgradient of SW-12 to 14.

2.2.3.7 Inorganics

Fence-averaged (Appendix 24) DO, pH, E_h, and conductivity profiles through Gate 1 on March 10, June 2, July 14, and September 4, 1997 (111, 195, 238, and 288 days after sources started) are shown in Figure 2.20. Results indicate that anaerobic conditions (DO < 1 mg/L and negative E_h measurements) generally existed through the length of the gate, with the strongest reducing conditions being observed in the iron environment. The only exception to this generality was observed during the July and September sampling events, when the average DO concentration at fence G108 approached 2 mg/L due to the influence of the ORC™. pH at the influent end of the gate was near neutral (pH 7 to 8), but quickly increased through the iron to values of approximately 9.5. A further pH increase, up to pH 11, was evident after contact with the ORC™ (downgradient of cassette C4 at early time, and downgradient of the alternate oxygen addition wells at late time). A gradual decline in pH was observed in the aquifer downgradient of the cassettes, possibly due to natural buffering, or as a result of mixing with water that bypassed the iron through the leak. Conductivity values decreased through the iron with a gradual increase observed along the length of the gate, possibly returning to background levels.

Figures 2.21 and 2.22 show the behavior of major inorganic constituents (cations and anions, respectively) through the length of the gate on days 111, 195, 238, and 288 (March 10, June 2, July 14, and September 4, 1997). All concentrations are fence-averaged; however, these results were quite consistent between individual points at a fence, particularly within the cassette system. From the figures it is evident that concentrations of calcium and sulfate and alkalinity decreased after contact with the

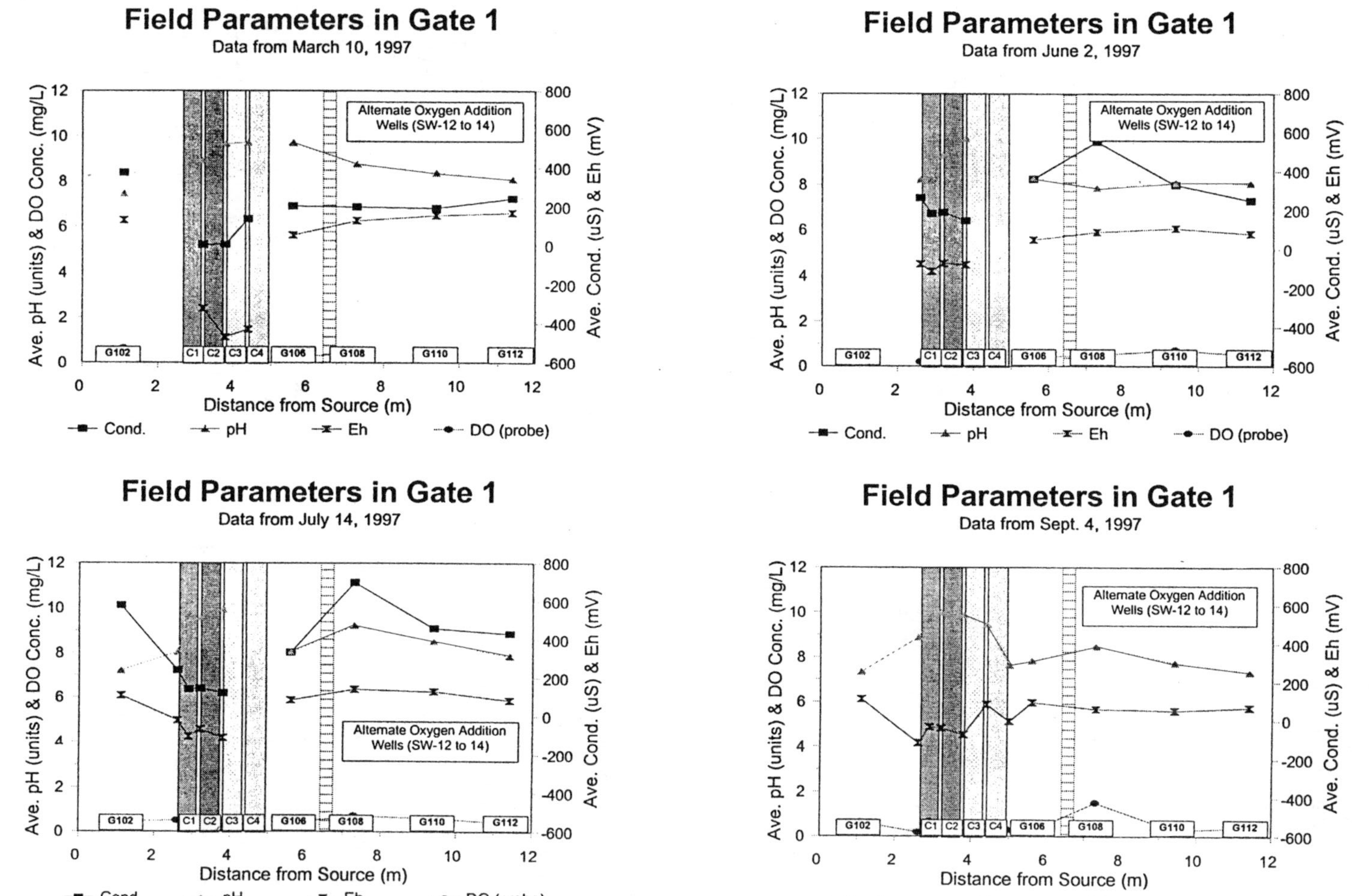

Figure 2.20 pH, DO, E_h, and conductivity profiles through Gate 1 (sampling on dates noted).

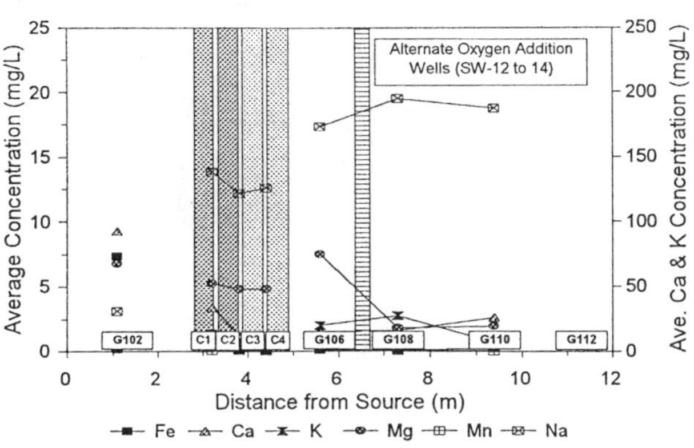

Cation Profile in Gate 1
Data from June 2, 1997
Alternate Oxygen Addition Wells (SW-12 to 14)
G102
C1 C2 C3 C4
G106
G108
G110
G112
Average Concentration (mg/L)
Ave. Ca & K Concentration (mg/L)
Distance from Source (m)
Fe Ca K Mg Mn Na

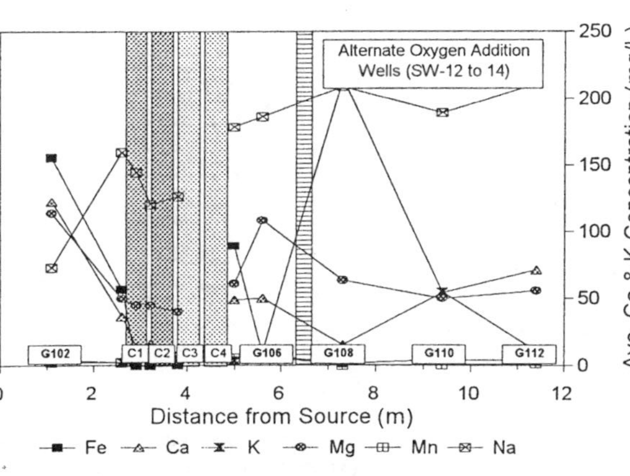

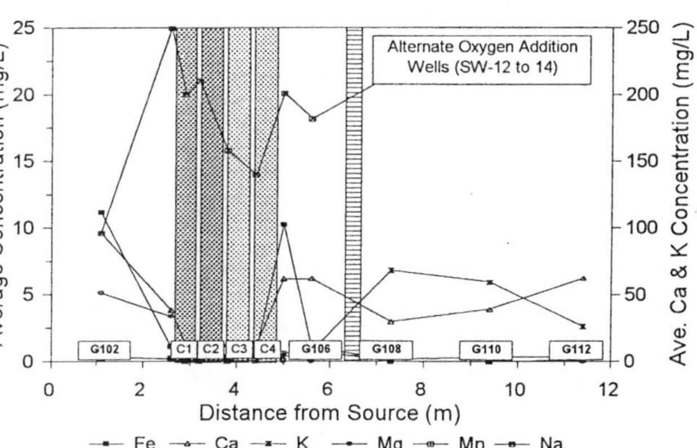

Figure 2.21 Cation profiles for Gate 1 (sampling on dates noted).

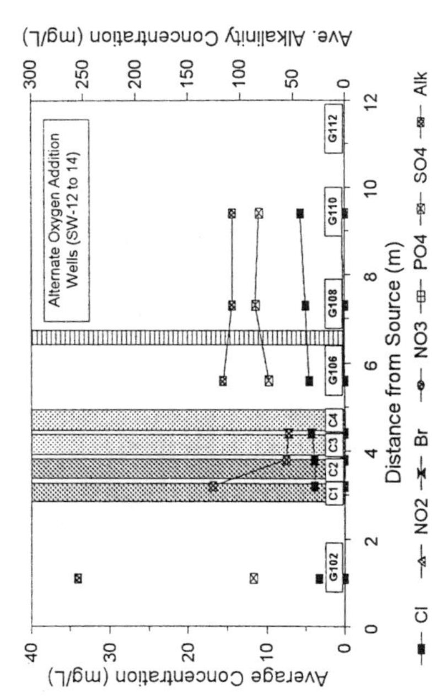

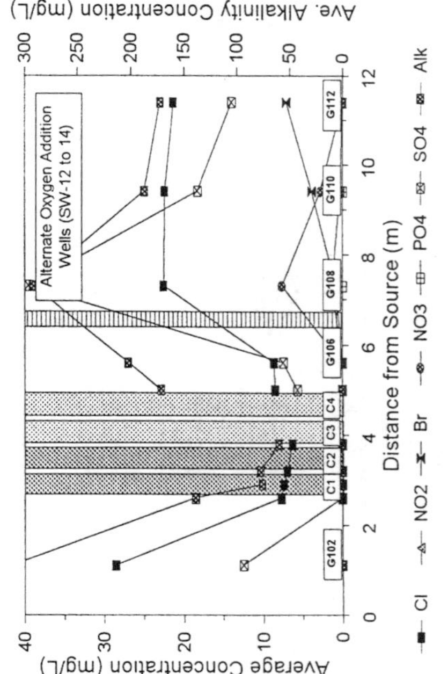

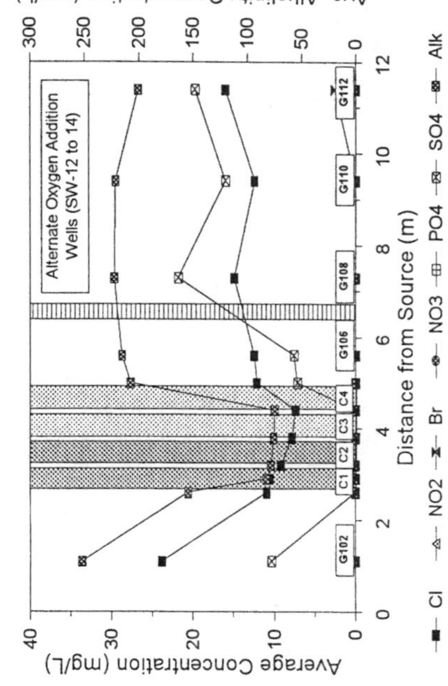

Figure 2.22 Anion profiles for Gate 1 (sampling on dates noted).

granular iron. The high pH (pH >9) of the emanating water may have caused the decrease in calcium and alkalinity due to $CaCO_3$ precipitation. Sulfate was possibly transformed to bisulfide in the environment caused by the iron (reducing conditions and pH >9). Although predicted by the reductive dechlorination theory, elevated concentrations of dissolved iron did not appear. Rather, dissolved iron concentrations also decreased through the granular iron, suggesting the formation of precipitates.

Downgradient of the cassettes, concentrations of calcium, sulfate, alkalinity, potassium, and magnesium increased to background or higher levels. Equilibration of water with aquifer materials, leading to the dissolution of aquifer minerals or surface coatings, is a possible explanation. Also, water bypassing the cassettes through the suspected leak (Appendix 14) could have contributed to the increases observed. Higher downgradient concentrations of calcium and alkalinity on later sampling dates suggested that the influence of the leak was becoming more pronounced, as previously discussed (Appendix 14). In the case of potassium and magnesium, increased dissolved concentrations were likely due to dissolution of ORC™ constituents. Manganese and phosphate concentrations were consistently low through the length of the gate. The lack of phosphate in the system possibly suggests that precipitate formation involving this compound was occurring or that microbial populations were efficient at removing any phosphate released from the ORC™. Concentrations of chloride tended to remain constant through the cassettes. Due to low concentrations of chlorinated compounds entering the cassettes, significant increases in chloride concentrations in this area were not expected, and a mass balance was not calculated. Elevated concentrations of bromide on the July and September sampling occasions were the result of tracer dilution testing in the alternate oxygen addition wells.

2.3 ADDITIONAL INVESTIGATIONS

Additional investigations were completed to investigate discovered or suspected limitations associated with the use of granular iron and ORC™ in sequential treatment systems. The first study presented deals with the suspected pH incompatibility problem between the two treatment zones and is described in Section 2.3.1. Investigations into the surface characteristics of the reactive media and the changes that occur over time are discussed in Section 2.3.2.

2.3.1 pH Tolerance of the ORC™

2.3.1.1 Problem Definition

ORC™ performance monitoring showed that no measurable increase in DO concentration occurred following the installation of the ORC™ socks in C4 (Appendix 13). Although the effective release of oxygen from the ORC™ given by the reaction $2MgO_2 + 2H_2O \rightarrow 2Mg(OH)_2 + O_2$ does not appear to depend on the initial pH of the groundwater system, preliminary field results indicated that high pH water emanating from the iron cassettes may have limited the desired reaction.

2.3.1.2 Study Objectives

The following questions related to the incompatibility of the iron pH system were addressed:

1. Is pH the condition controlling the performance of the ORC™?
2. If so, what is the desirable range in pH to activate the ORC™?
3. What alternatives might exist to reduce the pH of water to within the desired range, after its contact with the iron?
4. Can the iron be altered to eliminate the problem of high pH water?

2.3.1.3 Methodology

Laboratory batch experiments with ORC™ and water were completed to address the first two questions above. DO concentrations and pH were monitored over time following the addition of ORC™ (same formulation as used in the field; 50:50 mix of ORC™ powder and sand) to water (either deionized or iron-equilibrated water was used). To investigate the pH dependency of the DO release, various pH ranges were maintained through the drop wise addition of acid to the ORC™/water system. DO concentrations and pH were again monitored over time. Alternative methods to reduce the pH of the water contacting the ORC™ were investigated, including slow acid addition in the field, sparging using carbon dioxide, and the use of the natural buffering capacity of the aquifer materials. Also, the status of the iron technology with respect to preventing the release of high pH water was considered through a review of currently available literature. Details of these studies are provided in Appendix 15.

2.3.1.4 Results

Figure 2.23 shows the results of several ORC™ laboratory experiments and demonstrates the pH dependence of the oxygen-releasing reaction. Effective oxygen release from the ORC™ was achieved when pH values were maintained in the near-neutral range (7 to 8). DO concentrations of greater than 15 mg/L were obtained after only 10 minutes of contact between the water and ORC™. These results compare well to previous demonstrations of the product (Chapman et al., 1997; Smyth et al., 1995). At higher pH, increases in DO were considerably less. DO increased from 8 mg/L to approximately 11 mg/L when pH was maintained between 8 and 9, while for a pH range of 9 to 10, only a very slight increase in DO from 8 to 9 mg/L was observed. Substitution of low E_h, iron-equilibrated water for deionized water in laboratory experiments did not affect ORC™ performance at near-neutral pH. Thus, observed poor performance in the field was unrelated to the reduction potential (E_h) of the system.

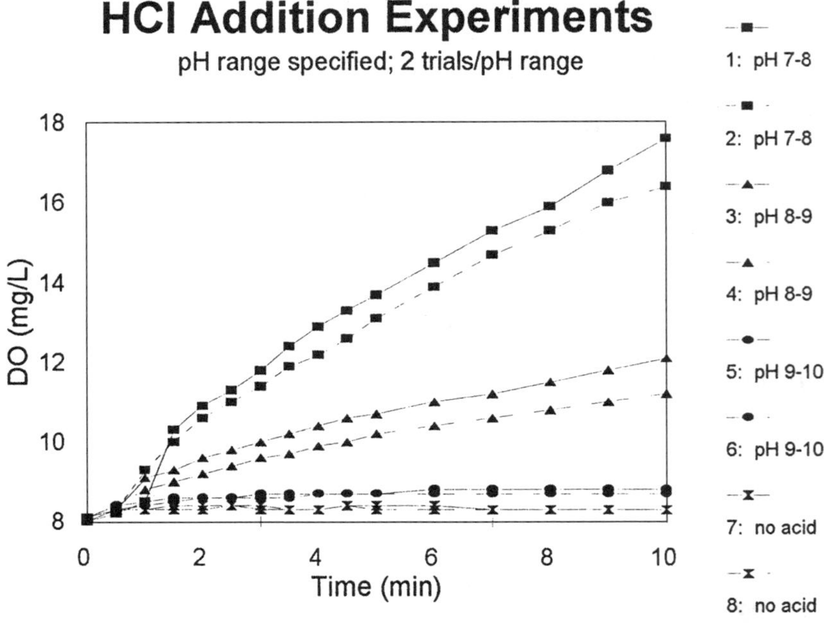

Figure 2.23 pH dependence of ORC™ reaction demonstrated in laboratory.

Various methods of reducing the pH of the groundwater emanating from the iron cassettes were considered including sparging with carbon dioxide, slow acid addition within the cassette system, the use of sulfide minerals mixed with the iron or following it in sequence, and the use of natural aquifer buffering. At the Alameda site, sparging with carbon dioxide was undertaken to reduce the pH of the water emanating from an iron barrier to create a favorable environment for microbial degradation (Chapter 5). At CFB Borden, mineral acid addition within the cassettes (between C3 and C4) successfully reduced the pH of the water emanating from the iron cassettes. However, effective oxygen release from the ORC™ was not achieved, possibly due to unfavorable system hydraulics. Mixing of sulfide minerals directly with granular iron to maintain a near-neutral pH within the iron system was demonstrated in the laboratory by Burris et al. (1995). Batch experiments using granular iron and pyrite to treat PCE and TCE were conducted, and over a 456-hour period, the system pH remained between 6.4 and 6.7. Use of sulfide minerals downgradient of a granular iron wall might also be effective at controlling the system pH. This alternative may be acceptable for field applications, although other side effects must be considered (e.g., the potential for increased concentrations of inorganic constituents, which may in turn cause increased precipitate formation and possible impacts to overall system performance). The natural buffering capacity of the aquifer materials downgradient of the cassette system was recognized in the early stage of this field demonstration, and the pH at fences G106 and G108 was near-neutral (Figure 2.24). Alternate oxygen addition wells (SW-12 to 14), located between these two fences, were adopted as ORC™ wells as of June 19, 1997. Use of these wells is discussed in Section 2.2.3.6 and Appendix 13.

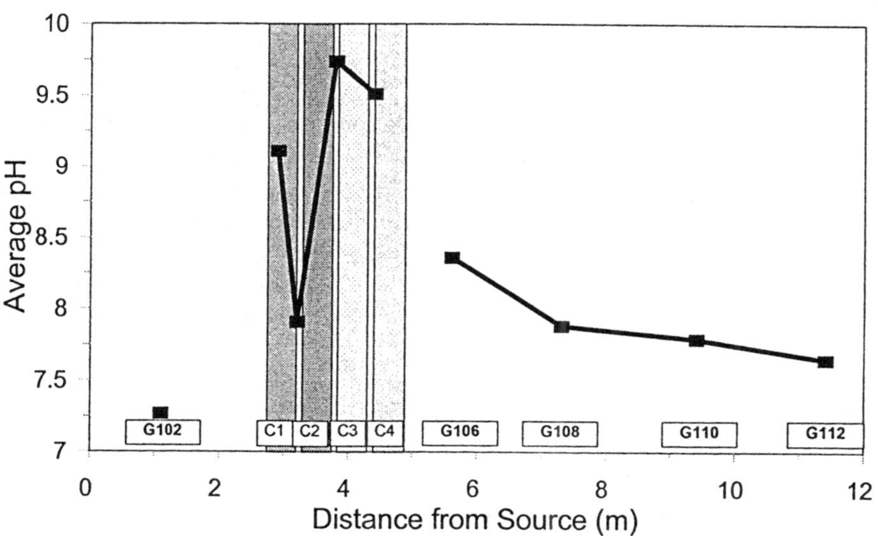

Figure 2.24 Natural pH buffering demonstrated.

Nickel-plated iron could provide a suitable alternative to the iron used in this study since it does not cause a pH increase upon contact with water (Gillham, personal communication). Laboratory studies have demonstrated accelerated degradation rates for chlorinated compounds in some cases; however, the technology is still in the early stages of development and is not ready for field use at this time.

2.3.2 Iron Surface Investigations

2.3.2.1 Problem Definition

Commercial-grade granular iron has been used in many previous experiments due to its availability and low cost. However, the presence on material surfaces of iron oxides, resulting from manufacturing, may reduce its reactivity. Also, the inactivation of the iron surface, or the formation of inorganic precipitates in an *in situ* iron wall, may limit its effective lifetime (Schuhmacher, 1995).

2.3.2.2 Study Objectives

The objectives of this study were

1. To investigate changes to the oxidized surface films that exist as a result of product manufacturing by comparing surface characteristics of granular iron samples prior to contact, and after equilibration, with Borden groundwater.
2. To investigate the formation of precipitates on iron surfaces after prolonged contact with organic contaminants.

2.3.2.3 Methodology

Cassette C1 (25% iron with 75% silica sand), cassette C2 (100% iron), cassette C3 (50% limestone screenings with 50% silica sand), and the natural sand aquifer were sampled. The commercial-grade granular iron used in this study was produced by Peerless Metal Powder and Abrasive (Detroit, MI). For brevity, only the results related to iron samples from cassette C2 are presented. "As received" granular iron samples are referred to as "background" samples. Samples collected after several months exposure to Borden groundwater (August 1996) are referred to as "exposed" samples, and samples obtained after contact with organic contaminants (September 3, 1997) are referred to as "final" samples.

Field samples of the reactive media were collected using a coring technique similar to that described by Starr and Ingleton (1992). Samples were obtained from the middle of the saturated aquifer, at a depth of approximately 2.7 m below ground. Care was taken to prevent exposure of the samples to atmospheric oxygen (cores were cut to the desired length and sealed immediately after removal from the subsurface; cores were stored and sample preparations completed in an anaerobic glove box; and samples were analyzed in water). Surface investigations were performed using a Renishaw 1000 Raman microscope system. A full description of the field sampling methods and surface analysis techniques, including sample handling and laboratory methods, is provided in Appendix 16.

2.3.2.4 Results

Good correlation between the Raman results from this study and available literature data is evident from Table 2.6, which provides a summary of observed and expected band positions for Raman vibrational modes. Raman spectra of the "as received" Peerless iron clearly indicated that both horizontal and vertical heterogeneity of each iron grain exist. The vertical heterogeneity was manifested by the presence of outer and inner surface oxide films. The outer surface layer consists of α-Fe_2O_3 and γ-Fe_2O_3 with randomly distributed islands of separated α and γ phases. Considerable variability between individual iron grains and on any particular grain was evident. These oxide films are visible to the naked eye (appear as rusty coating) and result from the manufacturing of the product, due to the high temperature treatment process that the iron undergoes. The presence of these oxides hinders the transfer of electrons at the interface (Pourbaix, 1973; Frankenthal and

Table 2.6 Raman Bands of Selected Iron Oxides in 200–800 cm^{-1} Region

Species	Raman Band Position (cm^{-1})									Ref.
α - Fe_2O_3	225	245	295		415	500	615			(1)
α - Fe_2O_3	227	245	293	298	414	501	612			(2)
α - Fe_2O_3	226	245	293		413	500	612			(3)
α - Fe_2O_3	224	244	291		409	499	609			this
	vs	vw	s		m	vw	w			work
γ - Fe_2O_3	265	300	345		395	515	645	670	715	(1)
γ - Fe_2O_3			350			505		660	710	(4)
γ - Fe_2O_3	263		350		380	505	650		740	(5)
γ - Fe_2O_3	266	290	342		380	502	640	668	714	this
	vw	vw	s, sh		s, sh	s	s, sh	s, sh	s, sh	work
Fe_3O_4		300	320		420	560	680			(6)
Fe_3O_4		294	319		415	540	669			(7)
Fe_3O_4						540	665			(1)
Fe_3O_4			310			540	669			this
			vb			w	vs			work
α - FeOOH	245	300	390	420	480	550	685			(1)
α - FeOOH		299	397		479	550	685			(8)
α - FeOOH	250	300	385		470	560				(5)
α - FeOOH		298	397	414	474	550				(2)
α - FeOOH	245	298	388	412	481	550	670			this
	m	s	s	m, sh	m	w	vb			work

Notes: A_1 modes and other lines of a high intensity are underlined.
Intensity statements and abbreviations: vs - very strong; s - strong; m - medium; w - weak; vw - very weak; vb - very broad; sh - shoulder. References: (1) Ohtsuka et al., 1986; (2) Thibeau, 1978; (3) Beattie and Gilson, 1970; (4) Boucherit et al., 1991; (5) Thierry et al., 1991; (6) Verble, 1974; (7) Odziemkowski et al., 1994; (8) Nauer et al., 1985.

Kruger, 1978) and therefore might slow the reductive dechlorination reaction (assuming that charge transfer at the iron surface is the rate-determining step in the reaction). In contrast, the inner layer, exposed by mechanical removal of the surface films, consists of magnetite (Fe_3O_4), which possesses less protective properties than the passive Fe(III) layers and allows the charge transfer to occur through it.

Figure 2.25 provides a comparison of Raman spectra data for exposed iron samples (shown as A and B in the figure) and for magnetite (C). It is evident from these spectra that, after equilibration, the γ-Fe_2O_3 disappeared from the iron surface, as had most of the α-Fe_2O_3 film (many of the α-Fe_2O_3 bands were no longer visible). However, complete removal of these films over the entire surface did not occur. Nonetheless, this behavior supports the autoreduction theory presented in the literature (Odziemkowski and Gillham, 1997). The formation of two new broad bands at 420 and 507 cm^{-1} indicates the presence of "green rust" (Boucherit and Hugot-Le Goff, 1992), formed as a result of the autoreduction process. Green rust is a thick unstable colloidal hydroxy salt layer, comprising Fe^{2+}, Fe^{3+}, hydroxyl ions, and interlayer hydroxyl or nonhydroxyl ions, and may include sulfates, halides, nitrates, and carbonates (Simard et al., 1997). Green rust allows charge transfer to occur through it.

The surface characteristics of the "final" samples were very similar to those observed for the "exposed" samples. Surface heterogeneity was evident, and identified compounds include "green rust," magnetite (Fe_3O_4), oxyhydroxides (FeOOH), and in certain spots α-Fe_2O_3 and γ-Fe_2O_3 films observed on the "background" samples. One spot of calcite was also identified on sample C2-3 obtained from the upgradient side of the cassette.

The presence of passive oxide films (consisting of α-Fe_2O_3 and γ-Fe_2O_3) indicates that for this particular commercial grade of granular iron, complete autoreduction of these films did not occur. The reactivity of the surface is expected to vary between commercially available granular iron materials (e.g., Master Builder, Fischer Scientific, Connelly) due to variations in the treatment

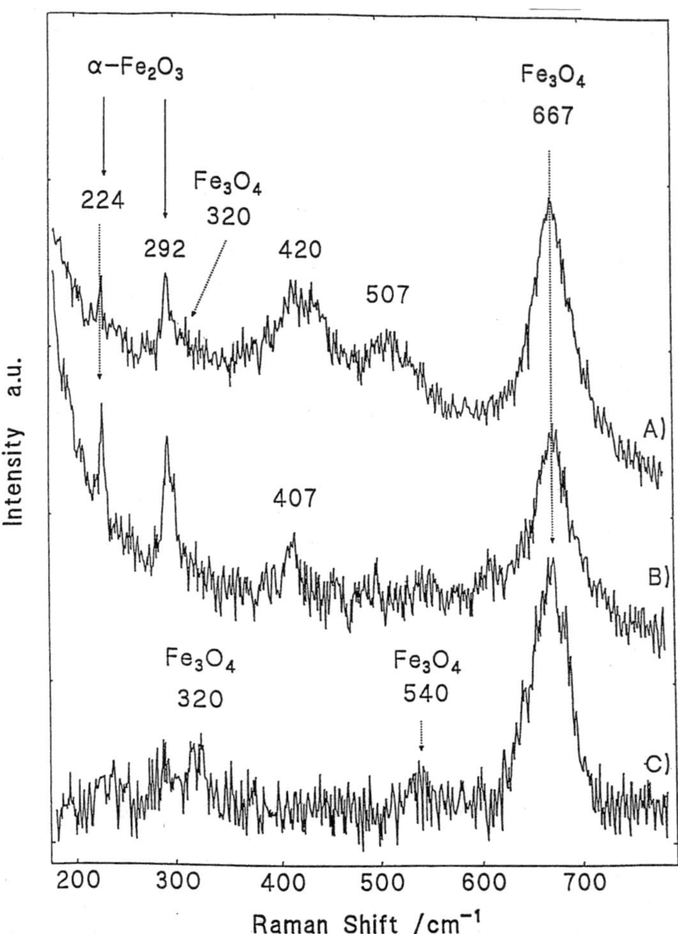

Figure 2.25 Comparison spectra of "exposed" iron sample (A & B) and magnetite (C).

processes to which the iron is subjected. Degradation rates for chlorinated compounds within the iron were shown to depend upon the surface area of the iron (Gillham and O'Hannesin, 1994). However, it is more likely only the unprotected portion of the surface that acts in the reductive dechlorination reaction (assuming that charge transfer at the surface is responsible for the reaction) and thus, only this percentage of the surface area should be considered in any half-life normalization calculations. Quantification of this percentage is difficult and subjective.

Magnetite and "green rust" were also observed on the metal surfaces, but they do not substantially hinder the transfer across the metal/solution interface. The presence of these compounds after long-term exposure to Borden groundwater and contaminants is not surprising. Iron oxyhydroxides found on the "final" samples occur due to contact with water and as a result of the autoreduction process. The formation of iron oxyhydroxide films on the iron surface will hinder the degradation process no matter what the origin of the reductive dechlorination reaction (i.e., what mechanism drives the degradation reaction).

Identification of only one spot containing carbonate minerals on the iron samples was encouraging with respect to the formation of precipitates on the iron surfaces, and the subsequent loss of hydraulic conductivity through the reactive wall. It was not surprising to identify this compound on the sample obtained on the upgradient side of the wall. Precipitate formation was typically observed, in the laboratory at least, at the influent end of columns containing iron (Schuhmacher,

1995). Thus, the upgradient portion of a reactive iron wall would be most susceptible to this deposition. Precipitate formation on samples from the upgradient side of cassette C1 would be expected; however, these samples were not analyzed as part of this work.

2.4 PERFORMANCE DATA EVALUATION

2.4.1 Mass Balance and Removal Efficiencies

2.4.1.1 Granular Iron

Carbon mass balances for PCE and CT through the cassettes were completed, and a thorough discussion of the calculations is provided in Appendix 12. For PCE, mass balances ranged from –5 to 48%, while for CT, they ranged from 18 to 28%. The mass balance for CT was complicated by the intrinsic degradation of this compound to form CF, upgradient of the iron cassettes.

A combination of factors was expected to contribute to poor mass balances even before they were attempted. To begin with, laboratory analyses were completed only for the expected breakdown products from the stepwise dechlorination of PCE and CT. However, other degradation pathways are known to exist (Roberts et al., 1996). Specific analyses for other possible partial degradation products (e.g., acetylene, acetate, etc.) were not conducted, and thus a perfect mass balance could not be expected. Sorption of chlorinated solvents in the granular iron has been demonstrated in the laboratory (Burris et al., 1995) and could be acting in the cassettes. Volatile losses could be another cause of the poor mass balances. Ethane and ethene are generally reactive compounds within a groundwater system; however, the presence of the open water zones and the mixing that was observed within them would be expected to enhance these losses. Maximum concentrations of these compounds were typically not measured at fence G103 (considered as the effluent concentration), suggesting that losses either due to volatilization or breakdown to carbon dioxide were occurring along the length of the groundwater flow path. Uncertainty with respect to the changing flow regime within the cassettes could also cause some problems, allowing longer residence times within the open water zones.

In general, the granular iron performed as expected and normalized degradation rates, for PCE at least, compared well with literature data. The observed rates were much greater than those reported by Jeffers et al. (1989) for intrinsic abiotic reactions, such as hydrolysis (half-lives of 8.7×10^{10} hour for PCE and 1.6×10^{7} hour for CF were reported).

A leak around the cassettes (details in Appendix 14) allowed contaminated water to bypass the iron, creating a PCE and CF plume in the aquifer from fences G106 to G112. This leak hindered the typically straightforward assessment of the ability of the granular iron to degrade chlorinated compounds. Nonetheless, a comparison of influent (at fence G1C0) and effluent concentrations (at fence G104) for the iron cassettes demonstrates the effectiveness of the iron technology. Table 2.7 provides this comparison.

Results indicate that typically greater than 93% of the chlorinated contaminants were removed using the granular iron technology. A slightly lower value (84%) was obtained for CT for the March 24, 1997 sampling event, although this low percentage is primarily an artifact of the low concentrations entering the system. Percent reduction values could not be determined for CT on the other sampling dates, since both influent and effluent concentrations were too low. DCM concentrations of between 18 and 25 μg/L persisted beyond the iron. The production and persistence of DCM through the iron was expected based on previous results (Gillham and O'Hannesin, 1994). At higher input contaminant concentrations the production of DCM should increase, although within the aerobic environment created downgradient of the granular iron this contaminant might biodegrade to concentrations below detection (Brunner et al., 1980). However, this behavior was not demonstrated in microcosm studies related to the Borden AATDF study (Appendix 11).

Table 2.7 Comparison of Influent and Effluent Concentrations for the Iron Cassettes

		Fence-Averaged Conc. (μg/L)		
		Influent	Effluent	Percent
Date	Compound	(G1C0)	(G104)	Reduction
March 24/97	PCE	71.4	**<MDL**	>93%
Day 125	CT	43.5	6.8	84%
16 PV	CF	87.2	<MDL	>94%
	DCM	**2.9**	18	N/A
June 2/97	PCE	47.3	**0.1**	>99%
Day 195	CT	**1.4**	**2.3**	N/A
28 PV	CF	85.6	**<MDL**	>94%
	DCM	6.1	**23.1**	N/A
July 14/97	PCE	38.1	**0.1**	>99%
Day 238	CT	**<MDL**	**0.9**	N/A
35 PV	CF	68.5	**<MDL**	>93%
	DCM	**12.6**	**25.0**	N/A

Notes: Day *x* refers to number of days since contaminant injection started (November 18, 1996). # PV refers to number of pore volumes, calculated based on a 6-day retention time within the iron cassettes (accounting for plume development and time for transport to the cassettes). Parameters that were not detected are shown as <MDL and were taken as 5 μg/L for the purposes of these calculations. Average concentration values below the LOQ (analyte specific) are shown in **bold**. For DCM, concentrations increased through the iron and percent reduction calculations are not appropriate (shown as N/A).

2.4.1.2 Aerobic Biodegradation

The success of any aerobic bioremediation effort depends directly upon the ability of the system to deliver oxygen to the microbial population. As the system was originally designed, ORC™ failed to fulfill this requirement. However, use of the ORC™ socks in the alternate oxygen addition wells seemed to overcome this problem, ultimately delivering at least an average of 3 mg/L of DO (measured at fence G108) to the groundwater environment.

A comparison of initial fence-averaged concentrations (measured on the day the ORC™ was installed, June 18, 1997) at fence G112 with the final fence-averaged concentrations (measured on October 21, 1997, 124 days after ORC™ installation) demonstrates the effectiveness of the enhanced bioremediation efforts undertaken (Table 2.8). Intrinsic losses between fences G106 and G112 are eliminated from percent reduction calculations using this method, assuming that input is steady state over this time period.

As discussed previously, complete aerobic degradation of toluene and CF was likely limited by the availability of oxygen within the aquifer. Nonetheless, losses of greater than 90% were observed

Table 2.8 Evaluation of the Effectiveness of Biodegradation Efforts

	Average Concentrations at Fence G112		Percent
Compound	June 18, 1997	October 21, 1997	Reduction
TOL	1026	95.5	91%
PCE	15.0	22.2	(48%)
CF	71.0	<MDL	>93%

Notes: All concentrations are given in μg/L. Average concentrations are fence-averaged (calculated as an arithmetic average). Parameters that were not detected are shown as <MDL and were taken as 5 μg/L for the purposes of these calculations. Percent reduction values shown in brackets are net gains.

for both of these compounds. Toluene degradation was likely an aerobic mineralization process, while the observed losses in CF might have occurred by cometabolism. For PCE, a net concentration increase was observed at fence G112 over time, since this compound had not yet reached steady state. Further, aerobic degradation of PCE was not expected (Mohn and Tiedje, 1992).

2.4.1.3 Overall System Performance

A direct comparison of Gate 1 performance and the control gate (Gate 2) demonstrates the effectiveness of the sequential iron/ORC™ system implemented at the AATDF site (Table 2.9). Fences G104 and G205 are located at approximately the same distance from the source and provide an effective point of comparison for chlorinated losses between the gates. At fence G104, located 3.8 m from the source, all chlorinated compounds, with the exception of DCM, were below detection. DCM concentrations of between 18 and 25 μg/L persisted beyond the iron; however, these decreased within the aquifer (at fence G106) to levels below the method detection limit (18 μg/L). Early data from fence G205 suggested no PCE degradation, while CT degraded to CF, which then moved unreacted through the aquifer (details in Chapter 3). CF concentrations at fence G205 appeared to decline at later time (October 21, 1997), indicating that further degradation may be occurring. Also, TCE and/or *c*DCE were detected at a few locations in both gates, suggesting that intrinsic PCE degradation was commencing. Apparent toluene losses between Gates 1 and 2 may be the result of sorption to the granular iron, volatile losses within the cassettes, or differences between input concentrations. Toluene was not expected to degrade within the iron.

A comparison between fences G112 and G212 provides an effective comparison between the overall performance of these two gates. At early time, a considerable improvement in performance of Gate 1 over Gate 2 (assessed at fences G112 and G212) was evident, particularly for the

Table 2.9 Comparison of Gates 1 and 2 Performance for All Contaminants

		Average Conc. (mγ/L)		Percent	Average Conc. (mγ/L)		Percent
Date	**Compound**	**G104**	**G205**	**Reduction**	**G112**	**G212**	**Reduction**
Mar 10/97	PCE	**<MDL**	1631	>99%	**<MDL**	63.6	>92%
	CT	**3.1**	103	97%	**<MDL**	**0.5**	N/A
	CF	**1.2**	209	>99%	41.9	93.7	55%
	DCM	**13.4**	**<MDL**	N/A	**<MDL**	**0.4**	N/A
	TOL	1350	5930	77%	**<MDL**	1000	>99%
June 2/97	PCE	**0.1**	79.4	>99%	**7.3**	113	94%
	CT	**2.3**	33.1	93%	**<MDL**	**<MDL**	N/A
	CF	**<MDL**	100.0	>95%	38.7	144	73%
	DCM	**23.1**	**<MDL**	N/A	**<MDL**	**<MDL**	N/A
	TOL	852	984	13%	1030	1730	41%
July 14/97	PCE	**<MDL**	98.9	>99%	17.7	270	93%
	CT	**<MDL**	**10.4**	91%	**<MDL**	**<MDL**	N/A
	CF	**<MDL**	190	95%	69.1	87.5	21%
	DCM	25.0	**<MDL**	N/A	**<MDL**	**<MDL**	N/A
	TOL	674	1590	58%	993	1220	19%
Oct. 21/97	PCE	**<MDL**	181	97%	22.2	84.0	74%
	CT	**<MDL**	**1.8**	N/A	**<MDL**	**<MDL**	N/A
	CF	**<MDL**	53.3	91%	**<MDL**	**<MDL**	N/A
	DCM	**17.9**	**<MDL**	N/A	**<MDL**	**<MDL**	N/A
	TOL	479	2140	78%	95.5	88.3	(8%)

Notes: Concentrations not detected during analyses are shown as <MDL and were taken as 5 μg/L for the purposes of these calculations. Average concentrations <LOQ (Appendix 6) are shown in **bold**. Percent reduction values shown in brackets indicate a net increase in compound concentration between Gate 1 and 2 values. N/A indicates that percent reduction calculations are not appropriate, since concentrations are low.

chlorinated compounds. Percent reduction values in the 80% to >90% range were calculated. Also, the gates were comparable in their treatment of toluene. Oxygen was not produced in the aerobic portion of Gate 1 until mid- to late-July 1997, and consequently no difference in biodegradation should have been expected between the two gates. At later time, the influence of the leak around the iron cassettes was greater than at early time, and concentrations of chlorinated compounds reaching the aerobic portion of Gate 1 increased. Concentrations of PCE at fence G112 increased with time; however, CF concentrations were declining, possibly due to aerobic, cometabolic processes. Significant losses in chloromethanes were observed at fence G212, and October 21, 1997 data showed no detectable chlorinated products of CT (concentrations of CF and DCM were both below the MDLs). The capability of the natural system to completely degrade CT and CF to nondetectable products (e.g., carbon dioxide) within an anaerobic environment was improving.

Losses of toluene in Gate 2 were ultimately comparable to those in Gate 1; however, the processes responsible for these losses were considerably different. In Gate 1, an aerobic environment was created (to some extent, at least) through the use of ORC™, and aerobic microbial populations were believed to be responsible for toluene degradation. Toluene in Gate 2 was likely used as an electron donor for anaerobic microbes (Chapter 3).

Although both gates ultimately provided similar performance, several factors suggest that the engineered technologies implemented in Gate 1 would provide more reliable and efficient site restoration:

- The degradation of chlorinated compounds using granular iron was limited by the existence of the leak between the sheet-piling wall and the outer cassette box. Nonetheless, the performance of the iron technology was as expected (normalized half-life values were comparable to literature values), and without the leak, the system should have continued to operate as effectively. The presence of toluene in the groundwater did not appear to hinder the degradation of chlorinated compounds within the iron.
- The stimulation of the natural microbial populations to degrade recalcitrant compounds such as PCE and CF is difficult and often requires considerable acclimation time. Degradation using the granular iron technology is virtually instantaneous. Complete degradation of the target compounds (PCE and CT) has been demonstrated previously, but a good understanding of the contaminant source and flow environment is required.
- Elevated concentrations of toluene may have served as a carbon substrate to drive Gate 2 further into the anaerobic regime and consequently created an environment conducive to the biodegradation of CF and possibly PCE. This process may be difficult to control and/or predict and may not apply for all sites.
- Aerobic toluene degradation has been shown to be rapid and complete (Allen-King et al., 1994). Provision of oxygen to the groundwater using ORC™ was demonstrated for Gate 1 once problems related to pH were overcome. Toluene degradation in Gate 2 occurs under bulk anaerobic conditions. Although this reaction has been observed, it is typically slower and more difficult to control than the aerobic process.

2.4.2 Design Limitations

2.4.2.1 Removable Cassette System

The removable cassette system is a convenient variation on the *in situ* Funnel-and-Gate theme, with flexibility to accommodate new and improved technologies, or variations in plume composition. Also, issues related to the effective lifetime of reactive media make the removable cassette system appealing, since lifting and refilling a cassette is expected to be a relatively simple operation. However, several issues that complicate treatment system design need to be addressed:

1. An effective seal of the cassettes within the Funnel-and-Gate is critical to the operation of the system.
2. The complexity of the groundwater flow system due to open water zones and varying hydraulic conductivities complicates degradation analyses.
3. The removable cassettes were designed with the capacity to be lifted and the reactive media replenished and/or replaced. Although repairs to leaky silicone pressure seals, involving the lifting of the cassettes, were conducted after the installation of the granular iron, no attempt to move the cassettes was made near the end of the project. Long-term exposure to groundwater could cause some rusting and corrosion that may hinder the removal of the cassettes for reuse. Future efforts in this regard might provide results useful for this technology.

The presence of a leak between the sheet-piling wall and the outer cassette box resulted in the formation of an unexpected chlorinated solvent plume in the aerobic portion of the system. It was possible to evaluate the performance of the iron technology based on monitoring data from within the cassettes themselves, but the overall success of the sequential treatment system requires demonstration that the leak can be repaired or circumvented in the future. A leak of this magnitude at a full-scale field demonstration would seriously compromise the results of any remediation effort.

The presence of open water zones in the cassette system created good lateral mixing of the contaminants and thus raises questions related to flow patterns in the system. Hydraulic modeling of the system might provide some insight into this issue. The open water environment of C4 into which the ORC™ socks were suspended was also believed to be a factor in the limited release of oxygen from the socks. Again, some modeling to support observations made might clarify this phenomenon.

2.4.2.2 Incompatibility between Iron and ORC™

The compatibility of iron-contacted groundwater and ORC™ was shown to be, at least to some extent, pH-dependent. Based on this idea, the need to reduce the pH of water emanating from the iron prior to its contact with ORC™ was recognized, and various methods of rectifying the problem were attempted. The presence of natural aquifer material between the two zones seemed to remedy the situation at this site. Ultimately though, good pH buffering capacity of the natural aquifer materials was important for the success of this design, and it was recognized that this natural ability to buffer the pH was limited. At Alameda, sparging with carbon dioxide was shown to provide effective pH control (Chapter 5), while at Borden mineral acid additions were undertaken. Any future installation of a sequential treatment system involving granular iron and ORC™ might consider using the natural pH buffering capacity of aquifer materials to minimize the pH problem. However, an engineered remedy might provide better long-term assurance of an appropriate solution.

2.4.2.3 High pH Downgradient of ORC™

The high pH resulting from the ORC™ reaction may reduce the ability of the natural microbial population to degrade the toluene, since many microbes prefer a near-neutral habitat. Previous field demonstrations of ORC™ have shown good degradation rates for petroleum hydrocarbons (Chapman et al., 1997), probably because the aquifers tested possessed an intrinsic capacity to buffer the pH to near neutrality. The Borden aquifer seems to provide a similar natural pH buffering capacity and no obvious limitations to the use of ORC™ as an oxygen source for aerobic biodegradation were observed.

2.4.2.4 Formation of Precipitates in Iron

In situ iron barriers should provide a long-term, low-maintenance alternative for the remediation of contaminated groundwater plumes. However, a potential limitation is the formation of inorganic precipitates that may inactivate iron surfaces or reduce the hydraulic conductivity of the system due to clogging of pores. Studies to date indicated that precipitate formation within the 100% iron cassette, C2, was minimal. Calcite was identified on the surface of only one sample from the upgradient side of cassette C2. It was recognized that precipitate formation would be most prevalent within the influent portion of the iron (i.e., the upgradient side of cassette C1); however, studies involving these samples were beyond the scope of this project. Based on the long-term performance of the granular wall installed at CFB Borden by O'Hannesin (1993), it is expected that precipitate formation within the iron cassettes will be minimal, and the iron technology is appropriate for use in groundwater environments similar to Borden (generally low alkalinity and total dissolved solids).

2.5 WASTE MANAGEMENT

All extracted water from the gates and waste collected during sampling was treated on site prior to discharge to the environment. Details of the aboveground treatment system are provided in Appendix 10. Upon final completion of the project, disposal of all contaminated materials (for example, source water, granular iron, ORC™ socks) was conducted according to appropriate guidelines.

2.6 CONCLUSIONS

The following conclusions can be drawn based on the field demonstration and additional investigations completed as part of this study.

The gate system provided adequate control of flow and the subsurface environment to assess the performance of the sequential treatment system. The leak between the sheet-piling wall and outer cassette box increased the complexity of data interpretation.

Effective degradation of most chlorinated components of the plume was achieved using granular iron. The only exception was DCM, which exited the iron environment at concentrations of between 18 and 25 μg/L. The formation and persistence of DCM through the iron cassettes were anticipated, based on previous studies.

The close proximity of the iron and ORC™ in the original system design did not permit the effective release of oxygen to the groundwater and, consequently, increased DO concentrations were not observed.

The release of oxygen from ORC™ was shown in the laboratory to be pH-dependent. DO concentrations increased from 8 to 15 mg/L within 10 minutes when pH was maintained between 7 and 8 through the use of dropwise acid addition. At higher pH ranges, oxygen release was significantly reduced (for pH = 8 to 9, maximum DO = 11 mg/L; for pH = 9 to 10, maximum DO = 9 mg/L).

Several alternatives for reducing the pH of water in the field were investigated. Slow acid addition was used to reduce the pH between C3 and C4. Also, pH buffering by natural aquifer materials created a near neutral pH environment approximately 2 m downgradient of the cassettes.

Installation of the ORC™ socks in the alternate oxygen addition wells resulted in elevated DO concentrations within the wells themselves (ranging from 20 to 50 mg/L with an average of approximately 30 mg/L over 4 months), and detectable increases in DO concentrations in the aquifer (at fence G108, DO levels up to 15 mg/L were recorded by late October 1997). Oxygen delivery to the aquifer was variable (measured at fence G108), possibly due to preferential pathways and heterogeneities within the natural sand aquifer. Elevated DO concentrations were not observed downgradient of fence G108, possibly due to rapid consumption.

Toluene degradation was observed within the aerobic portion of the system. Average concentrations were reduced from approximately 1500 μg/L at fence G106 to less than 200 μg/L at fence G112 by late September 1997 (after approximately two pore volumes had traversed the aerobic zone). Further toluene losses were observed by late October 1997 (the average concentration at fence G112 was 95 μg/L).

Cometabolic degradation of chloroform by toluene degraders or methanotrophs was suggested as a possible mechanism for observed concentration declines within the aerobic zone. PCE concentrations appeared to decline through the aerobic portion of the gate; however, this behavior was believed to be the result of slow plume development rather than degradation.

The identification of "green rust" on iron samples collected after several months' exposure to Borden groundwater supported the surface film autoreduction theory of Odziemkowski and Gillham (1997). Overall, precipitate formation in the iron cassettes was believed to be minimal.

2.7 RECOMMENDATIONS FOR FUTURE STUDIES

Several issues requiring further investigation resulted from this study; these are discussed below.

1. This work identified and demonstrated a pH incompatibility between iron-contacted groundwater and ORC™. More effective use of the ORC™ could be achieved with a better understanding of its limitations.
2. Modeling of the flow regime through the cassette system may provide some insight into the behavior of the ORC™ in the open water in cassette C4.
3. The removable cassettes were designed with the capacity to be lifted and the reactive media replenished and/or replaced. Although repairs to leaky silicone pressure seals, involving the lifting of the cassettes, were conducted after the installation of the granular iron, no attempt to move the cassettes was made near the end of the project. Long-term exposure to groundwater could cause some rusting and corrosion that may not permit the removal of the cassettes for reuse. Future efforts in this regard might provide some interesting results for the effective use of this technology.

Problems encountered during this study that could be avoided in future studies include the following:

1. Good seals need to be ensured between the cassette system and the sheet-piling walls. In May 1997, a similar cassette system was installed at another experimental site at CFB Borden. Considerable effort to ensure sound hydraulic seals was undertaken and a new method of creating these seals developed. Also, use of at least some percentage of bentonite in backfill material placed along the sides of the outer cassette box should help to slow migration of any water that passes through the corner seals.
2. Additional leak-detecting piezometers could be installed.
3. Until ORC™ limitations can be better defined, it is important to ensure in any coupled technology using ORC™ that the water contacting the ORC™ has a near-neutral pH.

REFERENCES

Agrawal, A. and P.G. Tratnyek, 1994. Abiotic Remediation of Nitro-Aromatic Groundwater Contaminants by Zero-Valent Iron. Presented at the 207th ACS National Meeting, San Diego, CA, March 13-18, 34(1): 492.

Allen-King, R.M., J.F. Barker, R.W. Gillham, and B.K. Jensen, 1994. Substrate- and Nutrient-Limited Toluene Biotransformation in Sandy Soil. *Environmental Toxicology and Chemistry*, 13(5): 693.

Austrins, C., 1997. Enhanced In-Situ Bioremediation of Hydrocarbon Contaminated Groundwater Using Oxygen and Ammonia Gas Injection in a Funnel and Gate Scheme. M. Sc. Thesis, Department of Earth Sciences, University of Waterloo, Waterloo, Ontario, Canada.

Ball, W.P., C.H. Buehler, T.C. Harmon, D.M. Mackay, and P.V. Roberts, 1990. Characterization of a Sandy Aquifer Material at the Grain Scale. *Journal of Contaminant Hydrology*, 5: 253.

Barker, J., C. Austrins, J. Gorman, J. Devlin, D. Smyth, and J. Cherry, 1995. Controlled In-situ Bioremediation of Groundwater. Published in the proceedings of the 5th Annual Symposium on Groundwater and Soil Remediation (GASReP), Toronto, Ontario, Oct. 2-6, 1995.

Barker, J.F., G.C. Patrick, and D. Major, 1987. Natural Attenuation of Aromatic Hydocarbons in Shallow Sand Aquifer. *Ground Water Monitoring Review*, Winter: 64.

Beattie, I.R., and T.R. Gilson, 1970. The Single-Crystal Raman Spectra of Nearly Opaque Minerals. Iron (III) Oxide and Chromium (III) Oxide. *Journal of the Chemical Society (A).* p. 980.

Bianchi-Mosquera, G.C., R.M. Allen-King, and D.M. Mackay, 1994. Enhanced Degradation of Dissolved Benzene and Toluene Using a Solid Oxygen-Releasing Compound. *Ground Water,* 14(1): 120.

Bonin, P.M.L., M.S. Odziemkowski, and R.W. Gillham, 1998. Influence of Chlorinated Solvents on Polarization and Corrosion Behaviour of Iron in Borate Buffer. *Corrosion Science*, 40(8): 1391–1409.

Borden, R.C., R.T. Goin, and C.M. Kao, 1997. Control of BTEX Migration Using a Biologically Enhanced Permeable Barrier. *Ground Water Monitoring and Remediation*, 17(1): 70-80.

Borden, R.C., C.A. Gomez, and M.T. Becker, 1995. Geochemical Indicators of Intrinsic Bioremediation. *Ground Water*, 33(2): 180.

Boucherit, N., A. Hugot-Le Goff, and S. Joiret, 1991. Raman Studies of Corrosion Films on Fe and Fe-6Mo in Pitting Conditions. *Corrosion Science*, 32: 497.

Boucherit, N., and A. Hugot-Le Goff, 1992. Localized Corrosion Processes in Iron and Steels Studied by *In Situ* Raman Spectroscopy. *Faraday Discussion,* 94: 17.

Brewster, M.L., A.P. Annan, J.P. Greenhouse, B.H. Kueper, G.R. Olhoeft, J.D. Redman, and K.A. Sander, 1995. Observed Migration of a Controlled DNAPL Release by Geophysical Methods. *Ground Water,* 33(6): 977.

Brunner, W., D. Staub, and T. Leisinger, 1980. Bacterial Degradation of Dichloromethane. *Appl. Environ. Microbiol.,* 40: 950-958.

Burris, D.R., T.J. Campbell, and V.S. Manoranjan, 1995. Sorption of Trichloroethylene and Tetrachloroethylene in a Batch Reactive Metallic Iron–Water System. *Environmental Science and Technology.*, 29(11): 2850.

Byerley, B.T., S.W. Chapman, D.A. Smyth, R.D. Wilson, and D.M. Mackay, 1996. Passive Oxygen Release for Enhancement of In-Situ Bioremediation. Presented at the I&EC Special Symposium, American Chemical Society, Birmingham, AL, September 1996.

Canadian Petroleum Products Institute (CPPI), 1995. *Bioremediation Handbook.* Prepared by Intera Information Technologies Ltd., Ottawa, Ontario.

Chapman, S.W., B.T. Byerley, D.J.A. Smyth, and D.M. Mackay, 1997. A Pilot Test of Passive Oxygen Release for Enhancement of In-Situ Bioremediation of BTEX-Contaminated Ground Water. *Ground Water Monitoring and Remediation,* Spring, 1997: 93.

Chapman, S.W., 1996. Spatial Variability and Oxygen Demand of a Petroleum Hydrocarbon Plume: Factors Affecting In-Situ Bioremediation. M.Sc. Thesis, University of Waterloo, Waterloo, Ontario, Canada.

Dean, K.J., W.F. Sherman, and G.R. Wilkinson, 1982. Temperature and Pressure Dependence of the Raman Active Modes of Vibration of α-quartz. *Spectrochimica Acta*, 38A(10): 1105.

Devlin, J.F., 1994. Enhanced In-Situ Biodegradation of CT and TCE Using a Permeable Wall Injection System. Ph. D. Thesis, Department of Earth Sciences, University of Waterloo, Waterloo, Ontario, pp. 633.

Focht, R., J. Vogan, and S.F. O'Hannesin, 1996. Field Applications of Reactive Iron Walls for In-Situ Degradation of Volatile Organic Compounds in Groundwater. *Remediation,* 6(3): 81.

Frankenthal, R.P., and J. Kruger (Eds.), 1978. *Passivity of Metals.* Electrochemical Society, Princeton, NJ.

Freeze, R.A. and J.A. Cherry, 1979. *Groundwater.* Prentice-Hall, Englewood Cliffs, NJ, p. 29.

Gillham, R.W., 1995. In-situ Treatment of Groundwater: Metal-Enhanced Degradation of Chlorinated Organic Contaminants. Published in the proceedings of the Recent Advances in Ground-Water Pollution Control and Remediation Conference, A NATO Advances Study Institute, Kemer, Antalya, Turkey, May 20-June 1, 1995.

Gillham, R.W. and S.F. O'Hannesin, 1994. Enhanced Degradation of Halogenated Aliphatics by Zero-Valent Iron. *Ground Water,* 32(6): 958.

Guiguer, N, T. Franz, J. Molson, and E. Frind, 1995. Flownet — User Guide. Waterloo Hydrogeologic and University of Waterloo.

Herman, R.G., C.E. Bogdan, A.J. Sommer, and D.R. Simpson, 1987. Discrimination among Carbonate Minerals by Raman Spectroscopy Using the Laser Microprobe. *Applied Spectroscopy,* 41(3): 437.

Hvorslev, M.J., 1951. Time Lag and Soil Permeability in Groundwater Observations. *U.S. Army Corps of Engineers Exp. Sta. Bull.,* 36, Vicksburg, MS.

Jeffers, P.M., L.M. Ward, L.M. Woytowitch, and N.L. Wolfe, 1989. Homogeneous Hydrolysis Rate Constants for Selected Chlorinated Methanes, Ethanes, Ethenes and Propanes. *Environmental Science and Technology,* 23(8): 965.

Little, C.D., A.V. Palumbo, S.E. Herbes, M.E. Lidstrom, R.L. Tyndall, and P.J. Gilmer, 1988. Trichloroethylene Biodegradation by a Methane-Oxidizing Bacterium. *Applied Environmental Microbiology,* 54: 951.

MacFarlane, D.S., J.A. Cherry, R.W. Gillham, and E.A. Sudicky, 1983. Migration of Contaminants in Groundwater at a Landfill: A Case Study, 1. Groundwater Flow and Plume Delineation. *Journal of Hydrology,* 63: 1.

Mackay, D.M. and J.A. Cherry, 1989. Groundwater Contamination: Pump-and-Treat Remediation. *Environmental Science and Technology,* 23(6): 630.

Mackay, D.M. D.L. Freyberg, P.V. Roberts, and J.A. Cherry, 1986. A Natural Gradient Experiment on Solute Transport in a Sand Aquifer, 1. Approach and Overview of Plume Movement. *Water Resources Research,* 22(13): 2017.

Matheson, L.J. and P.G. Tratnyek, 1994. Reductive Dehalogenation of Chlorinated Methanes by Iron Metal. *Environmental Science and Technology,* 28: 2045.

McClay, K., B.G. Fox, and R.J. Steffan, 1996. Chloroform Mineralization by Toluene-Oxidizing Bacteria. *Applied Environmental Microbiology,* 62(8): 2716.

Ministry of Environment and Energy of Ontario (MOEE), 1996. *Guideline for Use at Contaminated Sites in Ontario.* Queen's Printer for Ontario, 1996. No. ISBN 0-7778-4052-9.

Mohn, W.W. and J.M. Teidje, 1992. Microbial Reductive Dehalogenation. *Microbiological Reviews,* 56(3): 482.

National Research Council (NRC), 1993. *In-Situ Bioremediation - When Does it Work?* National Academy Press, Washington, DC.

Nauer, G., P. Strecha, N. Brinda-Konopik, and G. Lipstay, 1985. Spectroscopic and Thermoanalytical Characterization of Standard Substances for the Identification of Reaction Products of Iron Electrodes. *J. Therm. Anal.,* p. 813.

Nicholson, R.V., J.A. Cherry, and E.J. Reardon, 1983. Migration of Contaminants in Groundwater at a Landfill: A Case Study, 6. Hydrogeochemistry. *Journal of Hydrology,* 63: 131.

Norris, R.D., R.E. Hinchee, R. Brown, P.L. McCarty, J.T. Wilson, D.H. Kampbell, M. Reinhard, E.J. Bouwer, R.C. Borden, T.M. Vogel, J.M. Thomas, and C.H. Ward, 1994. *Handbook of Bioremediation,* Lewis Publishers, Boca Raton, FL.

Odziemkowski, M.S. and R.W. Gillham, 1997. Surface Redox Reactions on Commercial Grade Zero Valent Iron (Steel) and Their Influence on the Reductive Dechlorination of Solvent: Micro Raman Spectroscopic Studies. Presented at 213th ACS National Meeting, San Francisco, CA., American Chemical Society, Division of Environmental Chemistry, Preprints of Papers, 37(1): 177, April 1997.

Odziemkowski, M., J. Fils, and D.E. Irish, 1994. Raman Spectral and Electrochemical Studies of Surface Film Formation on Iron and Its Alloys with Carbon in $Na_2CO_3/NaHCO_3$ Solution with Reference to Stress Corrosion Cracking. *Electrochim. Acta,* 39: 2225.

O'Hannesin, S.F., 1993. A Field Demonstration of a Permeable Reactive Wall for the In-Situ Abiotic Degradation of Halogenated Aliphatic Organic Compounds. M. Sc. Thesis, University of Waterloo, Waterloo, Ontario, Canada.

Ohtsuka, T., K. Kubo, and N. Sata, 1986. Raman Spectroscopy of Thin Corrosion Films on Iron at 100 to 150°C in Air. *Corrosion,* 42: 476.

Orth, W.S. and R.W. Gillham, 1996. Dechlorination of Trichloroethene in Aqueous Solution using Fe^0. *Environmental Science and Technology,* 30(1): 66.

Pourbaix, M., 1973. *Lectures on Electrochemical Corrosion.* Plenum Press, New York.

Roberts, R.V., M.N. Gotz, and D.M. Mackay, 1986. A Natural Gradient Experiment on Solute Transport in a Sand Aquifer: III. Retardation Estimates and Mass Balances for Organic Solutes. *Water Resources Research,* 22(13): 2047.

Roberts, A.L., L.A. Totten, W.A. Arnold, D.R. Burris, and T.J. Campbell, 1996. Reductive Elimination of Chlorinated Ethylenes by Zero-Valent Metals. *Environmental Science and Technology,* 30(8): 2654.

Schuhmacher, T.T., 1995. Identification of Precipitates Formed on Zero-Valent Iron in Anaerobic Aqueous Solutions. M.Sc. Thesis, University of Waterloo, Waterloo, Ontario, Canada.

Scott, J.F. and S.P.S. Porto, 1967. Longitudinal and Transverse Optical Lattice Vibrations in Quartz. *Physical Review,* 161(3): 903.

Senzaki, T. and Y. Kumagai, 1988. Removal of Chlorinated Organics from Wastewater by Reduction Process: I. Treatment of 1,1,2,2-Tetrachloroethane with Iron Powder. *Kogyo Yosui*, 357: 2.

Simard, S., M. Odziemkowski, D.E. Irish, L. Brossard, and H. Ménard, 1997. In-Situ Micro-Raman Spectroscopic and Electrochemical Study of Pitting Corrosion of 1024 Mild Steel in Phosphate and Bicarbonate Solutions Containing Chloride and Sulfate. Submitted to *Journal Electrochemical Society*, May 1997.

Smyth, D.J.A., B.T. Byerley, S.W. Chapman, R.D. Wilson, and D.M. Mackay, 1995. Oxygen-Enhanced In-situ Biodegradation of Petroleum Hydrocarbons in Groundwater Using a Passive Interception System. Published in the proceedings from the 5th Annual Symposium on Groundwater and Soil Remediation (GASRep), Toronto, Ontario, Oct. 2-6, 1995.

Starr, R.C. and R.A. Ingleton, 1992. A New Method for Collecting Core Samples without a Drill Rig. *Groundwater Monitoring Review,* 12(4): 91.

Thibeau, R.J., C.W. Brown, and R.H. Hediersbach, 1978. Raman Spectra of Possible Corrosion Products of Iron. *Applied Spectroscopy*, 32: 532.

Thierry, D., D. Persson, C. Leygraf, N. Boucherit, and A. Hugot-Le Goff, 1991. Raman Spectroscopy and XPS Investigations of Anodic Corrosion Films Formed on Fe-Mo Alloys in Alkaline Solutions. *Corrosion Science*, 32: 273.

U.S. Environmental Protection Agency (US EPA), 1995. Labcert Bulletin, EPA-814 N 95001, August 1995.

University of Waterloo (UW), 1995. Revised Workplan for Passive and Semipassive Techniques for Groundwater Remediation at CFB Borden, submitted to AATDF, July 1995.

University of Waterloo (UW), 1997. Passive and Semipassive Techniques for Groundwater Remediation: Phase 1 Installation Report. Prepared for AATDF, Rice University.

Verble, J.L., 1974. Temperature-Dependent Light-Scattering Studies of the Verway Transition and Electronic Disorder in Magnetite. *Physical Review B,* 9(12): 5236.

Weber, E.J., 1996. Iron Mediated Reductive Transformations: Investigations of Reaction Mechanisms. *Environmental Science and Technology*, 30(2): 716.

Wilson, R.D. and D.M. Mackay, 1995. A Method for Passive Release of Solutes from an Unpumped Well. *Ground Water,* 33(6): 936.

Chapter 3

Intrinsic Remediation — Gate 2, CFB Borden

3.1 INTRODUCTION

Intrinsic remediation (IR) uses natural attenuation processes to reduce the environmental risk posed by groundwater contamination. Intrinsic remediation processes include dispersion, sorption, and abiotic and biological degradation. Evidence that intrinsic *bio*remediation (that subset of intrinsic remediation attributable to biological factors) is occurring at a site includes (1) lab-scale microcosms demonstrating loss, (2) field evidence of breakdown products, and (3) evidence that the plume is not growing over time (NRC, 1993).

In Gate 2, at the UW-AATDF site at CFB Borden, no engineered remediation was undertaken so that natural attenuation processes could be assessed. The background groundwater geochemistry indicated a moderately aerobic environment, typical of shallow, pristine sites. But, once the target organics were released to the aquifer, site conditions became reducing and could be considered generally anaerobic. If substantial intrinsic remediation occurred over the course of the project it was likely due to anaerobic processes.

As discussed in Chapter 1, in anaerobic environments, reductive dechlorination is a common biodegradation mechanism for chlorinated aliphatic compounds. The process involves the sequential replacement of chlorine atoms on the alkane or alkene molecule by hydrogen atoms. For this type of reaction to be thermodynamically favorable, the redox environment must be very reducing (i.e., excluding the presence of dissolved oxygen). The complete dechlorination of chlorinated VOCs requires the concerted efforts of a number of different organisms (fermentative, acetogenic, sulfate-reducing, and methanogenic bacteria). An ample supply of electron donor is also required for the dechlorinating organisms. The hydrogen produced by the metabolism of substrates such as methanol or toluene may serve as the direct substrate for microorganisms that use chlorinated solvent as the electron acceptor.

Numerous studies (e.g., Semprini et al., 1990; Bae et al., 1990; and Criddle et al., 1990) have shown IR of carbon tetrachloride (CT). Previous laboratory studies using Borden core material (Devlin, 1994) demonstrated that Borden microbial populations are capable of transforming CT. CT degradation is postulated to occur under denitrifying, sulfate-reducing, or methanogenic conditions. Degradation of CT is proposed to occur through more than one pathway, but the mechanism for pathway selection is unclear. Figure 3.1a shows several possible breakdown pathways. CT breakdown products that could be determined with the analytical methods employed in this project include carbon disulfide (CS_2), chloroform (CF), dichloromethane (DCM), chloromethane (CM), and methane.

Considerable laboratory and field research has shown that chlorinated ethenes can be biodegraded to nontoxic end products under appropriate anaerobic conditions. The degradation pathway for tetrachloroethene (PCE) is shown in Figure 3.1b. Freedman and Gossett (1989) and de Bruin

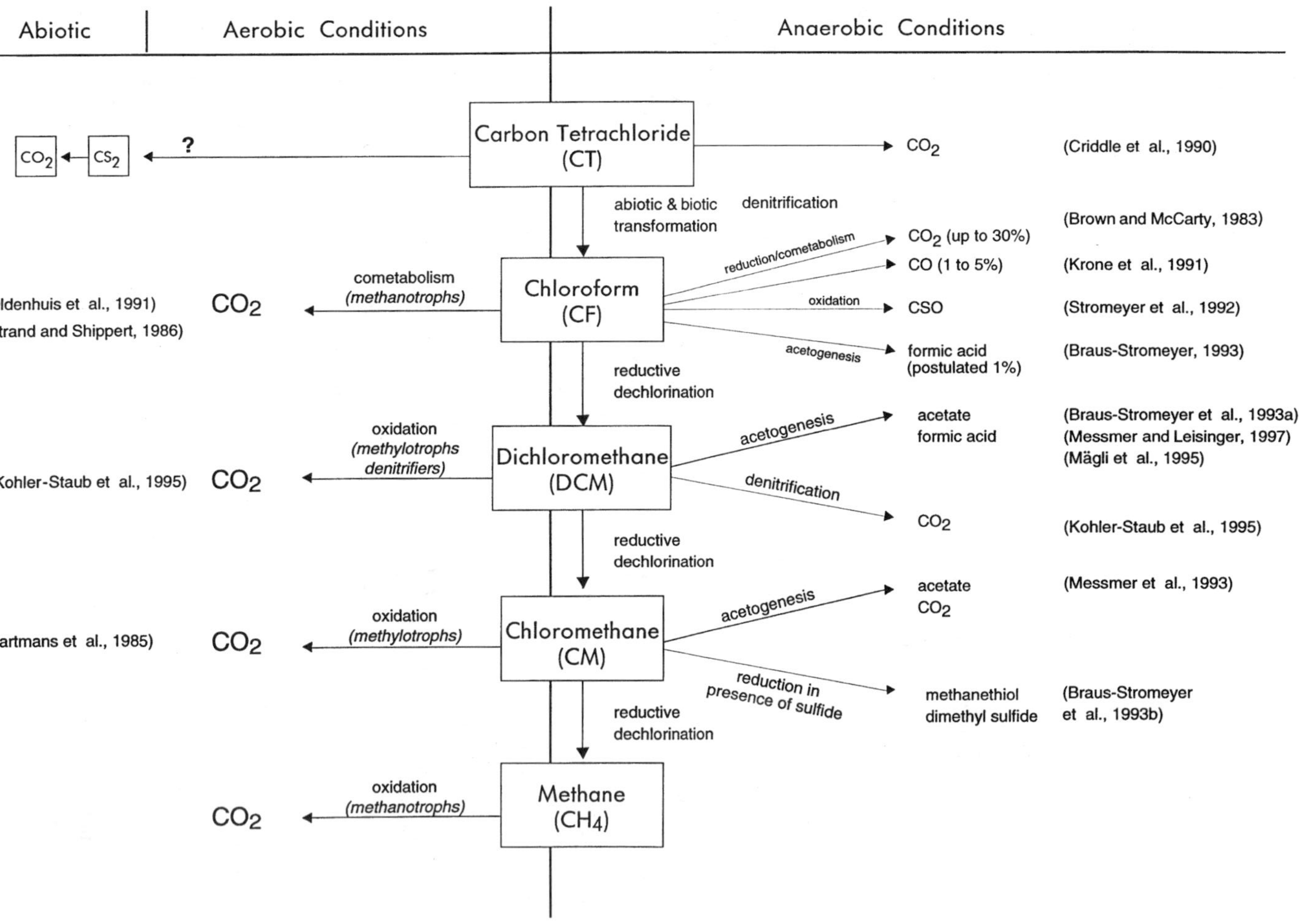

Figure 3.1a Pathways for the dechlorination of carbon tetrachloride, chloroform, and dichloromethane.

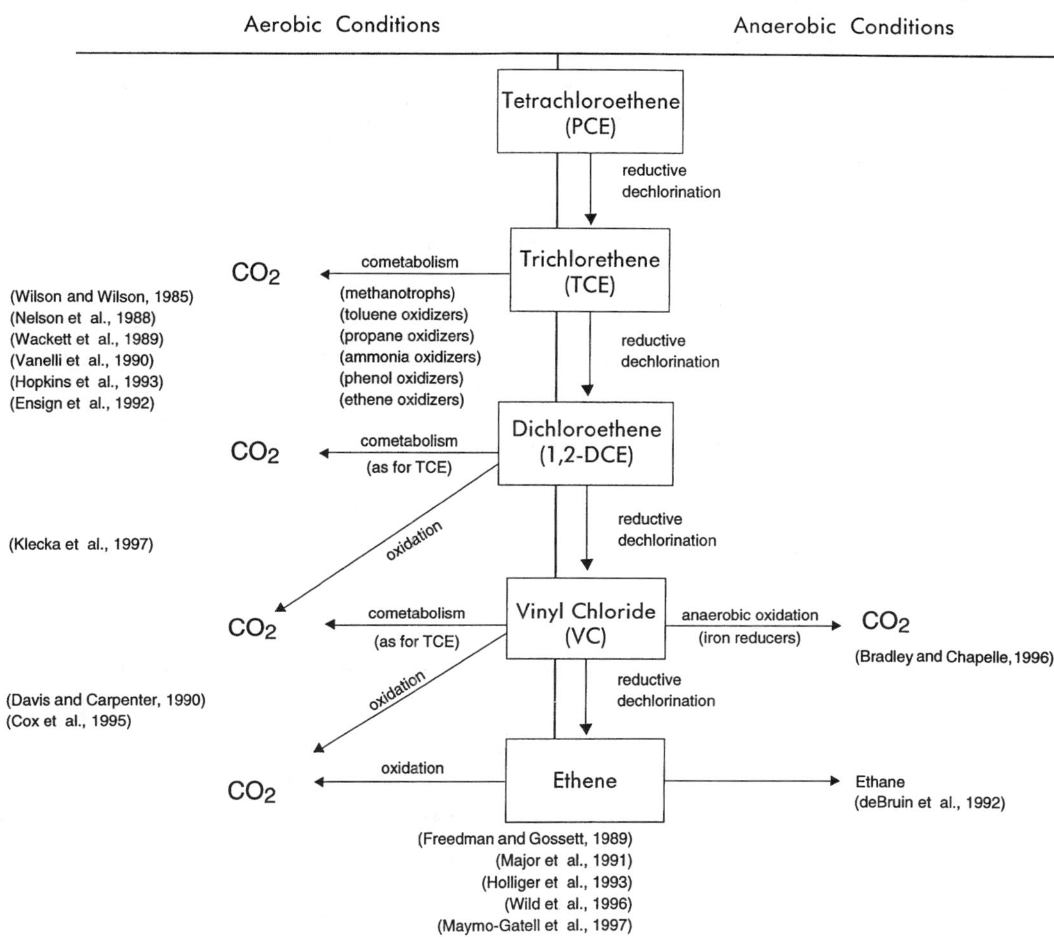

Figure 3.1b Pathways for the anaerobic and abiotic transformation of tetrachloroethene.

et al. (1992) have shown the complete conversion of PCE to ethene under methanogenic and acetogenic conditions in laboratory studies. Fiorenza et al. (1994), Lee et al. (1996), Major et al. (1991, 1995), and Wilson et al. (1994a, 1994b, 1995), among others, have reported the biodegradation of chlorinated ethenes to ethene under combinations of methanogenic, acetogenic, and sulfate-reducing conditions. As a result of these and other research studies the biodegradation of chlorinated aliphatic compounds has become a promising remedial technology.

PCE biodegradation to *c*DCE has been documented with Borden core material and groundwater in a 3-year laboratory microcosm study (Devlin, 1994). The results of this study were not conclusive, as replicates did not both show PCE transformation. They suggest that requisite microbial populations may exist at Borden, but are difficult to establish in laboratory conditions or the acclimation period is quite lengthy.

Toluene (TOL) was the aromatic compound investigated at the UW-AATDF Borden site. Previous studies conducted at Borden support biodegradation of TOL under denitrifying conditions (Barbaro et al., 1992). Unfortunately the breakdown products are not detected by the analytical methods used in this project, and evidence for biotransformation will be limited to evaluating TOL mass loss over the experimental period.

3.2 TECHNOLOGY PROCESS DESCRIPTION

Gate 2 at the UW-AATDF Borden experiment allowed monitoring of the extent and rate of degradation of the target organics without engineered treatments. The intent was to permit comparison of the engineered semipassive, *in situ* remediation treatment sequences installed in Gates 1 and 3 to the intrinsic processes occurring in Gate 2.

3.3 FIELD DEMONSTRATION

The controlled experiment was designed using sheet piling keyed into an underlying aquitard to construct a three-"gate" structure. Extraction wells located at the downgradient ends of the gates were operated to induce groundwater flow. A detailed description of construction activities and field installations was provided in the Phase 1 report (UW, 1997). The focus of this chapter is Gate 2, the "control" gate containing no engineered treatment sequences. The plan and section view of Gate 2 is shown in Figure 3.2.

The field demonstration at CFB Borden operated from May 1996 until October 31, 1997. The period of active release of target organics occurred between November 18, 1996 (day 0) and October 31, 1997 (day 347).

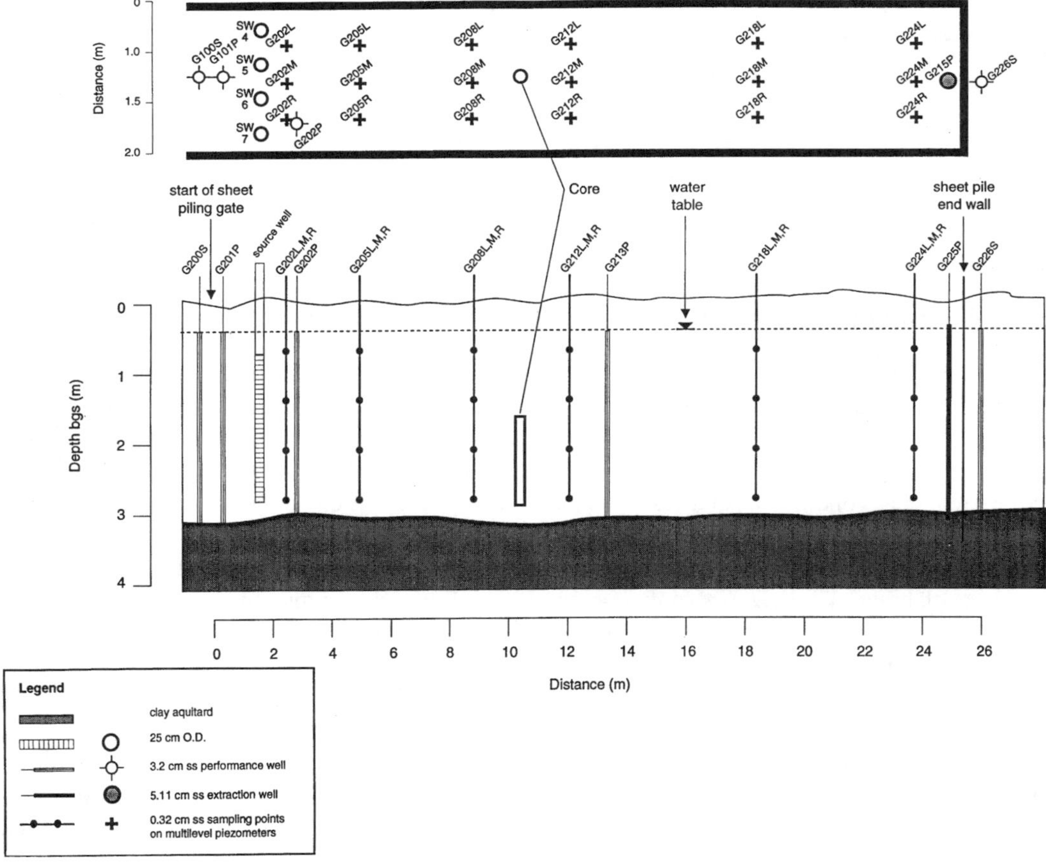

Figure 3.2 Plan and cross section of Gate 2.

3.3.1 Experimental Set-Up

General details of the Borden experimental site including site location, hydrogeology, and geochemistry are provided in Appendices 1 and 6. A chronology of field activities is provided as Appendix 2. Figure 3.2 shows a plan and cross section view of Gate 2, including the locations of the source wells, the groundwater monitoring network (consisting of multilevel piezometers and fully screened, stainless steel performance wells), and the groundwater extraction well.

Multilevel piezometer sampling points were located at approximately 1.2, 1.9, 2.6, and 3.3 m bgs. The shallowest point is referred to as 1 and the deepest is referred to as 4. For example, G202L-1 is the piezometer point in Gate 2, fence 202, located on the left side of the gate (in the direction of groundwater flow) and the first point (highest elevation) in the stalk of piezometers. The fence numbers refer to the approximate distance from the start of the sheet piling sides (not from the source wells, which are about 0.7 m inside the end of the sheet piling gates). Fully screened wells that are located within the sheet piling structure are indicated with a P, e.g., G202P. Those fully screened wells that are located outside of the sheet piling structure are referred to as S, surrounding, wells, e.g., G226S. A full description of the installation of the structure and monitoring points is given in the Phase I Installation Report, UW (1997).

3.3.2 Groundwater Sampling Program

The groundwater sampling program as described in the Work Plan (UW, 1995) was divided on the basis of the analytes being sampled. Groundwater samples were collected to track plume development and to assess potential intrinsic remediation. The following list of parameters were monitored within Gate 2:

1. VOCs, including the target organics CT, CF, PCE, and TOL
2. Dissolved hydrocarbon gases
3. Organic acids (formic, butyric, acetic, and propionic)
4. Inorganic parameters (anions, cations, alkalinity)
5. Bromide for tracer test
6. Conductivity
7. Dissolved oxygen
8. pH, E_h
9. Water level/head measurements

A description of each groundwater sampling program, the types of analyses completed, and the specific parameters analyzed, is provided in Appendix 4, as are sampling and field analysis methods. Laboratory analytical methods are detailed in Appendix 3. Quality assurance/quality control (QA/QC) procedures and results are given in Appendix 5. Timing for all sampling events is included in the Field Activities Chronology (Appendix 2).

A soil core was taken in July 1997 for microbial characterization. This core was collected as described in Appendix 4 at the location shown on Figure 3.2. Core material from about 1.8 to 3 m bgs was retained and transported to UW. At UW the cores were split and then used for microbial characterization.

3.4 RESULTS

Section 3.4.1 briefly discusses the background sampling results for Gate 2. Laboratory studies conducted to examine biotransformation of the target organics are highlighted in Section 3.4.2. The hydraulic performance of Gate 2 is given in Section 3.4.3. The development of a plume in Gate 2 is discussed in Section 3.4.4. Section 3.4.5 discusses the redox indicators and general geochemistry of groundwater, and Section 3.4.6 discusses the behavior of the target organics in Gate 2.

3.4.1 Background Groundwater Chemistry

Sampling to provide background geochemical conditions at the site was completed about 1 month after the extraction pumps were started. A complete description of background sampling activities and results is provided in Appendix 6. The groundwater exhibited neutral pH and was fresh, having moderate to low electrical conductivity. The groundwater was Ca-HCO_3 type, and its inorganic geochemistry was similar to other sites in the Borden aquifer (Nicholson et al., 1983; Mackay et al., 1986). Aerobic conditions were indicated by DO concentrations between 2 and 6 mg/L, and E_h measurements ranged from 50 to 300 mV. Several samples from near the extraction end of Gates 2 and 3 showed low (at or near the MDL) concentrations of PCE, likely resulting from a previous PCE release experiment at an adjacent site (Brewster et al., 1995). Dissolved hydrocarbon gas (DHG) sampling showed relatively low methane concentrations (up to 30 μg/L), but essentially undetectable concentrations of ethene, ethane, propane, and propene.

3.4.2 Laboratory Microcosm Studies

A lab microcosm experiment was completed using aquifer materials and site groundwater to investigate the potential for intrinsic bioremediation. Results indicated that PCE degradation was unlikely, but that TOL and CT were degradable in the aquifer under aerobic (TOL) and anaerobic (CT) conditions. Details and results of the microcosm experiment are given in Appendix 11.

3.4.3 Hydraulic Conditions

Water level data are provided in Appendix 8. Figure 3.3 shows three potentiometric surface plots of the water table in Gate 2, from June 29, 1996, November 13, 1996, and October 24, 1997. Measurements were taken in all accessible piezometers. It should be noted that the multilevel piezometers themselves were not used for water table readings, but the center stalk holding the multilevels was used, having their bottom few centimeters of PVC center stalk slotted and wrapped with Nytex screening material. The contouring shown in Figure 3.3 is somewhat irregular, likely reflecting the two different sources of data — fully screened wells and center stalks screened only at the bottom. Generally, the expected hydraulic gradient was achieved, and groundwater flowed to the extraction wells as designed.

The sheet piling was installed to hydraulically seal Gate 2 from Gates 1 and 3 on either side, and the potentiometric surface plots of all gates (see Appendix 8) indicates that each gate was hydraulically independent of the other gates.

Figure 3.4 shows water level fluctuations in fully screened wells 202P, 213P, and 225P. An overall variation of approximately 0.6 m was observed over the experiment. Lowest levels were observed from July to October, increasing over the winter to a peak in the spring. Analyses of contaminant degradation can be complicated by a fluctuating water table, since contaminants can be trapped in the vadose zone and can escape by volatilization, biodegrade, or reenter the groundwater in time of high water table. The maximum anticipated losses due to the above processes is 5% (Chapter 1).

Although the water level itself was quite variable over time, the gradient through the gate was relatively constant (i.e., the difference between 202P and 225P was relatively constant; see Figure 3.4). The imposed hydraulic gradient over the length of the gate was approximately 0.016 m/m.

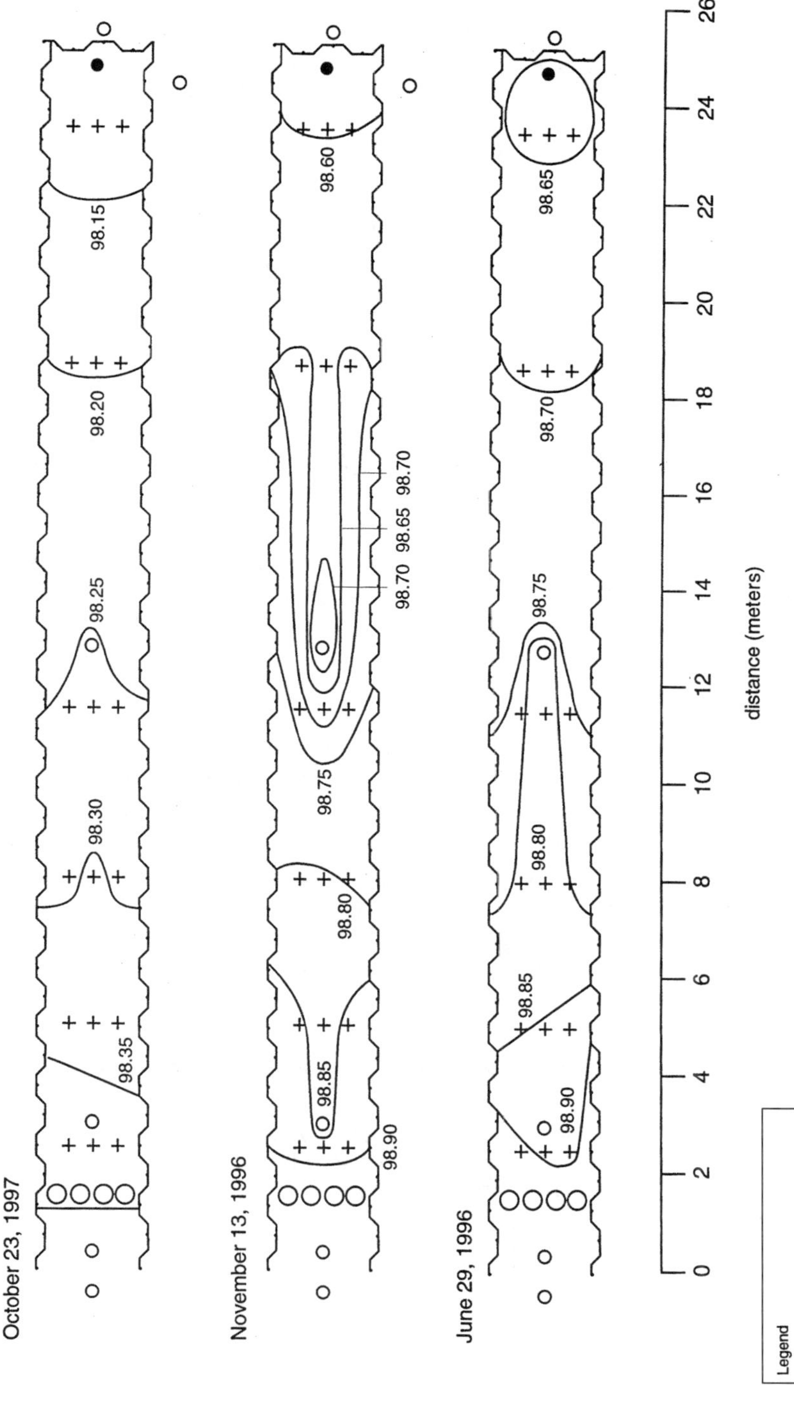

Figure 3.3 Potentiometric surface contours for Gate 2.

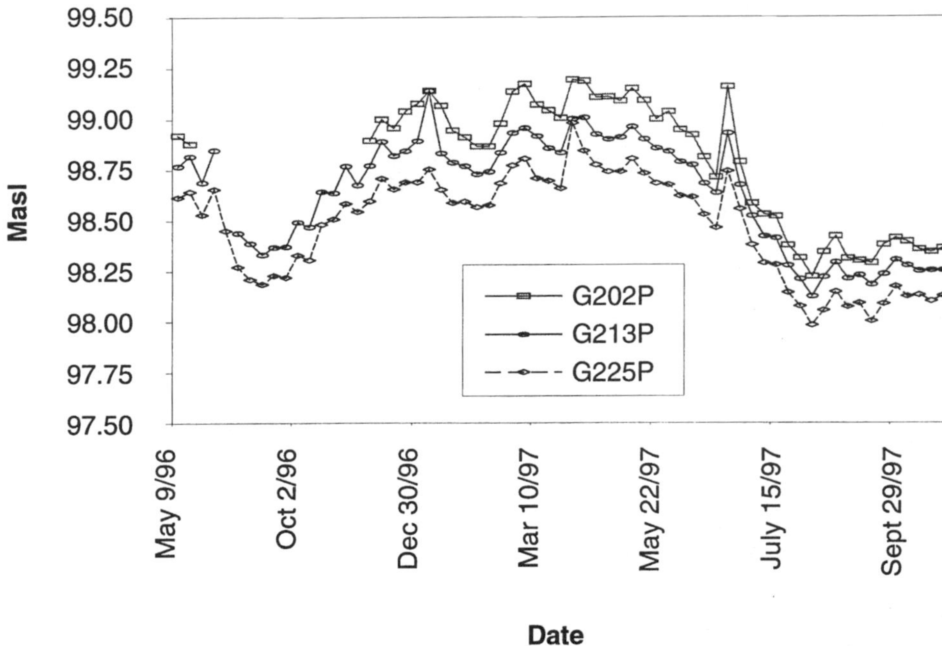

Figure 3.4 Fluctuations in the water level measured in wells 202P, 213P, and 225P.

In order to determine the groundwater velocity in Gate 2, a tracer test was completed prior to the release of VOCs to the aquifer. The background, setup, and operation of the tracer test are discussed in Appendix 7.

Calculation of typical tracer breakthrough curves required knowledge of the initial maximum concentration (C_o), which was not possible in this experiment, as the measured source well concentrations were not considered representative of the effective C_o in the aquifer. Sampling of the source wells indicated much higher concentrations of tracer (KBr) in the wells themselves compared to the nearby aquifer. Hence, alternate methods were employed. The Levenspiel method for determining the arrival time of the tracer pulse center of mass (Appendix 7) and the center of mass method (using the FORTRAN program, FENCE, see Appendix 24) were used to calculate velocities at each multilevel sampling point. The average velocities for groundwater from the source wells to various fences in Gate 2 are shown in Table 3.1.

Table 3.1 Calculated Groundwater Velocities From G202

Fence	Levenspiel Velocity (cm/d)	Center of Mass Velocity (cm/d)
G202	not determined	7.9
G205	9.0	9.5
G208	12.9	11.3
G212	12.1	10.4
G218	10.3	11.0
G224	10.1	9.9
Gate Average	10.9	10.0

Figure 3.5 (also Figure A7.3, located in Appendix 7) shows the mass of tracer that passed each point in the monitoring network and clearly shows that tracer was not strongly delivered (either by advection or dispersion) to the deepest points in the monitoring network.

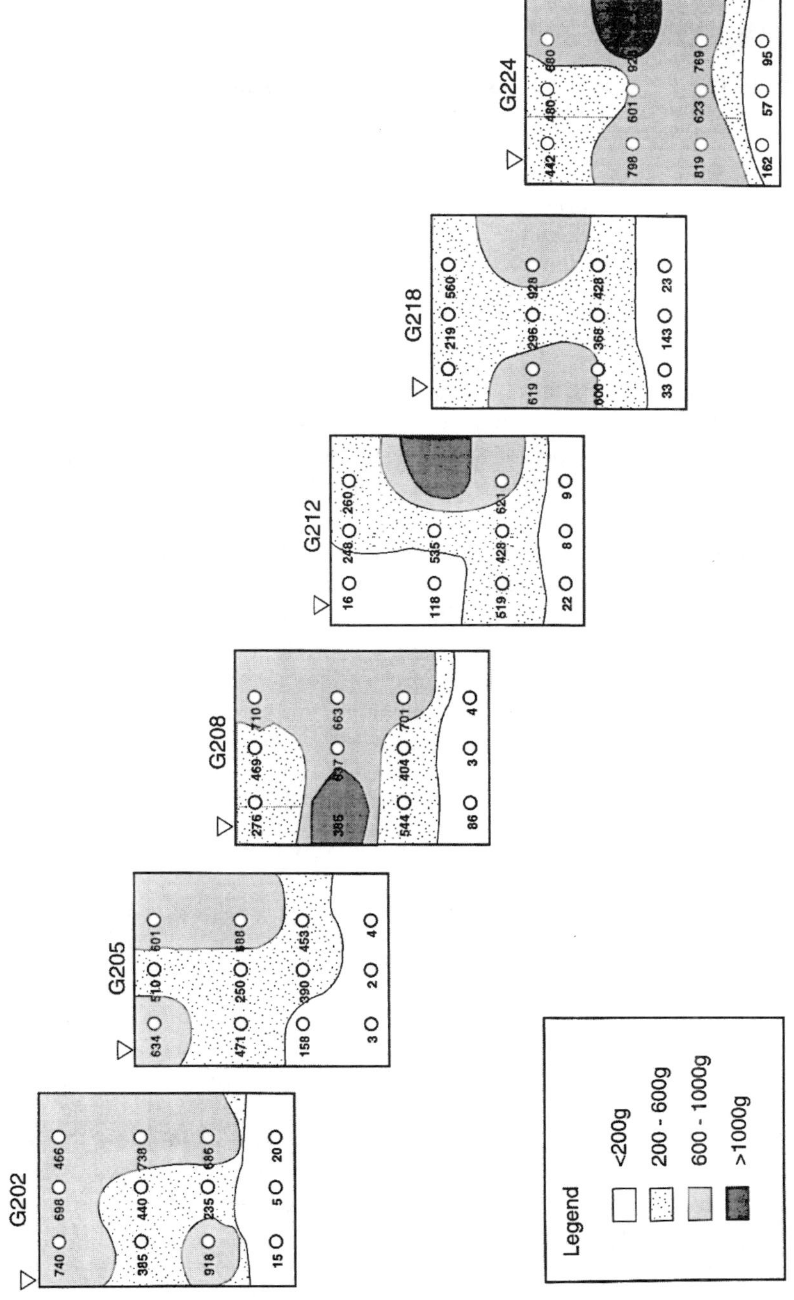

Figure 3.5 Tracer mass passing selected fences in Gate 2.

3.4.4 Plume Generation

A mixed organic plume consisting of PCE, CT, and TOL was created in Gate 2 from November 19, 1996 (day 0) until October 28, 1997 (day 344). The source was created in four large-diameter wells at the upgradient end of the gate (wells SW 4 to SW 8 in Figure 3.2) using semipassive, diffusive release emitters as described in Appendix 9. The target delivery concentrations to the aquifer and the actual concentrations delivered are shown in Table 3.2.

Table 3.2 Target vs. Observed Average Concentrations in Groundwater at Fence G202

Organic	Target Concentration (μg/L)	Average Concentration (μg/L) Day 70	Day 238
TOL	2,000–10,000	1,180	3,300
CT	1,000–2,000	32	1,133
PCE	1,000–2,000	61	381

The average TOL concentration was within the target range, but selected points (e.g., G202R-3) received much higher levels than the target, while others received almost no TOL (e.g., G202R-4). The PCE and CT took a much longer period to approach the target range. A pseudo-steady-state delivery of the target organics to the aquifer was not achieved, at least in individual points in fence G202.

The FENCE program was used to calculate the cumulative mass of VOCs that had passed fence G202 at days 112, 238, and 339. These are shown in Figure 3.6, and details are provided in Section 3.4.6. The linear increase in cumulative mass passing fence 202 over time, except for the increasing trend of CT, suggests that the overall input of mass to Gate 2 was reasonably steady, despite large fluctuations of input mass to particular depths. Interpretation of the fate of the target organics will consider the behavior of compound mass, rather than compound concentration, to take advantage of this apparently reasonably consistent overall mass input.

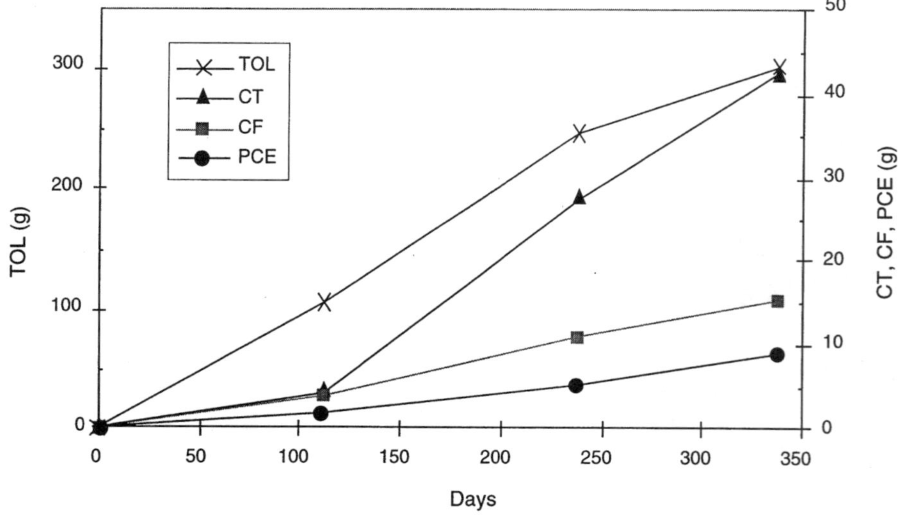

Figure 3.6 Cumulative mass of target organics passing fence G202 during the field experiment.

3.4.5 Redox and General Chemical Conditions

Section 3.4.1 outlined the background geochemical conditions in Gate 2. Conditions were moderately aerobic, with low methane concentrations, significant sulfate present, significant DO, and E_h values above +50 mV. After introduction of the target organics into Gate 2, conditions became anaerobic shortly after TOL had reached that segment of the aquifer. The E_h of the system became generally more negative, which is consistent with an anaerobic or microaerophilic environment. DO was reduced to <1 mg/L at most points, and methane concentrations were significantly higher than background levels in many locations, both consistent with an anaerobic environment. Sulfate was still present throughout the gate throughout the experiment, suggesting that sulfate reduction was not significant in Gate 2.

Dissolved hydrogen gas concentrations were measured in a few points in Gate 2 on about day 335. Chapelle et al. (1995) suggested that the H_2 concentration could be related to the predominant terminal electron-accepting microbial processes in the subsurface. This is discussed in Appendix 19, and the H_2 data gathered from Gate 2 is also presented in Table A19.1. A variety of anaerobic microbial conditions were suggested for groundwater in Gate 2 from the H_2 data, ranging from iron-reducing through sulfate-reducing to methanogenic.

While conditions in the groundwater had become anaerobic after addition of the target organics, a dominant, anaerobic microbial community was not clearly established. It is likely that the aerobic biodegradation of TOL (discussed in Section 3.4.6) was responsible for scavenging the DO and establishing anaerobic conditions in Gate 2 groundwater. While aerobic biodegradation of TOL was important in driving groundwater anaerobic, it was unlikely to have provided a major sink for TOL, given the few milligrams per liter of DO initially present in the groundwater in Gate 2 and the continuing addition of 2–4 mg/L TOL in DO-free groundwater at fence G202.

Fence G212 was monitored for organic acids (formic, acetic, butyric, and propionic) (Appendix 19). No formic, butyric, or propionic acids were detected, but on days 238 and 339 the levels of acetic acid at G212M-4 were 75 mg/L and 85 mg/L, respectively. Acetic acid is a common product of anaerobic metabolism, supporting the above conclusion that Gate 2 had become anaerobic.

3.4.6 Behavior of Target Organics

The area-weighted mass of target organics that had passed the monitoring points in fences G202, G208, G212, G218, and G224 up to and including days 112, 238, and 339 are depicted in Figures 3.7 to 3.10 for CT, CF, PCE, and TOL, respectively. The mass values are expressed in milligrams and were calculated using the FENCE program that is described in Appendix 24. If there were no increase in mass between two time periods, then the "total mass passing" the point does not increase. For example, in Figure 3.8 for G208 between days 238 and 339 at G208M1 the mass remains at 58 mg, indicating that at subsequent sampling periods beyond day 238 CF was detected at this point. Only fences where relatively complete data sets existed were used. Also, note that a fence was sampled only when the organics were likely to have been advected as far as that fence on the day of sampling. Table 3.3 summarizes the days when specific fences were sampled. The groundwater velocity was assumed to be 11.4 cm/day at all fences (see Figure A7.2 for actual distribution of velocities), and the porosity was assumed to be 0.33, typical of the Borden sand. The groundwater velocity of 11.4 cm/day, used for FENCE interpretation, was based on early estimates of the groundwater velocity, not those presented in Table 3.1. The average velocity differences are not expected to substantially affect interpretations. The cross-sectional area that was attributed to each fence varied according to the position of the water table. For this analysis an average water table was assumed to apply to all durations. Figure 3.11 shows the relative weighting factors that resulted from this assumption about the water table and from the spacing of monitoring points.

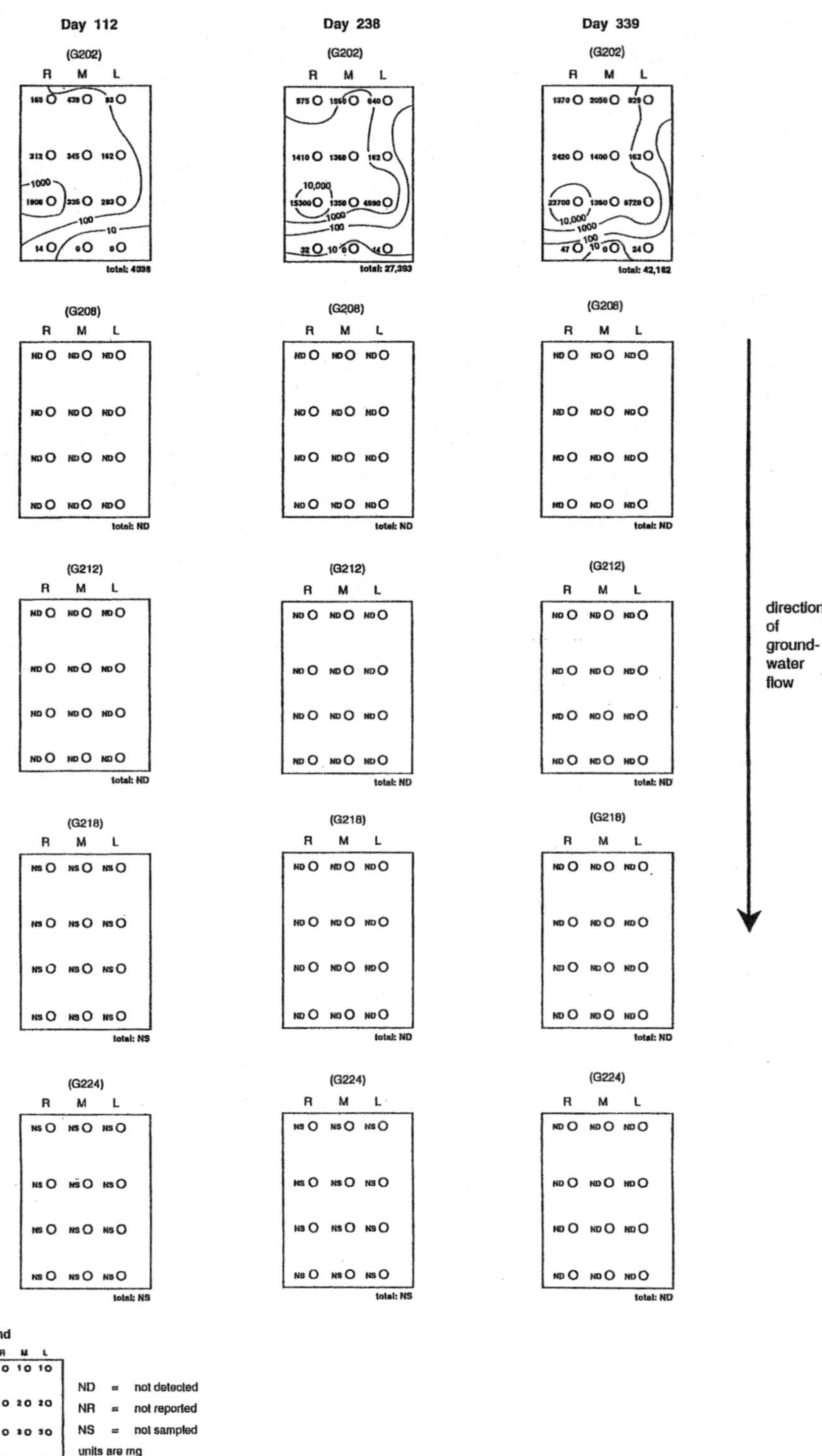

Figure 3.7 Total mass (mg) of CT that had passed fences G202, G208, G212, G218, and G224 on days 112, 238, and 339.

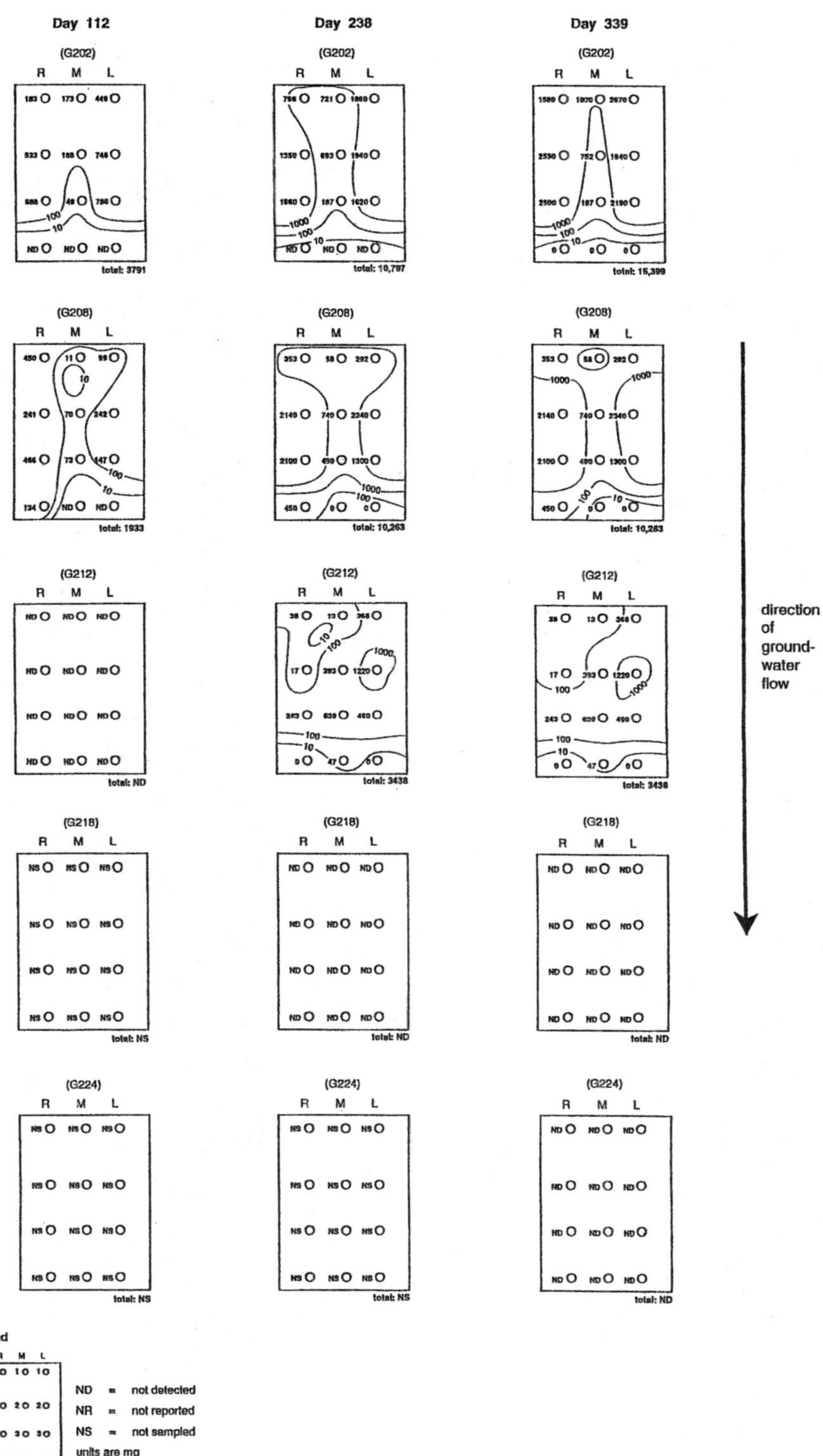

Figure 3.8 Total mass (mg) of CF that had passed fences G202, G208, G212, G218, and G224 on days 112, 238, and 339.

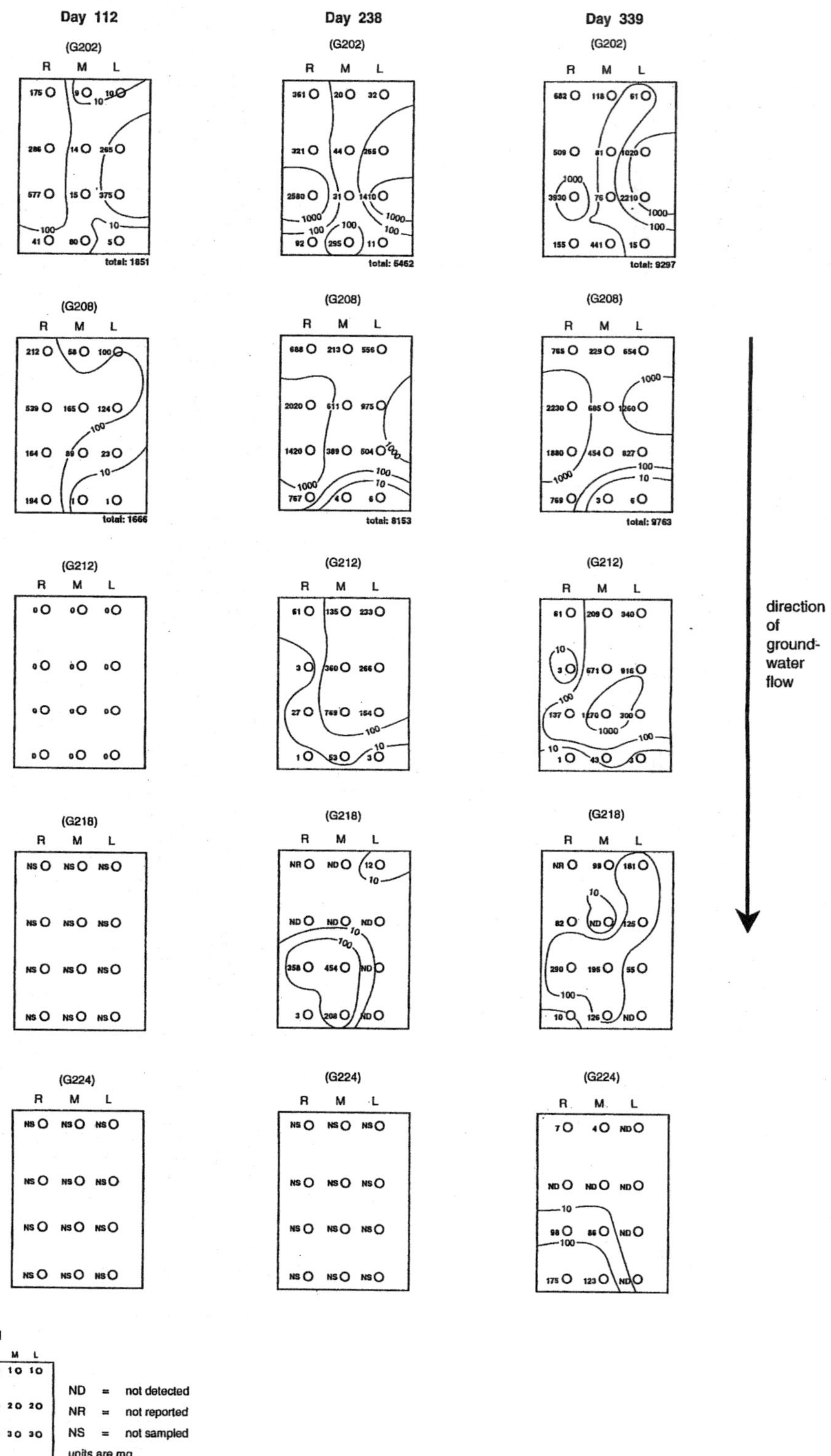

Figure 3.9 Total mass (mg) of PCE that had passed fences G202, G208, G212, G218, and G224 on days 112, 238, and 339.

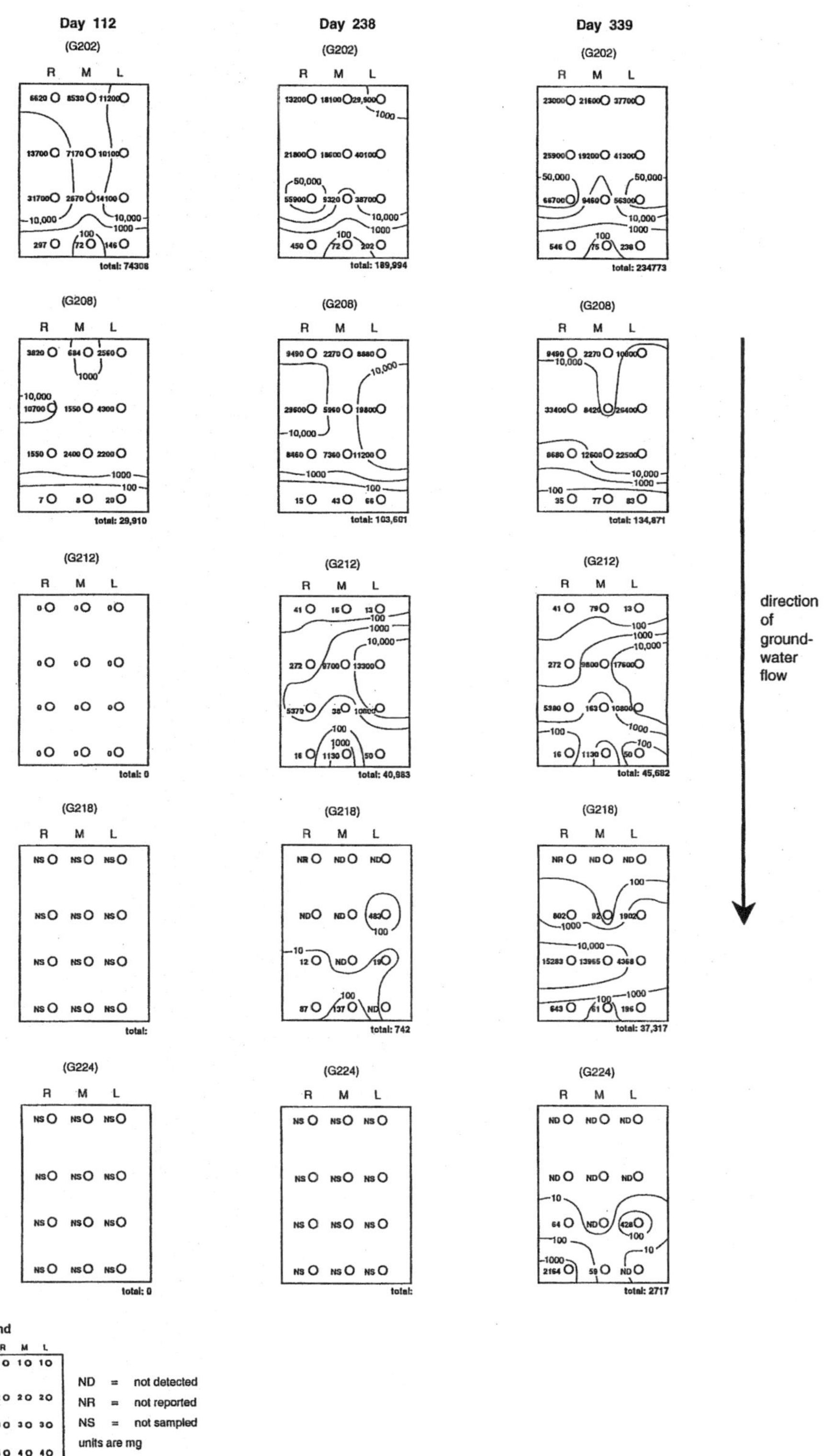

Figure 3.10 Total mass (mg) of TOL that had passed fences G202, G208, G212, G218, and G224 on days 112, 238, and 339.

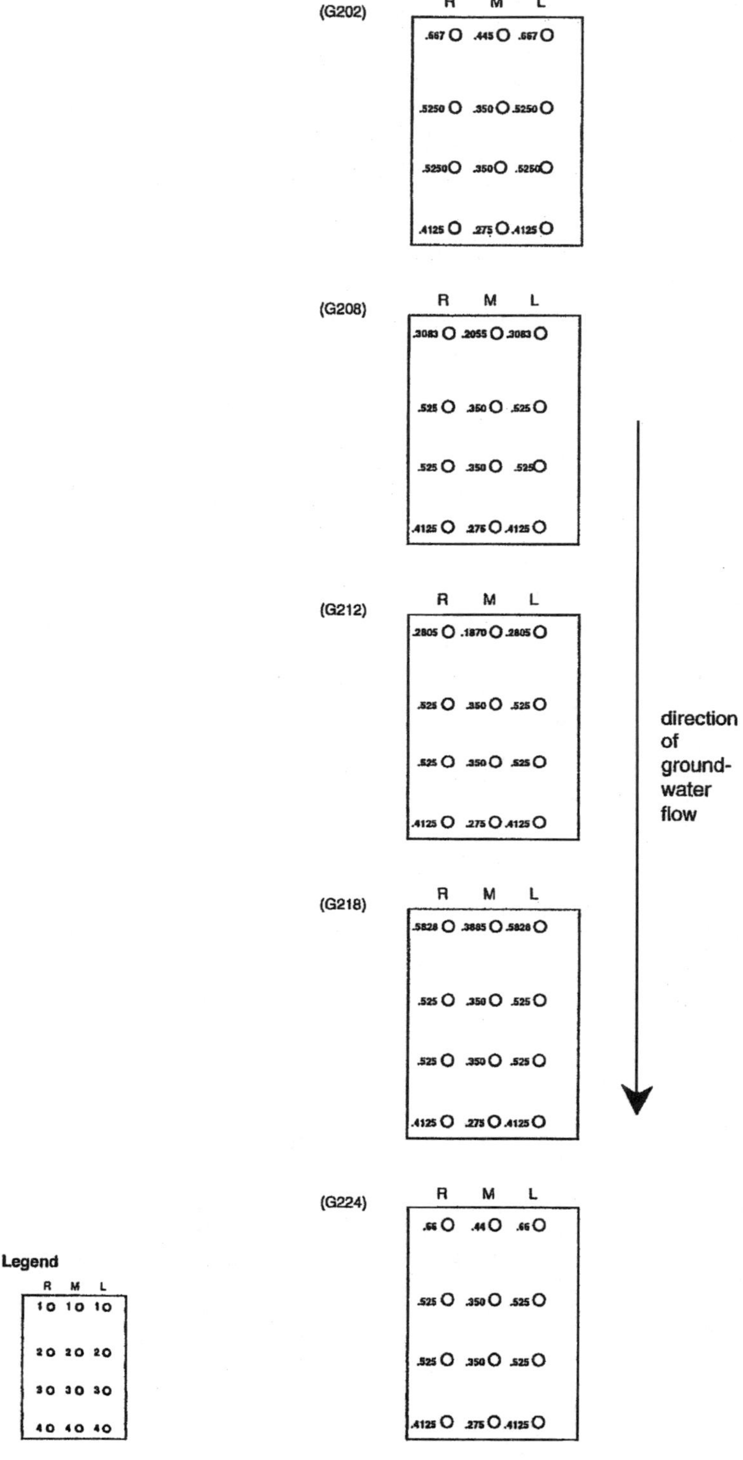

Figure 3.11 Relative weighting factors for each point in multilevel monitors used in the FENCE program.

Table 3.3 Days of Sampling Monitoring Points at Selected Fences in Gate 2

Fence	Day 14	28	42	70	112	196	238	339
G202	✓	✓	✓	✓	✓		✓	✓
G208				✓	✓			
G212					✓	✓	✓	✓
G218							✓	✓
G224								✓

CT was found at fence G202 at days 112, 238, and 339, but was not observed in significant mass downgradient at fences 208, 212, 218, or 224. CT was expected to advect at a retarded average linear groundwater velocity of about 6 cm/d (55% of groundwater velocity of 10.4 cm/day; see Table 3.4 for retardation estimate). Therefore, significant mass of CT was expected to pass G202 (0.9 m from source wells) within about 13 days. In fact, considerable CT was found on day 112 as expected. By day 238, significant mass of CT should have been found passing fence G212. This migration of CT was not observed (see Figure 3.7). In fact, CT was only found above the LOQ once beyond fence G205 during the experiment. The restricted distribution of CT and the appearance of CF, a reductive dechlorination product of CT, suggests that reductive dechlorination of CT had occurred intrinsically in Gate 2.

Table 3.4 Estimated Retardation Factors and Velocity for Organic Solutes in Gate 2

Compound	Retardation factor (R, dimensionless)	Solute velocity (cm/d)
PCE	2.7[a]	4
CT	1.8[b]	6
CF	1.26[b]	8
TOL	1.16[c]	9

[a] Pankow and Cherry, 1996.
[b] Devlin, 1994.
[c] Hubbard, 1992.

Considerable mass of CF passed fence G202 by day 112 (Figure 3.8), indicating that the reductive dechlorination of CT was occurring even upgradient of fence G202. CF was expected to advect at a retarded velocity of about 8 cm/day (see Table 3.4 for retardation estimate), so significant mass of CF was expected to have passed fence G208 by day 112, fence G218 by day 238, and also fence G224 by day 339. Figure 3.8 indicates the expectation for day 112 was met, but that no significant mass of CF passed fences G212, 218, and 224 even by the end of the experiment. This restricted distribution of CF suggests that it was being transformed intrinsically. Reductive dechlorination is the suspected reaction, given the anaerobic geochemical environment. A known product of CT reductive dechlorination, CS_2, was reported above the LOQ in six samples from points at G205M, but other breakdown products such as DCM or CM were not found above the MDL in Gate 2.

PCE was anticipated to advect with a retarded groundwater velocity of only about 4 cm/day (see retardation estimate in Table 3.4). By day 112, significant PCE mass should have passed fence 202, but not fence 208. By day 238, significant mass of PCE should have passed fence 208 but not fence 212, and by day 339, fence G212 but not G218. The more widespread distribution of PCE seen in Figure 3.9 suggests PCE may have been advecting slightly faster than this estimate, with, for example, significant mass apparently having passed fence G218 by day 339. This could be due to the early addition of PCE into the source well, SW-6, in Gate 2 from July to September 1996 as part of plume generation testing (see Appendix 9). Note that day 0 was taken as November

18, 1996. Unlike CT and CF, little restriction of the migration of PCE relative to its anticipated groundwater advection rate is evident from Figure 3.9. Also, typical products of reductive dechlorination of PCE (TCE, DCE, VC, ethene) were rarely detected in Gate 2. Only *c*DCE, in concentrations <30 μg/L, was detected, and then only on day 339 at five points. Therefore, reductive dechlorination of PCE appears to have been minimal in Gate 2 during the experiment.

TOL was expected to advect at about 85% of the groundwater velocity (10.4 cm/day) or at a retarded velocity of about 9 cm/day (see retardation estimate, Table 3.4). Significant concentrations appeared at fence G202 by day 14. Significant TOL concentrations were expected to have reached fences G208, G212, G218, and G224 by about day 70, 110, 170, and 225, respectively. Figure 3.10 shows that significant TOL had been passing fences G202 and G208 by day 112; also fence G212 and G218 by day 238 and fence G224 by day 339. The apparent appearance of TOL mass at the selected fences (Figure 3.10) was perhaps slightly later than anticipated from estimates of the retarded velocity of TOL, but the difference was not large. No inference of TOL mass loss can be derived from this analysis, but other analyses (below) indicate considerable TOL biodegradation occurred in Gate 2.

By day 339, all VOCs should have advected well beyond fence G218. The total mass that was estimated to have passed each fence by day 339 is shown in Table 3.5. The total mass was calculated using the FENCE program for groundwater concentrations on the days listed in Table 3.3. The mass passing Gate 202 was highest, in part because mass was passing this fence for the longest time. To assess potential mass loss of chemicals as they migrated through Gate 2, the average mass per day passing each fence during the experiment was calculated as the cumulative mass passing a fence divided by the number of days over which this mass was accumulating (see Figures 3.7 to 3.10). For example, fence G208 was sampled from days 70 to 339. Significant TOL mass was first found on day 70, so the period when TOL mass was accumulating at fence G208 was 269 days (339 – 70). Over that period, a calculated 135,000 mg TOL passed fence G208 and so the average daily mass of TOL passing fence 208 was 730 mg. The average daily mass that had passed each fence is also shown in Table 3.5.

Table 3.5 Cumulative Mass (mg) and the Average Daily Mass (mg/day) of CT, CF, PCE, and TOL Having Passed Selected Fences by Day 339

Fence	Chemical							
	CT		CF		PCE		TOL	
	(mg)	(mg/d)	(mg)	(mg/d)	(mg)	(mg/d)	(mg)	(mg/d)
G202	42200	130	15400	48	5500	17	23500	730
G208	ND	ND	10300	38	9800	37	135000	510
G212	ND	ND	3400	15	3900	17	45700	200
G218	ND	ND	ND	ND	1200	12	740	8
G224	ND	NA	ND	NA	490	NA	2700	NA

Note: ND: no significant mass detected; NA: not determined as duration was only 1 day.

This calculative approach is not particularly refined, mainly because the temporal variation in the input to Gate 2 was not closely defined by the limited sampling at fence 202. The reasonably linearly increasing cumulative mass passing fence 202 (Figure 3.6) over time, except perhaps for CT, suggested that large fluctuations of input concentrations to Gate 2 did not likely occur, and so this simple calculation was considered useful for assessing mass losses in Gate 2.

Figure 3.12 illustrates the apparent changes in the average daily flux of organics passing each fence up to day 339. CT essentially disappeared by about 7 m from the source wells. CF also decreased along the gate and no significant flux passed fence G212, in spite of the likely gain from CT dechlorination. PCE showed no real overall decline in daily mass flux moving down the gate, so it is likely that mass loss due to reductive dechlorination was minimal. The daily flux of TOL declined along the gate, indicating mass loss along the length of Gate 2.

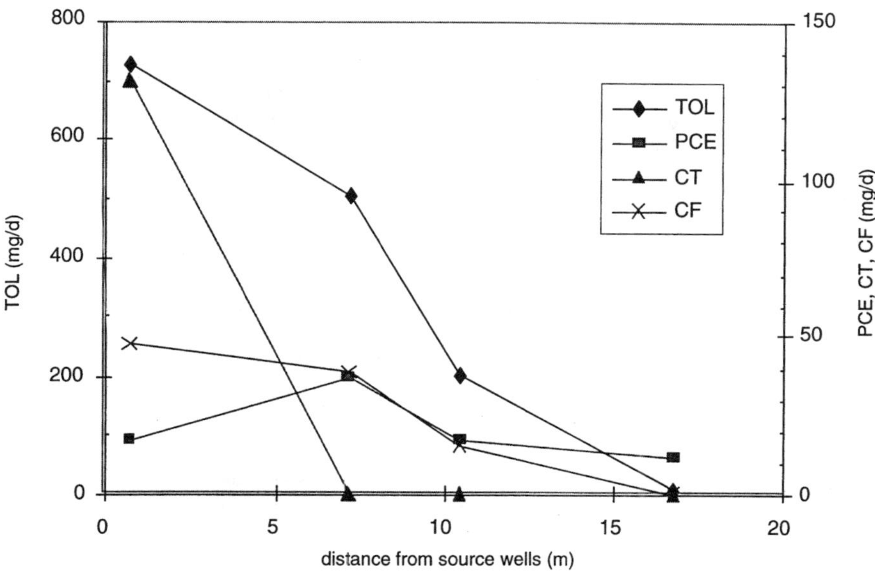

Figure 3.12 Average daily mass flux of target compounds at selected fences in Gate 2.

The declining daily flux of target organics passing downgradient fences on day 339 was examined to estimate an apparent biodegradation rate for each organic. A steady-state, first-order rate was assumed, and the reaction rate was determined by fitting mass flux profiles to a steady-state reaction equation:

$$J = J_0 \, e^{-kd/v} \tag{3.1}$$

where J is the flux (mg/day) at a fence beyond fence G202
J_0 is the flux (mg/day) at fence G202
k is the apparent first-order reaction rate (day^{-1})
d is the distance from fence G202 (m)
v is the retarded solute velocity (m/day), estimated from retardation information presented in Table 3.4

The average daily flux data (Table 3.5) were fitted to this equation, neglecting dispersion. This assumption was judged to be reasonable on the basis of agreement between profiles calculated using (3.1) and profiles using a steady-state solution of the advection–dispersion equation:

$$C = C_o \exp\left(\frac{v}{2D} - \left(\sqrt{\frac{k + \left(\frac{v^2}{4D}\right)}{D}}\right)x\right) \tag{3.2}$$

where D is the dispersion coefficient (m^2/day). Two values of D were considered, based on the dispersivities (α) ($D = \alpha v$) of 5 to 20 cm estimated from the tracer tests (see Appendix 7).

The best fit was obtained by minimizing the residual sum-of-squares between the actual and calculated data (using (3.1)) at the different fence locations. The data and best-fit lines are shown in Figure 3.13a and the apparent first-order rate constants and apparent half-lives are summarized in Table 3.6.

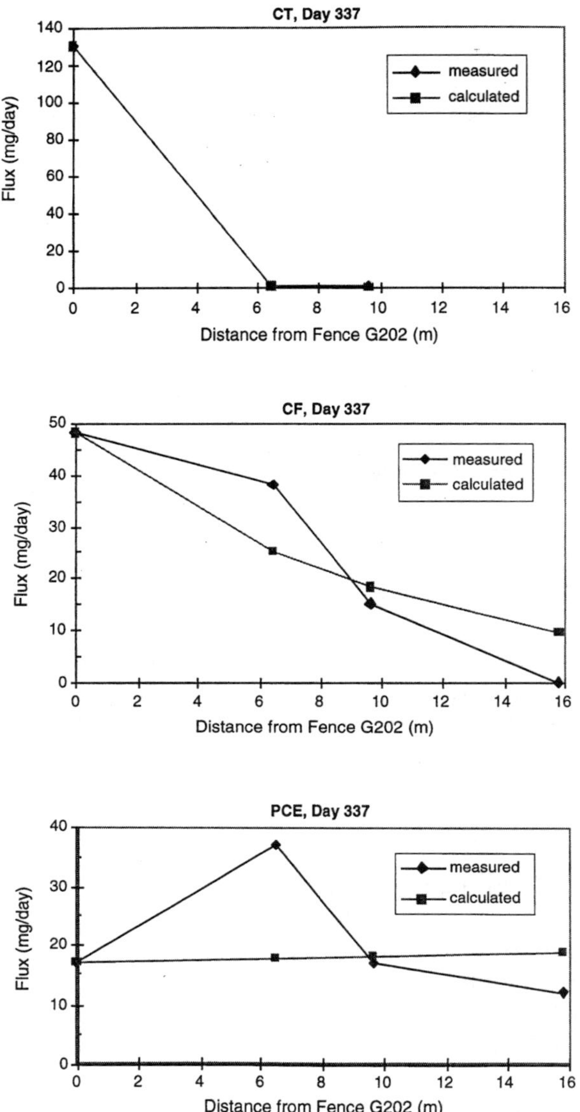

Figure 3.13a Measured and calculated (best-fit) average daily flux of target organics in selected fences in Gate 2.

The apparent rate constants and apparent half-lives shown in Table 3.6 confirm the very rapid transformation of CT, and the biotransformation of TOL and CF. PCE appears to have remained essentially untransformed. All of these transformations are considered to have occurred under the predominantly anaerobic conditions by the natural microbial community in Gate 2.

Pseudo-first-order rate constants for CT, CF, and TOL were estimated by fitting the profiles of maximum concentrations at each fence from the October 23, 1997 snapshot (day 339) to Equation (3.3) below:

$$C = C_o \exp\{-kx/v\} \tag{3.3}$$

where C_o is the input concentration to the profile (μg/L), i.e., the maximum concentration at G202,

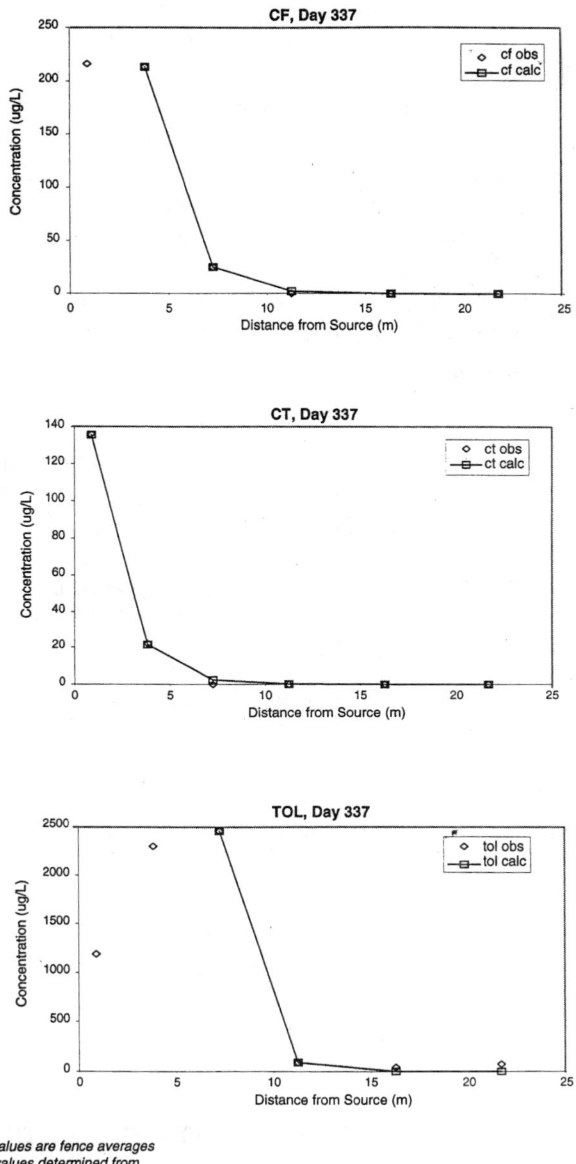

Figure 3.13b Observed and calculated concentrations of CF, CT, and TOL in Gate 2 on day 339.

k is the pseudo-first-order rate constant (day^{-1})

x is the distance from G202

v is the unretarded groundwater velocity (10 cm/day)

Maximum concentrations were used to provide a conservatively high estimate of k. This approach to estimating the kinetics assumes that the profile is at steady state, that dispersion can be ignored, and that the reaction occurs in the water phase only. In Gate 2 there was evidence for CT, CF, PCE, and TOL transformations. In the case of CT the reaction was underway before day 238, and this resulted in concentration distance profiles that were reasonably steady by the time of the day 339 snapshot. On this basis it was considered reasonable to use this profile to estimate pseudo-first-order rate constants. In the cases of TOL and CF, profiles were unlikely to be at steady state but maximum rate constants could be estimated. In the case of PCE, the late appearance of *c*DCE suggests some biotransformation, but the lack of steady state and the sporadic distribution of *c*DCE preclude the determination of a PCE pseudo-half-life. TCE was likely formed but never observed in any sampling events.

In order to calculate degradation rates for CF, assumptions in addition to the ones stated above had to be made. Since CF was formed from CT, the preferred equation for modeling its degradation would be one that incorporated the transformation sequence CT → CF → .... However, CF is not the only transformation product possible from CT, and determining the mass fraction that degraded along the CF pathway was not possible with the available data. Therefore, Equation (3.3) was used, again assuming that the concentrations measured at G202 were representative of inputs. Once again, the calculated rate constants are likely to be lower than the true ones because CF could have formed in the gate beyond the assumed input boundary. Fortunately, this error is likely to be reasonably small since the CF profile, like the CT profile, declined from input levels (216 μg/L) to near LOQs (G208 concentration average 25 μg/L, LOQ for CF = 25 μg/L) between the first two monitoring fences, G202 and G208.

Comparison of the apparent half-lives and the pseudo-first-order half-lives (Table 3.6) indicates that the CT degradation rate remained the same over the duration of the experiment. The CF and TOL pseudo-half-life improved with time, which may indicate a changing microbial population.

Table 3.6 Calculated Apparent First-Order Rate Constants and Half-Lives and Pseudo-First-Order Half-Lives for Organics Passing Selected Fences by Day 339

Compound	Apparent Rate Constant (day^{-1})	Apparent Half-Life[a] (days)	Pseudo-First-Order Half-Life[b] (days)
CT	0.05	13	11
CF	0.008	86	11
PCE	Undefined[c]	Undefined[c]	Undefined[d]
TOL	0.011	62	8.3

[a] Calculated from flux Equation (3.1).

[b] Calculated from maximum concentration Equation (3.3).

[c] The best-fit line had a positive slope, so no degradation indicated (see Figure 3.13).

[d] Not determined, not at steady state.

3.5 PERFORMANCE DATA EVALUATION

Section 3.4.6 demonstrated that without the augmentation of an engineered system, the natural aquifer populations in Gate 2 were capable of degrading CT, CF, and TOL and, to a far lesser degree, PCE. This degradation occurred under mainly anaerobic conditions.

CT transformation was complete by fence G205, with CF being completely attenuated by fence 208, about 7.3 m from the source wells. PCE appeared to be significantly attenuated, especially beyond about 7.3 m from the source wells, but most of this attenuation is likely due to retardation (reversible sorption) and dispersion near the front of the slowly advancing PCE plume. PCE was the most recalcitrant of the target organics delivered to Gate 2, with only very minor reductive dechlorination evident after about 300 days. TOL degradation was not complete by day 339, but considerable TOL had been apparently degraded in Gate 2 by that time, especially beyond fence G208, about 7.3 m from the source wells.

Figure 3.14 presents the area weighted average concentrations of target organics at selected fences in Gate 2 on day 339, the final sampling day. By day 339 (October 21, 1997), all target

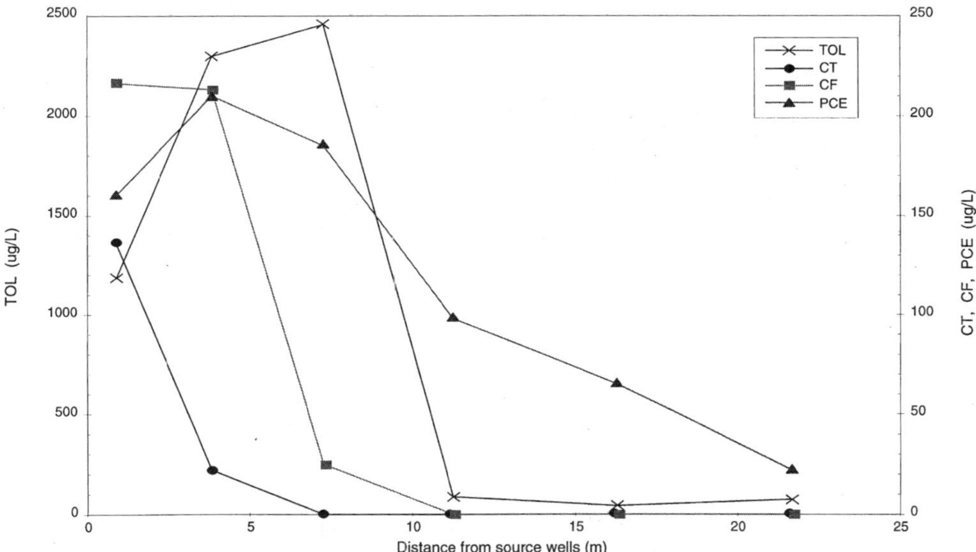

Figure 3.14 The area-weighted average concentrations of target organics at selected fences in Gate 2 on day 339.

compounds had undergone apparent attenuation in Gate 2, but biodegradation appeared to have been significant only for CT, CF, and TOL (discussion in Section 3.4.6). This sampling day represents the furthest advection of chemicals from the source wells, and the longest term for biodegradation to become established. PCE would still not have had sufficient time to fully penetrate the gate at near-source concentrations. With an estimated retarded velocity of only 4 cm/day, significant PCE concentrations (C/C_0 = 0.5) would be expected only about 13.5 m downgradient of the source wells (i.e., between fences G212 and G218). Thus, the lower mass of PCE apparent in fences G218 and G224 (Figure 3.14) likely represents retardation with little mass removal by biodegradation.

Table 3.6 presented the apparent degradation rates for organics considering their behavior in Gate 2 over most of the experiment. It is likely that natural biodegradation improved as populations evolved over the duration of the experiment, so the apparent biodegradation rates were recalculated by considering the data for the last sampling on day 339 (Figure 3.14).

If it is assumed that the profiles along the column represent a quasi-steady-state distribution brought about by biodegradation of a reasonably constant input of organic compounds, an apparent first-order biodegradation rate constant can be approximated. A first-order equation, identical to Equation (3.1), except that concentration rather than flux was considered, was fit to the average concentrations over distances for CT, CF, and TOL from Figure 3.14. Because PCE concentrations declined mainly due to retardation and little biotransformation of PCE to *c*DCE was evident, PCE was not considered in this analysis. The first-order equation was only fit to segments of the curves

Table 3.7 Calculated Apparent Degradation Rates for CT, CF, and TOL from Profiles in Gate 2 on Day 339

Compound	Apparent Rate Constant (day^{-1})	Apparent Half-Life (days)
CT	0.063	11
CF	0.063	11
TOL	0.083	8.3

when concentration decline along the gate was evident: only average concentrations from fences G202 to G208 were considered for CT, G205 to G218 for CF, and G208 to G224 for TOL.

This natural bioattenuation in Gate 2 (Figure 3.14) can be compared with the performance of the anaerobic/aerobic remedial system used in Gate 3, as judged from Figure 4.11. In both Gates, CT and CF were rapidly and completely attenuated by anaerobic biodegradation. Much more significant reductive dechlorination of PCE was evident in Gate 3 than in Gate 2 by the end of the experiment. TOL was well attenuated in both Gates, perhaps marginally better in Gate 3 than in Gate 2. The major difference was the improved remediation of PCE in Gate 3.

In Gate 2, significant biodegradation was very rapid for CT and appeared quickly for CF. Evidence for TOL biodegradation was slower to appear and the evidence of minor biodegradation of PCE appeared only after 300 days. This may reflect the time required for the microbial populations to adapt to the anaerobic conditions and to begin degrading the target organics.

By the end of the experiment, transformation of CT and CF was complete, and TOL transformation was well established. Apparent half-lives were short, ranging from 8.3 to 11 days. PCE was just beginning to be transformed.

3.6 WASTE MANAGEMENT

All groundwater that was extracted was treated using an aboveground treatment system, prior to surface discharge. Full details of the waste treatment train are given in Appendix 10. Any groundwater extracted from sampling or other investigative events was added to the pretreatment tank for appropriate treatment.

3.7 CONCLUSIONS

The microbial environment in Gate 2 evolved from mainly aerobic to mainly anaerobic, in part due to the input of the target organics and their initial utilization by aerobes to scavenge DO from the groundwater. Aerobic degradation of TOL was anticipated from laboratory experiments. Under the anaerobic conditions rapidly established in Gate 2, CT and CF were rapidly biodegraded with the appearance of only minor CS_2. PCE was not significantly biodegraded in Gate 2. TOL was significantly biodegraded in Gate 2, but persisted throughout the gate at the end of the experiment.

This experiment demonstrated the potential for anaerobic conditions to develop in mixed (hydrocarbons + chlorinated solvents) plumes that are conducive to reductive dechlorination of at least some chlorinated solvents (i.e., CT and CF). While considerable biodegradation of toluene also occurred anaerobically, it is less certain that other aromatics such as benzene would have continued to degrade once the oxygen had been utilized. Also, PCE was generally recalcitrant during the experiment, and the ultimate potential for PCE biodegradation in Gate 2 was not clear. While the potential for natural attenuation to remediate mixed plumes is recognized and the results from Gate 2 are encouraging, this field experiment did not succeed in attaining complete removal of all the target organics via intrinsic processes.

3.8 RECOMMENDATIONS FOR FUTURE STUDIES/NEEDS

In evaluating technologies for remediation it is useful to conduct field-scale controls to account for natural attenuation processes. In Gate 2, longer monitoring is recommended to evaluate whether the extent of the biodegradation of TOL will increase. Similarly, PCE biodegradation appeared to have been initiated only in the last month of the experiment, and its longer-term fate is of interest.

Additional, comprehensive monitoring would have improved the definition of source inputs, time, and distance required to attain significant mass removal of organics and evolving redox conditions. However, it is really the long-term performance of intrinsic bioremediation that is of interest. It is recommended that, rather than enhanced monitoring over short times, monitoring at a low frequency, but over many years, would provide a better assessment of natural bioremediation in a controlled field experiment. Clearly, an improved source system to provide more constant inputs without significant maintenance would be required to support such a long-term study.

REFERENCES

Bae, W., J.E. Oldencrantz, B.E. Rittmann, and A.J. Valocchi. 1990. Transformation kinetics of trace-level halogenated organic contaminants in a biologically active zone (BAZ) induced by nitrate injections. *Journal of Contaminant Hydrology*, 6: 53068.

Barbaro, J.R., J.F. Barker, L.A. Lemon, and C.I. Mayfield. 1992. Biotransformation of BTEX under anaerobic denitrifying conditions: field and laboratory observations. *J. of Contam. Hydrology*, 11: 245-272.

Bouwer, E.J. and P.L. McCarty. 1983. Transformation of 1- and 2-carbon halogenated aliphatic organic compounds under methanogenic conditions. *Appl. Environ. Microbiol.* 45(4): 1286-1294.

Bradley, P.M. and F.H. Chapelle. 1996. Anaerobic mineralization of vinyl chloride in Fe(III)-reducing aquifer sediments. *Environ. Sci. Technol.* 30: 2084-2086.

Braus-Stromeyer, S.A. 1993. Anaerobic Biodehalogenation of Chlorinated Methanes. ETH Zürich, Dissertation No. 10324.

Braus-Stromeyer, S.A., R. Hermann, A.M. Cook, and T. Leisinger. 1993a. Dichloromethane as the sole carbon source for an acetogenic mixed culture and isolation of a fermentative, dichloromethane-degrading bacterium. *Appl. Environ. Microbiol.* 59: 3790-3797.

Braus-Stromeyer, S.A., A.M. Cook, and T. Leisinger. 1993b. Biotransformation of chloromethane to methanethiol. *Environ. Sci. Technol.* 27: 1577-1579.

Brewster, M.L., A.P. Annan, J.P. Greenhouse, B.H. Kueper, B.H. Olhoeft, J.D. Redman, and K.A. Sander. 1995. Observed migration of a controlled DNAPL release by geophysical methods. *Ground Water*, 33(6): 977-987.

Chapelle, F.H., McMahon, P.B., Dubrovsky, N.M., Fujii, R.F., Oaksford, E.T., and Vroblesky, D.A. 1995. Deducing the distribution of terminal-electron-accepting processes in hydrologically diverse groundwater systems. *Water Resour. Res.* 31: 359-371.

Cox, E.E., E. Edwards, L. Lehmicke, and D.W. Major. 1995. Intrinsic biodegradation of trichloroethene and trichloroethane in a sequential anaerobic–aerobic aquifer. In: *Intrinsic Bioremediation*, R.E. Hinchee, J.T. Wilson, and D.C. Downey (Eds.). Battelle Press, Columbus, OH.

Criddle, C.S., J.T. DeWitt, D. Grbic-Galic, and P.L. McCarty. 1990. Transformation of carbon tetrachloride by *Pseudomonas* sp. strain KC under denitrification conditions. *Appl. Environ. Microbiol.* 56: 3240-3246.

Davis, J.W. and C.L. Carpenter. 1990. Aerobic biodegradation of vinyl chloride in groundwater samples. *Appl. Environ. Microbiol.* 56: 3878-3880.

deBruin, W.P., M.J.J. Kotterman, M.A. Posthumus, G. Schraa, and A.J.B. Zehnder. 1992. Complete biological reductive transformation of tetrachloroethene to ethane. *Appl. Environ. Microbiol.* 58: 1996-2000.

Devlin, J.F. 1994. Enhanced *In Situ* Biodegradation of Carbon Tetrachloride and Trichloroethene Using a Permeable Wall Injection System. Ph.D. dissertation, Department of Earth Sciences, University of Waterloo, Waterloo, Ontario, Canada.

Ensign, S.A., M.R. Hyman, and D.J. Arp. 1992. Cometabolic degradation of chlorinated alkenes by alkene monooxygenase in a propylene-grown *Xanthobacter* strain. *Appl. Environ. Microbiol.* 58: 3038-3046.

Fiorenza, S., E.L. Hockman, Jr., S. Szojka, R.M. Woeller, and J.W. Wigger. 1994. Natural anaerobic degradation of chlorinated solvents at a Canadian manufacturing plant. In: *Bioremediation of Chlorinated and Polycyclic Aromatic Hydrocarbon Compounds*, R. Hinchee, A. Leeson, L. Semprini, and S.K. Ong (Eds.), Lewis Publishers, Boca Raton, FL.

Freedman, D.L. and J.M. Gossett. 1989. Biological reductive dechlorination of tetrachloroethylene and trichloroethylene to ethylene under methanogenic conditions. *Appl. Environ. Microbiol.* 55: 2144-2151.

Hartmans, S. and J.A.M. de Bont. 1992. Aerobic vinyl chloride metabolism in *Mycobacterium aurum* L1. *Appl. Environ. Microbiol.* 58(4): 1220-1226.

Hartmans, S., J.A.M. de Bont, J. Tramper, and K.Ch.A.M. Luyben. 1985. Bacterial degradation of vinyl chloride. *Biotechn. Lett.* 7: 383-386.

Holliger, C., G. Schraa, A.J.M. Stams, and A.J.B. Zehnder. 1993. A highly purified enrichment culture couples the reductive dechlorination of tetrachloroethene to growth. *Appl. Environ. Microbiol.* 59: 2991-2997.

Hopkins, G.D., L. Semprini, and P.L. McCarty. 1993. Microcosm and *in situ* field studies of enhanced biotransformation of trichloroethylene by phenol-utilizing microorganisms. *Appl. Environ. Microbiol.* 59: 2277-2285.

Hubbard, C.E. 1992. Transport and Fate of Dissolved Methanol, Methyl-Tertiary-Butyl Ether and Monoaromatic Hydrocarbons in a Shallow Sand Aquifer. M.Sc. Thesis. Department of Earth Sciences, University of Waterloo, Waterloo, Ontario, Canada.

Klecka, E.L., N.J. Klier, R.J. West, J.W. Davis, D. Ellis, J.M. Odom, T.A. Ei, F.H. Chapelle, D.W. Major, and J. Salvo. 1997. Intrinsic bioremediation of chlorinated ethenes at Dover Air Force Base. In: *In Situ and On-Site Bioremediation, vol. 3.* Alleman, B.C. and A. Leeson (Eds.), Battelle Press, Columbus, OH.

Kohler-Staub, D., S. Frank, and T. Leisinger. 1995. Dichloromethane as the sole carbon source for *Hyphomicrobium* sp. Strain DM2 under denitrifcation conditions. *Biodegradation,* 6: 229-235.

Krone U.E., K. Laufer, R.K. Thauer, and H.P.C. Hogenkamp. 1989. Coenzyme F430 as a possible catalyst for the reductive dehalogenation of chlorinated C1-hydrocarbons in methanogenic bacteria. *Biochemistry,* 28: 10061-10065.

Krone U.E., R.K. Thauer, H.P.C. Hogenkamp, and K. Steinbach. 1991. Reductive formation of carbon monoxide from CCl4 and freons 11, 12 and 13 catalyzed by corrinoids. *Biochemistry,* 30: 2713-2719.

Lee, M.D., L.S. Sehayek, and T.D. Vandell. 1996. Intrinsic bioremediation of 1,2-dichloroethane. Symposium on Natural Attenuation of Chlorinated Organics in Ground Water. Dallas, September 11-13. EPA/540/R-96/509.

Mackay, D.M., D.L. Freyberg, P.V. Roberts, and J.A. Cherry. 1986. A natural gradient experiment on solute transport in a sand aquifer. 1. Approach and overview of plume movement. *Water Resources Research,* 22(13): 2017-2029.

Mägli, A., F.A. Rainey, and T. Leisinger. 1995. Acetogenesis from dichloromethane by a two-component mixed culture comprising a novel bacterium. *Appl. Environ. Microbiol.* 61: 2943-2949.

Major, D.W., E.E. Cox, E. Edwards, and P.W. Hare. 1995. Intrinsic dechlorination of trichloroethene to ethene in a bedrock aquifer. In: *Intrinsic Bioremediation*, R.E. Hinchee, J.T. Wilson, and D.C. Downey (Eds.), Battelle Press, Columbus, OH, pp. 197-203.

Major, D.W., E.H. Hodgins, and B.J. Butler. 1991. Field and laboratory evidence of *in situ* biotransformation of tetrachloroethene to ethene and ethane at a chemical transfer facility in North Toronto. In: *In Situ and On Site Bioreclamation*, R. Hinchee and R. Olfenbuttel (Eds.), Buttersworth-Heineman, Stoneham, MA.

Maymo-Gatell, X., J.M. Gossett, and S. H. Zinder. 1997. *Dehalococcus ethenogenes* strain 195: Ethene production from halogenated aliphatics. In: *In Situ and On-Site Bioremediation*, vol. 3, Alleman, B.C. and A. Leeson (Eds.), Battelle Press, Columbus, OH.

McCarty, P.L., L. Semprini, M.E. Dolan, T.C. Harmon, C. Tiedeman, and S.M. Gorelick. 1991. In situ methanotrophic bioremediation for contaminated groundwater at St. Joseph, Michigan. In: *On-Site Bioreclamation: Processes for Xenobiotic and Hydrocabon Treatment*, R. E. Hinchee and R. F. Olfenbuttel (Eds.), Butterworth Heinemann, Stoneham, MA, pp. 16-40.

Messmer, M. and T. Leisinger. 1997. Degradation of dichloromethane by *Dehalobacterium formicoaceticum.* Information on the Internet page of the Institute for Microbiology at the Swiss Federal Institute of Technology in Zürich, Switzerland.

Messmer, M., G. Wohlfarth, and G. Diekert. 1993. Methyl chloride metabolism of the strictly anaerobic, methyl chloride-utilizing homoacetogen strain MC. *Arch. Microbiol.* 160: 383-387.

National Research Council (NRC). 1993. *In Situ Bioremediation: When Does It Work*? National Academy Press, Washington, DC.

Nelson, M.J.K., S.O. Montgomery, and P.H. Pritchard. 1988. Trichloroethylene metabolism by microorganisms that degrade aromatic compounds. *Appl. Environ. Microbiol.* 54: 604-606.

Nicholson, R. V., J.A. Cherry, and E.J. Reardon. 1983. Migration of contaminant in groundwater at a landfill: A case study. *Journal of Hydrology*, 63: 131-176.

Oldenhuis, R., J.Y. Oedzes, J.J. van der Waarde, and D.B. Janssen. 1991. Kinetics of chlorinated hydrocarbon degradation by *Methylosinus trichorsporium* OB3b and toxicity of trichloroethylene. *Appl. Environ. Microbiol.* 57: 7-14.

Pankow, J.F. and J.A. Cherry. 1996. *Dense Chlorinated Solvents and Other DNAPLS in Groundwater.* Waterloo Press, Portland, OR.

Semprini, L., P.V. Roberts, G.D. Hopkins, and P.L. McCarty. 1990. A field evaluation of *in situ* biodegradation of chlorinated ethenes, 2. Results of biostimulation and biotransformation experiments. *Groundwater,* 28: 715-727.

Strand, S.E. and L. Shippert. 1986. Oxidation of chloroform in an aerobic soil exposed to natural gas. *Appl. Environ. Microbiol.* 52: 203-205.

Stromeyer, S.A., K. Stumpf, A.M. Cook, and T. Leisinger. 1992. Anaerobic degradation of tetrachloromethane by *Acetobacterium woodii*: Separation of dechlorinative activities in cell extracts and roles for vitamin B12 and other factors. *Biodegradation,* 3: 113-123.

University of Waterloo (UW). 1995. Revised Workplan for Passive and Semipassive Techniques for Groundwater Remediation at CFB Borden, submitted to AATDF, July 1995.

University of Waterloo (UW). 1997. Report on the Installation of the CFB Borden Experiments for Passive and Semipassive Techniques for Groundwater Remediation, submitted to AATDF, April 1997.

Vanellii, T., M. Logan, D.M. Arciero, and A. Hooper. 1990. Degradation of halogenated aliphatic compound by ammonia-oxidizing bacterium *Nitrosomonas eurpaea. Appl. Environ. Microbiol.* 56: 1169-1171.

Wackett, L.P., G.A. Brusseau, S.A. Householder, and R.S. Hanson. 1989. Survey of microbial oxygenases: trichloethylene degradation by propane-oxidizing bacteria. *Appl. Environ. Microbiol.* 55: 2960-2964.

Wild, A.P., W. Winklebauer, and T. Leisinger. 1996. Anaerobic dechlorination of trichloroethene, tetrachloroethene and 1,2-dichloroethane by an acetogenic mixed culture in a fixed-bed reactor. *Biodegradation,* 6: 309-318.

Wilson, J.T. and B. H. Wilson. 1985. Biotransformation of trichlorethylene in soil. *Appl. Environ. Microbiol.* 49: 242-243.

Wilson, J.T., J.W. Weaver, and D.H. Kampbell. 1994a. Intrinsic bioremediation of TCE in ground water at a NPL site in St. Joseph, Michigan. In: *Symposium on Bioremediation of Hazardous Wastes: Research, Development, and Field Evaluations.* U.S. Environmental Protection Agency EPA/600R-94/075, June 1994, San Francisco, CA.

Wilson, J.T., J.W. Weaver, and D.H. Kampbell. 1994b. Intrinsic bioremediation of TCE in ground water at an NPL site in St. Joseph, Michigan. In: *Symposium on Intrinsic Bioremediation of Groundwater,* U.S. Environmental Protection Agency EPA/540/R-94/515, August 1994, Denver, CO.

Wilson, J.T., D.H. Kampbell, J.W. Weaver, B. Wilson, T. Imbriglotta, and T. Ehlke. 1995. A review of intrinsic bioremedation of trichlorethylene in ground water at Picatinny Arsenal, New Jersey and St. Joseph, Michigan. In: *Symposium on Bioremediation of Hazardous Wastes: Research, Development, and Field Evaluations.* U.S. Environmental Protection Agency, EPA/600R-94/075, August 1995, Rye Brook, NY.

CHAPTER 4

Sequential Anaerobic/Aerobic Bioremediation — Gate 3 CFB Borden

4.1 TECHNOLOGY PROCESS DESCRIPTION

A pilot-scale, *in situ* bioremedial system was installed in Gate 3 to evaluate sequential anaerobic/aerobic bioremediation of groundwater containing dissolved chlorinated solvents and petroleum hydrocarbons. The *in situ* process operated in a semipassive manner with minimal operation and maintenance activity. This chapter describes the technology, and its design, implementation, and performance at the pilot scale, drawing on supporting design calculations and laboratory experiments, and specific field tests to document performance.

The conceptualized *in situ* treatment process is shown in Figure 4.1. The first treatment stage uses the addition of a carbon substrate, sodium benzoate, via a nutrient injection wall (NIW) to induce anaerobic conditions within the aquifer such that reductive dehalogenation of chlorinated solvents (i.e., CT and PCE) can occur. Sodium benzoate was added to cause *in situ* bacteria to utilize the available electron acceptors (O_2, SO_4^{2-}, etc.) to create highly reducing conditions downgradient of the NIW. In the second treatment stage, oxygen is introduced via a biosparge zone to promote aerobic biodegradation of toluene (TOL), chlorinated breakdown products remaining from the anaerobic system (e.g., DCE), and residual benzoate added as part of the anaerobic treatment. This sequence allows sections of the aquifer to become acclimatized to anaerobic and aerobic conditions, maximizing the potential to establish stable, consistent, *in situ* anaerobic and aerobic treatment for extended times.

4.1.1 Enhanced Anaerobic Bioremediation

The first stage in the treatment sequence is the induction of anaerobic conditions. These are achieved by periodic injection of a dissolved, labile organic carbon substrate using a nutrient injection wall (NIW, see Figure 4.1) as described by Devlin and Barker (1994, 1996). The NIW consists of filter sand filling the wall, and wells through which groundwater is extracted, amended with labile organic carbon and reinjected, essentially filling the NIW with nutrient-amended groundwater. The slug or pulse of amended groundwater advects downgradient of the NIW, while longitudinal dispersion mixes the pulse and nonamended groundwater before and after it. Pulse injection intervals (hereinafter referred to as "flush" intervals) are selected to create a downgradient segment of the aquifer continuously receiving a minimum level of organic carbon sufficient to support strongly anaerobic microbial activity. This is illustrated schematically in Figure 4.1. This continuously anaerobic zone would be the location of anaerobic reductive dechlorination of PCE and CT.

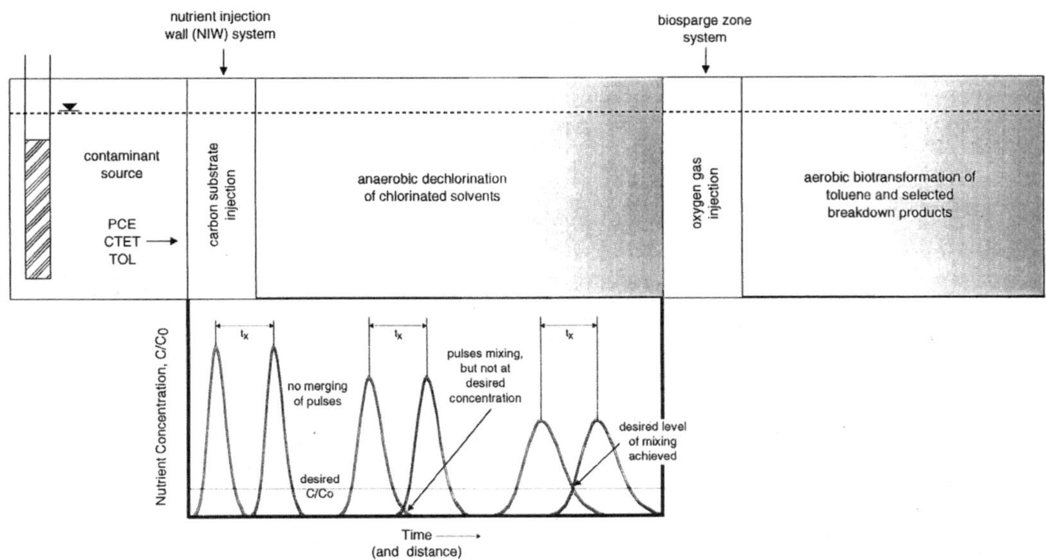

Figure 4.1 Conceptualized treatment sequence.

4.1.2 Enhanced Aerobic Bioremediation

The second stage in the treatment sequence is the induction of aerobic conditions to support *in situ* aerobic biodegradation of TOL, benzoate, and partial dechlorination products. This is achieved by periodically delivering oxygen gas across the bottom of a zone filled with pea gravel termed the "biosparge zone." The principal goal is to place as many residual oxygen bubbles into the pea gravel as possible while minimizing volatilization of contaminants. Subsequent to an oxygen addition event, this residual oxygen remains as a source of oxygen for groundwater migrating through the biosparge zone.

4.1.3 Coupling of Technologies

Incomplete dechlorination products, such as DCE and VC, that may be generated in the anaerobic treatment are more susceptible to aerobic biodegradation (McCarty and Semprini, 1994), as are petroleum hydrocarbons (e.g., Barker et al., 1987). In addition, aerobic cometabolism of DCE and VC may be promoted by methane, a common product of anaerobic metabolism. Thus, the coupling of anaerobic and aerobic treatment may enhance the performance of a simple anaerobic treatment for many CVOCs or a simple aerobic treatment system for petroleum hydrocarbons.

The interface between the strongly anaerobic zone and the biosparge zone to which O_2 is being added represents a major change in redox conditions. It was anticipated that adverse reactions, such as the oxidation of dissolved Fe(II) with precipitation of Fe(III) oxyhydroxides, could significantly reduce gate permeability and consume significant O_2. The biosparge zone is designed with a greater porosity to compensate for some mineral precipitation within it. The frequency of O_2 sparged into the biosparge gate can be tailored, to some extent, to satisfy the oxygen demand of Fe(II) entering the gate. Thus, the biosparge system is designed and can be operated to minimize the impact of mineral precipitation.

4.2 FIELD DEMONSTRATION

4.2.1 Experimental Design

General details of the Borden experimental site location, hydrogeology, and geochemistry are provided in Appendix 1. Section 4.1 described the sequential treatment system employed. A chronology of events is shown in Appendix 2. Construction was completed on December 21, 1996, initial tracer tests were conducted from May 15 to June 27, 1997, and source wells were established November 18, 1996 (termed "day 0") and continued to operate until October 28, 1997. Details of the construction and as-built drawings are included in UW (1997a).

The general layout of Gate 3 is shown in Figure 4.2. PCE, CT, and TOL were added to influent groundwater via source emitter wells (Appendix 9) and the concentrations established by sampling groundwater from the nine points in fence G302. The operation of the extraction well G325P caused groundwater to be drawn through the NIW, where an organic substrate, sodium benzoate, was added. Fences G304 and G306 were used to monitor the completeness of the nutrient additions. The extent of mixing and subsequent anaerobic bioremediation was determined by sampling

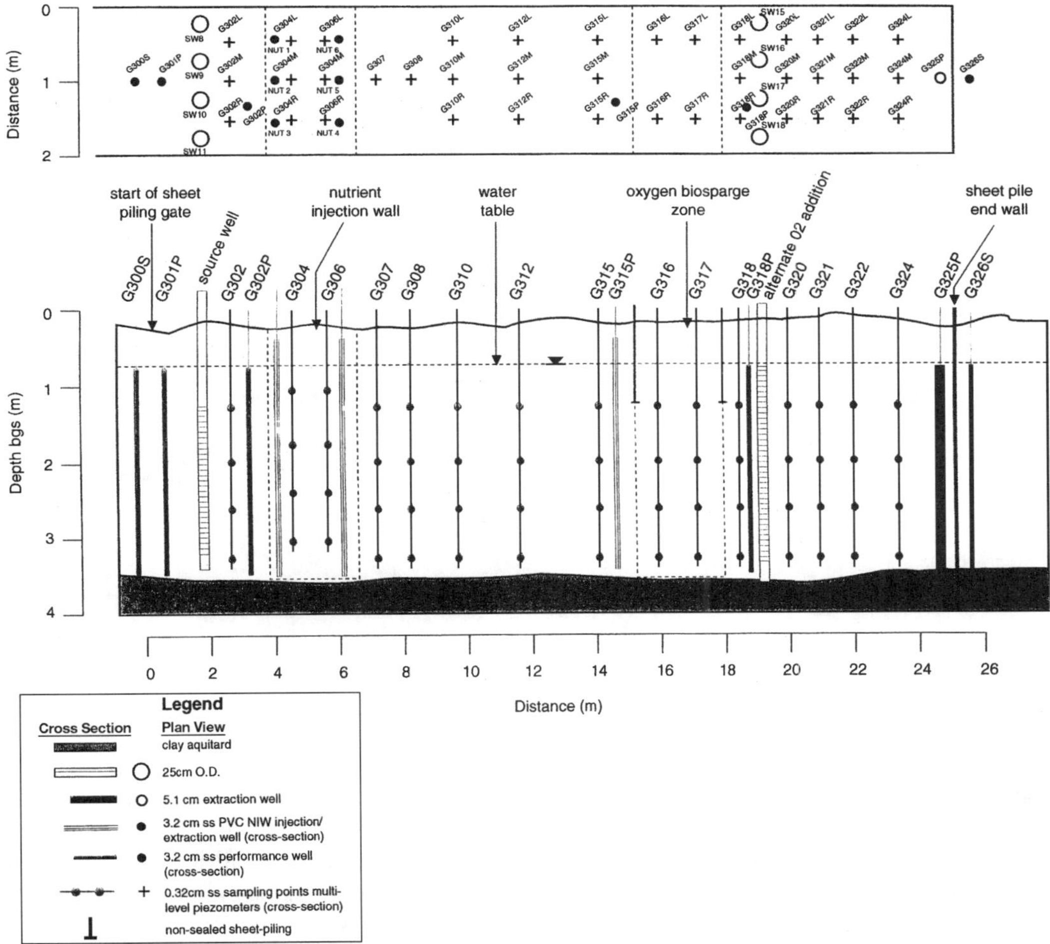

Figure 4.2 Plan view and cross section of Gate 3.

groundwater from fences G307 to G315. Then, the anaerobic groundwater was drawn through a biosparge zone where O_2 was added. The composition of gas in the headspace above the water table (potential offgas) was monitored. The alternate O_2 addition wells shown in Figure 4.2 were not used in this experiment. The extent of aerobic conditions in groundwaters was monitored in fences G316 and G324, as was the extent of bioremediation of organics and benzoate. Groundwater was extracted from well G325P at the downgradient end of the sheet-piling-isolated segment of the aquifer and treated before release on surface (see Appendix 10).

4.2.1.1 Anaerobic System

The first stage in the treatment sequence was the induction of anaerobic conditions in a permeable zone using an injection system (Figure 4.3). For the reasons discussed in Chapter 1, Section 1.4.2, benzoate was the organic nutrient injected in this pilot-scale test. The nutrient injection wall (NIW) consisted of coarse sand filling the wall, three 5.1-cm-diameter PVC extraction wells at the downgradient end, three identical reinjection wells at the upgradient end, a circulation pump, and an in-line nutrient amendment system comprising a water-driven proportional feeder pump and reservoir.

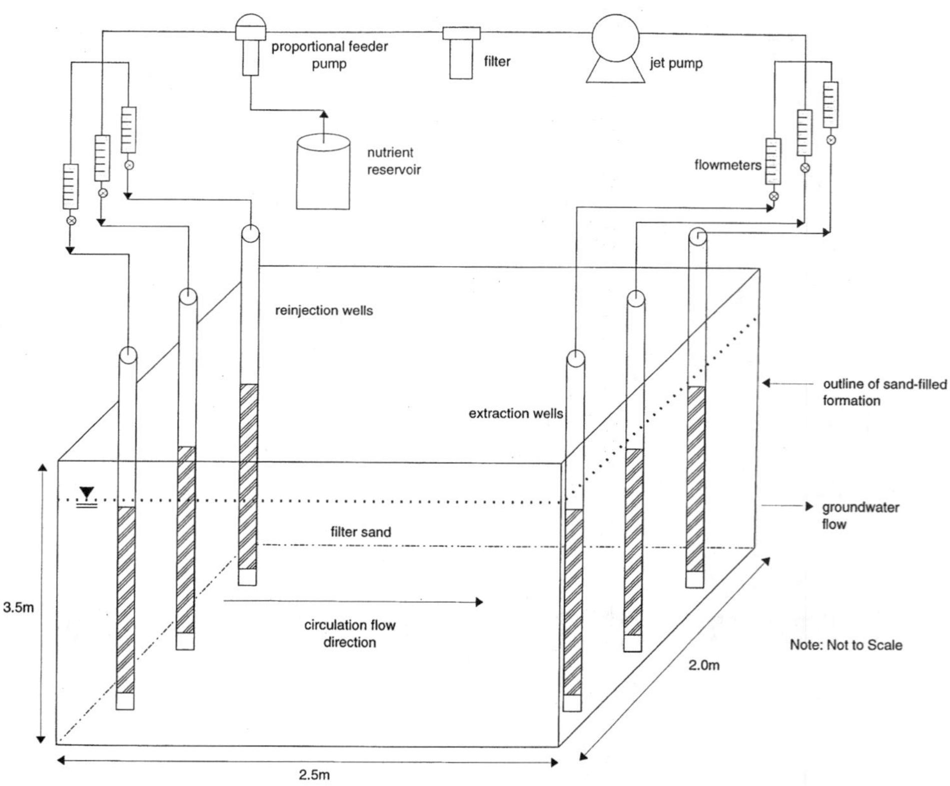

Figure 4.3 Longitudinal cross section of the NIW and schematic of the circulation/amendment system.

Intervals between nutrient injections (t_x, and hereinafter referred to as "flush" intervals) were set to create a zone within which a continuous supply of benzoate would be available (Figure 4.1). Fifteen flushes were done in the NIW from October 26, 1996 to September 25, 1997.

Operationally, nutrient injections were spaced 28 days apart to achieve a zone having a minimum concentration of 50 mg/L (designed concentration) sodium benzoate at a distance approximately 5 m downgradient of the NIW (see Appendix 17 for design modeling).

The flush operation was designed to extract, spike, and reinject the volume of groundwater within the 2.5 m by 3.5 m by 2.0 m NIW. Various scenarios were considered, but the protocol adopted required circulating about 4600 L of groundwater, within a 200-minute interval. Beginning December 13, 1997 (day 25) and for all subsequent nutrient flushes, selected points in piezometer G306M were monitored for conductivity to trace the arrival of the injected, amended groundwater at the extraction wells. Monitoring of points in the left and right piezometers was included beginning May 16, 1997 (day 177), to assess the thoroughness of solute distribution in the wall. Devlin (1994) showed such monitoring was a useful method to confirm the completeness of the flush. Typically, greater than 85% of the groundwater in the NIW was amended in each individual flush.

The groundwater was circulated within the wall via a jet pump (Home Plumber/McDougall model 312570) at a rate of approximately 23 L/min (6 U.S. gal/min). The extracted water was amended with a nutrient mixture via an automatic proportional feeder (APF; Autotrol Corp., model 2), designed to feed nutrient concentrate into the circulating groundwater at a rate proportional to flow. The feed ratio was set at 0.002, although it varied between 0.0018 and 0.002. Plumbing for the system was constructed from 2.5-cm and 3.75-cm ABS Polypipe tubing, and flexible 3.75-cm, high-pressure/vacuum line where necessary.

From September 26, 1996 to September 25, 1997, 15 nutrient flushes were completed. Initial objective sodium benzoate concentrations in the wall were 300 mg/L, which was subsequently lowered to approximately 180 mg/L beginning day 151, April 18, 1997 (see Appendix 17) when it became apparent that the anaerobic activity had been stimulated, but undesirably high residual benzoate concentrations were likely to reach the aerobic treatment zone. The estimated sodium benzoate concentration and the volume of amended groundwater for each flush are shown in Table 4.1.

Table 4.1 Summary of Sodium Benzoate, Mineral Salts, and Inorganic Tracers Introduced to the NIW in Each Flush

Day	Date	Flush	Volume (L)	Sodium Benzoate (mg/ L)	Mineral Salts, Total (mg/L)	KBr (mg/L)	KCl (mg/L)
	26/09, 02/10, & 07/10/96	1	15000[a]	106[a]			
	26,27/10/96	2	n/d	n/d			
14	02/12/96	3	5200	n/d			
25	13/12/96	4	n/d	n/d			
45	02/01/97	5	4500	n/d			
67	24/01/97	6	n/d	n/d			
95	21/02/97	7	5700	n/d			
123	21/03/97	8	4500[a]	300[a]			
151	18/04/97	9	6600	150		170	0
180	16/05/97	10	5000	190		0	240
207	13/06/97	11	5100	190	34	110	0
235	11/07/97	12	3900	190	34	110	0
262	08/08/97	13	2000	190	34	0	0
283	29/08/97	14	3600	180	33	0	0
311	25/09/97	15	2700	200	36	60	0

Notes: n/d: not determined. Mineral salts: about 5:5:5:1 of K_2HPO_4 : KH_2PO_4 : NH_4NO_3 : $MgSO_4 \cdot 7H_2O$.

[a] estimated

Beginning June 13, 1997 (day 207), inorganic nutrients (as a dilute mineral salts medium), based on the formulation of Mueller et al. (1991), were also injected to stimulate greater anaerobic activity. Table 4.1 indicates the average mineral salt concentrations achieved.

Inorganic tracers were added to the NIW with flushes 9 to 12 and 15 to allow for tracking of the benzoate pulses within the anaerobic zone, downgradient of the NIW. The concentration of tracers introduced is also given in Table 4.1.

4.2.1.2 Aerobic System

The second stage in the treatment sequence was the induction of aerobic conditions within a downgradient segment of the gate. Sparging of air into contaminated groundwater has been suggested as a remediation strategy to volatilize and/or biodegrade organics such as BTEX (Pankow et al., 1993; Johnson et al., 1993). Our approach was to sparge with small volumes of oxygen gas (O_2) to enhance *in situ* aerobic biodegradation while minimizing volatilization of organic contaminants. This was achieved by a biosparge system illustrated in Figure 4.4.

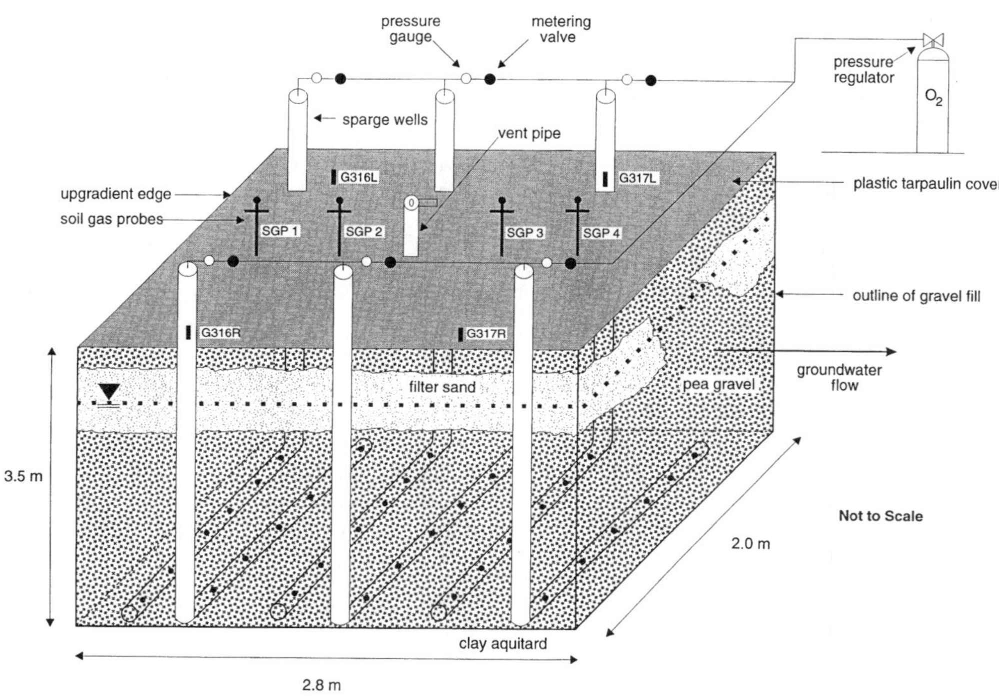

Figure 4.4 Schematic of the biosparge zone.

Delivery of oxygen gas was achieved via six L-shaped 5.1-cm-diameter PVC aeration wells, extending across the bottom of the biosparge gate. Each horizontal limb had vent holes located about every 5 cm. Four manifolded 0.63-cm OD polyethylene tubes extended to equally spaced lengths along the horizontal section of each well (Figure 4.5), with all vent wells fed by a single pressurized oxygen cylinder. Narrow diameter (0.64-cm OD and 0.3-cm ID) stainless steel soil gas probes fitted with Mininert valves monitored off gas concentrations immediately above the saturated zone (22 cm below ground surface).

The system was designed to deliver gaseous oxygen up to the residual gas phase capacity of the pea gravel (about 8.7% of the 46% porosity) within the gate (see Appendix 22). The principal goal was to place sufficient residual oxygen bubbles in the saturated pea gravel to support aerobic biodegradation of benzoate and TOL, while minimizing volatilization of contaminants.

Pea gravel was chosen as the medium to fill the gate, since laboratory studies by Ji et al. (1993) indicated that finer grained media perpetuated the formation of relatively few discrete air channels,

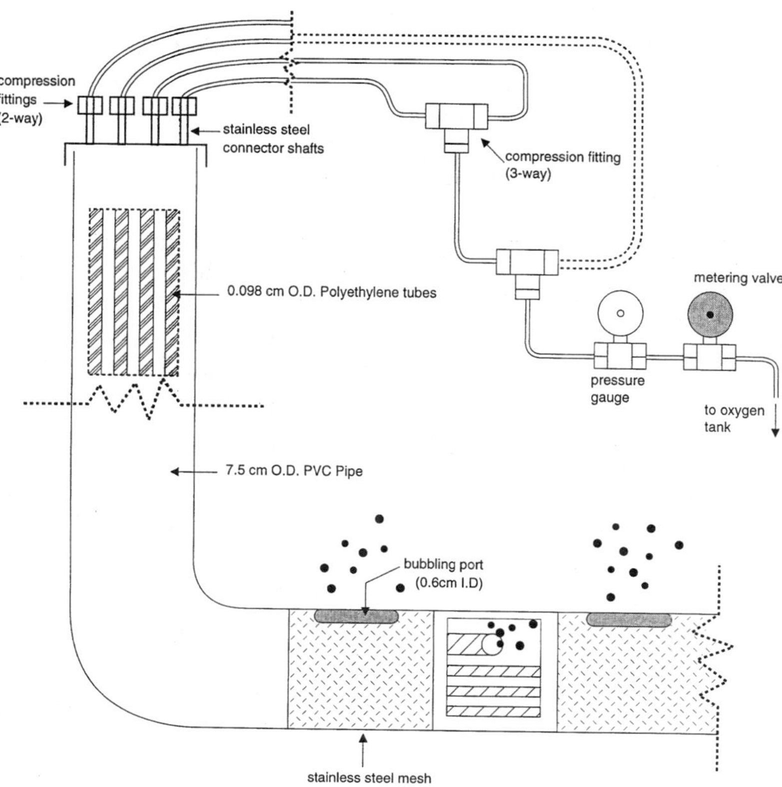

Figure 4.5 Schematic of the oxygen sparge wells.

which localized the residual oxygen bubbles along such channels. This would have resulted in poor overall transfer of oxygen from such bubbles to the advecting groundwater. The Ji et al. study indicated that, in pea gravel, gas flow would be through a greater volume of the material and that the widely distributed, residual, discrete bubbles would remain trapped in the media once gas delivery ceased. The broader distribution of residual bubbles was thought to allow for greater bubble–water contact, maximizing oxygen transfer to flowing groundwater. Sparging greater quantities of gas into the media would not have resulted in more oxygen bubbles being retained in the pea gravel, but would have generated additional off gas.

The unsaturated zone within the sparge gate had a wedge of more permeable pea gravel along the center line between less permeable coarse sand to "funnel" any volatilized contaminants toward the soil gas probes (see Figure 4.4). The greater moisture content of the finer material would also have tended to channel volatilized vapors toward a controlled outlet, minimizing possible short-circuiting of vapors up the sides of the sheet piling.

A tarp was installed over the top of the gate to minimize atmospheric contamination of soil gas samples when samples were drawn from the monitoring probes. Each soil gas probe (and multilevel piezometer) was sealed around the tarp using plumber's putty.

The frequency of O_2 sparging was calculated based on the BOD represented by 300 mg/L sodium benzoate and the estimation that about 0.4 m^3 of oxygen would be delivered each time (Appendix 22). Consequently, O_2 sparging was done initially on a twice-weekly basis from November 22, 1996 to July 18,1997 (day 242).

DO monitoring showed essentially no DO in groundwater in the biosparge zone, so either sufficient oxygen was being supplied and balanced by utilization, or insufficient oxygen was being supplied. The latter scenario was judged most likely as residual benzoate was found downgradient of the biosparge zone. For example, 25 to 30 mg/L of sodium benzoate was found in piezometers G318M and G324M in January 1997. Despite reductions in the concentration of benzoate introduced to the NIW beginning day 207 (June 13, 1997), up to 20 mg/L of sodium benzoate was still found at fence G318. In November 1997, a direct measurement revealed that only about 41 L (at STP), and not 390 L, of O_2 was being added during each 1 min sparge. This was far below the design objectives, and so insufficient O_2 addition was confirmed as the reason for the poor oxygenation of the biosparge zone up to day 242. During most of this time, operational problems with the NIW and lack of reductive dechlorination in the anaerobic segment of the gate had received most of the attention in Gate 3. In addition, TOL didn't enter the biosparge zone until late March 1997, so biosparge performance assessment was not possible until late in the project.

After day 242, O_2 sparging was increased to once each day, and as a result, an average 8 to 15 mg/L DO was typically observed in the biosparge zone groundwater. As noted above, each sparge event added about 40 L (at STP) of O_2 (about 57 g) to the biosparge zone. Sodium benzoate concentrations at fence G315, just upgradient of the biosparge zone, averaged less than 10 mg/L in October 1997 and this would exert a BOD of <1.29 g/day (groundwater flux of 0.04 m^3/day) in the biosparge gate. Also, an average of 160 mg/L acetic acid (BOD of 170 mg/L) was detected in the few samples taken from fence G315 in October 1997. This would exert an additional BOD of 6.8 g/day. Minor BOD as TOL and methane was likely less than 2 g/day. Under the daily, one minute O_2 sparge protocol, followed since day 242, about 57 g/day O_2 was being supplied, much more than required to meet the daily 11 g BOD entering the gate. Thus, it is reasonable to assume that the increased sparge frequency produced excess oxygen in the biosparge zone. It was only after day 242 that the biosparge zone was considered to be successfully operating, and the analyses of the biosparge gate remediation in Section 4.3.5 will concentrate only on the post-July 18, 1997 (day 242) data.

4.2.2 Sampling Program

The sampling program as described in the 1995 Work Plan (UW, 1995) included background groundwater chemical sampling, groundwater and soil gas sampling during the experimental phase, and some hydraulic head determinations during the experiment. The media sampled and analyses performed are listed in Table 4.2, with details of sampling methods provided in Appendix 4. Laboratory analytical methods are detailed in Appendix 3 and QA/QC procedures and results are given in Appendix 5. Groundwater sampling was conducted from May 22, 1996 to October 23, 1997 (day 339). Timing of sampling events is chronicled in Appendix 2.

Table 4.2 Media Sampled and Analyses Conducted in Gate 3

Parameter	Groundwater	Soil	Air
Volatile organic compounds (VOCs)	X		X (TOL and PCE)[a]
Dissolved hydrocarbon gases (DHGs)	X		
Sodium benzoate	X		
Major ions	X		
Alkalinity	X		
Dissolved oxygen	X		
Conductivity	X		
Redox potential	X		
pH	X		
Organic acids	X		
Microbiology		X	

[a] Within the unsaturated zone in the biosparge zone only.

4.3 RESULTS

The results of the field experiment are presented in the following sections. The background chemical conditions are described in Section 4.3.1. The hydraulic conditions for the test are presented in Section 4.3.2, including the distribution of hydraulic head and the groundwater velocity under pumping as estimated from tracer tests. The nature of the organic solute inputs to Gate 3 are discussed briefly in Section 4.3.3, and the results from the enhanced anaerobic and enhanced aerobic treatments are presented in Sections 4.3.4 and 4.3.5.

4.3.1 Background Conditions

Results of background sampling in the study site area during May 1996 are given in Appendix 6. The chemical characteristics of Gate 3 background groundwater are summarized in Table 4.3. The chemistry was similar to that found in Gates 1 and 2. The groundwater was fresh, Ca-bicarbonate type, with near-neutral pH. Low concentrations of DO and positive E_h values suggested aerobic conditions, but the presence of traces of methane was more characteristic of anaerobic conditions. Redox appeared heterogeneous, perhaps with methane being transported from upgradient anaerobic zones. Overall, the background groundwater is judged to be very weakly aerobic.

Table 4.3 Results of Background Sampling for Gate 3, May 1996

Parameter Analyzed	Average Concentrations (all depths included) G308	G321M
Field Parameters (units noted)		
pH	7.80	7.60
Eh (mV)	172	86
Conductivity (μs)	400	412
DO (mg/L)	3.9	2.9
Alkalinity (mg/L $CaCO_3$)	156	174
Inorganics (mg/L)		
Iron	n/d	n/d
Calcium	68.9	78.5
Potassium	1.3	1.8
Magnesium	5.5	5.8
Manganese	n/d	n/d
Sodium	17.9	9.7
Chloride	30.6	16.7
Bromide	n/d	n/d
Sulfate	17.6	16.5
Nitrate	n/d	n/d
Phosphate	n/d	n/d
Organics (μg/L)		
PCE	n/d	3.0
CT	n/d	n/d
Toluene	n/d	n/d
Methane	1.5	10.5
Ethane	n/d	0.8
Ethene	n/d	0.4

Note: n/d: not detected (below MDL).

In terms of typical potential electron acceptors, no nitrate was found, but 16 to 18 mg/L sulfate was found. Fe(III) is likely available in the aquifer solids, but no dissolved iron was detected.

Low concentrations of PCE (9.7 µg/L in piezometer G321M-2) would not have affected the demonstration, since it would have been flushed out prior to the introduction of organics into the source wells, and none was found upgradient of that point in May 1996.

4.3.2 Hydraulic Conditions

Water level measurements (corrected to sea level as datum) from June 1996 to October 1997 are given in Appendix 8. Figure 4.6 shows the distribution of hydraulic head for three dates. The head values generally decreased progressively along the gate toward the extraction well, indicating a hydraulic gradient in that direction as was intended. The head values in the biosparge zone on October 23, 1997 were likely affected by the sparging of oxygen and so could be anomalously elevated. The water level fluctuated seasonally as seen in Figure 2.8 for Gate 1. Such fluctuations likely had little impact on Gate 3 treatments, so they are not illustrated here (see Appendix 8).

Groundwater velocities from the source wells to each piezometer fence in Gate 3 are summarized in Appendix 7, with additional analysis of a later pulse tracer test given in Appendix 18. Results from the main tracer test (Appendix 7) indicated an average groundwater velocity up to fence G317 of 12 cm/day. This velocity was used in designing the benzoate pulse and oxygen addition protocols. Figure A7.4 (Appendix 7) illustrates updated estimates of the apparent groundwater velocity to various sampling points in Gate 3. While significant variations occur, there is no strong evidence for preferential flow paths or high velocity zones extending throughout a significant length of the gate.

An average groundwater velocity of 15 cm/day was determined from the nutrient pulse tracer test (Appendix 18). This test was conducted within the regularly scheduled sampling rounds, and is considered most representative of the hydraulic conditions within the experiment. Discrepancies between groundwater velocities determined from the main and nutrient pulse tracer tests (Appendices 7 and 18, respectively) are thought to be a result of variations in the water table (up to 1 m) over the duration of each test. This would have served to alter the cross-sectional area through which groundwater could flow. Under a relatively constant flux imposed by the pumping well, this fluctuation in the water table would have produced variation in the average linear groundwater velocity. In the following sections, a groundwater velocity of 15 cm/day is assumed for Gate 3.

4.3.3 Plume Generation

The mixed organic plume containing PCE, CT, and TOL was created in four source wells near the upgradient end of Gate 3 (see Figure 4.2). These wells were operated from November 18, 1996 (day 0) to October 28, 1997 (day 343). Concentrations of these organics in the source wells were variable and delivery to the aquifer was also variable. Details of contaminant plume generation are given in Appendix 9.

Concentration profiles at fence G302, 0.8 m downgradient of the source wells, indicated a high degree of spatial and temporal variability in contaminant input into the gate. At fence G302, toluene concentrations over time range from <MDL to 19,000 µg/L; CT from <MDL to 2100 µg/L; CF, a dechlorination product of CT, from <MDL to 700 µg/L; PCE from <MDL to 1200 µg/L. The presence of CF indicates some CT dechlorination had occurred intrinsically upgradient of fence G302. Figure 4.7 shows the distribution of TOL at fence G302 for day 339 to illustrate the spatial variability of input to Gate 3. The average concentrations at fence G302 over the ten sampling events are shown in Figure 4.8. Even for these averaged concentrations, temporal variability is large.

No monitoring was done at fence G302 from about day 140 to day 339. To indicate the variability in solute concentrations being delivered to the remedial systems during this time, average concentrations of PCE and TOL at fence G306, from day 154 to day 311, are also shown in Figure 4.8.

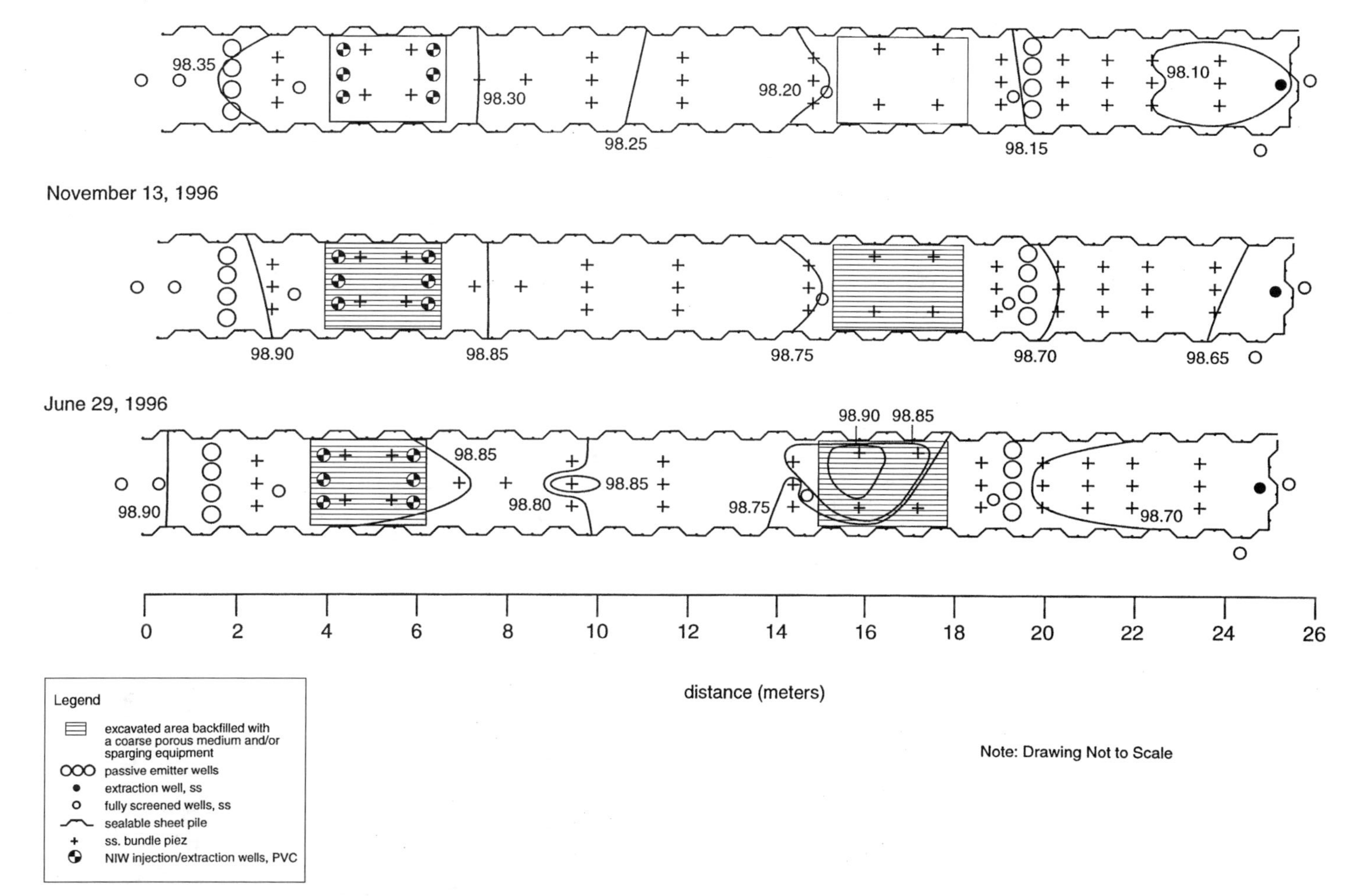

Figure 4.6 Distribution of hydraulic head in Gate 3.

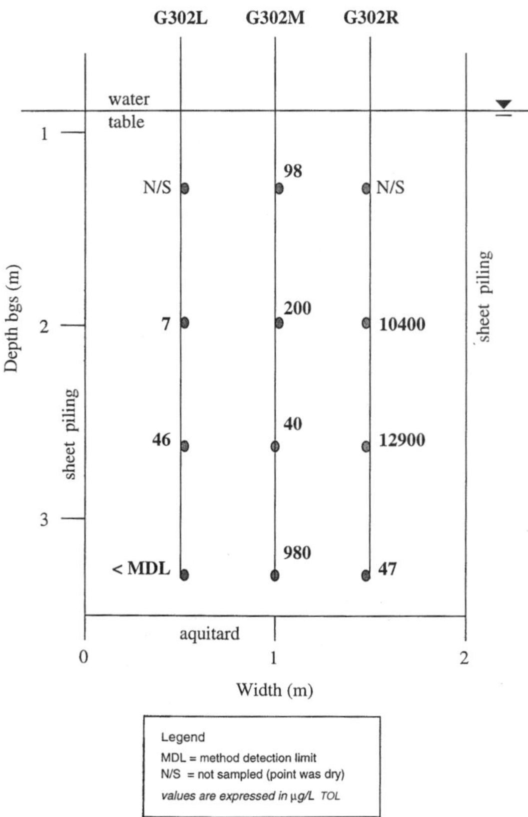

Figure 4.7 TOL concentrations at fence G302 for day 339 (October 23, 1997).

These average concentrations are discussed further in Section 4.3.4.1. The average concentrations at fence 306 show less temporal variability than at earlier times at fence 302, suggesting that the VOC concentrations being provided to the remedial systems in Gate 3, while still temporally variable, were more uniform after about day 140.

The highest VOC concentrations were often found in points G302L-2, -3, and G302R-1, -2, -3. These points also transmitted the highest tracer mass through fence G302 during the main tracer experiment (Figure A7.5, Appendix 7), but except for G302L-2, they do not represent higher-than-average groundwater velocity zones (Figure A7.5). The lack of direct correlation of concentrations and high velocity zones suggests that the spatial variability in organics distribution entering Gate 3 is likely due to a combination of aquifer heterogeneities and imperfectly developed source wells (see Appendix 9).

4.3.4 Anaerobic Treatment

4.3.4.1 Solute Delivery

The delivery of the designed mass of sodium benzoate, inorganic tracers, and other inorganic nutrients into the NIW was achieved on every occasion. Table 4.1 summarizes the concentration of each solute injected into the aquifer. During times of lower water table, smaller volumes of water were flushed, resulting in shorter flushing times (ranging down to 90 min) with the set pumping rate of 23 L/min.

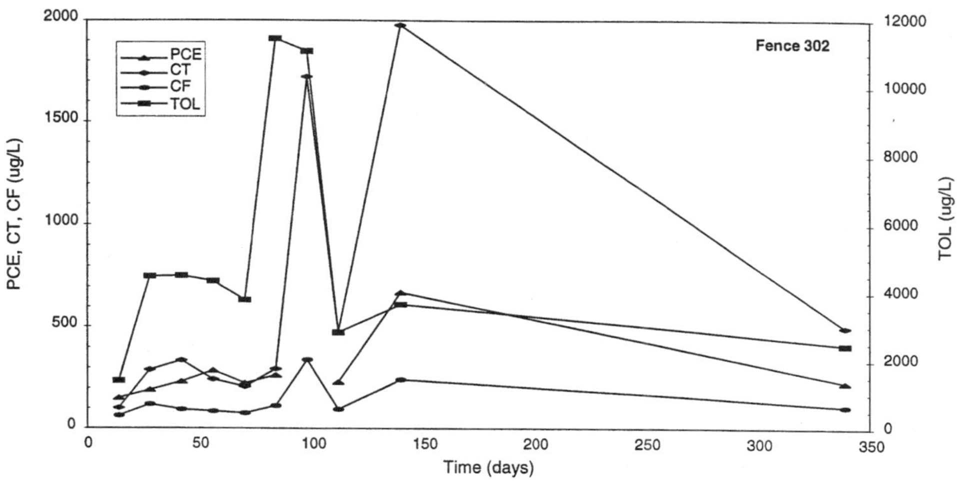

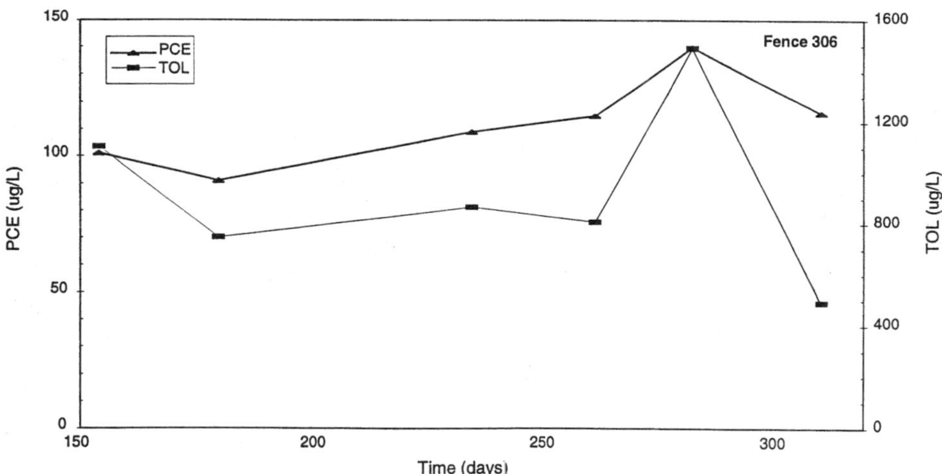

Figure 4.8 Area weighted average concentrations of VOCs at G302.

During each nutrient flush, from day 25 to 123, conductivity was monitored at points 2 and 3 in piezometer G306M, immediately upgradient of the NIW extraction wells. This was undertaken to assess when approximately one pore volume of the wall was amended with nutrient solution and flushing could cease. Plots of conductivity over time were examined and, in all cases, flushing appeared to be adequate, in that conductivity increased and stabilized before flushing was terminated. Representative plots are shown in Figure 4.9a.

On day 180, all monitoring points in fence G306 were sampled at regular intervals during the nutrient flush and groundwater analyzed for chloride (via an ion-specific electrode), which had been added to the nutrient concentrate on this occasion. Representative plots are shown in Figure 4.9b. In most points, the conductivity profiles showed the expected increase over time, suggesting the nutrient solution had swept through the NIW to those points. Maximum conductivity of 5000 to 6000 μS was achieved in 9 of the 12 points. Chloride response was, in some cases, very different from the conductivity response (for example, point G306M-4). Problems with the probe were

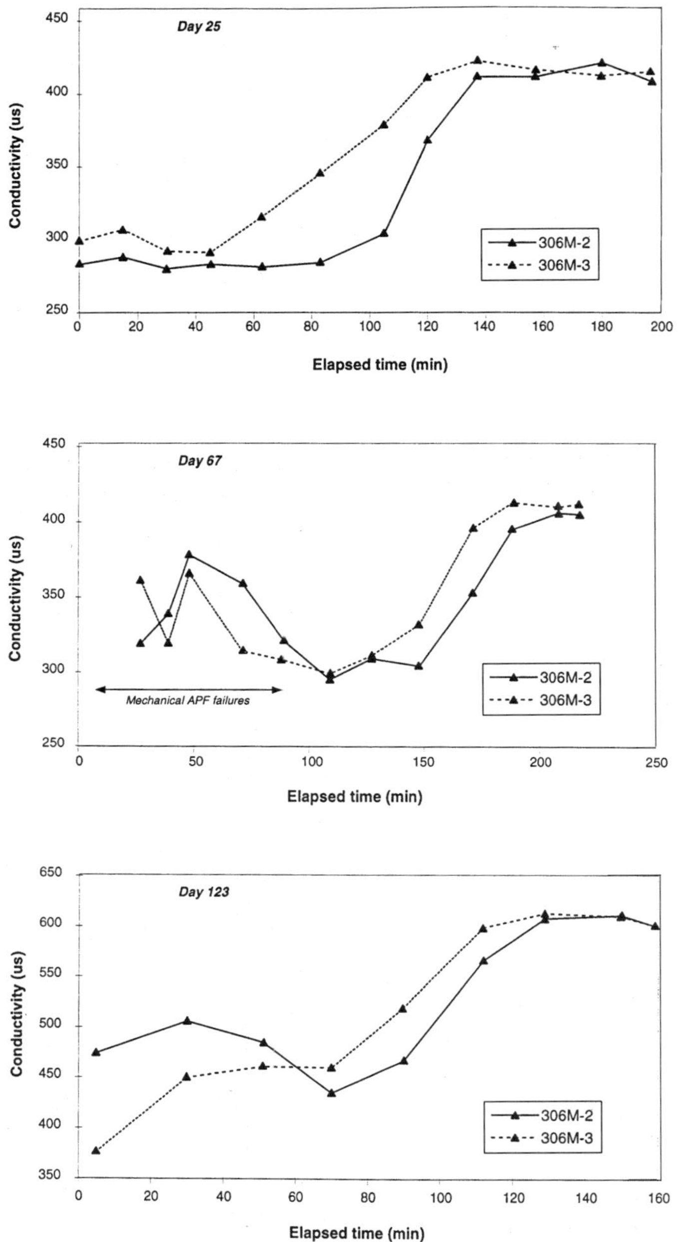

Figure 4.9a Typical conductivity changes in monitoring points G306M-2 and -3 during flushing of the NIW.

suspected, and so Cl^- analyses were discontinued for subsequent monitoring. The flushing protocol was deemed generally successful at replacing the groundwater within this zone with amended groundwater. For subsequent nutrient flushes, points G306L-2, G306L-4, G306M-2, G306M-3, G306R-2, and G306R-3 were monitored for conductivity to confirm adequate flushing.

On days 154, 180, 235, 262, 283, and 311, VOCs were measured in all (days 154, 180) or selected points in fence G306, in the NIW. It was anticipated that the flushing would have left the VOC concentrations reasonably uniform. Table 4.4 shows the average concentration, the standard deviation as a percentage of the mean concentration, and the range of concentrations for PCE and TOL for

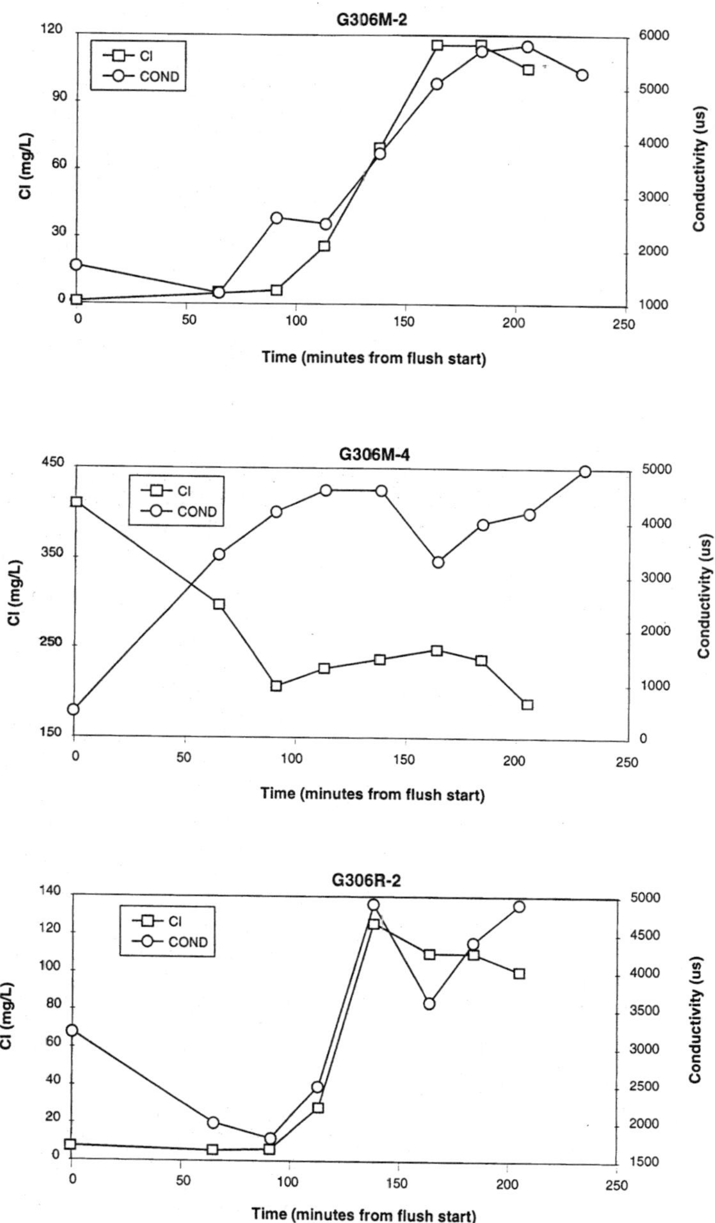

Figure 4.9b Conductivity and chloride changes in groundwater from selected points in fence G306, illustrating typical responses to flushing of the NIW.

these days, as well as similar information for day 154 and 180 from fence 302. In comparison with fence 302, the VOCs at fence 306, within the NIW, are much more homogeneous. The relative standard deviations were lower and the ranges smaller. In addition, the incomplete sampling (the lower *n* in Table 4.4) at fence 306 on days 235, 262, 283, and 311 probably generated average concentrations that adequately approximate the average concentration for fence 306, given the more homogeneous VOC distributions after flushing of the NIW. For example, the average PCE concentration for day 180 at fence 306, calculated using only the 6 points sampled on day 262, is 101 µg/L, which compares well to the average of 91 µg/L determined with all 12 points sampled on day 180.

Table 4.4 PCE and TOL Concentrations at Fences G302 and G306

Fence	Sampling Day	*n*	Average or Mean Concentration (mg/L)		Relative Standard Deviation (% of mean)		Range (mg/L)	
			PCE	TOL	PCE	TOL	PCE	TOL
G302	112	12	231	2838	126	109	1–796	17–7280
	339	10	242	2470	190	197	2–1120	<MDL–12880
G306	154	11	101	1104	24	26	69–148	716–1602
	180	12	91	748	17	25	65–114	449–897
	235	5	109	868	11	13	93–123	682–969
	262	6	115	809	19	33	90–144	549–1320
	283	5	140	1491	11	8.7	119–157	1336–1652
	311	6	116	492	24	21	73–140	293–571

4.3.4.2 Nutrient Pulse Mixing

To determine the extent to which consecutive nutrient pulses mixed downgradient, tracers were introduced with the April 18, 1997 (day 151) flush (Br^-) and with the May 16, 1997 (day 179) flush (Cl^-). Tracers were analyzed at fences G310 and G312 where overlap was desired and anticipated. Results (shown in Appendix 23) suggested sufficient overlap to assume continuous supply of sodium benzoate between fences G310 and G312. Nine of the 12 monitoring points in G312 had minimum concentrations of both ions in the range of 13 to 35 mg/L. Given background concentrations of bromide (Br^-) and chloride (Cl^-) in Gate 3 of below detection and 2 to 7 mg/L, respectively (fence G302), at least 5–25% of the water was amended with tracer. This translates into at least 10 to 50 mg/L sodium benzoate being present continuously. It was concluded that the pulsed injections established an adequate, continuous loading of generally >30 mg/L sodium benzoate over most of the gate cross section to groundwaters between fences G312 and G315.

4.3.4.3 Redox and General Geochemical Conditions

Benzoate and mineral nutrients (both from day 207) were injected in Gate 3 via the NIW to create very reducing conditions to encourage the development of a microbial community able to reductively dechlorinate PCE, CT, and CF. The development of such anaerobic conditions was assessed in Appendix 19, using such parameters as measured redox potential (E_h), and dissolved hydrogen (H_2), oxygen (DO), iron (Fe), sulfate (SO_4^{2-}), and methane (CH_4) in groundwater. Chapelle et al. (1995) suggest a hierarchy of certainty, with respect to redox level determinations, that is based on the number of agreeing lines of evidence. Thus, the interpretation considered all these redox-sensitive parameters.

It was concluded that anaerobic conditions, probably with sulfate reduction as the major electron-accepting process, were established between fences G310 and G315 before day 220 (late June to early July, 1997). By about day 300 (late September 1997) methanogenesis may have become the dominant electron-accepting process in this segment of Gate 3. Considerable acetate was found in these anaerobic groundwaters, indicating acetogens were active in the transformation of the benzoate added in the NIW. Thus, the strongly anaerobic conditions desired for microbial reductive dechlorination were induced downgradient of the NIW certainly by day 315 and perhaps less strongly anaerobic conditions by about day 220.

The pulsed injection of benzoate thus appears to have succeeded in creating the desired strongly reduced environment, but at least 7 months of pulsed additions were required. Mineral nutrients were included in the flushing solution starting on day 207. These could have reached the anaerobic zone beyond fence G312, 5 m downgradient of the NIW, by about day 240, assuming advection of inorganic nutrients at the groundwater velocity of 15 cm/day. Thus, these inorganic nutrients may have been largely responsible for the development of the strongly reducing conditions apparently established by day 315. Alternatively, it may simply have taken until day 315 for the *in situ* microbial population to attain the strongly reducing conditions and the mineral salts may have played only a minor role.

4.3.4.4 Volatile Organic Compounds (VOCs)

The assessment of the fate of VOCs in the anaerobic zone is divided into two periods of inferred, different degrees of reducing conditions: up to day 238 (July 14, 1997) and after day 238, starting with results of sampling on day 315. The pre-July 14 data will be examined first.

To assess the gross behavior of VOCs up to day 238, the average concentrations of CT, CF, PCE, and TOL at selected fences in Gate 3 for days 196 and 238 are presented in Figure 4.10. Note that the data for fence G306 are taken from day 180 and day 235. The biosparge zone was not yet producing continuous aerobic conditions in the biosparge zone by day 238, so fence G316 could still be within the anaerobic segment of Gate 3 for days 198 and 238.

CT was essentially completely attenuated before fence G310 throughout the initial 238 days of the experiment. Anaerobic reductive dechlorination to products including CF is suggested as the mechanism. CF was found at fence G302 by day 14, indicating its production from CT had begun between the source wells and fence G302. CF persisted through the anaerobic zone on day 196, but by day 238 CF, like CT, was essentially completely attenuated by fence G310. Only traces of other products of CT transformation such as CS_2 were detected at fence G302 up to day 98. After that time no dechlorination products beyond CF (DCM, CM) were found in Gate 3, so the further reductive dechlorination of CF did not result in measurable concentrations of dechlorination products.

Average PCE concentrations declined considerably throughout the anaerobic zone on both days 198 and 238. No dechlorination products (TCE, DCE, VC) were detected, so this general decline in PCE concentrations is attributed to either declining source concentrations or dispersive dilution. Figure 4.8 suggests that if any trend exists at fences G302 and G306, it is for increasing PCE concentrations over time, so a declining PCE source concentration does not seem to be indicated. A retarded solute velocity of about 5.5 cm/day was suggested for PCE (see retardation factor, Section 3.4.6), and so the midpoint of the PCE plume front could be just approaching the biosparge zone on day 196 and could be just into the biosparge zone by day 238. So the low average concentrations found at fences G312, G315, and G316 could reflect the arrival of the dispersed PCE plume front. In any event, no reductive dechlorination of PCE is evident up to day 238. TOL concentrations appear to be rising (day 196) or remaining steady (day 238) throughout the anaerobic zone (Figure 4.10). Average TOL concentrations at fence G302 do show considerable variation, with a "spike" of TOL about day 80 (Figure 4.8). Given an estimated velocity of about 13 cm/day for TOL, this spike would have been well through the anaerobic section by day 196, but other source "spikes" could be reflected in the high TOL found in the biosparge zone on both days (Figure 4.10). While anaerobic biodegradation of TOL is well documented (e.g., Acton and Barker, 1992) and it was seen in some anaerobic microcosms simulating unamended Borden conditions (Appendix 11), anaerobic microcosms amended with benzoate to mimic Gate 3 conditions did not demonstrate TOL biodegradation (Appendix 11, see Figure A11.12). The lack of significant TOL declines over the anaerobic segment in Gate 3 also suggests no significant TOL biodegradation had occurred anaerobically during the initial 238 days of the experiment.

By day 238, then, both CT and CF were attenuated by reductive dechlorination by fence G310, while PCE and TOL showed no compelling evidence of biodegradation.

Figure 4.10 indicates a different pattern for TOL for day 339, in part because data from fence 302 are included. The average TOL concentration dropped considerably between fences G302 and G310 (note: sampling at G306 was done on day 311) on day 339. This suggests considerable biodegradation of TOL could have been occurring in the unamended segment of Gate 3 before the

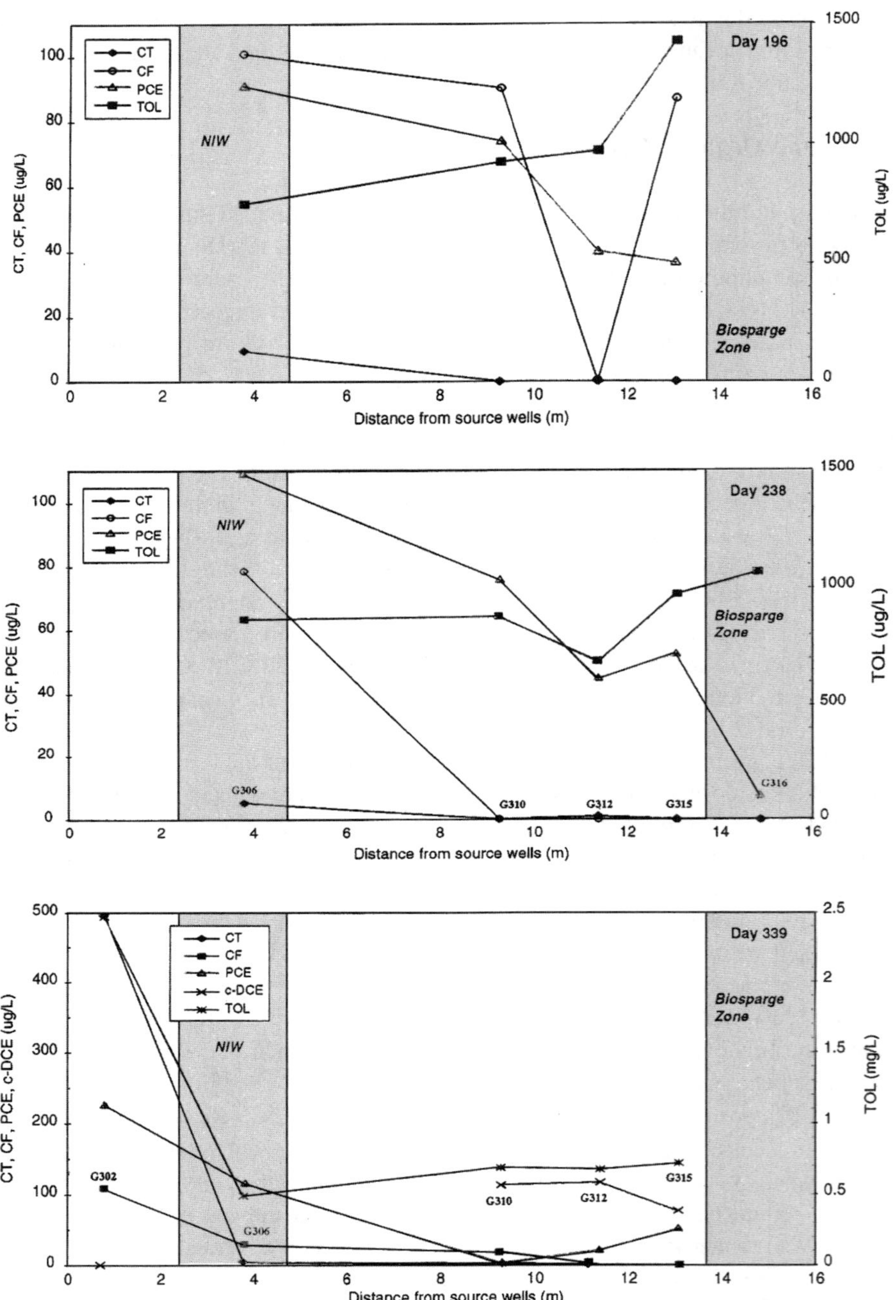

Figure 4.10 Average VOC concentrations at fences in the anaerobic segment of Gate 3 for days 196, 238, and 339.

NIW. Another possibility is TOL loss due to sorption and/or volatilization during nutrient injection and recirculation in the NIW. This could influence up to one third of groundwater passing the NIW, but the closed recirculation system should minimize volatilization and sorption is also likely to be minimal. Anaerobic biodegradation of TOL was seen in Gate 2 (Chapter 3) and in anaerobic microcosms not amended with benzoate (Appendix 11). However, little subsequent biodegradation was apparent downgradient of the NIW.

Figure 4.10 indicates that considerable reduction in average PCE concentration was occurring downgradient of the NIW and *c*DCE was appearing by day 339. On days 315 to 339, *c*DCE was found from fences G315 to G318. No *c*DCE was found at fence G306 after NIW flushing on days 235, 262, 283, or 311, so the reductive dechlorination of PCE to produce *c*DCE was likely not occurring in or upgradient of the NIW. From day 238 to 315, no sampling was done from fences G310 to G315, so the precise time when reductive dechlorination developed in Gate 3 is uncertain. Traces (<10 μg/L), usually <LOQ, of TCE were occasionally found in groundwater from fence G315 starting on day 238, so this could represent the onset of reductive dechlorination of PCE. By day 315, *c*DCE was found as far downgradient as fence G318. Assuming the *c*DCE formed in the anaerobic gate segment no further downgradient than G315 and that *c*DCE subsequently migrated with a retarded velocity of about 13 cm/day, then reductive dechlorination of PCE to *c*DCE must have begun before day 285 (i.e., about 30 days before day 315).

It appears that significant PCE (50–100 μg/L) arrived at fence G315 by day 196. *c*DCE was first encountered at fence G315 on day 316, so the onset of reductive dechlorination of PCE was within 120 days of PCE arrival. The estimate that *c*DCE could have been generated at fence G315 as early as day 285 (deduced from *c*DCE appearance at fence G318), suggests an onset of reductive dechlorination within about 90 days. Thus, the reductive dechlorination reaction appears to have been stimulated by the benzoate addition in Gate 3 within about 90 to 120 days of PCE arrival in the biostimulated zone.

The major PCE reductive dechlorination product appears to have been *c*DCE. The reductive dechlorination does not appear to have proceeded further than *c*DCE by day 339, since no VC and only traces of ethene were detected in groundwaters from Gate 3 at any time during the experiment.

In summary, CT was not found beyond fence G308 during the experiment, apparently having been completely dechlorinated by fence G310. This was also observed under intrinsic conditions in Gate 2. The major product of CT dechlorination was CF. CF appeared to have been completely dechlorinated in Gate 3 prior to fence G310 by day 238, perhaps marginally more rapidly than under the intrinsic conditions in Gate 2. TOL may have undergone some anaerobic biodegradation upgradient of the NIW, but little evidence for anaerobic biodegradation in the benzoate-stimulated zone was seen. Consequently, concentrations of about 1 mg/L TOL continued to leave the anaerobic zone of Gate 3. PCE only underwent reductive dechlorination in the benzoate-stimulated zone of Gate 3 after day 238 but likely prior to day 315, as evidenced by the appearance of *c*DCE by day 315. PCE persisted to the downgradient end of the anaerobic zone in Gate 3, being present at an average concentration of about 50 μg/L at fence G315 on the last sampling day, day 339. The major dechlorination product from PCE biodegradation was *c*DCE, which persisted into the aerobic segment of Gate 3.

4.3.4.5 Transformation Rates in the Anaerobic Zone

In the anaerobic zone, there was evidence of CT, CF, PCE, and TCE transformations. In the cases of CT and CF, the reactions were underway before July 14, 1997, and resulted in concentration–distance profiles that were reasonably steady by the time of the October 23rd snapshot. On this basis, it was considered reasonable to use those profiles to estimate pseudo-first-order rate constants for the disappearances of CT and CF, according to the method described below. In the cases of PCE and TCE, the evidence of transformation came late in the project with the appearance of TCE and *c*DCE, known dechlorination products of PCE. Additional evidence came from the decline in PCE concentrations at fence G315 between July 14 and October 23, 1997. The fact that PCE degradation occurred late in the project precluded a reasonable assumption of steady-state profile development in the anaerobic zone, and so no transformation rate was determined for this compound or its product, TCE.

Pseudo-first-order rate constants for CT and CF degradation were estimated by fitting the profiles of maximum concentrations at each fence from the October 23 snapshot (day 339) to the following equation:

$$C = \text{Co} \exp\{-kx/v\} \tag{4.1}$$

where Co is the input concentration to the profile (μg/L), i.e., the maximum concentration at G302, k is the pseudo-first-order rate constant (1/day), x is the distance from G302, and v is the unretarded groundwater velocity (15 cm/day).

Maximum concentrations were used to provide a conservatively high estimate of k. This approach to estimating the kinetics inherently assumes that the profile is at steady state, that dispersion can be ignored, and that the reaction occurs in the water phase only. The steady-state assumption has already been discussed. The assumption of negligible dispersion was evaluated by comparing calculations based on Equation (4.1) with those from a second equation that incorporated dispersion (provided by C. Neville, see Section 3.4.6). Negligible differences were found between the two solutions for dispersivity values typical of the Borden aquifer (1–10 cm). Equation (4.1) was preferred over the second solution, because it required fewer parameters for fitting. The assumption that reactions were limited to the solution phase is less certain, but does not affect the utility of the estimated rate constant as an indicator of effective removal rates, as long as the assumption is maintained in subsequent calculations involving k.

The fitting of the CT profile was straightforward, although the concentration declined between fences G302 and G306 to below the average LOQ (24 μg/L). As a result, the estimated rate constant may be lower than the true one since the available data were not able to define the steepness of the concentration descent along the profile. Nonetheless, the estimated rate constant was calculated to be 0.2 per day, corresponding to a pseudo-half-life of about 3.5 days. This rate of degradation compares favorably with that reported by Devlin (1994) for CT disappearance downgradient of a NIW in the Borden aquifer ($t_{1/2}$ = 3 – 7 days).

In order to calculate degradation rates for CF, assumptions in addition to the ones stated above had to be made. Since CF was formed from CT, the preferred equation for modeling its degradation would be one that incorporated the transformation sequence CT >> CF >> …. However, CF is not the only transformation product possible from CT, and determining the mass fraction that degraded along the CF pathway was not possible with the available data. Therefore, Equation (4.1) was used again assuming that the concentrations measured at G302 were representative of inputs. Once again, the calculated rate constants are likely to be lower than the true ones, because CF could have been formed in the gate beyond the assumed input boundary, G302. Fortunately, this error is likely to be reasonably small since the CF profile, like the CT profile, declined from input levels (450 μg/L) to near LOQs (maximum concentration at G306 = 52 μg/L; LOQ for CF = 25 μg/L) between the the first two monitoring fences, G302 and G306.

In the October 23 snapshot, a single anomalously high CF concentration (184 μg/L) was observed at fence G310. These data point deviated considerably from the trend in the other data collected on the same day, and are suspect. Two CF profiles were fitted, one including the point and the other excluding the point, to obtain a range of pseudo-first-order rate constants reflective of the minimal level of treatment gate performance. The estimated rate constants were 0.05 to 0.08 per day, corresponding to pseudo-half-lives of 15 and 9 days, respectively.

4.3.5 Aerobic Treatment

4.3.5.1 Oxygen Delivery

As indicated in Section 4.2.1.2, twice-weekly O_2 sparging began in November 1996. It wasn't until July 18, 1997 (day 242) that the sparging was increased to daily events. Thereafter, significant DO was found in most multilevel monitoring points in and downgradient of the biosparge zone. Table 4.5 presents the DO concentrations averaged over the points in each fence, for G315 to G324, measured from just before the increased sparging was initiated (day 238) until the end of monitoring

Table 4.5 Average Dissolved Oxygen in Selected Gate 3 Fences

	Average DO (mg/L)				
Fence	Day 238 July 14/97	Day 315 Sept. 30/97	Day 323 Oct. 7/97	Day 330 Oct. 14/97	Day 339 Oct. 23/97
G315	0.1	n/m	1.2	1.6	2.8
G316	0.9	7.8	11.5	10.6	14.4
G317	0.7	8.7	15.8	17.5	15.8
G318	0.2	6.2	4.3	5.2	7.2
G320	n/m	n/m	2.9	1.9	2.6
G324	n/m	n/m	2.8	2.1	2.6

Notes: G315 is immediately upgradient of the biosparge zone; G316 and 317 are in the biosparge zone; G318, G320, and G324 are downgradient of the biosparge zone. n/m: not measured.

in October 1997. Note that averages <2 to 3 mg/L may not reflect generally aerobic conditions. DO determinations by probe at <2 mg/L are difficult and averages of <2 or 3 mg/L often result from an occasional point having significant DO. Appendix 20 presents the specific DO concentrations.

Whereas <1 mg/L average DO was found in groundwater within the gate (G316 and G317) before the increased sparging on July 18, 1997, >7 mg/L average DO was observed thereafter. The detailed assessment of the delivery of oxygen to groundwater in the biosparge zone is presented in Appendix 20. Briefly, O_2 bubbles appeared over about 60% of the water surface when the pea gravel was excavated to the water table for the sparge test of December 19, 1996. This distribution was sufficiently broad to impart >2 mg/L DO to 22 of the 24 points within and downgradient of the biosparge zone on day 339. It was concluded that the daily sparging was sufficient to provide oxygen to the biosparge zone in excess of that required to support aerobic biodegradation of organics.

4.3.5.2 Off Gas Production and Composition

Appendix 20 also discusses the sampling of the headspace gas from the biosparge zone on September 30, and October 7 and 14, 1997. No TOL or PCE was detected. Detection limits were about 70 and 20 μg/L for TOL and PCE, respectively, using a portable field GC. Given these measurements and the fact that each sparge event added only about 40 L of gas to the system, the loss of PCE and TOL from the remedial system as off gas was minimal.

A calculation of the maximum loss as off gas for TOL, PCE, and *c*DCE is presented in Appendix 20. It was estimated that <1% of TOL, <5% of PCE, and <4% of *c*DCE entering the biosparge zone in groundwater was lost in off gas.

4.3.5.3 Biotransformations

Since adequate aeration of the biosparge zone was not achieved until July 18, 1997, the biotransformation of target organics was only assessed with data for July 14 (day 238) and later. Average concentrations of TOL, PCE, *c*DCE, and sodium benzoate for fences G315 to G324 are presented in Table 4.6.

A simple approach was employed to evaluate the apparent removal rate occurring within and downgradient of the biosparge zone of Gate 3. It was assumed that the snapshot samplings from day 238 to day 339 each represented a pseudo-steady-state profile, resulting from a constant input at fence G315, just before the biosparge zone, with a first-order reaction process resulting in the generally observed concentration declines through the biosparge zone and downgradient to fence G324. Each sampling date was expected to have a different reaction rate because the reaction processes, namely, aerobic biodegradation and volatilization, could evolve over the sampling period. The calculated first-order reaction rate represents the overall rate, resulting from a combination of

Table 4.6 Average Concentrations of Selected Organics for Fences G315 to G320, Day 238 to Day 339

		Average Concentration (TOL, PCE, cDCE μg/L; Na-benzoate mg/L)			
Date (Day)	**Fence**	**TOL**	**PCE**	***c*DCE**	**Na-benzoate**
July 14/97 (238) (in biosparge zone)	G315	970	52	0[a]	n/m
	G316	1070	6	0	15
	G317	340	2	0	8
	G318	470	1	0	5
Sept. 30/97 (315)	G315	710	56	14	7
	G316	780	5	16	25
	G317	560	5	11	3
	G318	180	4	12	1
Oct. 7/97 (323)	G315	740	61	28	2
	G316	780	3	21	7
	G317	540	0.5	9	16
	G318	230	1	10	1
	G320	140	1	7	3
	G324	120	0	1	3
Oct. 14/97 (330)	G315	740	59	48	10
	G316	750	4	25	11
	G317	540	0	11	2
	G318	290	0.4	8	5
	G320	100	0.9	8	0.5
	G324	80	0	2	0
Oct. 23/97 (339)	G315	720	51	76	n/m
	G316	710	5	30	4
	G317	480	1	17	0.5
	G318	170	0.8	9	0.3
	G320	150	2	10	0
	G324	40	1.2	3	0

Note: n/m: not measured.

[a] Zero mg/L concentration values were selected when no point in a fence contained >MDL concentrations.

volatilization and aerobic biodegradation. However, virtually all organics that volatilized must have subsequently biodegraded in the headspace, since <5% of the input concentrations could have been lost as off gas (Section 4.3.5.2).

The reaction rate was determined by fitting the profiles to a steady-state reaction equation:

$$C = C_0 \, e^{-xk/v} \tag{4.2}$$

where: C is the concentration at a fence beyond Gate G315

C_0 is the input fence averaged concentration (at Gate G315)

x is the distance (m) from fence to fence G315 (1.52, 2.43, 2.48, 5.58, and 9.57 m for fences 316, 317, 318, 320, and 324, respectively)

k is the apparent first-order reaction rate (day^{-1})

v is the average linear groundwater velocity, taken as 15 cm/day

The concentration data (Table 4.6) were fitted to this equation, neglecting dispersion and retardation (i.e., assuming steady state) and assuming a groundwater velocity of 15 cm/day through the biosparge system and downgradient aquifer. The best-fit line was obtained by minimizing the residual sum-of-squares between the actual and calculated data at the different fence locations. This was done using the Solver program in Excel 7.0.

Note that the biosparge zone was not sufficiently sparged to reliably maintain aerobic conditions on day 238. The DO was below 1 mg/L. After day 238, DO was usually above 1 mg/L throughout the aerobic treatment zone, from within the biosparge zone at G316 to the final downgradient monitor at G324, 8.4 m downgradient of the front of the biosparge zone (Table 4.5).

The apparent first-order reaction rate for TOL averaged about 0.023 day^{-1}, with an apparent half-life of 30 days. The PCE reaction rate averaged 0.21 day^{-1} ($t_{1/2}$ of about 3.3 days), while the apparent *c*DCE reaction rate averaged 0.045 day^{-1} ($t_{1/2}$ of about 15 days). These apparent rates likely represent the result of both volatilization (i.e., for all compounds) and aerobic biodegradation (for all compounds except PCE, which is not generally known to degrade aerobically). However, no TOL, PCE, or DCE was found in the headspace of the biosparge zone (Appendix 20), so it would appear that biodegradation in the groundwater or in the headspace was the major mass loss mechanism. Table 4.7 summarizes the apparent reaction rates obtained by this fitting.

Table 4.7 Summary of Apparent Reaction Rates (day^{-1}) and Half-Life (day) of TOL, Sodium Benzoate, PCE, and *c*DCE for the Biosparge Zone (G315–G324)

	TOL		Na-Benzoate		PCE		*c*DCE	
Day	*k* (day^{-1})	$t_{1/2}$ (day)	*k* (day^{-1})	$t_{1/2}$ (day)	*k* (day^{-1})	$t_{1/2}$ (day)	*k* (day^{-1})	$t_{1/2}$ (day)
238[a]	0.0247	28	0.067[b]	10[b]	0.181	3.8	ND	ND
315[a,c]	0.0205	33	NS	NS	0.188	3.7	0.0060	120
323	0.021	33	NS	NS	0.263	2.6	0.039	1812
330	0.020	35	0.088	7.9	0.230	3.0	0.0605	12
339	0.031	22	0.063	11	0.196	3.5	0.073	9.4

Notes: NS: no reasonable solution found because concentration increased downgradient. ND: not done, since compound was not found above LOQ.

[a] Averages only from fences G315–G318.

[b] Averages only from fences G316–G318.

[c] Dissolved oxygen began to appear consistently in biosparge zone.

The mass of PCE being removed was very low, with the average concentration input at G315 having been less than about 60 μg/L and the average PCE concentration having fallen below 10 μg/L before the first fence (G316) within the biosparge zone. Aerobic biodegradation of PCE is unlikely. Anaerobic biodegradation of PCE between fence G315 and the biosparge zone and volatilization within the biosparge zone were probably responsible for the evident loss of PCE. Volatilization alone was not likely responsible for the loss of PCE in groundwater downgradient of fence G315, since methane, a much more volatile organic, persisted through the biosparge zone (Figure 4.11).

The apparent *c*DCE reaction rate is difficult to relate to actual processes. The *c*DCE average input concentration at G315 was variable and low (<19 to 76 μg/L), but average concentrations near 10 μg/L generally persisted to fence G317 and often through fence G318, immediately downgradient of the biosparge zone. By fence G324, average concentrations fell below 3 μg/L, although point G324L-4 still had 23 μg/L on day 339. The *c*DCE apparent removal could have been due to both volatilization and aerobic biodegradation, but the lack of *c*DCE in the headspace suggests aerobic biodegradation was the major process. The presence of occasional *c*DCE concentrations in fence G324 even on day 339 suggests that the *c*DCE removal was not complete by the end of the field experiment.

The TOL removal in the biosparge zone was more significant, with the average concentration at the input fence, G315, of about 700–800 μg/L reduced to average concentrations of 170 to 290 μg/L immediately after the biosparge zone at fence G318 and further reduced to average concentrations of 40 to 120 μg/L by fence G324 (Table 4.6). With aerobic conditions becoming established after day 238, the apparent toluene reaction rate did not change much, with the apparent first-order

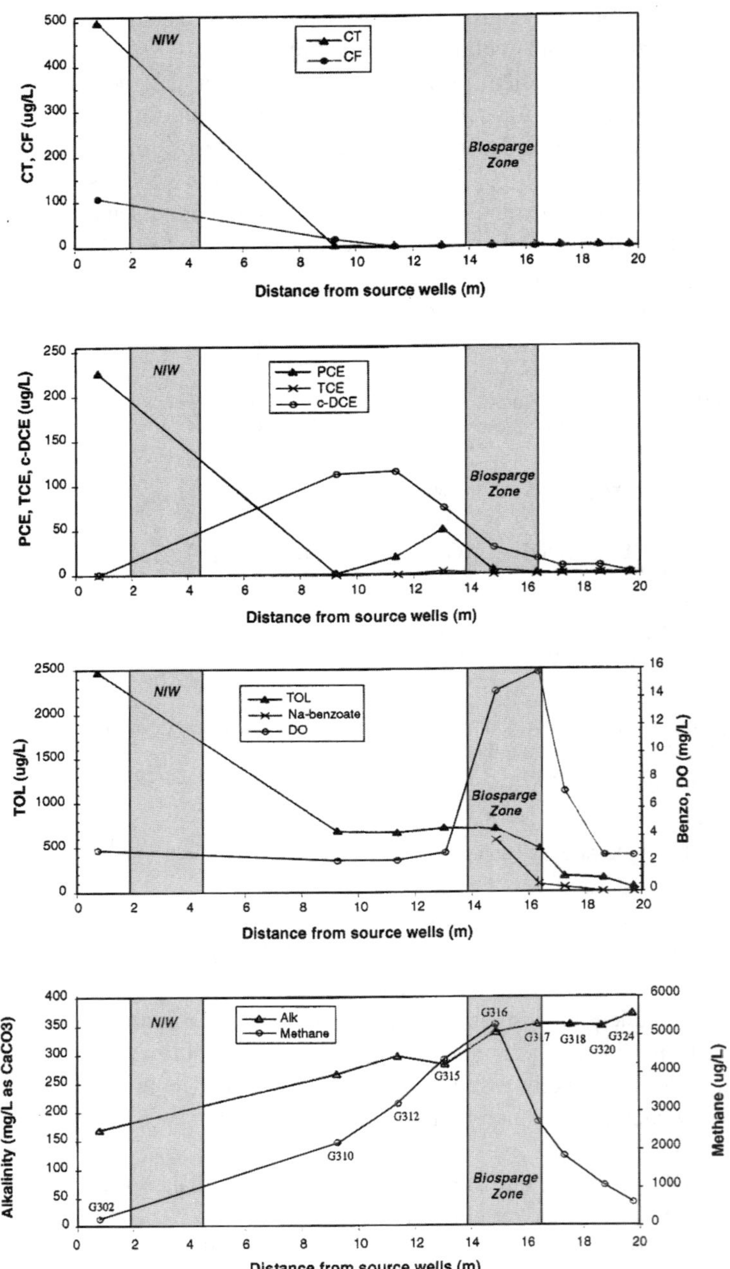

Figure 4.11 Distribution of average concentrations of various solutes in groundwater in Gate 3 on day 339.

half-life remaining at about 30 days. TOL may have been biodegraded aerobically in the biosparge zone on and perhaps prior to day 238, even though little excess DO was found in groundwater samples. This could account for the similar apparent half-life on day 238 and after, when excess O_2 was found in the biosparge zone.

The apparent half-life for aerobic biodegradation of TOL is often much less than 30 days in well-controlled laboratory studies. Appendix 11 points out that benzoate may be the preferred substrate when both benzoate and TOL are present in aerobic environments, as was the case at least in fence G316. This could explain why TOL concentration declines were not observed until after fence G316 (Table 4.6). This delay is reflected in the long (about 30-day) apparent half-life for TOL in the aerobic zone in Gate 3 (Table 4.7). There was apparently sufficient average DO in the groundwater at fence G324 to subsequently consume the toluene remaining on day 339, even the 380 μg/L at point G324L-4.

Considerable benzoate was degraded anaerobically with average concentrations at fence G315 less than 10 mg/L on days 315 and 322. Benzoate removal in the biosparge zone was also impressive. By day 339, benzoate was not found beyond fence G318, and only in G318M-4 and G318R-4 at <2 mg/L (see Appendix 20). The apparent reaction rate in the aerobic zone averaged about 0.07 day^{-1}, with an apparent half-life of about 10 days. This is within the range of half-life for benzoate calculated from breakthrough curve analysis in the anaerobic zone (see Appendix 18).

Overall the aerobic segment of Gate 3 functioned well after about day 238, with residual DO remaining in groundwater within and downgradient of the biosparge zone. Some of the least-oxygenated groundwater in fence G324, still containing TOL and/or *c*DCE on day 339, was from the deepest samples, indicating that oxygenation was still not optimal in the deepest groundwaters. Dechlorination products that exited the upgradient anaerobic treatment zone were adequately removed by fence G324, as were TOL and benzoate, when the average concentrations were considered. The major mass removal mechanism was concluded to be biodegradation.

4.4 PERFORMANCE SUMMARY

4.4.1 Benzoate and Mineral Salt Nutrient Delivery

The nutrient flush circulation/amendment system was successful in distributing the desired benzoate mass in the NIW. While the vertical and lateral distribution was not completely uniform, the benzoate was routinely added as designed and the desired, downgradient anaerobic zone was created and maintained. After day 207, mineral nutrients were included and these may have contributed to the development of the desired anaerobic conditions downgradient of the NIW.

4.4.2 Nutrient Pulse Mixing

Overlap of inorganic tracers at fence G312 produced as high or higher tracer concentrations than required to ensure continuous delivery of greater than 30 mg/L benzoate to *in situ* microorganisms between fences G312 and G315.

4.4.3 Geochemical Conditions and Microbial Habitat

Dissolved oxygen was typically low entering Gate 3. Decreasing sulfate concentrations, production of acetic acid and methane (see Figure 4.12), and the increased H_2 concentrations in groundwaters indicated that a strongly reducing zone with active acetogenesis and methanogenesis had been generated in the anaerobic zone downgradient of the NIW. Both methane and acetate concentrations increased between fences G310 and G315, especially over the period from day 238 to day 339, indicating that conditions had become even more conducive to microbes capable of reductively dechlorinating PCE, CTET, and CF.

Once O_2 sparging frequency was increased on day 242, the biosparge zone and downgradient aquifer in Gate 3 was generally aerobic, with only the deepest groundwater being poorly oxygenated (Appendix 20). Thus, the stable, aerobic environment required to support aerobic biodegradation of contaminants and benzoate was generated within and downgradient of the biosparge zone, except perhaps in the deepest groundwater. This would be expected to improve if sparging continued.

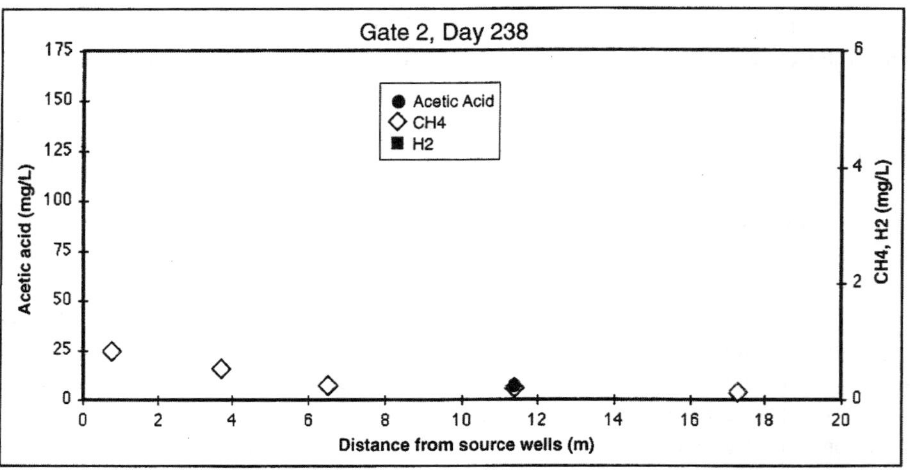

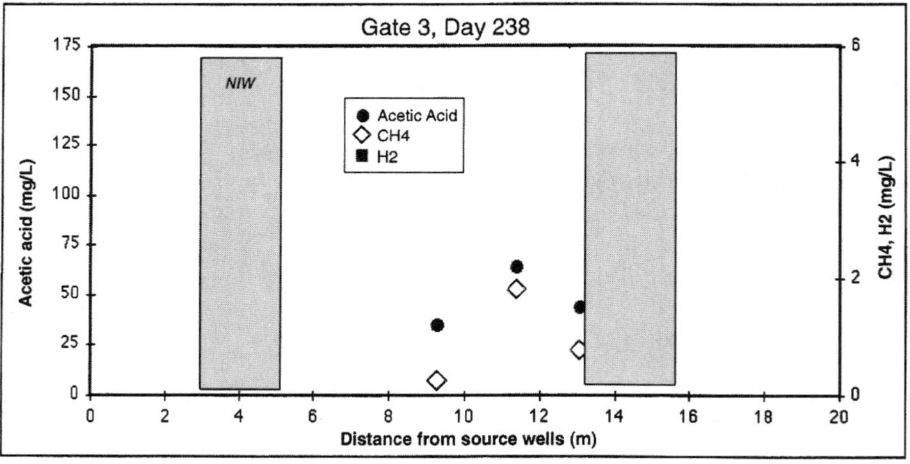

Figure 4.12a Average acetic acid, methane, and hydrogen concentrations at fences in Gates 2 and 3, day 238.

Overall, the NIW and biosparge zones were judged successful in creating both a strongly anaerobic and a strongly aerobic geochemical environment *in situ*.

4.4.4 Off Gas Production and Composition

Daily sparging with O_2 after day 242 was sufficient to maintain aerobic conditions in and downgradient of the biosparge zone. No PCE or TOL was detected in the biosparge zone headspace with daily sparging. A direct measurement indicated that each sparge event added only about 40 L of O_2 to the biosparge zone. Even if all this gas was vented as off gas, it would account for an insignificant flux of PCE, *c*DCE, or TOL from the system. The low concentrations and low volume of off gas suggests this technology could be applied even where air discharge is tightly regulated.

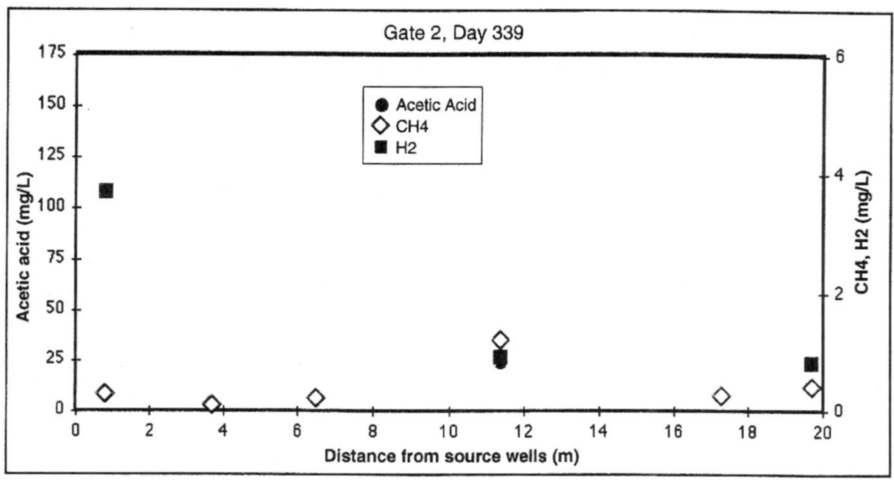

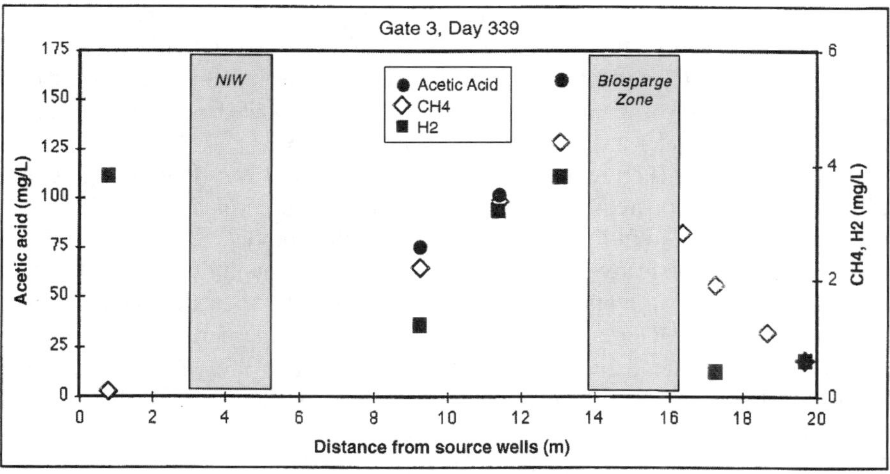

Figure 4.12b Average acetic acid, methane, and hydrogen concentrations at fences in Gates 2 and 3, day 339.

4.4.5 Organic Compounds

Since the performance of both the anaerobic and aerobic *in situ* treatments seemed to peak at the end of the field experiment, data from groundwater sampling on day 339 are considered to represent the best performance attained in Gate 3. The unweighted fence-average concentrations of CT, CF, PCE, *c*DCE, TOL, benzoate, and of the indicator parameters DO, methane, and alkalinity are presented in Figure 4.11.

Methane is an indicator of very anaerobic conditions. The increasing methane concentration through the anaerobic segment of Gate 3 attests to the success in creating the very anaerobic conditions desired to stimulate reductive dechlorinating microbial populations. It appears to be consumed in the aerobic segment of Gate 3. The alkalinity increased throughout the gate reflecting the production of CO_2 from the biodegradation and mineralization of, mainly, benzoate added in the NIW.

An increasing ratio between CF and CT concentrations between the source wells and the NIW is similar to the situation in Gate 2 and suggests rapid, intrinsic transformation of CT to CF in both gates. Variable CT concentrations in fence G302 made estimation of the apparent CT biotransformation rate difficult. However, taking the average concentration at fence G302 as 500 μg/L and at fence G310 as 1 μg/L (see Figure 4.11), a minimum apparent CT biotransformation rate of 9 μg/L/day can be calculated.

CF subsequently became the target contaminant to assess anaerobic system performance. CF was being produced from the dechlorination of CT and could itself be further dechlorinated. By day 339, CF was essentially completely attenuated by fence G312. Transport of CF mass into the gate was observed up to the end of the experiment (Figure 4.8), so declining source concentrations do not account for the disappearance of CF in Gate 3. If we assume the decline of concentrations at well G310M from a peak of 113 μg/L on day 70 to 15 μg/L on day 112, 42 days later, represents the mass loss rate, then an apparent biotransformation rate of 2.3 μg/L/day is suggested. This was in good agreement with a rate of 2.5 μg/L/day for the decline from day 112 to day 140 at G312.

CT and CF had been essentially removed from the groundwater within 10 m of the source wells. Field observations are supported by the microcosm results, where CT was rapidly degraded in conjunction with increasing concentrations of CF. Nominal concentrations of CS_2 were observed in microcosms, but only traces were found in groundwater in Gate 3. Apparently the anaerobic treatment produced complete attenuation of chlorinated methanes.

Throughout the experiment, fence-averaged PCE concentrations were declining along Gate 3. Mass loss may be inferred, but the lack of dechlorination products (TCE, *c*DCE, VC, and ethene) before day 315 indicates PCE-reductive dechlorination had not been stimulated up to day 238. This persistence was anticipated from microcosm results, which showed negligible decline of PCE concentrations over time under anaerobic conditions (Appendix 11).

In contrast, on day 339 (Figure 4.11) PCE had declined to a very low average concentration and *c*DCE had appeared. Apparently, the onset of reductive dechlorination of PCE in Gate 3 was between days 238 and 315. No further reductive dechlorination products were found. The anaerobic segment was likely not at a steady state, and the microbial community could still be in transition. Perhaps further dechlorination products would have been found if the experiment continued. Such an evolution was found by Major et al. (1991) with PCE-degrading microcosm experiments. Given the recent onset of PCE degradation, estimation of apparent half-life was not attempted. It does appear, however, that complete attenuation of PCE could be anticipated within the anaerobic segment of Gate 3.

On day 339, the average concentration of PCE at fence 315 was about 50 μg/L. By the next fence, G316, the average PCE concentration was reduced to about 5 μg/L. No PCE was found in the headspace gas in the biosparge zone, so at least some of this attenuation is likely through biodegradation. Since PCE is not normally biodegraded aerobically, it is most likely that anaerobic biodegradation between fence G315 and the biosparge zone contributed to the decline of PCE concentrations between fences G315 and G316,

The residual concentrations of *c*DCE exiting the anaerobic segment were well attenuated in the aerobic system, except in the deepest groundwater, which was less oxygenated. No *c*DCE was found in the headspace gas in the biosparge zone, so this attenuation is likely through biodegradation. The apparent half-life for *c*DCE degradation in the aerobic system was quite variable, but was approaching 10 days by the final sampling. It would appear that microbes in and downgradient of the biosparge system could aerobically biodegrade dechlorination products, such as *c*DCE and perhaps VC, if it had been produced.

TOL concentrations declined upgradient of the NIW, but not significantly in the benzoate-stimulated, anaerobic zone downgradient of the NIW. An average concentration of almost 1 mg/L was still entering the biosparge zone on day 339. TOL was significantly attenuated within the biosparge zone. In the downgradient aerobic segment almost all of the TOL had been biodegraded with only one deep point in fence G324, namely, G324L-4, having significant TOL (380 μg/L).

The apparent half-life was about 30 days (Table 4.7). As with the CVOCs, the biosparge zone headspace gas was not a significant sink for this TOL, so aerobic biodegradation is credited with bringing about the almost complete attenuation of TOL.

Sodium benzoate was degraded anaerobically with the production of acetate (see Figure 4.12 and Appendix 19 for acetate data). Still, significant concentrations persisted into the biosparge zone. This persisting benzoate could have been a preferred substrate relative to TOL and so the persistence of benzoate in the aerobic zone could have contributed to the persistence of TOL aerobically. The benzoate appeared to be completely attenuated in the aerobic segment of Gate 3. The apparent half-life for benzoate reaction was about 10 days (Table 4.7). Acetate was not monitored in the aerobic segment. Methane (Figure 4.12) was strongly attenuated in the aerobic segment of Gate 3, but almost 1 mg/L was still found at the downgradient end of the gate on day 339.

4.4.6 Comparison to Intrinsic Performance in Gate 2

In comparing the level of anaerobic conditions on day 238, the initial segment of Gate 2, up to about fence G212, appeared to be at least as reducing as the same segment of Gate 3, with considerable methane in fences G202 to G212. While conditions did not change significantly in Gate 2, the anaerobic segment of Gate 3 showed enhanced acetate production following the addition of benzoate, and increased levels of methane and H_2 indicating that conditions had become more reducing with increased activity of acetogens and methanogens. Such conditions were apparently conducive to microbes capable of reductively dechlorinating PCE after day 238.

The downgradient portion of Gate 2 remained anaerobic with some methane persisting (see Figure 4.11). In contrast, once O_2 sparging frequency was increased on day 242, the biosparge zone and downgradient aquifer in Gate 3 was generally aerobic.

Anaerobic transformation of CT mainly to CF was observed from very early times in both Gate 2 and Gate 3. On day 238, CF was found beyond fence G218, while it was not found in or downgradient of fence G310 in Gate 3. Thus, the benzoate additions in Gate 3 appear to have provided for earlier attenuation of CF than occurred naturally in Gate 2.

Up to day 315, there was no strong evidence for reductive dechlorination of PCE in either gate. On day 339, minor *c*DCE was found at fences G205 and G208 in Gate 2. From day 315 until 339, much more *c*DCE was found in groundwater in Gate 3, from fences G310 to G315, suggesting substantially more reductive dechlorination of PCE had occurred in Gate 3 than in Gate 2. Comparison on the last monitoring day 339 showed significant PCE persisting to fence G224 in Gate 2, while little PCE was found beyond fence G318 in Gate 3. Gate 3 performed much better than the intrinsic condition in Gate 2 after the desired microbial habitat became established in the anaerobic segment.

With *c*DCE being produced in Gate 3 and not significantly in Gate 2, there was concern that groundwater leaving Gate 3 could be of more environmental concern than *c*DCE-"free" groundwater leaving Gate 2. However, *c*DCE was sufficiently biodegraded in the aerobic segment of Gate 3. In fact, if reductive dechlorination of PCE increased over time in both gates, there is less likelihood that the dechlorination products such as *c*DCE would be attenuated in Gate 2 than in Gate 3, the latter having its aerobic treatment segment.

TOL was found throughout both gates by day 339, although considerable attenuation by dispersion and biodegradation likely occurred in both gates. At fences G224 and G324 on day 339, the fence average TOL concentrations were 100 and 40 μg/L, respectively. In Gate 3, there was sufficient DO for subsequent aerobic biodegradation of this TOL, and so rapid reduction of this 40 μg/L could have been expected in Gate 3. Greater persistence of TOL was expected in the anaerobic groundwater leaving Gate 2. However, the attenuation of TOL in both gates was significant and was perhaps only marginally better in Gate 3.

There was concern about the fate of chemicals added to Gate 3 (benzoate and the acetate and methane generated from biodegradation of benzoate). Judging by the data from fence G324, it appears that groundwater leaving Gate 3 would have had only a minor amount of CH_4 and no benzoate. No acetate data are available, but it was expected to also have biodegraded aerobically. If that is the case, there appeared to be sufficient O_2 in the groundwater exiting Gate 3 to permit aerobic biodegradation of residual CH_4 and acetate. In addition, the aerobic groundwater leaving Gate 3 was probably aesthetically superior to the anaerobic groundwater leaving Gate 2.

4.4.7 Adverse Interaction between Technologies and Design Limitations

One concern about coupling these *in situ* technologies (Section 4.1.3) was the potential for biofouling and subsequent reduction in gate permeability. There was no evidence of this problem from the field experiment, but a longer field trial may be required to evaluate this potential problem.

The addition of benzoate and mineral nutrients to stimulate anaerobic biodegradation was another concern, because the added BOD increased the O_2 required to be added in the biosparge zone. For example, sodium benzoate concentrations at G315M-3 were usually >100 mg/L before day 112, and after day 234 were <35 mg/L. After day 234, some of the decrease in benzoate entering the biosparge zone was likely made up by acetate (see data in Appendix 19, Table A19.2). In spite of this BOD loading, DO was found in groundwater throughout the biosparge zone and downgradient. Thus, the coupling of the two technologies in Gate 3 was successful, at least for the short term of this field experiment.

The delay in attaining aerobic conditions in the biosparge zone was mainly an operational limitation, overcome by increasing the O_2 addition frequency and perhaps aided by decreasing the benzoate addition in the NIW.

Unfortunately, steady-state performance capacity and longer-term operational limitations were not identified. Figure 4.11 summarizes the overall system performance attained on day 339. Average input concentrations of about 200 µg/L PCE, 600 µg/L CT plus CF, and 2.5 mg/L TOL had been reduced, on average, to <3 µg/L (PCE, CT, and CF) and to <40 µg/L (TOL) by the end of Gate 3. No PCE or TOL was detected in the off gas from the biosparge zone. The added benzoate had also been essentially completely removed, and overall groundwater chemistry was not adversely altered.

4.5 WASTE MANAGEMENT

Disposal of all contaminated groundwater was conducted successfully in accordance with the methods outlined in Appendix 10.

4.6 CONCLUSIONS

Groundwater moved through Gate 3 at an average velocity of about 12–15 cm/day, with a spatially heterogeneous distribution about this mean velocity. Solutes were added to this groundwater via source wells to produce average concentrations of about 230 µg/L PCE, 4300 µg/L TOL, 350 µg/L CT, and 100 µg/L CF in the first fence of multilevel monitoring piezometers, G302. The CF was produced by intrinsic reductive dechlorination of CT between the source wells and fence G302. Considerable spatial and temporal variability of solute concentrations at fence G302 and at downgradient fences was noted, likely reflecting problems with the source well system and the influence of aquifer heterogeneities.

Sufficient benzoate pulse mixing and overlap likely occurred between fences G312 and G315. This established anaerobic conditions which became particularly conducive to reductive dechlorinating microbes between days 238 and 315.

Complete transformation of CT and all associated breakdown products was achieved within the designated anaerobic zone by day 238.

Reductive dechlorination of PCE began after day 238 between the NIW and fence 315. The only product detected at significant concentrations in groundwater before the end of the experiment on day 339 was *c*DCE. The delay in the onset of reductive dechlorination from when PCE entered the anaerobic zone in significant concentrations (>LOQ) was estimated to be 90 to 140 days. This likely represents the time for adaptation of the *in situ* microbial population to generate the strongly anaerobic conditions utilizing benzoate and, later, mineral salts, added in the NIW. The reductive dechlorination was likely initiated or at least supported by the addition of mineral salts on days 207 and following. By day 339, most of the PCE had been dechlorinated within the anaerobic zone of Gate 3.

Sparging about 40 L of O_2 gas into the biosparge zone twice per week was not successful in developing aerobic conditions. When the sparging frequency was increased to once per day, the desired aerobic conditions were maintained in and downgradient of the biosparge zone. Essentially no volatile organics were lost as offgas from the headspace of the biosparge system. Thus, this system was very successful in adding sufficient O_2 to the groundwater passing through the biosparge zone while resulting in essentially no transfer of VOCs to the atmosphere.

Toluene may have undergone some intrinsic, anaerobic biodegradation in Gate 3, but significant toluene was biodegraded aerobically in and downgradient of the biosparge zone. PCE and *c*DCE were also reduced to <5 µg/L average concentrations by the end of the treatment system. Aerobic biodegradation also removed most of the benzoate and methane which entered the biosparge system.

Groundwater leaving Gate 3 at the end of the experiment (day 339) was considered completely remediated.

4.7 RECOMMENDATIONS FOR IMPLEMENTATION

Gate 3 was operating effectively for only the last about 100 days of the experiment, so not enough experience was gained to provide detailed recommendations for implementing this combination of technologies at other sites. This experiment was judged a successful demonstration of a concept that merits consideration for implementation at other sites.

The design approach for Gate 3 is highly recommended for future implementation. Benzoate appears to have facilitated reductive dechlorination of PCE after about 100 to 175 days. The absolute need for mineral nutrients was not demonstrated, but such nutrients may have been significant for the reductive dechlorination of PCE. The role of mineral salt nutrients deserves future evaluation to optimize the additions for creating *in situ* anaerobic conditions.

The carbon substrate concentration should be minimized so that it still drives the groundwater strongly anaerobic, but does not create an excessive BOD. The addition of benzoate, as in Gate 3, to attain steady concentrations of 20–50 mg/L appears effective and practical.

At another site with a less well-characterized aquifer, tracer tests as conducted in Gate 3 should be done to demonstrate that sufficient overlap of nutrient pulses is being attained. Adjustments to the flush schedule can be made, if necessary, following the tracer tests. Mineral salt nutrients should be included in at least some of the flushes.

For more passive, lower maintenance operations, a better nutrient flush/circulation system needs to be developed. This system should automatically initiate a flush and circulate the required volume of groundwater with the required amendments at a designed frequency. Flushing 1 to 1.5 NIW volumes should be sufficient.

Sparging with O_2 was very effective in creating the desired aerobic conditions in Gate 3. This system could also be automated to deliver the required volume of gas at a specific frequency. Using small volumes of gas, sparged only as frequently as needed to maintain aerobic conditions, produced no detectable VOCs in the off gas. The presence of the pea gravel seems to have facilitated the delivery of O_2 to groundwater by retaining O_2 bubbles in the system for continuing transfer to advecting groundwater. While other designs have been attempted in this project, the system employed in Gate 3 seems to have been the most effective, both technically and from a cost perspective.

Major permeability reductions were not evident in Gate 3, and the system was operating efficiently for a few months. Longer term operational problems could include biofouling, most likely in the aerobic system. For future application, hydraulic testing should be conducted across the anaerobic/aerobic interface to identify any permeability reduction. This testing should be considered after the dechlorinating and the aerobic microbial populations are both well established.

REFERENCES

Acton, D.W. and J.F. Barker. 1992. In situ biodegradation potential of aromatic hydrocarbons in anaerobic groundwaters. *Jour. Contam. Hydrol.*, 9: 325-352.

Barker, J.F., G.C. Patrick, and D. Major. 1987. Natural attenuation of aromatic hydrocarbons in a shallow sand aquifer. *Ground Water Monitoring and Remediation*, 7(1): 64-71.

Chapelle, F.H., P.B. McMahon, N.M. Dubrovsky, R.F. Fujii, E.T. Oaksford, and D.A. Vroblesky. 1995. Deducing the distribution of terminal-electron-accepting processes in hydrologically diverse groundwater systems. *Water Resources Research*, 31: 359-371.

Devlin, J.F. 1994. Enhanced *In Situ* Biodegradation of Carbon Tetrachloride and Trichloroethene Using a Permeable Wall Injection System. Ph. D. dissertation, Department of Earth Sciences, University of Waterloo, Waterloo, Ontario, Canada.

Devlin, J.F. and J.F. Barker. 1994. A semipassive nutrient injection scheme for enhanced *in situ* bioremediation. *Ground Water.* 32(3): 374-389.

Devlin, J.F. and J.F. Barker. 1996. Field investigation of nutrient pulse mixing in an *in situ* biostimulation experiment. *Water Resources Research.* 32(9): 2869- 2877.

Hubbard, C.E. 1992. Transport and Fate of Dissolved Methanol, Methyl-Tertiary-Butyl Ether and Monoaromatic Hydrocarbons in a Shallow Sand Aquifer. M.Sc. thesis, Department of Earth Sciences, University of Waterloo, Waterloo, Ontario, Canada.

Ji, W., A. Dahmani, D.P. Ahlfeld, J.D. Lin, and E. Hill, III. 1993. Laboratory study of air flow visualization. *Ground Water Monitoring and Remediation.* 13(4): 115-126.

Johnson, R.L., P.C. Johnson, D.B. McWhorter, R.E. Hinchee, and I. Goodman. 1993. An overview of *in situ* air sparging. *Groundwater Monitoring and Remediation*, 13(4): 127-135.

Major, D.W., Hodgins, E.W., and Butler, B.J. 1991. Field and laboratory evidence of *in situ* biotransformation of tetrachloroethene to ethene and ethane at a chemical transfer facility in North Toronto. In: Hinchee, R.E. and Olfenbuttel, R.F. (Eds.), *On-Site Bioremediation Processes for Xenobiotic and Hydrocarbon Treatments*, Butterworth-Heinmann, Boston, pp. 147-171.

McCarty, P.L. and L. Semprini. 1994. Groundwater treatment of chlorinated solvents. In: Norris, R. D. et al. (Eds.) *Handbook of Bioremediation*, Section 5, Lewis Publishers, Cheslea, MI, pp. 87-116.

Mueller, J.G., S.E. Lantz, B.O. Blattmann, and P.J. Chapman. 1991. Bench scale evaluation of alternative biological treatment processes for the remediation of pentachlorophenol and creosote contaminated materials: solid phase bioremediation. *Environmental Science and Technology.* 25: 1045-1055.

Pankow, J.F., R.L. Johnson, and J.A. Cherry. 1993. Air sparging in cutoff walls and trenches for control of plumes of volatile organic compounds (VOCs). *Ground Water.* 31(4): 654-663.

Pankow, J.F. and J. A. Cherry. 1996. Dense chlorinated solvents and other DNAPLS in groundwater. Waterloo Press, Portland, OR.

University of Waterloo (UW). 1995. Passive and Semipassive Techniques for Groundwater Remediation. Work Plan.

University of Waterloo (UW). 1997a. Passive and Semipassive Techniques for Groundwater Remediation. Phase 1 Installation Report.

Chapter 5

In Situ Sequential Treatment of a Mixed Organic Plume Using Granular Iron and Oxygen Sparging — NAS Alameda

5.1 TECHNOLOGY PROCESS DESCRIPTION

5.1.1 Overview

Selected technologies were applied to the pilot-scale remediation of a contaminant plume in the shallow, unconfined aquifer at Site 1, Naval Air Station (NAS) Alameda, California (Figures 1.3 and 5.1). The contaminants of interest were chlorinated volatile organic compounds (CVOCs), including trichloroethene (TCE), dichloroethene isomers (DCE), and vinyl chloride (VC), and monoaromatic hydrocarbons, specifically benzene, toluene, ethylbenzene, and xylenes (BTEX).

The selected technology applied at NAS Alameda was a pilot-scale Funnel-and-Gate, *in situ* remedial system. After system installation, its performance was evaluated by analyzing groundwater for contamination upgradient, within and in the effluent of the remedial gate. A control gate (no treatment zones) was also installed to assess intrinsic remediation for comparison to the remediation success in the remedial gate.

Remedial objectives (ROs) for the demonstration treatment system at Alameda were set to approximate California's maximum contaminant levels (MCLs). However, the remedial objectives for some contaminants (*c*DCE, VC, and benzene) were increased to concentration levels that are consistent with site-specific, risk management, and natural attenuation approaches. The remedial objectives proposed for the demonstration are listed in Table 5.1 along with California's MCLs for drinking water quality criteria and U.S. EPA Marine Ambient Water Quality Data (U.S.1331 EPA, 1992). The AATDF project would be considered successful if the ROs were attained, and/or if the capabilities and limitations of the technology were clearly defined.

5.1.2 Technology Description

The *in situ* technologies installed at NAS Alameda included (1) reductive dechlorination of chlorinated ethenes by contact with granular iron in a permeable reactive barrier, and (2) aerobic biodegradation of petroleum hydrocarbons and any remaining lesser-chlorinated compounds, stimulated by minimal O_2 and CO_2 addition via an *in situ* sparge system. The novel aspect of this technology was the combination of the anaerobic (granular iron) and aerobic (biosparge) treatment systems sequentially within a Funnel-and-Gate (Figure 5.2). The initial gate segment consisted of the granular iron media, and this was followed by the biosparge zone.

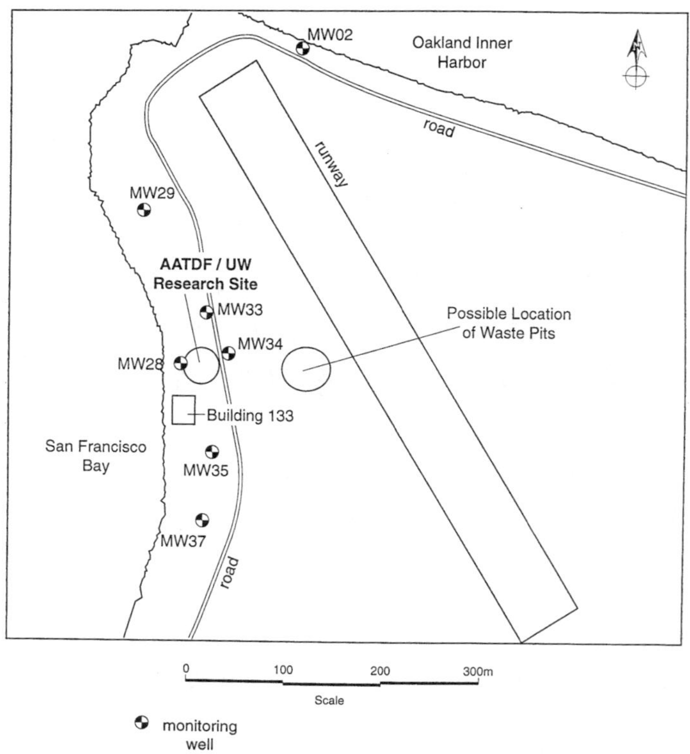

Figure 5.1 Alameda site location.

Table 5.1 AATDF and UW Remedial Objectives for Site 1, California MCLs and Marine Ambient Water Quality Criteria

Chemical Parameter	Proposed Remedial Objective (RO) (μg/L)	California MCL (μg/L)	Marine Ambient Water Quality Criteria (μg/L)
TCE	5	5	2000
1,1 DCE	6	6	224000
DCE	10	10	224000
*c*DCE	10	6	224000
VC	2	0.5	NA
Benzene	10	1	5100
Toluene	150	150	5000
Ethylbenzene	700	700	430
Xylenes, total	1750	1750	NA
pH	5–8.5[a]	NA	NA
Iron	500	NA	NA

Note: NA: not applicable.

[a] pH units.

For the purposes of experimentation, it was decided to operate the remedial gate under controlled groundwater flow conditions. Thus, the downgradient sheet piles were left in place to seal the gate and two groundwater extraction wells were placed at the most downgradient end. Groundwater was then drawn through the gate at controlled flow rates by operating the extraction wells at the specified pumping rate.

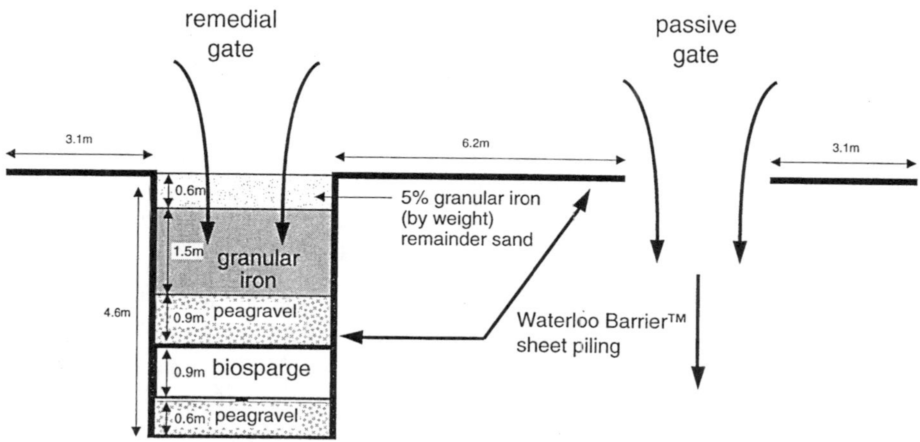

Figure 5.2 Schematic layout of the Funnel-and-Gate showing both the remedial and control gates.

The field demonstration was conducted in two phases, Operational Phase 1 (OP1) and Operational Phase 2 (OP2). Each phase was characterized by an induced groundwater extraction rate. In OP1, a high discharge rate (1.27 m^3/day) was established at the extraction wells and in OP2 a lower discharge rate (0.34 m^3/day) was established. OP1 was implemented to pump several pore volumes through the gate and to stress the system, i.e., study how the iron wall reacted to an accelerated flow velocity. OP2 was implemented to mimic passive conditions. Refer to Section 5.3.3 for a detailed explanation for each operational phase.

5.1.2.1 Granular Iron Treatment Technology

Laboratory and field tests have shown that under highly reducing conditions and in the presence of metallic surfaces, certain dissolved chlorinated organic compounds, e.g., PCE and TCE, in groundwater will degrade to nontoxic compounds such as ethene, ethane, and chloride. A detailed discussion of the iron metal technology is provided in Chapter 1. Basically, the process appears to be abiotic reductive dehalogenation, with the metal serving to lower the E_h in solution and serving as an electron source in the reaction (Gillham and O'Hannesin, 1994; Gillham, 1996). Using iron as the reactive metal, reaction half-lives (the time required to degrade one half of the original contaminant mass) are commonly several orders of magnitude lower than those measured under natural conditions. Because of these high rates of degradation, the low cost of the iron, the passive nature of the reaction, and because the compounds are degraded with production of minimal hazardous nonchlorinated organic by-products, the technology appears to have great promise for the remediation of contaminated groundwater. EnviroMetal Technologies Inc. (ETI), a private commercial firm, exclusively licenses this granular iron-based remediation technology. Refer to Chapters 1 and 2 for more detailed discussions.

5.1.2.2 Biosparge Technology

To promote the aerobic biodegradation of petroleum hydrocarbons and any remaining chlorinated compounds (DCE and/or VC), oxygen and carbon dioxide were added to the contaminated groundwater via coiled porous tubing, installed on the bottom of the biosparge zone. The oxygen was added to increase the DO levels to ~20 mg/L and the carbon dioxide was added to reduce the high pH levels (8–11) produced in the granular iron zone to a pH range of 7–8. The tubing emitted

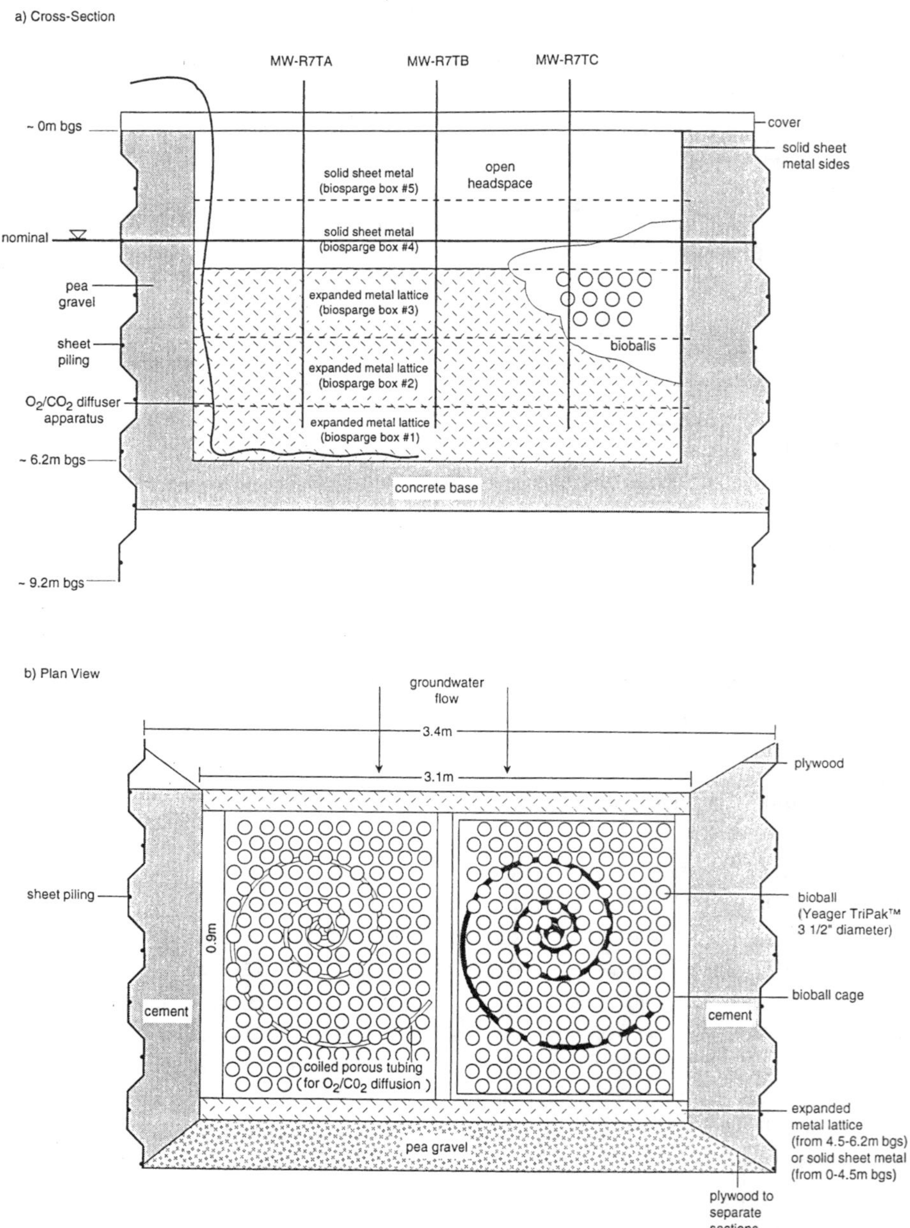

Figure 5.3 Cross section and plan view of the biosparge zone.

small oxygen and carbon dioxide bubbles into the water column packed with bacterial growth support material (Yeager™ TriPak, 5 cm diameter). Figure 5.3 presents a cross section and plan view of the biosparge zone. The sparged O_2 and CO_2 dissolve in the groundwater, aiding the biodegradation of the hydrocarbons and chlorinated compounds through the action of indigenous aerobic bacteria attached to the growth support material.

5.2 FIELD SITE DESCRIPTION

5.2.1 Site 1 History

Site 1 is located in the northwestern part of NAS Alameda, on the northwestern tip of Alameda Island. It is situated on artificial fill hydraulically placed on top of natural bay mud estuarine deposits (silts and clays). Historical information indicates that filling occurred in the late 1930s (PRC, 1993).

Starting in the 1940s, cleaning solvents and waste petroleum hydrocarbons were disposed in unlined waste pits excavated in the fill soil at Site 1. Small quantities of solid waste were also disposed in the waste pits. One historical disposal area, defined from aerial photographs (PRC, 1993), labeled as "Waste Pits" in Figure 5.1, is apparently the source of a plume of dissolved chlorinated organics and petroleum hydrocarbons that flows westward to San Francisco Bay. This dissolved plume is the location of this demonstration project.

5.2.2 Site 1 Hydrogeology

Site 1 overlies a 120- to 150-m-thick sequence of quaternary unconsolidated sediments that unconformably overlie Jurassic/Cretaceous Franciscan bedrock (Rogers and Figuers, 1991). Typical cross sections (PRC, 1993) indicate that a sandy artificial fill occurs from the ground surface to a depth of approximately 6 m. The sandy fill rests on top of Holocene bay mud (silt and clay), which was the sea floor prior to placement of the fill. The bay mud unit is approximately 4.6 to 6 m thick in the western portion of Site 1.

The sediments beneath Site 1 are subdivided into two aquifers. The deeper one consists of undivided Pliocene/Pleistocene coarse-grained terrestrial deposits at a depth greater than 61 m bgs (PRC, 1993). The shallower aquifer, which occurs within 30 m of the ground surface, consists of two separate water-bearing zones. The shallowest water-bearing zone consists of the sandy artificial fill, and hosts the AATDF demonstration. Groundwater within this unit is unconfined, with a water table that fluctuates seasonally. The second water-bearing zone occurs within sandy sediments below the Holocene bay mud unit, from a depth of approximately 11 to 23 m. Groundwater within this unit is semiconfined (PRC, 1993).

In the conceptual hydrogeologic model of the site, shallow groundwater is recharged through unpaved surface areas at NAS Alameda and flows radially outward to the Bay (PRC, 1993). As the fresh groundwater nears the edge of the island, it mixes with saline water in the Bay. The flow system is complicated by the internal structure of the artificial fill and by tidal effects.

Alameda Island is surrounded on two sides by the San Francisco Bay; tidal effects in the shallow aquifer at Site 1 are possible. A comprehensive study of the hydraulic effects of the tides was performed in April 1992 (PRC, 1993). As would be expected, the most pronounced water level changes occurred in wells located near the edge of the Bay. It was found that the tidal fluctuations were relatively small in the uppermost artificial fill water-bearing zone, typically <5 cm in the area of the demonstration.

Calculating the mean groundwater flow direction in aquifers affected by tidal fluctuations requires filtering of the water-level data. Average groundwater elevation contours for the uppermost artificial fill water-bearing zone were calculated in April 1992 using the method of Serfes (PRC, 1993). Once the corrected water level data were contoured, it was found that the averaged groundwater elevation mimicked the shape of the northern part of Alameda Island, with a higher elevation in the interior of the island and a lower elevation along the margin of the Bay. This supports the conceptual model of groundwater flowing radially from the center of the island outward to the Bay. Groundwater flow in the Site 1 area is estimated to be primarily horizontal due to the presence of the underlying bay mud aquitard and the location of groundwater recharge and discharge areas. The average groundwater gradient near Site 1 measured in April 1992 was approximately 1.8×10^{-3}. Given a hydraulic conductivity value of 2.2×10^{-5} m/s and an assumed effective porosity of 20% (perhaps to reflect zones of finer material), the average groundwater velocity in April 1992 was 1.7 cm/day (approximately 7 m/year) (PRC, 1993).

Samples collected from groundwater monitoring wells within the artificial fill aquifer during the 1993 Solid Waste Assessment Test (SWAT) investigation contained total dissolved solids (TDS) ranging from 800 to 8,700 mg/L. This is much lower than TDS concentrations within the same hydrogeologic unit further south in Site 2. The groundwater at Site 1 is generally fresh and Na (Ca, Mg)-Cl or Na (Ca, Mg)-HCO_3 type. The relatively fresh water likely reflects the direct recharge by infiltration or precipitation (PRC, 1993). Iron concentration, locally up to 10 mg/L, and methane occurrence reflect the anaerobic nature of the groundwater.

5.2.3 Demonstration Project Site Conditions

The AATDF Demonstration Project Site is located in the southwestern part of Site 1, adjacent to San Francisco Bay (see Figure 5.1). This area was selected for the experiment because of its hydrogeologic characteristics and contaminant blend and distribution. The demonstration site is located within a small contaminant plume that apparently emanates from former waste disposal pits near the runway and flows westward before discharging to San Francisco Bay. The contaminant plume was previously identified during an earlier investigation of groundwater contamination at NAS Alameda. To further delineate the groundwater plume, two separate hydrogeologic investigations were completed to evaluate, in three dimensions, the subsurface conditions at the Demonstration Project Site (see UW, 1996a and UW, 1996b).

The artificial fill in the vicinity of the Demonstration Project Site is generally silty sand to sand (SM to SP in the Unified Soil Classification System), containing approximately 5% to 15% fines (i.e., soil having a grain size smaller than a number 200 sieve). However, the very upper portion of the unit, to a depth of approximately 1.2 m, contains more silt-size sediment, up to 40% of the sample. Also, a thin (approximately 0.6-m thick), dark gray sandy clayey silt layer was penetrated in all three exploratory borings in July 1996, from a depth of approximately 1.2 to 2.1 m bgs. The base of the sandy artificial fill occurs at the contact with the underlying Bay Mud clay, at a depth of approximately 7.0 m bgs.

Unconfined groundwater was encountered at a depth of approximately 1.2 to 2.1 m bgs during the July 1996 investigation (UW, 1996a). From these data, groundwater contours were drawn using stabilized groundwater elevations from the three newly installed monitoring wells, and it was found groundwater flow direction is toward the west, consistent with the Site 1 regional groundwater flow direction.

Using the hydraulic gradient calculated from the July 1996 investigation (UW, 1996a), the mean hydraulic conductivity obtained from the slug tests (2×10^{-5} m/s), and an assumed effective porosity of 30%, the groundwater seepage velocity was estimated to be 5.2×10^{-2} m/day (5 cm/day). A 30% porosity value was used, as it was felt to be more representative for a sandy aquifer than the value of 20% used by PRC (1993). Note, however, that the calculated groundwater gradient may be significantly influenced by the tidal cycle.

In the area of the field demonstration, groundwater contained up to 210 mg/L *c*DCE, 26 mg/L VC, and <1 mg/L concentrations of other solvents such as TCE, 1,1-DCE, and DCE. The highest concentrations of chlorinated volatile organic compounds (CVOCs) occurred within the upper portion of the artificial fill water-bearing zone, approximately 2.8–3.7 m bgs below ground surface. Since groundwater flow within the unit is primarily horizontal, this suggests that the source of the contamination is relatively shallow. Based on the data, contamination appears to be more extensive in the shallow part of the aquifer, but is still present at a depth of 6.0 m (Figure 5.4). Data can be found in UW (1996a) and UW (1996b).

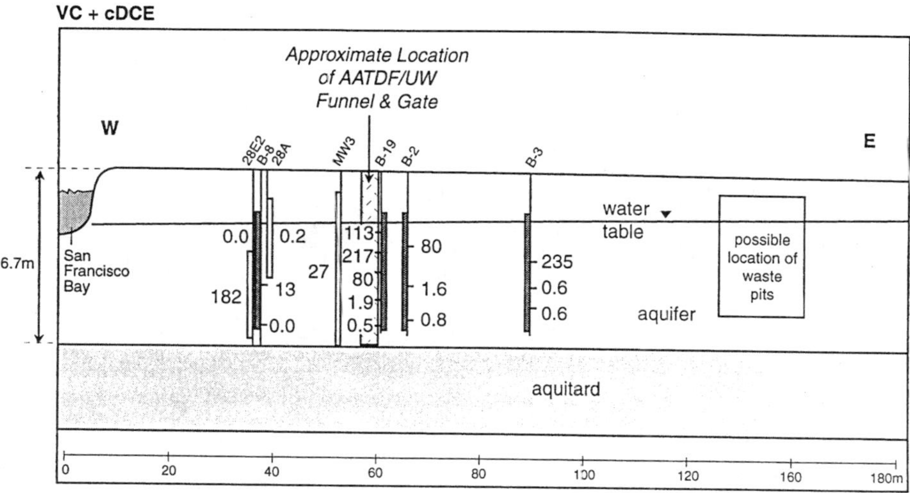

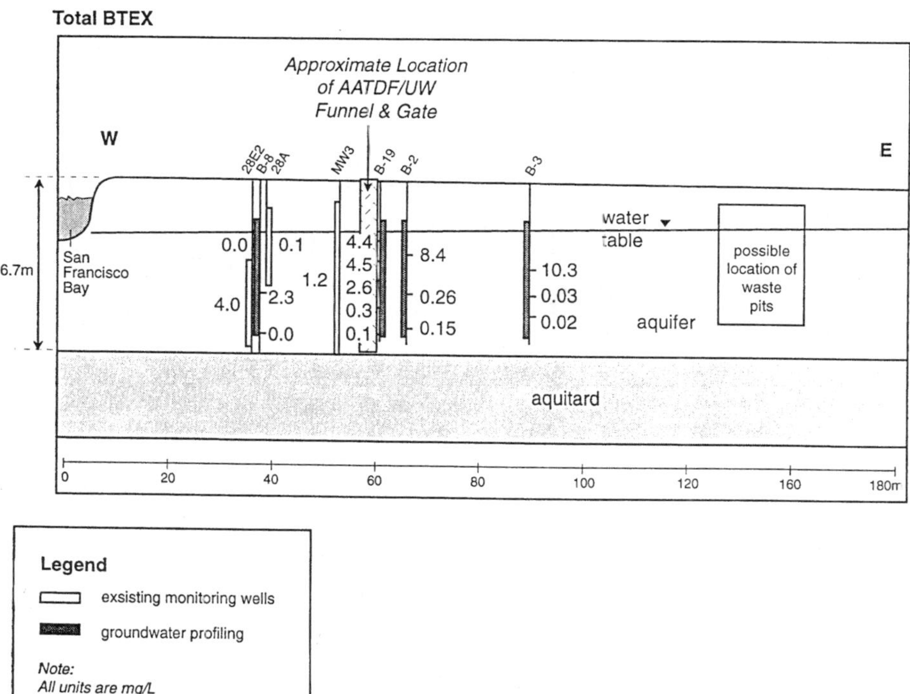

Figure 5.4 Total *c*DCE and VC and total BTEX concentrations collected from both Site 1 investigations, July 1996 and October 1996.

Moderate concentrations of BTEX were detected in many of the groundwater samples collected during both the July and October 1996 investigations. Toluene concentrations ranged from nondetect up to 7 mg/L, benzene up to 0.3 mg/L, and concentrations of ethylbenzene and xylenes were <1 mg/L. As with the CVOCs, the highest concentrations of BTEX occurred within the upper portion of the aquifer (Figure 5.4). Data can be found in UW (1996a) and UW (1996b).

The pilot-scale treatment gate was positioned to encounter the highest concentrations of CVOCs and BTEX in the shallow plume, in a location free of buildings, roads, etc. (see Figure 5.5).

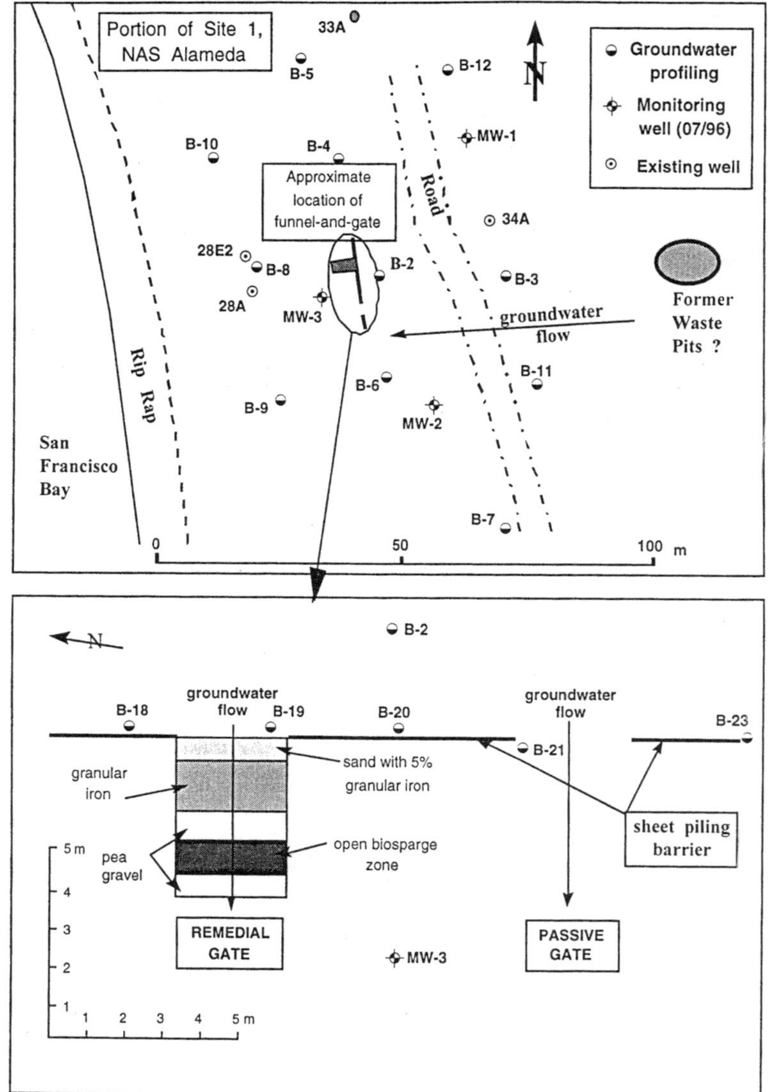

Figure 5.5 Location of the Funnel-and-Gate based on data collected from the two site investigations.

5.3 FIELD DEMONSTRATION

5.3.1 Description and Design

A Funnel-and-Gate was designed with two gates. One was instrumented with the remedial systems, while the other was left undisturbed, to serve as a control for the documentation of natural (intrinsic) remediation processes. Figure 5.2 shows a schematic layout of the pilot-scale system. The Revised Work Plan (UW, 1996b) and the Installation Report (UW, 1997c) detail the final design and installation of the Funnel-and-Gate.

The remedial gate was approximately 4.5 m long, 3 m wide, and 6 m deep. The concept of the remedial gate was to funnel the contaminated groundwater first through a 0.6-m (2-ft) permeable zone packed with a mixture of sand and granular iron, approximately 3–5% by weight of iron. The next zone consisted of 1.5 m (5 ft) of 100% granular iron followed by 0.9 m (3 ft) of pea gravel. Effluent from the granular iron treatment was directed to the 0.9-m (3-ft) biosparge zone, where aerobic biodegradation was enhanced through oxygen and carbon dioxide amendments. The final section contained 0.61 m (2 ft) of pea gravel to facilitate groundwater extraction and serve as the final compliance monitoring zone.

There were a number of areas that required further investigation before the Funnel-and-Gate design was finalized. These areas included estimating the flow rate through the gate and the required residence time within the granular iron wall as well as defining the operational requirements of the biosparge zone (specifically to determine the frequency of oxygen and carbon dioxide additions that would create conditions suitable for aerobic biodegradation).

5.3.1.1 Flow Rate through the Gate

An initial set of numerical simulations was completed prior to installation to aid in the design of the Funnel-and-Gate configuration (see UW, 1996b for complete details). To ensure adequate residence time within the granular iron, design simulations were conducted to produce conservative estimates of groundwater flow through the gate. On the basis of these simulations the Funnel-and-Gate configuration that was chosen had an estimated flow rate of about 0.34 m^3/day (12 ft^3/day) through the remedial gate. This flow rate was then used as the extraction well discharge rate within the remedial gate for OP2.

An additional set of numerical simulations was conducted in July 1997, after the Funnel-and-Gate had been installed, to provide insight into potential scale-up configurations. At the same time, the numerical model for the chosen gate design was updated to more closely resemble the actual Funnel-and-Gate configuration and existing hydraulic gradient. The basic simulations were run with a gradient of 0.002 to represent the daily mean average gradient (i.e., tidally filtered), and also a gradient of 0.005 to represent the maximum gradient within a tidal cycle. Figure 5.6 shows the computed volumetric gate discharge for the remedial and control gates vs. the simulated natural gradient. The design simulations showed that the groundwater velocity within the remedial gate (with its sheet piling sides) would be different than the control gate. Also, while the flow rate through the remedial gate remained fixed by the extraction wells, the flow rate through the control

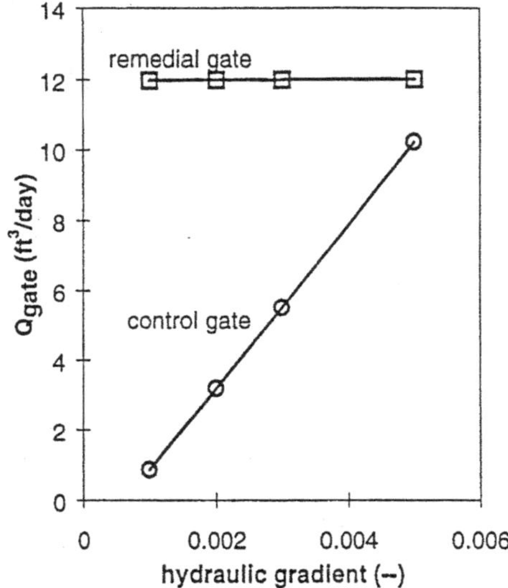

Figure 5.6 Gate volumetric discharge vs. natural hydraulic gradient.

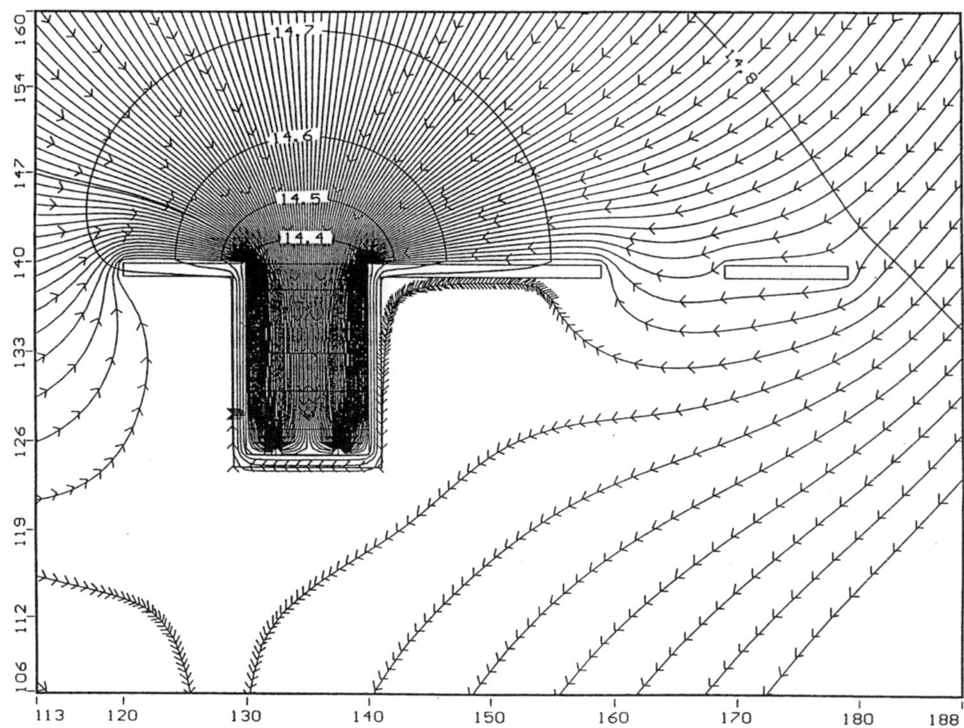

Figure 5.7a Flow paths from the faster pumping rate, OP1 (1.27 m³/day). Modeling was with Visual MODFLOW, V. 2.50 © 1995-1997, Waterloo Hydrogeologic software.

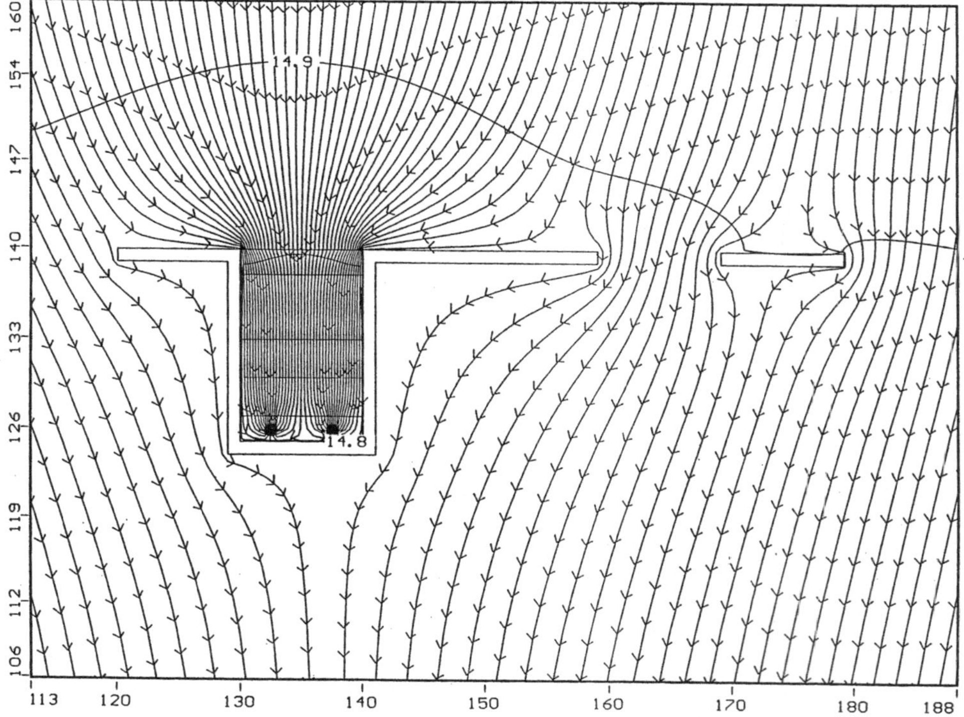

Figure 5.7b Flow paths from the slower pumping rate, OP2 (0.34 m³/day). Modeling was with Visual MODFLOW, V. 2.50 © 1995-1997, Waterloo Hydrogeologic software .

gate could have varied as the natural hydraulic gradient changed over time (seasonal variation). The severity of the gradient variations and associated hydraulic complexities were such that simple comparisons of the remedial and control gate performances were not justified, and so no such comparisons have been attempted here.

Further simulations showed the effect on the groundwater flow paths when the extraction rate was changed. Figure 5.7 shows a plan view of the flowpaths with a total extraction well discharge rate of 1.27 m^3/day (45 ft^3/d) (Figure 5.7a) and 0.34 m^3/day (12 ft^3/d) (Figure 5.7b). Figure 5.7a, in particular, shows that the flow through the control gate could have been greatly influenced by the extraction in the remedial gate. This further complicated the interpretation of concentration changes along the length of the control gate especially during OP1, when the higher extraction rate applied.

Other simulations indicated that "passive" conditions within the remedial gate (based on a natural gradient of 0.002) could be duplicated using a total extraction well discharge rate of about 0.13 m^3/d (4.7 ft^3/d). However, such a low flow rate would not have exchanged sufficient pore volumes through the gate during the time available to be assured of representative reaction rates, particularly in the granular iron. Therefore, the higher flow rate of 0.34 m^3/day was used to approximate "passive" conditions.

5.3.1.2 Treatability Study for the Granular Iron

A bench-scale treatability study was completed to estimate the time required for treatment of the CVOCs. This study, conducted at the University of Waterloo, consisted of passing groundwater collected from MW-3 (fully screened well from Site 1, NAS Alameda) through the granular iron material (Peerless Metal Powders and Abrasive), selected for the field demonstration (see UW, 1996b for complete details on the treatability study). Table 5.2 shows the measured groundwater concentrations used in the column study and half-lives calculated for these CVOCs in the treatability study. These half-lives suggest that a residence time in the iron zone of between 7 and 8 days is required for treatment of the CVOCs to desired levels.

Table 5.2 Measured Concentrations Used in the Column Study and the Resultant Calculated Half-Lives

Chemical Parameter	Concentration (μg/L)	Measured Half-Life (hours)
TCE	10.0	1.5
cis-1,2 DCE	32,154	12.1
VC	25,625	11.2

Note: Concentration of water used in column study.

At the estimated passive groundwater flow rate of 6.1 cm/day (refer to Table 5.3 for OP2, following) through the granular iron, the flow-through thickness of the iron wall should be at least 0.5 m. For a velocity of 20 cm/day (assumed fast flow during OP1), a flow-through thickness of 1.6 m would be required. These thicknesses coincide relatively well with those presented in Tratnyek et al. (1997).

It was decided that a slightly higher groundwater velocity of 22 cm/day would be used, increasing the flow-through thickness of the granular iron wall to 1.76 m. Since the chlorinated solvent plume was not uniform and CVOCs were encountered at one location in concentrations even higher than those given in Table 5.2 (see UW, 1996a), it was recognized that this thickness might not be adequate locally. However, since mixing and some dilution was expected within the remedial gate and in the iron zone, the potential for breakthrough was expected to be significant only during the short phase of accelerated pumping. As a cost-effective design was desired, it was decided to implement only a 1.5-m flow-through thickness for the granular iron zone. The small portion of granular iron (3–5% by weight) incorporated into the upgradient sand zone would provide additional treatment of CVOCs.

5.3.1.3 Biosparge Zone

The design objective of the biosparge zone was to provide sufficient oxygen to groundwater for aerobic biodegradation of BTEX and any remaining CVOCs to the remedial objectives given in Table 5.1. Several designs were considered, ranging from slotted vertical caissons to gravel-filled zones to open excavation. The design selected was a series of five steel boxes, stacked one on top of the other. Each box was 0.9 m wide by 3.1 m long and 1.2 m high. The bottom three boxes were open to groundwater flow due to upgradient and downgradient perforated sheet metal sides (expanded metal lattice, 1/4 in. centered holes on gauge 11 steel). The top two boxes had solid sheet metal on the up- and downgradient ends which rendered these boxes impermeable. Lying on the bottom of the biosparge zone was a coiled porous tube that was used to deliver either oxygen or carbon dioxide to the groundwater. Rigid, plastic bacterial growth support medium (Yeager™ TriPaks) were placed in cages, to keep the medium submerged in the water column. Refer to Figure 5.3 for a cross-sectional and plan view of the biosparge zone. There was a cover over the biosparge unit, to limit and capture any off gas. The space overlying the water table but below the biosparge cover was packed with plastic growth support medium serving as a headspace bioreactor, where excess sparged O_2, perhaps containing traces of VOCs, could reside for a period of time sufficient to degrade the VOCs.

As the effluent from the granular iron zone was anaerobic and would have a high pH (8–11) (see Chapter 2), both oxygen and carbon dioxide gases were sparged into the biosparge zone to increase the DO levels and to decrease the pH, respectively. During OP1, the remedial gate was pumped at a high flow rate (1.27 m^3/day), and oxygen and carbon dioxide were sparged into the biosparge zone on several occasions to determined whether conditions suitable for aerobic biodegradation (DO >5 mg/L, pH ~7) could be achieved (refer to Appendix 28 for sparging details). The frequency and length of each sparge event was originally based on the flux of BOD entering the biosparge. Early BOD data showed a very low oxygen demand (~0.01 mg/L) in the biosparge zone. This could be due to low influent concentrations of BTEX and CVOCs entering the biosparge zone. As a result of the low BOD values, a variety of low volume sparge patterns were attempted but none established continuously aerobic conditions in the biosparge zone. Therefore, aerobic conditions were not constantly maintained during OP1.

After mid-June 1997, it was decided to increase O_2 addition in the biosparge zone. Due to logistical problems with the field demonstration, efforts to create these suitable conditions were limited prior to this time. To maintain consistent levels of DO and pH, automated timers were installed on the oxygen and carbon dioxide cylinders so that sparging events would occur on a regular basis. It was determined that oxygen should be sparged at a delivery pressure of 20 psi, 6 times in 24 hours with each event lasting 20 minutes. Carbon dioxide was only sparged once every 3 weeks for 10 minutes at a 10 psi delivery pressure. This sparging pattern maintained the desired conditions in the biosparge zone. Refer to Appendix 28 for complete details on the development of these sparging patterns.

5.3.2 Construction of the Remedial Gate

Once the design of the Funnel-and-Gate was finalized, construction of the gates began. Refer to UW (1997c) for complete details on the construction activities. The pilot-scale design, as shown in Figure 5.8, consisted of a rectangular "box" (to become the remedial gate) 3.0 m wide by 4.6 m in the direction of groundwater flow, flanked by about 3.0 m of impermeable sheet piling on either side of the remedial and control gates. Therefore, 6.0 m (edge to edge) of impermeable sheet piling separated the remedial gate from the control gate.

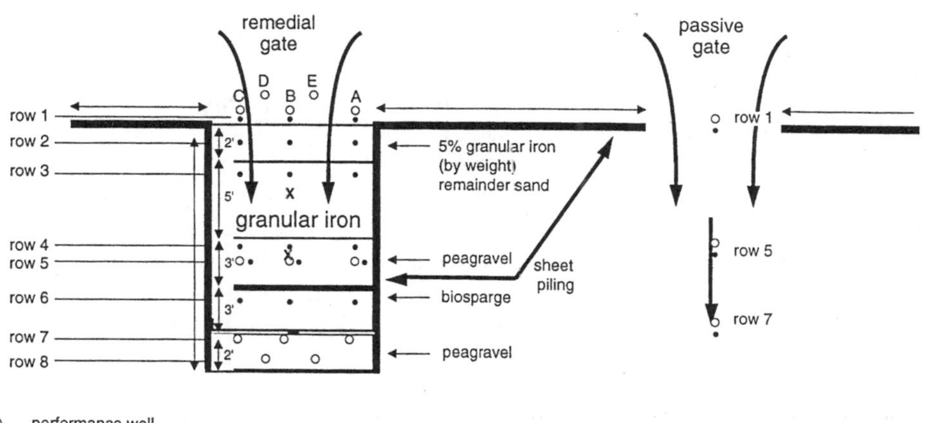

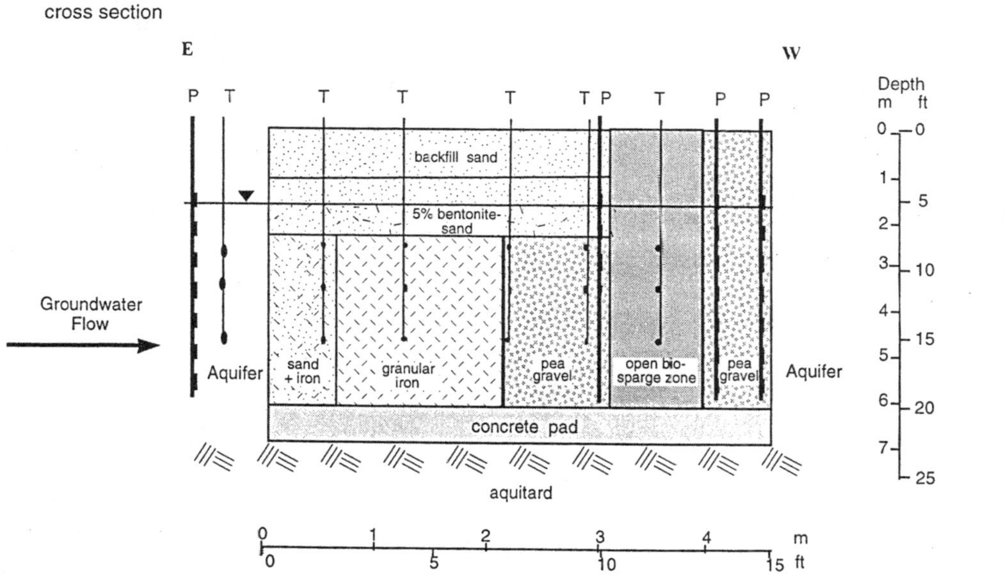

Figure 5.8 Locations of the groundwater monitoring wells and a cross section of the remedial gate only.

The sections of the remedial gate (from the upgradient end) are as follows (see Figure 5.8):

1. Sand with 5% (by weight) granular iron was 0.6 m thick and served to initiate the reductive dechlorination process and separate the 100% iron wall from the native material.
2. Granular iron section was 1.5 m thick and reductively dechlorinated the chlorinated ethenes.
3. First pea gravel section was 0.9 m thick and separated the anaerobic and aerobic technologies.
4. Biosparge zone was 0.9 m thick and was used to aerobically biodegrade BTEX and remaining chlorinated compounds such as *c*DCE or VC.
5. Second pea gravel section was 0.6 m thick and housed the extraction wells.

The "box" was constructed of sheet piles driven to a depth of about 8.5 m bgs (2.4 m into the bay mud aquitard). Native unconsolidated material was excavated to a depth of 6.7 m bgs, and a 0.6-m-thick cement floor was poured on the bottom of the excavation. The installation of the cement floor served to prevent underlying material from invading the treatment zone, to support the weight of the iron, to provide a bottom for the sparging system, and to prevent the introduction of contaminants into the gate via upward flow. The monitoring wells and the first two perforated biosparge boxes were lowered into the installation. The three granular sections — sand, iron, and pea gravel — were then backfilled simultaneously to around 2.4 m bgs. The reason for backfilling to this depth was to ensure that the water table would never fall below the top of the iron surface as the grains would become oxidized, rendering them unreactive. According to PRC (1993), the water table fluctuates between 0.9–2.1 m bgs. The biosparge zone top was protected and remained empty. The third perforated biosparge box was then lowered into the gate, and backfilling around it was completed to 2.4 m bgs. The two solid biosparge boxes were installed to surface, keeping the biosparge zone free of fill. A geotextile liner was then placed on top of the first three granular sections, followed by a 100% bentonite layer (5 cm thick) and then fill mixed with 5% bentonite was added (1.2 m thick). These waterproof liners and impermeable sections were added to force the groundwater down and through the iron wall. Finally, 1.2 m of clean sand was added up to ground surface. The last section immediately following the biosparge zone was filled with pea gravel from the cement floor at 6.1 m bgs up to grade. Refer to the Alameda Installation Report (UW,1997c) for complete details on the design, construction, and installation of the remedial and control gates.

There were two types of wells installed: (1) 5-cm ID stainless steel, fully screened wells and (2) 0.3-cm ID stainless steel, three-point multilevel wells (Figure 5.8). For complete details on well installation, size of well, and survey information, refer to the Alameda Installation Report (UW, 1997c). The wells in the remedial gate were aligned in rows 1 through 8 (from east to west). In each row, there were three wells placed across the width of the gate (from north to south). There were either fully screened wells (P wells) or multilevel piezometers (T wells) or a combination of the P and T wells in each row. The position of the three wells across the width of the gate were designated A, B, or C, with A being the most southern location and C being the most northern location (refer to Figure 5.8).

The explanation for the well identification is as follows:

R1PA and R5TC-3.7

where: R1 or R5= Row #
P = Performance well (fully screened well)
T = Treatment well (multilevel piezometer)
A or C = Location across the width of the gate
3.7 = Depth of piezometer (m bgs)

5.3.3 Operational Phases of the Experiment

The field demonstration was conducted in two operational phases, OP1 and OP2, which were based on different groundwater flow rates: a higher flow rate, OP1 (1.27 m^3/day), and a lower flow rate, OP2 (0.34 m^3/day). Velocities within each of the remedial gate sections were calculated based on pumping rates and the specific porosity. The following equation was used to calculate the velocities:

$$V = \frac{Q}{A \times n} \tag{5.1}$$

where: Q = extraction rate (m^3/day)
A = cross-sectional area of the flow (m^2)
n = porosity

Table 5.3 presents the assumed properties and calculated groundwater velocities in each section of the remedial gate for both OP1 and OP2.

Table 5.3 Groundwater Velocities in the Remedial Gate for OP1 and OP2

Gate Section	Porosity	OP1 — Velocity (m/day)	OP2 — Velocity (m/day)
Sand/Iron	0.3	0.38	0.10
Granular Iron	0.5[a]	0.23	0.06
Biosparge Zone	0.9[b]	0.13	0.03
Pea Gravel	0.4[c]	0.29	0.08

[a] ETI, personal communication.

[b] Assumed.

[c] Katic, personal communication.

5.3.3.1 *Operating Phase 1 (OP1)*

Once the upgradient sheet piling had been removed, the extraction pumps were set to operate at a rate of 1.27 m^3/day. Continuous pumping was started on February 3, 1997 and continued to April 21, 1997. Throughout OP1, wells in both gates were monitored for organic, inorganics, field parameters, BOD, and water level measurements (refer to Appendix 26 for complete details on sampling protocols). Also, O_2 and CO_2 were periodically sparged into the biosparge zone to study whether the DO and pH levels could be controlled (see Appendix 28 for sparging details). On April 21, 1997, OP1 was completed and the pumping rate was decreased to 0.34 m^3/day. Approximately 11.6 pore volumes had been drawn through the granular iron zone and 3.7 pore volumes had been extracted from the total gate.

5.3.3.2 *Operating Phase 2 (OP2)*

Operating Phase 2 began on April 21, 1997 by reducing the groundwater flow rate through the remedial gate to 0.34 m^3/day. This flow rate represents a slower flow rate than OP1 but is still somewhat above the average ambient flow rate across the site. The treatment and control gates were sampled for the same analytes established in OP1, but at a reduced frequency consistent with the slower groundwater velocity. Although the addition of oxygen and carbon dioxide was automated by flow timers, the preferred sparging pattern was not established until mid-August. The automation allowed the biosparge zone to be regularly sparged, creating the required conditions for aerobic biodegradation. Complete details on how this sparging protocol was developed can be found in Appendix 28.

Once the sparging protocol was established, an intensive sampling plan for the evaluation of the aerobic treatment system was developed and these data were used to calculate the degradation rates observed in the biosparge zone (refer to Appendix 29 for complete details). The amount of contaminant loss via volatilization during oxygen sparging was determined by sampling the headspace gas on three occasions.

By December 31, 1997 approximately 10.1 pore volumes had passed through the granular iron and 3.1 pore volumes had passed through the entire gate. Discussions continued at the onset of OP3 for pumping the remedial gate under a different flow rate, which would be included as part of the Remedial Investigation/Feasibility Study ongoing at the Alameda, NAS.

5.3.4 Media Sampled

5.3.4.1 Air

The local regulatory body, Bay Area Air Quality Management District (BAAQMD), had set limits on certain contaminant releases: 0.007 lb/day of VC and 0.02 lb/day of benzene. As it was unlikely that this project would exceed these limits, the demonstration project was exempt from strict regulatory monitoring requirements. The biosparge zone was designed to aerobically degrade volatilized contaminants in the headspace. The hypothesis was that all the volatilized contaminants would degrade in the headspace and not escape to the atmosphere. However, it was anticipated that during a sparge event some volatile contaminants might escape, hence an air sampling protocol was adopted. Also, off gas from the biosparge zone was directed through an activated carbon unit to trap any escaping VOCs.

All air sampling was done in the headspace of the biosparge zone using a Summa canister. Sampling procedures for headspace analysis can be found in Appendix 26. All air sampling events are listed in Appendix 26, Table A26.3.

5.3.4.2 Groundwater

Groundwater sampling for OP1 and OP2 was divided into the following categories:

1. Background sampling (during OP1 only)
2. Performance sampling
3. Treatment sampling
4. Biosparge sampling
5. Tracer test sampling

A brief description outlining the rationale for each of the listed groundwater sampling categories can be found in Appendix 26 along with the specific sampling methods. Quality assurance/quality control practices are described in Appendix 5, and analytical methods can be found in Appendix 3. All groundwater sampling events for OP1 and OP2 are listed in Appendix 26.

5.3.5 Description of the Chemical Data

Chemical data gathered in this project were classified as "Performance," P, or "Treatment," T. The type of chemical data collected was consistent throughout the two operational phases of the experiment.

Performance data were used to assess the overall performance of the system with respect to the remedial objectives, emphasizing the fate of target organics and degradation products, and to study any geochemical changes of the influent groundwater that occurred due to the two remedial technologies. These data represented the vertically integrated concentrations from fully screened wells. Chemical data included target organics, degradation products, metals, anions, alkalinity, and field parameters such as DO and pH. Sampling methods for Performance chemical data are given in Appendix 26. The locations of all the Performance wells are shown in Figure 5.8.

Treatment data were used to assess the specific processes operating within the granular iron wall and the biosparge unit. Chemical data included target organics, degradation products, dissolved hydrocarbon gases, and field parameters such as DO and pH. Monitoring was conducted in three-point, multilevel wells from row 1, upgradient of the gates, to row 6 in the biosparge zone. In rows 7 and 8, downgradient of the biosparge zone, only fully screened wells were installed, and so treatment data were obtained from these. Sampling methods for treatment chemical data are summarized in Appendix 26. The locations of the treatment monitoring wells are shown Figure 5.8.

5.3.6 Results

5.3.6.1 Background Sampling

Background sampling occurred on January 23, 1997, approximately 5 weeks after the upgradient sheet piles had been removed but prior to the activation of extraction wells. All the fully screened wells in the remedial gate were sampled and samples were analyzed at Intertek Testing Services (San Jose, CA) for the target organics and degradation products. Only VC, *c*DCE, and toluene were reported. TCE, 1,1-DCE, *t*DCE, benzene, ethylbenzene and xylenes (*p-/m-* and *o-*) concentrations were all below the method detection limits (Appendix 5). A summary of the analytical results is listed in Table 5.4.

Table 5.4 Analytical Results of the Background Sampling Event, January 23, 1997

Well	Vinyl Choride (μg/L)	*c*DCE (μg/L)	Toluene (μg/L)
R1PA	5600	56000	< 2500
R1PB	5600	34000	1400
R1PC	10000	35000	1600
R5PA	160	890	< 50
R5PB	180	950	< 25
R5PC	180	1200	< 50
R7PA	510	2700	< 100
R7PB	570	3000	< 120
R7PC	550	3000	< 100
R8PA	440	2400	< 100
R8PB	450	2600	< 100

Note: < preceding a number indicates the method detection limit.

Groundwater from row 1 wells R1PA, R1PB, and R1PC exhibited high levels of contamination and was regarded as representative of influent groundwater composition. However, groundwater from wells in rows 5, 7 and 8, which are downgradient of the granular iron wall, also contained *c*DCE and VC, above the ROs and MCLs.

5.3.6.2 Water Level Measurements

Figure 5.9 shows the water table fluctuations over time in the remedial gate during OP1 and OP2. The imposed hydraulic gradient in the remedial gate during OP1 (flow of 1.27 m^3/day) averaged between 0.03 and 0.04. During OP2 (flow of 0.34 m^3/day), the induced hydraulic gradient averaged between 0.01 and 0.02. As an artificial gradient was imposed, the resultant groundwater velocities in each of the gate sections were primarily dependent on the flow rate and porosity values.

5.3.6.3 Groundwater Velocities in the Remedial Gate

During OP2, velocities through selected sections in the remedial gate were estimated using two different methods, a tracer test and two In Situ Permeable Flow Sensor (ISPFS) probes. The results of the tracer test will be discussed first, followed by the ISPFS results, and then a comparison of the two tests with the velocity inferred from the pumping rates will be presented.

Tracer Test

A tracer test was conducted in the remedial gate from May 7 through June 16, 1997, during OP2, to assess groundwater velocities throughout the various reactive media and to define any preferential flow pathways. Potassium bromide was used as the conservative tracer. The tracer was

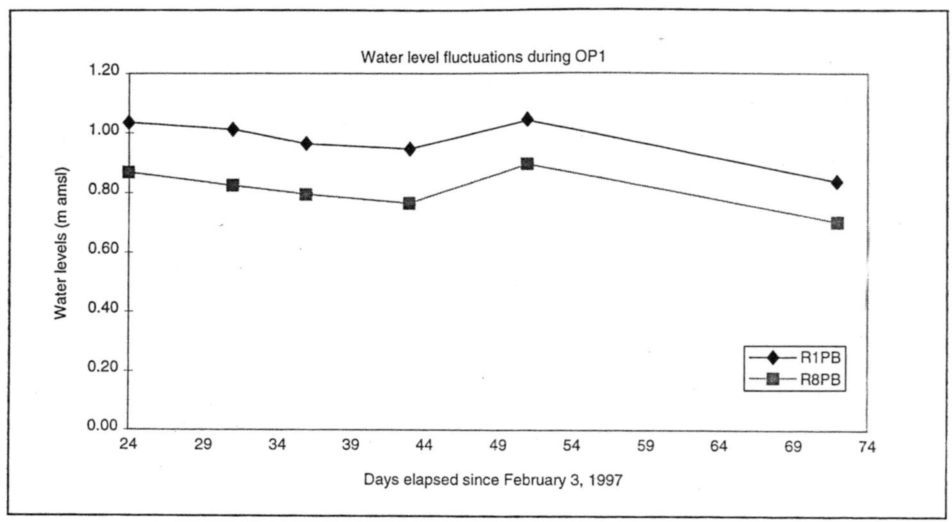

OP1: Pumps were turned on February 3, 1997, extracting at 1.27 m^3/day

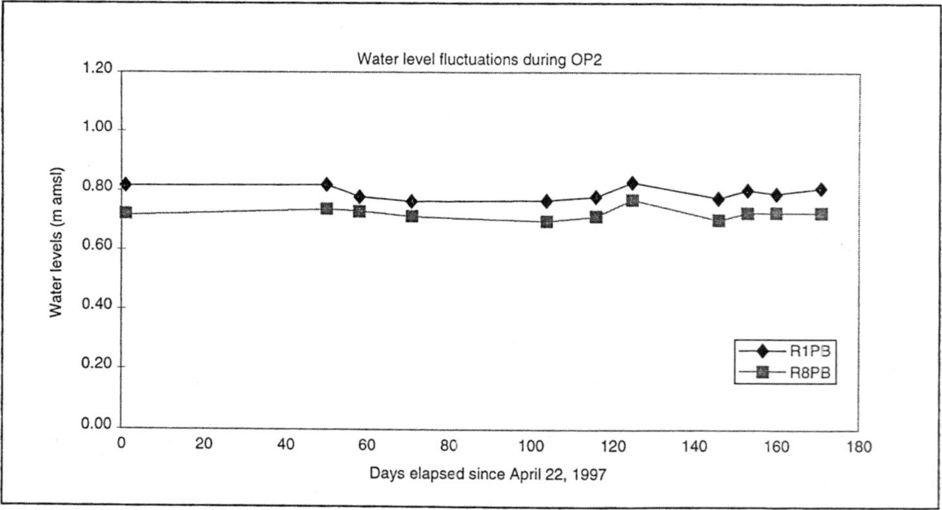

OP2: Flow rate was reduced on April 22, 1997 to 0.34 m^3/day

Figure 5.9 Water level fluctuations over time at R1PB and R8PB in the remedial gate.

injected in R1PB and drawn toward R1PC, which was operated as an extraction well. Refer to Appendix 27 for complete details. Table 5.5 lists the calculated velocities at the various sampling points obtained from the tracer test.

Porosity differences in the various gate materials (i.e., sand/iron, granular iron, and pea gravel) may explain the differences in the average velocities calculated for the different sampling points. Therefore, velocities calculated for the R1TB sampling points may reflect an average velocity through native materials immediately upgradient of the gate, while R2TB sampling point velocities would be an average of native and sand/iron velocities. Velocity estimates for R3TB sampling points would be an average of velocities from the native material, sand/iron, and the granular iron velocity. The bromide tracer velocity also varied with depth. For example, higher concentrations in R1TB arrived at approximately 200 and 400 hours at 3.7 and 4.9 m bgs, respectively. As the

Table 5.5 Average Groundwater Velocities Along Centerline (Column B wells) in the Remedial Gate as Determined by a Tracer Test

Sampling Point	Distance from Injection Well (m)	Location of Sampling Point	Travel Time (days)	Average Groundwater Velocity (m/day)
R1TB-2.8	0.60	Native Aquifer	3	0.24
R1TB-3.7	0.60	Native Aquifer	8	0.06
R1TB-4.9	0.60	Native Aquifer	18	0.03
R2TB-2.8	0.96	Sand/Iron	—	—
R2TB-3.7	0.96	Sand/Iron	24	0.03
R2TB-4.9	0.96	Sand/Iron	18	0.06
R3TB-2.8	1.60	Granular iron	—	—
R3TB-3.7	1.60	Granular iron	21	0.09
R3TB-4.9	1.60	Granular iron	25	0.06

Note: Multilevel sampling point blocked, no data available.

tracer progressed toward R2TB, higher concentrations arrived at approximately 575 and 400 hours at 3.7 and 4.9 m bgs, respectively. This could be the result of preferential flow within the gate materials, an artifact of the tracer injection, and/or the result of some degree of density driven flow.

In Situ Permeable Flow Sensors

In Situ Permeable Flow Sensors (ISPFS) directly measure 3D groundwater flow velocity in saturated, unconsolidated formations. Darcy velocities in the range of 3×10^{-7} cm/s to 3×10^{-4} cm/s can be accurately measured without the need to measure the hydraulic conductivity of the medium (Ballard, 1996). Two ISPFSs were installed within the remedial gate, one in the granular iron zone and the second in the first pea gravel zone, downgradient of the iron. Of the two ISPFS deployed, only the probe placed in the iron zone could measure groundwater flow velocity magnitude and direction. In the pea gravel zone, the relatively high permeability of the porous medium permitted the thermal gradients. Heat, produced by the ISPFS, produced significant convection gradients that were reported by the probe as velocity gradients. An erroneously high vertical gradient was reported from the probe; hence the groundwater flow velocity could not be measured in this zone. Consequently, only the results from the ISPFS installed in the iron zone are discussed here.

Horizontal groundwater velocity (measured within the iron zone, after May 9, 1997) using the ISPFS (0.09 to 0.14 m/day) was similar to that calculated using gate dimensions, estimated iron porosity, and pumping rates during the same time period (0.06 m/day; refer to Table 5.3 for OP2).

Vertical groundwater flow velocity, as measured by the iron zone ISPFS (0.45 to 0.60 m/day upwards), was higher than that anticipated for groundwater moving through the iron portion of the gate. Although vertical flow within this zone cannot be ruled out, it is likely that the velocity measured by the ISPFS was erroneously high as a result of convection produced by the probe heater.

Groundwater within the iron zone of the gate was indicated to be moving at approximately 25 degrees north, relative to the longitudinal axis of the gate. Although it was anticipated that groundwater would flow parallel to the longitudinal gate axis, it is possible that variations in materials near the iron zone ISPFS resulted in a preferential flow direction at an angle relative to the long axis.

Comparison of the Velocity Measurements

Groundwater velocity through the remedial gate during OP2 was estimated using three different techniques (i.e., a tracer test, ISPFSs, and based on the pumping rate). Tracer velocity estimates, based on tracer breakthrough from sampling points, averaged 0.04 m/day in the sand/iron section and 0.07 m/day in the granular iron section. Groundwater velocities based on the pumping rate were 0.10 m/day in the sand/iron section and 0.06 m/day in the granular iron section. Velocities in the granular iron, at least, were comparable. According to the ISPFSs, the horizontal groundwater velocity within the granular iron zone ranged from approximately 0.09 to 0.14 m/day. While this is in reasonable agreement with the velocities estimated by the tracer tests and pumping rates, the anomalous horizontal flow direction and the vertical flow indicated by the ISPFS results suggest the ISPFS was not providing reliable groundwater velocity results. The agreement of the ISPFS velocity with that produced by the other techniques is considered fortuitous. The groundwater velocity derived from the pumping rates will be used in subsequent considerations.

5.3.6.4 Field Parameters

Figures 5.10 and 5.11 illustrate typical "snapshots" of the pH and DO profiles through the remedial gate during OP1 and OP2, respectively. Data measured from the multilevels in column B (rows 1, 2, 3, 4, 5, and 6) were arithmetically averaged over the three depths as there was little depth discrete variation. pH and DO measurements were also taken from the fully screened wells (rows 7 and 8) along column B (refer to Figure 5.8 for location of column B).

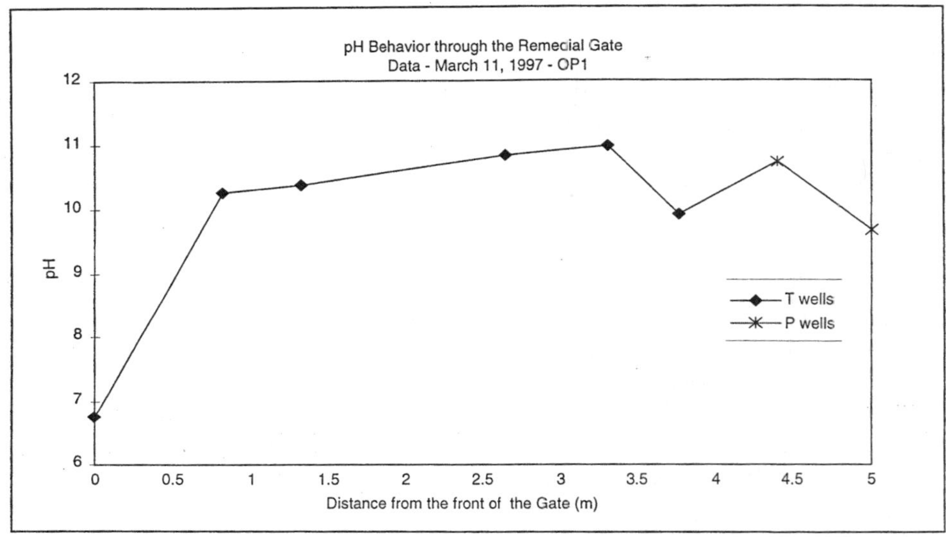

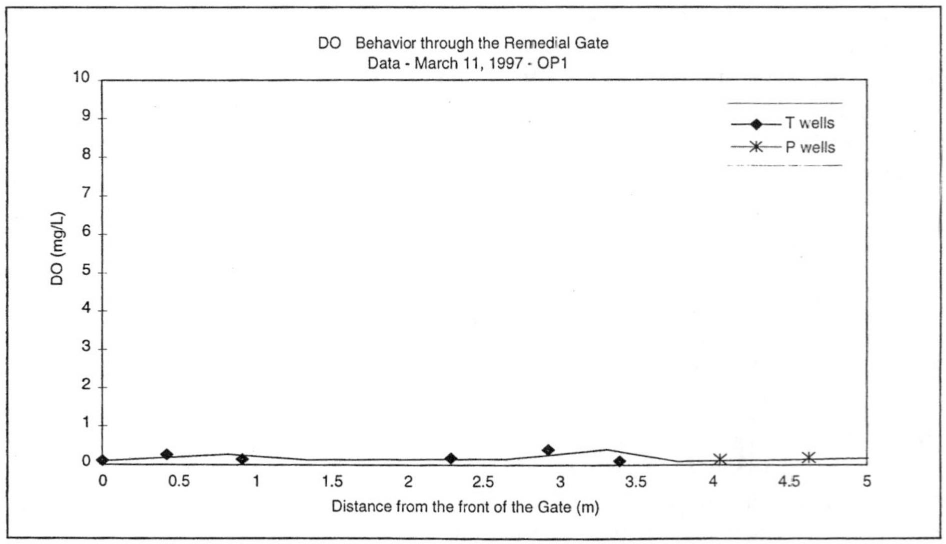

Figure 5.10 pH and dissolved oxygen (DO) through the remedial gate during OP1.

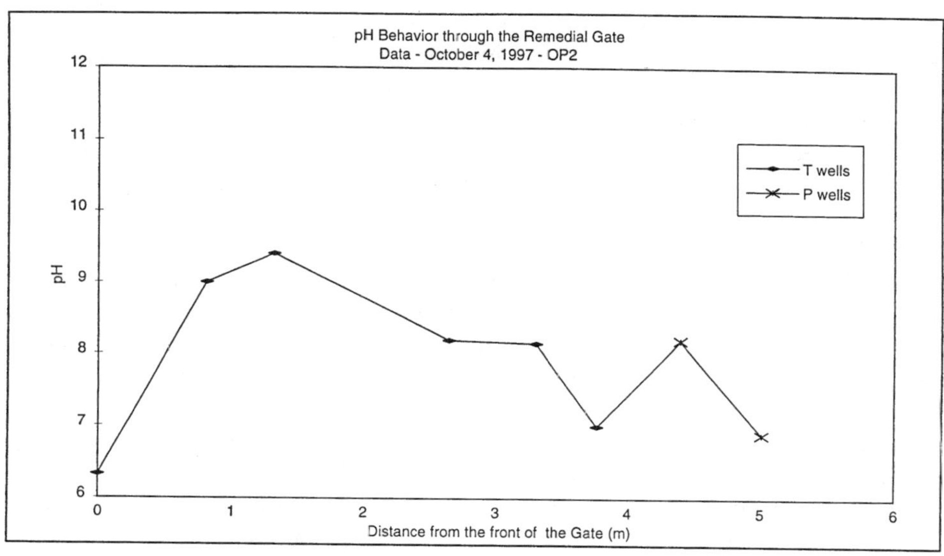

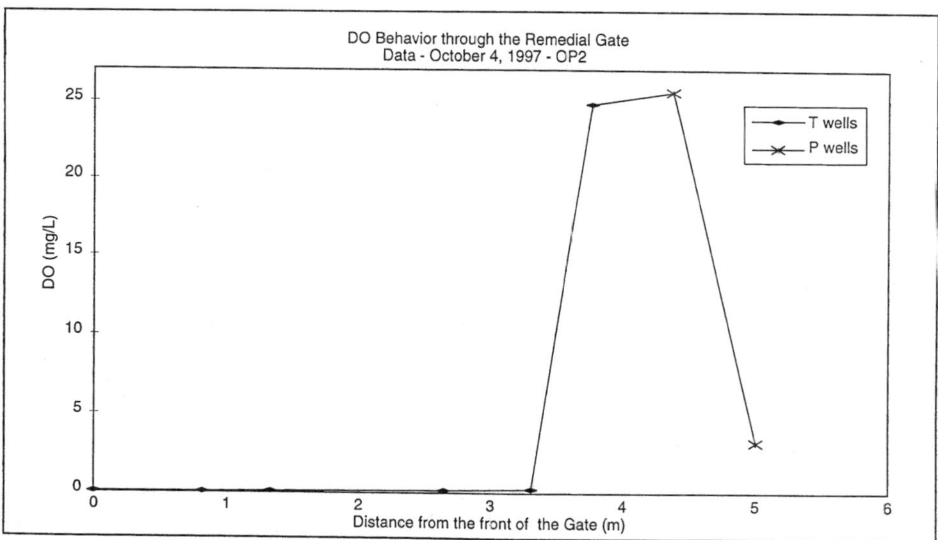

Figure 5.11 pH and dissolved oxygen (DO) through the remedial gate during OP2.

The natural pH of the groundwater, as measured in row 1, was, on average, 6.5 for both OP1 and OP2. As the groundwater passed through the sand/iron section, the pH increased to between 8 and 10, and once downgradient of the granular iron wall, the pH ranged between 8 and 11. During OP1, when the biosparge gate was not regularly sparged with carbon dioxide, the pH remained high throughout the entire gate length (Figure 5.10[top]). During OP2 (Figure 5.11[top]), carbon dioxide was regularly sparged into the biosparge zone and this successfully decreased the pH of the water in the biosparge zone to levels around 6 to 8. However, in general, the pH levels in the remedial gate downgradient of the iron zone were lower in OP2 than in OP1. The cause of this decline in pH was not resolved, but it is possible that the carbon dioxide was diffusing upgradient into the pea gravel and neutralizing the groundwater as far as row 4. This is reasonable as there was detectable DO in row 4 (up to 5.4 mg/L in R4TC-4.9 as of January 1998), which may explain the loss of BTEX compounds from rows 4 to 5.

The DO in the influent groundwater (row 1) for OP1 and OP2 was close to the MDL. During OP1, with no regular oxygen additions into the biosparge zone, the DO levels remained close to the MDL throughout the entire length of the remedial gate (Figure 5.10[bottom]). During OP2, with regular additions of oxygen, the DO levels in and downgradient of the biosparge zone (rows 6, 7, and 8) increased (Figure 5.11[bottom]). The DO levels were judged sufficient to support aerobic biodegradation in and downgradient of the biosparge zone after about July 1997.

5.3.6.5 Inorganic Data

Figures 5.12 and 5.13 illustrate two snapshots of cations, alkalinity, and anions throughout the remedial gate sampled during OP1 (April 18, 1997) and OP2 (September 30, 1997), respectively. With respect to the cations, it was evident that calcium, potassium, iron, magnesium, and manganese concentrations decreased along the length of the gate, whereas sodium increased along the same path. The decrease in concentration of these species was probably due to the formation of precipitates and/or cation exchange. The increase in sodium concentration may be due to a cation exchange phenomenon, i.e., sodium being exchanged with iron or calcium, causing an increase in aqueous concentrations of sodium and a subsequent decrease in the aqueous concentrations of iron and calcium. Alkalinity also decreased throughout the gate, probably due to the formation of carbonate precipitates. These results are slightly different than what was observed in the bench-scale iron treatability study where little change in sodium, magnesium, and potassium concentrations resulted from contact with the granular iron, although a decrease in manganese was observed. Concentrations of both calcium and alkalinity decreased with passage through the column, consistent with the field observations.

With respect to the anions, Figures 5.12 and 5.13 show that chloride increased from Row 1 to Row 5 (i.e., before and after the iron wall). This increase was probably due to the production of chloride ions from the reductive dechlorination of the chlorinated ethenes. Note, however, that the groundwater was slightly saline, and chloride concentrations in row 1 fluctuated over time. During OP1 and OP2, an increase in sulfate concentrations was observed through the gate. Bromide concentrations did not change during OP1 but increased for OP2. This increase in bromide concentration might be a reflection of a variable, saline environment and/or excess concentrations due to the tracer experiment. Nitrite, nitrate, and phosphate influent concentrations were consistently below method detection levels (0.5, 1.50, and 2.50 mg/L, respectively).

Table 5.6 lists the influent and effluent concentrations of the major inorganic species for both the treatability study and the remedial gate during the snapshots in OP1 and OP2 (Figures 5.12 and 5.13). Two separate sets of inorganic data from the column test are presented. Nitrate, nitrite, and phosphate were not included in the table, as these concentrations were all below method detection limits in both the remedial gate and the column study. Bromide was not included, as it was not analyzed in the column study.

To summarize, the alkalinity and most of the cation concentrations decreased through the gate with the exception of sodium, which increased through the gate. The decrease in cation concentrations may be due to carbonate precipitation, i.e., $CaCO_3$ initiated by the high pH waters and the increase in sodium may be a result of a cation exchange process that may have been adding sodium to the groundwater. No increase in dissolved iron was observed even though dissolved iron was potentially introduced into solution by corrosion of the granular iron. The reduction in dissolved iron concentrations through the gate suggests formation of iron carbonates, iron hydroxides, and/or oxyhydroxides. Similar inorganic trends, except for the increase in sodium, were observed in Gate 1 at the Borden field experiment. In Gate 1, the concentration of sodium did not seem to change throughout the length of the gate, suggesting that a unique geochemical process was occurring at Alameda that was adding sodium to the groundwater.

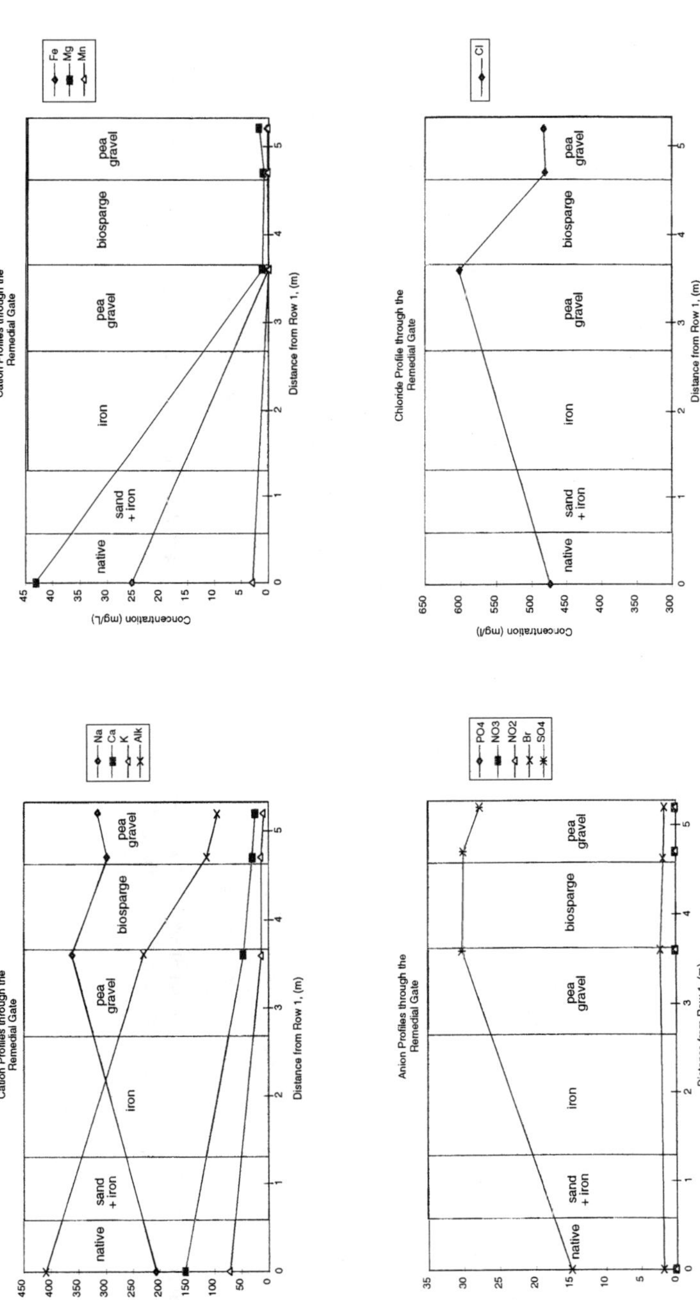

Figure 5.12 Cation, anion, and alkalinity profiles through the remedial gate sampled on April 18 1997. Data, collected from Performance wells, represent column B.

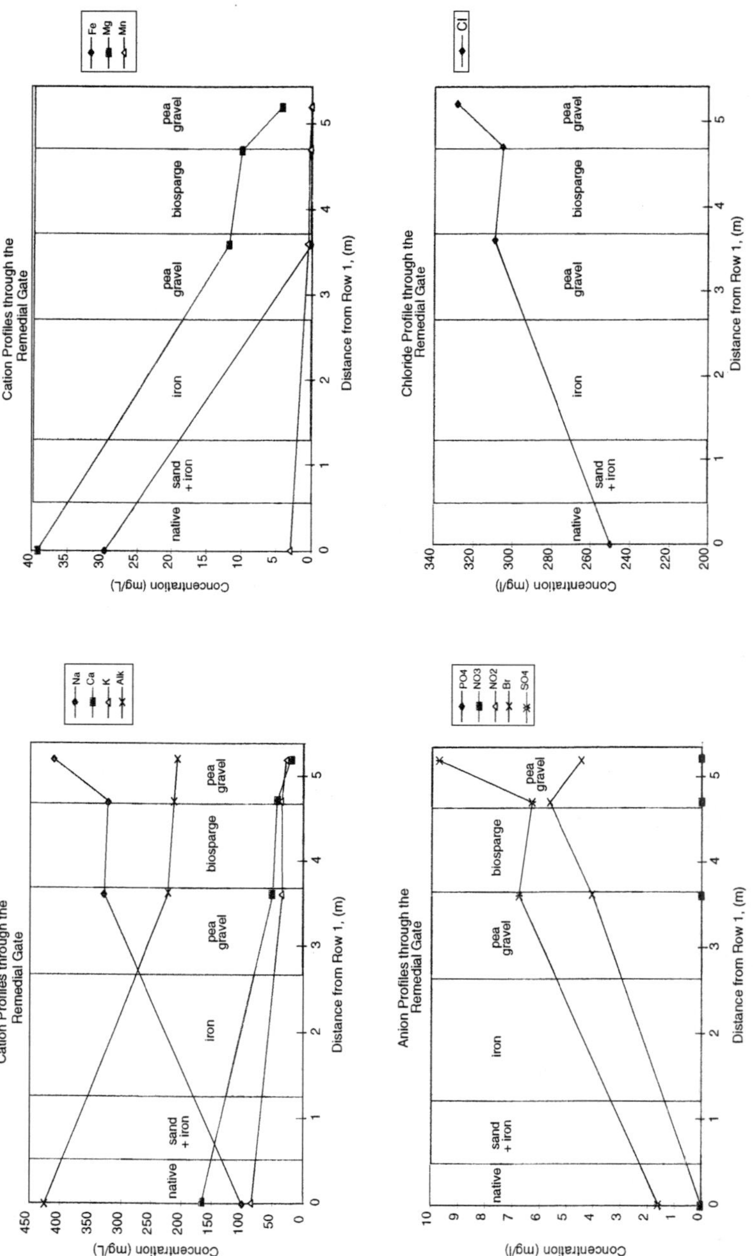

Figure 5.13 Cation, anion, and alkalinity profiles through the remedial gate sampled on September 30 1997. Data, collected from the Performance wells, represent column B.

Table 5.6 Comparison of the Inorganic Concentrations from the Iron Treatability Study and the Remedial Gate for OP1 and OP2

Inorganic Parameter	Iron T.S.[a] Influent	Iron T.S.[a] Effluent	OP1 Influent R1PB	OP1 Effluent R8PB	OP2 Influent R1PB	OP2 Effluent R8PB
Iron	5.4	0.6	26	0.08	29.7	1.17
	2.1	0.3				
Sodium	111	109	275	494	101	412
	110	108				
Magnesium	30	25	42.6	0.35	39.2	4.31
	29	25				
Calcium	142	18	139	70.8	167	22.1
	115	19				
Potassium	27	27	76.7	18.4	84.5	29
	26	25				
Manganese	1.8	0.4	2.71	0.01	2.91	0.03
	1.8	0.3				
Chloride	223	233	528	536	250	328
	224	238				
Sulfate	3	ND	14.9	27.9	1.59	9.70
	3	ND				
Alkalinity	371	156	424	218	425	209
	368	162				

Notes: ND means not detected, all units are expressed in mg/L.

[a] Results from the Treatability Study (T.S.) are from 2 samples.

5.3.6.6 Organic Data

Although OP1 data were initially examined, it was concluded that the pumping rate was too high for complete reductive dechlorination to occur (recall also that aerobic biodegradation in the biosparge zone was not attempted or studied during OP1). Therefore, only OP2 data will be analyzed and discussed. Treatment of the organics within the iron segment of the gate will be discussed first, then analysis of data in the biosparge zone will be presented.

Granular Iron Zone

Figure 5.14 illustrates the VC concentrations (sampled from fully screened wells) throughout the remedial gate on June 16, 1997, July 31, 1997, and September 30, 1997 (all during OP2). Performance data were analyzed and plotted to represent the three columns of the gate, A, B and C (refer to Figure 5.8 for locations of columns). A decrease in the influent VC concentrations through reductive dechlorination in the granular iron wall is illustrated from the first two points on Figure 5.14.

The influent VC concentrations were relatively steady over time, whereas the concentrations detected downgradient of the granular iron wall decreased over time (Figure 5.14). This trend was also observed with *c*DCE over the same sampling events (refer to Appendix 30 for a figure illustrating this trend). The influent concentrations for the other chlorinated ethenes (PCE, TCE, 1,1-DCE, and *t*DCE) were not as steady. Effluent concentrations downgradient of the granular iron wall were all below method detection limits (refer to Appendix 30 for figures illustrating the behavior of the other chlorinated ethenes).

The above approach examined vertically averaged concentrations from fully screened wells through the remedial gate for a specific snapshot in time. Figure 5.14 compares average concentrations from a volume of water at one well to average concentrations further downgradient at a different well. Although uncertainty exists, general degradation trends can be inferred from these data, assuming steady-state conditions exist. However, this uncertainty is even greater when considering the multilevel well data, because a very specific flow path from one point to the next must be assumed. This is not a reasonable assumption, so the treatment data were analyzed by calculating

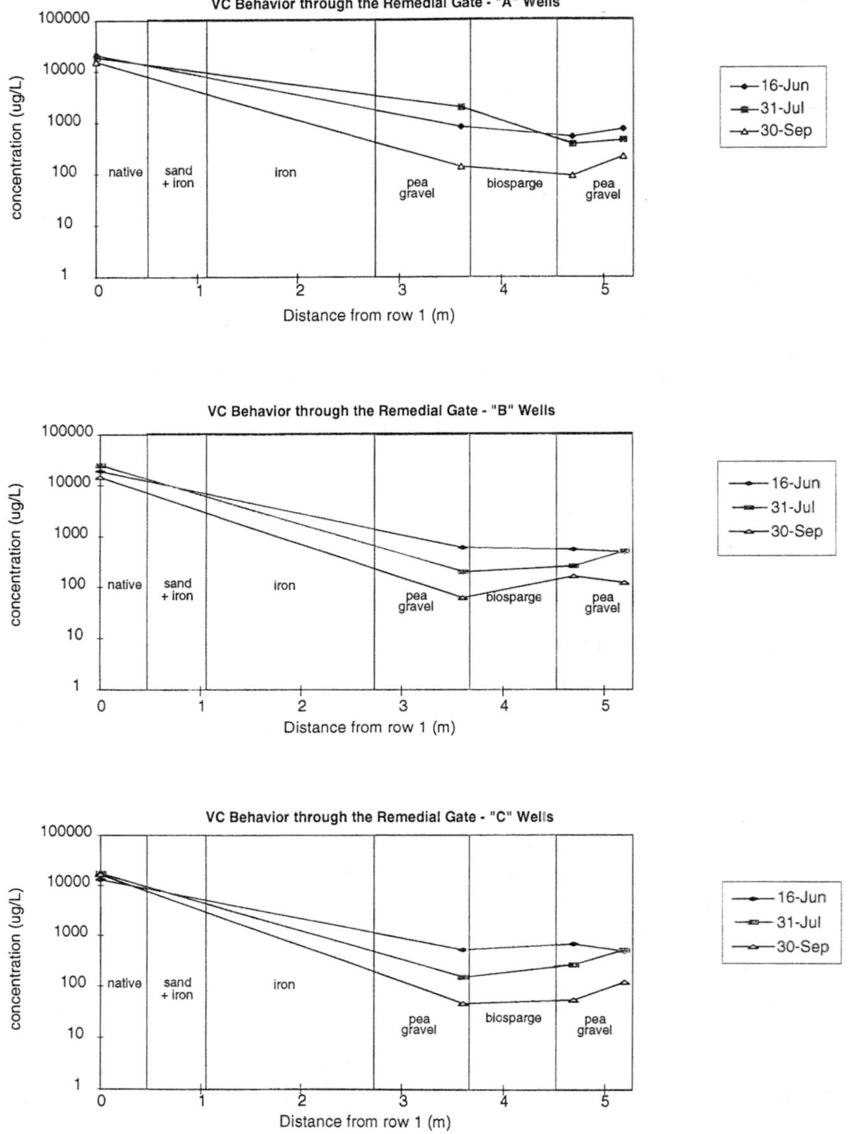

Figure 5.14 VC behavior through the remedial gate, sampled on June 16, July 31 and September 30 1997. Data represents fully screened wells from Columns A, B, and C.

the total mass that passed rows 1, 3, and 4 over a 5-month time period. The total mass passing rows 1 and 3 over the 4-month period from April 1997 to July 1997 was compared to the total mass passing row 4 over the 4-month period from May 1997 to August 1997. The time offset of 1 month accounted for the approximate travel time between rows 1 and 4 (note: there was no sampling completed in August, but when July 31, 1997 data were compared to September 30, 1997 data, they were similar in magnitude, hence assuming similar values for August was felt reasonable). The total mass passing a row of multilevel wells over a period of time was calculated using the FORTRAN program, FENCE (refer to Appendix 24 for complete details on the calculations). A decrease in the integrated mass at row 4 compared to the integrated mass at row 1 was assumed to be due to degradation that occurred within the granular iron section, between rows 1 and 4.

Degradation rates were calculated for TCE, *c*DCE, *t*DCE, and 1,1-DCE using an analytical solution to the chain decay kinetic equation, assuming first-order kinetics applied. A satisfactory analytical solution for VC could not be obtained, so a numerical solution was used to solve for VC degradation rate. Analytical equations have the advantage of providing exact solutions. Numerical solutions are based on approximations and are not exact. In order to gain confidence in the rate constants estimated from the numerical solution, in particular that for VC, the time increment (Δt) was reduced until the rate constants for TCE and the DCE isomers estimated numerically were within about 2.5% of those obtained analytically (refer to Appendix 29 for the equations and assumptions). These degradation rates were calculated by using the solver option in Excel 5.0 and allowing the calculated masses to be fit to the FENCE-generated masses. There were only three data points representing the three FENCE-generated masses from rows 1, 3, and 4. The first data point represents row 1, second data point represents row 3, and the last data point represents row 4. These data points were plotted and then the degradation rate constants were calculated by using the analytical and numerical solutions.

The data points being fit did not generally lie on an ideal first-order curve. So a range of rate constants was obtained by solving the analytical and numerical solutions in two ways. The first method involved fitting the solution to the first and last data points, ignoring the middle data points. Since the calculated masses at row 3 were lower than the masses at row 4, this method calculated rate constants that were the minimum of the range, and hence the most conservative (Figures 5.15a and 5.15b for the analytical and numerical solution, respectively). The second method involved fitting all three data points, a procedure that was thought to provide a more realistic estimate of the rate constants, since all the data were considered (Figures 5.16a and 5.16b for the analytical and numerical solution, respectively). Table 5.7 lists the degradation rates and half-lives calculated from the two different fitting methods for both the analytical and numerical solutions.

Table 5.7 Degradation Rate Constants and Half-Lives Calculated from the Analytical and Numerical Solutions

	Analytical		Numerical	
Contaminant	**Fitting first and last point**	**Fitting all three points**	**Fitting first and last point**	**Fitting all three points**
TCE				
Degradation rate Constant (1/day)	0.231	0.599	0.238	0.614
Half-life (days)	3.0	1.1	2.9	1.1
***c*DCE**				
Degradation rate Constant (1/day)	0.097	0.255	0.098	0.238
Half-life (days)	7.1	2.7	7.1	2.9
***t*DCE**				
Degradation rate Constant (1/day)	0.151	0.331	0.152	0.335
Half-life (days)	4.6	2.1	4.5	2.1
1,1-DCE				
Degradation rate Constant (1/day)	0.129	0.315	0.130	0.319
Half-life (days)	5.3	2.2	5.3	2.2
VC				
Degradation rate Constant (1/day)	NA	NA	0.404	0.609
Half-life (days)	NA	NA	1.7	1.1

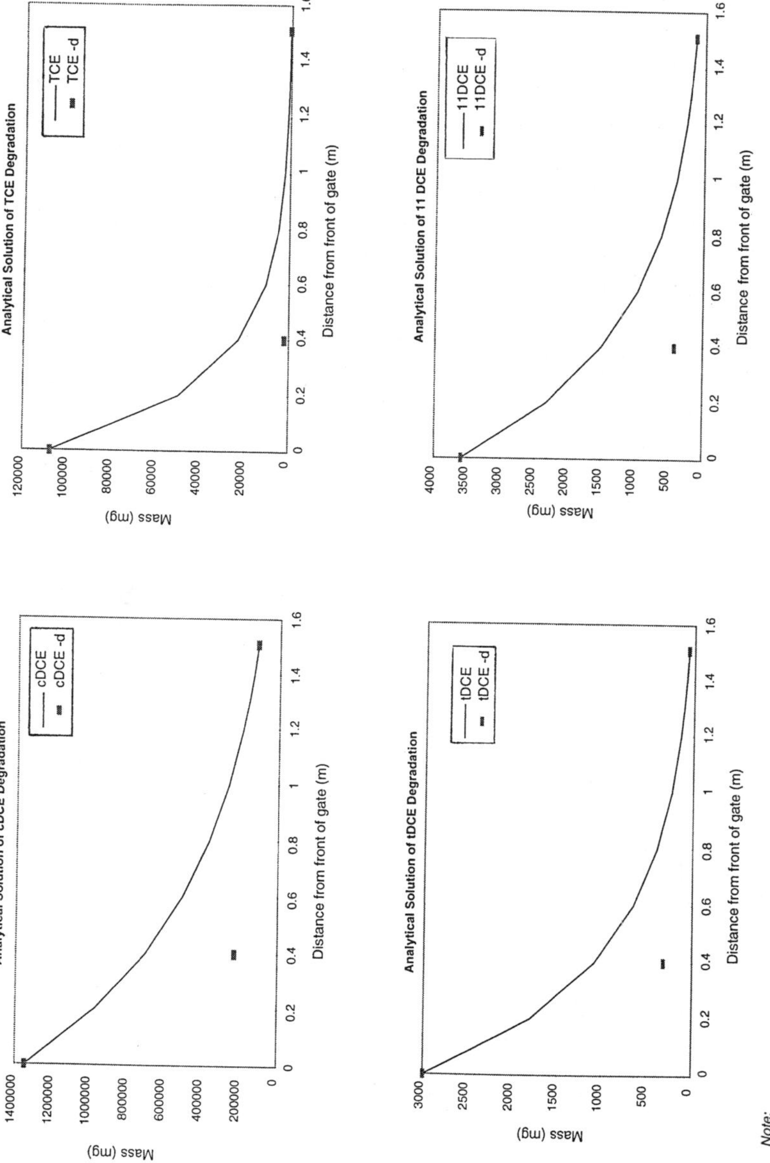

Note:
Plots represent fitting the first (Row 1) and last (Row 4) points only (middle (Row 3) points are shown but not fitted).

Figure 5.15a Analytical solution for the reductive dechlorination of the chlorinated ethenes.

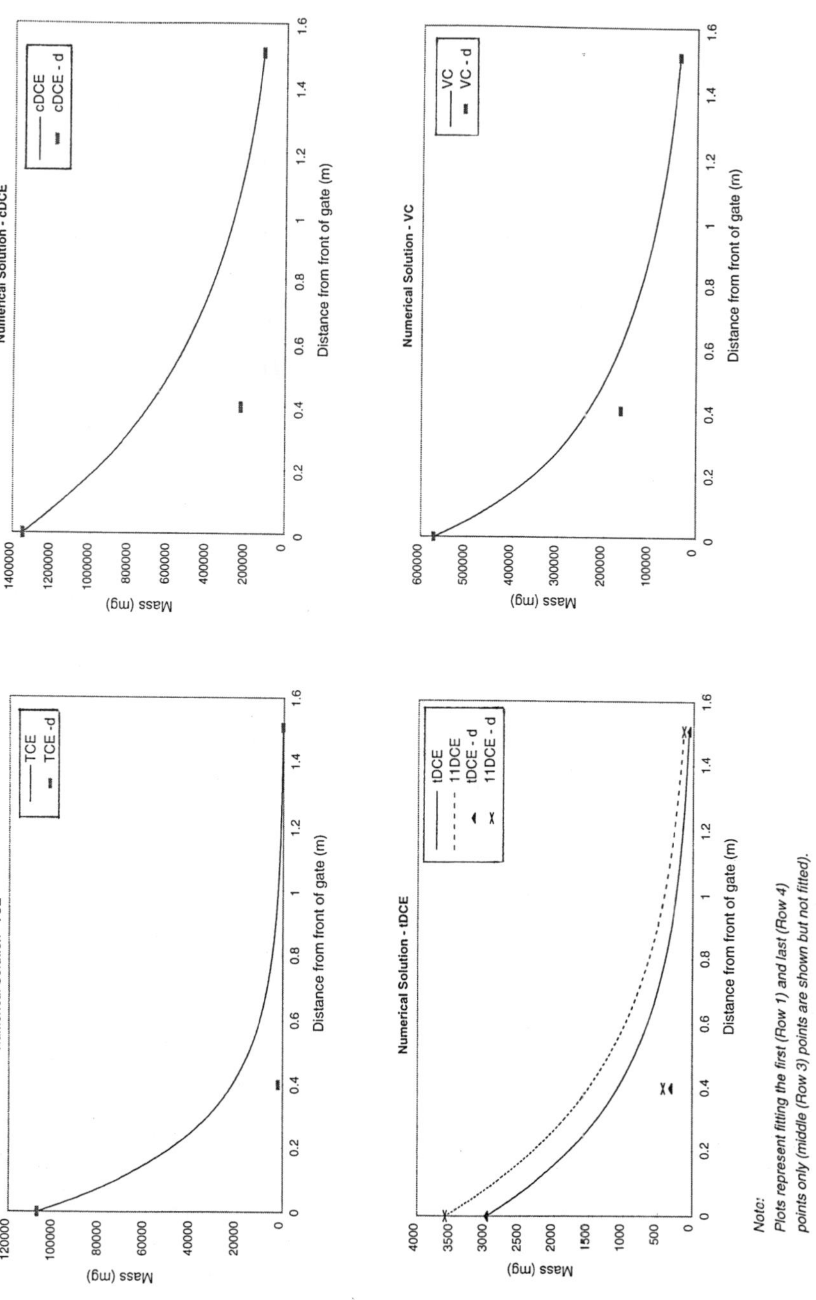

Note:
Plots represent fitting the first (Row 1) and last (Row 4) points only (middle (Row 3) points are shown but not fitted).

Figure 5.15b Numerical solution for the reductive dechlorination of the chlorinated ethenes.

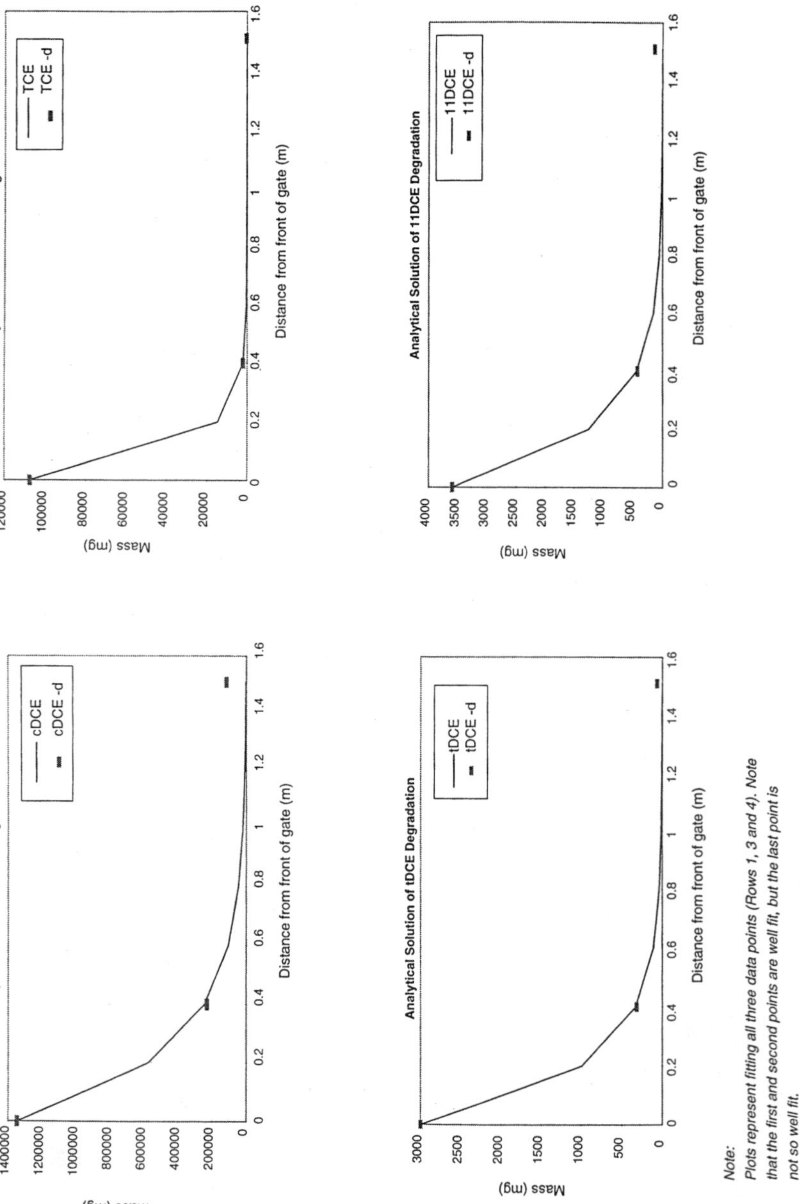

Note:
Plots represent fitting all three data points (Rows 1, 3 and 4). Note that the first and second points are well fit, but the last point is not so well fit.

Figure 5.16a Analytical solution for the reductive dechlorination of the chlorinated ethenes.

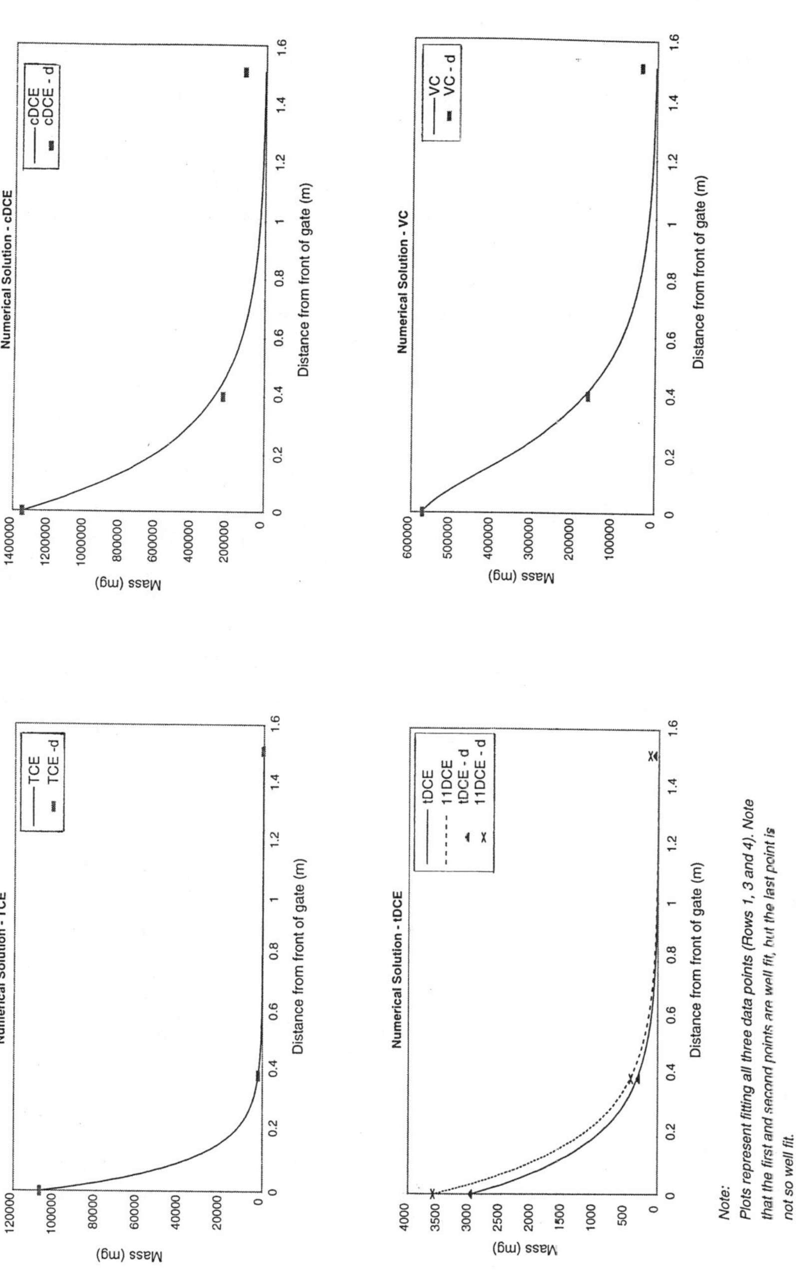

Note:
Plots represent fitting all three data points (Rows 1, 3 and 4). Note that the first and second points are well fit, but the last point is not so well fit.

Figure 5.16b Numerical solution for the reductive dechlorination of the chlorinated ethenes.

A possible explanation of the imperfect first-order behavior of the data is that "hot" spots within the plume that were not detected in row 1 could have penetrated the granular iron wall without complete degradation, and then been intersected by row 4 monitors. This would have resulted in apparent increases in contamination at row 4 and may have caused the degradation rates to seem slower than they actually were in the granular iron wall. Based on this possibility, it seems reasonable that fitting all three data points may represent a more realistic view of the performance of the iron wall.

Figure 5.17 illustrates the behavior of toluene through the remedial gate on June 16, 1997, July 31, 1997, and September 30, 1997. The influent concentrations vary from nondetectable up to 4 mg/L, and the effluent concentrations averaged between nondetectable and 20 μg/L, suggesting near-complete attenuation within the iron wall (recall the RO for toluene was 150 μg/L). As toluene is not subject to reduction by granular iron, the attenuation is probably due to sorption to the iron. A similar characteristic was observed in the iron of Gate 1 in CFB, Borden, Canada (see Chapter 2). Eventually, the toluene is expected to break through the granular iron, so the apparent sorption is not considered a useful long-term remediation process.

Appendix 30 contains figures illustrating the behavior of the other BTEX compounds. In general, the influent concentrations of benzene, ethylbenzene, and the xylenes varied, but the compounds were all below detection downgradient of the iron wall, suggesting that they are also subject to pronounced retardation within the granular iron medium.

Biosparge Zone

Preliminary observations can also be made about aerobic treatment based on the trends from Figure 5.14. Starting in mid-June 1997, a sparging protocol for oxygen and carbon dioxide was implemented (refer to Appendix 28 for complete details) that maintained suitable and consistent aerobic conditions that would enhance aerobic biodegradation of the BTEX compounds and any remaining chlorinated ethenes. BTEX was expected to biodegrade and, according to Davis et al. (1990), VC also degrades under aerobic conditions. In addition, volatilization of the VOCs was also expected to reduce their concentration as groundwater passed through the biosparge system. Hence, the July 31, 1997 and September 30, 1997 data might have been expected to show a decrease in VC concentrations between rows 5 and 8, bracketing the biosparge zone. There does not seem to have been such a decrease in concentration across the biosparge zone for any individual sampling event. However, there does seem to be a decrease in VOC concentrations in the biosparge zone over time (Figure 5.14).

If the biosparge zone were visualized as a mixed bioreactor, where the variable influent concentrations are homogenized, then it would seem reasonable to expect little variation in organic concentrations across the biosparge zone at any one time, but instead concentrations should decline, perhaps to a steady state, over time.

Modeling degradation through the biosparge zone was undertaken by presenting it as a large, completely mixed bioreactor that accounted for instantaneous partitioning of VOCs into the sparge bubbles, as well as storage in the unsaturated headspace. The equation was developed from the mass balance as follows: the mass in the bioreactor at any time, Mass_t, equals the mass that was present at the previous time step, Mass_{t-1}, plus whatever mass entered the reactor over the intervening time increment, Δt, due to advection from the granular iron zone, Mass_{in}, minus the mass that was removed from the reactor by the process of advection, Mass_{out}, volatilization, Mass_{vol}, and biodegradation Mass_{bio}.

$$\text{Mass}_t = \text{Mass}_{t-1} + \text{Mass}_{\text{in}} - \text{Mass}_{\text{out}} - \text{Mass}_{\text{bio}} - \text{Mass}_{\text{vol}} \tag{5.2}$$

The mass in the bioreactor at time t can be subdivided into mass in the headspace (Mass_{at}) and mass in the water (Mass_{wt}):

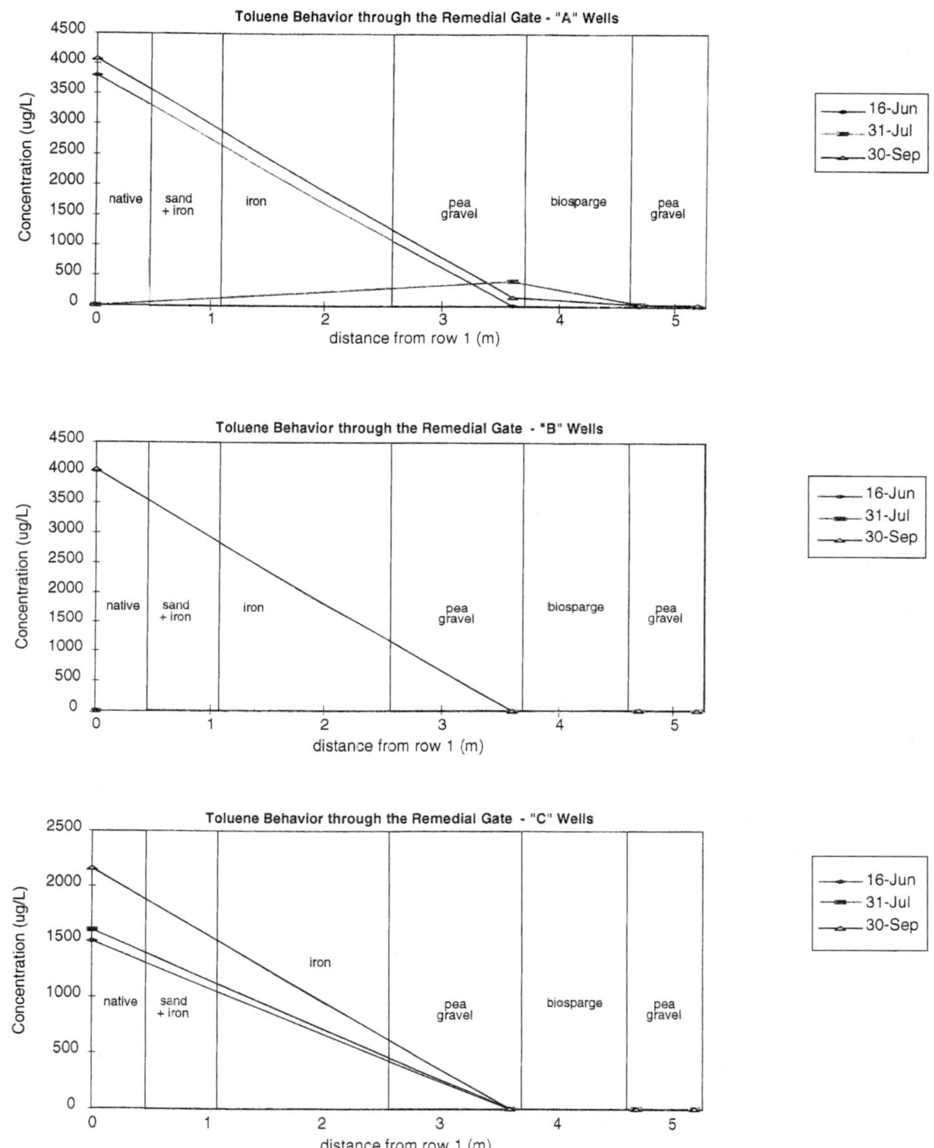

Figure 5.17 Toluene behavior through the remedial gate sampled on June 16, July 31, and September 30 1997. Data represents fully screened wells from columns A, B, and C.

$$\text{Mass}_{wt} + \text{Mass}_{at} = \text{Mass}_{wt-1} + \text{Mass}_{at-1} + \text{Mass}_{\text{in}} - \text{Mass}_{\text{out}} - \text{Mass}_{\text{bio}} - \text{Mass}_{\text{vol}} \quad (5.3)$$

Mass entering the system can be calculated from the influent concentration and the pumping rate times the time over which the pumping occurred:

$$\text{Mass}_{\text{in}} = C_{\text{in}} \times Q \times \Delta t \quad (5.4)$$

Mass leaving the system can be calculated from the pumping rate times the concentration in the bioreactor at the earlier time times the time interval:

$$\text{Mass}_{\text{out}} = C_{t-1} \times Q \times \Delta t \tag{5.5}$$

Mass in the system at any time is calculated from the concentrations of contaminants in the water and gas phase, and the volumes of those phases:

$$\text{Mass}_w = C_{wt} \times V_w \tag{5.6}$$

$$\text{Mass}_a = C_{at} \times V_a \tag{5.7}$$

Henry's Law provides a means of reducing the number of concentration variables, since

$$C_a = H \times C_w \tag{5.8}$$

Employing Henry's Law, we find

$$\text{Mass}_a = H \times C_w \times V_a \tag{5.9}$$

When sparging occurs, the mass stripped is the mass that partitions into the sparge bubbles. If this is assumed to occur instantaneously, then using Henry's Law to express all concentrations as water concentrations,

$$\text{Mass}_{\text{vol}} = Q_a \times H \times C_{wt-1} \times \Delta t \tag{5.10}$$

Biodegradation is assumed to occur by first-order kinetics; therefore, the rate of decline in concentration is equal to the rate constant times the concentration, or

$$\frac{\Delta C}{\Delta t} = -K_1 \times C \tag{5.11}$$

Rewriting this in terms of mass,

$$\Delta\text{Mass} = \text{Mass}_{\text{bio}} = -K_1 \times C_{wt-1} \times \Delta t \times V_w \tag{5.12}$$

Now, assembling the parts of the mass balance equation, we obtain

$$\begin{aligned}(C_{wt} \times V_w) + (H \times C_{wt} \times V_a) = {} & (C_{wt-1} \times V_w) + (H \times C_{wt-1} \times V_a) + (C_{in} \times Q \times \Delta t) \\ & - (C_{wt-1} \times Q \times \Delta t) - (Q_a \times H \times C_{wt-1} \times \Delta t) - (K_1 \times C_{wt-1} \times \Delta t \times V_w)\end{aligned} \tag{5.13}$$

From this equation, it can be shown that

$$\frac{dC}{dt} = \frac{Q \times C_{\text{in}}}{(V_w \times V_a \times H)} - \left(\frac{Q + K_1 \times V_w + Q_a \times H}{(V_w + V_a \times H)}\right) \times C \tag{5.14}$$

The differential equation, when solved, yields

$$C = \frac{Q_w \times C_{in}}{Q_w + K_1 \times V_w + Q_a \times H} + \left(C_o - \frac{Q_w \times C_{in}}{Q_w + K_1 V_w + Q_a \times H}\right) \exp\left(-\left(\frac{Q_w + K_1 \times V_w + Q_a \times H}{V_w + V_a \times H}\right)t\right) \quad (5.15)$$

The concentrations used were collected during the biosparge sampling events (Appendix 29 contains all the data for these calculations). These data were used because they represent groundwater that had been sufficiently aerated and with reasonable pH (7–8). Row 6 (biosparge zone) concentrations of *c*DCE and VC were calculated as area-weighted averages then plotted over time (Figures 5.18[top] and 5.18[bottom] for *c*DCE and VC, respectively). None of the BTEX compounds were analyzed because concentrations downgradient of the iron wall were at or below the LOQ.

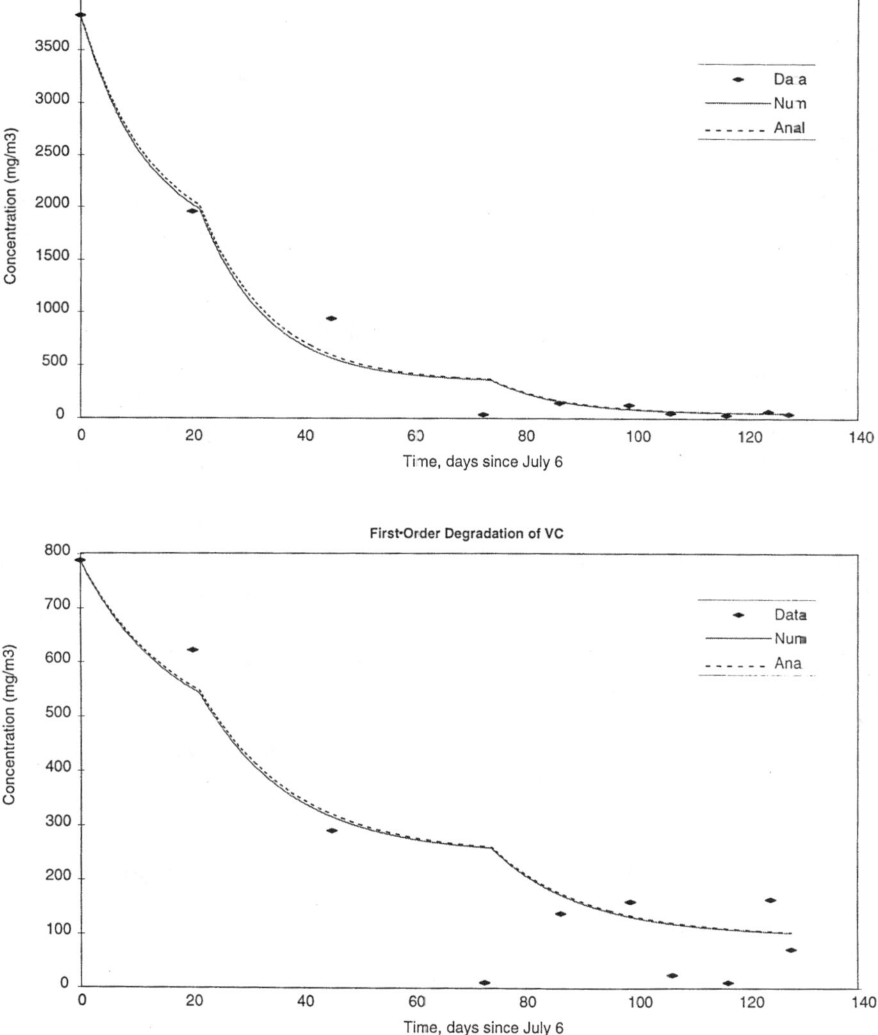

Figure 5.18 Analytical and numerical solution for the declining concentrations of *c*DCE and VC in the biosparge zone over time.

Mass removal via both biodegradation and volatilization can be described by first-order kinetics. Zero-order kinetics were also applied to biodegradation, but the data were poorly fit. Pseudo-first-order rate constants or biodegradation rates were calculated for TCE, *c*DCE, and VC using the analytical solution above (Equation 5.15) and the numerical solution (Equation 5.13). Refer to Appendix 29 for the equations used and the assumptions made for each of these equations. The rate constants were calculated using the solver option in Excel 5.0 as previously described (Section 5.3.6.6). The pseudo-first-order rate constants were calculated using an oxygen flow rate of 4.7 L/min (measured in the field) (0.564 m^3/day).

Figure 5.18 illustrates the analytical and numerical fit to the data for *c*DCE (Figure 5.18[top]) and VC (Figure 5.18[bottom]). The first-order degradation constant for *c*DCE using an oxygen flow rate of 4.7 L/min was calculated to be 0.045 1/day, which corresponds to a half-life of 15 days. In the case of VC, the first-order degradation constant was estimated to be 0.0097 1/day, corresponding to a half-life of 71 days. Concentration values below the LOQ for VC (22 μg/L) were not considered in this estimate; hence, the values presented may be conservatively high. Values below the LOQ for *c*DCE (7 μg/L) were also not included in the calculation.

5.4 PERFORMANCE DATA EVALUATION

5.4.1 Granular Iron

The field pseudo-half-lives for the chlorinated organics were 3 to 15 times higher than those predicted from the column study (see Table 6.4). The differences observed were quite reasonable considering the inherent differences between the controlled input to the column in the lab and the complex input to the gate in the field. The main discrepancy probably stems from concentration differences that existed in the two tests. For *c*DCE, half-lives calculated from the column study were based on input concentrations of 32 mg/L, whereas field-based half-lives were calculated using time-integrated data (including concentrations as high as 220 mg/L). Nonetheless, the flow-through thickness of the iron wall was based on an accelerated velocity (OP1), hence breakthrough during OP2 should have been minimal. Since there was significant breakthrough, it was obvious that the half-lives calculated from the column study were too short. Breakthrough could also have resulted from surface saturation of the iron. The dechlorination reaction is known to be surface-related; hence as the solute concentrations increased, the possibility of surface saturation would also have increased. As the surface becomes saturated, the pseudo-first-order rate constants would be expected to decrease, which would increase the chance for breakthrough. Nevertheless, based on the total solute mass passing rows 1 and 4 from April to August 1997, there was still >91% removal for each chlorinated organic (Table 5.8).

Table 5.8 Summary of Percent Removal Remediation within Granular Iron Wall During OP2

Contaminant	Row 1 Mass (mg)	Row 4 Mass (mg)	% Mass Removed
TCE	107,500	278	99.7
1, 1-DCE	3595	137	96.1
*c*DCE	1,347,000	116,500	91.3
*t*DCE	2997	66	97.8
VC	571,900	37,180	93.5

From the multilevel data, it was determined that the plume was spatially and temporally variable, hence loading to the granular iron wall was not uniform. The high, nonuniform loading on the granular iron wall may explain the breakthrough at specific areas in the wall (i.e., R4TA-2.8 and

R4TA-3.7). After observing breakthrough from these and other multilevel monitors along column A, it was concluded that a "hot spot," representing the core of the plume, was located near the southern wall, column A, of the gate.

5.4.2 Biosparge Zone

During OP1, none of the BTEX compounds were detected in the groundwater downgradient of the granular iron wall, as BTEX compounds were retarded in the wall due to sorption. During the early part of OP2 (April 1997), benzene began to break through above the maximum contaminant level (MCL) (1 μg/L) and later on (June 1997), toluene began to breakthrough below the MCL (150 μg/L). However, benzene and toluene were normally only detected in row 4, and sometimes row 5, but rarely past row 5. Perhaps they were aerobically degraded in the pea gravel or volatilized from row 5, as DO as high as 6 mg/L was detected in R5TB-3.7. Ethylbenzene and the xylenes had not yet broken through the granular iron wall at the time of writing this document.

To estimate the percentage of mass that was biodegraded versus volatilized, the analytical solution, Equation (5.15), was solved for steady-state conditions. Under steady-state conditions,

$$C_{wt} = \frac{Q \times C_{\text{in}}}{Q + (K_1 \times V_w) + (Q_a \times H)} \tag{5.16}$$

where C_{wt} = concentration in the biosparge water (mg/m^3)
Q = extraction rate (0.34 m^3/day)
C_{in} = concentration from row 4 (influent, mg/m^3)
K_1 = pseudo-first-order degradation rate constant (*c*DCE = 0.0454 1/day, VC = 0.0097 1/day)
V_w = volume of water (10.83 m^3)
Q_A = flow rate of air (0.564 m^3/day)
H = Henry's Law constant (*c*DCE = 0.312, VC = 0.964)

Henry's Law constant is taken from Gossett (1987) with units of m^3·atm·mol^{-1}. The Henry's Law constant was then divided by RT (with units m^3·atm·mol^{-1}) to make the constant dimensionless. Since Equation (5.16) is nonlinear in $k * V_w$ and $Q_A * H$, contributions to mass removal cannot be calculated as simple additive percentages. Instead, percent removals were estimated using a four-step procedure involving the calculation of

1. Biodegradation and volatilization
2. Volatilization only ($k = 0$)
3. Biodegradation only ($Q_A = 0$)
4. Neither biodegradation nor volatilization ($k = 0$, $Q_A = 0$)

The results of this estimation for *c*DCE and for VC are outlined in Tables 5.9 and 5.10, respectively.

To evaluate the contribution of mass loss from each process (biodegradation and volatilization), a ratio of the percent loss was taken. For example, 66% of *c*DCE was removed with both biodegradation and volatilization (scenario 1); 34% with only the volatilization term active (scenario 2), and 59% with only the biodegradation term active (scenario 3). Since 34% + 59% ≠ 66%, nonadditivity is demonstrated. As an approximation of the contribution to mass removal from the two sinks, it was assumed that 34%/(34% + 59%) = 36% of the 66% total was due to volatilization and 59%/(34% + 59%) = 63% of the total was due to biodegradation. Therefore, for *c*DCE, biodegradation accounted for about 65% of the mass removal and about 35% was removed via volatilization. With respect to VC, the same calculations suggested biodegradation only removed

Table 5.9 Percent *c*DCE Removal from the Biosparge Zone

	Scenario	Influent Conc., Row 4 (μg/L)	Biosparge Conc., Row 6 (μg/L)	% Removal
1	$k = 0.0454$ $Q_A = 0.564$	142	47.91	66
2	$k = 0$ $Q_A = 0.564$	142	93.57	34
3	$k = 0.0454$ $Q_A = 0$	142	58.05	59
4	$k = 0$ $Q_A = 0$	142	142	0

Note: Influent concentrations were sampled on September 30, 1997. Effluent concentrations were calculated using Equation (5.2).

Table 5.10 Percent VC Removal from the Biosparge Zone

	Scenario	Influent Conc., Row 4 (μg/L)	Biosparge Conc., Row 6 (μg/L)	% Removal
1	$k = 0.0454$ $Q_A = 0.564$	283	97.31	66
2	$k = 0$ $Q_A = 0.564$	283	108.88	62
3	$k = 0.0454$ $Q_A = 0$	283	216.20	24
4	$k = 0$ $Q_A = 0$	283	283	0

Note: Influent concentrations were sampled on September 30, 1997. Effluent concentrations were calculated using Equation (5.2).

30% of the mass and volatilization removed 70%. Laboratory microcosm results (Appendix 11) supported biodegradation of *c*DCE and VC, but found that VC was readily biotransformed to levels below detection, whereas *c*DCE was only partially biotransformed. Mass removal via biodegradation could probably be improved if a larger active biomass were established in the biosparge zone.

As predicted, *c*DCE and VC were found in the headspace of the biosparge zone. It was found that headspace concentrations decreased over time, which agrees with the decrease in contaminant concentrations measured from the water column (Table 5.11). This decrease may be related to an increase in the amount of aerobic biodegradation occurring in the biosparge zone.

To evaluate the previous estimate of the mass loss due to volatilization, potential headspace concentrations were calculated using Henry's Law (Equation 5.8, Section 3.6.6.2) and water concentrations measured in the biosparge zone and then compared to the measured headspace concentrations (Appendix 26). Table 5.11 summarizes the measured and calculated values.

Table 5.11 Calculated and Measured Headspace Gas Concentrations for *c*DCE and VC

Date	*c*DCE Concentrations			VC Concentrations		
	C_{water} Meas.	C_{air} Meas.	C_{air} Calc.	C_{water} Meas.	C_{air} Meas.	C_{air} Calc.
July 14, 1997	336.7	18.7	105.1	247.2	62.3	238.3
October 2, 1997	47.9	0.30	15.0	97.3	0.36	93.8

Notes: All units are μg/L. Meas. means measured concentration; Calc. means calculated concentration.

Assuming that the headspace gas had equilibrated with the water in the biosparge zone, significantly higher concentrations should have been present in the headspace than were found (Table 5.11). Some possible explanations for lower concentrations detected in the headspace:

1. Equilibrium between the aqueous and air phase had not been established during sparging.
2. There was aerobic biodegradation of the volatilized contaminants within the headspace.
3. Leaks in the biosparge lid allowed the gas to escape.

As it is not possible to quantify how much biodegradation was occurring in the headspace, the above predictions of mass loss due to volatilization must be considered preliminary.

5.4.3 Overall Performance

The overall performance of the remedial gate during OP2 was assessed by comparing the average concentrations entering the remedial gate (as measured in fully screened wells in row 1) to averaged concentrations leaving the gate (average of the two fully screened wells in row 8) for a particular sampling event. A percent removal was then calculated based on the differences in concentrations (Table 5.12).

Table 5.12 Bulk Performance of the Remedial Gate Based on Percent Removal

Contaminant	% Removal 16-Jun-97	% Removal 31-Jul-97	% Removal 30-Sep-97
TCE	100	NA	100
1,1-DCE	NA	NA	100
*c*DCE	95.7	98.5	99.8
*t*DCE	NA	NA	100
VC	96.7	97.6	99.0
Benzene	NA	NA	100
Toluene	100	NA	99.7
Ethylbenzenes	NA	NA	100
Xylenes (O)	*NA	*NA	100
Xylenes (P/M)	*NA	*NA	100

Notes: NA: a percent loss could not be calculated because concentrations in rows 1 and 8 were both <MDL. *NA: on those sampling days, only total xylene was analyzed, but due to input concentrations below MDL, a percent loss could not be calculated.

Assessing the overall performance of the system based on fully screened wells might suggest a more positive performance of the gate than if the performance was judged using multilevel monitors. Fully screened wells produce vertically averaged solute concentrations, which are likely less than the maximum concentration that could be found using numerous depth-discrete sampling points. The use of data from the different types of wells requires a clear statement of whether the remedial objectives are for bulk groundwater quality or for maximum concentrations in a potentially small volume of groundwater. However, since the biosparge zone appeared to homogenize concentrations from row 6 through to row 8, multilevel monitors might report similar values to that of the fully screened wells, suggesting that the overall performance may be reliably assessed for either objective using row 8 wells.

Based on the high percent removals, the performance of the remedial gate was judged excellent. Remediation of *c*DCE and VC within the granular iron was incomplete, probably because of their localized high influent concentrations. With respect to BTEX, an accurate assessment of its aerobic biodegradation could not be inferred, because compounds detected in the biosparge zone were at or below their LOQ. Although Table 5.12 suggests that there was excellent removal of the BTEX compounds, most of the mass was removed via sorption within the iron wall. The field demonstration was not of sufficient duration to allow the BTEX compounds to break through the iron wall.

The overall performance of this field demonstration was also assessed relative to the remedial objectives that were set at the beginning of the project. Table 5.13 lists the remedial objectives for the project and the contaminant concentrations sampled from R8PA and R8PB on the last sampling, September 30, 1997.

Table 5.13 Comparison of Remedial Objectives and Effluent Concentrations Detected in R8PA and R8PB

Compound	Remedial Objective (μg/L)	Effluent in R8PA (μg/L)	Effluent in R8PB (μg/L)
TCE	5	<5	<5
1,1 DCE	6	<5	<5
*c*DCE	10	136[a]	131[a]
*t*DCE	10	<5	<5
VC	2	217[a]	121[a]
Benzene	10	<5	<5
Toluene	150	21.8	<5
Ethylbenzene	700	<5	<5
Xylenes (Total)	1,750	<5	<5
pH	5–8.5	6.89	8.78[a]
Iron	500	70	170

[a] Value exceeded remedial objective.

Vinyl chloride and *c*DCE were the only organic contaminants that did not meet the remedial objectives. Possible reasons were previously discussed. Even though these compounds broke through the iron wall, the percent removal for VC was 99% and 99.8% for *c*DCE. Removal for these compounds may improve if their aerobic biodegradation rates increase in the biosparge zone. The pH measured in R8PB was also above the remedial objectives, but this is not a concern as the pH in the biosparge zone was consistently within the desired pH range. It was not uncommon for the field parameters to vary when they were sampled in the fully screened wells, i.e., pH in row 8 would fluctuate from 6 to 9.

On January 12, 1998, the remedial gate was sampled for the last time (because these data were collected so late, the degradation rates were not recalculated). It was discovered that the peak influent concentrations (row 1) for *c*DCE and VC had decreased significantly since the last sampling event in September 1997. Table 5.14 compares data from rows 1, 4, and 6 for the last two sampling events.

Table 5.14 Analytical Results for the Remedial Gate Sampled on September 30 1997 and January 12, 1998 for *c*DCE, VC, and Toluene

	*c*DCE (μg/L)		VC (μg/L)		Toluene (μg/L)	
Well	Sept. 30	Jan. 12	Sept. 30	Jan. 12	Sept. 30	Jan. 12
R1TA-2.8	70,990	25,950	49,670	19,300	4654	4686
R1TA-3.7	169,600	384.1	21,700	550.1	6273	2548
R4TA-2.8	747.7	<MDL	2179.0	18.75	774.1	43.57
R4TA-3.7	100.7	<MDL	138.2	20.18	89.50	16.99
R6TA-2.8	51.25	<MDL	32.34	<MDL	3.1	<MDL
R6TA-3.7	49.65	<MDL	21.25	<MDL	<MDL	<MDL

Note: <MDL means less than method detection limit.

For the low influent concentrations, complete removal is evident by row 6 (January 12, 1998 data). These latest row 6 concentrations meet the ROs listed in Table 5.13. Although these low influent concentrations were not often measured in the groundwater over the length of the field demonstration, they might represent a level of contamination that the remedial gate can effectively treat, whereas at higher concentration, i.e., 220 mg/L of *c*DCE and 40 mg/L of VC, it was found that the gate could not completely treat these levels.

This field demonstration has shown that treatment using granular iron and a biosparge system in series can reduce high concentrations of chlorinated ethenes (remediation of BTEX compounds was not assessed due to retardation within the iron wall). Although the remedial objectives for all the contaminants were not met (*c*DCE and VC) at high influent concentrations, the mass removal and/or overall performance of the system was >90% for all contaminants, suggesting that this particular combination of treatment technologies is a viable option for plume control. With influent *c*DCE concentrations of <26 mg/L, the ROs were met, while with influent *c*DCE concentrations around 170 mg/L, the ROs were not met.

5.4.4 Adverse Interaction *Between* Technologies and Design Limitations

The experience in Gate 1 of the Borden experiment (Chapter 2) highlighted the potential for high pH to be produced from the granular iron zone. While ORC™ was not used at the Alameda demonstration, high pH (above pH 8.5 and perhaps even above pH 9) was also anticipated to reduce microbial success in the biosparge zone. The periodic sparging of CO_2 seemed to have overcome this problem. Sparging CO_2 is practical, given the active, daily sparging of O_2. It would be practical to automate the system, periodically switching the sparge gas to CO_2. At Alameda, this solution was efficient and cost-effective.

The granular iron wall had not succeeded in treating *c*DCE nor VC to the remedial objectives. This appeared to be due to localized higher-than-anticipated concentrations of these CVOCs entering the granular iron zone, but it could also have been the result of poor performance of the granular iron at the locally high (>200 mg/L) concentrations of *c*DCE entering the remedial gate. Laboratory experiments were conducted as part of Morkin's M.Sc. research (1998) to evaluate the latter possibility.

Further monitoring may find that the aerobic biodegradation of *c*DCE and VC improves over time, and the failure to meet remedial objectives within the granular iron treatment may be supplemented by sufficient aerobic biodegradation to meet the ROs by the end of the remedial gate.

Continuous aerobic conditions in the biosparge zone were not achieved until O_2 sparging was increased well above the initially designed rate. This problem was also noted in Gate 3 at Borden (Chapter 4). It is not clear what the limitation was at Alameda. In 1998, O_2 sparging will be gradually reduced in both frequency and duration of each sparging event in order to arrive at a more passive, but effective, sparging protocol.

5.5 WASTE MANAGEMENT

All extracted water from the remedial gate and any wastewater generated from sampling was treated on site using activated carbon drums. Details of the aboveground treatment system are provided in Appendix 31.

5.6 CONCLUSIONS

The following conclusions are based on the field demonstration. During OP2, groundwater velocities were measured via the pumping rate, a tracer test, and ISPFS. Tracer velocity estimates in the sand/iron section were 4 and 7 cm/day in the granular iron section, which agreed fairly well with the pumping rate velocities of 10 cm/day in the sand/iron section and 6 cm/day in the iron section. ISPFS velocity in the iron zone was 9 to 14 cm/day, but was considered unreliable due to the anomalous horizontal flow direction and vertical flow measurements.

Assessment of the multilevel data shows excellent degradation (>91%) of the chlorinated organics using the granular iron at high influent concentrations. At lower influent concentrations (January 1998 data), almost complete degradation (>99%) was observed.

The pseudo-half-lives calculated from the field data were 3 to 15 times larger then those predicted from the column study.

From the multilevel data, it was determined that the plume was spatially and temporally variable, hence loading on the iron wall was not uniform. The nonuniform loading on the iron wall may explain the breakthrough in specific areas in the wall, i.e., R4TA-2.8 and R4TA-3.7, at the higher influent concentrations. After observing breakthrough from these and other sampling points along column A, it is concluded that a "hot spot," representing the core of the plume, was located near column A of the gate.

Oxygen sparging at intervals of six times in 24 hours for 20 minutes at a delivery pressure of 20 psi was capable of establishing aerobic conditions in a bioball-packed *in situ* bioreactor. Carbon dioxide sparging intervals of once per month for 10 minutes at a delivery pressure of 20 psi were sufficient to neutralize the elevated pH of the groundwater that has passed through the granular iron porous medium.

The biosparge zone supported aerobic biodegradation of VC and *c*DCE, although complete attenuation did not occur. The modeling suggests that mass removal for *c*DCE was predominantly by biodegradation, and with respect to VC, the dominant process was volatilization, although some biodegradation was predicted. Laboratory microcosm results supported biodegradation of *c*DCE and VC, but found that VC was readily biotransformed to levels below detection, whereas *c*DCE was only partially biotransformed.

Using granular iron with a biosparge zone in series has been shown to effectively treat a high concentration mixed plume to near MCL levels, suggesting that this combination is a viable technology.

5.7 RECOMMENDATIONS FOR FUTURE STUDIES

The following recommendations for future studies can be made:

1. To prevent heterogeneous loading of contaminants on the granular iron surfaces, a mixing zone upgradient of the iron barrier should be installed to partially homogenize concentrations within the groundwater, i.e., an open water mixing zone similar to that installed in Gate 1 at Borden (refer to Chapter 2 of this report). By homogenizing the contaminant concentrations, the "hot spots" of the plume can be diluted, resulting in a more even loading to the treatment wall and a reduced risk of contaminant breakthrough.
2. Because the plume proved to be so spatially variable, it is imperative that the initial site characterization be done in sufficient detail that high concentration regions of the plume can be identified. To obtain the needed level of detail, multilevel sampling or profiling should be used. Sampling from fully screened wells is likely to obscure plume hot spots because of mixing and dilution in the well. Accurate location of the core of the plume is critical for correct positioning of the Funnel-and-Gate.
3. The required flow-through thickness of the iron wall should be estimated on the basis of a bench-scale treatability study using groundwater collected from the plume. If the groundwater collected does not contain contaminant concentrations representative of "hot spots," the flow-through thickness may be underdesigned. Therefore, prior to conducting treatability studies, care should be taken to ensure that conservative concentrations are used. This may mean spiking groundwater for the treatability study to raise concentrations to maximum values found at the site.

4. Aerobic microcosm studies should be conducted as part of the overall treatability study to assess the biodegradability of the target organics present in the groundwater. Additional microcosms could then be prepared to evaluate whether inorganic nutrients are a limiting factor for successful aerobic biodegradation.
5. Sparging rates in the biosparge zone should be minimized to reduce the volatilization of the contaminants from the water column. Once the required populations of aerobes have been established, the sparge rate should then be reduced to the minimum level required to maintain aerobic conditions that still support biodegradation.
6. Monitoring the current system at Site 1 (at the current flow rate) should continue to assess long-term treatment trends within the granular iron wall and the biosparge zone. Because the influent concentrations decreased as of the last sampling event, reasons for this variation should be addressed. If it is found that the influent concentrations increase again allowing the chlorinated compounds to breakthrough the iron wall, the flow rate should be reduced. Should BTEX begin to breakthrough the granular iron wall, its potential for biodegradation within the biosparge zone should then be assessed.

REFERENCES

Anderson, R.L. 1987, *Practical Statistics for Analytical Chemists*, van Nostrand Reinhold, New York.

Ballard, S. 1996. The *in situ* permeable flow sensors: a ground-water flow velocity meter. *Ground Water*, 34(2): 231-240.

Barker, J.F., Smyth, D., and Cherry, J.A. 1994. Controlled *in situ* groundwater treatment. *Proc. 1994 NATO/CCMS Pilot Study Meeting*, Oxford, UK, September 1994.

Bowles, M. et al. 1995. In-situ remediation of hydrocarbon contaminated groundwater in low hydraulic conductivity media using trench and gate technology. Proc. Groundwater & Soils Remediation Program 5th Annual Conf., Toronto, October 2–6, 1995.

Cherry, J.A., Feenstra, S., and Mackay, D.M. 1996. Concepts for the remediation of sites contaminated with dense non-aqueous phase liquids (DNAPLs). *Dense Chlorinated Solvents*, Pankow, J.F. and Cherry, J.A. (Eds.), Waterloo Press, Guelph, ON.

Davis, J.W. and Carpenter, C.L. 1990. Aerobic biodegradation of vinyl chloride in groundwater samples. *Applied and Environmental Microbiology.* 56(12): 3878-3880.

Devlin, J.F. 1996. A method to assess analytical uncertainties over large concentration ranges with reference to volatile organics in water. *Ground Water Monitoring & Remediation*, 16(3): 179-185.

Focht, R., Vogan, J., and O'Hannesin, S. 1996. Field application of reactive iron walls for *in situ* degradation of volatile organic compounds in groundwater. *Remediation*, 6(3): 81-94.

Gillham, R.W. 1996. *In situ* treatment of groundwater: Metal-enhanced degradation of chlorinated organic contaminants. *Advances in Groundwater Pollution Control and Remediation.* Aral, M.M. (Ed.), Kluwer Academic, Netherlands, pp. 249-274.

Gillham, R.W. and O'Hannesin, S.F. 1994. Enhanced degradation of halogenated aliphatics by zero-valent iron. *Ground Water*, 32(6): 958-967.

Gorman, J.K. 1995. Bioremediation of Hydrocarbon Contaminated Soil and Groundwater Using an *in situ* Funnel-and-Gate System. M.Sc. project, University of Waterloo.

Gossett, J. 1987. Measurement of Henry's Law constants for C1 and C2 chlorinated hydrocarbons. *Environmental Science and Technology.* 21: 202-208.

Granger, D. Nutrient Addition Techniques for Funnel-and-Treatment Gates Systems. M.Sc. thesis, University of Waterloo, in progress.

Katic, D. 1996. Laboratory determinations of residual gas phase content after air sparging in a coarse-grained porous medium based on physical displacement of fluids. Personal communication.

Morkin, M. Sequential In-situ Treatment of a Mixed Organic Plume Using Granular Iron and Biosparging. M.Sc. thesis, University of Waterloo, Waterloo, Ontario, Canada.

O'Hannesin, S. 1993. A Field Demonstration of a Permeable Reaction Wall for the *in situ* Abiotic Degradation of Halogenated Aliphatic Organic Compounds. M.Sc. thesis, University of Waterloo.

PRC, 1993. Solid Waste Assessment Test (SWAT) Report, NAS Alameda, Alameda, CA.

PRC, 1996. Site 1 and Site 2 Radiation Survey Report, NAS Alameda, Alameda, CA.

Rogers, J.D. and Figuers, S.H. 1991. Engineering Geologic Site Characterization of the Greater Oakland-Alameda Area, Alameda and San Francisco Counties, California. Final Report to National Science Foundation, Grant No. BCS-9003785.

Snedecor, G.W. and Cochran, W.G. 1989. *Statistical Methods,* 8th ed., Iowa State University Press, Ames, IA.

Starr, R.C. and Cherry, J.A. 1994. *In situ* remediation of contaminated ground water: the Funnel-and-Gate system. *Ground Water*, 32(3): 465-476.

Taylor, J.K. 1987. *Quality Assurance of Chemical Measurements*, Lewis Publishers, Boca Raton, FL.

Tratnyek, P.G., Johnson, T.L., Soherer, M.M., Eykholt, G.R. 1997. Remediating groundwater with zero valent metals: chemical considerations in barrier design. *Ground Water Monitoring and Remediation*, 17(4): 108-114.

University of Waterloo (UW). 1996a. Workplan for Semipassive Groundwater Remediation Demonstration Project.

University of Waterloo (UW). 1996b. Revised Workplan for Semipassive Groundwater Remediation Demonstration Project.

University of Waterloo (UW). 1997c. Installation Report for Alameda, Funnel-and-Gate.

U.S. Environmental Protection Agency (U.S. EPA). 1992. Water Quality Standards; Establishment of Numeric Criteria for Priority Toxic Pollutants. *Federal Register*, December 22, Part 2. 40 CFR Part 131.

CHAPTER 6

Part I — Conclusions

6.1 REVIEW OF OBJECTIVES

As stated in Chapter 1, the objectives of this project were divided along geographic lines, depending on the field site where the various phases of the work were completed. At the Borden site, the goals were to compare the performances of the various treatment methods utilized, to determine the lowest achievable concentrations under the conditions of the test, and to quantify performance related parameters for the design of the Alameda experiment. The objectives of the Alameda site experiment were to achieve the remedial objectives for the target organic contaminants (risk reduction) and dissolved iron (to deal with aesthetic concerns and potential precipitation of oxides and hydroxides in the aerobic part of the gate) and to maintain near-neutral pH water (to foster the development of biological activity downgradient). A further objective was to provide clear guidance and advice for the full-scale application of the remedial technologies at Alameda or other sites. The latter objective is addressed primarily in Part II of this volume. However, selected recommendations are presented here as judged appropriate for the sake of clarity and the completeness of this report. In the following sections, each of these objectives is discussed with reference to the data and interpretations presented in the earlier chapters of this report.

6.2 THE BORDEN EXPERIMENT

6.2.1 Comparison of Treatment Methods

The treatment methods tested in this project included two pairs of engineered sequential anaerobic–aerobic technologies contained within sheet piling gates. In the first gate, Gate 1, the treatment system comprised reductive dechlorination of chlorinated solvents on granular iron in a removable cassette system, followed by aerobic biodegradation of toluene and other oxidizable organics, using ORC™ as a source of dissolved oxygen. In Gate 3, a second treatment system consisted of anaerobic biodegradation of chlorinated solvents using a NIW to supply labile organic carbon (benzoate) and nutrients, followed by aerobic biodegradation of toluene and other oxidizable compounds, using a biosparge zone to supply dissolved oxygen. Between these two gates was a control gate, Gate 2, in which natural attenuative processes were permitted to occur without any engineered enhancements.

6.2.1.1 Comparison of Redox Environments in Gates 1, 2, and 3

The water entering the gates was not uniformly aerobic or anaerobic. Bulk groundwater samples prior to activating the source wells generally contained detectable DO (~1 mg/L), but it was common for DO levels below 0.5 mg/L to be encountered at various multilevel sampling locations. However, after source well activation, bulk samples collected in the gates were almost always anaerobic (<0.6 mg/L). In addition, dissolved methane was detectable in shallow groundwater in all three gates, indicating active methanogenesis in the upgradient aquifer. In spite of the apparent methanogenesis going on nearby, the redox environment of the experimental site was somewhere between aerobic and sulfate reduction, based on the low but detectable DO concentrations and consistently detectable background sulfate (approximately 20 mg/L) in the groundwater, and hydrogen measurements. Background nitrate was below detection limits. These conditions are believed to have prevailed in the control gate throughout the project, at least until the final inorganic sampling snapshot on July 14, 1997 (day 238).

The groundwater in Gate 1, upon contacting the granular iron, experienced a significant decline in the redox state. As one would expect on the basis of thermodynamic considerations, the presence of zero valent iron lowers the reduction potential of the system considerably below that achievable by methanogens. This electron-rich environment is largely responsible for driving the reductive dechlorination of PCE and its products, producing $Fe^{2+}(aq)$ as a by-product. Despite the highly reducing quality of the water and Fe^{2+} production, the concentration of dissolved iron did not increase above 2–3 mg/L. These low dissolved iron concentrations are attributable, in part, to the pH of the water, which increased from 7 to about 10 within the granular iron cassettes, likely causing precipitation of iron hydroxide. Surface analysis of precipitates on the granular iron, by Raman spectroscopy, supports the hypothesis that the Fe^{3+}, as the hydroxide, is subsequently autoreduced on the surface to produce a magnetite phase and green rust (refer to Section 2.3.2, Chapter 2). In this way the granular iron resists rapid passivation by hydroxide coatings.

Contrary to previous experience, the ORC™ in this experiment, while installed in the open water cassettes, did not release sufficient oxygen to raise the DO level above 2 mg/L, a concentration too low to support the aerobic biological activity needed to treat the organic load. It was subsequently determined through laboratory studies that the ORC™ used in this project does not release substantial amounts of molecular oxygen in water with a pH exceeding 9. Attempts to correct the pH inhibition of ORC™, by lowering the pH of the water in the ORC™ cassettes, failed. However, the interference was overcome by installing ORC™ in large diameter, fully screened, PVC wells (the alternate oxygen addition wells) located about 1.3 m downgradient of the cassette system. This success was partly due to the pH buffering capacity of the aquifer material, and may also have been due to a flow regime in the wells more conducive to maintaining contact between near-neutral pH water and the socks than that in the cassettes. Further work is needed to evaluate this possibility. In spite of the improvement in the performance of the ORC™ over the duration of the experiment, aerobic conditions (DO > 2 mg/L) were not uniformly maintained throughout the gate, downgradient of the alternate oxygen addition wells. The non-uniform aerobic condition resulted from a combination of a heterogeneous velocity distribution in the aquifer, which limited oxygen transport from the wells at certain locations and the natural chemical/biological oxygen demand of the aquifer. It is believed that aerobic conditions would have eventually prevailed downgradient of the alternate oxygen addition wells if the ORC™ socks had been maintained there for a longer period of time.

In Gate 3, sulfate reduction was stimulated within a few weeks of initiating the benzoate additions. Evidence of acetogenesis and methanogenesis was also obtained downgradient of the NIW after July 1997. These conditions were confirmed by direct sampling of sulfate, acetate, and methane, and were also inferred by hydrogen concentrations in the aquifer. It was the objective of the biosparge zone to return the aquifer to an aerobic state. However, initial difficulties were experienced in overcoming the oxygen demand imposed by the benzoate and, to a lesser extent, the toluene loading. In an effort to correct this problem, benzoate injection concentrations were decreased from 300 to 180 mg/L, and the sparging interval was increased from semiweekly to daily. Following these corrective measures, aerobic conditions were maintained within and downgradient of the biosparge. It is anticipated that with time the natural chemical oxygen demand of the aquifer would become exhausted, leading to the formation of a more extensive aerobic zone. Nevertheless, following the adjustments to the benzoate input concentrations and the sparging frequency, benzoate degradation appeared to be complete with sufficient oxygen remaining in the groundwater to fuel complete methane, acetate, and toluene biodegradation.

6.2.1.2 Comparison of Treatment of Target Organics in Gates 1, 2, and 3

On November 19, 1996, the source wells were activated. By July 14, 1997 (day 238), approximately one pore volume of groundwater had been flushed through Gates 2 and 3, and approximately two pore volumes had been flushed through Gate 1. A snapshot sampling, conducted on July 14, 1997 (Figure 6.1), reveals that CT was the most reactive of the target organics (PCE, CT, and TOL), having penetrated no further than about 7 m into any of the gates. CT underwent a transformation that limited its progress through the gates, but required no engineered treatment systems to occur (maximum concentrations at each multilevel monitor in Gate 1 are shown in Figure 6.1; Figures 6.2 to 6.6 also show maximum concentrations). One hundred days later, on October 21, 1997 (day 336), the distribution of CT had changed very little, suggesting a pseudo-steady-state profile had developed.

At the time of the July 14 sampling snapshot, it was found that CF, a known dechlorination product of CT, was extensively distributed throughout both Gates 1 and 2. (Note: the extensive distribution of CF in Gate 1 occurred due to a leak in the seals surrounding the cassette assembly. Also, traces of CS_2 were detected in some locations early in the experiment but the production of this compound did not continue throughout the project.)

The picture was very different for CF, which by day 238 had migrated throughout Gate 1 and most, if not all, of Gate 2, but was absent at quantifiable levels in Gate 3 (Figure 6.2). Data collected earlier in Gate 3 indicated that CF had, in fact, been transported at least as far as the biosparge zone, but was subsequently attenuated, presumably due to anaerobic biodegradation (see Figure 4.10, Chapter 4). By day 238, the attenuation was complete (down to a concentration of 25 µg/L, the LOQ for CF; refer to Appendix 6). One hundred days later, the distribution of CF in Gate 3 was essentially unchanged, suggesting that a pseudo-steady-state profile had been established. However, the CF distributions in Gates 1 and 2 changed considerably between days 238 and 339, resulting in a reduced maximum penetration distance of only about 6 to 7 m by the October date. CF was attenuated to concentrations below MDLs in water that passed through the granular iron cassettes. This is discussed in more detail in Chapter 2. Subsequently, CF was also apparently degraded downgradient of the alternate oxygen addition wells in Gate 1, following the ORC™ installation. The mechanism for this apparent aerobic degradation of CF has not been established but cometabolism, with TOL as the primary substrate, is a possible explanation. No further dechlorination products of CF, such as DCM or chloromethane, were detected in the groundwater, except in the cassettes of Gate 1, where the granular iron converted CF to DCM but was apparently unable to further dechlorinate that product (DCM). Given that there are no plausible abiotic mechanisms to explain CF attenuation in Gates 2 and 3, these data demonstrate that the aquifer was capable of biodegrading CT and all its dechlorination products to below LOQs without engineered treatment systems. It has also been demonstrated that the lag time for complete destruction of the chlorinated methanes was reduced as a result of the benzoate and nutrient additions to the aquifer in Gate 3.

It is hypothesized that the TOL, supplied to the aquifer as a target organic, served as an electron donor and supported the anaerobic biological activity that ultimately dechlorinated CF in all three

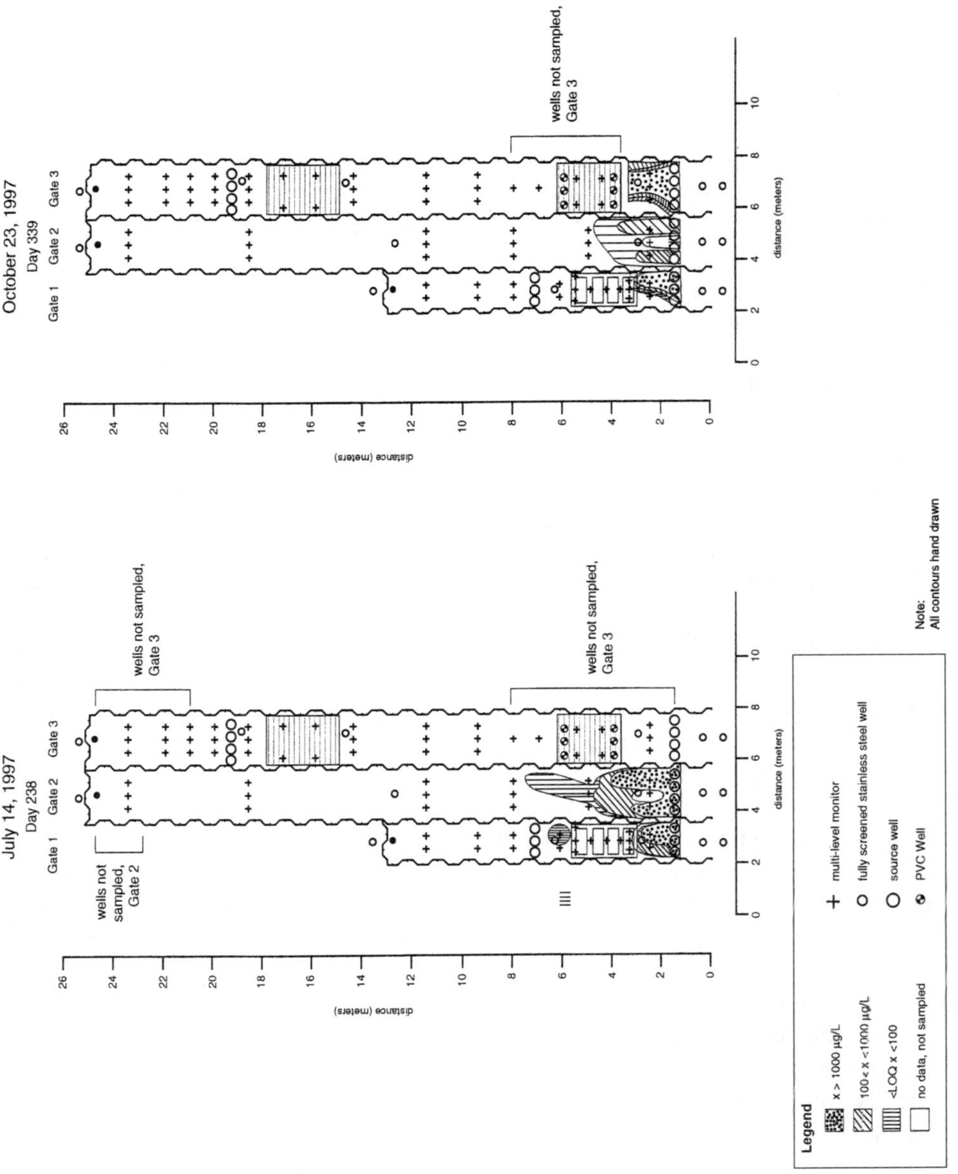

Figure 6.1 Plan view of CT progress through Gates 1, 2, and 3 on July 14, 1997 and October 23, 1997.

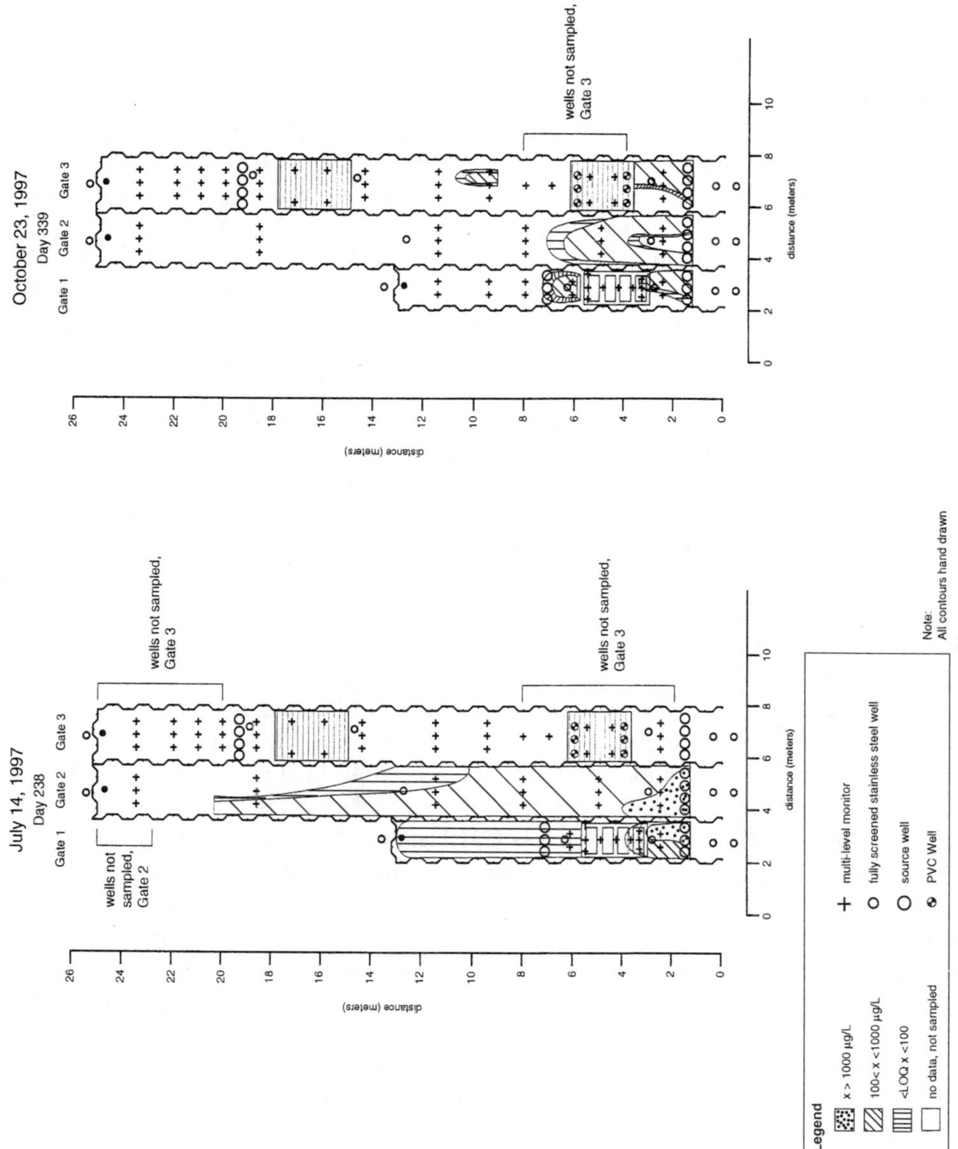

Figure 6.2 Plan view of CF progress through Gates 1, 2, and 3 on July 14, 1997 and October 23, 1997.

gates. Thus, at the conclusion of the Borden experiment it was shown that chlorinated methane compounds could be effectively treated by passive and semipassive *in situ* bioremediation. Under the conditions of this test, the presence of reduced organic carbon, as TOL or benzoate, appears to have driven intrinsic biodegradation at a rate sufficiently rapid to render further engineered enhancements unnecessary. Given that some accumulation of DCM was evident in the granular iron cassettes but not in the bioactive zones of the aquifer material, bioremediation would seem to be the preferred treatment method for handling carbon tetrachloride and its derivatives, under the conditions of this experiment.

At the time of the July 14 snapshot sampling (Figure 6.3), there was no evidence of PCE degradation except in the granular iron cassettes. As expected, the chlorinated ethene compounds in the cassettes were reduced to nonchlorinated products, their concentrations decreasing to below the LOQs. Nevertheless, PCE was transported throughout Gate 1, presumably through the leaky seals mentioned above. Its progress through Gates 2 and 3 was consistent with that expected for a retarded but otherwise nonreactive solute. By day 238, the PCE front had penetrated further into Gate 2 than Gate 3 because it was introduced to Gate 2 before November 19, 1996 (day 1) as part of the source well prototype testing completed in the summer of 1996 (UW, 1997). Additional evidence for the lack of biodegradation was the absence of detectable dechlorination products of PCE (e.g., TCE, *c*DCE, VC) in any part of the treatment gates outside the granular iron cassettes.

By the October 23, 1997 snapshot sampling, a great deal had changed. In Gate 1, the short-circuiting of water around the cassette assembly had become critical, and no further assessment of granular iron performance was possible. The change in the hydraulics of this system are suggestive of clogging in the cassettes, although no evidence of this was found in cores of the cassette media. It is hypothesized that if any clogging occurred, it was concentrated on or near the cassette screens. Further work is needed to confirm this possibility.

By the time of the October snapshot, PCE dechlorination products, including TCE (Figure 6.4) and primarily *c*DCE (Figure 6.5), were detectable in all of the gates. The occurrence of *c*DCE in Gate 1, outside the cassettes, and in Gate 2 is clear evidence of intrinsic biodegradation of PCE. As was the case for CF biodegradation, it is believed that the presence of TOL in the contaminant mixture may have contributed to the relatively short lag time leading up to PCE degradation (cf. a lag time difference between Gates 2 and 3 of less than 100 days, Gate 3 exhibiting the lower lag time), although this could not be confirmed in laboratory batch studies undertaken concurrently with the field experiment. However, the highest concentrations of *c*DCE, by far, were measured in Gate 3 between fences G310 and G315, corresponding to the anaerobic mixing zone of the benzoate pulses. This is strong evidence that the nutrient and benzoate additions accelerated the PCE degradation process significantly. In fact, PCE degradation was sufficiently rapid that movement of the contaminant, at concentrations above the LOQ, was not observed beyond fence G316 between days 238 and 339 of the experiment. TCE was only detected at concentrations near the LOQ (7–15 μg/L) and showed no signs of developing a plume that extended beyond the anaerobic zone.

The distribution of *c*DCE in Gate 2 was limited to the zone bounded by fences G208 and G212 (approximately 5–11 m from source). It remains unclear how much of this zone contained micro-organisms that were actually involved in the transformation process, and therefore how much of the zone was contributing to the remediation. However, in Gate 3, the *c*DCE distribution was more extensive, stretching from fence G310 to the end of the gate, G324 (none was detected at G302, next to the source wells, but no monitoring points were sampled between G302 and G310 in the October 21 snapshot). Concentrations exceeding 100 μg/L were limited to the benzoate mixing zone, ending at G315 (about 14 m from the source wells), which almost exactly corresponds to the maximum extent of the PCE contamination. This suggests that the *c*DCE was actively being produced from PCE over most of the anaerobic benzoate mixing zone. The data also suggest that this degradation occurred with minimal net TCE production. VC was not identified in any of the

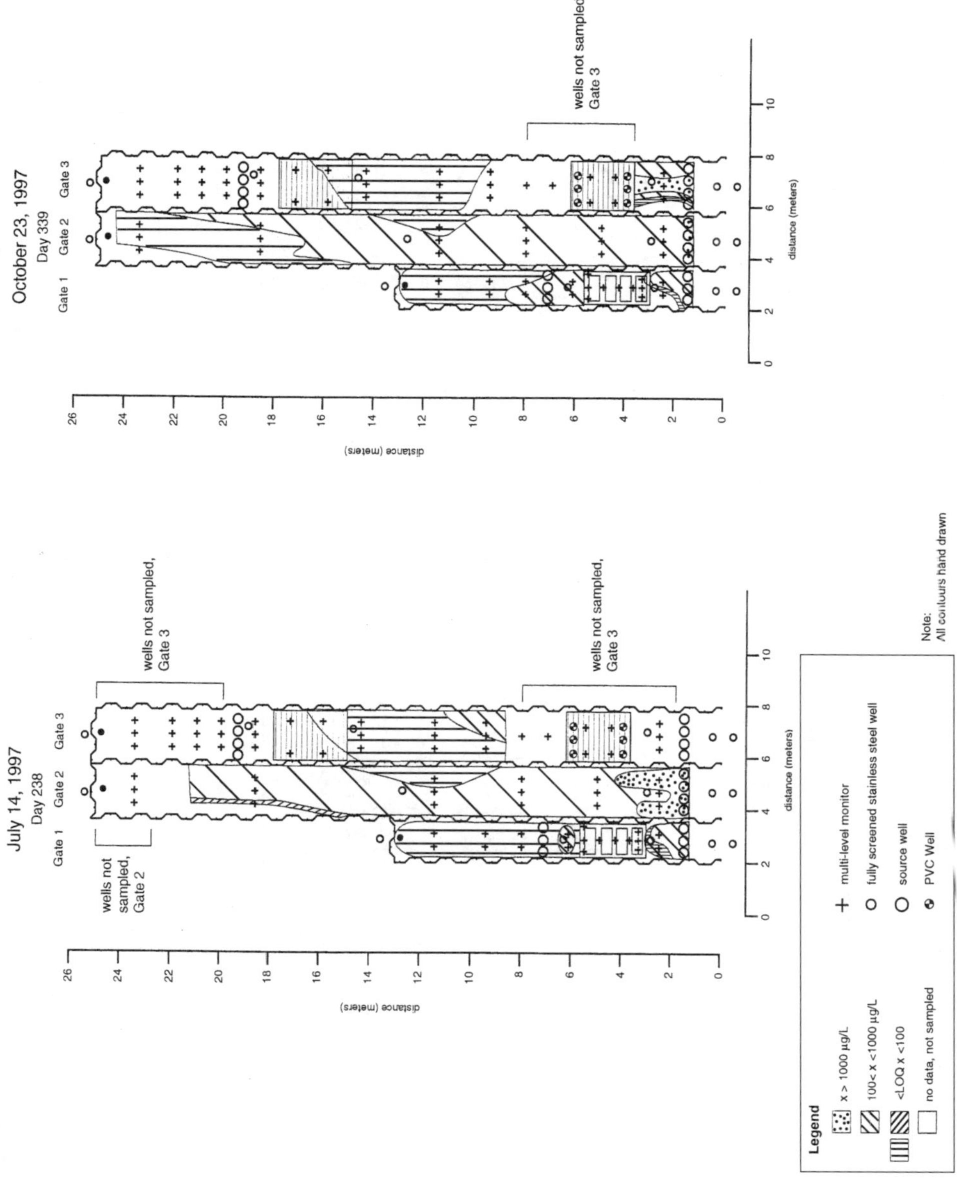

Figure 6.3 Plan view of PCE progress through Gates 1, 2, and 3 on July 14, 1997 and October 23, 1997.

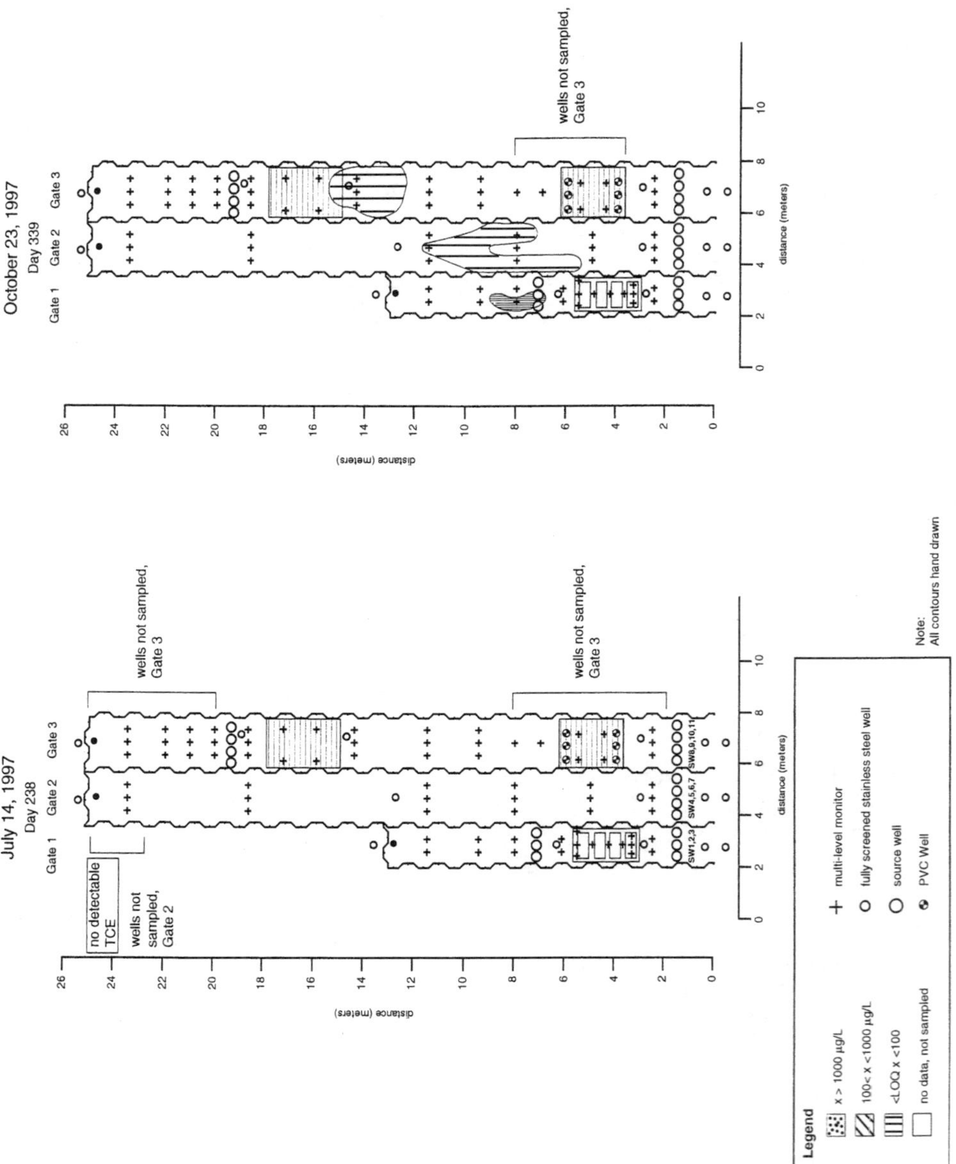

Figure 6.4 Plan view of TCE progress through Gates 1, 2, and 3 on July 14, 1997 and October 23, 1997.

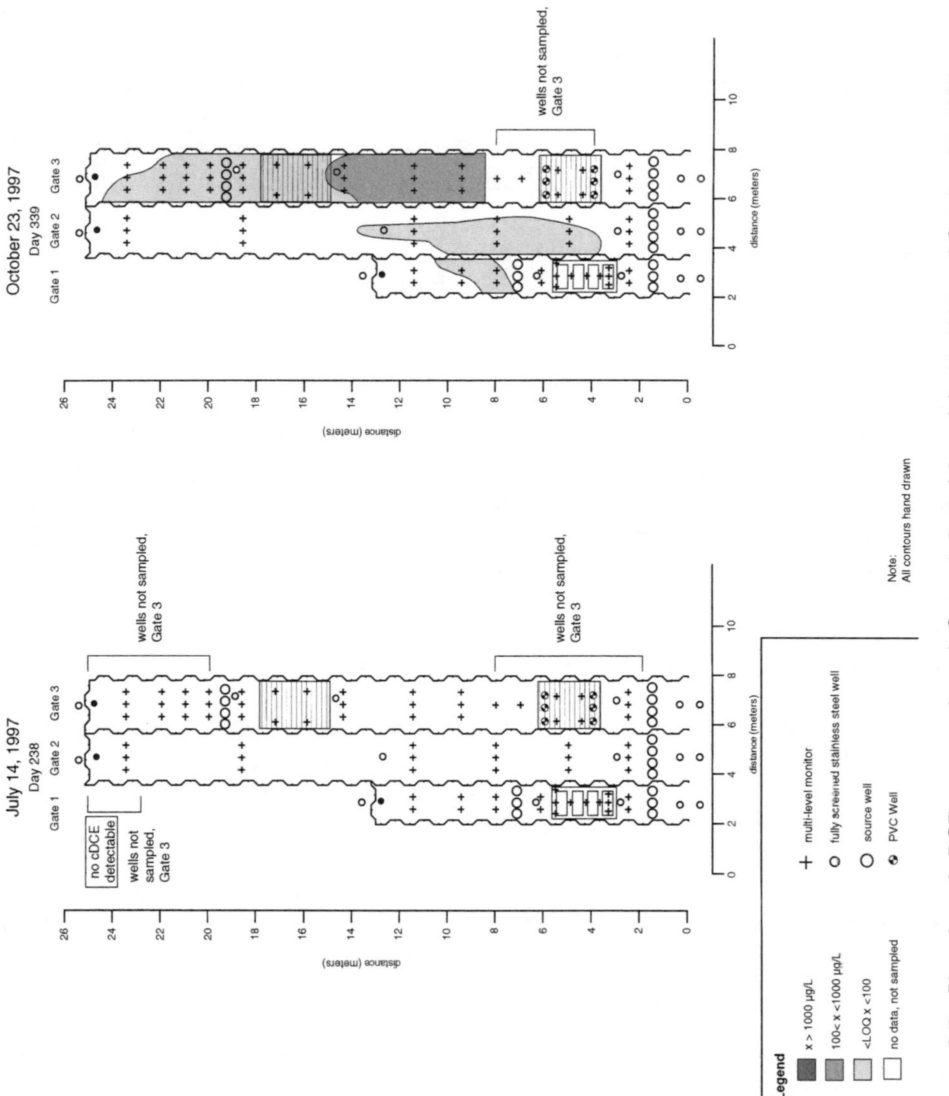

Figure 6.5 Plan view of *c*DCE progress through Gates 1, 2, and 3 on July 14, 1997 and October 23, 1997.

groundwater samples collected over the project, and the completely dechlorinated products ethene and ethane almost never exceeded 1 μg/L in concentration, so the available evidence suggests that the anaerobic biodegradation of PCE did not proceed beyond *c*DCE (in Gate 2 and 3).

The possibility that *c*DCE might not be degradable within the gates was anticipated at the outset of the experiment. It was hoped that this compound (and VC if it had appeared) could be removed from the groundwater by aerobic biodegradation within or downgradient of the biosparge zone. Unfortunately, due to the late appearance of *c*DCE in the gates, no laboratory microcosm tests with Borden aquifer material were conducted to evaluate the aerobic biodegradation of *c*DCE or VC. However, aerobic biodegradation experiments involving these compounds were performed as part of the Alameda phase of the work, and demonstrated the following order of disappearance rate TOL > VC > *c*DCE. All three contaminants were shown to aerobically degrade. The variation in disappearance rates suggests preferential degradation of some compounds over others. This was demonstrated in laboratory microcosms prepared with aquifer material from the Borden site where it was shown benzoate was preferred over TOL (Appendix 15). Such preferential utilization or degradation could, in part, explain the sporadic low concentrations of *c*DCE and TOL that persisted in the aerobic water at the end of Gate 3.

The results of the anaerobic treatment comparison point to granular iron as being the preferred method for the treatment of chlorinated ethene compounds, since it was shown to completely degrade PCE and its dechlorination products. However, the performance of the bioremediation gate was very encouraging and indicates that the NIW approach is worthy of further development.

Throughout the first 238 days of the experiment, there was little or no evidence that TOL was being treated in any of the gates. This was largely the result of the limited background oxygen available and of unsuccessful initial attempts at creating aerobic zones within Gates 1 and 3; the original design relied on the aerobic biodegradation of TOL in all three gates. The design objective was most nearly met in Gate 1 following the installation of ORC™ in the alternate oxygen addition wells (SW12, 13, and 14), located downgradient of the cassette assembly, on June 19, 1997 (day 212). As discussed previously (Section 6.2.1.1), aerobic conditions were created and sustained over part of the aquifer cross section downgradient of the ORC™ wells. This resulted in a noticeable decline in TOL concentrations (average concentrations along fence G106 declined from ~1500 to <500 μg/L), particularly at G108R where the oxygen delivery was most pronounced (Figure 6.6; compare day 238 and day 339 at G108R).

A comparison of the Gate 2 (the control gate) snapshots on days 238 and 339 reveals TOL concentration reductions not explainable by source strength variability (Figure 6.6). Comparisons in breakthrough masses at various fences in Gate 2 indicate that toluene attenuation is considerably more pronounced than PCE attenuation, even with retardation taken into account (refer to Section 3.4.6, Chapter 3). Since TOL is known to biodegrade anaerobically (albeit at a lower rate than aerobic biodegradation), it seems plausible that this process is responsible for the observed concentration reductions. There is some suggestion of this in Gate 3 also, although it is much less evident from a comparison of the snapshots. Benzoate utilization upgradient of the biosparge zone in Gate 3 may have reduced the rate of anaerobic TOL consumption in that area, in a fashion consistent with the results of the microcosm study.

The evidence for aerobic TOL degradation in Gate 3 is encouraging, with reductions of maximum concentrations in excess of 1000 μg/L entering the biosparge zone to an average concentration of about 40 μg/L at the final fence. It should be noted that in a few locations (including one at fence G324), the TOL concentrations remained in excess of 300 μg/L. However, the sampling points concerned were located near the base of the aquifer and may have received relatively little aeration. It is considered encouraging that with daily sparging in the biosparge zone sufficient dissolved oxygen was introduced to the gate to theoretically fuel the biodegradation of all the benzoate, acetate, methane, and toluene. Further work is needed to fine-tune the biosparge zone and the benzoate flux (for anaerobic biodegradation) so that these two technologies can work together more efficiently. Further work is also needed to evaluate the

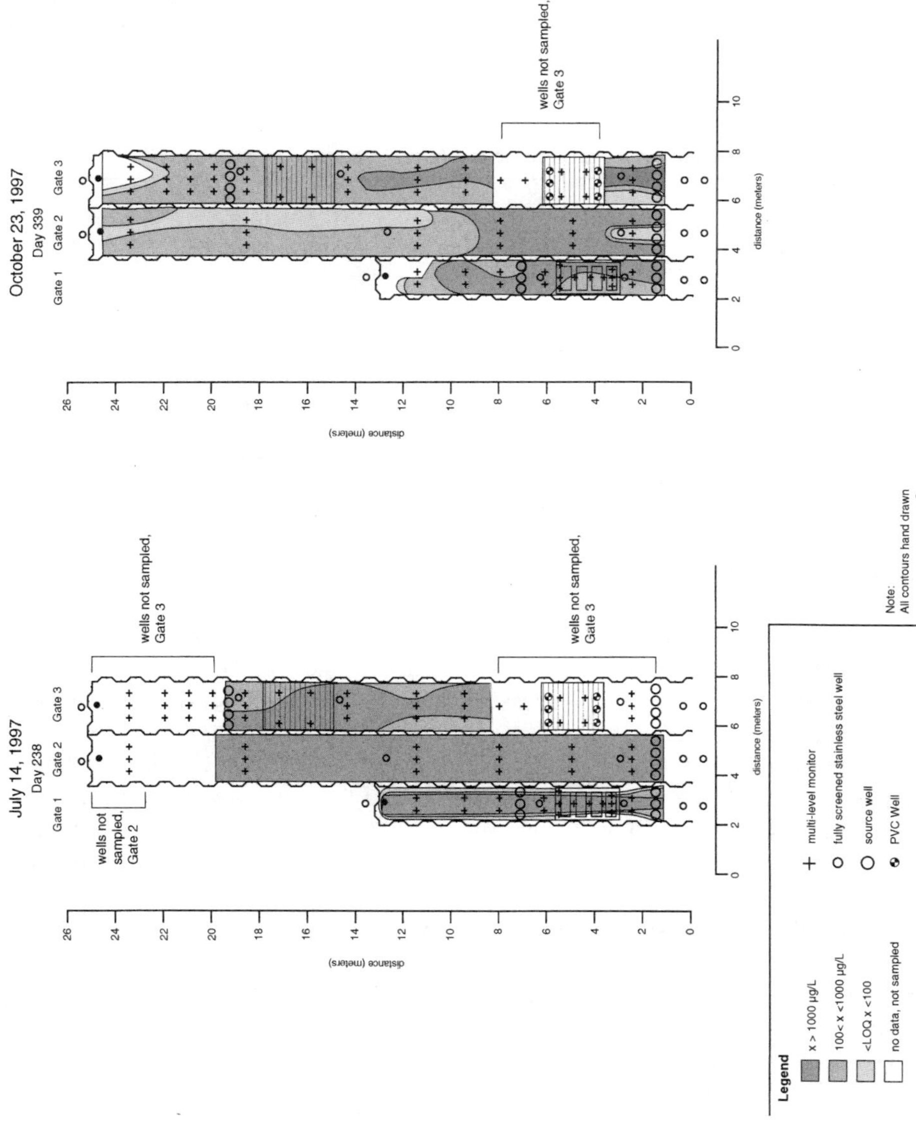

Figure 6.6 Plan view of TOL progress through Gates 1, 2, and 3 on July 14, 1997 and October 23, 1997.

long-term sustainability of the anaerobic and aerobic zones in close proximity. Nevertheless, these preliminary results are very encouraging and demonstrate that coupled anaerobic–aerobic *in situ* treatment systems are feasible.

The treatability of the various compounds encountered over the course of this project is summarized in Table 6.1. The granular iron was found to be an effective medium for the treatment of chlorinated ethenes and, to a lesser extent, chlorinated methanes. The reverse was true of the anaerobic biodegradation treatment. Toluene apparently underwent anaerobic biodegradation intrinsically, but not sufficiently quickly to be remediated within the available space. Aerobic degradation of TOL and *c*DCE in Gate 3 was also indicated, although the observed removal rates included volatilization. The available data suggest that the mass losses due to volatilization were small, but the data were insufficient to accurately quantify this contribution. There were insufficient data to evaluate the aerobic biodegradation of DCM.

Table 6.1 Summary of Target Compound (and Products) Degradability

Compound	Granular Iron	Anaerobic (Bio)degradation	Aerobic Biodegradation
CT	Y	Y	N
CF	Y	Y	Y
DCM	N	Y	?
PCE	Y	Y	N
TCE	Y	Y	N
*c*DCE	Y	N	?
TOL	N	Y	Y

Notes: Y = degradation confirmed or inferred, N = degradation not observed,? = insufficient data for assessment.

6.2.2 Operational Aspects of Gates 1, 2, and 3

The treatment of the chlorinated organic chemicals began immediately upon contact of contaminated groundwater with the granular iron in the cassette system of Gate 1. Apart from the intrinsic transformation of CT to CF (and to a lesser extent CS_2), the reactions in the granular iron cassettes were the only indication of reductive dechlorination for several months into the experiment. By March 1997 (approximately 100 days after activating the source wells) it was apparent that CF was biodegrading in the benzoate mixing zone of Gate 3, and the same process began in Gate 2 after about 250 days. PCE degradation did not become apparent (on the basis of *c*DCE appearance at G315L) in Gate 3 until late September 1997 (approximately 300 days into the experiment) and in Gate 2 for a month or so after that. This amounted to a lag time in Gate 3 of less than 175 days (the minimum being about 40 days) based on the difference between arrival times of PCE in the anaerobic zone and the first appearance of *c*DCE. The effective lag times in Gate 2 were on the order of twice as long, given that *c*DCE first appeared within 5 m of the sources wells and 30 days after its appearance in Gate 3. With the onset of reductive dechlorination in Gates 2 and 3 the remediation of CT-related compounds was complete, but that of PCE and its transformation products was not. However, *c*DCE, the most abundant product of PCE degradation, was apparently treatable by aerobic biodegradation in Gate 3. By fence G324, the final fence in the gate, *c*DCE was below MDLs at 9 out of 11 functioning sampling points. The remaining two points, 324L-4 and 324M-3, corresponded to concentrations of 23 and 10 μg/L, respectively. Thus, although complete restoration of the groundwater was not achieved, there is compelling evidence for the attenuation of the chlorinated ethenes in Gate 3. In contrast, transformations of *c*DCE in Gate 2 were not observed. While the results of the Gate 3 experiment are very promising and clearly demonstrate the viability of semipassive bioremediation of chlorinated compounds, the lag times and incomplete degradation are obvious disadvantages compared with granular iron.

The use of removable cassettes to support the granular iron and ORC™ in this experiment was a novel aspect to the research. The potential advantages of this system included the flexibility to add or delete cassettes containing any reactive material that proved to be limiting or overdesigned, as well as replacing cassettes containing exhausted material, all without the trouble or expense of re-excavation. In practice, two severe limitations were encountered. First, the bentonite seals intended to prevent flow around the cassette assembly, next to the sheet piles, were defective. This resulted in a small amount of flow around the cassette assembly in the early part of the experiment that worsened as the experiment progressed until, near the end of the experiment, virtually all the flow in Gate 1 appeared to pass through the leak. Since there was no evidence of clogging within the porous media in the cassettes (based on coring), it is hypothesized that minor clogging of the upgradient screen may have been the cause of the diverted flow. Since this experiment was initiated, improved seal designs have been developed and implemented at the Borden site, as part of other projects. Therefore, this operational difficulty is not considered to be a serious one vis-a-vis the overall efficacy of the technology.

The second limitation pertains to the poor performance of the ORC™ when it was installed in open cassettes downgradient of the granular iron. Laboratory data strongly indicated a pH sensitivity in the ORC™ used; the release of molecular oxygen was severely hindered at high pHs (greater than pH 9). Unfortunately, water flowing out of the granular iron cassettes was characterized by a pH of 9.5 or more, so a pronounced interference resulted. Attempts were made to rectify the problem by regularly adding mineral acids to the ORC™ cassette to maintain a low pH. However, these attempts failed because, we believe, the open water in the cassette was insufficiently mixed to ensure that the low pH water consistently contacted the ORC™ socks. It is hypothesized that high-pH halos developed around the socks in response to the normal ORC™ reactions, and that the flow through the cassette was too slow to displace these halos. The superior performance of ORC™ in the alternate oxygen addition wells, where flow velocities would be expected to be higher, due to a narrow annulus between the socks and the well screen, supports this view.

Consideration was given to placing the ORC™ upgradient of the granular iron, but this was dismissed on the basis of insufficient space for aerobic biodegradation to occur, rapid accumulation of oxides (hence loss of reactivity) on the granular iron surfaces, and the poor aesthetic quality of the water leaving the system if downgradient aeration was not implemented. Ultimately, a solution was found by moving the ORC™ from the cassettes to large diameter wells located 1.3 m downgradient. In the short-term, the pH problem was resolved by relying on the buffering capacity of the aquifer to maintain neutral conditions and, as alluded to above, the improved hydraulics of the wells assured proper functioning of the ORC™. Over the long term, pH control might require engineered intervention, since the buffering capacity of the aquifer is most likely finite.

The operation of the NIW in Gate 3 proceeded without any serious difficulties. Occasional problems were experienced with the proportional feeder pumps used to administer the benzoate/nutrient/tracer concentrate. This resulted in a few pulses containing less solute mass than planned. The difficulties were attributable to premature pump wear by fine sand particles in the water passing through the system. More exhaustive well development is recommended to alleviate this problem in future systems.

The biosparge zone is believed to have performed satisfactorily in its delivery of oxygen to the groundwater. However, the sparging frequency had to be adjusted from the original design interval of 1 minute, twice weekly to 1 minute daily to maintain aerobic conditions within the sparge zone. In vertical laboratory columns, the 1 minute sparge duration was determined sufficient to emplace oxygen bubbles at residual saturation throughout the gravel medium. On the basis of bubble breakout patterns at the water table, which were examined prior to initiating the experiment, it was estimated that 60% of the biosparge gravel actually received oxygen bubbles (Chapter 4, Section 4.3.5.1). Great care had to be taken not to apply too much pressure to the sparging lines since overpressurizing the lines can lead to the creation of preferential pathways for the bubbles through the gravel, potentially restricting the bubble distribution, and limiting oxygen delivery to the groundwater. DO profiles collected from multilevels in the gravel indicate that oxygen delivery occurred throughout the sparge zone, although not with perfect uniformity. At the time of completion of the project, additional work was needed to optimize the sparging duration and frequency, as well as the benzoate injection concentrations, so that the oxygen demand in the groundwater entering the biosparge zone could be minimized.

In all three gates, contaminants were introduced using passive diffusion emitter devices installed in large diameter wells. This kind of delivery system could be adapted for general remedial use by employing it to introduce desirable substances, such as oxygen, to aquifers instead of contaminants, so some comment about the operational aspects of this technology is warranted. The solutes reaching the aquifer from the source wells (SW) were found to be more heterogeneously distributed (vertically) than had been anticipated. Also, the concentrations of solutes in the SWs were generally higher than those measured at sampling points in the aquifer nearby. Although these discrepancies were, in part, unavoidable due to aquifer heterogeneities and they did not compromise the project in any significant way, minimizing them is clearly a desirable goal. One strategy for achieving this goal was revealed by the tracer studies conducted at the alternate oxygen addition wells. These tests clearly demonstrated the pronounced effect that well development has on passive solute delivery. It is therefore of utmost importance that wells installed for passive solute release be thoroughly developed prior to use, even when they are installed in highly permeable material such as sand.

The snapshot profiles (Figures 6.1 to 6.6) reveal that intrinsic remediation (Gate 2) was an ineffective strategy for treating PCE (and probably its transformation products, although there are insufficient data to make this determination with confidence) or TOL within the gate area. In contrast, PCE was attenuated to below LOQ concentrations by granular iron in Gate 1, and also by anaerobic bioremediation in Gate 3. The Gate 1 system maintained concentrations of TCE, *c*DCE, and VC below the LOQs, except where flow bypassed the cassettes. There was also strong evidence for the onset of aerobic TOL biodegradation in Gate 1. In Gate 3, *c*DCE and TOL persisted at low concentrations (<25 μg/L) in sporadic locations near the end of the gate. Also, a TOL concentration of 381 μg/L occurred at one location on the final fence (324L-4) in a deep, poorly aerated zone. Nevertheless, TOL and *c*DCE appeared to have undergone aerobic biodegradation, when the data are considered overall. Furthermore, in spite of any limitations that might exist, there is unambiguous evidence that the treatment systems tested in this project accelerated the remedial processes compared with the control gate. Additional work is needed to refine various aspects of the technologies tested, but sequential, semipassive groundwater remediation has been shown to be a viable alternative for the cleanup of mixed contaminant plumes.

6.2.3 Lowest Achieved Concentrations of Contaminants

In order to assess the lowest achievable concentrations of contaminants by the various treatment systems, comparisons were made between the two active remediation gates (Gates 1 and 3) and the control (Gate 2). To distinguish between intrinsic processes and engineered remediation, the comparisons were made at locations where contaminants were present in the control gate but absent in the treatment gates (Table 6.2). Also, to ensure that the comparisons were conservative, the most upgradient locations meeting the above criterion were used for each compound. For example, at the end of the experiment in Gate 3 the most upgradient fence where PCE was not detected was G318. This compares with fence G218 in Gate 2 where PCE was present at 188 μg/L. These are the two entries reported in Table 6.2. In the cases where contaminant concentrations were not reduced to nondetectable levels within the Gate 1 or Gate 3 structures, the values reported in Table 6.2 are those from the most downgradient fences where comparisons with Gate 2 were reasonable (i.e., a common distance from source wells). Maximum concentrations occurring at multilevel points on these fences are reported since minimum concentrations may reflect anomalies due to flow effects rather than remediation. Concentrations that fell between the reported MDL and the LOQ (Appendix 5) are reported as trace quantities. It was not possible to assess certain compounds (TCE, *c*DCE) by this comparison since their lowest concentrations may have been more a function of physical mixing processes such as dispersion or dilution (in Gate 2 especially) rather than degradation. This also applies to DCM in Gate 1. The fact that all these compounds were produced as degradation products makes comparisons with Gate 2 difficult, since their first appearance in the gates occurred at different distances from the source wells. However, it is possible to say that of the four degradation products that formed in the gates (CF, DCM, TCE, and *c*DCE) only *c*DCE reached the final sampling fence (G324), and that was at a maximum concentration of only 22 μg/L.

It is apparent from an examination of Table 6.2 that the treatment gates performed extremely well, reducing concentrations of most target contaminants to levels below LOQs, and in some cases below MDLs. The notable exception to this was TOL which, ironically, is a compound known to be easily biodegradable. The apparent persistence of TOL is a reflection of the limited DO availability that was achieved in the treatment gates, particularly at depth in Gate 3. It is believed that with additional time to overcome aquifer reduction capacities, the desired aerobic conditions and the degradation of TOL could be achieved.

Table 6.2 Comparison of Target Organic Concentrations in Gates 1, 2, and 3

Compound	Fence	Gate 1	Gate 2	Gate 3
CT	G104/205	ND	t	—
	G312/212	—	ND	ND
CF	G104/205	ND	238	—
	G312/212	—	ND	ND
DCM		—	—	—
	G312/212	—	ND	ND
PCE	G104/205	ND	322	—
	G318/218	—	188	t
TOL	G108/208	248	7,900	—
	G324/224	—	892	380

Notes: t = trace, ND = not detected, — = data not available. Units are μg/L.

6.2.4 Performance Parameters

First-order kinetics were assumed for the degradation kinetics of all the contaminants considered. This convention was adopted, in part, because the processes studied (reactions on granular iron, biodegradation) are often pseudo-first order in nature, and partly for ease of comparison. It was further assumed that all reactions occurred in the aqueous phase, both in the estimation of the rate constants (and half-lives) and in the calculation of percent mass removed, below. Handling the data in this way leads to entirely empirical degradation rates; details of the processes and mechanisms involved are ignored. For example, although sorption to aquifer solids is known to occur, and in some cases may accelerate reaction rates (e.g., CT to CF), the need for explicit retardation factor corrections to the rate constants is eliminated by treating the transformations as equivalent aqueous phase reactions. Degradation rates are simply compared as empirically derived pseudo-half-lives. The data used to calculate the pseudo-half-lives in Gates 2 and the anaerobic section of Gate 3 consisted of maximum concentrations at the monitoring fences on day 339, and hence are considered conservative. The data used to estimate half-lives in the iron cassettes, Gate 1, and the aerobic section of Gate 3 were fence averages. The aerobic section of Gate 3 was averaged for these calculations because zones of poor oxygen delivery at the base of the aquifer resulted in uncharacteristically limited aerobic degradation at depth. The use of maximum concentrations from these regions of the aquifer would have resulted in a misleadingly poor assessment of the overall biosparge zone performance.

The effectiveness of a treatment method can be gauged by comparing the pseudo-half-lives to the residence times of the contaminants in the reactive zones, and estimating the percent mass removal that can be expected within these zones (Table 6.3).

Table 6.3 Summary of Performance Parameters

Compound	Zone	Res. Time (days)	Pseudo- $t_{1/2}$ (days)	% Mass Removal
CT	Iron	6	—	—
	Intrinsic	130	13, 11	99.9-100
	Anaer. Bio	70	3.5	100
CF	Iron	6	0.7	99.7
	Intrinsic	130, 120	86, 11	70[a], 100
	Anaer. Bio	70	9-15	96-99.5
PCE	Iron	6	0.4	100
	Intrinsic	130	—	NR
	Anaer. Bio	70	—	—
TOL	Aer. ORC™	30	—	—
	Intrinsic	130, 93	62, 8.3	64[a], 100
	Biosparge	45	30, 22	65[a], 76

Notes: — = not determined, NR = no significant removal.

[a] Based on pseudo-first-order rate constants (see Chapter 3 or 4) representing project-averaged degradation rates. Percent removals indicated by snapshot data from the October 23 sampling, given as the second number, were greater than this, possibly due to changes in microbial communities.

$$\%MR = \left(1 - \exp\left[\frac{-(0.693 \cdot t_r)}{t_{1/2}}\right]\right) \cdot 100\% \tag{6.1}$$

where %MR = percent mass removed, t_r = residence time, $t_{1/2}$ = pseudo-half-life.

The residence times listed in Table 6.3 are those calculated for nonsorbing solutes migrating through the reactive zones with a groundwater velocity of about 15 cm/day for Gates 2 (conservatively high) and 3, and 20 cm/day through the cassettes in Gate 1. In the case of Gate 2 ("intrinsic" in Table 6.3), two residence times, half-lives, and percent mass removal estimates are given for some compounds. These represent project-averaged and end-of-project performances, respectively. Pseudo-half-lives were determined as discussed in Chapters 2, 3, and 4. Examination of Table 6.3 reveals that removal of the chlorinated target contaminants (CT, PCE) and at least one breakdown product (CF) was successfully accomplished in Gates 1 and 3. Intrinsic processes were also sufficient to control CT and CF levels, although they were less successful with PCE (no pseudo-half-life was determined due to incomplete breakthrough at the final monitoring fence, G224). Half-lives for PCE degradation in Gate 3 and TOL degradation in the aerobic section of Gate 1 were also omitted from Table 6.3, because the rates of removal could not reasonably be assumed to be at steady state at the end of the experiment. In Gate 3, with PCE not at steady state, *c*DCE was not at steady state and therefore has been omitted from Table 6-3. In spite of the difficulties associated with quantifying the PCE removal rates in the anaerobic zone of Gate 3, the data show virtually complete PCE removal there. The available data also suggest aerobic removal of TOL in

Gate 1 and *c*DCE in Gate 3. Attempts were made to quantify TOL and *c*DCE removal in Chapters 2 and 4, respectively. However, while encouraging, these results are considered too preliminary to be included in the performance assessment (Table 6.3).

The limited success of establishing active aerobic zones is reflected in the lower TOL mass removal percentages in the biosparge zone. However, as discussed previously, the aerobic zones in both Gates 1 and 3 were in the process of becoming established when the project came to an end. Also, the deeper sections of the aquifer in Gate 3 were susceptible to oxygen limitations due to the limited ability of the biosparge system, as designed for this experiment, to aerate the groundwater uniformly at depth. Therefore, the performance parameters reported in Table 6.3, for TOL in the biosparge zone, are not considered reflective of what the technology is capable of achieving. At the end of the experiment, TOL removal appears to have been occurring quite rapidly in Gate 2, probably as an anaerobic process. Further work is needed to evaluate this development in detail.

6.3 THE ALAMEDA EXPERIMENT

6.3.1 System Performance and Performance Parameters

The Alameda experiment drew upon the preliminary results of the Borden work to design the sequential treatment system most likely to achieve the remedial objectives (the actual performance with respect to those objectives is discussed in the following section). On the basis of complete chlorinated ethene degradation with no lag time for the onset of the reactions, granular iron was selected as the preferred anaerobic treatment method. Peerless was chosen as the supplier for the granular iron to maintain consistency with the Borden experiment. In order to minimize expenses, and to avoid complex hydraulics as much as possible, removable cassettes were not utilized.

With the decision made to use granular iron for anaerobic treatment, it remained to select the aerobic treatment method. Ultimately, it was decided that biosparging was the best choice for introducing DO to the gate, because the preliminary Borden experience demonstrated a pH-related interference between iron-contacted water and the ORC™. Laboratory investigations into the distribution of bubbles in a gravel-filled biosparge zone (UW, 1997) indicated a strong sensitivity to delivery pressure; too low a pressure would not deliver bubbles to the porous medium, while too high a pressure created irreversible pathways through which bubbles would pass preferentially in subsequent sparge events. The difficulties and risks associated with implementing a gravel-filled biosparge zone in the field had not yet been investigated at Borden, so the decision was made to maintain as open a biosparge zone as possible, to ensure that no preferential pathways developed. The medium adopted to support the bacterial populations comprised Yeager TriPak™, 5-cm-diameter PVC bioballs. Unlike gravel, this material is inherently less dense than water, so the balls could not be emplaced throughout the entire depth of the biosparge zone; the lower few meters were left as open water. The extreme openness of the biosparge system, together with the frequent mixing of the groundwater during sparging, made the Alameda biosparge system resemble a mixed bioreactor more than a porous media gate. Data interpretation was carried out accordingly.

Preliminary site investigations of the shallow aquifer (consisting of an artificial fill unit, see Section 5.2.2, Chapter 5, this report) at Alameda indicated that groundwater velocities were on the order of 1.7 cm/day toward the Bay, and that the organic contaminant plume was characterized by elevated concentrations of *c*DCE, VC, and TOL (peak values were 218, 16, and 9 mg/L, respectively; cf. July 26, 1996 sampling results, UW, 1996) and lower concentrations of TCE, 1,1-DCE, *t*DCE, benzene, ethylbenzene, and the xylenes (all <1 mg/L). Bulk water samples collected for the purposes of the treatability study contained *c*DCE at 32 mg/L and VC at 26 mg/L; other contaminants were present with concentrations less than 1 mg/L each. It was believed that these levels were generally representative of the plume. Degradation rates and half-lives for reactions on granular iron were estimated from column tests with this water, and the permeable iron barrier was designed to treat this level of contamination.

The treatability study indicated that for the highest anticipated flow rate (1.27 m^3/day in Operating Phase 1 (OP1)), an iron wall thickness of 1.52 m was adequate to treat the contaminant loading. As an added precaution, the wall was augmented with a 0.45-m-thick reactive barrier comprising a 3–5% iron/sand mixture at the influent end of the gate. Velocity estimates for flow through the permeable barrier in OP2, in which the flow rate was 0.34 m^3/day, were on the order of 6–9 cm/day (based on a tracer study, Table 5.6). Similar velocities were estimated from Darcy's Law calculations and ISPFS measurements (although the reliability of the latter determinations were considered questionable), corresponding to residence times of 21 to 25 days. Residence times of 17 days were used in the performance parameter calculations (Table 6.4) to reflect time spent in the granular iron alone, using the highest velocity estimates (most conservative) from Table 5.5, Chapter 5. Performance parameters for contaminant degradation in the iron wall were determined by fitting 4-month integrated mass loadings (during OP2) at three fences, located before, within, and beyond the wall, to a first-order kinetic model (Table 5.8 and Table 6.4). Percent mass removals for the estimated residence times of contaminants in the wall were calculated as previously described (Section 6.2.3). Also shown in Table 6.4 are the half-lives determined from the treatability study involving granular iron packed columns.

Table 6.4 Summary of Performance Parameters for the Alameda OP2

Compound	Treatment Zone	Treatability $t_{1/2}$ (days)	Residence Time (days)	Pseudo-$t_{1/2}$ (days)[a]	% Mass Removed
TCE	Iron	0.06	17	1.1–2.9	98–100
*t*DCE	Iron	—	17	2.1–4.5	93–99
1,1-DCE	Iron	—	17	2.2–5.3	89–99
*c*DCE	Iron	0.5	17	2.9–7.1	81–98
	Biosparge	—	32	15	77
VC	Iron	0.47	17	1.1–1.7	99.9–100
	Biosparge	—	32	70	27

Note: — = data not available.

[a] half-lives taken as numerically fit values (see Table 5.8).

The field pseudo-half-lives for organics degradable on granular iron were three to ten times longer than those for the same compounds in the lab (and for those obtained for PCE and CF in the Borden field trial). As a result, complete mass removal of the contaminants within the iron wall was not realized, although >90% removal was achieved in all but one case (*c*DCE). The cause of the discrepancy between lab and field half-lives is partially attributable to the different methods used to derive them; treatability half-lives were estimated from steady-state column profiles, while the field values came from 4-month integrated mass profiles. Nevertheless, the differences observed are larger than expected even accounting for these distinctive estimation methods. A more important contributor to the discrepancy is likely the variability and magnitude of concentrations in the plume. It was discovered, during the preliminary sampling program, that *c*DCE existed in some parts of the plume at concentrations in excess of 200 mg/L, considerably more than the 32 mg/L used in the treatability test. Two issues arise as a result of this realization: first, the large difference between the concentrations encountered during the site characterization study and those obtained for the treatability study indicates a highly heterogeneous distribution of contaminant concentrations in the plume. It is possible that unexpectedly high concentrations of any or all of the contaminants (e.g., *c*DCE at or above 200 mg/L) existed in other, undetected parts of the plume and were transported into the gate. However, since the gate was designed for OP1 flow rates, there was a significant factor of safety built into the design, and enough iron was installed in the gate to treat several hundred milligrams per liter of any of the chlorinated organics at the lower OP2 flow rates, assuming the validity of the treatability study rate constants. This leads to the second issue: the possibility that the rate constants calculated from the treatability study are not applicable to the degradation of such high concentrations of chlorinated compounds. Since the reaction is known to be surface related, the possibility of surface saturation increases as solute concentrations rise. As the surface becomes saturated, the pseudo-first-order rate constants would be expected to decline, possibly leading to partial breakthrough of contaminants from the wall, as was observed.

To overcome the problem of surface saturation, two strategies are currently feasible: first, increase the amount of available reactive surface or, second, decrease the concentration of incoming reactive solute. The first option requires that the wall thickness be increased. This can only be done economically if the concentrations in the plume are well known, i.e., detailed plume characterization is essential, probably by either multilevel monitoring or profiling techniques (Zemo et al., 1994). The second option requires that the plume be diluted prior to entering the wall. A possible method of accomplishing this, while maintaining the passive character of the treatment system, is to precede the reactive barrier with an open water zone, similar to the space between the cassettes in Gate 1 of the Borden experiment. The Borden experience indicated that this kind of pretreatment structure would do much to homogenize the concentrations passing from the aquifer into the wall, thereby reducing the maximum concentrations and the risk of breakthrough due to surface saturation.

The biosparge gate design for implementation at Alameda possessed an inherently higher risk of stripping volatiles from groundwater than did the Borden design. Specifically, this risk resulted from more frequent sparging (three times daily compared to once daily at Borden) and longer sparging times (12 to 15 minutes compared to 1 minute at Borden). The increased sparging was necessitated by the open biosparge structure that exhibited a low residual saturation of oxygen bubbles. To take this into consideration, the data were interpreted using a mixed bioreactor model that accounted for instantaneous partitioning of VOCs into the sparge bubbles, as well as storage in the unsaturated headspace. First-order kinetics were assumed to describe the biodegradation process (a zero-order model was also used for comparison, but found to describe the data very poorly), and the rate constants were estimated for *c*DCE and VC, the only two contaminants present in sufficient concentration to make the analysis possible. As can be seen in Table 6.4, the apparent degradation rates of *c*DCE and VC were insufficient to effect complete removal within the biosparge zone. Based on the average residence times, it is estimated that about 30 to 80% of the mass entering the aerobic zone was degraded. Laboratory microcosm studies qualitatively supported this finding by demonstrating that the Alameda aerobes were capable of degrading both *c*DCE and VC. The field site removal rates could be improved if a larger active biomass were established in the biosparge zone. The available data do not clearly show that a steady-state population had developed at the time of writing, so it is possible that the biosparge zone was capable of supporting, and was in the process of growing, a larger population up to the time of the most recent sampling. With time, mass removal rates in the biosparge gate might increase to levels required to meet the remedial objectives of the project. Further work is needed to determine how best to accelerate the required microbial growth in the event that the existing growth rates are found to be inadequate.

A limitation of the coupling of granular iron with aerobic biodegradation at Borden was a pH interference between iron-contacted water and ORC™. Although ORC™ was not used at Alameda, there was still concern that the alkaline water, flowing out of the permeable iron wall, might prevent the establishment of a robust aerobic microbial population in the biosparge zone. To neutralize the alkaline water, carbon dioxide gas periodically sparged into the biosparge zone, in addition to oxygen. This measure proved to be successful at maintaining the desired pH range (7 to 8). The above analysis of degradation behavior indicates that the two technologies did function well in sequence. No other interferences related to the coupling of anaerobic and aerobic processes were noted over the course of the experiment.

6.3.2 Remedial Objectives

The remedial objectives (ROs) for the project were established in terms of the concentrations of contaminants in the effluent from the treatment gate. These were reviewed and discussed in Chapter 5, Section 5.1.1. In most cases, the objectives were met throughout the project. However, during periods of high (several hundred milligrams per liter) inputs of *c*DCE, this compound and VC were not treated to the objectives, with average effluent concentrations of about 169 μg/L (RO = 10 μg/L) and 134 μg/L (RO = 2 μg/L), respectively (Table 6.5). In spite of the breakthrough of these two compounds, the data indicate that they were both subject to more than 90% mass removal from the groundwater by reaction on the granular iron, and an additional 20 to 80% mass removal from the iron effluent by aerobic biodegradation in the biosparge zone. Given the extremely high concentrations of these compounds entering the system (10 to 200 mg/L), the results obtained are considered excellent. Furthermore, subsequent monitoring of the system (data to be supplied under separate cover) reflecting performance during periods of low (50 mg/L) *c*DCE loading indicates that all remedial objectives can be met, as designed.

Table 6.5 Comparison of Remedial Objectives and Breakthrough Concentrations at Fence 8 on September 30, 1997

Compound	Remedial Objectives (μg/L)	Fence 8PA (μg/L)	Fence 8PB (μg/L)
TCE	5	<5	<5
1,1-DCE	6	<5	<5
*t*DCE	10	<5	<5
*c*DCE	10	136[a]	131[a]
VC	2	217[a]	121[a]
Benzene	10	<5	<5
Toluene	150	21.8	<5
Ethylbenzene	700	<5	<5
Xylenes, total	1750	<5	<5
pH	5–8.5	6.89	8.78[a]
Iron	500	70	170

[a] Value exceeded remedial objective on September 30, 1997, following *c*DCE loadings in excess of 200 mg/L, but did not exceed remedial objectives in January 1998, with *c*DCE input levels of 50 mg/L.

The pH at row 8 was slightly in excess of the RO, but this was not considered a problem since most of the pH measurements made in the biosparge zone were within the desired range and the microbial activity appears to have been minimally affected by any such fluctuations.

The assessment of contaminant distributions in the plume entering the treatment gate has been shown to be an important step in the site characterization leading up to the design phase of a field installation. In such an assessment, depth-discrete multilevel sampling is required to identify regions of excessively high contaminant concentrations. The same issues arise in the assessment of the gate performance vis-a-vis the ROs. In the Borden work (Section 6.2), samples from multilevel monitors were considered, resulting in the most critical assessment possible. The wells in row 8 at Alameda were fully screened, and therefore yield samples with vertically averaged solute concentrations. The question arises as to whether this method of sampling biases the results in favor of meeting the ROs when in fact ROs have not been achieved at some discrete depths. However, in the case of the Alameda project, this issue was less contentious than it might first appear. The high degree of mixing that occurs in the biosparge zone largely eliminates any significant vertical variation in solute concentration, so that the samples collected from fully screened wells and multilevel monitors in downgradient pea gravel would be expected to yield very similar levels of contamination. This issue should be considered carefully when gates constructed with only porous media-based treatment systems are employed.

6.4 SELECTED RECOMMENDATIONS FOR TECHNOLOGY APPLICATION

Recommendations concerning each of the technologies tested in the Borden experiment are provided in Chapters 2 to 4; selected recommendations concerning the Alameda project are provided in Chapter 5, and details are provided as part of the Technology Evaluation Report (BEAK International, Inc.), Part II of this volume. A summary of the major recommendations discussed in this Technology Demonstration Report are provided below.

6.4.1 Passive Delivery of Solutes to Aquifers from Wells

1. Wells must be installed and thoroughly developed prior to attempting to use them for solute delivery. This recommendation applies to wells installed in both well sorted and poorly sorted aquifers.
2. Downhole assemblies should consist only of polymer materials. No connections or fittings should be installed below the top of casing. This measure will minimize the possibility of leakage, resulting in overloads of solutes to the groundwater in the well.
3. Density driven flow is likely to be more pronounced in the standing water within the well than it would be if the solute were introduced directly to the porous medium. Care must be taken to consider the likelihood and consequences of increased solute flux from the bottoms of the source wells due to density effects.

6.4.2 Granular Iron in Removable Cassettes Coupled to ORC™ for Oxygen Delivery

1. Cassette assemblies should either be redesigned to eliminate corner seals, or installed with water-tight corner seals between them and the gate walls. It is recommended that if bentonite is chosen to establish these seals, it should be emplaced with forms to eliminate the possibility of invasion by aquifer material. The material beside the cassette assembly, between the corner seals, should also be of low permeability as an added precaution against leaks. In some cases, a mixture of bentonite and native material could serve this purpose.
2. With the exception of the first cassette, volumes of standing water between cassettes should be minimized to reduce the possibility of volatile losses to the atmosphere and to simplify data interpretation. A small volume of open water prior to the first cassette containing granular iron can promote the partial homogenization of the concentrations entering the cassette. This could have the beneficial effect of more evenly distributing the reaction over the iron surface, prolonging the reactive material lifetime.
3. ORC™ performs acceptably when it is installed in wells, and in contact with near-neutral pH groundwater. ORC™ is not recommended for use in open water cassettes due to the possible formation of high pH halos around the socks, causing poor oxygen release performance. This is particularly the case when the ORC™ cassette is located immediately downgradient of granular iron, since iron-contacted water is characteristically at about pH 10.

6.4.3 Anaerobic and Aerobic Bioremediation Using An NIW and a Biosparge Wall

1. All fully screened wells for injection and extraction in the NIW should be developed until the water produced is consistently clear, to avoid unnecessary wear of pump parts.
2. Benzoate was shown to be a substrate suitable for the support of reductively dechlorinating micro-organisms. However, it is recommended that additions of benzoate be supplemented with inorganic nutrients, such as nitrogen, phosphorus, and/or trace metals (Appendix 11), to accelerate the onset of the desired transformations.
3. Under the conditions of the Borden experiment, a mixing zone, downgradient of the NIW, of not less than 10 m is required for the development of chlorinated ethene degrading populations. A mixing zone of 5 m is sufficient for the bioremediation of chlorinated methanes, provided that a labile carbon source, such as toluene, is a co-contaminant in the plume. A longer mixing zone may be required if the only reduced carbon substrate is the added one (e.g., benzoate).

4. Oxygen sparging frequencies can initially be determined on the basis of BOD estimates from the anticipated organic loading to the biosparge zone. However, these levels should be closely monitored in the field and adjustments made to both the organic loadings (if possible) and the sparging frequencies and duration, as required.

6.4.4 Granular Iron Coupled to a Biosparge Reactor

1. Detailed plume characterization based on multilevel monitoring or profiling is recommended to fully understand the nature and geometry of the loading to the granular iron. This could be of great importance in cases where solutes are present in such high concentrations that surface saturation effects might become an issue.
2. Treatability studies should be conducted with site groundwater containing peak concentrations of contaminants to best predict the likely degradation rates. If samples of such groundwater are difficult to collect directly, then the samples that are collected should be spiked with the necessary solutes to achieve the peak concentrations.
3. Microcosm studies should be conducted as part of the overall treatability study to assess the aerobic biodegradability of target organics at the site.
4. Sparging rates in the biosparge zone should be minimized to limit stripping of volatile contaminants from the groundwater. The optimization of the sparging rate should occur only after the required populations of aerobes have become established.

6.5 CONCLUDING REMARKS

This project has shown, at two field sites, that the treatment of mixed contaminant plumes by *in situ*, semipassive remedial technologies is a viable option. The sequential operation of anaerobic and aerobic treatment technologies was anticipated to be challenging from the standpoint of geochemical and microbiological interferences. In practice, these kinds of interferences proved to be less problematic than first thought. During the experiment, the only case of permeability loss due to biofouling or precipitate formation occurred in the cassette system and the severity of the problem was magnified by the failure of seals between the cassette assembly and the sheet piles. It remains uncertain if the problem would have been noticed at all if not for the faulty seals. Further work is needed to evaluate the longer term behavior of this kind of treatment system, to see if the anticipated interferences become more important with time, and to evaluate the sensitivity of the system to differing native geochemical environments.

The advantages of removable cassettes containing granular iron over permanently installed iron walls were not explored as part of this project. The granular iron life expectancy, based on field tests conducted elsewhere, and in progress, greatly exceeded the duration of this experiment, so cassette replacement was not required. However, some of the disadvantages of removable cassettes were apparent in the project. They included greater expense and difficulty in installation compared to cassette-free systems; complex contaminant transport between porous media and open water sections; relatively high possibility for volatilization and possibly inappropriate hydraulics for ORC™ functioning. Further work is needed to determine if the flexibility afforded by the cassette design outweighs these disadvantages over the long term. Over the short term, the use of cassettes is probably difficult to justify.

Anaerobic bioremediation was shown to be a viable alternative for the removal of CT and PCE and many of their chlorinated transformation products. Comparisons between the progress of the chlorinated compounds in the control gate at Borden (Gate 2) with the biodegradation gate (Gate 3) indicated that the chloromethanes were intrinsically degradable in the presence of toluene, also a target organic. However, Gate 3 outperformed the control gate in two respects: first, the lag times leading up to the onset of reductive dechlorination were slightly shorter and, second, there was evidence that PCE and possibly TCE degraded more rapidly in Gate 3 than in Gate 2. It should be noted that the lag times experienced in the Borden experiment were at least partially due to the pristine nature of the aquifer before the experiment. At sites where contamination has existed for some extended time, acclimation of subsurface microorganisms would be expected and the lag times would likely be shorter or absent altogether. Additional work is required to investigate this general aspect of the technology performance as well as the continued development of TCE, *c*DCE, and VC degraders in the anaerobic treatment zone at Borden, specifically.

The success of the anaerobic treatment zone in dechlorinating PCE to *c*DCE is encouraging both scientifically and from the standpoint of technology development. However, it also raises some practical concerns. The concerns comes from the perception that *c*DCE is less desirable in the groundwater than the original contaminant, PCE. However, the aerobic biodegradation of *c*DCE was demonstrated at the Alameda site and appeared to be an active process at Borden as well. Further work is needed to determine the factors limiting *c*DCE degradation in the aerobic zones and to optimize this process. Given the uncertainties that remain in the ultimate performance of the anaerobic biodegradation zone, granular iron is the preferred choice for treating shallow plumes of dissolved chlorinated ethenes and a variety of other halogenated compounds. However, for plumes at great depth, or for the treatment of chemicals not reactive in the presence of granular iron, the anaerobic bioremediation gate may still have application.

Historically, aerobic biodegradation of petroleum hydrocarbons has been the most successful example of engineered bioremediation. It is therefore ironic that in this work the anaerobic processes appeared to be more complete than the aerobic ones. However, this result is more a function of the limited time available for the fine-tuning and evaluation of the aerobic systems tested. Since the anaerobic processes were the first in the treatment sequence, considerably more time was available to identify and deal with problems in these zones than in the aerobic process zones. At the end of the project, the most important limitations for aerobic biodegradation were (1) overcoming the reduction capacity of the native aquifer material so that an extended aerobic zone could develop, and (2) overcoming dissolved organic carbon loading, particularly of benzoate in Gate 3, so that oxygen could be available for the biodegradation of target organics (TOL, *c*DCE). Limitation (1) above is thought to be one of finite duration. Further work is needed to evaluate how long that time interval is likely to be. Limitation (2) requires additional work to optimize the delivery of benzoate to the anaerobic zone, supporting the anaerobic bacteria, while minimizing its impact on the dissolved oxygen supply in the aerobic zone.

The hydrogeologic conditions and plume encountered at the Alameda site were far removed from the "ideal" aquifer and contamination scenario created at Borden. Nevertheless, by the time of the September sampling, the performance of the sequential anaerobic–aerobic treatment system was sufficient to reduce the concentrations of all but two contaminants (*c*DCE and VC) to the ROs for the project. In spite of not meeting the ROs, the Alameda treatment gate removed 98 to 99% of the mass of these two compounds over a few meters of flow path. The failure of the system to reach the ROs for these compounds is partially due to unexpectedly high loadings of *c*DCE that may have resulted in near surface saturation of the iron medium. Once input concentrations diminished (January 1998), all the chlorinated compounds were treated to below ROs in the permeable iron barrier. Furthermore, it remains possible that with increased biomass buildup in the biosparge zone, the ROs may yet be achievable even during periods of extraordinarily high loadings. These results underline the main conclusion of this report: that the treatment of mixed contaminant plumes by sequential, *in situ* anaerobic–aerobic treatment technologies is a viable option for controlling groundwater contamination.

REFERENCES

University of Waterloo (UW). 1995. Revised Workplan for Passive and Semipassive Techniques for Groundwater Remediation at CFB Borden. Submitted to AATDF, July 1995.

University of Waterloo (UW). 1997. Passive and semipassive techniques for groundwater remediation: Phase 1 Installation Report. Prepared for AATDF, for Environmental Technology, U.S. Department of Defense DOD-AATDF Administration, Rice University. Submitted April 1997.

Zemo, D.A., Pierce, Y.G., Gallinatti, J.D. 1994. Cone penetrometer testing and discrete-depth ground water sampling techniques: a cost-effective method of site characterization in a multiple-aquifer setting. *Ground Water Monitoring and Remediation*, 14(4): 176-182.

PART II

Technology Design and Evaluation

Michaye McMaster
Geosyntec Consultants, Inc.

Robert Focht
EnviroMetal Technologies

Evan Cox
Geosyntec Consultants, Inc.

John Vogan
EnviroMetal Technologies

Executive Summary

The University of Waterloo (UW) was contracted by Rice University under Advanced Applied Technology Demonstration Facility (AATDF) to develop and demonstrate passive and semipassive techniques for remediation of groundwater containing mixtures of chlorinated solvents and petroleum hydrocarbons. These organic chemicals are commonly present in groundwater at U.S. Department of Defense (DOD) facilities and are frequently present in mixed plumes. Current remediation technologies for groundwater containing these chemicals is costly, and frequently results in transfer of chemicals to other media (e.g., air) or locations (e.g., landfill) rather than destruction of the chemicals. The goal of the research was to assess the performance of innovative passive and semi-passive remediation technologies and provide guidance and advice for their full-scale use at DOD facilities. Such technologies are potentially less costly and result in compound degradation rather than transfer.

Field demonstrations were conducted by UW at two sites, Canadian Forces Base (CFB) Borden in Ontario, Canada and Site 1 at Naval Air Station (NAS) Alameda, California. The Borden site investigation was designed to evaluate four *in situ* treatment technologies for remediating mixed plumes, and to provide data to select the most appropriate technologies to implement at an existing plume at a DOD facility. Results of the Borden demonstrations provided data supporting the coupled *in situ* treatment technologies. A sequenced permeable reactive barrier (SPRB) consisting of a zero-valent metal component and an aerobic biosparge component was judged to be the best configuration of those investigated. This configuration was applied to the Alameda site and a pilot-scale SPRB was installed at the Alameda site to treat a portion of an existing mixed plume. Results of the Alameda demonstration provided strong evidence to support the effectiveness and cost competitiveness of a combined zero-valent metal-bioremediation SPRB. More than 90% of the mass of chlorinated VOCs was removed by reaction with the granular iron. An additional 20 to 80% of the mass present in the iron effluent was removed by aerobic biodegradation in the biosparge zone. Given the extremely high concentrations of these compounds entering the system (10 to 200 mg/L), the amount of mass removed during treatment (92 to 98%) indicates excellent performance of the SPRB system.

BEAK International, Inc. was retained by UW and Rice University to provide technical expertise and to prepare an evaluation of SPRB technology. This report evaluates SPRBs for *in situ* treatment systems for groundwater remediation.

Two hypothetical full-scale SPRBs were designed and costs were estimated for Alameda to evaluate what level of performance could be expected of a full-scale SPRB, and to generate cost information for the technology. It was assumed that a Funnel-and-Gate configuration, as used in the pilot, would be used in the hypothetical full-scale system. Two remediation scenarios were evaluated: Case A consisted of capturing a relatively narrow, high concentration core area of the plume inside the 1000-µg/L VOC contour (plume width of about 80 ft); Case B consisted of capturing groundwater containing more than 10 µg/L of cDCE (plume width of about 350 ft).

The costs of the two SPRB systems were compared to costs of pump-and-treat (P&T) systems designed to capture and treat the same plume dimensions. Both cumulative cost and net present value (NPV) calculations over a 30-year period were used to compare the SPRBs to P&T. For Case

A, P&T cumulative costs exceed SPRB costs after 10 years of operation. For Case B, P&T cumulative costs exceed SPRB costs after 16 years of operation. The Case A SPRB has a NPV that is $104,000 less than P&T. The Case B P&T has a NPV that is $242,000 less than the SPRB. The capital costs of SPRBs have a large influence on long-term cost estimates.

The various design assumptions made with respect to construction method, input concentrations, regulatory criteria, and frequency of Operation, Monitoring, & Maintenance (OM&M) have a significant effect on SPRB costs and therefore the cost competitiveness of SPRBs with conventional alternatives. For example, cost competitiveness of SPRBs would increase over that observed at Alameda if different construction methods were employed. Other indirect factors such as land-lease costs and local public perception could also influence the preferential selection of SPRB technology at the specific site.

SPRB technology could be applicable at many DOD facilities. For example, DOD estimates that more than 8300 sites on more than 1500 installations will require remediation of contaminated materials (U.S. EPA, 1997). Of these sites, groundwater contamination has been reported for at least 2280 sites. Halogenated and nonhalogenated VOCs are the most prevalent groundwater contaminants, detected at 74% of all DOD sites having known groundwater contamination (more than 1680 sites). The estimated cost to complete remediation work at contaminated DOD facilities is more than $28.6 billion, much of which will be devoted to groundwater remediation.

Significant incentive exists for the development of innovative and cost-effective remediation technologies for organic chemicals in groundwater. In particular, incentive exists to develop passive and semipassive *in situ* technologies that can provide lower long-term cost as compared to conventional remediation systems (e.g., P&T). A variety of potential SPRB configurations exist for simultaneous remediation of several organic chemicals in groundwater plumes. The most cost-effective configuration for a given mixed plume will be dependent on the target chemicals to be remediated and site-specific conditions.

CHAPTER 7

Introduction to the Technology Evaluation

This report has been prepared for the University of Waterloo (UW) Department of Earth Sciences and the DOD-AATDF project administered through Rice University. UW was contracted by Rice to conduct an investigation of various passive and semipassive techniques for remediation of mixed contaminant groundwater plumes. A work plan (UW, 1995) outlined the goals and objectives of the research program. One deliverable of the original work plan was a Technology Evaluation Report (TER), prepared by the project Principal Investigator, to assess technology transfer and applicability of the technologies studied. While the research was in progress, the DOD-AATDF staff indicated that the TER should be subcontracted to a science and engineering company familiar with the focus of the research program. This company would be contracted to complete the TER according to the DOD-AATDF guidelines. Accordingly, UW retained EnviroMetal Technologies, Inc. (ETI) and BEAK International, Inc. to

- Provide technical expertise to select and implement the pilot-scale experiment at a UW-AATDF research site (NAS Alameda);
- Use pilot-scale data to design and cost a hypothetical full-scale system for the NAS Alameda site (Hypothetical Site); and
- To prepare the TER for sequenced permeable reactive barrier (SPRB) technologies.

The UW-AATDF research was completed at two sites, Canadian Forces Base (CFB) Borden and Site 1 at Naval Air Station (NAS) Alameda. Installation reports for each field site and a Final Technical Document (FTD) summarizing the results of the entire experiment were submitted to Rice (UW, 1997a, 1997c, 1998).

The technical approach and background are presented in Section 7.1. Section 7.2 provides a brief summary of the technologies demonstrated in the UW-AATDF research program. Chapter 8 describes the application of a SPRB at a hypothetical site. Chapter 9 presents a measurement procedures and monitoring program plan. Chapter 10 compares costs and performance of both the hypothetical SPRB and a hypothetical groundwater extraction and treatment (pump-and-treat [P&T]) system. Chapter 11 discusses the applicability of the SPRB technology. Report references are provided in Chapter 12. A list of acronyms and abbreviations used in this report is provided at the beginning of this volume.

7.1 BACKGROUND

Chlorinated solvents are among the most frequently detected contaminants in groundwater. These include the chlorinated ethenes (e.g., tetrachloroethene, also known as perchloroethene (PCE), trichloroethene (TCE), the dichloroethene isomers (*cis*-1,2- (*c*DCE), *trans*-1,2- (*t*DCE), and

1,1-DCE), and vinyl chloride (VC)) and chlorinated methanes (e.g., carbon tetrachloride (CT), chloroform (CF), dichloromethane (DCM, also known as methylene chloride), and chloromethane (CM)). Petroleum-derived aromatic compounds, such as benzene, toluene, ethylbenzene, and the xylene isomers (*ortho-*, *meta-*, *para-*), collectively termed "BTEX," are also common groundwater contaminants. The occurrence of both chlorinated solvents and petroleum hydrocarbons (here termed "mixed plumes") is not uncommon. The DOD estimates that more than 1680 groundwater impacted sites may exist where more than one contaminant class is present (EPA, 1997).

The geochemical conditions required for *in situ* treatment of chlorinated solvents are often quite different from those required for the treatment of petroleum-derived aromatic compounds. Dissolved BTEX compounds are quite rapidly biodegraded under oxidizing or aerobic conditions. Less-chlorinated compounds (e.g., VC, CM, and DCM) are also biodegradable under aerobic conditions, while highly chlorinated solvents usually degrade more rapidly under reducing (anaerobic) conditions. Therefore, treatment of these two classes of contaminants in one mixed plume requires fundamentally different treatment processes. Sequential redox treatment is a suitable technology to treat both chemical groups (i.e., chlorinated solvents and BTEX) in a mixed plume.

Experience has shown that many contaminated sites cannot be quickly remediated through excavation and/or the operation of conventional pump-and-treat systems (P&T). This has led to the development of new strategies for the cleanup of contaminated sites, particularly long-term remedial strategies focusing on risk reduction rather than complete restoration. In particular, many emerging innovative technologies are designed to control plume size, rather than treat source areas. To minimize expenditures, a long-term plume control technology should have low operation and maintenance costs. Therefore, researchers have been asked if plume control could be attained with passive or semipassive *in situ* treatment systems (e.g., permeable reactive barriers, PRBs). Over the past 5 years, considerable experience has been gained with such technologies, both at the lab scale and in the field (Section 7.2). Technologies emphasized in the UW-AATDF project destroy contaminants *in situ* using reductive dechlorination processes (i.e., granular iron and microbial processes) and enhanced aerobic biodegradation processes (i.e., addition of oxygen to promote biodegradation). Technologies that do not destroy the contaminants, but instead immobilize them by processes such as sorption, were not considered.

7.2 SUMMARY OF THE TECHNOLOGY DEMONSTRATION

Many types of mixed contaminant plumes exist at DOD facilities, (e.g., chlorinated VOCs, nonchlorinated VOCs, chlorobenzenes, PAHs, and energetics); however, chlorinated VOCs and BTEX plumes are among the most common. Therefore, the UW-AATDF research was confined to plumes that contained mixtures of chlorinated solvents (e.g., chlorinated methanes and ethenes) and petroleum hydrocarbons (BTEX). *In situ*, semipassive technologies known to successfully destroy these classes of contaminant were reviewed. Several independently developed technologies were selected and combined to form the sequential treatment systems evaluated in the UW-AATDF research.

The overall treatment strategy was to have the contaminants encounter the reductive dechlorination treatment sections first, followed by the aerobic treatment sections. This order of encounter was selected because the reductive dechlorination treatments were known to produce transformation products that could then be degraded aerobically (e.g., PCE → VC; CT → DCM). Also, aerobic water exiting the sequential treatment system would be of a generally higher aesthetic quality than anaerobic water.

The project was divided between two field sites: Canadian Forces Base Borden (the Borden site), located near Barrie, Ontario, and NAS Alameda (the Alameda site), located near Oakland, California (Figure 7.1). The Borden site investigation was designed to evaluate four *in situ* treatment technologies for remediating mixed plumes, and to provide the data required to select the most

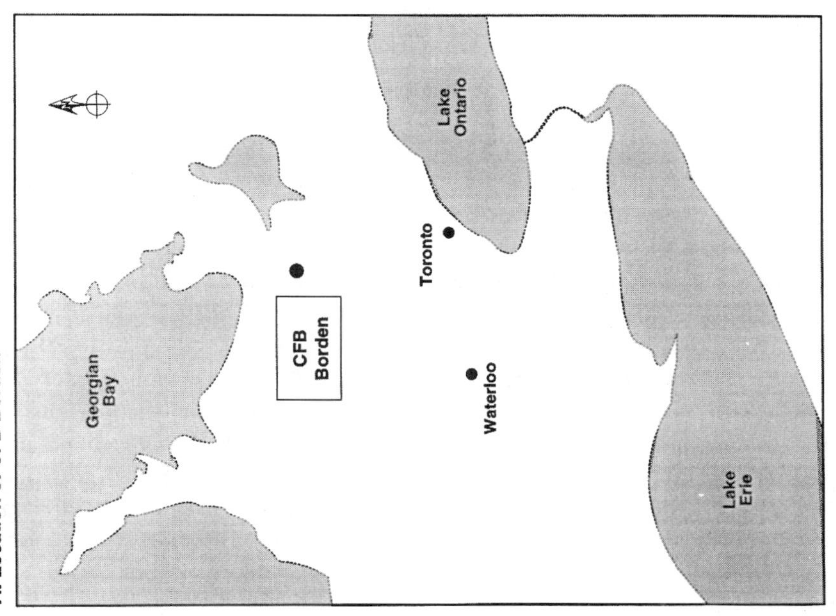

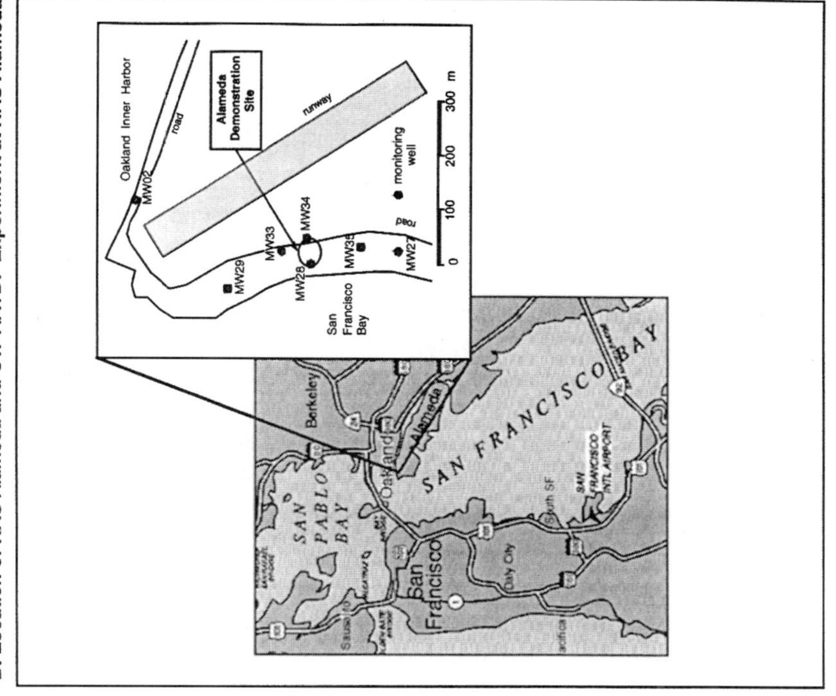

Figure 7.1 Research sites at CFB Borden, Canada and NAS Alameda, California.

appropriate pair of technologies (anaerobic and aerobic) to implement at a DOD site that contained a mixed plume. The DOD site selected was NAS Alameda. At the Alameda site, one set of technologies was implemented and evaluated at the pilot scale in an existing, mixed plume. A detailed description of the research completed at both Borden and Alameda is presented in Part I of this volume. A summary is provided below of research objectives, site descriptions, and results obtained for the investigations conducted at both the Borden site (Section 7.2.1) and the Alameda site (Section 7.2.2).

7.2.1 The Borden Experiment

The experiment at the Borden site was designed, constructed, installed, and tested in 1995 and 1996. Release of chemicals to the systems began in November 1996 and continued through October 1997.

7.2.1.1 Objectives of the Borden Experiment

The specific goals of the CFB Borden experiment, as stated in the Work Plan (UW, 1995), were to

1. Compare the performances of two anaerobic treatment options for removal of dissolved CT and PCE from groundwater, and two oxygen delivery methods for aerobic biodegradation of TOL in groundwater.
2. Determine the lowest solute concentrations achievable by each treatment technology under the conditions of the test.
3. Quantify performance related parameters for use in the experimental design at the DOD facility.

While the first two objectives were met, the third objective was not realized due to a compressed time schedule. The Alameda experiment was designed and installed at about the same time the Borden experiment was beginning. Nevertheless, some preliminary information collected from the Borden experiment was used in the design of the Alameda experiment.

7.2.1.2 Site Description of CFB Borden

The Borden experiment was conducted in a shallow sand aquifer located on the CFB Borden test site, in Ontario, Canada. Details of the site geology, hydrogeology, and aquifer characteristics can be found in the published literature (Egboka et al., 1983; Mackay et al., 1986; Sudicky, 1986; Ball et al., 1990). The research site area is characterized by sandy material overlying a till aquitard that is situated at about 3.5 to 4 m bgs (Brewster et al., 1995). Groundwater flow is generally to the west at a rate of about 10 cm/day, and the water table is shallow (<1 m bgs). The groundwater has a near-neutral pH and contains dissolved calcium (50 to 160 mg/L), sulfate (3 to 20 mg/L), chloride (2 to 50 mg/L), and bicarbonate (150 to 770 mg/L). Although the groundwater usually contains low levels of dissolved oxygen (DO), <2 mg/L, there were indications that portions of the aquifer were mildly reducing in character. For instance, both nitrate and phosphate were not present at detectable concentrations. Low concentrations of PCE (<10 μg/L) were detected in two wells. The presence of PCE was likely related to previous experiments conducted with PCE in the immediate area.

7.2.1.3 Experimental Design

The technologies selected for evaluation at CFB Borden included two methods for the reduction of chlorinated solvents:

- A PRB containing granular iron for reductive dechlorination of the selected chlorinated solvents
- A permeable nutrient injection wall (NIW) for the stimulation of anaerobic biodegradation of chlorinated solvents.

These technologies were coupled with one of two methods of stimulating aerobic biodegradation of BTEX and the less-chlorinated solvent compounds:

- Oxygen Releaseing Compound (ORC™) socks, supplied by Regenesis, Inc., and
- An oxygen biosparge gate.

These treatment technologies were installed in two sheet-pile "alleys" (gates) created within the shallow sand aquifer at Borden. The gates were up to 25 m in length, 1 to 2 m wide, and 4 m deep (Figure 7.2). The granular iron technology within a removable cassette system was combined with the oxygenation technology using ORC™ in Gate 1. In Gate 3 the NIW was combined with an oxygen biosparging zone. Gate 2 served as a control, with only natural attenuation occurring. Each gate design included upgradient source wells for contaminant release and downgradient pumping wells. This configuration provided the ability to control the groundwater flow direction and velocity through each gate as well as control the influent contaminant concentrations. Extensive multilevel monitoring of groundwater provided the data required to assess the remedial progress of each technology and the interactions between the technologies. After installation and initial testing, the *in situ* treatment of CT, PCE, and TOL, and selected transformation products (e.g., CF and *c*DCE) were followed for a period of 339 days.

7.2.1.4 Borden Site Results

The performance of each sequenced treatment system and each treatment within the sequenced system were evaluated independently. A complete description of the results and conclusions is contained in Part I. A summary of the target organic compounds and daughter products that were observed to have undergone transformation is listed in Table 7.1. Reported residence times, pseudo-half-lives, and percentage mass removals for each of the technologies studied are listed in Table 7.2. First-order kinetics were assumed for degradation of all contaminants. It was further assumed that all reactions occurred in the aqueous phase. In manipulating the data in this manner, the half-life calculated is an empirical degradation rate, and accordingly is referred to as a "pseudo-half-life." The results of each sequenced treatment system are highlighted in the following sections.

Table 7.1 Summary of Chemicals Degraded at Borden and Alameda

Compound	Granular Iron	Anaerobic Biodegradation	Aerobic Biodegradation
CT	Y	Y	N
CF	Y	Y	N
DCM	N	Y	?
PCE	Y	Y[a]	N
TCE	Y	Y	N
*c*DCE	Y	N	Y[b]
TOL	N	Y	Y

Notes: Y - degradation confirmed; N - degradation not observed; ? - insufficient data for assessment.

[a] Inferred by the presence of *c*DCE.

[b] Unknown if degradation was cometabolic or aerobic oxidation.

Table 7.2 Summary of Performance Parameters for CFB Borden Experiments

Compound	Technology	Residence Time (Days)	Pseudo Half-Life (days)	% Mass Removal
CT	Granular Iron (Gate 1)	6	ND	ND
	Intrinsic (Gate 2)	130	13, 11	99.9-100
	Anaerobic Bio (Gate 3)	70	3.5	100
CF	Granular Iron (Gate 1)	6	0.7	99.7
	Intrinsic (Gate 2)	130, 120[a]	86, 11	70,[b] 100
	Anaerobic Bio (Gate 3)	70	9 to 15	96 – 99.5
PCE	Granular Iron (Gate 1)	6	0.4	100
	Intrinsic (Gate 2)	130	ND	NR
	Anaerobic Bio (Gate 3)	70	ND	ND
TOL	Aerobic ORC (Gate 1)	30	ND	ND
	Intrinsic (Gate 2)	130, 93	62, 8.3	64,[b] 100
	Biosparge (Gate 3)	45	30, 22	65,[b] 76

Notes: ND - not determined; NR - no significant removal.

[a] Two values are residence times computed by two slightly different methods.

[b] Likely improving at end of experiment,

Borden, Gate 1: Granular Iron and ORC™ — Gate 1 at CFB Borden consisted of a removable cassette system that contained both granular iron and ORC™ technologies in close proximity (Figure 7.2). The input concentrations of chemicals entering Gate 1 were variable, typically ranging as follows:

CT:	200–700 μg/L
CF:	60 –90 μg/L
TOL:	3000–6000 μg/L
PCE:	200–400 μg/L

Reduction of the chlorinated organic chemicals began immediately upon contact with the granular iron in the cassette system of Gate 1. Complete dechlorination of the chlorinated ethenes was noted in Gate 1 (Table 7.2). A half-life for CT in granular iron could not be determined due to the low concentrations entering the granular iron cassettes. CF was the primary chlorinated methane to contact the granular iron. The CF was quickly and effectively transformed to DCM (Table 7.1).

Little aerobic biodegradation occurred in Gate 1, while the ORC™ was in the cassette system because oxygen was not released from the ORC™ socks. Laboratory data strongly indicated that the ORC™ was sensitive to pH; release of molecular oxygen was inhibited at high pH (greater than pH 9). Unfortunately, water flowing out of the granular iron cassettes had a pH of 9.5 or more, so a pronounced interference resulted. Attempts were made to rectify the problem by regularly adding mineral acids to the ORC™ cassette to maintain a low pH. However, inadequate mixing of groundwater in the cassette with the added acids prevented lower pH water from consistently contacting the ORC™ socks. Additionally, the ORC™ itself increases the pH of the water when the ORC™ reaction occurs. It is hypothesized that high pH halos developed around the socks in response to the ORC™ reactions and that the groundwater flow through the cassette was not sufficient to displace these halos.

The problem was rectified by moving the ORC™ socks from the cassette system and placing them in contingency wells (25 cm OD) located 1.3 m downgradient of the cassette system. DO concentrations in the aquifer downgradient of the contingency wells ranged from <1 to >20 mg/L. The variability was attributed to the presence of preferential flow paths; this finding wass supported by tracer test results. In the short term, the pH problem was resolved by relying on the buffering capacity of the aquifer to maintain neutral conditions. The improved hydraulics of the wells (vs.

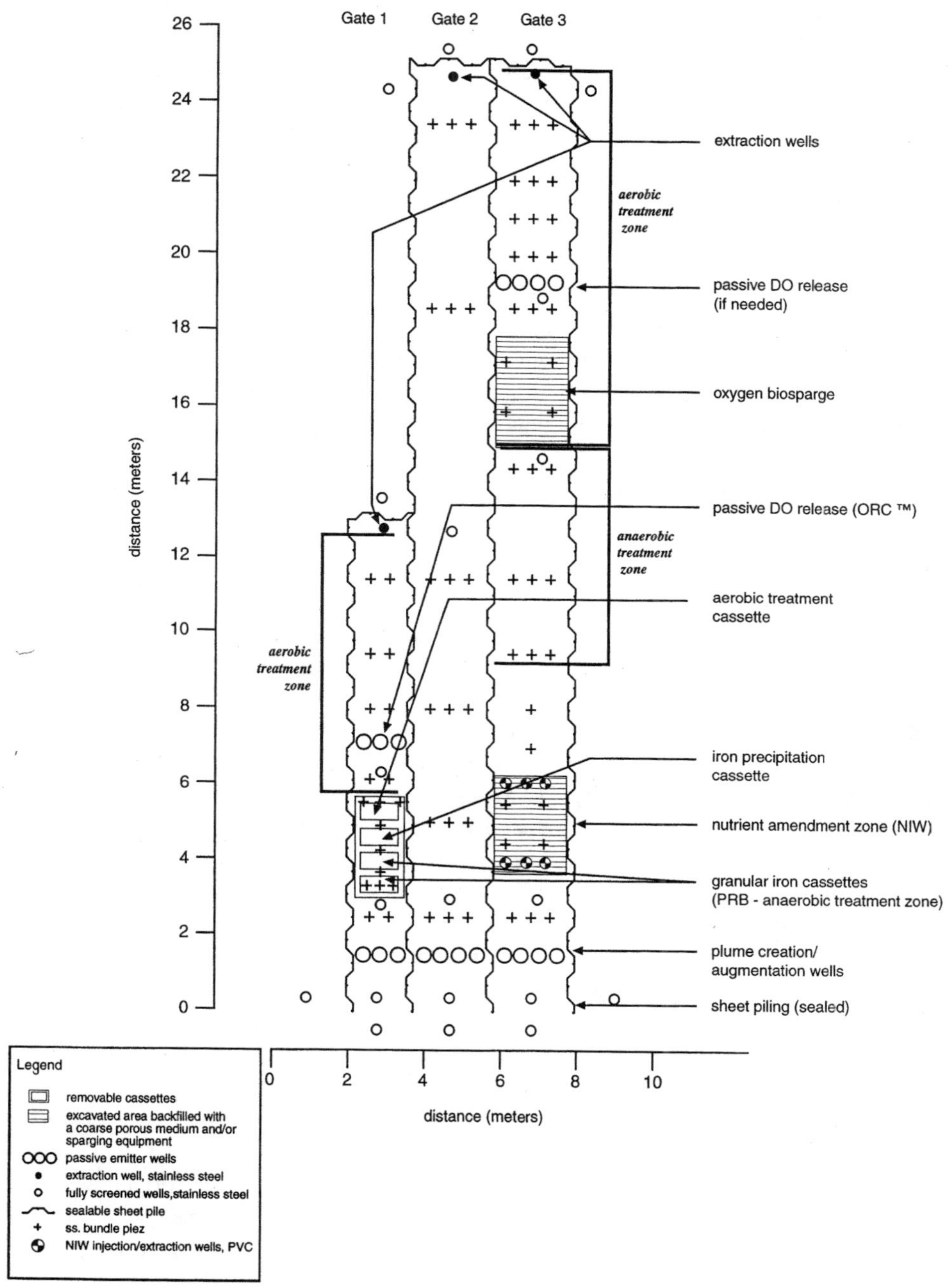

Figure 7.2 Experimental design for the Borden experiment.

the cassette) assured proper functioning of the ORC™. Over time, pH control might require engineered intervention, since the buffering capacity of the aquifer is likely finite.

Borden, Gate 2: Natural Attenuation — Table 7.2 summarizes the results from Gate 2 of the Borden experiment. The evaluation of intrinsic remediation concluded that biodegradation of CT and CF was complete about 8 m downgradient of the source input. Quantifiable TOL loss was slow to appear, and based on the redox conditions this loss was likely through anaerobic biodegradation.

A TOL half-life of 8.3 days was calculated using data from the later stages of the experiment. At the end of the experiment, TOL was still observed in many monitoring points, but at levels lower than if no biotransformation processes had occurred (i.e., only transport of TOL through the gate). PCE degradation was very slow to occur; after 337 days, *c*DCE was observed at several wells. These results indicate that the natural microbial populations were able to transform the chemicals that were introduced to the aquifer. Lag (or acclimation) periods were as follows (in order of shortest lag to longest lag time): CT < CF and TOL < PCE. Relative to Gates 1 and 3, the observed degradations of CF, CT, and PCE were slower. TOL degradation rates appeared to be about the same in both Gates 2 and 3. A TOL degradation rate in Gate 1 was not compiled due to insufficient data, making a comparison to Gate 2 or 3 not possible.

Borden, Gate 3: NIW and Oxygen Biosparge — Gate 3 at the Borden Site coupled a NIW (Devlin, 1994; Devlin and Barker, 1996) with an aerobic biosparge zone. Benzoate was added to the groundwater system via the NIW to stimulate anaerobic biodegradation. The plume entering the remediation zone of Gate 3 generally contained

CT:	100–500 μg/L
CF:	100–300 μg/L
PCE:	200–700 μg/L
TOL:	500–2000 μg/L

Table 7.2 summarizes the results from Gate 3 at Borden. Within 238 days after activating the source wells, CT and CF were completely removed in the anaerobic zone (see Figure 7.2). PCE degradation did not begin until approximately 300 days into the experiment. PCE was presumably dechlorinated via TCE (detected at concentrations below the LOQ) to *c*DCE, which was observed to accumulate in the anaerobic zone. VC was not detected. While the results of the Gate 3 experiment were promising, the considerable lag times (300 days for PCE to begin to disappear) and incomplete chlorinated solvent degradation (*c*DCE persistence from PCE parent) were disadvantages of anaerobic biodegradation compared with granular iron. At other sites where the microbial populations are acclimated to the contaminants, lag times may be considerably shorter.

The aerobic biosparge zone performed satisfactorily in its delivery of oxygen to the groundwater. At the time the experiment was completed, *c*DCE and TOL persisted through the anaerobic zone. TOL and *c*DCE were essentially completely attenuated within the aerobic zone (see Figure 7.2). Thus, the biosparge treatment system was more successful in stimulating biodegradation of TOL and the breakdown products of reductive dechlorination than was the ORC™ in Gate 1.

7.2.1.5 Borden Site Conclusions

The *in situ* treatment technologies coupled for the Borden experiment performed well, reducing the concentrations of most contaminants to levels below LOQs, and in some cases below MDLs. The noticeable exception to this trend was TOL. The general persistence of TOL is a reflection of the limited DO availability that was achieved in the aerobic treatment gates. It is believed that, with additional time to overcome aquifer reduction capacities, the desired aerobic conditions could be created and TOL degradation would quickly proceed.

7.2.2 The Alameda Experiment

In late 1995, a number of U.S. DOD sites having mixed contaminant plumes were screened to assess suitability for the second experiment. The suitability of five DOD sites for demonstration of this technology was evaluated in terms of 12 criteria, including site hydrogeology, contaminant concentrations, site access, and facility support (UW, 1997a). Site 1 at NAS Alameda, in Alameda, CA, was selected as the DOD facility for the pilot-scale SPRB demonstration.

7.2.2.1 Objectives of the Alameda Experiment

The U.S. DOD site was a pilot-scale installation of the best remedial technologies identified through the Borden experiments. The pilot-scale demonstration was designed to treat a portion of an existing plume at a DOD facility. The NAS Alameda Site was selected and the remedial objectives (ROs) for the *in situ* SPRB were generally set to California Maximum Contaminant Levels (MCLs) (see Table 7.3).

Table 7.3 Proposed Remedial Objectives for Site 1, NAS Alameda; California MCLs; and Marine Ambient Water Quality Criteria

Chemical Parameter	Proposed Remedial Objective for Alameda SPRB	California MCL	Marine Ambient Water Quality Criteria
TCE	5	5	2,000
1,1-DCE	6	6	224,000
*t*DCE	10	10	224,000
*c*DCE	10	6	224,000
VC	2	0.5	NR
Benzene	10	1	5,100
Toluene	150	150	5,000
Ethylbenzene	150	700	430
Xylenes, total	150	1,750	NR
Iron	0.5 mg/L	NR	NR
pH	5 – 8.5 (pH units)	NR	NR

Notes: All units are expressed as μg/L unless otherwise noted. NR – no criteria reported.

7.2.2.2 Description of Site 1, NAS Alameda

As part of the site selection process, a subsurface assessment of the NAS Alameda site was conducted at Site 1 in July 1996. An additional assessment was conducted in October 1996, after the site was selected to collect the data required to optimize the location of the pilot-scale SPRB (UW, 1996; UW, 1997b).

At Site 1, NAS Alameda, an apparently narrow plume of chlorinated solvents (*c*DCE, VC, and minor TCE) and petroleum hydrocarbons (mainly toluene and benzene) was selected for the demonstration. The plume occurs in an unconfined aquifer that is comprised of sandy silt, underlain by a clay-rich aquitard at a depth of about 6.5 m. The water table is about 1.5 m below ground surface. The groundwater flows laterally at a rate of 1–3 cm/day toward San Francisco Bay. A minor tidal influence has been noted at the demonstration site, which is located about 30 m from the Bay (PRC, 1993).

The shallow groundwater is generally fresh water (Na(Ca, Mg)–Cl or Na (Ca, Mg)–HCO_3 type) but somewhat brackish at depth (Cl up to 1200 mg/L at a depth of about 5 m), reflecting the influence of the adjacent saltwater bay. Dissolved iron concentrations, locally up to 10 mg/L, and the presence of methane reflect the reducing character of the plume. In the area of the demonstration, groundwater contains the following chemicals at the concentrations listed:

*c*DCE:	~220,000 μg/L
VC:	~26,000 μg/L
Toluene:	~8,000 μg/L
Benzene:	~300 μg/L

Other chlorinated solvents such as TCE, 1,1-DCE, and *t*DCE, and the petroleum hydrocarbons ethylbenzene and xylene (all isomers), were generally present at concentrations of less than 1000 μg/L.

7.2.2.3 Experimental Design

The shallow, watertable aquifer at Site 1, NAS Alameda was characterized (see Section 7.2.2.2), and a location for the SPRB demonstration selected. Modeling was completed, using VisualMod-Flow™ (Waterloo Hydrogeologic Software) to assist in the design of the pilot-scale test (UW, 1996). A control gate, containing no treatment sequences, was also installed. The evaluation of the pilot-scale SPRB continues and the hydraulics of groundwater capture by the Funnel-and-Gate is currently being evaluated in an ongoing research project (funded by Naval Facilities Engineering Service Center, Port Hueneme, CA).

The sequenced technologies selected for implementation at NAS Alameda were granular iron for reductive dechlorination of chlorinated solvents followed by biosparging for biodegradation of petroleum hydrocarbons and less chlorinated solvents (Figure 7.3). The granular iron proved reliable in Gate 1 at Borden and was selected over the NIW, partly due to the very high solvent concentrations present at the Alameda site. The biosparge system tested in Gate 3 at Borden was judged to have outperformed the ORC™ in the cassette system in Gate 1 at Borden, mainly due to the apparent inhibition of oxygen release from ORC™ in the presence of high pH water generated in the granular iron zone. At NAS Alameda, a pea gravel zone separates the granular iron and the biosparge zone, and acts as a transition zone. An open biosparge gate was constructed and filled with a biomass support medium, Yeager TriPak™ (5-cm-diameter PVC bioballs placed in cages to keep them submerged). The extreme openness of the biosparge gate and the frequent mixing of the groundwater during sparging events made the system operate like a continuously mixed bioreactor. Periodic sparging of carbon dioxide gas was selected as a means to control the high pH water emanating from the granular iron. The sparging apparatus used to deliver oxygen was also used to deliver carbon dioxide on an as-needed basis.

The two technologies were installed in a Funnel-and-Gate system, within an approximately 3.3 m wide by 4.6 m long treatment gate (UW, 1997b). Sealable joint sheet piling (Waterloo Barrier™) funnels were installed to a depth of about 10 m, 3.5 m below the top of the aquitard. The gate was excavated down to the aquitard (6.5 m depth below ground surface).

Results of a treatability study conducted by ETI indicated that approximately 1.5 m of granular iron (flow-through thickness) was sufficient to treat the solvents (mainly *c*DCE, VC, and lesser concentrations of TCE) in the bulk of the contaminated groundwater. It was recognized that small volumes of more highly concentrated groundwater might not be treated to the remedial objective concentrations. The biosparge technology, designed to promote complete aerobic biodegradation of hydrocarbons (mainly TOL and benzene at the Alameda site), was also expected to support aerobic biodegradation of the *c*DCE (cometabolism) and VC (mineralization) that passed through the iron gate, such that remedial objectives could be achieved.

Treatment performance was assessed at two different groundwater flux rates. Initially, contaminated groundwater was drawn through the gate at a rate of 1.12 m^3/day. This phase of the experiment, termed Operating Phase 1 (OP1), was intended to stress the SPRB so that the upper limits of the granular iron system treatability could be investigated. In the second phase of the experiment, termed Operating Phase 2 (OP2), the flow rate was reduced to 0.34 m^3/day. The OP2 flow rate is considered reflective of what the ambient groundwater velocity through the SPRB will be once the downgradient sheet piling is removed (June 1998). Only OP2 data are discussed in the following section (7.2.2.4).

The biosparge zone operating procedure was not fully established until about June 1997 (after OP1 and during OP2), when the rate of oxygen sparging was increased to six 20-minute pulses per day. Carbon dioxide was sparged monthly or as needed to maintain the groundwater pH below

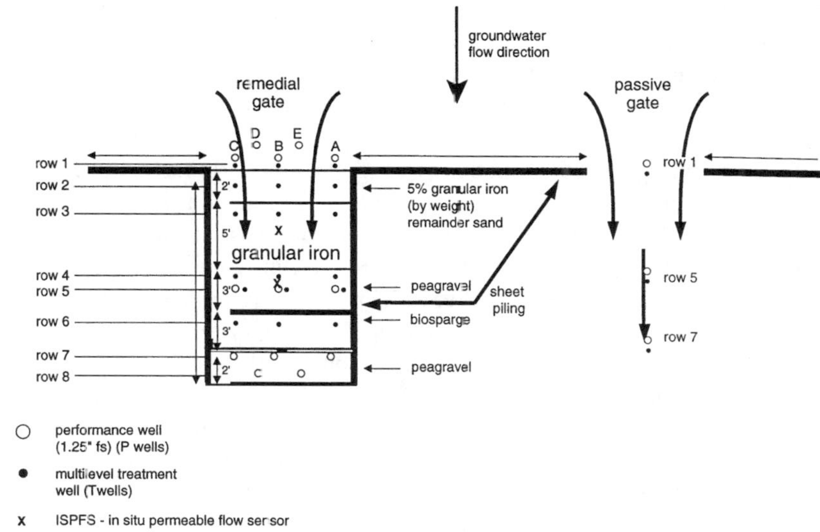

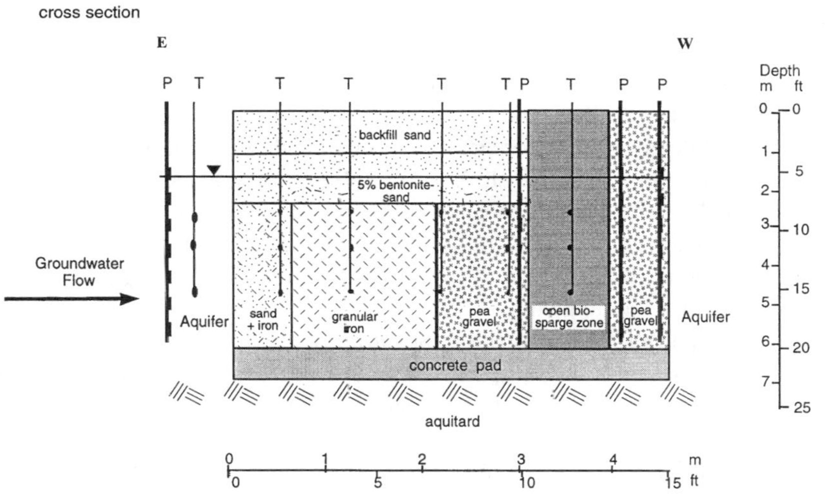

Fgiure 7.3 Locations of the groundwater monitoring wells and a cross section of the remedial gates.

about 8.0 in the biosparge zone. Major monitoring events were completed regularly (e.g., about every 60 days) during OP2. Additional monitoring to define aspects of the granular iron performance and the biosparge zone performance was conducted as needed.

7.2.2.4 Results and Conclusions of OP-2 at Alameda

OP1 operated for a period of about 75 days. At the end of OP1, approximately 11.6 pore volumes had passed through the iron PRB and about 3.7 pore volumes had been extracted from the entire gate (UW, 1998). During OP1, breakthrough of *c*DCE and VC was observed. During most of OP1, the biosparge zone was not operating consistently and BTEX had not yet appeared in the biosparge zone, so an assessment of biosparge performance was not attempted.

OP2 operation began immediately after OP1. Since OP1 had produced breakthrough, chlorinated solvents (i.e., *c*DCE and VC) were present at the downgradient section of the iron PRB. A second transition period, due to change in flow conditions, occurred. From June 1997 until September 1997 the remedial objectives (ROs) for *c*DCE (10 μg/L) and VC (2 μg/L) were not met. In general the ROs for all other compounds were generally met.

Contaminant concentrations entering the SPRB were extremely variable both in space and time. The dominant organic compounds entering the SPRB were VC and *c*DCE, with concentrations in the tens and hundreds of milligrams per liter, respectively. BTEX concentrations entering the gate were generally less than 10 mg/L in total. Table 7.4 provides a summary of data from selected fully screened wells during OP2 (see Section 5.3.2 for well nomenclature). By December 1997, approximately 10 pore volumes had passed through the iron PRB and 3.1 pore volumes through the whole gate (UW, 1998). Recent data (e.g., January and April 1998) indicate that ROs are being met at row 8. Table 7.5 provides a summary of performance parameters for the Alameda experiment, up to and including the January 1998 data.

In spite of the breakthrough of *c*DCE and VC during 1997, the data indicate that they were both subject to more than 90% mass removal from the groundwater by reaction with the granular iron. An additional 20–80% mass removal from the iron effluent was observed in the biosparge zone. Given the extremely high concentrations of these compounds entering the system (10 to 200 mg/L), the results obtained are considered excellent (UW, 1998).

BTEX compounds were highly retarded in the granular iron, and only sporadic and low (<200 μg/L) concentrations were detected in the monitoring points immediately downgradient of the iron zone. Therefore, the performance of the aerobic biosparge zone for BTEX biodegradation could not be adequately assessed. However, based on scientific literature and lab microcosm experiments using Alameda material, there is no reason to believe that milligrams-per-liter quantities of BTEX would not have been biodegraded in the biosparge zone.

The pseudo-half-lives in the biosparge zone (see Table 7.5) were derived by treating the biosparge zone as a completely mixed bioreactor with input and output of groundwater occurring over a number of sampling rounds in mid- to late-1997. Based on the average residence times, it is estimated that about 80–90% of the chlorinated solvent mass entering the aerobic zone was removed, with much of this being attributed to biodegradation (Table 7.5). Laboratory microcosm studies supported this finding by demonstrating that the indigenous microorganisms at Alameda were capable of aerobically degrading both *c*DCE (likely via cometabolism) and VC. Volatilization is known to have also occurred. The field site removal rates could likely be improved if a larger active biomass were established in the biosparge zone. The degradation behavior observed in the biosparge zone indicates that the two technologies functioned well in sequence.

A potential limitation of the coupling of granular iron with aerobic biodegradation at Alameda that was identified at Borden was that the high pH water exiting the permeable iron wall might prevent the establishment of a robust aerobic microbial population in the biosparge zone. To neutralize the alkaline water, carbon dioxide gas was periodically sparged into the biosparge zone (in addition to oxygen). This measure proved to be successful at maintaining the desired pH range (7 to 8). No other interference related to the coupling of the anaerobic and aerobic process was noted over the course of the experiment.

7.2.3 Conclusions of the UW-AATDF Experiments

The treatability of the various compounds encountered over the course of this project is summarized in Table 7.1. The granular iron (Gate 1 and Alameda) was found to be an effective medium for the treatment of chlorinated ethenes and, to a lesser extent, chlorinated methanes. Granular iron coupled with ORC™ (Gate 1) in close proximity did not function ideally. Granular iron coupled with biosparge (Alameda) functioned acceptably during the first year of operation.

Anaerobic biodegradation (Gates 2 and 3) was effective in dechlorinating CT and CF. Intrinsic biodegradation of TOL (Gates 1, 2, and 3) was noted under anaerobic conditions. Anaerobic dechlorination of PCE was observed with *c*DCE being produced as the main end product.

Table 7.4 Results of Analyses for Selected Volatile Organic Compounds in Groundwater Samples During Operating Phase 2, Pilot Scale SPRB, Site 1 NAS Alameda, California

MONITORING WELL SAMPLE DATE	R1PA Sep-97	R1PA Jan-98	R1PA Apr-98	R1PB 30-Sep-97	R1PB 12-Jan-98	R1PB Apr-98	R1PC 30-Sep-97	R1PC 12-Jan-98	R1PC Apr-98
Volatile Organic Compounds									
Benzene	146	75	161	156	<2	127	112	<2	172
1,1-Dichloroethene	372	<4	276	516	32	195	68.3	<4	102
cis-1,2-Dichloroethene	99,600	53,000	110,200	117,000	33,000	100,400	15,300	19,000	44,870
trans-1,2-Dichloroethene	536	<4	165	796	<2	144	88.8	<4	88.8
Ethylbenzene	111	<2	103	124	<2	88	70.2	<2	108
Trichloroethene	36.3	8	69	24.7	6	41	<5	<3	19
Toluene	4,080	3,500	5031	4,050	2,200	3762	2,160	3,800	3,666
Vinyl Chloride	15,700	11,000	26,900	14,500	11,000	24,400	16,300	3,900	31,020
m,p-Xylene	435	<7	369	491	<7	278	235	<7	343
o-Xylene	201	<4	172	224	<4	129	102	<4	135

MONITORING WELL SAMPLE DATE	R1PD 30-Sep-97	R1PD 12-Jan-98	R1PD Apr-98	R1PE 30-Sep-97	R1PE 12-Jan-98	R1PE Apr-98	R5PA 30-Sep-97	R5PA 12-Jan-98	R5PA Apr-98	R5PB Sep-98	R5PB 12-Jan-98
Volatile Organic Compounds											
Benzene	184	<2	108	156	<2	118	<2	12	41	<2	<2
1,1-Dichloroethene	499	<4	53	298	<4	42	<4	<4	<4	<4	<4
cis-1,2-Dichloroethene	111,000	45,000	20,600	64,800	2,500	14,800	55	48	83	56	17
trans-1,2-Dichloroethene	735	<4	80	313	<4	80	4	<4	<4	<4	<4
Ethylbenzene	111	<2	74	79.6	<2	118	<2	<2	960	<2	<2
Trichloroethene	42.6	<3	16	33.7	<3	8.8	<5	<3	<3	<3	<3
Toluene	4,240	4,600	1,696	3,320	2,800	2,000	143	200	750	7	4
Vinyl Chloride	22,100	13,000	21,200	18,600	1,400	22,000	138	54	160	62	2.3
m,p-Xylene	415	<7	210	270	<7	<7	<7	<7	<7	<7	<7
o-Xylene	189	<4	95	116	<4	<4	<4	<4	<4	<4	<4

continued

Table 7.4 continued Results of Analyses for Selected Volatile Organic Compounds in Groundwater Samples During Operating Phase 2, Pilot Scale SPRB, Site 1 NAS Alameda, California

MONITORING WELL SAMPLE DATE	R5PB Apr-98	R5PC Sep-97	R5PC 12-Jan-98	R5PC Apr-98	R8PA Sep-97	R8PA Jan-98	R8PA Apr-98	R8PB Sep-97	R8PB Jan-98	R8PB Apr-98
<u>Volatile Organic Compounds</u>										
Benzene	<2	<2	<2	22	<2	<2	<2	<2	<2	<2
1,1-Dichloroethene	<4	<4	<4	<4	<4	<4	<4	<4	<4	<4
cis-1,2-Dichloroethene	15.8J	92	81	<8	136	4	<8	131	6	<8
trans-1,2-Dichloroethene	<4	<4	<4	<5	<4	<4	<4	<4	<4	<4
Ethylbenzene	<2	<2	<2	<2	<2	<2	<2	<2	<2	<2
Trichloroethene	<3	<3	<3	<3	<3	<3	<3	<3	<3	<3
Toluene	41	<5	4	111	22	<5	14.3J	<5	<5	<5
Vinyl Chloride	24	47	30	30	217	<12	<12	121	2.4	<12
m,p-Xylene	<7	<7	<7	<7	<7	<7	<7	<7	<7	<7
o-Xylene	<4	<4	<4	<3	<4	<4	<4	<4	<4	<4

Notes:

All concentrations in μg/L.

< — Analyte was not detected at a concentration greater than the associated quantitation limit.

J — Estimated value.

Table 7.5 Summary of Performance Parameters for the Alameda Experiment

Compound	Treatment Zone	Treatability $t_{1/2}$ (days)[a]	Residence Time (days)	Pseudo $t_{1/2}$ (days)	% Mass Removed
TCE	Iron	0.06	17	1.1 to 2.9	98 to 100
*t*DCE	Iron	—	17	2.4 to 4.5	93 to 99
1,1-DCE	Iron	—	17	2.2 to 5.3	89 to 99
*c*DCE	Iron	0.5	17	2.9 to 7.1	81 to 98
	Biosparge	—	32	15	77
VC	Iron	0.47	17	1.1 to 4.7	99.9 to 100
	Biosparge	—	32	70	27

Note: — data not available.

[a] $t_{1/2}$:half-life.

Source: UW (1998).

Aerobic degradation of TOL (Gates 1 and 3) was also demonstrated. Insufficient data (Gates 1, 2, and 3) were collected to evaluate the aerobic degradation of DCM, but some aerobic biodegradation (Gate 3) of *c*DCE was observed, possibly as a result of cometabolism in the presence of toluene or methane. Aerobic biodegradation of *c*DCE and VC was observed in the Alameda experiment. Degradation of BTEX compounds in the biosparge gate at Alameda was not assessed due to retardation in the granular iron.

In general, the complexities of coupling the anaerobic/aerobic technologies selected here were not manifested and the SPRB systems performed extremely well under the conditions investigated. Long-term performance assessments should be undertaken to ensure that fouling does not become an issue.

CHAPTER 8

Application of a Sequential Treatment System at a Hypothetical Site

In light of the potential advantages of passive *in situ* "permeable reactive barriers" (PRBs) as a means of remediating groundwater, several documents have been prepared describing the general approach to implementing these systems. These reports include a design protocol document prepared for the USAF (Battelle, 1997) and a report on emplacement technologies and monitoring techniques for PRBs prepared for the U.S. DOE (MSE Technology Applications, 1996). A guidance document that will discuss PRB design and installation methods for specific organic and inorganic contaminants is currently being prepared by the U.S. EPA. Several other protocols describing specific *in situ* biological treatment processes have been prepared for the USAF and others (e.g., ITRC, 1997).

The objective of this section of the TER is to illustrate how a **full-scale** SPRB combining two (or more) treatment processes in sequence for the treatment of organic contaminants would be designed and implemented at a particular site (the "hypothetical site"). Rather than repeat general "background" information regarding PRB design referred to above, this chapter focuses on specific issues related to the design and implementation of a SPRB to treat a mixed organic plume. The design of this type of PRB may be significantly different than that of a PRB containing a single treatment zone and/or those designed to treat a single "group" of contaminants (the use of granular iron to treat chlorinated ethenes, for example).

8.1 HYPOTHETICAL SITE

8.1.1 Site Location

The combined chlorinated solvent and BTEX plume at Site 1 at NAS Alameda, where the UW-AATDF pilot-scale demonstration is located, was chosen to serve as the hypothetical location for the full-scale installation. The mixture of contaminants present at Alameda occurs in shallow plumes at several DOD facilities, so the strategy outlined in this report for implementation may be readily transferable to several other sites. Moreover, the cost estimates presented may also serve as a useful approach for implementing/costing this remedy at other sites. The detailed modeling and site characterization effort completed as part of the demonstration phase (Section 7.2) facilitated the design of the hypothetical full-scale system. Data collected prior to and during the design and installation of the pilot-scale phase were used to obtain accurate quotations from contractors familiar with the site. The comparison of practical "full-scale" costs to the costs incurred in the R&D demonstration was facilitated by using this approach.

8.1.2 Plume Dimensions

The plume at Alameda contains a very narrow and very highly concentrated core as delineated in the depth-discrete sampling site investigations. Site-specific ROs have not been defined for the site. The discharge/receptor for this plume is the San Francisco Bay, which would imply that the Marine Ambient Guidelines would be appropriate water quality criteria. The Marine Ambient Guidelines for the contaminants identified in the plume were shown in Table 7.3. Realistically, it is unlikely that these guidelines would be selected as the site ROs. The California Drinking Water MCLs were also presented in Table 7.3. Given that it is unlikely that the groundwater at the site will ever be used for consumptive purposes, these guidelines may not appropriately reflect possible site-specific ROs. Therefore, it was decided that two scenarios, using two different ROs, would be investigated.

Case A would be designed to remediate just the core of the plume, and would be reflective of more lenient ROs (i.e., capturing a plume represented by the 1000-µg/L total VOC contour). In Case A the SPRB is designed to capture the relatively narrow, high concentration core area of the plume inside the 1000-µg/L VOC contour, which is about 80 ft (25 m) wide. Figure 8.1 shows a plan view schematic of the Case A scenario at the site. Geochemical and geological characteristics of the aquifer were determined during recent site investigations (Section 7.2.2). The Case A system would be considered small for most DOD plumes.

Case B would capture a much larger portion of the plume (i.e., capturing a plume represented by the 10-µg/L *c*DCE contour) to reflect more stringent ROs (MCLs). This system (Figure 8.2) would be about 350 ft (107 m) wide at the face of the plume. Costs for Case B may be more representative of plumes requiring treatment at other DOD facilities.

For both Case A and Case B the plume was assumed to persist throughout the vertical saturated thickness of the shallow aquifer (about 18 ft (5.5 m)), and assumed to be bounded by a clay aquitard at a depth of 22 ft (6.8 m) below ground surface.

8.2 GENERAL SYSTEM CONFIGURATION

The configuration of the *in situ* SPRB developed for the hypothetical full-scale application at Alameda was the same as that demonstrated at the pilot-scale, i.e., a granular iron zone followed by a biosparge zone. The granular iron zone would degrade the chlorinated solvents and would be followed by a biosparge zone which would promote degradation of BTEX compounds. The advantages of using a biosparge zone versus ORC™ or O_2 emitter tubes were described in Section 7.2. However, the use of ambient air vs. oxygen gas (for the pilot-scale) was considered in the biosparge zone in the full-scale design to minimize treatment costs. Two methods for introducing air to the biosparge zone were also considered: the use of a gas delivery system at the base of a high permeability zone (similar to the demonstration), and the introduction of the gas via wells installed in either an artificial high permeability zone or in the native aquifer sediments.

The residence time required for treatment, the anticipated groundwater velocity through the SPRB, and the required upgradient "capture zone" were used to determine the dimensions of an SPRB needed to achieve the ROs for a given contaminant plume. Two basic configurations were considered for full-scale application of the SPRB, a laterally continuous PRB and a configuration described in the literature as Funnel-and-Gate (Starr and Cherry, 1993).

In a Funnel-and-Gate configuration, low permeability funnels installed perpendicular to the groundwater flow path direct groundwater toward the SPRB or treatment gates. By providing a "focused" treatment area the Funnel-and-Gate configuration may be more appropriate if several forms of passive or semipassive treatment using different types of reactive materials are to be employed at the site. The Funnel-and-Gate configuration may be particularly advantageous if the materials need to be replaced or replenished periodically over the life of the system. These focused

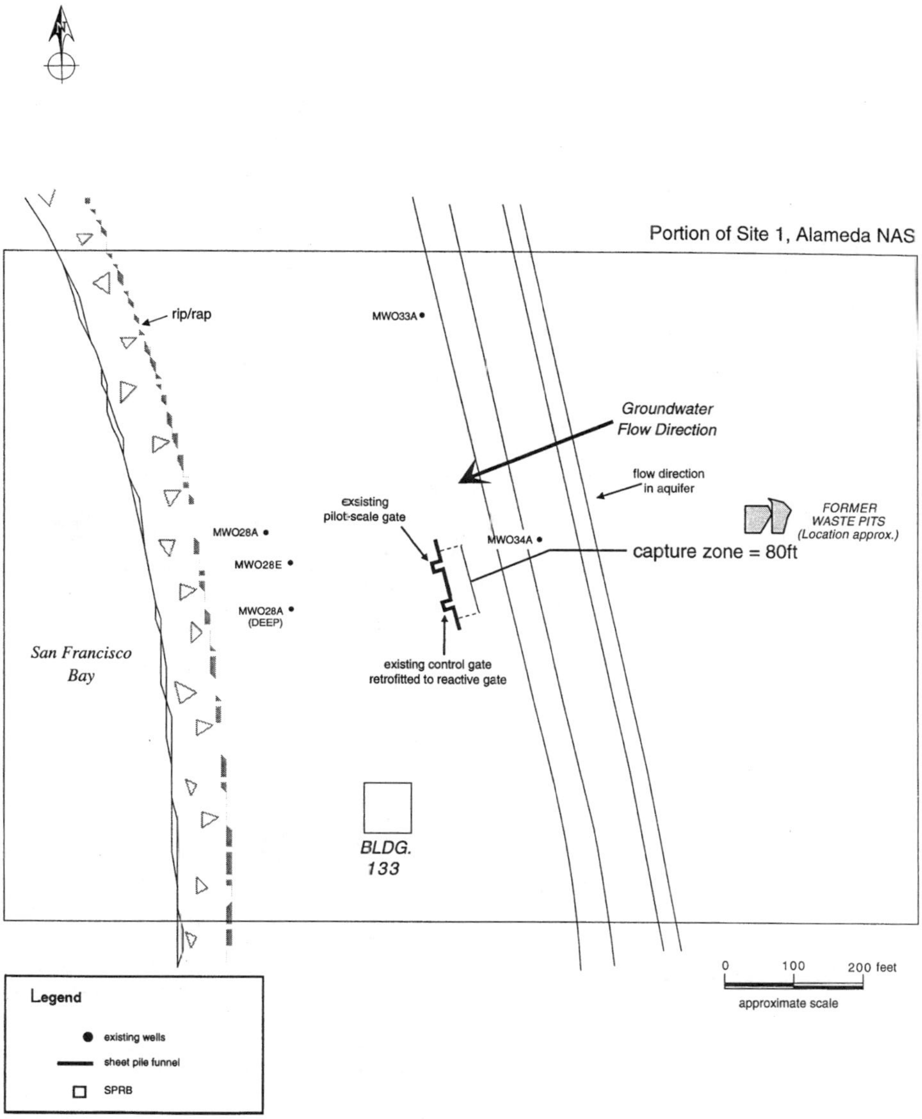

Figure 8.1 Schematic of Alameda full-scale system — Case A.

treatment areas will also facilitate monitoring the performance of the system. A Funnel-and-Gate system seemed particularly appropriate for the Alameda examples for several reasons, including the following:

1. Air sparging could be a focused treatment when in a gate design.
2. A potential drawback to the use of a Funnel-and-Gate system, underflow of contaminated water, was prevented at the site by the presence of a low permeability clay layer at the base of the system.
3. Completion of the pilot-scale Funnel-and-Gate system at the site provided a sound basis for estimating the costs of a larger Funnel-and-Gate system.

The final decision on SPRB configuration for any particular site would be made based on consideration of the site-specific factors listed in Table 8.1, in the context of the proposed sequenced treatment processes.

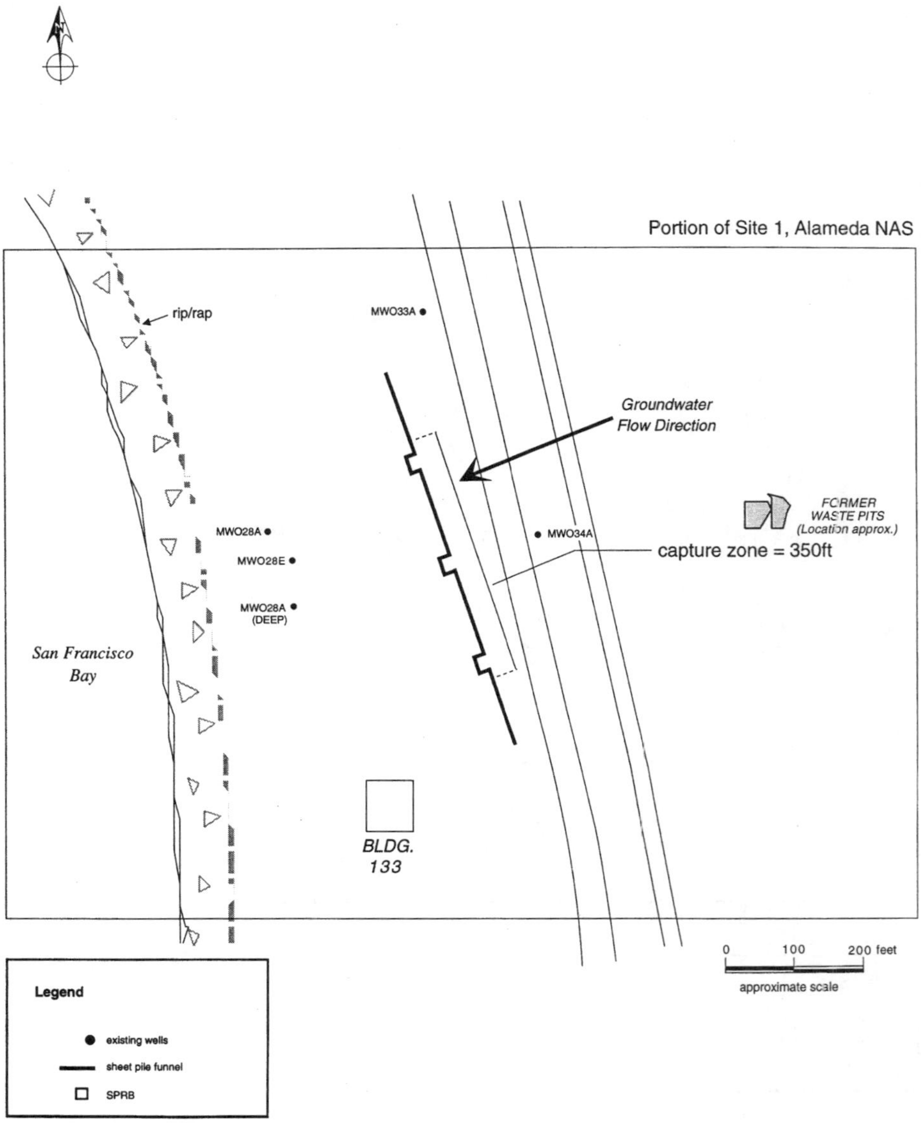

Figure 8.2 Schematic of Alameda full-scale system — Case B.

Table 8.1 Site Characterization Needs for Sequenced Permeable Barrier Systems

Aquifer Properties	Groundwater Qualities
Hydraulic conductivity	VOC concentration
Gradient	Major inorganic chemistry
Groundwater velocity	Background nutrient levels
Depth to confining unit	Plume width
Geology of confining unit	Plume depth
Geotechnical data	Location of downgradient receptors

8.3 ENGINEERING DESIGN

Design of a SPRB typically involves establishing treatment objectives, collecting site characterization data, completing treatability studies, and performing groundwater flow modeling. These data are needed to determine the appropriate dimensions of the SPRB. The application of a specific conventional engineering safety factor to SPRB design is difficult due to the nature of the degradation processes involved (often involving first-order kinetics) and the uncertainty in determining design parameters such as groundwater flow velocity, and variations in influent concentrations. The design approach outlined in this section is as follows: determine what possible maximums may occur in input design parameters (chemical concentration, groundwater flow rate), gauge the effect of these maximums on system costs and performance, and arrive at reasonable design compromises (if necessary) that still ensure regulatory compliance over the widest range of circumstances possible.

8.3.1 Treatment Objectives

The regulatory framework of the project should be evaluated in terms of identifying where possible, the treatment goals (desired concentrations) for groundwater exiting the SPRB; these should be related to the location of downgradient receptors and the ability of natural attenuation/intrinsic bioremediation processes to further remediate the plume downgradient of the SPRB (if required).

8.3.2 Site Characterization

As with any technology, the initial step in the design is to review existing site data to determine what chemicals are present and their concentrations, with a goal of developing a "treatment hypothesis" that identifies candidate treatment processes and their possible order of sequencing. This review also identifies data gaps that prevent refinement of this treatment hypothesis. Site data which require evaluation include the following:

- Depth and fluctuations of the water table;
- Depth to the top and bottom of the contaminated zone;
- Lateral boundaries of the contaminated zone;
- Flow direction;
- Groundwater velocity;
- Lateral and vertical distribution of organic compounds;
- Geotechnical data as they pertain to "constructability issues"; and
- Inorganic chemical constituents; in particular, the background levels of potential nutrient sources for bioremediation (e.g., NO_3^-, PO_4^{3-}, SO_4^{2-}).

One of the unique issues differentiating the design of PRBs or SPRBs from conventional pump-and treat (P&T) approaches is the need to obtain more data related to the spatial variation in contaminant concentration. P&T system design is often based on the "average" contaminant concentration requiring treatment, because groundwater from different wells is typically mixed prior to entering the treatment system. Mixing of groundwater may not occur upgradient to a PRB; thus PRBs are typically designed using anticipated maximum influent concentrations. To date, field data have shown little mixing/dispersion occurring upgradient of PRBs, and it is currently impractical, in most environments, to consider incorporating an upgradient mixing zone into the design of such systems. Therefore, determining spatial variation in contaminant concentration where the PRB will be installed is a necessary part of the site characterization process.

Ascertaining groundwater velocity is very important in the design of PRBs and SPRBs. For PRBs, the thickness of the treatment zone will be proportional to the background groundwater velocity (both the hydraulic gradient and conductivity, i.e., velocity) in the aquifer, and therefore this parameter becomes integral to the design. In P&T designs, the aquifer hydraulic conductivity (K), not the actual groundwater velocity, directly affects pumping well yields and capture zones and is therefore an important design parameter for these systems.

Site characterization efforts to fill the "data gaps" could consist of a variety of groundwater and soil analyses. These are summarized in Table 8.1. Appropriate sampling and analytical methods for site characterization are described in detail in many of the protocol documents identified in Section 8.1, as well as in a variety of other DOD, EPA, and ASTM publications.

For both Case A and Case B at Alameda NAS it was assumed that the plume was well defined. No additional field programs would need to be undertaken for design of the full-scale hypothetical system. Table 8.2 presents the site background and characteristics of Site 1 at NAS Alameda.

Table 8.2 Site Characteristics of Site 1, NAS Alameda, California and Treatment Systems

Site Background:

Historical Activity That Generated Contamination: SIC Code: 9711A (Public Administration, National Security, Miscellaneous)

Management Practices That Contributed to Contamination: Unlined Disposal Pits

Site Characteristics:

Media Treated: Groundwater (*in situ*)

Contaminants Treated: Volatiles — Halogenated (TCE, DCE, and VC) and Volatiles — Nonhalogenated Petroleum Hydrocarbons (BTEX)

Treatment System:

Primary Treatment System: Passive Treatment Walls (granular iron) and Bioremediation

Supplemental Treatment System: None

8.3.3 Treatability Studies

The need for treatability studies and their scope can vary depending on the contaminants present, their concentrations, and the treatment sequence envisioned for these compounds. The purpose of the treatability study is to test and refine the "treatment hypothesis" developed as a result of the site characterization effort.

A key aspect of the design of successful treatability studies is to ensure that representative water samples are tested. Reaction rates need to be determined in conditions as close as possible to the field, so that accurate estimates of required field-scale residence times can be established. The tests should be conducted using groundwater from the subject plume, as the degradation rates of some organics in certain materials may be greatly influenced by the chemical composition of groundwater. If the treatment material to be tested is to be placed in the downgradient part of a SPRB, then the water used in the tests of these materials should represent the effluent chemistry of the upgradient treatment zone. Among the key changes occurring in groundwater in sequence that could affect the performance of downstream processes (i.e., bioremediation following granular iron treatment) are

- Changes in pH and redox;
- Removal, conversion, or addition of key inorganic constituents (SO_4^{2-}, NO_3^-, etc.);
- Production of organic breakdown products not originally present in groundwater entering the SPRB due to reactions occurring in the first/upgradient treatment zone; and
- Inorganic changes resulting from mineral precipitation within the treatment materials.

For many plumes of chlorinated solvents, granular iron may be envisioned as part of the SPRB system. Laboratory column tests may be required prior to full-scale design. Batch tests with iron generally do not provide reliable rate data for design of field installations (Gillham, 1996); batch tests commonly give higher half-lives than dynamic column tests, and interpretation of results can be complicated by the fact that at early time, sorption and degradation can be occurring simultaneously. Instead, laboratory column tests are more useful in determining chemical degradation rates under conditions that more closely approximate the operating conditions anticipated in the field. In some instances, degradation rates for chlorinated ethenes (TCE, for example) may be obtained from existing data collected from other sites.

For SPRBs involving biological treatment of petroleum hydrocarbons or BTEX compounds by biological processes that are enhanced by oxygen or nutrient addition, existing literature values for biodegradation rates (under conditions where the presence of the amendment is not limiting) can be used for design purposes. Treatability studies in such cases could be limited to batch microcosm studies. Such studies would be used to confirm that biodegradation is measurable under the anticipated field conditions. Column tests may be of use when examining pulse nutrient injection, where a nutrient source is periodically injected into the treatment zone.

For organic compounds such as 1,2-dichloroethane (1,2-DCA) and DCM, where less is known of their biodegradation behavior, more detailed treatability studies (again in the form of microcosm tests) might be required. Such tests would investigate a variety of amendments to examine the robustness of the degradation processes and also determine relative rates of reaction using various amendments. As a means of identifying amendments for testing, historical aqueous geochemical data can be reviewed to examine the rate of microbial utilization of background levels of amendments present in the aquifer.

Limitations of treatability studies need to be recognized and accounted for when the study results are used as input parameters into the field design. Possible limitations include the following:

- Minimization or elimination of an inhibitory site condition;
- Establishment of a nonnative microbial population in the reactive materials due to laboratory contamination (this may be significant in longer-term laboratory studies);
- The effect of laboratory temperatures on the observed reaction rate;
- The influence of inorganic mineral precipitation, if it occurs, on reactive material performance may also be related to the temperature of the system; and
- Sampling and transport may alter the chemistry of the tested groundwater from that occurring *in situ*.

Usually, these limitations can be accommodated by building safety factors into the full-scale design of the SPRB.

Treatability Studies for Cases A and B — Treatability studies for both Case A and Case B would be of similar design. For the purposes of this exercise, it is assumed that the treatability studies for both cases are the same. The granular iron design treatability studies completed for the hypothetical full-scale system at Alameda would be similar to those performed for the pilot test. A mixture of 50% iron and 50% aquifer material would be tested in column studies in addition to a column of 100% iron, to evaluate the use of a lower percentage of iron in treatment gates at the margins of the plume, where influent *c*DCE and VC concentrations are lower. The difference in effluent chemistry between 50 and 100% iron treatments could potentially cause differences on the efficiency of downgradient microbial treatment.

BTEX degradation rates obtained from previous studies and the pilot-scale test would be used to calculate the input of ambient air needed to promote aerobic biodegradation. There would be no need for additional microcosm studies.

8.3.4 Groundwater Flow Modeling

Groundwater flow models are used to examine the effect of PRBs on existing groundwater flow patterns in the vicinity of the proposed installation. Of the numerous models available for this purpose, most include a particle tracking routine to calculate groundwater pathlines and travel times. The travel time in each part of a SPRB needs to be determined using the hydraulic properties of each specific treatment material. The width of the upgradient aquifer captured and treated by the PRB (the capture zone) is also evaluated. Previous modeling studies have shown that the velocity in a gate is insensitive to the gate length in the direction parallel to groundwater flow.

Several articles have been published describing the predicted effects of PRBs on existing flow patterns in contaminant plumes. The initial work by Starr and Cherry, examining Funnel-and-Gate configurations, was published in 1994. These two-dimensional simulations were extended to examine vertical flow patterns by Shikaze et al. in 1996. Austrins (1997) describes the modeling completed as part of a biotreatment gate in western Canada.

Design Modeling for Hypothetical Cases — In many instances, an existing groundwater flow model from a site can be modified to include a PRB. However, no such model existed for the Alameda site. Based on the presence of a low permeability confining layer at depth, a two-dimensional (2-D) plan-view model was considered sufficient for design purposes.

As described in Section 7.2, a 2-D model was developed for evaluation of the pilot-scale system. This model was expanded to evaluate both full-scale scenarios proposed herein (Appendix 33). As noted previously, each gate contains an iron treatment zone followed by a biosparge zone. The candidate *in situ* materials (i.e., granular iron, pea gravel, and biosparge zone) in the gates were assigned hydraulic properties based on previous laboratory and field determinations (UW, 1998).

For Case A, designed to capture an 80-ft (24-m)-wide plume, it was assumed that the control gate of the existing pilot-scale system would be replaced by a second treatment gate. Additional sheet pile would be added to create the desired capture zone (Figure 8.1).

Results of the Case A simulations performed by the University of Waterloo are presented in Appendix 33. These results are used together with projected residence time requirements to determine the size (i.e., flow-through dimensions) of each of the two reactive zones required in each SPRB gate. The key design assumptions used to determine treatment zone dimensions are shown in Table 8.3.

Table 8.3 Summary of Design Assumptions for SPRB Case A and B

	Assumed Value	
Parameter	**Case A**	**Case B**
Plume width	80 ft (24 m)	350 ft (107 m)
Plume depth	22 ft (6.8 m)	22 ft (6.8 m)
Saturated thickness	18 ft (5.5 m)	18 ft (5.5 m)
Gate velocity	0.11 ft/day (3.4 cm/day)	0.14 ft/day (4.3 cm/day)
Influent concentration — CVOCs	See Table 8.4	See Table 8.6
Influent concentration — Total BTEX	10 mg/L	1 mg/L

For Case B, it was assumed a larger SPRB constructed to capture a 350-ft (107-m)-wide plume would not make use of the existing pilot (Figure 8.2). Rather, a three-gate Funnel-and-Gate configuration was modeled. The central gate would treat the high concentrations in the core of the plume and would be separated by lengths of low permeability funnel from the two outer gates, which would be designed to treat the lower concentrations of VOCs observed at the plume margins.

Results of the Case B simulations performed by the University of Waterloo are presented in Appendix 33. These results are used together with projected residence time requirements to determine the size (i.e., flow-through dimensions) of each of the two reactive zones required in each of the three SPRB gates. The key design assumptions used to determine treatment zone dimensions are shown in Table 8.3.

8.3.5 Determination of SPRB Dimensions

The dimensions of the SPRB are determined by the groundwater modeling and required residence time. Residence time requirements are based on projected influent concentrations, desired treatment goals, and degradation rates determined during the treatability study phase. Influent concentrations used in the design are based on groundwater chemistry data collected along the line of installation.

Degradation of most of the organic contaminants under consideration usually follow first-order kinetics, and thus required residence times can be determined from the measured degradation half-life. Using half-lives and influent concentrations as input parameters, a simple calculation can be completed to predict the residence time required to reach the desired treatment level. Useful models are available that will accommodate simultaneous degradation of several compounds and will also accommodate the simultaneous formation and degradation of breakdown products (Figure 8.3). This is of particular importance when evaluating chlorinated solvent degradation (e.g., the production of *c*DCE and VC from TCE).

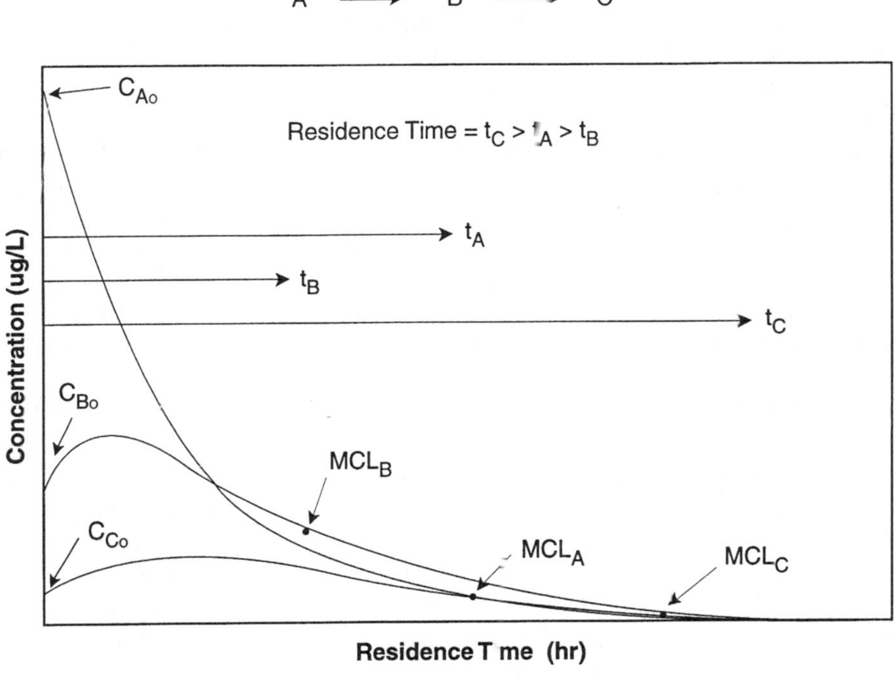

Figure 8.3 Conceptual residence time calculation for VOC removal.

In a SPRB, the residence times needed for each treatment process are added together to calculate the total residence time needed. The residence time can then be integrated with the predicted groundwater velocity from the design modeling effort to determine the required SPRB dimensions (i.e., the dimension of each part of the SPR parallel to the flow path). These dimensions become the core of the design developed for the field application.

8.3.5.1 SPRB Dimensions — Case A

Expansion of the pilot-scale system to capture the core of the plume (Case A) would include the installation of a second 10-ft (3-m)-wide treatment gate in the control gate of the pilot-scale system, and the addition of 40 ft (12 m) of funnel section on each side of the pilot-scale system (Figure 8.1). Groundwater modeling showed that this configuration could capture and treat about an 80-ft (24-m)-wide plume. Groundwater flow velocities through the two gates would be about 0.11 ft/day (3.4 cm/day), but could range as high as 0.29 ft/day (8.8 cm/day), depending on background aquifer gradients.

Treatment Zone Dimensions Needed for Chlorinated Compound Degradation — Influent concentrations of *c*DCE and VC from four different data sets gathered during the pilot-scale were used to assess the residence time needed for degradation of the VOCs present in the plume. Both the apparent degradation rates (half-lives) determined from the pilot-scale trial and those measured during the bench-scale test conducted using groundwater from the site were examined to evaluate the potential range in residence time that could be required to treat the observed chemicals.

The residence times listed in Table 8.4 were used as input to a first-order sequential degradation model similar to the one illustrated in Figure 8.3. The degradation model accounts for simultaneous production and degradation of VOCs. Pilot-scale half-lives of 40 hr were used for both *c*DCE and VC in the first simulation. The second set of simulations used half-lives of 24 and 22 hr for *c*DCE and VC, respectively. To account for field effects (mainly lower field temperature versus laboratory) these half-lives are a factor of two greater than those measured in the bench-scale test. Although available field data are relatively sparse, they indicate that this decrease in degradation rates may occur. Based on previous laboratory studies conducted at UW, *c*DCE is calculated to degrade to 2% VC and 98% ethene and ethane at each step of the simultaneous model runs. The required residence time was also estimated for two sets of effluent quality criteria: the proposed site-specific remedial objectives, and California MCL standards. These input parameters and resulting residence times are summarized in Table 8.4. As shown in Table 8.4, the required residence time in the iron zone could be as low as 11 days or as high as 31 days depending on the assumed influent concentrations, half-lives, and regulatory criteria.

Using the maximum estimated residence time of 31 days and a flow velocity of 0.11 ft/day (3.4 cm/day), an iron thickness of 3.4 ft (1 m) would be required. The required thickness would increase to 9 ft (2.7 m) at the higher flow velocity of 0.29 ft/day (8.8 cm/day), and would decrease to about 1.2 ft (0.6 m) if the shortest residence time of 11 days and the low velocity (0.11 ft/day) are used. Given the groundwater flow conditions at the site, a minimum thickness of 3.4 ft (1 m) should be used in the second gate. Further, it was assumed (arbitrarily) that this minimum thickness would be increased by 50% (to 5.1 ft) to accommodate higher flow velocities that might occur at certain times of the year. Using an average saturated thickness of about 18 ft (5.5 m) and a gate width of 10 ft (3 m), the volume of iron can be calculated as follows:

$$\begin{aligned}\text{Volume of Iron} &= \text{width of treatment zone} \times \text{saturated height} \\ &\quad \times \text{thickness in the direction of flow} \\ &= (10\text{ ft} \times 18\text{ ft} \times 5.1\text{ ft}) \\ &= 918\text{ ft}^3\ (26\text{ m}^3)\end{aligned}$$

Table 8.4 Granular Iron Design Parameters and Residence Time Estimates — Case A Expanded Pilot Configuration

VOC Concentrations from Historical Data	Assumed Influent Concentration (μg/L)		Design Parameters: Using Apparent Field-Scale Half-Lives (*c*DCE HL of 40 hours and VC HL of 40 hours)				Design Parameters: Using Scaled Laboratory Half-Lives (*c*DCE HL of 22 hours and VC HL of 24 hours)			
			PRO[a]		MCL[b]		PRO[a]		MCL[b]	
	*c*DCE	VC	T[e]	R.T.[f]	T[e]	R.T.[f]	T[e]	R.T.[f]	T[e]	R.T.[f]
BH03-1[c] (1996)	219,000	16,700	3.1–3.9	28	3.4–4.3	31	1.7–2.1	15	1.7–2.4	17
MW-3[d] (1996)	22,400	4,800	2.2–2.8	20	2.2–3.2	23	1.2–1.5	11	1.2–1.8	13
19 March 1997 Pilot Influent Concentrations	62,000	9,600	2.4–3.1	22	2.4–3.5	25	1.4–1.8	13	1.4–2.0	14
17 June 1997 Pilot Influent Concentrations	110,000	21,000	2.6–3.4	24	2.6–3.8	27	1.5–2.0	14	1.5–2.1	15

[a] PRO — Residence time (in days) based on proposed (site-specific) remedial objectives: *c*DCE = 10 μg/L and VC = 2 μg/L.

[b] MCL — Residence time (in days) based on California MCL: *c*DCE = 6 μg/L and VC = 0.5 μg/L.

[c] Depth-discrete sampling point from July 1996 site investigation.

[d] Temporary monitoring well installed in July 1996 site investigation (UW, 1996). Well was removed during construction of pilot-scale system.

[e] T — thickness (in ft) of granular iron zone needed. First value is for low velocity (0.11 ft/day) and second value is for higher velocity (0.14 ft/day).

[f] R.T. — Residence time (days).

Assuming a bulk density for the iron of 0.08 ton/ft^3 (2.6 t/m^3), about 73 tons (66 t) of iron material would be required.

Treatment Zone Dimension Needed for BTEX Compound Degradation — The design objective of the biosparge zone is to provide sufficient oxygen to groundwater to promote aerobic degradation of BTEX, and/or any remaining less chlorinated solvents (e.g., DCE or VC) released from the upgradient granular iron zone. The open biosparge zone design used for the pilot-scale is shown in Figure 8.4. There are a number of possible means of suppling oxygen to groundwater, but, for the purposes of this report, a system similar to the one currently in place at the pilot-scale test was used.

A variety of flow thicknesses in the flow direction were considered for the biosparge zone and the residence times for these under potential groundwater fluxes and velocities are shown in Table 8.5. Based on the data from Table 8.5, a 3-ft (0.9-m)-thick zone was selected for the pilot-scale. This zone provided both reasonable residence times and acceptable construction costs. Unfortunately, breakthrough of BTEX compounds to the biosparge zone during pilot-scale testing was not substantial; therefore the FTD (UW, 1998) could not assess the performance of the pilot-scale system in treating BTEX. For this report it will be assumed that all design parameters selected for the pilot-scale were valid.

Table 8.5 Residence Times Calculated for Four Groundwater Velocities and Five Biosparge Zone Volumes – Alameda Pilot-Scale Test

		Groundwater Velocity (ft/day) [Groundwater Flux (ft^3/day)]			
		0.2	0.27	0.65	1
Biosparge Zone Thickness (ft)	Biosparge Zone Volume (ft^3)[a]	[9.0]	[12.2]	[29.3]	[45]
		Residence Time (days)[b]			
2	300	33	25	10	7
3	450	50	37	15	10
4	600	67	49	20	13
5	750	83	61	26	17
6	900	100	74	31	20

[a] Based on a zone that is 10 ft wide and has a saturated thickness of 15 ft.

[b] Residence time calculated as biosparge zone volume divided by groundwater flux.

Source: Based on UW, 1996 (NAS Alameda Work Plan).

Appendix 34 provides background information concerning the determination of BTEX loading to the sparge zone. It was assumed that an equal fraction of each of the components of BTEX was present. Theoretically, for every mole of BTEX only 1.59 moles of oxygen are required (see Appendix 34). Conservatively, this number was increased to 6 moles of oxygen supplied for each mole of BTEX to account for other biochemical oxygen demand (BOD) and chemical oxygen demand (COD) processes. Therefore, if the influent water to the biosparge zone contained 7 mg/L of total BTEX, then up to 15 mg/L of oxygen would need to be added.

At the high groundwater velocity (0.29 ft/day), the residence time in the expanded pilot-scale system (Case A) would be about 10 days. At the lower groundwater velocity (0.11 ft/day), the residence time would be more than 30 days with a zone thickness of 3 ft. Based on this finding, no further increase in biosparge gate size would be necessary for the expanded pilot-scale.

During pilot-scale testing at NAS Alameda, the sparge frequency was determined based on the residence times of water in the sparge zone (UW, 1996). The initial frequency of sparging was once per day for about 10 minutes. This frequency was insufficient to maintain constant aerobic conditions. Numerous investigations into the sparge frequency concluded that the frequency required to maintain sufficient dissolved oxygen (DO) was six times a day for about 15 minutes each interval. This frequency resulted in a time-weighted average (TWA) DO that ranged from 25 to 30 mg/L. If the interval was only three times a day for 15 minutes, the TWA DO ranged from 10 to 15 mg/L.

Determination of sparging frequency in the full-scale design was based on the BTEX loading to the biosparge zone. In the core of the plume the BTEX concentrations range from <1 to 7 mg/L; elsewhere BTEX concentrations generally ranged from not detected to ~1 mg/L. To degrade a total BTEX load of 10 mg/L, a conservative estimate of the core plume concentrations, up to 21 mg/L of DO would be required (assuming 2.1 g of oxygen is needed for each gram of BTEX). Correspondingly, only 2 mg/L of DO is needed if the BTEX load is 1 mg/L (see Appendix 34). The final sparging frequency should be determined through field testing during the start-up and troubleshooting period.

8.3.5.2 SPRB Dimensions — Case B

As shown in Figure 8.3, a full-scale SPRB to treat the entire plume to the "mapped" lateral extent of the 10 µg/L *c*DCE contour might consist of three 25-ft (7.6-m) gates separated by 125 ft (38 m) of funnel section, with an additional 125 ft (38 m) of funnel on either end of the system. Based on groundwater modeling, the 575-ft (175-m)-long system could capture and treat a plume about 350 ft (107 m) wide. Each gate would have about the same capture width (117 ft (36 m)), and thus the same flow velocity of about 0.14 ft/day (4.3 cm/day) would occur in each gate. However, periodic flow velocities as high as 0.38 ft/day (11.6 cm/day) may occur depending on the background aquifer gradient.

Treatment Dimensions for Chlorinated Compound Degradation — This design assumed the central gate would be positioned to treat concentrations in the narrow core of the plume. The residence times determined in the previous section for the second gate of the pilot-scale expansion would therefore apply to this gate. Based on existing plume data, it was assumed that the outer two gates would only be required to degrade a maximum total VOC concentration of 1000 µg/L. Influent concentrations of 750 and 250 µg/L for *c*DCE and VC, respectively, were selected for these gates based on the ratio of these compounds observed in the field. Simulations were run using two sets of degradation rates and regulatory criteria, as for the previous scenario. Table 8.6 lists the required residence times for these gates (a maximum of 31 days in the central gate and 16 days in the outer gates).

Given these estimated residence times, an iron flow-through thickness of 4.3 ft (1.3 m) and 2.2 ft (0.7 m) would be required in the core gate and two outer gates, respectively, at a flow velocity of 0.14 ft/day (4.3 cm/day). As in Case A, these thicknesses were increased by 50% to accommodate higher flow rates. Assuming an average saturated thickness of about 18 ft (5.5 m) and gate widths of 25 ft (7.6 m), the volume of iron can be calculated as follows:

$$\begin{aligned}\text{Volume of Iron} &= \text{number of gates of a given length} \times \text{length of treatment zone} \times \\ &\quad\ \text{saturated height x thickness in the direction of flow} \\ &= 1 \times (25\ \text{ft} \times 18\ \text{ft} \times 6.5\ \text{ft}) + 2 \times (25\ \text{ft} \times 18\ \text{ft} \times 3.3\ \text{ft}) \\ &= 2925\ \text{ft}^3 + 2970\ \text{ft}^3 \\ &= 5895\ \text{ft}^3\ (167\ \text{m}^3)\end{aligned}$$

This equates to about 472 tons (434 t) of iron.

Treatment Zone Dimensions for BTEX Compound Degradation — In the Case B Funnel-and-Gate configuration, the BTEX and/or less-chlorinated CVOC loading in the peripheral gates is expected to be minimal (based on current site characterization data), so the biosparge would not

Table 8.6 Granular Iron Design Parameters and Required Residence Time Estimates – Case B

	Concentration (μg/L)		Residence Time (days)							
			Using Apparent Field-Scale Half-Lives (*c*DCE HL of 40 hours and VC HL of 40 hours)				Using Scaled Laboratory Half-Lives (*c*DCE HL of 22 hours and VC HL of 24 hours)			
			PRO[a]		MCL[b]		PRO[a]		MCL[b]	
	*c*DCE	VC	T[d]	R.T.[e]	T[d]	R.T.[e]	T[d]	R.T.[e]	T[d]	R.T.[e]
Gate in Core of Plume	see Table 8.4		2.2–3.9	20–28	2.5–4.3	23–31	1.2–2.1	11–15	1.4–2.4	13–17
Gates Outside of Plume Core	750	250	1.3	12	1.8	16	0.8	7	1.0	9

[a] PRO — Residence time (in days) based on proposed (site-specific) remedial objectives: *c*DCE = 10 μg/L and VC = 2 μg/L.

[b] MCL — Residence time (in days) based on California MCL: *c*DCE = 6 μg/L and VC = 0.5 μg/L.

[c] See Figure 8.2.

[d] T — Thickness (in ft) of granular iron zone needed. First value is for low velocity (0.11 ft/day) and second value is for higher velocity (0.14 ft/day).

[e] R.T. — Residence Time (days).

need to be increased in size from the Case A scenario. In fact a reduction in the sparging frequency would be expected due to reduced BOD/organic compound loading. In the central gate, the biosparge zone would not be increased in thickness, as the residence time should be adequate, but the sparging frequency may have to be increased to provide extra DO for degradation.

8.4 INSTALLATION

Construction of a SPRB usually involves installation of the funnel, excavation of native material to install the SPRB, and installation of the treatment media and monitoring networks. There are different methods for the installation of funnels (cutoff walls) and the PRBs. Possible construction methods and the general construction sequence are summarized in the following sections.

8.4.1 Possible Construction Methods

A variety of excavation techniques have been used to construct most of the shallow (40 ft bgs or less) PRBs installed to date. The use of a sheet pile trench box (a box of rectangular sheet piling is driven into the ground and native soils excavated) allowed expeditious and accurate placement of the iron materials and also facilitated construction of the biosparge gate at the Alameda pilot (UW, 1997b). Injection methods are being considered for emplacement of granular iron treatment zones to greater depths, but these appear less amenable to the installation of biologically active systems.

Often, the choice of construction method (and material) for the funnel section is dependent on the geotechnical characteristics of the aquifer materials and funnel costs (including associated soil disposal costs). The presence of cobbles and related difficult installation conditions may preclude the use of sheet piling at some sites, whereas excavation of heavily contaminated soils (and related disposal costs) may preclude the use of certain types of slurry walls. For the Alameda full-scale examples, sealable joint sheet pile (Waterloo Barrier™) was used in the cost estimate for the funnel sections.

8.4.2 Construction Sequence

The construction sequence for installation of the SPRB is similar to that followed for other subsurface works. The design engineer and the contractor would attend a project design meeting at the site to discuss the requirements of the design and plans for the development of construction procedures. The contractor would prepare detailed drawings and detailed written construction procedures for the installation of the system based on the design requirements. The contractor would also prepare a health and safety plan for construction activities to be conducted at the site. This plan would comply with U.S. Occupational Safety and Health Administration (OSHA) and other requirements. The design and construction procedures would incorporate the installation of performance monitoring equipment. When considering treatment gate construction, other factors common to subsurface construction procedures need to be considered, including those listed below:

1. The need for dewatering during excavation of gate sections.
2. Means and costs of groundwater and soil disposal. Every effort should be made to minimize the waste volumes generated, and the selection of construction method should involve consideration of this issue (Battelle, 1996).
3. Health and safety concerns. The only "hazard" associated with granular iron used to date has been nuisance dust. There may be material handling issues that arise depending on a liquid or gas amendment selected for a biologically mediated treatment zone. Entry of construction personnel into gate excavations, if necessary, does represent a significant health and safety issue.
4. Disruption of site activities.

In addition, particular care is needed when considering the connection between the low permeability funnel sections and permeable gate sections in order to avoid bypass of contaminated groundwater around the gate sections.

Following regulatory approval of the design and receipt of necessary construction permits, installation of the system would proceed. The major activities and requirements of the installation task would include the following:

- Mobilization of all equipment required for the installation of the system to the site;
- Completion of all necessary site preparation work required prior to initiating construction, including grading, setup of storage areas, and setup of decontamination zones in accordance with the site health and safety plan;
- Installation of the Funnel-and-Gate sections of the system as per the approved drawings and construction procedures;
- Disposal of any generated solid or liquid construction "wastes" using approved methods;
- Site restoration and cleanup to agreed-upon conditions following the completion of the site construction activities;
- Decontamination and demobilization of all equipment and materials and supplies; and
- A detailed location and elevation survey of the installed treatment system components and monitoring wells.

A final report would be prepared following system installation, including as-built detailed drawings for the installation and a location and elevation survey of each component. These drawings would include the exact locations and elevations of any monitoring wells installed as part of the system. The report would also update the detailed written construction procedures for the installation of the system incorporating any changes made during the installation process.

8.5 OPERATIONS AND MAINTENANCE ISSUES

The two major issues affecting the operations and maintenance (O&M) requirements for PRBs and SPRBs are the loss of system reactivity over time and possible permeability decreases in the reactive section of the system. The former could lead to breakthrough of contaminants on the downgradient side of the PRB, while the latter could cause contaminated groundwater to bypass beneath or around the system. Currently, only periodic maintenance (once every few years) is envisioned for iron PRBs, while much more frequent maintenance (semiannually or annually) may be required for biosparge systems in PRBs.

A major factor affecting O&M costs is the possible need for periodic rejuvenation of iron wall sections affected by precipitates. The iron material itself should last for several years to decades. The precipitates (if significant) will form in a narrow zone at the upgradient aquifer/iron interface. Rejuvenation therefore could be as simple as agitating the upgradient face of the iron with an auger to restore the permeability of this material. No significant precipitates were observed in cores collected from an *in situ* reactive wall at a University of Waterloo test site 2 and 4 years after its installation. This wall performed consistently for over 5 years (O'Hannesin and Gillham, 1998). Additionally, there is no evidence of biofouling or plugging in cores obtained from commercial *in situ* granular iron PRBs operating for over 2 years (Vogan et al., 1998).

For biosparge or nutrient addition systems, biofouling of the gas/nutrient delivery system may occur. As noted in Section 7.2, the oxygen injection system at Alameda was not cleaned during the first year of operation. The O&M procedures for these zones could involve their physical removal and cleaning, or the periodic injection of a chemical solution designed to remove/deplete the fouling material.

The interface between the upgradient granular iron treatment zone and downgradient biosparge zone may be another location of potential microbial growth. The effects of such growth may be minimized during design by ensuring that a high conductivity/high porosity interface (e.g., pea gravel) exists in this area.

Chapter 9

Measurement Procedures and Monitoring Program

The compliance monitoring well network established for a SPRB is dependent on the site-specific characteristics of the installation, which will determine the number and spacing of wells and their screen intervals (corresponding to the vertical extent of the treatment zone). The following are general considerations:

1. Monitoring wells should be placed both within the treatment zone and as close to the downgradient edge of the system as possible.
2. If possible, a set of wells should be placed along the line of groundwater flow direction upgradient of, within, and immediately downgradient of the system.
3. Wells placed in the aquifer downgradient of the system may continue to exhibit VOC concentrations for some time following system installation due to desorption of VOCs from the aquifer. Concentration vs. time profiles using data from these wells may show characteristic tailing over time.
4. Wells whose screens extend throughout the full vertical extent of the SPRB will give a vertically averaged indication of wall performance. The appropriateness of such wells should be considered at the time of SPRB installation.
5. Groundwater monitoring should include VOCs present at the site and breakdown products resulting from the SPRB.
6. Appropriate location of monitoring points to confirm biological treatment will depend on the biodegradation mechanism being utilized. For example, the effects of a biosparge or nutrient addition zone may not be seen until the augmented groundwater moves into downgradient geologic materials (either artificially emplaced sand/gravel or the aquifer).
7. Because the various treatment zones can be relatively thin at some sites, it may be quite easy to pull water in from different zones during well development and sampling (e.g., from the iron into an upgradient gravel zone). Low flow sampling techniques and purge volumes should be taken into account for the possibility of sampling artifacts when interpreting field results.
8. Monitoring frequency may depend on the groundwater flow velocity in the SPRB and the time needed for the treatment systems to reach steady state. In most instances, quarterly VOC monitoring should be sufficient to gauge system performance over the first year. Field parameters that can serve as useful secondary indicators of performance include pH, E_h, specific conductance, DO, BOD, and alkalinity. These parameters and groundwater temperature can be measured concurrently with the VOC sampling.
9. Water level monitoring on a monthly basis (at most) is adequate for the first year. Water level monitoring can be reduced to quarterly or semiannually thereafter.
10. Inorganic parameters may be sampled on a semiannual basis to gauge the amount of mineral precipitation occurring in the reactive zone. Not all wells sampled for VOCs need to be sampled for these parameters. Probably two or three upgradient and downgradient wells and two or three wells in the treatment zone will suffice. If inorganic nutrients source(s) are used the list may need to be revised to include nutrients and conversion products. Parameters of interest include:

SO_4^{2-}	Ca^{2+}
Cl^-	Mg^{2+}
Alkalinity (HCO_3^-)	K^+
NO_3^-	Na^+
PO_4^{3-} (if present)	Fe^{2+}
TDS	Mn^{2+}

11. This list of organics could be augmented by any inorganic nutrient source and/or its conversion products added in a biotreatment zone. The use of specific nutrient sources may dictate more frequent monitoring (e.g., quarterly) of nutrient/product concentrations in specific wells.
12. Although not necessary for compliance, other wells can be placed at various distances within the treatment zone to enable calculation of VOC degradation rates, as opposed to having only two data points (upgradient and downgradient) for this purpose. These wells could conceivably provide data that could warn of the need for rejuvenation/corrective action (i.e., the detection of unexpectedly high VOC concentrations at a given point within the treatment zone) and allow for rejuvenation of the treatment zone prior to VOCs passing untreated though the SPRB.
13. Wells could be placed in clean zones parallel and adjacent to the end of the SPRB, and at the upgradient edge of the gate sections. These wells could be monitored over time for VOCs. Occurrence of VOCs in these wells may indicate that contaminated groundwater is bypassing the system.
14. Specific microbial analyses may be conducted to gauge the performance of the biotreatment zones. These may include conventional biomass determinations, general lipid biomass determinations to gauge the size of microbial populations, and/or phospholipid fatty acid (PLFA) analyses to better characterize the microbial populations present in the biotreatment zone.
15. Measurement of concentrations of dissolved gases (e.g., methane, ethane, ethene, and carbon dioxide) may also be used to gauge the performance of biological treatment processes.
16. Other measurements may be of use in gauging hydraulic performance of a SPRB. Hydraulic conductivity (slug) tests and groundwater velocity measurements (*in situ* meters, tracer tests) could be completed in selected wells. Coring of the treatment sections should not be considered as part of a routine program in commercial installations.

9.1 MONITORING PLAN — SPRB CASE A

We have assumed that quarterly sampling for the first 2 years would be required by the regulatory agencies. If the system is performing as designed, it is reasonable to assume that the program would be scaled back following this initial period (i.e., twice per year).

The monitoring network envisioned for the Case A SPRB at Alameda is shown in Figure 9.1. Although several wells would be installed as part of system construction, the center line of wells in each gate plus one well at each end of the outer funnel section would be the focus of the monitoring program.

Aquifer monitoring wells located upgradient and downgradient of each gate would serve as the key compliance monitoring wells (i.e., to gauge how influent and effluent concentrations meet the design criteria). These wells would be monitored quarterly for the first 2 years, together with wells on either end of the system. Concentrations in wells on either end of the system would be monitored specifically to ensure that higher VOC concentrations indicating bypass of contaminated groundwater around the system do not occur over time.

As shown in Tables 9.1 and 9.2, the compliance well data would be supplemented by quarterly results for VOCs from the remaining wells along the center line of each gate, which will enable the performance of each treatment process to be quantified. Semiannual sampling for major cations, anions, and biological parameters will also assist in documenting system performance.

Water levels collected quarterly will allow the hydraulics of the system to be evaluated on a seasonal basis. Anomalous groundwater elevations, especially from wells on either side of the funnel section, could be used to judge the potential for bypass around or under the system.

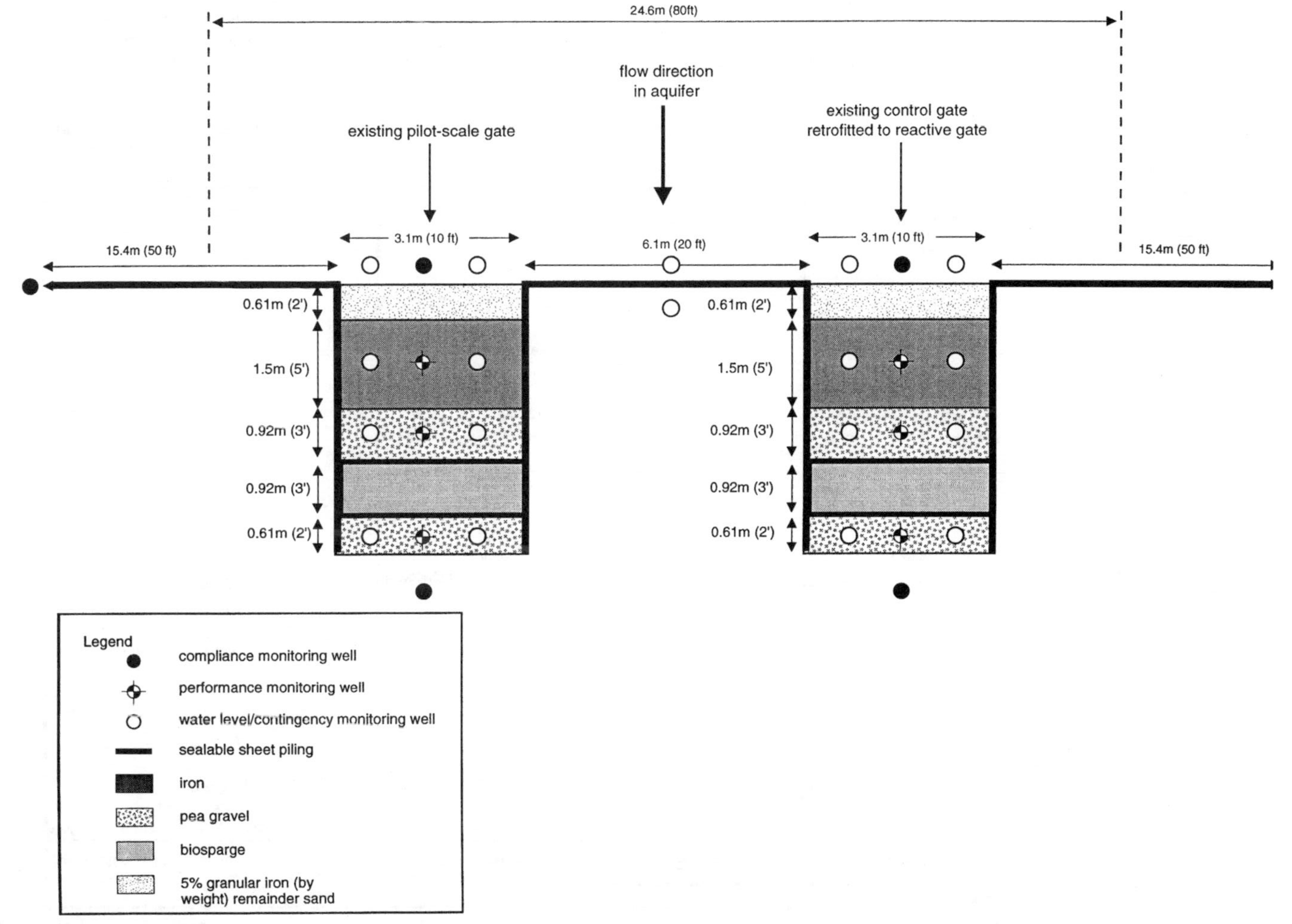

Figure 9.1 Monitoring network — Alameda Case A.

Table 9.1 SPRB Monitoring Program, Alameda Case A — Expanded Pilot Configuration

Parameter	Year 1 - 2		Year 3 - 30	
	Location	Frequency	Location	Frequency
Volatile Organic Compounds PCE Benzene TCE Ethylbenzene cDCE Toluene VC Xylenes Chlorobenzene	Five wells in center row of each gate and four wells located at the end and middle of funnel sections	Quarterly	Five wells in center row of each gate and six wells located at the end and middle of funnel sections.	Semiannually
Major Cations & Anions Calcium Chloride Iron (Total) Nitrate Magnesium Phosphate Manganese Sulfate Potassium Sodium Alkalinity TDS	Five wells in center row of each gate	Semiannually	Five wells in center row of each gate.	Annually
Biological DHG BOD Oxygenates VFA	Three downgradient wells in center row of each gate	Semiannually	Three downgradient wells in center row of each gate.	Annually
Field pH Temperature ORP (E_h) DO	Five wells in center row of each gate	Quarterly	Five wells in center row of each gate.	Semiannually
Hydraulic Performance Groundwater Elevation	All Wells	Quarterly	All Wells	Semiannually
Groundwater Velocity	Gate Wells	As Required	Gate Wells	As Required
Hydraulic Conductivity	Iron Wells	As Required	Iron Wells	As Required

PCE = Tetrachloroethene TCE = Trichloroethene cDCE = cis-1,2-Dichloroethene VC = Vinyl Chloride TDS = Total Dissolved Solids
DHG = Dissolved Hydrogen Gas BOD=Biochemical Oxygen Demand VFA = Volatile Fatty Acids ORP = Oxygen Reducing Potential DO = Dissolved Oxygen

Notes:

a Performance monitoring of VOCs includes the middle row of wells in each gate located in the iron and pea gravel sections as well as the two wells in the middle of the funnel sections; refer to Figure 9.1.

b Compliance monitoring of VOCs includes the upgrandient and downgradient aquifer wells from each gate as well as the two wells at either end of the treatment system; refer to Figure 9.2.

Table 9.2 SPRB Monitoring Program, Alameda Case B — 350 ft Full-Scale Configuration

Parameter	Year 1 - 2		Year 3 - 30	
	Location	Frequency	Location	Frequency
Volatile Organic Compounds PCE, TCE, cDCE, VC, Benzene, Ethylbenzene, Toluene, Xylenes, Chlorobenzene	Five wells in center row of each gate and six wells located at the end and middle of funnel sections	Quarterly	Five wells in center row of each gate and six wells located at the end and middle of funnel sections	Semiannually
Major Cations & Anions Calcium, Iron (Total), Magnesium, Manganese, Potassium, Sodium, Chloride, Nitrate, Phosphate, Sulfate, Alkalinity, TDS	Five wells in center row of each gate	Semiannually	Five wells in center row of each gate	Annually
Biological DHG, Oxygenates, BOD, VFA	Three downgradient wells in center row of each gate	Semiannually	Three downgradient wells in center row of each gate	Annually
Field pH, ORP (E_h), Temperature, DO	Five wells in center row of each gate	Quarterly	Five wells in center row of each gate	Semiannually
Hydraulic Performance Groundwater Elevation	All Wells	Quarterly	All Wells	Semiannually
Groundwater Velocity	Gate Wells	As Required	Gate Wells	As Required
Hydraulic Conductivity	Iron Wells	As Required	Iron Wells	As Required

PCE = Tetrachloroethene TCE = Trichloroethene cDCE = cis-1,2-Dichloroethene VC = Vinyl Chloride TDS = Total Dissolved Solids
DHG = Dissolved Hydrogen Gas BOD=Biochemical Oxygen Demand VFA = Volatile Fatty Acids ORP = Oxygen Reducing Potential DO = Dissolved Oxygen

Notes:

a Performance monitoring of VOCs includes the middle row of wells in each gate located in the iron and pea gravel sections as well as the four wells in the middle of the funnel sections; refer to Figure 9.1.

b Compliance monitoring of VOCs includes the upgrandient and downgradient aquifer wells from each gate as well as the two wells at either end of the treatment system; refer to Figure 9.2.

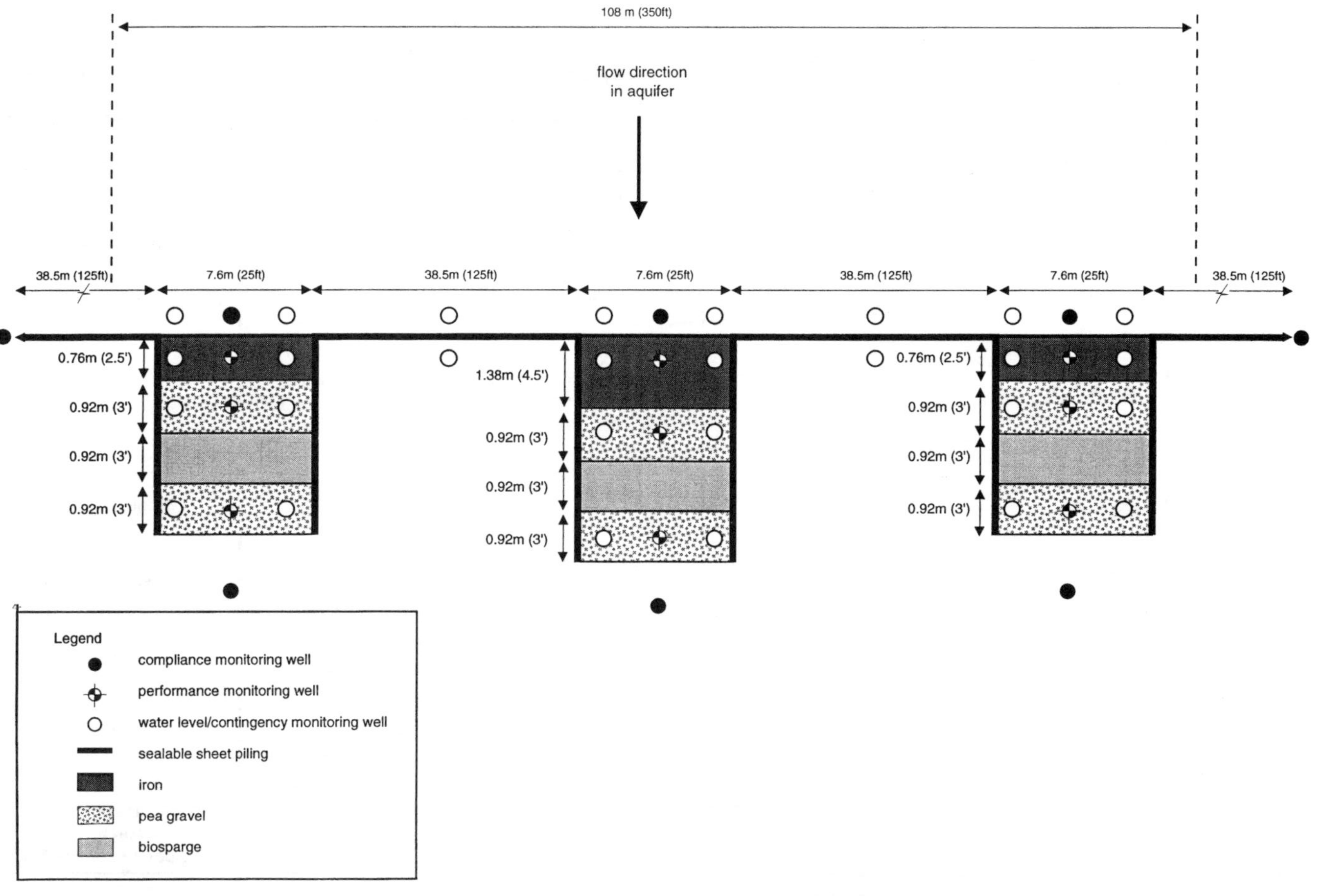

Figure 9.2 Monitoring network — Alameda Case B.

9.2 FULL-SCALE MONITORING PLAN — SPRB CASE B

The monitoring plan for Case B is identical to that designed for Case A (Figure 9.2). The addition of one additional gate and an additional funnel section for Case B will increase the number of wells. An increase in the number of wells results in an increase in the number of samples, even if the same general program as Case A is followed.

CHAPTER 10

Cost and Performance

The following chapter presents the costs of the two full-scale SPRB systems described in Chapter 8 for NAS Alameda. To facilitate comparison of the SPRB to conventional treatment alternatives, cost estimates were also developed for two P&T systems. The Case A P&T captures and treats the 80-ft (24-m)-wide plume (i.e., for comparison to the Case A SPRB), and the Case B P&T is a larger system that captures and treats the 350-ft (107-m)-wide plume (i.e., for comparison to the Case B SPRB).

Capital costs for each of the SPRB and P&T treatment systems are presented in Section 10.1. OM&M costs for the SPRB and P&T systems are presented in Section 10.2. Section 10.3 presents the comparison of the SPRB and P&T costs from both cumulative cost and net present value (NPV) perspectives. Section 10.4 discusses how design input parameters such as VOC concentrations, groundwater flux, and regulatory criteria could affect these cost estimates. Section 10.5 discusses other site-specific factors that can affect SPRB or P&T system costs, and summarizes the costs and performance of the SPRB and P&T systems for Case A and B.

10.1 CAPITAL COST ESTIMATES

10.1.1 SPRB Capital Cost Estimate

Capital cost estimates for the two Alameda full-scale system alternatives were developed based on the conceptual designs presented in Chapter 8. C^3 Environmental (Breslau, Ontario) supplied construction cost estimates for both scenarios. C^3 Environmental was involved in construction of the Alameda pilot-scale system, and therefore their cost estimates are considered realistic.

Iron costs (delivered to the site) were assumed to be \$430/t (\$475/ton), based on recent quotations from iron suppliers. Soil and groundwater disposal costs were based on those incurred during the pilot-scale construction program. The SPRB capital cost estimate assumes that sealable joint sheet piling could be used for the funnel sections; full-scale estimates were based on costs incurred installing the sealable joint sheet piling in the pilot-scale study. Costs for licensing, royalties, etc. are not included in these estimates, as they are considered system-specific, and may or may not be applicable depending on the treatment processes chosen.

Capital costs for the two systems are included in Table 10.1. For Case A the total capital costs were estimated to be \$697,500. For Case B the total capital costs were estimated to be \$1,031,500.

10.1.2 Pump-and-Treat System Capital Cost Estimate

The site-specific numerical model of groundwater flow developed by University of Waterloo (Appendix 33) was used to evaluate the groundwater extraction rates that would be required in the

Table 10.1 Summary of Captial Cost Estimates

Cost Category		Funnel and Gate SPRB					Pump & Treat	
Interagency WBS #	Construction Cost Elements	Case A (80 ft plume)	Case A- MCL (80 ft plume)	Case B (350 ft plume)	Case B -Slurry (350 ft plume)	Case B- MCL (350 ft plume)	Case A (80 ft plume)	Case B (350 ft plume)
Before Treatment Cost Elements								
33 01 - Both[a]	Mobilization	10,000	10,000	20,000	20,000	20,000	1,000	1,000
33 01 - Both	Site Preparation & Access	2,000	2,000	7,500	7,500	7,500	2,000	7,500
33 03 - SPRB	Funnel Wall Installation	43,000	43,000	195,000	82,500	195,000	NA	NA
33 03 - SPRB	Gate Construction	84,000	84,000	235,000	235,000	235,000	NA	NA
33 02 - Both	Aquifer Monitoring Wells	4,000	4,000	9,000	9,000	9,000	3,500	5,000
33 03 - P&T	GTS Installation	NA	NA	NA	NA	NA	10,000	10,000
Treatment Cost Elements								
33 12 - SPRB	Granular Iron Material	35,000	25,500	224,000	224,000	103,000	NA	NA
33 11 - SPRB	Biosparge Construction	2,500	2,500	15,000	15,000	15,000	NA	NA
33 13 - Both	Dewatering/Groundwater Treatment	3,000	3,000	18,000	18,000	18,000	NA	NA
33 13 - P&T	Extraction Pumps, Wells/Well Development	NA	NA	NA	NA	NA	34,000	53,000
33 13 - P&T	GTS Installation	NA	NA	NA	NA	NA	35,000	35,000
After Treatment Costs								
33 18 - P&T	Groundwater Reinjection	NA	NA	NA	NA	NA	5,000	8,000
33 19 - Both	Soil Disposal	10,000	10,000	60,000	60,000	60,000	1,000	1,000
33 21 - Both	Demobilization	6,000	6,000	10,000	10,000	10,000	1,000	1,000
Sub-Total		**$199,500**	**$190,000**	**$793,500**	**$681,000**	**$672,500**	**$92,500**	**$121,500**
Other Costs								
	Design, work plans, meetings, treatability study, construction oversight, reporting, etc. (assumed to be 30% of construction)	60,000	57,000	238,000	204,500	202,000	28,000	36,500
	Pilot Cost	438,000	438,000					
TOTAL CAPITAL COSTS		**$697,500**	**$685,000**	**$1,031,500**	**$885,500**	**$874,500**	**$120,500**	**$158,000**
Net Present Value Calculation (includes OM&M from Table 10.2)								
Discount Rate: 8%		0.08	0.08	0.08	0.08	0.08	0.08	0.08
NET PRESENT VALUE		**$1,093,000**	**$1,073,500**	**$1,511,000**	**$1,354,000**	**$1,365,000**	**$1,197,000**	**$1,268,000**

NA = not applicable.

different pump-and-treat systems; one designed to capture the 80 ft (24 m) wide plume and the second to capture the 350-ft (107-m)-wide plume (i.e., the same upgradient capture zones as for the Case A and B SPRB *in situ* alternatives). A two-well system was selected for the 80-ft plume (Figure 10.1), and a three-well system (Figure 10.2) for the 350-ft-wide plume. Based on the model results, a combined extraction rate of 0.4 gpm (1.5 L/min) was predicted for the two-well system, and an extraction rate of 2.6 gpm (9.8 L/min) for the three-well system.

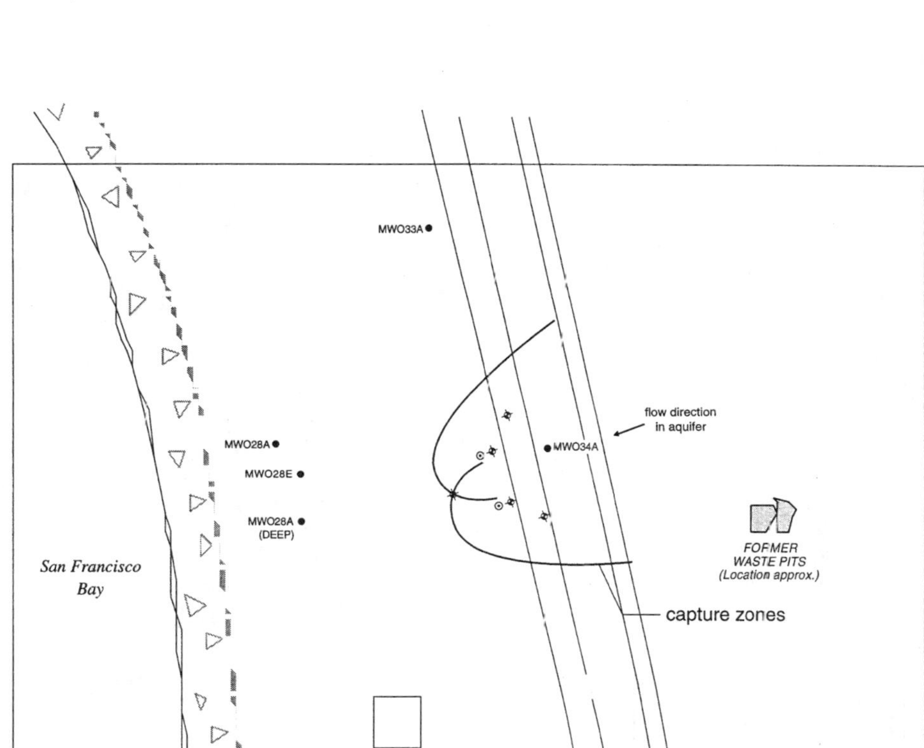

Figure 10.1 Schematic of Alameda pump-and-treat alternative for 80-ft plume.

The *c*DCE and VC concentrations for both Case A and Case B requiring aboveground treatment were assumed to be 100 and 11 mg/L, respectively. It was assumed that treatment would be achieved via air stripping, with the off gas treated using activated carbon. Other input assumptions and a general description of the system are provided in Appendix 35. Cost estimates for the two systems are included in Table 10.1. For Case A the capital costs were estimated to be \$120,500, while for Case B the total capital costs were estimated at \$158,000.

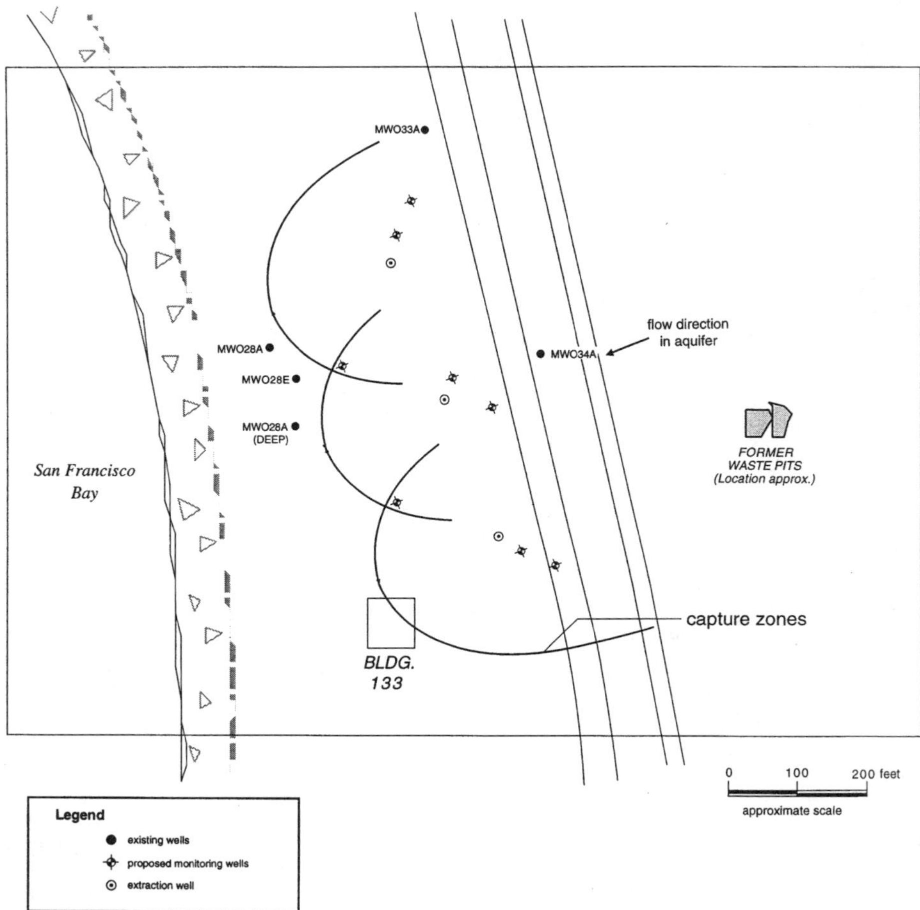

Figure 10.2 Schematic of Alameda pump-and-treat alternative for 350-ft plume.

10.2 OPERATIONS, MAINTENANCE, AND MONITORING COSTS

The OM&M costs for both technologies are technology specific. Operation and maintenance tasks are summarized in Chapter 9 for the SPRB systems and in Appendix 35 for the P&T systems. Compliance and performance monitoring of the plume was considered for each technology (see Tables 9.1, 9.2, and Appendix 35). The same reporting requirements were applied to both technologies. For the first 2 years, reports would be produced quarterly; for subsequent years (3 to 30) reporting would be semiannually. OM&M costs for both SPRB and P&T are reviewed in the following sections.

10.2.1 OM&M for SPRB

The SPRB monitoring program (Chapter 9) was developed according to the assumptions outlined in Section 8.6. The breakdowns of sampling costs for the Case A and B SPRB are presented in Table 10.2 and 10.3, respectively. Groundwater monitoring costs for the two full-scale SPRB scenarios were prepared in accordance with the monitoring programs described in Chapter 9. Compliance and performance monitoring were included the monitoring program; these costs are summarized in Table 10.4. With respect to system maintenance cost, it was assumed that periodic scarification of the iron zones in the gate sections, to increase permeability lost by formation of mineral precipitates (Section 8.5), would occur once every 7 years, at a cost of $7,000 for Case A and $14,000 for Case B, for each event. These costs are based on recent contractor cost estimates for performing this procedure. The second part of routine OM&M involves cleaning the biosparge system; it was assumed that this would occur once annually, at a cost of $1,000 and $2,000 per year for Cases A and B, respectively. For the first 2 years, compliance and performance monitoring costs would be incurred quarterly. For the remaining 28 years, compliance and performance monitoring costs would be incurred semiannually. Annual OM&M costs for Case A were $59,600 for each of years 1 and 2 and $30,800 for each of years 3 to 30. For Case B, annual OM&M for the first 2 years was $71,300 annually and for the remaining 28 years $37,700 annually.

10.2.2 OM&M for P&T

The assumptions used to derive the OM&M costs for the P&T alternatives are presented in Appendix 35. The costs are summarized in Table 10.4. For the first 2 years, compliance monitoring costs would be incurred quarterly; for the remaining 28 years, costs would be incurred semiannually. Performance monitoring of the groundwater treatment system would be conducted on a monthly basis, with two samples being taken per event. Operator labor for the P&T systems is comprised of twice-weekly site visits to review system operation. Annual OM&M costs for Case A were $116,000 for each of years 1 and 2 and $91,800 for each of years 3 to 30. For Case B, annual OM&M for the first 2 years was $120,200 annually and $94,500 annually for the remaining 28 years.

10.3 COMPARISON OF SPRB COSTS TO P&T COSTS

Two methods were used to compare SPRB costs to those of the pump-and-treat alternative; cumulative costs incurred over a 30-year period, and NPV costs calculated using the costs provided above and an 8% discount rate. NPV costs for the SPRB and P&T systems in Case A and B are included in Table 10.1. Both accounting methods have been used by others to compare the costs of single treatment PRB technology to that of conventional alternatives. All costs presented are for a 30-year period. A 30-year time frame was considered reasonable, as this time period is used as a costing basis under several federal and state legislative programs.

10.3.1 Cumulative Costs

Case A — The SPRB capital costs and yearly OM&M costs for Case A were summed to generate a 30-year cumulative cost of $1,677,100. P&T capital costs and yearly OM&M costs for Case A were summed to generate a 30-year cumulative cost of $2,922,170. Figure 10.3 presents the Case A cost curves. After 10 years of operation, the P&T cumulative costs exceed SPRB costs (see cross-over on Figure 10.3 between the two base case scenarios).

Table 10.2 SPRB Monitoring Costs, Alameda Case A — Expanded Pilot Configuration

Parameter	Year 1 - 2					Year 3 - 30				
	Location	Frequency	Number of Samples	Cost/Sample	Total Cost	Location	Frequency	Number of Samples	Cost/Sample	Total Cost
Volatile Organic Compounds PCE Benzene TCE Ethylbenzene cDCE Toluene VC Xylenes Chlorobenzene	Five wells in center row of each gate and four wells located at the end and middle of funnel sections.	Quarterly	**Performance**[a] 4/yr x (3 gate wells x 2 gates + 2 funnel wells) = 32/yr	150	4,800	Five wells in center row of each gate and six wells located at the end and middle of funnel sections.	Semi-annually	**Performance**[a] 2/yr x (3 gate wells x 2 gates + 2 funnel wells) = 16/yr	150	2,400
			Compliance[b] 4/yr x (2 gate wells x 2 gates + 2 funnel wells) = 24/yr	150	3,600			**Compliance**[b] 2/yr x (2 gate wells x 2 gates + 2 funnel wells) = 12/yr	150	1,800
Major Cations & Anions Calcium Chloride Iron (Total) Nitrate Magnesium Phosphate Manganese Sulfate Potassium Alkalinity Sodium TDS	Five wells in center row of each gate.	Semi-annually	2/yr x (5 gate wells x 2 gates) = 20/yr	50	1,000	Five wells in center row of each gate.	Annually	1/yr x (5 gate wells x 2 gates) = 10/yr	50	500
Biological DHG BOD Oxygenates VFA	Three downgradient wells in center row of each gate.	Semi-annually	2/yr x (3 gate wells x 2 gates) = 12/yr	150	1,800	Three downgradient wells in center row of each gate.	Annually	1/yr x (3 gate wells x 2 gates) = 6/yr	150	900
Field pH Temperature ORP (Eh) DO	Five wells in center row of each gate.	Quarterly	4/yr x (5 gate wells x 2 gates) = 40/yr	Included in sampling labor costs[c].		Five wells in center row of each gate.	Semi-annually	2/yr x (5 gate wells x 2 gates) = 20/yr	Included in sampling labor costs[c].	
Hydraulic Performance Groundwater Elevation	All Wells	Quarterly	4/yr x 29 wells = 116/yr	Included in sampling labor costs[c].		All Wells	Semi-annually	2/yr x 29 wells = 58/yr	Included in sampling labor costs[c].	
Groundwater Velocity	Gate Wells	As Required				Gate Wells	As Required			
Hydraulic Conductivity	Iron Wells	As Required				Iron Wells	As Required			

PCE = Tetrachloroethene TCE = Trichloroethene cDCE = cis-1,2-Dichloroethene VC = Vinyl Chloride TDS = Total Dissolved Solids
DHG = Dissolved Hydrogen Gas BOD = Biological Oxygen Demand VFA = Volatile Fatty Acids ORP = Oxygen Reducing Potential DO = Dissolved Oxygen

Notes:

a Performance monitoring of VOCs includes the middle row of wells in each gate located in the iron and peagravel sections as well as the two wells in the middle of the funnel sections, refer to **Figure 9.1**.
b Compliance monitoring of VOCs includes the upgrandient and downgradient aquifer wells from each gate as well as the two wells at either end of the treatment system, refer to **Figure 9.2**.
c Sampling labour costs for years 1 and 2 are estimated to be \$6,300/yr, and \$3,100/yr for years 3 to 30.

Table 10.3 SPRB Monitoring Costs, Alameda Case B — 350 ft Full-Scale Configuration

Parameter	Year 1 - 2					Year 3 - 30				
	Location	Frequency	Number of Samples	Cost/Sample	Total Cost	Location	Frequency	Number of Samples	Cost/Sample	Total Cost
Volatile Organic Compounds PCE, TCE, cDCE, VC, Benzene, Ethylbenzene, Toluene, Xylenes, Chlorobenzene	Five wells in center row of each gate and six wells located at the end and middle of funnel sections.	Quarterly	**Performance**[a] 4/yr x (3 gate wells x 3 gates + 4 funnel wells) = 52/yr	150	7,800	Five wells in center row of each gate and six wells located at the end and middle of funnel sections.	Semi-annually	**Performance**[a] 2/yr x (3 gate wells x 3 gates + 4 funnel wells) = 26/yr	150	3,900
			Compliance[b] 4/yr x (2 gate wells x 3 gates + 2 funnel wells) = 32/yr	150	4,800			**Compliance**[b] 2/yr x (2 gate wells x 3 gates + 2 funnel wells) = 16/yr	150	2,400
Major Cations & Anions Calcium, Iron (Total), Magnesium, Manganese, Potassium, Sodium, Chloride, Nitrate, Phosphate, Sulfate, Alkalinity, TDS	Five wells in center row of each gate.	Semi-annually	2/yr x (5 gate wells x 3 gates) = 30/yr	50	1,500	Five wells in center row of each gate.	Annually	1/yr x (5 gate wells x 3 gates) = 15/yr	50	750
Biological DHG, Oxygenates, BOD, VFA	Three downgradient wells in center row of each gate.	Semi-annually	2/yr x (3 gate wells x 3 gates) = 18/yr	150	2,700	Three downgradient wells in center row of each gate.	Annually	1/yr x (3 gate wells x 3 gates) = 9/yr	150	1,350
Field pH, ORP (Eh), Temperature, DO	Five wells in center row of each gate.	Quarterly	4/yr x (5 gate wells x 3 gates) = 60/yr	Included in sampling labor costs[c].		Five wells in center row of each gate.	Semi-annually	2/yr x (5 gate wells x 3 gates) = 30/yr	Included in sampling labor costs[c].	
Hydraulic Performance Groundwater Elevation	All Wells	Quarterly	4/yr x 45 wells = 180/yr	Included in sampling labor costs[c].		All Wells	Semi-annually	2/yr x 45 wells = 90/yr	Included in sampling labor costs[c].	
Groundwater Velocity	Gate Wells	As Required				Gate Wells	As Required			
Hydraulic Conductivity	Iron Wells	As Required				Iron Wells	As Required			

PCE = Tetrachloroethene TCE = Trichloroethene cDCE = cis-1,2-Dichloroethene VC = Vinyl Chloride TDS = Total Dissolved Solids
DHG = Dissolved Hydrogen Gas BOD=Biochemical Oxygen Demand VFA = Volatile Fatty Acids ORP = Oxygen Reducing Potential DO = Dissolved Oxygen

Notes:

a Performance monitoring of VOCs includes the middle row of wells in each gate located in the iron and peagravel sections as well as the four wells in the middle of the funnel sections, refer to **Figure 9.1.**

b Compliance monitoring of VOCs includes the upgrandient and downgradient aquifer wells from each gate as well as the two wells at either end of the treatment system, refer to **Figure 9.2.**

c Sampling labour costs for years 1 and 2 are estimated to be $10,400/yr, and $5,200/yr for years 3 to 30.

Table 10.4 Summary of Operation, Maintenance, and Monitoring Cost Estimates

Cost Category	SPRB Funnel-and-Gate		Pump & Treat	
	Case A (80 ft Plume)	Case B (350 ft Plume)	Case A (80 ft Plume)	Case B (350 ft Plume)
Monitoring				
Compliance Monitoring[b]	1,800 (3,600)[a]	2,400 (4,800)[a]	2,700 (5,400)[a]	3,600 (7,200)[a]
Performance Monitoring[c]	3,800 (7,600)[a]	6,000 (12,000)[a]	5,700	5,700
Sampling Labor	3,100 (6,300)[a]	5,200 (10,400)[a]	3,400 (4,900)[a]	3,900 (6,000)[a]
Operator Labor			41,200	41,200
Electrical	100	100	1,000	1,100
Carbon Consumption			10,700	10,700
Maintenance				
Iron Zone Maintenance[d]	1,000	2,000		
Biosparge Zone Maintenance	1,000	2,000		
Equipment Maintenance			3,700	4,200
Other Costs				
Groundwater Discharge			3,400	4,100
Reporting	20,000 (40,000)[a]	20,000 (40,000)[a]	20,000 (40,000)[a]	20,000 (40,000)[a]
Total Annual O&M Cost	$30,800 ($59,600)[a]	$37,700 ($71,300)[a]	$91,800 ($116,000)[a]	$94,500 ($120,200)[a]

Notes:

a Costs for year 1 and 2. Compliance monitoring is 4 times/year for Years 1&2 then 2 times/year for Years 3-30.

b Compliance monitoring includes costs for compliance VOC analyses. For SPRB see Table 9.1 and 9.2 for each Case. For Pump and Treat see Appendix 35.

c Performance monitoring includes costs for all analyses excluding compliance VOCs. For SPRB see Table 9.1 and 9.2 for ea For Pump and Treat VOC analyses to assess performance of the Off-Gas Treatment. See Appendix 35 for details.

d Iron maintenance, which is assumed to occur every seven years, is divided by seven to obtain an annual maintenance cost

Case B — The SPRB capital costs and yearly OM&M costs for Case B were summed to generate a 30-year cumulative cost of $2,225,700. P&T capital costs and yearly OM&M costs for Case B were assumed to generate a 30-year cumulative cost of $3,045,100. Figure 10.4 presents the Case B cost curves. After 16 years of operation, the P&T cumulative costs exceed SPRB cumulative costs (see crossover on Figure 10.4 between the two base case scenarios).

10.3.2 Net Present Value

Case A — The NPV costs of the SPRB and the P&T systems calculated using the capital (Section 10.1) and OM&M (Section 10.2) costs presented earlier are $1,093,000 and $1,197,000, respectively. The SPRB has an NPV that is $104,000 less than the P&T.

Case B — The NPV costs of the SPRB and the P&T systems calculated using the capital (Section 10.1) and OM&M (Section 10.2) costs presented earlier are $1,510,000 and $1,268,000, respectively. The P&T has an NPV that is $242,000 less than the SPRB.

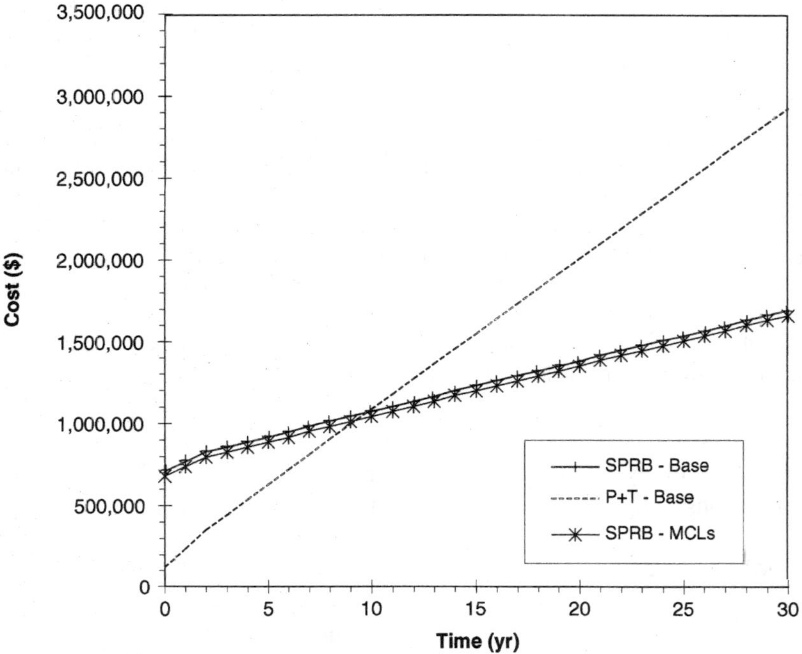

Figure 10.3 Thirty-year cost comparisons for the Case A Funnel-and-Gates and pump-and-treat.

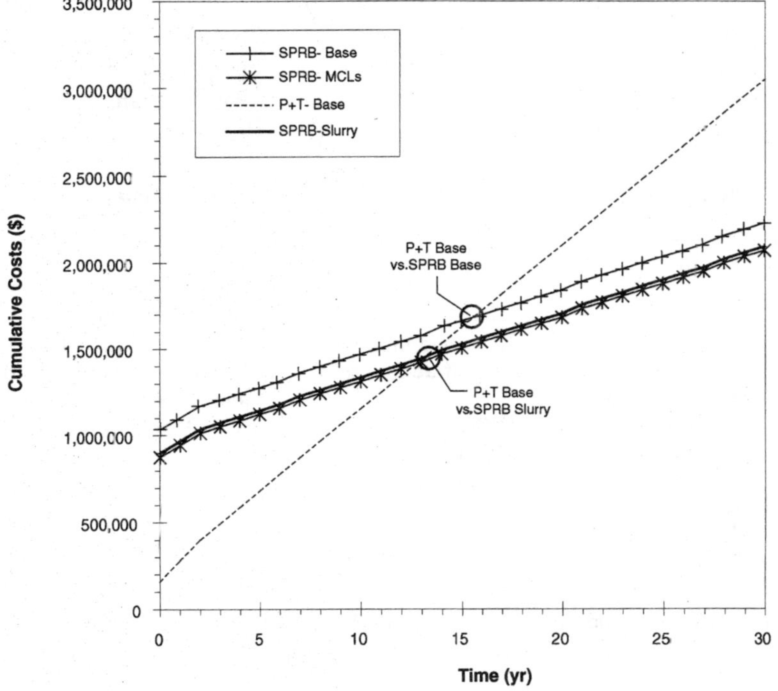

Figure 10.4 Thirty-year cost comparisons for the Case B Funnel-and-Gates and pump-and-treat.

10.4 FACTORS AFFECTING COST COMPETITIVENESS OF *IN SITU* TREATMENT

Several economic aspects of the SPRB technology are apparent from the cumulative cost curves (Figures 10.3 and 10.4):

- The capital costs of the P&T systems are significantly lower than those of either SPRB.
- The capital cost of *in situ* treatment (SPRB) is significantly affected by the size of plume to be captured. This is less significant for the P&T system.
- Lower annual OM&M costs result in less cumulative costs for the smaller *in situ* treatment (Case A) than the comparable P&T system after about 10 years of operation. Because of higher initial (capital) costs associated with building the larger SPRB, this "crossover" does not occur until year 16 for Case B.
- Case A SPRB NPV costs were lower than the comparable P&T case. Case B SPRB NPV costs were higher than the comparable P&T system costs.

As noted above, the capital costs of the SPRB at Alameda significantly influence the cost competitiveness of this technology relative to the pump-and-treat alternative. The following discussion addresses how various design parameters can affect these cost comparisons.

10.4.1 Effluent Criteria

At many sites, natural attenuation factors and distance to receptors could be considered when setting effluent limits on groundwater exiting the *in situ* treatment system. Design simulations were reexamined to calculate the residence time needed to reach a total loading of 100 μg/L of chlorinated solvents and the effect of these revised effluent criteria on iron zone residence times (Table 10.5 and 10.6). Maximum residence times decreased to 18 days in the plume core and 5 days in the outer gates of Case B. The gate thickness in each case can therefore be reduced according to these lower residence time requirements, as shown in Table 10.7, leading to a corresponding decrease in capital (iron) costs. The effects of these revised effluent criteria on the 30-year cost comparisons between SPRBs and P&Ts for Case A and Case B are shown graphically in Figures 10.3 and 10.4 (see curves labeled SPRB-MCLs). Note that the costs of the P&T systems do not change, as it is not likely that effluent that contains 100 μg/L of CVOC would be allowed.

Table 10.5 Granular Iron Design Parameters and Residence Time Estimates — Case A, Treatment to Total VOCs to 100 μg/L

VOC Concentrations from Historical Data	Assumed Influent Concentration (μg/L)		Predicted Residence Time (day) to Reach 100 μg/L	
	cDCE	VC	Using Apparent Field-Scale Half-Lives	Using Scaled Laboratory Half-Lives
BH03-1[a] (1996)	219,000	16,700	18	11
MW-3[b] (1996)	22,400	4,800	13	8
19 March 1997 Pilot Influent Concentrations	62,000	9,600	16	9
17 June 1997 Pilot Influent Concentrations	110,000	21,000	17	10

[a] Depth-discrete sampling point from July 1996 site investigation (UW, 1996).

[b] Temporary monitoring well installed in July 1996 site investigation (UW, 1996). Removal during construction of pilot-scale system.

Table 10.6 Design Parameters and Residence Time — Full-Scale System to Treat Total VOCs to 100 μg/L

Source for VOC Concentrations	Assumed Influent Concentration (μg/L) cDCE	VC	Predicted Residence Time to Reach 100 μg/L Total VOCs: Using Apparent Field-Scale Half-Lives	Using Scaled Laboratory Half-Lives
Gates in core of plume	See Table 10.5		13–18	8–11
Other gates (Case B)	750	250	5	3

Table 10.7 Summary of Effect of Residence Time Requirements on Treatment Cost

Scenario (Design Thickness (ft))	Gate Thickness if Influent VOC = TCE (ft)	Cost Reduction	Gate Thickness if Effluent Criteria = 100 μg/L (ft)	Cost Reduction
Additional gate, Case A (5.1)	3.9 (1.2 m)	$9,000	3.0 (0.9 m)	$15,000
Centre gate, Case B (6.5)	5 (1.5 m)	$26,000	3.8 (1.2 m)	$46,000
Outer gates, Case B (3.3)	2.1 (0.6 m)	$41,000	1.1 (0.3 m)	$75,000
Total Cost reduction, Case B	—	$67,000	—	$121,000

10.4.2 Construction Methods

The base case designs incorporate the use of sealed joint sheet piling for the funnel sections of the Funnel-and-Gate treatment system. Slurry walls represent a less expensive alternative for funnel construction, providing that site-specific soil conditions permit their use. The small amount of extra funnel needed in Case A (about 80 ft) likely means that slurry wall construction would not be cost-competitive with sheet piling. The cost of the 500 ft of sealed joint sheet piling funnel sections required for Case B, according to the C^3 estimate, is about $195,000. Assuming a unit cost for slurry walls of $7.50/ft^2, slurry wall funnel construction costs would decrease to about $82,500, lowering SPRB capital costs to $885,500. As shown in Figure 10.4 (i.e., lined marked as SPRB-Slurry), this results in the "crossover" of costs at 13 years compared to P&T. This change in construction method is equivalent to a cost savings of about 3 years over the base case comparison (i.e., crossover occurred at 16 years in base case comparison).

10.4.3 Dimensions of Treatment Zone for Chlorinated Solvent Removal

The type and concentration of influent VOCs, in particular the chlorinated compounds, will affect the residence time requirements in the iron treatment zone of the SPRB, and therefore the dimensions and total cost of the SPRB. If TCE were the dominant VOC, then the amount of granular iron required would decrease, since the TCE half-life is less than the *c*DCE or VC half-lives. This decrease in iron requirements would reduce capital costs. On the other hand, the SPRB dimensions would not need to be expanded if VC was the predominant contaminant, rather than *c*DCE, because VC and *c*DCE are degraded in the SPRB at roughly the same rate.

For the P&T, the higher levels of TCE would decrease carbon consumption rates and therefore reduce the OM&M component of the costs of this system. Conversely, the occurrence of higher levels of VC (vs. *c*DCE) would increase carbon consumption in the P&T system, resulting in a general increase in the OM&M costs associated with the P&T system.

10.4.4 Different Configurations for the Biosparge Zone

The pilot-scale biosparge system served as the design basis for the biological component of the hypothetical full-scale SPRB. One alternative for the biosparge zone is to have this zone backfilled with pea gravel rather than using the open sparge zone designed for the pilot-scale. This configuration could still use the diffuser apparatus at the base of the gate (installed after excavation is complete) and have the sparge interval designed accordingly. The cost of materials would be decreased for both Case A and Case B. As well, it may not be necessary in the Case B scenario to install open biosparge zones in the outer gates because the levels of BTEX and other contaminants in these areas are minimal at this site. Instead, a few wells with sparge-type diffusers might be able to effect the necessary treatment. Also, the requirement for oxygen supply was conservatively estimated.

10.4.5 Effect of Changing Design Criteria on P&T Costs

The P&T alternatives designed and costed for Cases A and B have the potential to accommodate increases in groundwater extraction rates and increases in groundwater contaminant concentration. The groundwater treatment system specified for both Case A and B has a hydraulic capacity of 5 gpm (19 L/min) as compared with the design flow for Case A of 0.4 gpm (1.5 L/min) and 2.6 gpm (9.8 L/min) for Case B. If an extraction rate that is greater than 5 gpm is required, increased capital must be invested for larger pumps, air strippers, blowers, and potentially larger GPC vessels.

An increase in flow rate or contaminant concentration, or a combined increase in both, will result in higher contaminant mass flux into the GTS. The consumption of GPC will increase proportionally with the increase in contaminant mass flux, which in turn will result in a proportional increase in annual GPC replacement costs, thus increasing the operating cost of the system.

Changes in effluent criteria will not affect capital or operating cost of the system due to the high contaminant removal rate the GTS can provide.

10.5 SUMMARY

As shown in the previous sections, the costs of applying SPRB technology at NAS Alameda are competitive with the cost of P&T alternatives. This is encouraging because the applicability of the SPRB technology to other mixed plumes has only begun to be evaluated, and in many cases SPRBs may turn out to be less costly (compared to P&T) than the two cases examined here. Existing iron PRBs installed to date have shown that this technology is a cost-effective alternative to P&T in many situations.

There are a number of key factors that significantly affect the cost of the hypothetical full-scale SPRBs that were specifically designed and costed for the Alameda site. For instance:

- To ensure pricing accuracy, the construction methods costed for the hypothetical full-scale system at Alameda (sealable joint sheet pile) were identical to those employed in the smaller, pilot-scale system (accuracy is ensured because the construction method was tested during the installation of the pilot-scale at Alameda). While this choice yielded accurate cost estimates, the actual construction method that would be selected for a much larger scale system (100 ft wide) could be significantly different than for a small-scale system (25 ft wide). New cost estimates presented in Section 10.4.2 indicate that savings of 15% of the capital cost (and 10% of the NPV) could be realized by using slurry wall technology to create the funnel sections. These findings indicate that the full-scale SPRBs that would be designed to treat the larger plumes more often found at other sites could offer cost advantages over traditional technologies.
- The very high concentrations of *c*DCE and VC (two compounds that are treated effectively by air stripping/GAC but have relatively long half-lives in contact with granular iron) that required

treatment at Alameda resulted in the cost of the SPRB being high compared to the cost of a P&T system. In many other cases, the concentrations of these compounds will be much lower, and the SPRB technology will be more cost-competitive.

- The combination of a low hydraulic gradient and moderate hydraulic conductivity observed at the Alameda site ensures that a large capture zone can be achieved with a relatively low pumping rate. This means that the P&T system will be more cost-effective in this instance. At many DOD sites the hydraulic conductivity is lower, and hydraulic gradients are higher such that a higher extraction rate is required to achieve the desired capture. In these cases, SPRBs may be more cost-effective than traditional technologies.

Several factors not considered in the cost comparison presented above could influence the costs of both SPRBs and P&T systems at other sites. Some of these factors are

- Permitting costs for disposal of treated groundwater, if applicable, will increase the OM&M cost of a P&T system.
- The presence of a highly heterogeneous aquifer may decrease the effectiveness of a P&T system.
- The general public may be more accepting of treatment technologies such as SPRBs that result in destruction of the chemical of concern (vs. P&T which just transfers the chemical to another medium, often GAC, for transport to another facility).
- The reduction or elimination of aboveground structures (associated with P&T, but not generally employed in SPRBs) may enhance the usefulness or value of the site, or allow the property to be transacted.

In summary, while costs of the hypothetical SPRBs that were designed for the Alameda site were comparable to the costs of P&T systems, several site-specific characteristics of the Alameda site that are not shared by many other DOD sites may have caused the costs to be relatively high. SPRBs are likely to be even more cost-competitive at other DOD sites. The potential cost-effectiveness of an SPRB over traditional technology needs to be carefully evaluated on a site-specific basis.

Chapter 11

Applicability of SPRB Technology

According to the U.S. EPA's Markets and Technology Trends Report, DOD estimates that more than 8300 sites on more than 1500 installations will require remediation of contaminated materials (EPA, 1997). Of these sites, data identifying the type of matrix requiring cleanup (e.g., groundwater, surface water, soil, and/or sediment) were available for only 3212 sites. Groundwater is the most frequently contaminated matrix at these sites, with contamination reported for 71% of the sites (2280 sites). A breakdown of the major organic contaminant groups indicates that VOCs (halogenated and nonhalogenated) are the most prevalent groundwater contaminants, occurring at 74% of all DOD sites having known contaminated groundwater (more than 1680 sites).

The estimated cost to complete remediation work at contaminated DOD facilities is more than $28.6 billion. Based on the numbers summarized above, it is expected that remediation of groundwater containing VOCs and other organic chemicals will represent the most significant portion of this remediation cost estimate. Significant incentive exists for DOD to develop innovative and cost-effective remediation technologies for organic chemicals in groundwater. In particular, incentive exists to develop passive and semipassive *in situ* technologies that can provide lower long-term cost as compared to conventional remediation systems (e.g., P&T).

Of the technologies that could be potentially coupled in SPRB applications, reactive/destructive technologies show the most promise, as compared to sorptive technologies such as activated carbon. Zero-valent metal refers to a wide range of remedial technologies that are based on metal compounds, of which granular iron is one type. Zero-valent metals and bioremediation are reactive/destructive technologies that are well suited for SPRB application: they are robust, cost effective for long-term applications, and, when combined, can address a wide range of organic contaminants in groundwater. Reference to SPRB technology in the remainder of this section is meant to imply combined zero-valent metal and bioremediation applications.

The following sections summarize the range of target chemicals that can potentially be remediated in groundwater using SPRB technology, present information regarding potential SPRB configurations, and list some of the limitations associated with the use of innovative technologies and SPRB technology for site remediation.

11.1 TARGET CHEMICALS

SPRB technology is applicable to a wide variety of chemicals present in groundwater environments and is particularly valuable for remediation of groundwater plumes containing mixtures of chemicals. It is expected that zero-valent metal technologies will be most appropriate for halogenated VOCs, while bioremediation will be most appropriate for the other chemicals in the plume. Table 11.1 lists a range of chemicals treatable with SPRB technology. The list

Table 11.1 Treatment Alternatives for Common Organic Contaminants

Contaminant Class	Granular Iron	Enhanced Aerobic Biodegradation (Oxygen)	Enhanced Anaerobic Biodegradation (Electron Acceptor Addition)	Enhanced Anaerobic Biodegradation (Electron Donor Addition)
		Halogenated VOCs		
Chlorinated Solvents	Yes[a]	Yes[d]	Yes[d]	Yes[a]
		Nonhalogenated and Semihalogenated VOCs		
Petroleum Hydrocarbons	No	Yes[a]	Yes[a]	NA
PAH Compounds	No	Yes[a]	Yesd	NA
BTEX Compounds	No	Yes[a]	Yes[a]	NA
Chlorobenzenes	Possibly[c]	Possibly[c]	Yes[a]	NA
Chlorophenols	Possibly[c]	Possibly[c]	Yes[d]	NA
		Other Chemicals		
Propellants — Perchlorate	Possibly[c]	No	No	Yes[b]
Propellants — NDMA	Yes[b]	Possibly[c]	Possibly[c]	Possibly[c]
Pesticides (e.g., Diisopropyl Methylphosphonate)	Possibly[c]	Possibly[c]	Possibly[c]	Possibly[c]

Notes: NA — not applicable.

[a] Successful tests at field scale.

[b] Field trials are ongoing.

[c] Laboratory-scale studies are ongoing, no field scale tests, and chemical specific.

[d] Chemical specific.

includes halogenated VOCs, nonhalogenated VOCs (including BTEX), halogenated and nonhalogenated semi-VOCs (SVOCs), organic pesticides and herbicides, and explosives or propellants. Remediation of these general groups of chemicals in SPRB systems is discussed in the following sections.

Halogenated VOCs — Halogenated VOCs can often be present in groundwater at sites requiring cleanup, with TCE being the most frequently detected halogenated VOC. Natural (intrinsic) biodegradation of TCE will result in the production of *c*DCE, VC, and ethene at a significant number of these sites. All of the halogenated VOCs related to TCE can typically be remediated using zero-valent metal technologies in stand-alone applications. However, numerous other halogenated compounds cannot be treated by zero-valent metal technology alone; therefore, their treatment may require a coupled approach. For example, dechlorination of CT and CF produces DCM, which is not dechlorinated by zero-valent metal. However, DCM easily biodegrades under both anaerobic and aerobic conditions; therefore, an SPRB application that couples zero-valent metal and bioremediation may be appropriate and cost effective for CT plumes. Similarly, 1,2-DCA (used as a nitroplasticizer in propellant manufacturing at DOD-related facilities) is not dechlorinated by zero-valent metal applications but is easily biodegraded under both anaerobic and aerobic conditions. SPRB technology may therefore be applicable for sites having mixed halogenated VOC plumes containing 1,2-DCA (as well as TCE and/or 1,1,1-TCA).

Nonhalogenated VOCs — Nonhalogenated VOCs might also be present in groundwater at many sites requiring cleanup. BTEX comprises the most significant portion of nonhalogenated

VOCs observed; ketones and other aromatics make up the balance. Given historic use and disposal practices, nonhalogenated VOCs are likely to be present together with halogenated VOCs, creating a significant number of mixed groundwater plumes at DOD sites. These plumes would be good candidates for SPRB technology. Several bioremediation applications exist for remediation of the nonhalogenated components of these plumes. For example, in the case of BTEX or ketones, aerobic bioremediation could be installed either upstream or downstream of a zero-valent metal treatment system. In the case of TEX compounds, anaerobic bioremediation through nitrate addition may also be appropriately applied either upstream or downstream of a zero-valent metal treatment system.

Nonhalogenated and Halogenated Semi-VOCs — Nonhalogenated SVOCs (e.g., PAHs) and halogenated SVOCs (e.g., chlorobenzenes) can also be found at contaminated sites. While much of this SVOC contamination is likely related to soil (e.g., for PAHs), potential exists for these chemicals to be present in groundwater.

Most SVOCs, either halogenated or not, can be biodegraded aerobically. In contrast, only selected SVOCs such as chlorophenols and chlorobenzenes can be biodegraded under anaerobic conditions. Therefore, anaerobic–aerobic SPRB configurations (see Section 11.2) would be expected to be applicable for remediation of mixed groundwater plumes containing a mixture of these compounds.

Propellants and Explosives — Explosives and propellants may also be present at some sites. Two common chemicals associated with propellants that can be present in groundwater are

- *n*-Nitroso dimethylamine (NDMA), a reaction product of unsymmetrical dimethyl hydrazine, a component of liquid rocket fuels; and
- Perchlorate (an inorganic anion), used as a solid rocket fuel.

Both chemicals are known to cause adverse health effects and therefore have low regulatory standards in groundwater. Currently, remediation of NDMA in groundwater is accomplished by conventional groundwater extraction and treatment (destruction) by UV oxidation. These remediation systems are costly to operate and maintain and often have short life cycles (10–15 year recapitalization periods). Current remedial options for perchlorate are extremely limited. For both compounds, treatment using SPRB technology represents an effective remediation alternative. For example, NDMA can be transformed by zero-valent metals to biodegradable compounds such as dimethylamine (DMA) and ammonia. Therefore, coupled zero-valent metal and bioremediation SPRBs may be widely applicable for sites contaminated by NDMA and halogenated VOCs. Similarly, perchlorate has been known to biodegrade under anaerobic reducing conditions. Bioremediation is thought to be one of the few technologies that may be effective at remediating perchlorate (Renner, 1998).

Pesticides, Herbicides, and Other Chemicals. Significant promise exists for use of SPRB technology for *in situ* remediation of groundwater containing mixed plumes of halogenated VOCs and complex compounds such as organophosphates (e.g., diisopropyl methylphosphonate; DIMP), pesticides, and herbicides. While these compounds are typically associated more with soil than groundwater contamination, their presence in groundwater has been reported at some facilities. Many common groundwater remediation technologies do not adequately treat these complex organic compounds due to their physical properties (e.g., low volatility for stripping applications), but most are biodegradable under appropriately enhanced conditions. Evaluation of SPRB technology for these types of mixtures is ongoing, and the applicability of the technology for new chemicals is being evaluated.

11.2 TREATMENT CONFIGURATIONS

A variety of potential SPRB configurations exist for simultaneous remediation of several organic chemicals in groundwater plumes. The most cost-effective configuration for a given site will depend on the target chemicals to be remediated and site-specific conditions. The NAS Alameda and CFB Borden demonstrations evaluated several anaerobic–aerobic SPRB configurations to treat chlorinated solvents and BTEX in groundwater. In addition to these configurations, several other configurations may have been applicable, including anaerobic–anaerobic applications and aerobic-anaerobic applications.

The first consideration in designing a SPRB is the redox environment required for treatment of the chemical components. Table 11.2 lists the redox conditions under which a variety of common chemicals degrade. This information (or its equivalent) can be used to determine appropriate sequencing of the SPRB components. An overall consideration is how to work with existing redox and geochemical conditions at a site to minimize the amount of engineering required to manipulate redox conditions. The following sections discuss several common SPRB configurations involving zero-valent metals and bioremediation to remediate mixed plumes containing chlorinated solvents and other chemicals.

Table 11.2 Comparative Analysis of Costs to Implement Treatment Alternatives in Sequenced Treatment Zones

	Granular Iron and Enhanced Aerobic Biodegradation (O_2)[a]	Granular Iron and Enhanced Aerobic Biodegradation (Nutrient)	Granular Iron and Enhanced Anaerobic Biodegradation (Nutrient)
Site Characterization Costs	—	I	I
Treatability Study			
Complexity/Cost	—	D, N	D, N
Design Costs	—	N	N
Construction Costs	—	D	D
Monitoring Costs	—	I	I
Ease of Regulatory Approval	—	D	D

Notes: I – increased cost relative to base case. N – cost approximately the same as base case. D – decreased cost relative to base case.

[a] SPRB Case B as presented in the text.

Zero-Valent Metal — Aerobic Biodegradation Configuration — The SPRB configuration that will be evaluated most commonly to treat mixed plumes of halogenated VOCs and other organic compounds is zero-valent metal followed by aerobic biodegradation. As was indicated in Table 11.1, many groundwater contaminants can be bioremediated aerobically, including: BTEX and other nonhalogenated VOCs (e.g., ketones); halogenated VOCs such as DCM and 1,2-DCA that are not treated by zero-valent metal applications; most halogenated SVOCs (e.g., chlorobenzenes, chlorophenols) and nonhalogenated SVOCs (e.g., PAHs); and propellant-related products such as NDMA. Sequencing the anaerobic process (zero-valent metal) upgradient of the aerobic bioremediation process is additionally advantageous over the reverse configuration (aerobic bioremediation then zero-valent metal) in that the aerobic bioremediation process will effectively "polish" the groundwater that has passed through the zero-valent metal zone, removing the traces of lesser-chlorinated VOCs that may exit the zero-valent metal wall. The relatively rapid VOC aerobic degradation rates lead to aerobic treatment that is faster than anaerobic treatment. However, to some extent the zero-valent metal component and the aerobic bioremediation component of the sequenced system will be at odds in terms of oxygen demand/delivery. Engineering of the system will need to address the potential interactions of the reactive processes to optimize remediation of all target contaminants. All things considered, anaerobic–aerobic SPRB configurations should be widely applicable for long-term passive/semipassive groundwater remediation.

Figure 11.1a illustrates remediation of groundwater containing mixed organics using an anaerobic–aerobic SPRB configuration. The configuration consists of the metal component placed in advance (hydraulically upgradient) of the aerobic bioremediation component. Oxygen required for the aerobic biodegradation reactions would be supplied through common delivery mechanisms such as air sparging (as tested for the NAS Alameda demonstration), passive oxygen release using ORC™ or equivalent (as tested at CFB Borden), peroxide addition, or ozonation.

Aerobic Biodegradation — Zero-Valent Metal — Depending on site-specific redox and chemistry conditions, it may be advantageous to reverse the sequence of treatment presented above, applying aerobic biodegradation processes first, then zero-valent metal downgradient as shown in Figure 11.1b. A benefit of this configuration may be reduced oxygen demand as compared to anaerobic–aerobic systems, which may improve the efficiency and reduce the cost relative to having an aerobic bioremediation component downstream of the zero-valent metal wall. However, this configuration does not allow for polishing of low concentrations of dechlorination products (e.g., VC) that may exit the zero-valent component or for improvement of secondary water quality parameters. In addition, the system must be engineered such that added oxygen is consumed prior to entering the zero-valent metal component to minimize formation of iron hydroxides on the upstream face of the metal barrier. Use of aerobic–anaerobic configurations are expected to be dependent on site-specific conditions. However, given that most of the common contaminants at DOD sites can be aerobically biodegraded, then applicability of aerobic–anaerobic applications should be high.

Zero-Valent Metal — Anaerobic Biodegradation — Perhaps the most straightforward SPRB configuration consists of a sequenced zero-valent metal anaerobic biodegradation application where halogenated VOCs are treated through reductive dechlorination using a zero-valent metal reactive barrier, while other organics are bioremediated under anaerobic conditions created by the zero-valent metal reactive barrier and the addition of selected nutrients. As indicated in Table 11.1, many common groundwater contaminants can be bioremediated under anaerobic nitrate-reducing and/or sulfate-reducing conditions, including: TEX compounds and possibly benzene; halogenated VOCs such as DCM and 1,2-DCA that are not treated by zero-valent metal applications; selected halogenated SVOCs (e.g., chlorobenzenes, chlorophenols); and selected propellants such as perchlorate. In these cases, anaerobic bioremediation may be promoted by either the addition of electron acceptors such as nitrate or sulfate or by the addition of electron donors such as alcohols (e.g., ethanol), sugars (e.g., glucose), or organic acids (e.g., acetate). The use of anaerobic–anaerobic SPRB configurations should be widely applicable for long-term passive/semipassive groundwater remediation at DOD sites, particularly for sites contaminated with halogenated VOCs and propellants, and for sites where the groundwater is already anaerobic, with a low redox potential.

Figure 11.1c illustrates remediation of groundwater containing mixed organics using a zero-valent-metal anaerobic bioremediation SPRB configuration. The configuration consists of the zero-valent metal component placed in advance of the bioremediation component; however, the order of these components could be reversed depending on site conditions (e.g., target chemicals, natural redox conditions, microbial activity). Given that the zero-valent metal technology creates highly reducing conditions in groundwater, it will generally be advantageous to place the anaerobic bioremediation component downstream of the metal component to reduce the effort required to further manipulate redox conditions in the aquifer.

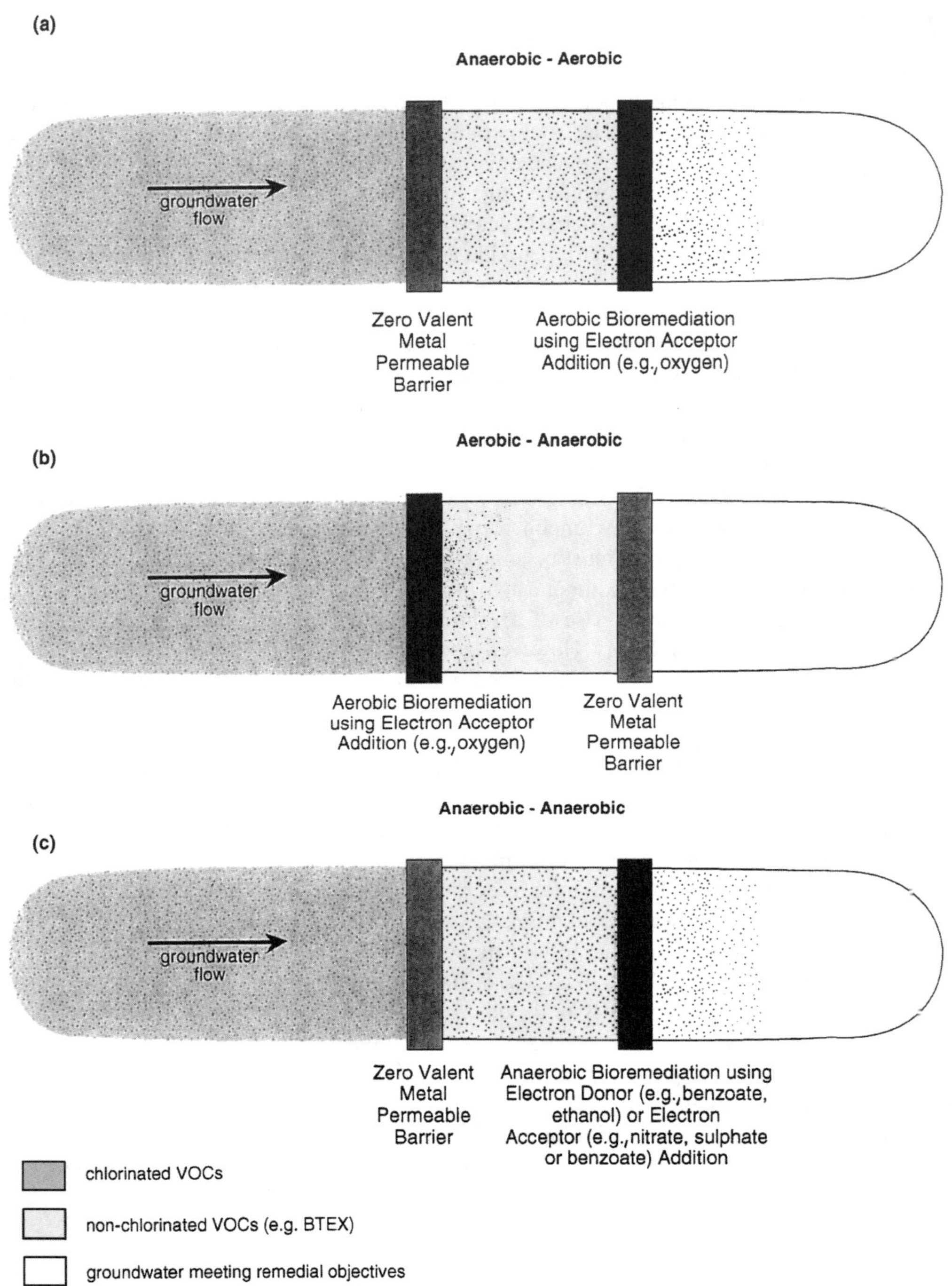

Figure 11.1 Potential SPRB configurations for remediation of mixed plumes.

11.3 LIMITATIONS

The development and use of innovative technologies such as SPRB technology face several limitations that can be grouped into general categories that are applicable to all innovative technologies and a SPRB-specific category. These potential limitation categories are discussed below.

General limitations that apply to the development and use of innovative technologies include the following:

- Cost of development and testing. The costs associated with successful development, implementation, monitoring, and validation of innovative technologies typically exceed equivalent costs for conventional technologies. Potential long-term cost advantages of the innovative technology may be overlooked because of the up-front costs.
- Cost uncertainty for implementation and long-term operation. The lack of cost history on the technology may make stakeholders uncertain or hesitant to select the technology for application. In addition, some organizations may prefer lump-sum or fixed-price contracts. These types of contracts are often not appropriate for small vendors of innovative technologies.
- Regulatory and public perception. Depending on stakeholder involvement, active remediation technologies that utilize more complex engineered systems may be preferred over innovative passive and/or semipassive remediation technologies.

Limitations that are specific to the development and implementation of SPRB technology include the following:

- Current construction methods limit cost-effective installation of SPRB systems to depths of approximately 50 ft below ground surface (although construction methods for deeper installations are rapidly evolving).
- Application of SPRB systems in fracture-dominated flow environments (clays or bedrock) is difficult because of the nature of the flow system.
- Groundwater may contain one or more chemicals that are not treatable by either zero-valent metal or bioremediation, or reaction rates are too slow to allow feasible implementation of the technology.
- Chemicals (e.g., CF if at very high concentration) that may inhibit biological processes may be present at the site, or conditions that may limit the effectiveness of the bioremediation component of the SPRB.
- The lack of long-term (public domain) performance data impedes site owner and regulatory approval.

As indicated above, installation and operation methods are evolving for PRB systems and bioremediation applications and therefore the applicability of SPRB technology should continue to increase.

References for Part II, Technology Design and Evaluation

Austrins, C. 1997. Enhanced In-Situ Bioremediation of Hydrocarbon Contaminated Groundwater Using Oxygen and Ammonia Gas Injection in a Funnel-and-Gate Scheme. M.Sc., University of Waterloo, Waterloo, Ontario, Canada.

Ball, W.P., Buehler, C.H., Harmon, T.C., Mackay, D.M., and Roberts, P.V. 1990. Characterisation of a sandy aquifer material at the grain scale. *Journal of Contaminant Hydrology*, 5: 253-295.

Battelle, Inc. 1997. Design guidance for application of permeable barriers to remediate dissolved chlorinated solvents. Prepared for First Lieutenant D. O'Sullivan. AL/EQ Tyndall Air Force Base, Florida, 32403. Contract No. F08637-95-D-6004 DO 5501.

Brewster, M.L., Annan, A.P., Greenhouse, J.P., Kueper, B.H., Olhoeft, G.R., Redman, J.D., and Sander, K.A. 1995. Observed migration of a controlled DNAPL release by geophysical methods. *Ground* Water, 33(6): 977-987.

Devlin, J.F. 1994. Enhanced In-Situ Biodegradation of Carbon Tetrachloride and Trichloroethene Using a Permeable Wall Injection System. Ph.D. thesis, Department of Earth Sciences, University of Waterloo, Waterloo, Ontario, Canada.

Devlin, J.F., Barker, J.F. 1996. Field investigation of nutrient pulse mixing in an in-situ biostimulation experiment. *Water Resources Research*, 32(9): 2869-2877.

Egboka, B.C.E., Cherry, J.A., Farvolden, R.N., and Frind, E.O. 1983. Migration of contaminants in groundwater at a landfill: a case study. *Journal of Hydrology*, 63: 51–80.

Focht, R., Vogan, J., and O'Hannesin, S. 1996. Field application of reactive iron walls for in-situ degradation of volatile organic compounds in groundwater. *Remediation*, Summer, 1996.

Gillham, 1996. In-situ treatment of groundwater: metal-enhanced degradation of chlorinated organic contaminants, Advances in *Groundwater Pollution Control Remediation*. M.M. Aral (Ed.), Kluwer Academic Publishers, Netherlands, pp. 249-274.

Interstate Technology and Regulatory Cooperation. 1997. Design Guidance for Application of Permeable Barrier Walls to Remediate Dissolved Chlorinated Solvents. http://www.sso.org/ecos/itrc/reports/pbw.htm.

MSE Technology Applications, Inc. 1996. Analysis of Technologies for the Emplacement and Performance Assessment of Subsurface Reactive Barriers for DNAPL Containment. MSE Technologies, Butte, MT.

Mackay, D.M., Freyberg, D.L., Roberts, P.V. 1986. A natural gradient experiment on solute transport in a sand aquifer 1. Approach and overview of plume movement. *Water Resources Research*, 2(13): 2017-2029.

O'Hannesin, S.F. and Gillham, R.W. 1998. Long-term performance of an in situ "iron wall" for remediation of VOCs. *Ground Water*, 36(1): 164-170.

PRC. 1993. Solid Waste Assessment Test (SWAT) Report, NAS Alameda, Alameda, California. Private consultant's report.

Renner, R. 1998. Perchlorate-tainted wells spur government action. *Env. Sci. Technol.* May: 210A.

Shikaze, S., Austrins, C.D., Smyth, D.J.A., Cherry, J.A., Barker, J.F., and Sudicky, E.A. 1995. The hydraulics of a Funnel-and-Gate system: a three-dimensional analysis. Solutions '95. International Association of Hydrogeologists Congress, June 5 to 7, Edmonton, Alberta.

Starr, R.C. and Cherry, J.A. 1994. In-situ remediation of contaminated groundwater: the Funnel-and-Gate system. *Ground Water*, 32(3): 465-476.

Sudicky, E.A. 1986. A natural gradient experiment on solute transport in a sand aquifer: spatial variability of hydraulic conductivity and its role in the dispersion process. *Water Resources Research*, 22(13): 2069-2082.

U.S. Environmental Protection Agency (EPA). 1997. *Cleaning Up the Nation's Waste Sites: Markets and Technology Trends*. 1996 ed. EPA 542-R-96-005.

University of Waterloo (UW). 1995. Revised Workplan for Passive and Semipassive Techniques for Groundwater Remediation at CFB Borden, submitted to AATDF, July 1995.

University of Waterloo (UW). 1996. Revised workplan for semipassive groundwater remediation demonstration project at Site 1, Alameda Navel Air Station, California. Submitted to AATDF, December, 1996.

University of Waterloo (UW). 1997a. Summary of potential DOD sites for demonstration of the passive and semipassive, in-situ groundwater remedial system. Submitted to AATDF, May 21, 1997.

University of Waterloo (UW). 1997b. Report on the installation of the pilot-scale, in-situ groundwater treatment system at NAS Alameda, Alameda, CA. Submitted to AATDF, November 1997.

University of Waterloo (UW). 1998. Final technical document. Submitted to AATDF, March 1998.

Vogan, J.L., Butler, B.J., Odziemski, M.S., Friday, G., and Gillham, R.W. 1998. Inorganic and biological evaluation of cores from permeable iron reactive barriers. Designing and Applying Treatment Technologies: Proc. of the First Int. Conf. on Remediation of Chlorinated and Recalcitrant Compounds, May 18-21, 1998, Monterey, CA. Wickramanayake, G.B. and Hinchee, R.E. (Eds.), Battelle Press, Columbus, OH. 1(6): 163-168.

APPENDIX 1

Borden Site Description

A1.1. GENERAL STUDY LOCATION

The UW-AATDF Borden study was conducted in an unconfined sand aquifer at Canadian Forces Base (CFB) Borden, located about 60 km northwest of Toronto, Ontario, as shown in Figure A1.1. CFB Borden is a Canadian military base used mainly for training purposes. The University of Waterloo (UW) established an experimental site at Borden in 1978. This site is occasionally used for controlled experiments in which industrial chemicals are released carefully into the subsurface environment, and detailed monitoring is used to characterize their migration in either the vadose or saturated zones of the surficial aquifer.

An abandoned sand quarry that forms a broad local depression served as the location for most early experimentation by UW researchers. More recent research has been conducted in an elevated grassy area on the western side of the sand pit. Military installations and residential subdivisions of CFB Borden border the western and northern sides of these research areas. A municipal landfill, operated from 1940 to 1976, is located approximately 350 m to the south of the sand quarry. This landfill and the associated contaminant plume that migrates northward under the sand pit were the original focus of work at the site (MacFarlane et al., 1983). The wooded areas to the east and north of the sand pit contain mainly oak, pine, birch, and scrub brush. Seasonal marshes occupy the low lying parts of these wooded areas. A small northward-flowing creek flows through the woods to the east of the sand pit.

A second distinct research site known as the sheet piling cell area was established in 1990. Installations in this area are shown in Figure A1.2. It is located approximately 1 km northeast of the sand pit and less than 100 m from the small creek, with access by gravel road. This site covers approximately 2.5 hectares and is completely surrounded by woods. The UW-AATDF experiment is being undertaken in a recently deforested portion of this research area.

Topography of the area is flat to undulating, with local elevations ranging from 222 m above sea level (asl) in the sand pit to approximately 210 m asl near the sheet pile cell area. Climatic data collected just north of the base until 1970 indicate that mean daily temperature ranges from 12.4 to 0.6°C, with extremes from +37°C to -41°C. Mean total precipitation averaged 59 cm of rain and 241 cm of snow. A typical year has 171 days of frost, 88 days of rain, and 56 days of snow (MacFarlane et al., 1983).

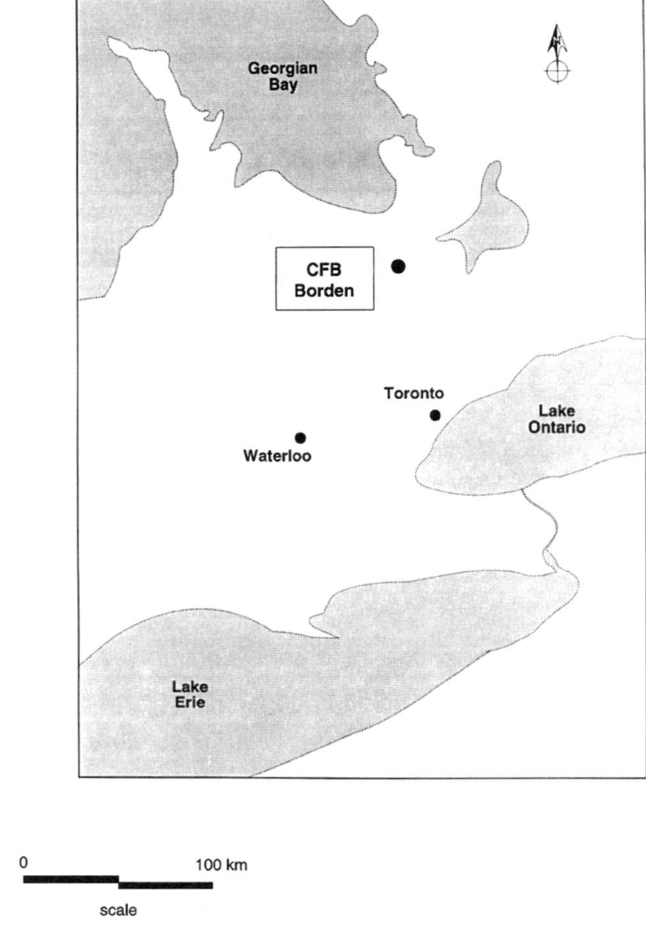

Figure A1.1 Regional location and site map of CFB Borden.

A1.2 THE UW-AATDF SITE

To locate the site, four cores and five 2.5-cm monitoring wells were installed prior to construction; the locations of each are shown in Figure A1.2. The cores were used to determine the approximate depth to clay in the area and the monitoring wells were installed to determine the approximate groundwater flow direction. The footprint position of the site was determined by the gradient measured in the monitoring wells.

The UW-AATDF Borden site consisted of three gates to control groundwater flow in the natural sand aquifer (Figure A1.3 shows the footprint of the site). Sealable-joint sheet piling, keyed into the underlying clay layer, formed the walls of the gates, which were aligned along the predominant local groundwater flow direction. The upgradient ends of the gates were open to allow capture of the natural groundwater flow, while downgradient ends were closed. The hydraulic gradient within the system was controlled by pumping from fully screened, 5-cm OD stainless steel extraction wells located at the downgradient end, at a rate sufficient to achieve a groundwater flow velocity comparable to the natural rate (10 to 15 cm/day). The extracted groundwater was collected and treated on site (see Appendix 10 for details).

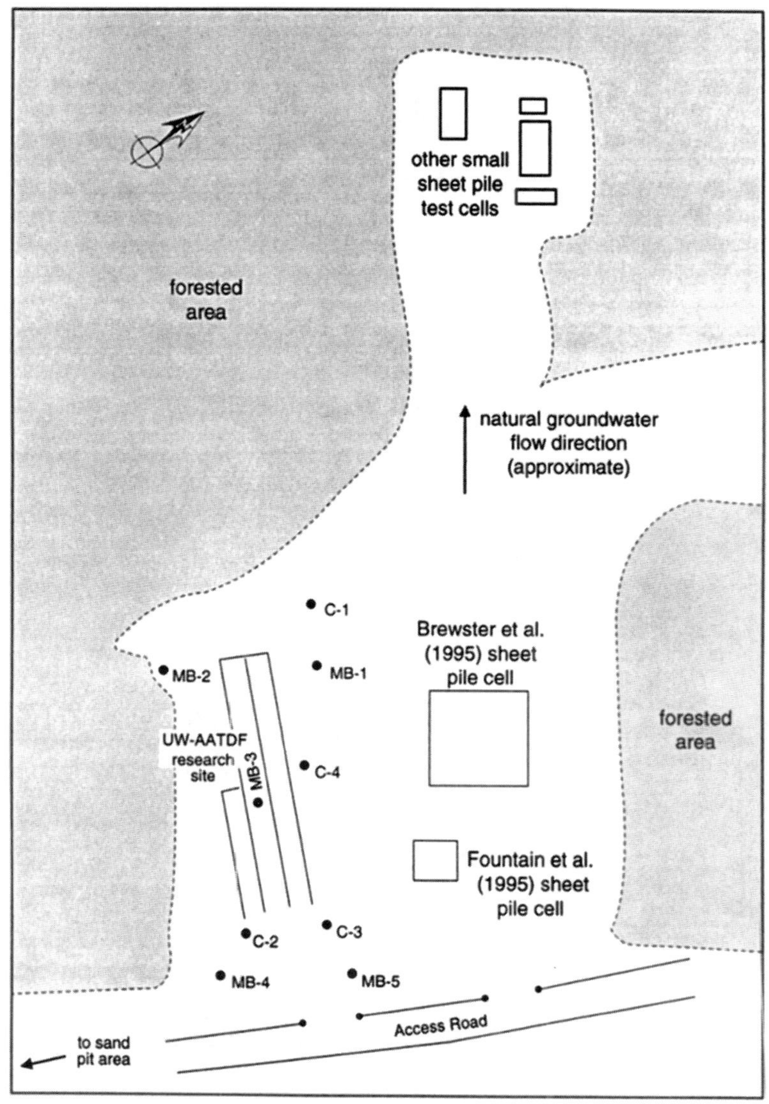

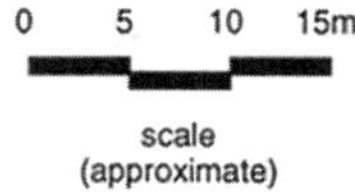

Figure A1.2 Plan view of test site and nearby research installations.

The site was covered to create a controlled environment. The frame was covered by an opaque tarp that reflects sunlight but allows some heat to penetrate. This structure prevented uncontrolled groundwater recharge to the site and permitted suitable conditions for the operation of the site over the winter months.

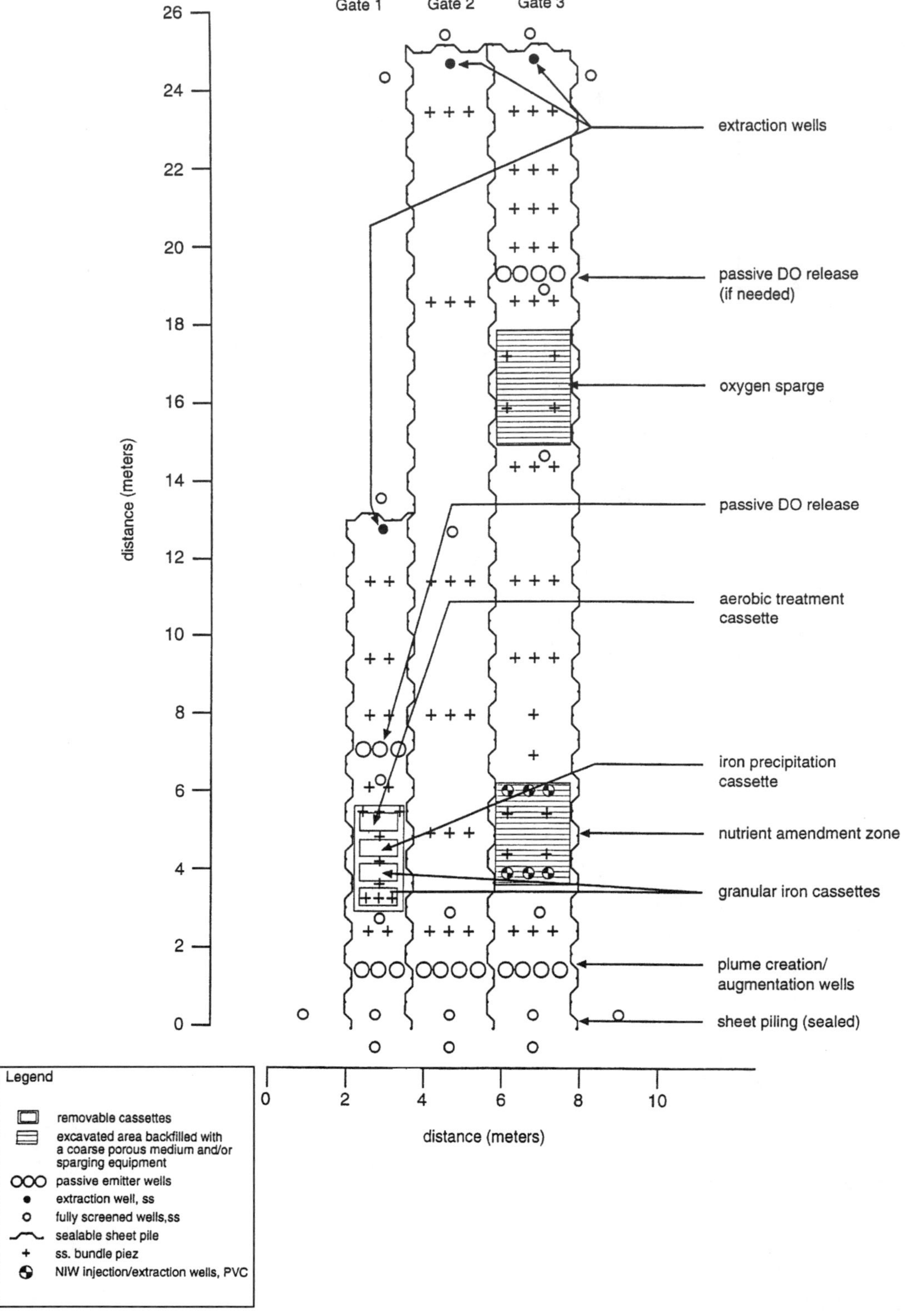

Figure A1.3 Diagram of experimental gates and instrumentation.

The site was heavily instrumented for groundwater sampling. Multilevel piezometers consisted of four 0.3-cm OD stainless steel tubes strapped to a central 1.25-cm OD PVC stalk. Sampling tubes extended to four equally spaced depths of approximately 1.2, 1.9, 2.6, and 3.3 m in the aquifer, providing capabilities for discrete lateral and vertical monitoring of the subsurface. A fine stainless steel screen was attached to the tip of each sampling tube and prevented fouling or clogging. Conventional monitoring wells fully screened across the saturated zone (0.6 to 3.7 m bgs) and constructed of 3.1-cm OD stainless steel were installed at selected locations throughout the site. These wells were sampled regularly to assess performance, using U.S. EPA methods. A row of eleven 25-cm OD PVC source wells screened between 0.9 and 3.1 m bgs was located at the open end of the gates, while an additional seven of these wells were installed further downgradient in Gates 1 and 3. Figure A1.4 shows the site monitoring network.

A1.3 PHYSICAL HYDROGEOLOGY

At the Borden site, surficial sand overlies a clay deposit. The sand aquifer was deposited in a prograding foreshore beach environment in glacial Lake Algonquin approximately 12,000 years ago (Bolha, 1986). The surficial sand creates an unconfined aquifer ranging in thickness from 9 m in the sand pit to about 4 m thick at the sheet pile cell area. The surficial aquifer at Borden is composed of clean, well-sorted, fine- to medium-grained sand. Although quite homogeneous relative to many aquifers of similar origin, cores and excavations reveal distinct bedding features that can significantly control solute transport processes. These include primarily horizontal and parallel structures, with some convolute and cross-bedding features. Texture of individual strata and laminae range from silt to coarse-grained sand with occasional pebbles. Mineralogical studies of bulk samples from the sand pit indicate that the aquifer material consists of 58% quartz, 19% feldspars, 14% carbonates, 7% amphiboles, and 2% chlorite. Median grain size in cores from the sand pit range from 0.07 to 0.69 mm, with a very low clay size fraction (Mackay et al., 1986).

At the sheet pile research area, a clayey silt layer underlies the surficial sand at a depth of 4.5 (±0.5) m. This clayey silt layer is considered to be an effective aquitard and serves as the bottom of the hydrogeologic system relevant to this investigation. This unit is composed of 13 m of clay and silty clay, interbedded with thin sand laminations and sand lenses.

The groundwater flow system in the shallow, unconfined Borden aquifer has been described in several previous studies. Average water table depth at the site is approximately 0.5 m bgs, with a seasonal fluctuation on the order of 1.0 m. Water table elevations are highest from March to June in response to spring snowmelt and rains. The water table gradually declines through the summer, then recovers in the fall months with increased rainfall.

Mean values for porosity and bulk density of the aquifer material in the sand pit area are 0.33 and 1.81 g/cm^3, respectively. Hydraulic conductivities estimated by slug tests have yielded a range from 5×10^{-3} to 1×10^{-2} cm/s, with a mean value on the order of 7×10^{-3} cm/s. The best estimate of a yearly average horizontal hydraulic gradient in the vicinity of the sand pit area is 0.0043, with a range from 0.0035 to 0.0054. No detectable vertical hydraulic gradients have been observed in the surficial aquifer. Natural groundwater flow velocity is estimated at between 5 and 15 cm/day based on these horizontal hydraulic gradients and natural gradient tracer test data (MacFarlane et al., 1983; Mackay et al., 1986).

A1.4 GROUNDWATER GEOCHEMISTRY

The background geochemistry of Borden aquifer groundwater in the site is presented in Appendix 6. At the site there likely exist localized anaerobic zones within the generally aerobic aquifer. This is based on the levels and presence of sulfate, absence of nitrate, and presence of oxygen.

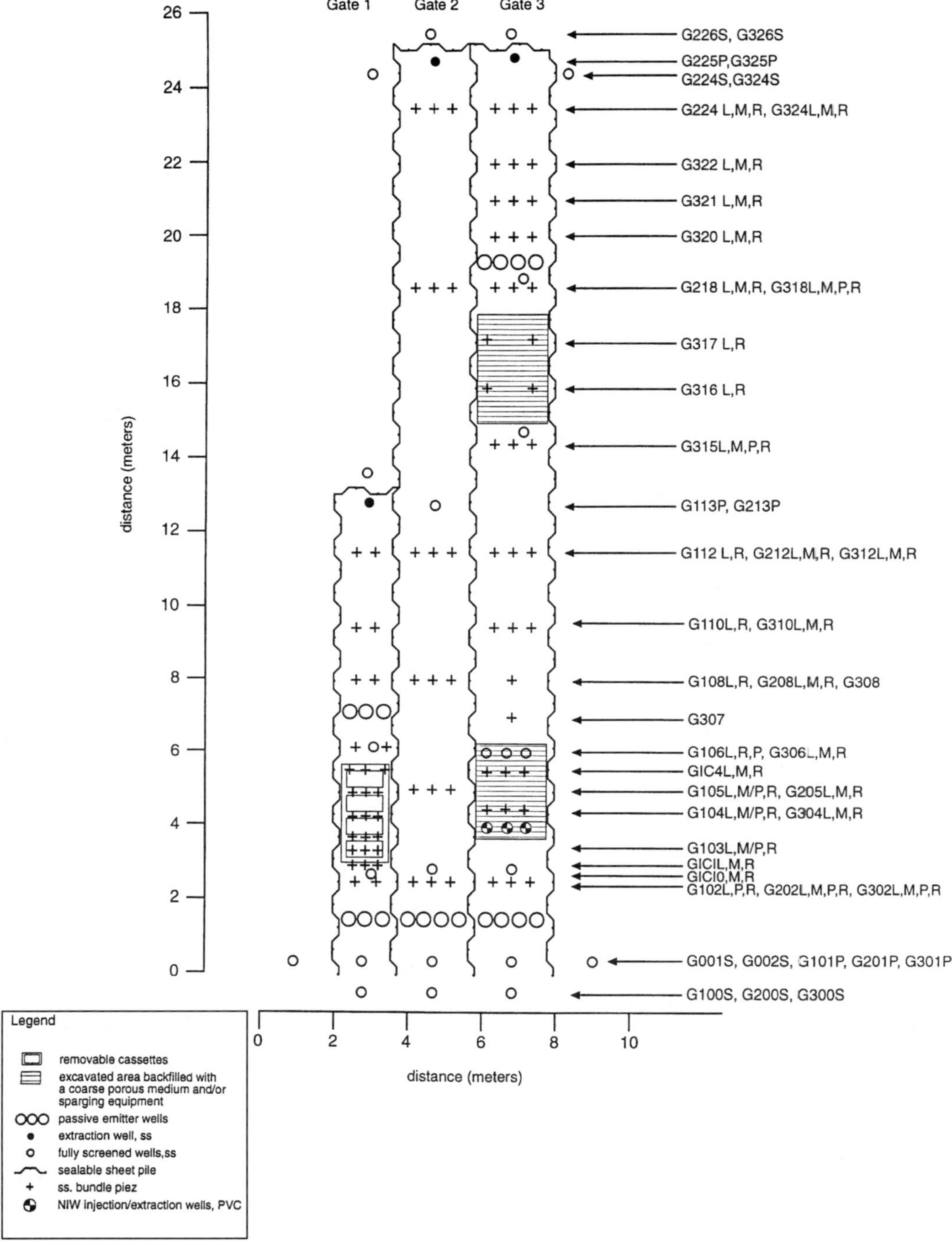

Figure A1.4 Well designations.

The presence of calcium carbonate in the aquifer solids results in high calcium and alkalinity of the groundwater. Groundwater temperature may range locally from 6 to 15°C, depending on the season and depth, but is generally 6 to 8°C with a variation of less than 2°C.

A1.5 GROUNDWATER CONTAMINATION NEAR THE UW-AATDF SITE

The surficial Borden aquifer has no significant value as a water resource for human consumption, due to contamination from various sources, including an old landfill, petroleum leakage, and road salt application (Cherry et al., 1995). Although the landfill leachate sometimes affects experiments carried out in the sand pit, the UW-AATDF study site is located near the eastern edge of the landfill plume and appears unaffected based on the sampling completed in 1996–1997. Potential for VOC contamination from this source is negligible, since no VOCs have been detectable in the leachate in previous studies.

Previous experiments conducted in the sheet pile test cell area have involved nonaqueous phase liquids including PCE, TCE, and CF. Major installations for these tests are shown in Figure A1.2. The experiments described by Brewster et al. (1995), Fountain et al. (1995), and others were conducted within sheet pile enclosures. These sheet pile cells prevented the migration of contaminants out of the immediate experimental area in all but one case (Brewster et al., 1995). Trace concentrations of PCE were detected in the background sampling event, for the current experiment, in location G224M-1 and G321M-2 and are attributed this PCE leak from the earlier experiments.

A1.6 REFERENCES

Bolha, J. 1986. A Sedimentological Investigation of a Progradational Foreshore Sequence: CFB Borden. Unpublished M.Sc. thesis, University of Waterloo, Waterloo, Ontario, Canada.

Brewster, M.L., Annan, A.P., Greenhouse, J.P., Kueper, B.H., Olhoeft, G.R., Redman, J.D., and Sander, K.A. 1995. Observed migration of a controlled DNAPL release by geophysical methods. *Ground Water*, 33(6): 977-987.

Cherry, J.A., Barker, J.F., Feenstra, S., Gillham, R.W., Mackay, D.M., and Smyth, D.J. 1995. The Borden site for groundwater contamination experiments: 1978-1995. Published in the proceedings of *In Situ Subsurface Remediation: Research and Strategies*, Stuttgart, Germany, September 24-25, 1995.

Fountain, J.C., Starr, R.C., Middleton, T., Beikirch, M., Taylor, C., and Hodge, D. 1996. A controlled field test of surfactant-enhanced aquifer remediation. *Ground Water*, 34(5): 910-916.

MacFarlane, D.S., Cherry, J.A., Gillham, R.W., and Sudicky, E.A. 1983. Migration of contaminants in groundwater at a landfill: a case study, 1. Groundwater flow and plume delineation. *Journal of Hydrology*, 63: 1-29.

Mackay, D.M., Freyberg, D.L., Roberts, P.V., and Cherry, J.A. 1986. A natural gradient experiment on solute transport in a sand aquifer, 1. Approach and overview of plume movement. *Water Resources Research*, 22(13): 2017-2029.

APPENDIX 2

Chronology of Borden Activities

The following table provides a detailed chronology of all activities related to the experimental phase of the UW-AATDF Borden project.

Date	Event
	1995
Oct and Nov	Site cleared of vegetation; preliminary site investigation
Dec 5 to 22	Major construction activities (sheet piling, outer cassette box installed)
	1996
Jan 2 to Mar 14	Major construction activities continued (cassettes, nutrient flushing, and biosparge zones installed)
Feb 27 to Mar 5	Source well installations
Apr 9 to 12	Piezometer, ss well installations
Apr 23	Iron installed in C1 and C2 (Gate 1)
Apr 26 and 30	Started extraction wells in all gates
Apr 28	Background alkalinity sampling in Gates 2 and 3
late April	Installation of multilevels G316 and G317
May 6	Problem w/ cassette seals identified
May 7	Extraction pumping started in Gates 2 and 3
May 15	Tracer test started in Gates 2 and 3
May 22, 23, and 27	Background sampling in all gates
May 29	Cassettes (C1 and C2) lifted and seals replaced
June 6	Site surveyed (elevations and distances)
June 10	Resampling for DHGs (background); washed limestone installed in C3 Gate 1); tracer test started in Gate 1
June 13	Background alkalinity sampling in Gate 1
June 15	Tracer input to Gates 2 and 3 ended (after 30 days)
June 19	Effluent from extraction wells sampled for VOCs
June 19 to 31	Conducted laboratory column sparging experiments
June 20	Sampling points installed in cassettes
June 25	G1CK piezometers installed (Gate 1)
June 29	Water levels measured
July 3	Tracer input diluted in Gate 1
July 8 to 10	Inorganic background sampling in Gate 1
July 9	Source prototype started in SW-6
July 11	Water levels measured
July 17	Treatment system started
July 23	Source wells stirred and CT added; water levels measured
July 31	Tracer input to Gate 1 ended (after 51 days)
Aug 1	Source drums stirred; cores collected outside Gate 3
Aug 2 to 16	Intermittent pumping of source wells to remove tracer input
Aug 7	Source drums stirred; water levels measured; completion of input plumbing for biosparge gate (Gate 3); first field test of biosparge gate (Gate 3)

Date	Event
Aug 15	Water levels measured
Aug 19	First observation of precipitates in Gate 1 extraction line
Aug 22	Source wells "developed" (excluding SW-6); CT and PCE added to source drums
Aug 23	Source drums mixed
Aug 26	Construction of permeable wall circulation system completed (Gate 3)
Aug 27	Inorganic monitoring of Gate 1 (background sampling); water levels measured
Aug 28	S. Fiorenza visit; second and third field sparge tests
Aug 29	Solid sampling for precipitate assessment (Gate 1)
Sept 2	Water levels measured
Sept 6	Toluene added to source drums; 3-day tracer test in Gate 3 started
Sept 9 to 20	Raman analyses conducted on Aug 29 samples
Sept 12	Source prototype test completed (SW-6)
Sept 13	SW-6 "developed"
Sept 18	Water levels measured
Sept 25	Field comparison of DO analyses methods; water levels measured
Oct 1	Initial microbial core collection (outside Gate 3)
Oct 2	Water levels measured; first nutrient flush (Gate 3); first Gate 3 field parameter data collected
Oct 7	First nutrient flush continued; first nutrient (benzoate) samples collected
Oct 8	ORC™ socks installed in C4 (Gate 1)
Oct 8 to Dec 19	ORC™ performance monitoring program
Oct 9	Top of cassettes sealed (Gate 1); end of 3-day tracer test (Gate 3)
Oct 16	Water levels measured
Oct 17	Sources started in all gates
Oct 21	Leak discovered in SW setup
Oct 23	Water levels measured
Oct 24 to 27	Source leak testing
Oct 26 and 27	Nutrient flush #2
Oct 27	Source leak detected in Gate 1 (SW-1)
Oct 30	SW-1 purged 4 well volumes to remove leaked source water; water levels measured
Nov 1	VOC sampling of discharged water
Nov 8 to 20	Winterized site
Nov 12	ORC™ socks removed from C4 (Gate 1)
Nov 13	Water levels measured
Nov 15	New ORC™ replaced in C4 and G1C4 piezometers installed (Gate1)
Nov 19	Float sensors installed in source drums; source restarted
Nov 21	Preliminary piezometer fence distances measured in all 3 gates; water levels measured
Nov 22	1.5 L DI water added to CT drum; first delivery of oxygen gas to biosparge gate (Gate 3)
Nov 29	Water levels measured; nutrient flush #3 (Gate 3)
Dec 2	Source wells stirred; plume development monitoring; water levels measured; nutrient flush #3 continued; fourth field sparge test; sparging interval set at 3 times/week
Dec 5	1 L DI water added to toluene drum
Dec 13	Water levels measured; nutrient flush #4 (Gate 3)
Dec 16	Major sampling in all gates; water levels measured
Dec 19	Fifth field sparge test; sparging interval set to 2 times/week
Dec 19 to 22 and 27	Construction of unsaturated zone; installation of soil gas monitoring apparatus in biosparge gate (Gate 3)
Dec 27	Water levels measured
Dec 30	Plume development monitoring; 1 L DI water added to CT drum; 2 L added to PCE drum (leak in cabinet discovered); water levels measured
1997	
Jan 2	Nutrient flush #5 (Gate 3)
Jan 6	External laboratory sampling (U.S. EPA wells); water levels measured
Jan 10	Source drums stirred
Jan 13	Plume development monitoring; water levels measured
Jan 15	500 mL DI water added to CT drum
Jan 17	Leak in CT source suspected

Date	Event
Jan 17 to 24	Leak testing of source wells conducted
Jan 20	Water levels measured
Jan 21	500 mL DI water added to toluene drum
Jan 24	1 L DI water added to CT drum; CT system restarted; source drums stirred; nutrient flush #6 (Gate 3)
Jan 27	Major sampling in all gates; water levels measured
Jan 28	Lab analysis of benzoate/conductivity correlation
Jan 31 to Mar 13	ORC™ laboratory experiments completed
Feb 1	Source drums stirred; tracer sampling ended in Gate 1
Feb 3	External laboratory sampling (U.S. EPA wells); water levels measured; field parameters measured in Gate 3
Feb 5	U.S. EPA wells G001S, G002S, and G106P installed
Feb 7	Source drums stirred
Feb 10	Plume development monitoring; water levels measured; external laboratory sampling of wells G001S, G002S, and G106P; 500 mL DI water added to CT drum
Feb 14	Source drums stirred; test of nutrient flush system (Gate 3)
Feb 20	Water levels measured
Feb 21	Toluene source disengaged temporarily since input concentrations too high; nutrient flush #7 (Gate 3)
Feb 24	Plume development monitoring; water levels measured; distances measured in Gate 3
Feb 28	Source drums stirred
late Feb	Permeable wall circulation system replumbed (Gate 3)
Mar 3	External laboratory sampling (U.S. EPA wells); water levels measured; initiated pretests of permeable wall circulation and amendment system on a regular basis
Mar 10	Major sampling in all gates; water levels measured; 1 L DI water added to CT drum
Mar 17	Source drums stirred; water levels measured; completed modifications to permeable wall circulation system (Gate 3)
Mar 18 to Apr 18	Acid addition experiment in field (Gate 1)
Mar 21	Sampling points installed upgradient of C1 (Gate 1); nutrient flush #8 (Gate 3)
Mar 24	Detailed VOC sampling in cassettes (to assess degradation in iron); water levels measured
Mar 27	Source drums stirred
Mar 31	External laboratory sampling (U.S. EPA wells); water levels measured; 1 L DI water added to CT drum
Apr 4	Source drums stirred; external laboratory sampling of G213P; initiated detailed nutrient pulse characterization plan in anaerobic zone of Gate 3
Apr 7	Plume development monitoring; water levels measured; toluene circulation system restarted
Apr 11	Source drums stirred; 1 L DI water added to PCE drum
Apr 15	Water levels measured; 1 L DI water added to toluene and CT drums; ORC™ socks removed from C4; new socks suspended in C4 (Gate 1)
Apr 18	Source drums stirred; nutrient flush #9 (Gate 3)
Apr 21	Water levels measured
Apr 25	Source drums stirred; iron cylinder added to site water treatment system
Apr 25 to May 13	Site de-winterized
Apr 28	External laboratory sampling (U.S. EPA wells)
Apr 29	1 L DI water added to toluene drum
Apr 30	Water levels measured
May 1	Source drums stirred
May 5	Water levels measured
May 7	1.5 L DI water added to CT drum; leaks at joints in lines identified and fixed
May 9	Source drums stirred
May 12	ORC™ socks removed from C4 (Gate 1); well dilution test of SW-12 to 14 prior to development
May 13	Water levels measured
May 14	UW short-course tours
May 15 and 16	500 mL DI water added to CT drum; field parameters measured in Gates 1 and 2 (2 pts per fence)
May 16	Nutrient flush #10 (Gate 3)
May 21 and 22	Well development of SW-12 to 14 (Gate 1)

Date	Event
May 22	Stickups and distances measured for all wells; source drums stirred; water levels measured
May 26	VOC sampling of source wells and effluent lines (discharged water); well dilution test started again (Gate 1)
May 27	Water levels measured
May 28	ORC™ socks shipped for oxygen "lockup" analyses
May 29	Source drums stirred
May 30	Ended main tracer test in Gate 2
June 2 and 3	Major sampling for all gates; water levels measured
June 5	Source drums stirred
June 10	Water levels measured
June 11	Source circulation system set on timer (3 hr "on"; 9 hr "off")
June 12	Source drums stirred; further development of SW-13 and 14 (Gate 1)
June 13	Well dilution test for SW-12 to 14 (Gate 1); nutrient flush #11 (Gate 3)
June 16	Water levels measured
June 18	VOC sampling for discharged water; plume development monitoring (Gate 1)
June 19	Source drums stirred; ORC™ installed in SW-12 to 14, complete with monitoring points (Gate 1)
June 19 to present	ORC™ performance monitoring program started (Gate 1)
June 20 to 21	Completed 24-hr sparge/DO monitoring test in biosparge gate (Gate 3)
June 23	External laboratory sampling (U.S. EPA wells); water levels measured
June 25	Field parameters measured in Gates 1 and 2 (2 pts per fence)
June 26	Initiated anaerobic zone monitoring program in Gate 3
June 27	Source drums stirred; ended main tracer test in Gate 3
July 4	Water levels measured; source drums stirred; VOC sampling of discharged water; source wells and drums sampled; 500 mL each of CT, PCE, and toluene added to source drums; bio-cores collected; well G318P installed
July 7	Anaerobic monitoring in Gate 3 completed
July 8	Water levels measured
July 9	Stirred drums; source wells and drums sampled for VOCs; U.S. EPA well G318P purged of 30 L
July 11	Nutrient flush #12 (Gate 3)
July 14 and 15	Major sampling in all gates; water levels measured
July 18	Biosparge oxygen delivery automated; sparge frequency increased to once daily; source drums stirred
July 21	initiated DO monitoring in aerobic section of Gate 3
July 23	Water levels measured
July 24	Source drums stirred
July 28	Water levels measured
Aug 5	Biosparge system monitoring apparatus sealed (Gate 3), water levels recorded
Aug 8	Nutrient flush #13 (Gate 3)
Aug 11	Source drums stirred
Aug 18	Water levels measured; oxygen tank for biosparge system replaced (Gate 3)
Aug 20	Source drums stirred
Aug 26	Water levels measured
Aug 29	Nutrient flush #14 (Gate 3)
Sept 3	Solid sampling for surface analyses (Gate 1)
Sept 4	Major sampling (Gate 1 and SWs and effluent lines from treatment system); water levels measured
Sept 5	Raman analyses of Gate 1 samples commenced
Sept 8	Solid sampling in Gate 1 completed; started tracer test in cassettes (Gate 1); source drums stirred; water levels measured
Sept 15	Source drums stirred
Sept 16	Water levels measured
Sept 22	Pump maintenance completed; water levels measured; source drums stirred
Sept 24	Tracer test in cassettes abandoned (Gate 1)
Sept 25	Nutrient flush #15 (Gate 3)
Sept 29	VOC and field parameter sampling in aerobic portion of Gate 1; chloride tracer injected in cassettes and tracer test restarted (Gate 1); water levels measured
Sept 30	VOC, nutrient, and field parameter sampling in aerobic portion of Gate 3; field GC work on unsaturated zone in biosparge gate (Gate 3)

Date	Event
Oct 7	VOC, nutrient, and field parameter sampling in aerobic portion of Gate 3; field GC work on unsaturated zone in biosparge gate (Gate 3); source drums stirred; water levels measured
Oct 13 to 18	Hydrogen sampling by UMass and UW personnel
Oct 14	VOC, nutrient, and field parameter sampling in aerobic portion of Gate 3; field GC work on unsaturated zone in biosparge gate (Gate 3); water levels measured in Gate 3
Oct 16	Chloride tracer test in cassettes abandoned (Gate 1); source drums stirred; water levels measured
Oct 21	VOC, DHG, and field parameters sampled in Gates 1 and 2; external laboratory sampling (U.S. EPA wells); field alkalinity analyses (Gate 3); organic acid samples collected (Gates 2 and 3); water levels measured
Oct 23	VOC, DHG, nutrient, and field parameter sampling in Gate 3; water collected from G104M-2 for sulfide laboratory experiments; water levels measured in all piezometers
Oct 27	VOC samples collected from source drums and treatment system effluent; plume source injection stopped; drained lines and began dismantling source apparatus; effluent lines winterized
Oct 29	ORC™ performance monitoring program completed (Gate 1)
Oct 31	Gate 3 aerobic section DO monitoring halted
Nov 6	Volumetric oxygen flow rate check for biosparge system (Gate 3); oxygen delivery system dismantled (Gate 3); extraction pumps in Gates 2 and 3 shut "off"; cores collected from bentonite corner seals between sheet-piling wall and outer cassette box (Gate 1; further related work beyond the scope of this project)
Dec 6	Extraction pumps in Gates 1 shut "off"; pumped source wells to remove contaminant mass in source wells

Appendix 3

Laboratory Analytical Procedures

A variety of different laboratories and procedures were used to analyze the samples collected over the duration of the UW-AATDF project. The following lists the laboratories and the type of analyses each performed.

University of Waterloo, Organic Geochemistry Laboratory

1. Volatile Organic Compounds (VOCs)
2. Dissolved Hydrocarbon Gases (DHGs)
3. Benzoate
4. Carbon Dioxide and Oxygen
5. Biological Oxygen Demand (BOD)
6. Dissolved Organic Carbon (DOC)

Barringer Labs (formerly MDS), Mississauga, Ontario

1. Volatile Organic Compounds (VOCs — U.S. EPA standards)
2. Organic Acids (acetic, butyric, formic, and propionic acids)

Water Technology, Inc., Burlington, Ontario

1. Alkalinity in Water
2. Ammonia
3. Anions
4. Dissolved Metals (cations)

InterTek Testing Services (formerly Inchcape Testing Services), San Jose, California

1. Volatile Organic Compounds (VOCs)
2. Cam-17 Metals
3. Toxicity Characteristic Leaching Procedure (TCLP), Lead

In the following pages methods of analysis and each laboratory's QA/QC measures are listed for the compounds identified above, with the exception of Intertek Testing Services.

UNIVERSITY OF WATERLOO ORGANIC GEOCHEMISTRY LABORATORY

Volatile Organic Compound Analysis

CRITICAL PARAMETERS: Toluene, Benzene, *p* + *m*-Xylene, *o*-Xylene, Tetrachloroethene, Trichloroethene, *cis*- and *trans*-Dichloroethene, Vinyl Chloride, Carbon Tetrachloride, Chloroform, Dichloromethane, Chloromethane, and Carbon Disulfide

Automated Headspace Procedure

SAMPLE PREPARATION: Samples are collected in 40-mL glass EPA sample vials and stored in a 4°C refrigerator until analyzed. At the time of analysis, sample vials are uncapped and 14.3 mL of sample is transferred to a 22-mL autosampler vial using a 30-mL glass syringe. The autosampler vial is then sealed with a Teflon-faced septum and aluminum crimp seal. A second autosampler vial is also filled to capacity and crimp-sealed. This latter subsample serves as a backup for analysis (if required) or as an in-house check (if required). Any backup samples are analyzed within the holding time limits (14 days, when preserved, 7 days without preservative). Sample vials to be analyzed are then placed in the autosampler carousel.

Dilution of the sample may be necessary if the analyte concentration in the sample exceeds the calibration curve concentration. In this case, organic free water is used as the diluent.

Calibration standards are prepared by filling the same 22-mL autosampler vials with organic-free water and then removing 8 mL from the total volume. Vials are quickly spiked with appropriate hydrocarbon standards (dissolved in methanol) and sealed. Standards are run at 23–30% level (or 2–3 for every 10 samples analyzed). Different ranges of standards are also processed, and typically this range is 5–6 levels of calibration standards.

Blanks are run every 20–30 samples. Spiked samples are not usually completed.

GAS CHROMATOGRAPHIC (GC) ANALYSIS: The prepared samples and calibration standards are run on a Hewlett Packard 5890 gas chromatograph equipped with a split injection port, capillary column, PID (photoionization detector), and a Varian Genesis headspace autosampler. Peak areas are measured by Varian Star 4.0 integration software and linear regressions are completed using a program for multisegmented calibration curves (Devlin, 1996). An external standard method of calibration is used.

Detection limits for these compounds are found to be 2–15 mg/L using the EPA procedure for method detection limit (MDL). The MDLs for the AATDF project are provided in Appendix 5. But in general the MDLs for the chlorinated ethenes (e.g., PCE, TCE, DCE, and VC) and hydrocarbons (e.g., BTEX) are lower than those MDLs for the chlorinated methanes (e.g., CT, CF, and DCM).

GC CONDITIONS:

Column:	DB-VRX 30 m × 0.32 mm I.D., 1.8 m film thickness
Carrier:	Helium at 3.5 mL/min
Oven:	Temperature Program: Initial temperature 35°C, hold 5 min, then ramp at 10°C/min to 110°C, hold for 4 min
Injector:	Split 12:1, 150°C
Detector:	PID (11.7 eV), 65°C; helium makeup gas at 30 mL/min

HEADSPACE ANALYZER CONDITIONS:

Vial volume:	22.3 mL
Water volume:	14.3 mL

Platen temperature:	70°C
Sample equilibrium time:	30.0 min
Mixing time:	4.0 min
Pressurization time:	1.0 min
Loop fill time:	0.3 min
Loop equilibrium time:	0.2 min
Loop size:	1.0 mL
Injection time:	0.3 min
Valve temperature:	165°C
Line temperature:	165°C

REFERENCE:

Devlin, J.F. 1996. A method to assess analytical uncertainties over large concentration ranges with reference to volatile organics in water. *Ground Water Monit. Remed.* 16(3): 179-185.

Dissolved Hydrocarbon Gases

PARAMETERS: Methane, Ethene, Ethane, Propene, Propane, Isobutane, 1-Butene, n-Butane

SAMPLE PREPARATION: Water samples are collected in screw-cap glass containers fitted with Teflon-lined septa and stored at 4°C for less than 1 week until analyzed. A 15-mL aliquot is withdrawn from the sample bottle into a 30-mL glass syringe followed by 13 mL of helium. The syringe is shaken and allowed to equilibrate for 3 hr. A 5-mL aliquot of the gas phase is injected for chromatographic analysis.

GAS CHROMATOGRAPHIC ANALYSIS: The gas samples are analyzed with a Hewlett Packard 5840A gas chromatograph equipped with a flame ionization detector and a 2-mL sample loop. The gas chromatograph is calibrated in an external standard mode using several concentrations of analyzed gas mixtures (purchased from Praxair). The concentrations of the dissolved hydrocarbons in the original water samples are determined using the Ideal Gas Law, Henry's Law, and solubility coefficients. The detection limits range from 0.1 to 0.5 mg/L depending on the specific dissolved hydrocarbon. Blanks are run each day and when compounds are detected, these amounts are subtracted from the reported values.

GC CONDITIONS:

Column:	Megabore GS-Q, 30 m
Carrier:	Helium at 12 mL/min
Oven:	Isothermal, 100°C
Injector:	100°C
Detector:	200°C

QUALITY ASSURANCE DATA:

COMPOUND	MDL (mg/L)
Methane	0.1
Ethene	0.1
Ethane	0.1
Propene	0.1
Propane	0.1
Isobutane	0.5
1-Butene	0.5
n-Butane	0.5

Analysis of Benzoate By HPLC

SAMPLE BOTTLE PREPARATION: Bottles and other glassware are soaked in a commercial alkaline cleaning solution for several hours, then rinsed with deionized water, dilute nitric acid; and again with deionized water. The bottles are dried overnight in a 90°C oven.

SAMPLE COLLECTION AND HANDLING: Samples are collected in 20-mL borosilicate glass screw-cap vials and treated with concentrated formic acid (5 mL per 4-mL sample) to preserve the sample and to convert benzoate to benzoic acid. During transport the samples are stored on ice and frozen when received at the lab. Prior to analysis, samples are thawed and filtered using a stainless steel swinney syringe filter holder containing a 13-mm Millipore 7 HV filter (0.45 mm pore size). Aliquots (150 mL) of filtered sample are injected on-column via a 50-mL sample loop. All samples and standards are equilibrated to room temperature (ca. 22°C) before injection.

REAGENTS: The following reagents are used: acetonitrile, nanopure water, sodium acetate, sodium benzoate, formic acid, and concentrated acetic acid.

CHROMATOGRAPHIC ANALYSIS: The analysis is performed on a Gilson HPLC (high-pressure liquid chromatograph) equipped with a Gilson Holochrome UV detector, Gilson 302 Pump, and Gilson 802B Manometric Module. A guard column is installed to protect the analytical column. The mobile phase (20% acetonitrile/80% sodium acetate buffer, pH 4.3) is continually degassed with helium (flow rate of 0.59 mL/min) during analysis.

The HPLC is calibrated by the external standard method. Two stock standards of 1000-mg/L sodium benzoate are prepared independently and diluted to several concentrations in water with formic acid addition. These standards are used to establish response, linearity, and detection limits. At least three different concentrations of standards, with three replicates each, are analyzed and the data subjected to linear regression (through zero). The resulting regression equation is used to determine unknown sample concentrations. A matrix standard was also prepared using Borden groundwater, and diluted to the same degree as the unknowns to rule out matrix effects.

HPLC CONDITIONS:

Analytical column:	200 mm Chromsep Reverse Phase column (ID = 3 mm) packed with ChromSpher C18 (particle size 5 mm)
Guard column:	Pelliguard LC-18 cartridge, 20 mm in length
Injector:	Rheodyne 7161 with a 50-mL sample loop
UV Detector:	Range = 0.05, wavelength = 254 nm
Mobile phase:	20:80, acetonitrile/0.02 M sodium acetate, pH 4.3 w/acetic acid

QUALITY ASSURANCE DATA: Table 1 shows the method detection limit (MDL) for the analytical procedure. MDL is defined as the minimum concentration of substance that can be identified, measured, and reported with 99% confidence that the analyte is greater then zero.

TABLE 1:

COMPOUND	MDL (mg/L)
Benzoic Acid	2.0

Carbon Dioxide and Oxygen Analysis

INTRODUCTION: The Fisher/Hamilton Gas Partitioner, Model 29, may be used for the analysis of a number of gases, including carbon dioxide and oxygen. The instrument is designed for emplacement of two chromatographic columns in series with a detector at the end of each column. This arrangement permits the analysis of widely different types of gases in a single sample. Specific column materials are chosen so that each column will separate a different set of gases in a mixed sample. Some components of the mixture are separated in the first column and measured with the first detector; others move through the first column and are resolved in the second column, and then detected.

SAMPLE SYRINGE PREPARATION: Glass syringes to be used for samples or calibration gases are soaked in a commercial alkaline cleaning solution for several hours, then rinsed with deionized water, dilute nitric acid, and more deionized water.

SAMPLE COLLECTION AND HANDLING: Gas was obtained directly from microcosm head-spaces by syringe. The barrel and plunger of the syringe are first wetted with organic-free water prior to sample collection to ensure an airtight seal. The wetted syringe is fitted with a Mininert™ valve, then filled and emptied at least three times with sample gas in order to obtain a representative sample. The valve is closed to seal the sample in the syringe.

Samples are normally analyzed immediately after collection. If this proves impossible, the sealed syringes are placed in containers under water and stored at 4°C. The samples are equilibrated to room temperature before analysis. A 1-mL sample loop is used to introduce gas samples into the carrier gas stream with a precision of ~0.3%.

GAS PARTITIONER ANALYSIS: With the Gas Partitioner setup used, carbon dioxide is retarded in column 1, while oxygen and nitrogen rapidly pass through and are detected as a composite peak at the first detector. Carbon dioxide is eluted from column 1 and detected at the first detector, while oxygen and nitrogen separate in column 2 and are measured individually at the second detector. Carbon dioxide is permanently absorbed as it enters column 2, and never reaches the second detector.

Calibration is by the external standard method, using commercially obtained, certified gas mixtures (three concentrations in triplicate) in the expected range of the collected samples. When the chromatogram of the standard mixture has been obtained, the peak height of each component is measured from the actual baseline of the peak. The obtained calibration data are subjected to linear regression analysis and the resultant equation is used to determine unknown sample concentrations.

At least 3 samples of a calibration mixture are run prior to the analysis of unknown samples and also after every 10 unknown samples, to ensure the gas partitioner is operating in a consistent manner. Sample peak heights are measured, and the concentration of the unknown components determined using the linear regression equation.

GAS PARTITIONER CONDITIONS:

Column 1:	30% di-2-ethylhexyl-sebacate on 60-80 mesh Chromosorb-P
Column 2:	Molecular Sieve 13X
Carrier:	Helium at 20 mL/min
Temperature:	Room temperature
Detectors:	Thermal conductivity

QUALITY ASSURANCE DATA: The method detection limit for this procedure has not yet been determined according to EPA protocol. However, under normal operating conditions, the following concentrations in a 1-mL gas sample can be detected:

Carbon dioxide	0.589 mg/L
Oxygen	7.14 mg/L
Nitrogen	30 mg/L
Methane	30 mg/L
Carbon monoxide	50 mg/L
Hydrogen	150 mg/L

Some of this information was obtained from "Respiratory gas analysis with the Fisher/Hamilton Gas Partitioner," Bulletin No. 65-12A. Instrument Division, Fisher Scientific, Pittsburgh, PA.

Biochemical Oxygen Demand

SAMPLE HANDLING: Since the sampling site was very distant from the laboratory, samples for biochemical oxygen demand (BOD) analysis could not be tested immediately, but were maintained at 4°C during transport and storage, and analysis begun within 5 days of sample collection. Many received samples were highly alkaline and were adjusted to a pH of 6–8 with 6N HCl prior to BOD analysis. Some received samples contained levels of dissolved oxygen (DO) above those expected of water in equilibrium with the atmosphere, because pure oxygen was used in the Alameda biosparge system. These samples were not purged to adjust to ambient DO level at 20°C, as the Standard Method for BOD describes, since this would cause loss of volatile organic compounds. Consequently, loss of oxygen during incubation of such samples may be observed.

BOD ANALYSIS: Analysis of BOD was performed using the Standard Method 5210 for BOD_5 (Eaton et al., 1995), which was modified somewhat (see below; Chapman et al., submitted) to optimize analysis of samples containing volatile organic contaminants. The test measures the oxygen required for the biological degradation of organic material (carbonaceous demand) and the oxygen required for the biochemical degradation of inorganic material (sulfides, ferrous iron, and nitrogenous demand).

In summary, the method consists of making several dilutions of the sample within 300-mL glass BOD bottles (Wheaton). Dilutions of 5, 10, 25, 50, and 75% were typically used. Dilutions are made with a diluent which has been aerated, and contains inorganic nutrients (nitrogen, phosphorus, trace elements) to permit microbial metabolism. All BOD bottles are also inoculated with 2 mL of microbial seed culture. The DO is measured initially, and following 5 days of incubation at 20°C using a dissolved oxygen probe (MI-730 Oxygen Electrode and Meter, Microelectrodes, Inc.) that was standardized with 2% sodium sulfide (zero standard) and air-saturated distilled water, according to the manufacturers instructions. The BOD of each sample is computed from the difference between initial and final DO.

MODIFICATIONS TO STANDARD METHOD: For this work an adapted microbial population (microbial seed culture) was developed. The culture was initiated in the laboratory on September 26, 1996 from approximately 750 mL of groundwater (MW-3) and 200 g of aquifer material (AWA101 Boring B-14, depth 8.5 to 9 ft, date July 26, 1996) obtained from the Alameda site. The seed culture was maintained in an aerated 2-L flask which was incubated at room temperature in the dark. The culture was fed with small amounts of either contaminated Alameda groundwater, gasoline-contacted groundwater (aqueous phase from 1:10 mixture of gasoline and deionized water), or neat BTEX, several times per week. Inorganic nutrients were added monthly (0.6 mL of each of the four stock inorganic solutions used to prepare the BOD diluent).

In addition to the dilution water blanks, seed controls, and glucose–glutamic-acid checks required in Standard Method 5210 (Eaton et al., 1995), a check consisting of gasoline-saturated water was used with each set of samples to specifically verify the ability of the seed culture to degrade petroleum hydrocarbons.

References:

Eaton, A.D., L.S. Clesceri, and A.E. Greenberg. (Eds.). 1995. *Standard Methods for the Examination of Water and Wastewater*, 19th ed. American Public Health Association, American Water Works Association Water Environment Federation, Washington, DC.

Chapman, S.W., M.R. Vandergriendt, B.J. Butler, and D.M. Mackay. 1997. Measurement of oxygen demand of petroleum hydrocarbon-contaminated groundwater. Accepted for publication in *Bioremediation J.*

Nonpurgeable Dissolved Organic Carbon

A Dohrmann DC 190 TOC/DOC analyzer was used to determine nonpurgeable dissolved organic carbon (NDOC) content of Alameda groundwater. Through catalytic oxidation (a platinum catalyst maintained at 680°C), organic material in the sample is completely oxidized to CO_2 and H_2O. An oxygen flow of 200 cc/min sweeps the CO_2-containing steam through a condenser and into a gas/liquid separator, which traps most of the H_2O. Final removal of H_2O is accomplished by a dehumidifier operating at 0–10°C. The dried CO_2-containing gas is then passed through a halogen scrubber and a CO_2-specific nondispersive infrared (NDIR) detector for peak quantification. Inorganic carbon (IC) in the sample is determined by introducing sample into the IC reactor, which contains room-temperature acidified water. In this acidic environment, all forms of IC are purged from the solution as CO_2. The CO_2-containing oxygen then flows through the dehumidifier and onto the NDIR detector, where it is quantified. The amount of inorganic carbon in the sample is subtracted from the total carbon (TC) value, to obtain the amount of organic carbon in a sample. To obtain dissolved organic carbon (DOC) measurements rather than total organic carbon (TOC) measurements, particulates are removed from the liquid before carbon analyses are conducted.

SAMPLE COLLECTION AND HANDLING: Samples were collected in 20-mL glass scintillation vials, acidified with 4 drops of phosphoric acid, and stored on ice or at 4°C until analyzed. The samples were filtered (0.45-μm filters) before analysis to remove particulates, and purged for 6 min to remove volatile compounds, since the intent of the analysis was to quantify DOC other than the target contaminants in the Alameda groundwater.

PREPARATION OF STANDARDS AND CALIBRATION: Standards are prepared using potassium biphthalate in organic-free water acidified to pH 2 with phosphoric acid. A stock solution of 1000 mg C/L is prepared by adding 0.2126 g $KC_8H_5O_4$ to 100 mL organic-free water. The stock is then further diluted in organic-free water, as appropriate. Calibration is normally over the range of 500 to 10 mg C/L, but this may vary depending upon the samples analyzed. The method detection limit is 1 mg C/L.

Before calibration, water is injected several times (200-mL injection volume) to obtain a blank value. Then the calibration is begun. The standard is injected three times to obtain a response factor. The response is then checked by running the standard again, and also by running several checks of lower concentration. A linear regression is then performed, using the calibration standard and the check data to obtain a standard curve. Values for unknown samples are obtained from the linear regression equation.

DOHRMANN DC 190 SETUP:

Gas flow	200 mL/min
Furnace temperature	680°C
Injection volume	200 mL
IC chamber	1/3 full with 10% phosphoric acid
Inlet	injection by syringe is selected
Method	either a TC (1 injection) or a TOC (2 injections) analysis is selected, as appropriate

INJECTION OF SAMPLES: Samples are injected in the same manner as standard solutions. A sample volume of 200 mL is used. For DOC analyses, the Analyzer is set for a TOC analysis, and two injections of each sample are performed in sequence. The first is through the standard injection port, the second through the IC injection port. When analyzing a number of samples, calibration checks are run after every 10 samples.

BARRINGER LABS, MISSISSAUGA, ONTARIO (FORMERLY MDS LABORATORIES)

MDS Environmental Services Limited

ANALYTICAL METHOD SOP SUMMARY

	VML-24E-WT	[E]NABLED/[D]ISABLED: E	SAMPLE TYPE: Water
COVER PAGE DESCRIPTION:	Volatiles, Priority Pollutant, Extended List(SIM Mode)		
INTERNAL EPL DESCRIPTION:	Volatiles, Extended List(MIBK etc) - SIM Mode		

METHOD SUMMARY:	Ambient temperature Purge & Trap/capillary GC/MS analysis of water for volatile organics. Internal std. quantitation with surrogate standard recoveries.		
KEYWORDS:	VOLATILES, WATER, GC/MS, MPM16 A/S		
DEPARTMENT:	O	WORKCENTER: VOC	WORKSTATION: VMS
INSTRUMENTATION:	Hewlett-Packard 5890II GC, O.I.4460A P&T, MPM16 A/S, Chemserver D/S		

METHOD NUMBERS(COVER PAGE):	FILENAMES:
U.S. EPA Method No. SW-846 8260	EPL8260A

METHOD NUMBERS(INTERNAL USE):	FILENAMES:

STORAGE METHOD: 4 C	
MINIMUM VOLUME REQUIRED: 40 ml	PRICE(1-4): 285 PRICE(5 AND UP): 230
DILUTION FACTOR:	PRESERVATION CODE: G CONTAINER: 5
MAXIMUM HOLD TIME: 6 days	PRESERVATION CODE: CONTAINER:

PROCESS STEP TIMES (MINUTES):		
	PREPARATION:	120
	ANALYSIS:	45
	SUPERVISOR APPROVAL:	0
	REPORT:	0
	TOTAL TIME:	165

SAMPLE PROCESSING

STEP 1:	Tune GC/MS to meet performance and BFB spectra com pliance.
STEP 2:	Calibration - fortify 10.0ml of reagent water with the specified levels of CAL AND SURROGATE stds.
STEP 3:	Analysis - 10.0ml of sample is fortified with the specified level of SURROGATE std.
STEP 4:	Reanalysis of the appropriate dilution for samples that exceed the calibration range.
AQUISITION PROGRAM:	vocus.m vocus.s = SEQUENCE
QUANT PROGRAM:	vocus.m (on Chemserver)

MDS Environmental Services Limited

ANALYTICAL METHOD SOP SUMMARY

TESTCODE:	VML-24E-WT	[E]NABLED/[D]ISABLED: E	SAMPLE TYPE: Water
COVER PAGE DESCRIPTION:	Volatiles, Priority Pollutant, Extended List(SIM Mode)		
INTERNAL EPL DESCRIPTION:	Volatiles, Extended List(MIBK etc) - SIM Mode		

STANDARDS:	STANDARD IDENTIFICATION:	CONCENTRATION:	UNITS:	VOLUME ADDED:
FORTIFICATION:	vocus - 10 - Sep/95	20.0	NG/SPRG	2.0 – UL
SURROGATE:	NA			–
STD. REF. MATERIAL:	PMX - 110	20.0	NG/SPRG	2.0 – UL
INTERNAL STANDARD:	SURR + INT	10.0	NG/SPRG	1.0 – UL
CALIBRATION STD 1:	vocus - 10 - Sep/95	5.0	NG/SPRG	0.5 – UL
CALIBRATION STD 2:	vocus - 10 - Sep/95	10.0	NG/SPRG	1.0 – UL
CALIBRATION STD 3:	vocus - 10 - Sep/95	20.0	NG/SPRG	2.0 – UL
CALIBRATION STD 4:	vocus - 10 - Sep/95	40.0	NG/SPRG	4.0 – UL
CALIBRATION STD 5:	vocus - 10 - Sep/95	100	NG/SPRG	10 – UL

QUALITY CONTROL RECORDS:	INSTRUMENT PERFORMANCE:	
	Q1 - METHOD BLANK:	BLANK
	Q2 - REAGENT SRM:	VOC_SRM
	Q3 - MATRIX SRM:	VOC_SRM
	Q4 - REPLICATE SRM:	
	OTHER:	

WORKSHEET/DCI INFORMATION:	WORKSHEET FORMAT: 4 - Volatile Organics	
	AUTOMATIC WORKSHEET GENERATION (Y/N): N	[S]INGLE/[M]ULTIPLE PROJECTS: S
	DCI TYPE (AUT/MAN/GST/TXT): AUT	DCI Q3 NAME: VOC_SRM

WGT. CAPTURE INFORMATION:	REQUIRES WEIGHING (Y/N):	WEIGH Q3 & Q4 (Y/N):	WEIGH Q4 ONLY (Y/N):
	RECOMMENDED CONTAINER:		
	TARGET WEIGHT:	LOW:	HIGH:

MDS Environmental Services Limited

ANALYTICAL METHOD SOP SUMMARY

TEST CODE **VML-24E-WT** Volatiles, Priority Pollutant, Extended List(SIM Mode) PAGE: 1 OF 2

ANALYTE	CAS#	MDL	LOQ	UNITS	DISPOSAL LIMIT	Q2 HIGH & LOW	Q2 FACTOR	BLANK HIGH	SPIKE ACTUAL	SPIKE LOW	SPIKE HIGH
Benzene	71-43-2	0.1	0.2	ug/L	0	70 — 130	1	0.4	20.0	12.0	26.0
Bromodichloromethane	75-27-4	0.1	0.2	ug/L	0	70 — 130	1	0.4	20.0	12.0	26.0
Bromoform	75-25-2	0.1	0.2	ug/L	0	70 — 130	1	0.4	20.0	12.0	26.0
Bromomethane	74-83-9	0.8	1.6	ug/L	0	50 — 150	0.5	3.2	40.0	20.0	60.0
Carbon Tetrachloride	56-23-5	0.1	0.2	ug/L	0	70 — 130	1	0.4	40.0	28.0	52.0
Chlorobenzene	108-90-7	0.1	0.2	ug/L	0	70 — 130	1	0.4	20.0	12.0	26.0
Chloroethane	75-00-3	0.8	1.6	ug/L	0	50 — 150	0.5	3.2	100.0	60.0	180.0
Chloroform	67-66-3	0.1	0.2	ug/L	0	70 — 130	1	0.4	20.0	12.0	26.0
Chloromethane	74-87-3	0.8	1.6	ug/L	0	50 — 150	0.5	3.2	40.0	20.0	60.0
cis 13 Dichloropropene	542-75-6	0.2	0.4	ug/L	0	70 — 130	1	0.8	20.0	12.0	26.0
Dibromochloromethane	124-48-1	0.1	0.2	ug/L	0	70 — 130	1	0.4	20.0	12.0	26.0
13-Dichlorobenzene	541-73-1	0.1	0.2	ug/L	0	70 — 130	1	0.4	20.0	12.0	26.0
12-Dichlorobenzene	95-50-1	0.1	0.2	ug/L	0	70 — 130	1	0.4	20.0	12.0	26.0
14-Dichlorobenzene	106-46-7	0.1	0.2	ug/L	0	70 — 130	1	0.4	20.0	12.0	26.0
11-Dichloroethane	75-34-3	0.1	0.2	ug/L	0	70 — 130	1	0.4	20.0	12.0	26.0
12-Dichloroethane	107-06-2	0.1	0.2	ug/L	0	70 — 130	1	0.4	20.0	12.0	26.0
11-Dichloroethene	75-35-4	0.2	0.4	ug/L	0	70 — 130	1	0.8	20.0	12.0	26.0
12-Dichloropropane	594-20-7	0.1	0.2	ug/L	0	70 — 130	1	0.4	20.0	12.0	26.0
Ethyl Benzene	100-41-4	0.1	0.2	ug/L	0	70 — 130	1	0.4	20.0	12.0	26.0
12-Dibromoethane	106-93-4	0.1	0.2	ug/L	0	70 — 130	1	0.4	20.0	12.0	26.0
Dichloromethane	75-09-2	0.3	0.6	ug/L	0	70 — 130	1	1.2	20.0	12.0	26.0
mp Xylenes	106423/108383	0.1	0.2	ug/L	0	70 — 130	0.5	0.4	40.0	24.0	52.0
o Xylene	95-47-6	0.1	0.2	ug/L	0	70 — 130	1	0.4	20.0	12.0	26.0
Styrene	100-42-5	0.1	0.2	ug/L	0	70 — 130	1	0.4	20.0	12.0	26.0
1122-Tetrachloroethane	79-34-5	0.1	0.2	ug/L	0	70 — 130	1	0.4	20.0	12.0	26.0
Tetrachloroethene	127-18-4	0.1	0.2	ug/L	0	70 — 130	1	0.4	20.0	12.0	26.0
Toluene	108-88-3	0.1	0.2	ug/L	0	70 — 130	1	0.4	20.0	12.0	26.0
trans 12 Dichloroethene	156-60-5	0.2	0.4	ug/L	0	70 — 130	1	0.8	20.0	12.0	26.0
trans 13 Dichloropropene	78-88-6	0.4	0.8	ug/L	0	70 — 130	1	1.6	20.0	12.0	26.0
111-Trichloroethane	71-55-6	0.1	0.2	ug/L	0	70 — 130	1	0.4	20.0	12.0	26.0
112-Trichloroethane	79-00-5	0.1	0.2	ug/L	0	70 — 130	1	0.4	20.0	12.0	26.0
Trichloroethene	79-01-06	0.1	0.2	ug/L	0	70 — 130	1	0.4	20.0	12.0	26.0

MDS Environmental Services Limited

ANALYTICAL METHOD SOP SUMMARY

TEST CODE	VML-24E-WT	Volatiles, Priority Pollutant, Extended List(SIM Mode)	PAGE: 2 OF 2

ANALYTE	CAS#	MDL	LOQ	UNITS	DISPOSAL LIMIT	Q2 HIGH & LOW	Q2 FACTOR	BLANK HIGH	SPIKE ACTUAL	SPIKE LOW	SPIKE HIGH
Trichlorofluoromethane	353-54-8	0.8	1.6	ug/L	0	50 — 150	0.5	3.2	40.0	20.0	60.0
Vinyl Chloride	9003-22-9	0.25	0.5	ug/L	0	50 — 150	0.5	1.0	40.0	20.0	60.0
cis 12 Dichloroethene	156-59-2	0.2	0.4	ug/L	0	70 — 130	1	0.8	20.0	12.0	26.0
Carbon Disulphide	75-15-0	0.1	0.2	ug/L	0	70 — 130	0.5	0.4	40.0	28.0	52.0
2-Chloroethyl Vinyl Ether	110-75-8	0.2	0.4	ug/L	0	70 — 130	0.5	0.8	40.0	28.0	52.0
Vinyl Acetate	108-05-4	10.0	20.0	ug/L	0	40 — 150	0.1	40.0	200.0	80.0	300.0
1112-Tetrachloroethane	630-20-6	0.1	0.2	ug/L	0	70 — 130	1	0.4	20.0	12.0	26.0
Acrolein	107-02-8	10.0	20.0	ug/L	0	40 — 150	0.1	40.0	200.0	80.0	300.0
Acrylonitrile	107-13-1	5.0	10.0	ug/L	0	40 — 150	0.1	20.0	200.0	80.0	300.0
Acetone	67-64-1	5.0	10.0	ug/L	0	40 — 150	0.1	40.0	200.0	80.0	300.0
2-Butanone	78-93-3	10.0	20.0	ug/L	0	40 — 150	0.1	40.0	200.0	80.0	300.0
4-Methyl-2-Pentanone	108-10-1	5.0	10.0	ug/L	0	40 — 150	0.1	20.0	200.0	80.0	300.0
2-Hexanone	591-78-6	5.0	10.0	ug/L	0	40 — 150	0.1	20.0	200.0	80.0	300.0
Methyl-t-butyl Ether	1634-04-4	1.0	5.0	ug/L	0	70 — 130	1	0.8	20.0	12.0	26.0
4-BrFlbenzene-Surrogate	460-00-47-sur	na	na	ug/L	0	70 — 130	1	na	na	na	na
Toluene-d8	108-88-3-d8	na	na	ug/L	0	70 — 130	1	na	na	na	na
Dibromofluoromethane - Surrogate	75-61-6-sur	na	na	ug/L	0	70 — 130	1	na	na	na	na
						—					
						—					
						—					
						—					
						—					
						—					
						—					
						—					
						—					
						—					
						—					
						—					
						—					
						—					
						—					

MDS Environmental Services Limited

ANALYTICAL METHOD SOP SUMMARY

TESTCODE:	OA4-02-WT	[E]NABLED/[D]ISABLED:	E	SAMPLE TYPE:	Water
COVER PAGE DESCRIPTION:	Organic Acid Scan(Formic, Acetic, Propionic and Butyric)				
INTERNAL EPL DESCRIPTION:	Organic Acid Scan(Formic, Acetic, Propionic and Butyric)				
METHOD SUMMARY:	Determination of Organic acids in water by Ion-Exclusion Chromatography.				
KEYWORDS:					
DEPARTMENT:	I	WORKCENTER:	AC	WORKSTATION:	MAN
INSTRUMENTATION:	Dionex Ion Chromatograph, Series 4500i				

METHOD NUMBERS(COVER PAGE):	FILENAMES:
Ministry of Labour 1991(Modification)	
(Reference - Molson Breweries Technical	
Services Centre Method 1992)	

METHOD NUMBERS(INTERNAL USE):	FILENAMES:

STORAGE METHOD:	4 C				
MINIMUM VOLUME REQUIRED:	40 ml	PRICE(1-4):	70	PRICE(5 AND UP):	50
DILUTION FACTOR:		PRESERVATION CODE:	C	CONTAINER:	4
MAXIMUM HOLD TIME:	7 days	PRESERVATION CODE:		CONTAINER:	

PROCESS STEP TIMES (MINUTES):		
	PREPARATION:	5
	ANALYSIS:	12
	SUPERVISOR APPROVAL:	0
	REPORT:	0
	TOTAL TIME:	17

SAMPLE PROCESSING

STEP 1:	
STEP 2:	
STEP 3:	
STEP 4:	
AQUISITION PROGRAM:	
QUANT PROGRAM:	

MDS Environmental Services Limited

ANALYTICAL METHOD SOP SUMMARY

TESTCODE:	OA4-02-WT	[E]NABLED/[D]ISABLED: E	SAMPLE TYPE: Water
COVER PAGE DESCRIPTION:	Organic Acid Scan(Formic, Acetic, Propionic and Butyric)		
INTERNAL EPL DESCRIPTION:	Organic Acid Scan(Formic, Acetic, Propionic and Butyric)		

STANDARDS:	STANDARD IDENTIFICATION:	CONCENTRATION:	UNITS:	VOLUME ADDED:
FORTIFICATION:				–
SURROGATE:				–
STD. REF. MATERIAL:				–
INTERNAL STANDARD:				–
CALIBRATION STD 1:				–
CALIBRATION STD 2:				–
CALIBRATION STD 3:				–
CALIBRATION STD 4:				–
CALIBRATION STD 5:				–

QUALITY CONTROL RECORDS:	INSTRUMENT PERFORMANCE:	
	Q1 - METHOD BLANK:	BLANK
	Q2 - REAGENT SRM:	ORG_ACID_SPK
	Q3 - MATRIX SRM:	ORG_ACID_SPK
	Q4 - REPLICATE SRM:	
	OTHER:	

WORKSHEET/DCI INFORMATION:	WORKSHEET FORMAT: 1 - Metal/Conventional Analysis	
	AUTOMATIC WORKSHEET GENERATION (Y/N): Y	[S]INGLE/[M]ULTIPLE PROJECTS: S
	DCI TYPE (AUT/MAN/GST/TXT): MAN	DCI Q3 NAME: ORG_ACID_SPK

WGT. CAPTURE INFORMATION:	REQUIRES WEIGHING (Y/N):	WEIGH Q3 & Q4 (Y/N):	WEIGH Q4 ONLY (Y/N):
	RECOMMENDED CONTAINER:		
	TARGET WEIGHT:	LOW:	HIGH:

MDS Environmental Services Limited

ANALYTICAL METHOD SOP SUMMARY

TEST CODE	OA4-02-WT	Organic Acid Scan(Formic, Acetic, Propionic and Butyric)	PAGE: OF

ANALYTE	CAS#	MDL	LOQ	UNITS	DISPOSAL LIMIT	Q2 HIGH & LOW	Q2 FACTOR	BLANK HIGH	SPIKE ACTUAL	SPIKE LOW	SPIKE HIGH
Formic Acid	64-18-6	0.2	0.2	mg/L	0	85 — 115	50	0.4	5.0	3.0	7.0
Acetic Acid	64-19-7	0.3	0.3	mg/L	0	85 — 115	28.6	0.6	8.8	5.28	12.3
Propionic Acid	79-09-4	0.7	0.7	mg/L	0	85 — 115	14.3	1.4	17.5	10.5	24.5
Butyric Acid	107-92-6	1.0	1.0	mg/L	0	85 — 115	10.0	2.0	25.0	15.0	35.0
						—					
						—					
						—					
						—					
						—					
						—					
						—					
						—					
						—					
						—					
						—					
						—					
						—					
						—					
						—					
						—					
						—					
						—					
						—					
						—					
						—					
						—					
						—					
						—					
						—					
						—					
						—					
						—					

WATER TECHNOLOGY, INC., BURLINGTON, ONTARIO

Method for the Determination of Alkalinity in Water

Part 1

4. GLASSWARE PREPARATION

4.1 All glassware used in this procedure is to be cleaned with a detergent, rinsed thoroughly and drained before use. To ensure that the glassware is free of analyte, the glassware should be rinsed with deionized water prior to use.

5. CONTAMINATION CONTROL

5.1 To prevent contamination of the standard solution and any cross-contamination between subsequent samples, the electrodes should be rinsed thoroughly with deionized water and blotted dry after removal from any solution.

6. INTERFERENCES

6.1 Soaps, oily matter, suspended solids, or precipitates may coat the glass electrode and cause a sluggish response. Allow additional time between titrant additions to let electrode come to equilibrium or clean the electrodes occasionally. Do not filter, dilute concentrate, or alter sample.

7. REAGENTS, WATER AND STANDARDS

7.1 Sodium carbonate solution, approximately 0.02N: Dry 2 to 3 g primary standard Na_2CO_3 at 250°C for 4 hours and cool in a desiccator. Weigh 1.0 ± 0.2 g (to the nearest mg), transfer to a 1-L volumetric flask, fill flask to the mark with distilled water, and dissolve and mix reagent. Do not keep longer than 1 week.

7.2 Standard sulfuric acid, 0.1N: Dilute 2.8 mL concentrated sulfuric acid to 1000 mL with deionized water.

7.3 Standard sulfuric acid, 0.02N: Dilute 200 mL 0.1000N standard acid to 1000 mL with deionized water.

7.4 Whatman buffer solutions at pH 4, 7, and 10.

8. APPARATUS

8.1 Orion Research Model 701A Digital Ionalyzer
8.2 Borosilicate glass buret, 50 mL
8.3 Magnetic stirrer and stirbars
8.4 Volumetric pipets
8.5 Polyethylene or glass beakers

9. SAMPLING AND SAMPLE SIZE

9.1 No sample preservation is required.
9.2 A minimum 100 mL of sample is required for alkalinity measurement.
9.3 Samples may be submitted in either plastic or glass bottle.
9.4 Samples must be stored at 4°C and analyzed within 4 days of submission.

10. SAMPLE PREPARATION/EXTRACTION PROCEDURE

10.1 No special preparation is required.
10.2 Warm chilled samples to room temperature prior to alkalinity measurement.

11. INSTRUMENT OPERATION/MEASUREMENT PROCEDURE

11.1 Rinse electrodes with deionized water and blot dry with tissue.
11.2 Measure room temperature and adjust temperature dial on meter.
11.3 Adjust dial on meter to measure pH.

12. CALIBRATION PROCEDURE

12.1 Calibrate meter by submersing electrode in three separate buffer solutions of pH 4, 7, and 10 while solutions are being mechanically stirred. Adjust pH meter to read the appropriate pH. Electrodes must be rinsed with deionized water and blotted dry with tissue between each analysis of buffer solution to prevent any cross-contamination of the solutions.
12.2 Standardize sulfuric acid solution by potentiometric titration of x mL of 0.02N Na_2CO_3 to pH 4.5 on a weekly basis. Determine normality of acid by the use of the following equation:

$$\text{Normality of acid (N)} = (A * B) / (53 * C)$$

where A = g Na_2CO_3 weighed into 1-L flask
B = mL Na_2CO_3 solution taken for titration
C = mL acid used

12.3 Check daily alkalinity of standard Na_2CO_3 solution by titrating 10 mL of the solution with standardized sulfuric acid solution. Calculate alkalinity of Na_2CO_3 by use of the following equation:

$$\text{Alkalinity, mg } CaCO_3/L = (A * N * 50000)/\text{volume of sample (mL)}$$

where A = mL standard acid used
N = normality of standard acid

Calculate percent difference from true value as follows:

$$\% \text{ Difference} = \text{measured alkalinity} * 1000/1060 \text{ mg/L}$$

13. SAMPLE ANALYSIS PROCEDURE

13.1 Determine the appropriate end point pH. Prepare sample and titration assembly. Titrate to the end points pH. As the end point is approached make smaller additions of acid and be sure that pH equilibrium is reached before adding more titrant.
13.2 Quality control samples with their LIMS codes include:

13.2.1 Blanks — analysis of deionized water to ensure that there is no contamination due to laboratory procedure. (METH_BLANK)

13.2.2 Duplicates — a replicate analysis of an homogeneous sample to show method precision. (DUPLICATE)

13.2.3 References — a certified reference material or a sample with known alkalinity to show method accuracy. (REFERENCE)

13.2.4 Spikes— a replicate sample spiked with a known amount of standard solution to show both method precision and accuracy as well as detecting interferences. (SPIKE)

14. RUN FORMAT/CONTROL LIMITS

14.1 For each batch of 16 or less samples, a blank, duplicate, reference, and standard is analyzed.
14.2 After every ten samples analyzed, the pH meter calibration is verified by checking the pH of the three buffer solutions.
14.3 Control limits consist of upper and lower warning and control limits. If a warning limit is reached, you must ensure that the system is stable and is not going out of control. If a control limit is exceeded, you must abandon the run, troubleshoot the system to find and correct the problem, and carry out the analytical run again.

	UWL	LWL	UCL	LCL
DUP	2.36%	0%	3.21%	0%
REF	106%	89.6%	110%	85.6%
SPIKE	100%	99.6%	101%	99.4%

PART 2 — SCIENTIFIC BACKGROUND FOR A METHOD

1. DEFINITIONS

Alkalinity — a measure of the acid-neutralizing capacity of water. It is the sum of all titratable bases. Alkalinity is a measure of an aggregate property of water and can be interpreted in terms of specific substances only when the chemical composition of the sample is known.

2. PARAMETERS MEASURED

2.1 Sample data are reported in mg $CaCO_3$/L.

3. EQUIPMENT, REAGENTS

See Part 1 — Sections 7 and 8.

4. POLLUTION PREVENTION PROCEDURES

No hazardous chemical is available as the result of this analysis.

5. METHOD PERFORMANCE SPECIFICATIONS

5.1 The MDL for alkalinity is 3 mg/L. This is based on an average METH_BLANK of 1.22 mg/L and eight replicates of 20 mg/L standard which give the results of:

20 mg/L
20
21
20
20
21
20
20 for a standard deviation of 0.4629. MDL = 1.22 + 3 * 0.46929 = 2.6 rounded 3.0 mg/L.

5.2 Other performance specs included:

REF recovery = 97.6% S = 4.0
SPIKE recovery = 100% S = 0.233
DUP.RPD = 0.65% S = 0.853

5.3 This lab is certified for alkalinity test by CAEAL.

6. SAMPLE COLLECTION, PRESERVATION, STORAGE, AND HANDLING

6.1 A minimum 100 mL of sample is required for alkalinity measurement.
6.2 Samples may be submitted in either plastic or glass bottle.

7. SAMPLE PRETREATMENTS

7.1 No sample preparation is necessary.

8. CALCULATIONS/FLAGS/DIGITS

8.1 Alkalinity, mg $CaCO_3$/L = $A * N *$ 50000/volume of sample (mL)

where A = mL standard acid used
N = normality of standard acid

8.2 Data are generally reported to three significant digits, but with a maximum of two digits after the decimal point. Typical reported data would look like 0.33, 1.33, 5.33, 15.3, 103, 1030, etc.
8.3 Data below the MDL are reported using the "t and w" system, where "t" is a measurable result below the calculated MDL and "w" represents no instrument response to the sample. Typical data reported below the MDL would look like 0.05t, or w if the MDL is 0.10 mg/L.

9. QUALITY VALIDATION

9.1 Quality validation requires that the QC_TYPE samples (Section 13.2) are completed and that the results of those samples be reasonable and fall within acceptable ranges.
9.2 The MDL must be reasonable compared to sample size and dilutions and data below MDL should have appropriate flags.
9.3 Data should be examined for consistency with historical customer data if possible.

10. CORRECTIVE ACTIONS/INTERVENTIONS

10.1 Upon sample arrival, it is the responsibility of the analyst to ensure that any special instructions given by the customer regarding sample preparation and/or analysis are clearly understood and performed. If there is any doubt as to sample handling or the method of sample preparation the analyst is to seek advise from his/her supervisor or contact the client personally for clarification.
10.2 If duplicate samples submitted by the client do not appear similar (i.e., color or suspended solid content appear different), the analyst performing the analysis shall make a record in his/her lab notebook and the observation is to be noted in the sample comment section of the final report.
10.3 If the method blank contains a concentration above the method detection limit for the analyte but is less than 5% of the average concentrations of that analyte in the samples, the method blank shall be reported as analyzed. If the blank concentrations contribute to more than 5% of the concentration

of the analyte, the blank must be reanalyzed. If it falls below this acceptable level the second time analyzed the results may be reported. If the high blank reoccurs, it will be necessary to investigate further the source of the high blank by checking the glassware cleaning and rinsing procedure, reagent chemical purity, etc. If the source of the contamination can be found and rectified within a relatively short period of time, it may be necessary to obtain more sample and reanalyze the entire batch of samples affected. If additional sample is unavailable the original set of data and its corresponding high blank shall be recorded and flagged in the sample comment section of the data report.

10.4 Upon analysis, if the duplicate results of any batch falls outside of the acceptable control limits, as defined by previous analyses of similar or like samples, the duplicate samples shall be rerun. If the precision falls within acceptable limits the second time, these data shall be reported to the client in the final report. If a duplicate result from the second set and one of the results of the first set fall within acceptable precision limits, they shall be reported as duplicates in the final report. If the duplicate precision remains out of the control limits, both of the original sets of duplicate data shall be reported to the client and the data flagged in the sample comment section of the final report. When there is insufficient sample for reanalysis and it is feasible to do so, contact the client for additional sample. If there is no sample left, the original data and its duplicate shall be reported and the data flagged in the sample comment section of the final report. If the duplicate precision is poor but the absolute concentration of the analytes fall close to the method detection limit of the analyte in question, the results may be reported and an explanation given in the sample comment section of the final report.

10.5 Upon analysis, if the spike recoveries fall outside of acceptable control limits as defined by previous analyses of similar or like samples, the sample shall be rerun. If the spike recoveries fall within acceptable limits for the second time, these data shall be reported to the client in the final report. If the spike recoveries remain out of control after rerunning the samples, additional sample must be obtained and the sample and its spike reanalyzed. If the second set of sample and spiked sample fall within spike recovery limits, they shall be reported in the final report. If the spike recovery remains out of the control limits, while the duplicate precision of the original unspiked sample is acceptable, the original sample shall be reported and the source of the poor spike recovery shall be investigated. When there is insufficient sample and it is feasible to do so, contact the client for additional sample. If there is no sample left, the original data and its spiked sample recoveries shall be reported and the data flagged in the sample comment section of the final report.

10.6 If an external reference material falls outside of acceptable control limits as defined by previous analyses of the same material, the sample shall be rerun. If the recovery of the reference standard falls within acceptable limits the second time, these data shall be recorded. If the reference recovery remains out of control after rerunning of the samples, additional sample must be obtained and the sample reanalyzed. If the sample falls within acceptable accuracy limits, it shall be reported. If the reference recovery remains out of the control limits, the entire data set must be rejected.

11. PROCEDURE FOR LEGAL SAMPLES

11.1 Legal samples have a Chain of Custody form which must be signed by the person who has the samples at any time. Legal samples must be in a locked fridge or fume hood at all times when they are not attended by the person with custody of the samples. After analysis the samples should be signed back to the customer with the chain of custody form. It is important that all lab notes, books, output, QC samples, and anything else related to the legal case be legible. defensible, and available for the customers when they ask for it.

12. DISPOSAL OF SAMPLES

12.1 Liquid samples are disposed of down the drain with copious amounts of water.

13. REFERENCES

13.1 Standard Methods for the Examination of Water and Wastewater, 18th ed., Arnold E. Greenberg, Lenore S. Clesceri, and Andrew D. Eaton, Eds., 1992.

Alkalinity Relationships

Result of Titration	Hydroxide Alkalinity as $CaCO_3$	Carbonate Alkalinity as $CaCO_3$	Bicarbonate Concentration as $CaCO_3$
P = 0	0	0	T
P < 1/2T	0	2P	T – 2P
P = 1/2T	0	2P	0
P > 1/2T	2P – T	2(T – P)	0
P = T	T	0	0

METHOD FOR THE DETERMINATION OF AMMONIA IN WATER

PART 1 — STANDARD OPERATING PROCEDURE

1. SCOPE AND APPLICATION

1.1 Ammonia is present naturally in surface and wastewaters. Its concentration generally is low in groundwaters because it adsorbs to soil particles and clays and is not leached readily from soils. It is produced largely by deamination of organic nitrogen-containing compounds and by hydrolysis of urea. At some water treatment plants ammonia is added to react with chlorine to form a combined chlorine residual.

1.2 In the chlorination of wastewater effluents containing ammonia, virtually no free residual chlorine is obtained until the ammonia has been oxidized. Rather, the chlorine reacts with ammonia to form mono- and dichloramines. Ammonia concentrations encountered in water vary from less than 10 mg/L ammonia nitrogen in some natural surface and groundwaters to more than 30 mg/L in some wastewaters.

2. SUMMARY/PRINCIPLE OF METHOD

2.1 This automated procedure for the determination of ammonia utilizes the Berthelot reaction, in which the formation of a blue-colored compound believed to be closely related to indophenol occurs when the solution of an ammonium salt is added to sodium phenoxide, followed by the addition of sodium hypochlorite. A solution of EDTA is added to the sample stream to eliminate the precipitation of the hydroxides of calcium and magnesium. Sodium nitroprusside is added to intensify the blue color.

2.2 Ammonia concentration is measured colorimetrically by means of the Technicon (TRAACS 800) analyzer containing a 660-nm filter.

3. SAFETY CONSIDERATIONS

3.1 This method does not address all safety issues associated with its use. A reference file of material safety data sheets (MSDSs) is available in Central Registry.

4. GLASSWARE PREPARATION

4.1 All glassware used in this procedure is to be cleaned with hot water, rinsed thoroughly, and drained before use.

4.2 The Technicon analyzer is cleaned by pumping it through with a solution of 5N H_2SO_4.
4.3 Sample cups for the analyzer should be rinsed with 10% sulfuric acid, followed by deionized water and finally by an aliquot of the sample itself.

5. CONTAMINATION CONTROL

5.1 Brij-35 contains phosphorus as an impurity. Cleanse system well if phosphorus is to be measured afterward.

6. INTERFERENCES

6.1 Sample turbidity may interfere with NH_3 analysis, therefore all samples should be filtered prior to analysis.

7. REAGENTS, WATER, AND STANDARDS

7.1 Unless otherwise specified, all chemicals should be of ACS grade or equivalent. Milli-Q or distilled water can be used.
7.2 Alkaline Phenol: Add 83 g liquified phenol (approx. 90% (C_6H_5OH)) to about 800 mL of deionized water. While cooling under tap water or in an ice bath, slowly with swirling add 96.0 g of sodium hydroxide (50% w/w solution (NaOH)). Cool to room temperature, dilute to 1 L with deionized water and mix thoroughly. Store in an amber glass container. CAUTION: This material is corrosive. Stability: 2 weeks.
7.3 Sodium Hypochlorite Solution: Dilute 86 mL of sodium hypochlorite (5% solution (NaOCl)) commercial grade) to 100 mL with distilled water and mix thoroughly. Commercially available bleach is 5.25% active. If used, only 82 mL is needed. Dilute to 100 mL with deionized water and mix thoroughly. Prepare fresh weekly.
7.4 Sodium Nitroprusside Solution: Dissolve 1.1 g of sodium nitroprusside [$Na_2Fe(CN)_5NO_2H_2O$] in about 600 mL of deionized water. Dilute to 1 L with deionized water and mix thoroughly. Store in an amber container. Stability: 1 month.
7.5 Disodium EDTA: Dissolve approximately 1.0 g of 50% w/w sodium hydroxide and 41.0 g of disodium EDTA in about 800 mL of deionized water. Dilute to 1 L. Add 3 mL of Brij-35 and mix well. (Brij-35 is a Registered Trademark of Atlas Chemical Industries, Inc.)
7.6 Stock Standard A, 100 mg/L N: In a 1-L volumetric flask containing about 800 mL of deionized water dissolve 1 mL of chloroform. Add 0.4717 g of ammonium sulfate ($(NH_4)_2SO_4$) and swirl to dissolve. Dilute to 1 L with distilled water and mix thoroughly.
7.7 Stock Standard B, 20 mg/L N: Dilute 20.0 mL of stock standard A to 100 mL with deionized water and mix thoroughly. Prepare fresh daily.
7.8 Working Standard Solutions: To prepare working standard solutions of 0.6, 1.8, and 3.0 mg/L N, transfer aliquots of 3.0, 9.0, and 15.0 mL of Stock Standard B, respectively, into individual 100-mL volumetric flasks and dilute to volume with deionized water and mix thoroughly. Working standard solutions must be prepared fresh daily.

8. APPARATUS

8.1 Volumetric flasks; 100, 1000 mL
8.2 TRAACS 800 Continuous Flow Analytical System

9. SAMPLING AND SAMPLE SIZE

9.1 A minimum of 50 mL of sample is required for NH_3 analysis.

9.2 Preserve sample by the addition of 2 mL/L concentrated sulfuric acid and store at 4°C.
9.3 When ammonia is done with NO_2, NO_3, in 1–2 days, addition of acid is not necessary.

10. SAMPLE PREPARATION

10.1 Sample turbidity should be removed by filtration prior to analysis.
10.2 No other sample preparation required.

11. INSTRUMENT OPERATION/MEASUREMENT PROCEDURE

11.1 Turn on computer, monitor, printer, air, autosampler, and the Traacs consoles (main and slave consoles).
11.2 Set up an edit file for an instrument run.
11.3 Set up samples and standards in tray on analyzer as set out in edit file.
11.4 Follow directions for reagent setup as described in Appendix 1.
11.5 Follow directions for base and gain adjustment as described in Appendix 1.

12. CALIBRATION PROCEDURE

12.1 A three-level calibration curve is created by the results of the analysis of three standards of 0.6, 1.8, and 3.0 ppm run at the beginning of an automated sequence. A full set of calibration standards is rerun at the end of every sequence to ensure instrument stability.

13. SAMPLE ANALYSIS PROCEDURE

13.1 Samples are transferred to sample cups which have been previously acid washed and dried. Sample dilution may be necessary if concentration of NH_3 is found to be outside of calibrated range. Dilutions of samples are made with deionized water.
13.2 Quality control samples include:
- 13.2.1 Blanks — analysis of deionized water to ensure that there is no contamination due to laboratory procedure.
- 13.2.2 Duplicates — a replicate analysis of an homogeneous sample to show method precision.
- 13.2.3 Spikes — a replicate sample spiked with a known amount of stock standard solution to show both method precision and accuracy as well as check for interferences.
- 13.2.4 References — an EPA-certified reference material to show method accuracy.

14. RUN FORMAT/CONTROL LIMITS

14.1 A typical sample run is as follows:

P	3.0 mg/L	NH_3 as N	Calibration
C	0.6		
C	1.8		
C	3.0		
H	3.0		
L	0.6		
L	0.6		
	METH_BLK		
	METH_BLK		

METH_BLK
REF
REF
Sample
Sample
Sample
Sample
DUP
SPIKE
Sample
Sample
Sample

This continues with 20% of total number of analysis being QC samples.

Sample
Sample
Sample
CHECK_STD
REF
METH_BLK
METH_BLK
H 3.0
L 0.6
L 0.6
G 1.8
END

14.2 Control limits consist of upper and lower warning and control limits. If a warning limit is reached, you must ensure that the system is stable and is not going out of control. If a control limit is exceeded, you must abandon the run, troubleshoot the system to find and correct the problem, and carry out the analytical run again.

	UWL	LWL	UCL	LCL	
DUP	11.7%	0	15.7%	0	(RPD)
REF	112%	87.8%	119%	81.6%	
SPIKE	105%	84.8%	110%	79.7%	

PART 2 — SCIENTIFIC BACKGROUND FOR A METHOD

1. DEFINITIONS

2. PARAMETERS MEASURED

2.1 Sample data are reported in mg/L NH_3 as N.

3. EQUIPMENT, REAGENTS, SUPPLIES SPECIFICATIONS

3.1 See Part 1, Sections 7 and 8.
3.2 Suggested suppliers: TRAACS

Pulse Instrumentation Ltd.
433 Birch Crescent
Saskatoon, Sask
S7N 2K2
306-249-3000 (Phone)
306-249-3082 (Fax)

Folio Instruments Inc.
262 Manitou Drive, Unit 3
Kitchener, Ontario
N2C 1L3
519-748-4612 (Phone)
519-748-1535 (Fax)

3.3 Suggested suppliers: Chemicals

Caledon Laboratories: certified reagent grade chemicals
Mallinckrodt:
J.T. Baker Chemical: Baker Grade chemicals

4. POLLUTION PREVENTION PROCEDURES

The wastes containing phenols are collected and disposed of through the system operated by CCIW stores. Unused portion of samples are returned to the customer.

5. METHOD PERFORMANCE SPECIFICATIONS

5.1 The MDL for NH_3 is 0.1 mg/L. This is based on an average METH_BLK of 0.05 mg/L and eight replicates of 0.5 mg/L standard which gave results of

0.49 mg/L
0.50
0.49
0.49
0.49
0.49
0.50
0.49 for a standard deviation of 0.0046 mg/L. MDL = 0.05 + 3 * 0.0046 = 0.064 rounded to 0.1 mg/L.

5.2 Other performance specs include:

Duplicate RPD = 3.63% S = 3.86
REF recovery = 100% S = 5.69
SPIKE recovery = 94.8% S = 4.89

6. SAMPLE COLLECTION, PRESERVATION, STORAGE AND HANDLING

6.1 Samples are preserved by the addition of 2 mL/L concentrated sulfuric acid and stored at 4°C.

7. SAMPLE PRETREATMENTS

7.1 No sample preparation is required.

8. CALCULATIONS/FLAGS/DIGITS

8.1 Prepare standard curves using the instrument software by plotting peak heights of standard processed through the manifold against NH_3 concentration in standards. Compute sample ammonia concentration by comparing sample peak height with standard curve. This is carried out by TRAACS computer.
8.2 Data are generally reported to three significant digits, but with a maximum of two digits after the decimal point. Typical reported data would look like 0.33, 1.33, 5.33, 15.3, 103, 1030, etc.
8.3 Data below the MDL are reported using the "t and w" system where "t" is a measurable result below the calculated MDL and "w" represents no instrument response to the sample. Typical data reported below the MDL would look like 0.05t, or w if the MDL is 0.10 mg/L.
8.4 The reported MDL on data reports must be corrected for dilution of the sample and sample size. The MDL in LIMS is for a typical sample, not necessarily for each actual sample analyzed.

9. QUALITY VALIDATION

9.1 Quality validation requires that the QC_TYPE samples (Section 13.2) are completed and that the results of those samples be reasonable and fall within acceptable range.
9.2 All calibration data must be valid and properly documented.
9.3 The MDL must be reasonable compared to sample size and dilutions and data below MDL should have appropriate flags.
9.4 Data should be examined for consistency with historical customer data if possible.
9.5 If any high samples were encountered, carry over to the next sample needed to be examined.

10. CORRECTIVE ACTIONS/INTERVENTIONS

10.1 Upon sample arrival, it is the responsibility of the analyst to ensure that any special instructions given by the customer regarding sample preparation and/or analysis are clearly understood and performed. If there is any doubt as to sample handling or the method of sample preparation the analyst is to seek advise from his/her supervisor or contact the client personally for clarification.
10.2 If duplicate samples submitted by the client do not appear similar (i.e., color or suspended solid content appear different), the analyst performing the analysis shall make a record in his/her lab notebook and the observation is to be noted in the sample comment section of the final report.
10.3 If the method blank contains a concentration above the method detection limit for the analyte but is less than 5% of the average concentrations of that analyte in the samples, the method blank shall be reported as analyzed. If the blank concentrations contribute to more than 5% of the concentration of the analyte, the blank must be reanalyzed. If it falls below this acceptable level the second time analyzed the results may be reported. If the high blank reoccurs, it will be necessary to investigate further the source of the high blank by checking the glassware cleaning and rinsing procedure, reagent chemical purity, etc. If the source of the contamination can be found and rectified within a relatively short period of time, it may be necessary to obtain more sample and reanalyze the entire batch of samples affected. If additional sample is unavailable the original set of data and its corresponding high blank shall be recorded and flagged in the sample comment section of the data report.

10.4 Upon analysis, if the duplicate results of any batch falls outside of the acceptable control limits, as defined by previous analyses of similar or like samples, the duplicate samples shall be rerun. If the precision falls within acceptable limits the second time, these data shall be reported to the client in the final report. If a duplicate result from the second set and one of the results of the first set fall within acceptable precision limits, they shall be reported as duplicates in the final report. If the duplicate precision remains out of the control limits, both of the original sets of duplicate data shall be reported to the client and the data flagged in the sample comment section of the final report. When there is insufficient sample for reanalysis and it is feasible to do so, contact the client for additional sample. If there is no sample left, the original data and its duplicate shall be reported and the data flagged in the sample comment section of the final report. If the duplicate precision is poor but the absolute concentration of the analytes fall close to the method detection limit of the analyte in question, the results may be reported and an explanation given in the sample comment section of the final report.

10.5 Upon analysis, if the spike recoveries fall outside of acceptable control limits as defined by previous analyses of similar or like samples, the sample shall be rerun on the instrument. If the spike recoveries fall within acceptable limits for the second time, these data shall be reported to the client in the final report. If the spike recoveries remain out of control after rerunning the samples, additional sample must be obtained and the sample and its spike reanalyzed. If the second set of sample and spiked sample fall within spike recovery limits, they shall be reported in the final report. If the spike recovery remains out of the control limits, while the duplicate precision of the original unspiked sample is acceptable, the original sample shall be reported and the source of the poor spike recovery shall be investigated. When there is insufficient sample and it is feasible to do so, contact the client for additional sample. If there is no sample left, the original data and its spiked sample recoveries shall be reported and the data flagged in the sample comment section of the final report.

10.6 If an external reference material falls outside of acceptable control limits as defined by previous analyses of the same material, the sample shall be rerun on the instrument. If the recovery of the reference standard falls within acceptable limits the second time, these data shall be recorded. If the reference recovery remains out of control after rerunning of the samples, additional sample must be obtained and the sample reanalyzed. If the sample falls within acceptable accuracy limits, it shall be reported. If the reference recovery remains out of the control limits, the entire data set must be rejected.

11. PROCEDURE FOR LEGAL SAMPLES

11.1 Legal samples have a Chain of Custody form which must be signed by the person who has the samples at any time. Legal samples must be in a locked fridge or fume hood at all times when they are not attended by the person with custody of the samples. After analysis the samples should be signed back to the customer with the chain of custody form. It is important that all lab notes, books, output, QC samples, and anything else related to the legal case be legible, defensible, and available for the customers when they ask for it.

12. DISPOSAL OF SAMPLES

12.1 Waste containing phenols is collected and disposed of through CCIW stores.

13. REFERENCES

13.1 *Standard Methods for the Examination of Water and Wastewater*, 18th ed., Arnold E. Greenberg, Lenore S. Clesceri, and Andrew D. Eaton, Eds., 1992.

13.2 Technicon TRAACS 800 Method Industrial Manual No. 787-86T, Feb. 1986.

METHOD FOR THE DETERMINATION OF ANIONS IN WATER

PART 1 — STANDARD OPERATING PROCEDURE

1. SCOPE AND APPLICATION

1.1 Determination of the common anions such as bromide, chloride, fluoride, nitrate, nitrite, phosphate, sulfate, and thiosulfate often is desirable to characterize a water and/or to assess the need for specific treatment. Ion chromatography provides a single instrumental technique that may be used for their rapid, sequential measurement. It eliminates the need to use hazardous reagents and it effectively distinguishes among the halides (Br^-, Cl^-, and F^-) and the oxy-ions (SO_3^{2-}, SO_4^{2-}, or NO_2^-, NO_3^-).

1.2 This method is applicable, after filtration to remove particles larger than 2.0 μm, to surface, ground-, and wastewaters as well as drinking water. Some industrial process waters, such as boiler water and cooling water, also may be analyzed by this method.

2. SUMMARY/PRINCIPLE OF METHOD

2.1 A small portion of a filtered, homogeneous, aqueous sample or a sample containing no particles larger than 0.45 μm is injected into an ion chromatograph. The sample merges with the eluent stream and is pumped through the ion chromatographic system. Anions are separated on the basis of their affinity for the active sites of the column packing material. Conductivity detector readings (either peak area or peak height) are used to compute concentrations.

3. SAFETY CONSIDERATIONS

3.1 This method does not address all safety issues associated with its use. A reference file of material safety data sheets (MSDSs) is available in Central Registry.

4. GLASSWARE PREPARATION

4.1 All glassware used in this procedure is to be cleaned with a phosphate-free detergent, rinsed thoroughly, and drained before use.

5. CONTAMINATION CONTROL

5.1 Because method sensitivity is high, avoid contamination by reagent water and equipment. Determine any background or interference due to the matrix when adding the QC sample into any matrix other than reagent water.

6. INTERFERENCES

6.1 Any two species that have similar retention times can be considered to interfere with each other. This method has potential coelution interference between short-chain acids and fluoride and chloride. Solid-phase extraction cartridges can be used to retain organic acids and pass inorganic anions. The interference-free solution then can be introduced into the ion chromatograph for separation.

6.2 A high concentration of any one ion also interferes with the resolution, and sometimes retention, of others. Sample dilution overcomes many interferences.

6.3 Chlorate and bromide coelute under the specified conditions. Determine whether other anions in the sample coelute with the anions of interest.

6.4 Best separation is achieved with sample pH between 5 and 9. When samples are injected the eluent pH will seldom change unless the sample pH is very low. Raise sample pH by adding a small amount of hydroxide salt to enable the eluent to control pH.
6.5 When analyzing for anions in drinking water and other environmental samples, the presence of high levels of carbonate, interfering metals (such as magnesium and calcium), and excess base are often encountered. In an untreated sample, these may cause disturbances in the chromatographic baseline. The presence of high levels of carbonate will result in large carbonate peak which may interfere with the quantitation of chloride. In extreme cases carbonate may overload the column and cause loss of early eluting analytes. Alkaline earth metals, such as magnesium and calcium, may cause severe interferences in the early part of the chromatogram.
6.6 When using a borate/gluconate eluent, high pH samples may result in a pH system peak in the middle portion of the chromatogram, which can interfere with chlorate, bromide, and nitrate analysis. In many of these samples, the analyte levels are too low for dilution of the sample to be practical. Dilution would cause analyte levels to drop below the detection limits for direct injection.

7. REAGENTS, WATER, AND STANDARDS

7.1 Reagent water: Deionized water of 18 megohom-cm resistivity containing no particles larger than 0.20 μm.
7.2 Borate/gluconate concentrate: To a 1-L volumetric flask add 16 g sodium gluconate, 18 g boric acid, and 25 g sodium tetraborate decahydrate. Add approximately 500 mL Milli-Q (deionized) water and mix thoroughly until dissolved, then add 250 mL glycerin. Fill the flask to the 1-L mark with Milli-Q water and mix thoroughly. Concentrate may be stored refrigerated for up to 6 months before replacement.
7.3 Borate/gluconate eluent (pH 8.5, conductivity 270 ms): Place approximately 500 mL of Milli-Q water into a 1-L volumetric flask and add 20 mL of borate/gluconate concentrate, 20 mL n-butanol, and 120 mL acetonitrile. Fill the flask to the mark with Milli-Q water and mix thoroughly. Filter through a 0.22-μm Durapore membrane (GVWP).
7.4 Concentrated Standards: Prepare separate 1000 ppm standard concentrates by diluting each of the following salts, dried to a constant weight at 105°C, to 100 mL with Milli-Q deionized water. Store in plastic bottles in a refrigerator; these solutions are stable for at least 1 month. Verify stability.

Cl^-	0.165 g NaCl (ACS) ± 0.001 g
NO_2^-	0.150 g $NaNO_2^-$ (ACS) ± 0.001 g
ClO_3^-	0.108 g $NaClO_3^-$ (ACS) ± 0.001 g
Br^-	0.149 g KBr (ACS) ± 0.001 g
NO_3^-	0.137 g $NaNO_3$ (ACS) ± 0.001 g
SO_4^{2-}	0.148 g Na_2SO_4 (ACS) ± 0.001 g
PO_4^{3-}	0.223 g K_3PO_4 (ACS) ± 0.001 g
$S_2O_3^{2-}$	0.141 g $Na_2S_2O_3$ (ACS) ± 0.001 g

7.5 Working Standard (10 ppm): Pipet 1 mL of each standard concentrated into a 100-mL volumetric flask and dilute with Milli-Q deionized water. Prepare a fresh working standard weekly.

8. APPARATUS

8.1 Waters Ion Chromatographic system, equipped with a Waters Series 590 Pump, a WISP U6K Rheodyne Sample Injector, an IC-Pak A HR Column, a 430 Conductivity Detector, and a Spectra-Physics SP4290 integrator.
8.2 Volumetric flasks; 1000 mL, 100 mL

9. SAMPLING AND SAMPLE SIZE

9.1 A minimum of 50 mL of sample is required for anion analysis.

10. SAMPLE PREPARATION/EXTRACTION PROCEDURE

10.1 MilliTrap™ H^+ Membrane Cartridges are used in pretreatment of the sample prior to analysis. The MilliTrap™ H^+ cartridge is a hand-held, multiple-use, disposable device that removes cation and carbonate interferences from many matrices such as drinking, surface, or wastewater samples.

11. INSTRUMENT OPERATION/MEASUREMENT PROCEDURE

11.1 System equilibration: Set up ion chromatograph in accordance with manufacturer's directions. Install guard and separator columns and begin pumping eluent until a stable base line is achieved. The background conductivity of the eluent solution is 270 µs ± 10%.

11.2 Determine retention time for each anion by injecting a standard solution containing only the anion of interest and noting the time required for a peak to appear. Retention times vary with operating conditions and with anion concentration.

12. CALIBRATION PROCEDURE

12.1 When the standards are at room temperature, filter them through a 0.45-µm cartridge filter, place about 2 mL of the standard solution in a plastic cup, and place it on the tray of the autosampler. Then adjust the injection volume to 50 µL. Push the run button and record the peak heights of the standard components on the integrator. Calibrate the instrument after feeding the appropriate data to the integrator as described in the operation manual.

13. SAMPLE ANALYSIS PROCEDURE

13.1 After warming up to room temperature and filtration through a 0.45-µm cartridge filter, place about 2 mL of the sample in a plastic cup and place it on the tray of the autosampler (keep the injection volume 50 µL as the calibration standard). Push the run button and record the peaks of the sample components on the integration. At the end of the run you will obtain a printout of data of the unknown sample.

13.2 Quality control samples include:

13.2.1 Blanks — analysis of deionized water to ensure that there is no contamination due to laboratory procedure.

13.2.2 Duplicates — a replicate analysis of an homogeneous sample to show method precision.

13.2.3 Spikes — a replicate sample spiked with a known amount of stock standard solution to show both method precision and accuracy and check for any interferences.

13.2.4 References — an EPA-certified reference material to show method accuracy.

14. RUN FORMAT/CONTROL LIMITS

14.1 A typical sample run is as follows:

calibration standard
check standard
blank
samples

----
duplicate
reference
spike

14.2 Control limits consist of upper and lower warning and control limits. If a warning limit is reached, you must ensure that the system is stable and is not going out of control. If a control limit is exceeded, you must abandon the run, troubleshoot the system to find and correct the problem, and carry out the analytical run again.

		UWL	LWL	UCL	LCL
	Dup	6.32%	0%	8.38%	0%
Cl	Ref	112%	91.3%	117%	86.1%
	Spike	107%	94.9%	110%	91.9%
	Dup	15.4%	0%	20.7%	0%
Br	Ref	111%	91.7%	116%	86.8%
	Spike	112%	93.2%	117%	88.5%
	Dup	9.62%	0%	12.7%	0%
NO_2	Ref	108%	95.0%	111%	91.7%
	Spike	118%	89.1%	125%	81.8%
	Dup	14.2%	0%	18.7%	0%
NO_3	Ref	118%	86.8%	125%	79.1%
	Spike	115%	85.5%	122%	78.1%
	Dup	15.2%	0%	20.5%	0%
ClO_3	Ref	105%	94.7%	108%	92.1%
	Spike	115%	79.7%	124%	70.8%
	Dup	14.2%	0%	18.2%	0%
PO_4	Ref	111%	88.2%	117%	82.4%
	Spike	112%	81.9%	120%	74.3%
	Dup	14.7%	0%	19.8%	0%
SO_4	Ref	115%	84.4%	123%	76.6%
	Spike	108%	92.7%	111%	89.0%
	Dup	8.71%	0%	11.4%	0%
S_2O_3	Ref	106%	89.3%	110%	85.1%
	Spike	122%	87.1%	131%	78.4%

PART 2 — SCIENTIFIC BACKGROUND FOR A METHOD

1. DEFINITIONS

2. PARAMETERS MEASURED

2.1 The following is a list of the anions analyzed in this method.

Cl^- (chloride)	NO_2^- (nitrate)
ClO_3^- (chlorate)	Br^- (bromide)
NO_3^- (nitrate)	SO_4^{2-} (sulfate)
$PO_4$3– (phosphate)	$S_2O_3^{2-}$ (thiosulfate)

3. EQUIPMENT, REAGENTS, SUPPLIES SPECIFICATIONS

3.1 See Part 1 — Section 7 and 8.
3.2 Suggested supplier:

Water Ltd.
3687 Nashua Drive
Mississauga, ON L4V 1V5
Tel: 1-800-252-4752
Fax: 1-905-678-9237

4. POLLUTION PREVENTION PROCEDURES

4.1 There are no harmful chemicals from this analysis.

5. METHOD PERFORMANCE SPECIFICATIONS

5.1 The MDL of these anions are as follows:

Cl	0.5 mg/L
NO_2	0.5 mg/L
Br	1.0 mg/L
NO_3	1.5 mg/L
ClO_3	1.5 mg/L
SO_4	2.0 mg/L
PO_4	2.5 mg/L
S_2O_3	5.5 mg/L

5.2 Other performance specifications include:

Cl	Dup. RPD = 2.18%	S = 2.07%
	Ref. Recovery = 102%	S = 5.22%
	Spike Recovery = 101%	S = 2.99%
Br	Dup. RPD = 4.89%	S = 5.27%
	Ref. Recovery = 101%	S = 4.83%
	Spike Recovery = 103%	S = 4.71%
NO_2	Dup. RPD = 3.51%	S = 3.05%
	Ref. Recovery = 102%	S = 3.28%
	Spike Recovery = 104%	S = 7.27%
NO_3	Dup. RPD = 5.17%	S = 4.52%
	Ref. Recovery = 102%	S = 7.71%
	Spike Recovery = 100%	S = 7.35%
ClO_3	Dup. RPD = 4.71%	S = 5.26%
	Ref. Recovery = 99.9%	S = 2.62%
	Spike Recovery = 97.6%	S = 8.94%

PO_4	Dup. RPD = 6.08%	S = 4.05%
	Ref. Recovery = 99.8%	S = 5.78%
	Spike Recovery = 97.2%	S = 7.64%
SO_4	Dup. RPD = 4.34%	S = 5.17%
	Ref. Recovery = 99.9%	S = 7.76%
	Spike Recovery = 100%	S = 3.71%
S_2O_3	Dup. RPD = 3.32%	S = 2.70%
	Ref. Recovery = 97.7%	S = 4.20%
	Spike Recovery = 105%	S = 8.76%

6. SAMPLE COLLECTION, PRESERVATION, STORAGE, AND HANDLING

6.1 For short-term preservation for 1 to 2 days store at 4°C.

7. SAMPLE PRETREATMENTS

7.1 No sample pretreatment is necessary.

8. CALCULATIONS/FLAGS/DIGITS

8.1 Prepare standard curves using the instrument software by plotting peak heights of standards processed through the manifold against anion concentration in standards. Compute sample anion concentration by comparing sample peak height with standard curve.

8.2 Data are generally reported to three significant digits, but with a maximum of 2 digits after the decimal point. Typical reported data would look like 0.33, 1.33, 5.33, 15.3, 103, 1030, etc.

8.3 Data below the MDL are reported using the "t and w" system where "t" is a measurable result below the calculated MDL and "w" represents no instrument response to the sample. Typical data reported below the MDL would look like 0.05t, or w if the MDL is 0.10 mg/L.

8.4 The reported MDL on data reports must be corrected for dilution of the sample and sample size. The MDL in LIMS is for a typical sample, not necessarily for each actual sample analyzed.

9. QUALITY VALIDATION

9.1 Quality validation requires that the QC_TYPE samples (Section 13.2) are completed and that the results of those samples be reasonable and fall within acceptable ranges.

9.2 All calibration data must be valid and properly documented.

9.3 The MDL must be reasonable compared to sample size and dilutions and data below MDL should have appropriate flags.

9.4 Data should be examined for consistency with historical customer data if possible.

9.5 If any high samples are encountered, carry over to the next sample needed to be examined.

10. CORRECTIVE ACTIONS/INTERVENTION

10.1 Upon sample arrival, it is the responsibility of the analyst to ensure that any special instructions given by the customer regarding sample preparation and/or analysis are clearly understood and performed. If there is any doubt as to sample handling or the method of sample preparation the analyst is to seek advise from his/her supervisor or contact the client personally for clarification.

10.2 If duplicate samples submitted by the client do not appear similar (i.e., color or suspended solid content appear different), the analyst performing the analysis shall make a record in his/her lab notebook and the observation is to be noted in the sample comment section of the final report.
10.3 If the method blank contains a concentration above the method detection limit for the analyte but is less than 5% of the average concentrations of that analyte in the samples, the method blank shall be reported as analyzed. If the blank concentrations contribute to more than 5% of the concentration of the analyte, the blank must be reanalyzed. If it falls below this acceptable level the second time analyzed the results may be reported. If the high blank reoccurs, it will be necessary to investigate further the source of the high blank by checking the glassware cleaning and rinsing procedure, reagent chemical purity, etc. If the source of the contamination can be found and rectified within a relatively short period of time, it may be necessary to obtain more sample and reanalyze the entire batch of samples affected. If additional sample is unavailable the original set of data and its corresponding high blank shall be recorded and flagged in the sample comment section of the data report.
10.4 Upon analysis, if the duplicate results of any batch falls outside of the acceptable control limits, as defined by previous analyses of similar or like samples, the duplicate samples shall be rerun on the instrument. If the precision falls within acceptable limits the second time, these data shall be reported to the client in the final report. If a duplicate result from the second set and one of the results of the first set fall within acceptable precision limits, they shall be reported as duplicates in the final report. If the duplicate precision remains out of the control limits, both of the original sets of duplicate data shall be reported to the client and the data flagged in the sample comment section of the final report. When there is insufficient sample for reanalysis and it is feasible to do so, contact the client for additional sample. If there is no sample left, the original data and its duplicate shall be reported and the data flagged in the sample comment section of the final report. If the duplicate precision is poor but the absolute concentration of the analytes fall close to the method detection limit of the analyte in question, the results may be reported and an explanation given in the sample comment section of the final report.
10.5 Upon analysis, if the spike recoveries fall outside of acceptable control limits as defined by previous analyses of similar or like samples, the sample shall be rerun on the instrument. If the spike recoveries fall within acceptable limits for the second time, these data shall be reported to the client in the final report. If the spike recoveries remain out of control after rerunning the samples, additional sample must be obtained and the sample and its spike reanalyzed. If the second set of samples and spiked sample fall within spike recovery limits, they shall be reported in the final report. If the spike recovery remains out of the control limits, while the duplicate precision of the original unspiked sample is acceptable, the original sample shall be reported and the source of the poor spike recovery shall be investigated. When there is insufficient sample and it is feasible to do so, contact the client for additional sample. If there is no sample left, the original data and its spiked sample recoveries shall be reported and the data flagged in the sample comment section of the final report.
10.6 If external reference material falls outside of acceptable control limits as defined by previous analyses of the same material, the sample shall be rerun. If the recovery of the reference standard falls within acceptable limits the second time, these data shall be recorded. If the reference recovery remains out of control after rerunning of the samples, additional sample must be obtained and the sample reanalyzed. If the sample falls within acceptable accuracy limits, it shall be reported. If the reference recovery remains out of the control limits, the entire data set must be rejected.

11. PROCEDURE FOR LEGAL SAMPLES

11.1 Legal samples have a Chain of Custody form which must be signed by the person who has the samples at any time. Legal samples must be in a locked fridge or fume hood at all times when they are not attended by the person with custody of the samples. After analysis the samples should be signed back to the customer with the chain of custody form. It is important that all lab notes, books, output, QC samples, and anything else related to the legal case be legible, defensible, and available for the customers when they ask for it.

12. DISPOSAL OF SAMPLES/REAGENTS

12.1 Liquid samples are disposed of down the drain with copious amounts of water.

13. REFERENCES

13.1 *Standard Methods for the Examination of Water and Wastewater*, 19th ed., Arnold E. Greenberg, Lenore S. Clesceri, and Andrew D. Eaton, Eds., 1995.
13.2 Waters Ion Analysis Method A-119.

Determination of Metals (Total and Dissolved) in Water by Inductively Coupled Plasma (ICP)

PART 1 — STANDARD OPERATING PROCEDURE

1. SCOPE AND APPLICATION

1.1 The effects of metals in water and wastewater range from beneficial through troublesome to dangerously toxic. Some metals are essential, others may adversely affect water consumers, wastewater treatment systems, and receiving waters. Some metals may be either beneficial or toxic depending on concentration.
1.2 Inductively coupled plasma (ICP) techniques are applicable over a broad linear range and are especially sensitive for refractory elements.

2. SUMMARY/PRINCIPLE OF METHOD

2.1 Total Metals: A suitable aliquot of sample is digested by the addition of HCl, HNO_3, and H_2O_2 and concentrating down to a known volume. Digestion is necessary to break down organic matter and to convert metal associated with particulates in a sample to a dissolved form that can be determined by inductively coupled plasma spectroscopy (ICP). The sample is then diluted to an appropriate final volume.
2.2 Dissolved Metals: The sample is first filtered through a 0.45-mm filter and preserved to a pH < 2 with HNO_3. A suitable aliquot sample is then analyzed without further digestion.
2.3 The sample is pumped into the ICP, where a sample aerosol is generated in an appropriate nebulizer and spray chamber and is carried into the plasma. The plasma consists of a flowing stream of argon gas ionized by an applied radio frequency field. It is inductively coupled to the ionized gas by a water-cooled coil surrounding a quartz "torch" that supports and confines it.
2.4 The high temperature of the plasma excites atomic emission efficiently and the ionization of a high percentage of atoms produces ionic emission spectra. The efficient excitation provided by the ICP results in low detection limits for many elements. The light emitted from the ICP is focused onto the entrance slit of a monochromator that effects dispersion. A precisely aligned exit slit is used to isolate a portion of the emission spectrum for intensity measurement using a photomultiplier tube.

3. SAFETY CONSIDERATIONS

3.1 This method does not address all safety issues associated with its use. A reference file of material safety data sheets (MSDSs) for reagents used in this method is available in Central Registry.

3.2 CAUTION: Many metal salts are extremely toxic and may be fatal if swallowed. Wash hands thoroughly after handling.
3.3 The inorganic acids used for sample preparation are hazardous. Proper safety techniques should be observed and adequate ventilation used when handling these reagents.

4. GLASSWARE PREPARATION

4.1 All glassware used in this procedure is to be cleaned with 10% HNO_3, rinsed thoroughly with deionized water, and drained before use.

5. CONTAMINATION CONTROL

5.1 Avoid introducing contaminating metals from container, distilled water, or membrane filters. Some plastic caps or cap liners may introduce metal contamination; for example, zinc has been found in black Bakelite-type screw caps as well as in many rubber and plastic products, and cadmium has been found in plastic pipet tips. Lead is a ubiquitous contaminant in urban air and dust. Laboratory gloves coated with talc or other powders may also be a source of metals contamination.
5.2 Thoroughly clean sample containers with a metal-free non-ionic detergent solution, rinse with tap water, soak in acid, and then rinse with metal-free water. For quartz, Teflon, or glass materials, use 1:1 HNO_3, 1:1 HCl, or aqua regia (3 parts HCl/1 part HNO_3) for soaking. For plastic material, use 1:1 HNO_3 or 1:1 HCl. Reliable soaking conditions are 24 hours at 70°C. Chromic acid or chromium-free substitutes may be used to remove organic deposits from container, but rinse containers thoroughly with water to remove traces of chromium. Do not use chromic acid for plastic containers or if chromium is to be determined. Always use Milli-Q water or equivalent in analysis and reagent preparation.
5.3 For analysis of microgram-per-liter concentrations of metals, airborne contaminants in the form of volatile compounds, dust, soot, and aerosols present in laboratory air may become significant. To avoid contamination use "clean laboratory" facilities such as commercially available laminar-flow clean-air benches or custom-designed work stations and analyze blanks that reflect the complete procedure.
5.4 Precharging of sample containers with preservatives may lead to leaching of trace metals from sample containers. Precharging is to be avoided where possible.

6. INTERFERENCES

6.1 Spectral interferences — Light emission from spectral sources other than the element of interest may contribute to apparent net signal intensity. Sources of spectral interference include direct spectral line overlaps, broadened wings of intense spectral lines, ion-atom recombination continuum emission, molecular band emission, and stray (scattered) light from the emission of elements at high concentrations. Avoid line overlaps by selecting alternate analytical wavelengths. Avoid or minimize other spectral interference by judicious choice of background correction positions. A wavelength scan of the element line region is useful for detecting potential spectral interferences and for selecting positions for background correction. Make corrections for residual spectral interference using empirically determined correction factors in conjunction with the computer software supplied by the spectrometer manufacturer. Do this by analyzing single-element solutions of 100 mg/L concentration and noting for each element channel the apparent concentration from the interfering substance that is greater than the element's instrument detection limit.
6.2 Physical interferences — Physical interferences are effects associated with sample nebulization and transport processes. Changes in the physical properties of samples, such as viscosity and surface tension, can cause significant error. This usually occurs when sample containing more than 10% (by volume) acid or more than 1500 mg dissolved solids per liter are analyzed using calibration

standards containing ≤ 5% acid. Viscous samples should be handled as solids and weighed out with final results reported as mg/g. If physical interference is present, compensate for it by sample dilution, by using matrix-matched calibration standards, or by applying the method of standard addition.

6.3 Chemical interferences are caused by molecular compound formation, ionization effects, and thermochemical effects associated with sample vaporization and atomization in the plasma. Normally these effects are not pronounced and can be minimized by careful selection of operating conditions (incident power, plasma observation position, etc.). Chemical interferences are highly dependent on sample matrix and element of interest. As with physical interferences, compensate for them by using matrix-matched standards or by standard addition.

7. REAGENTS, WATER AND STANDARDS

7.1 Use reagents that are of reagent grade or equivalent. Use deionized water for preparing all calibration standards, reagents, and for dilution.

7.2 Hydrochloric acid, HCl; concentrated and 1:1

7.3 Nitric acid, HNO_3.

7.4 Hydrogen peroxide, 30% H_2O_2.

7.5 Standard stock solutions: Delta Standards: MES-WAST 1012 – 100 ± 0.5% μg/mL in 5% HNO_3 contains calcium (Ca), sodium (Na), chromium (Cr), magnesium (Mg), silicon (Si), cobalt (Co), manganese (Mn), copper (Cu), nickel (Ni), iron (Fe), potassium (K), vanadium (V), cadmium (Cd), lead (Pb), and zinc (Zn) from the CAEAL certification parameter list. The standard also contains aluminum (Al), lithium (Li), barium (Ba), strontium (Sr), boron (B), thalium (Tl), and beryllium (Be).

7.6 Working standard: 10 μg/mL standard prepared from standard stock solution by 1:10 dilution in deionized water used for calibration purposes.

7.7 Addition stock standard solutions for metals not contained in Delta MES-WAST 10/2 stock standard above obtained individually from Delta @ 1000 ± 3 μg/mL.

7.8 Reference stock standards:

- 7.8.1 Reference Standard for continuous calibration verification. 40 L of standard is prepared in 2% HNO_3. The standard contains Al, Ba, Be, Bi, Ca, Cd, Co, Cr, Cu, Fe, K, Li, Mg, Mn, Na, Ni, Pb, Ti, V, and Zn @ 5 μg/mL. This reference is analyzed with each batch of samples.
- 7.8.2 SRM 1643c, U.S. Geological Survey, branch of Quality Assurance
- 7.8.3 U.S. EPA Water Pollution Quality Control Sample — Trace Metals I
- 7.8.4 U.S. EPA Quality Control Sample for Mineral Analyses

8. APPARATUS

8.1 Thermo Jarrell Ash - AtomScan 16 sequential inductively coupled plasma emission spectrometer equipped with autosampler.

8.2 Hot plate

8.3 Glass beakers (NALGENE beakers used for trace metals analyses)

8.4 Watch glasses

8.5 Graduated cylinders, various sizes

8.6 Volumetric flasks, various sizes

8.7 Teflon or quartz beakers for Ag and/or B analyses

8.8 Glass pipets, Class A, various sizes

8.9 Eppendorf type pipets, adjustable to 100 to 1000 μL

9. SAMPLING AND SAMPLE SIZE

9.1 A minimum of 250 mL of sample is required for metals analysis.

10. SAMPLE PREPARATION/EXTRACTION PROCEDURE

10.1 SAMPLE DIGESTION

10.1.1 No digestion is performed if dissolved metals are to be determined. The sample is filtered through a 0.45-mm membrane filter after sampling, preserved with 5% HNO_3 and run on the ICP. Dilutions if necessary are done with 5% HNO_3.

10.1.2 For all metals except Ag and B a suitable aliquot, typically 100 mL, of a well-mixed sample is transferred to a beaker. Carefully add 10 mL of concentrated HCl, 3 mL of concentrated HNO_3, 3 mL of H_2O_2 (30%), and 3 boiling chips to the beaker place on a hot plate.

10.1.3 For Ag and B proceed as with 10.1.2 but do not add HCl. HCl can lead to a AgCl precipitate and the loss of BCl_3 vapors.

10.2 Evaporate the sample down to approximately 20 mL. Once sample digestion is complete, allow the beaker to cool to room temperature and transfer the contents including the rinsings (excluding the boiling chips) to a volumetric flask. For low level trace metal work, the digested sample is left at a final volume of 25 mL for ICP analysis. Higher level samples are diluted to an appropriate final volume with 5% HNO_3.

11. INSTRUMENT OPERATION/MEASUREMENT PROCEDURE

11.1 Using the Thermo Jarrell Ash - AtomScan 16 Spectrometer Operators Manual as a guide, set up the ICP and optimize for the samples to be run.

11.2 The elements commonly analyzed are read at the specified wavelengths found in Table A3.1. These wavelengths may vary from instrument to instrument depending on sample type, but are typically the most sensitive as determined by the instrument software.

12. CALIBRATION PROCEDURE

12.1 A three-point calibration line is created for each individual element from the results of the analysis of a high-level 10-μg/mL standard, a standard at < 10 times the method detection limit and a machine blank. A machine blank is a 2% solution of environmental-grade nitric acid in deionized water used as the machine rinsing solution.

12.2 Calibration checks are run after every five samples and if they are found to differ by more than 10% from the calibration standard, a restandardization is completed.

13. SAMPLE ANALYSIS PROCEDURE

13.1 Quality control samples include:

13.1.1 Method Blank — analysis of deionized water taken through digestion step to check for any contamination or interferences from digestion procedure.

13.1.2 Duplicates — a replicate analysis of an homogeneous sample to show method precision.

13.1.3 Spikes — a replicate sample spiked with a known amount of stock standard solution to show both method precision and accuracy as well as any interferences.

13.1.4 References — a certified reference material to show method accuracy.

13.2 Quality control samples are incorporated into every batch of samples run and must make up at least 20% of the total number of samples analyzed.

14. RUN FORMAT/CONTROL LIMITS

14.1 Each run begins with the analysis of a high-level standard (10 μg/mL) and a machine blank. High level check standards are analyzed every five samples and if found to differ by more than 10% from the calibration standard, a restandardization is completed.

A typical run format for metals analysis, once calibration has been completed, is as follows:

Sequence	Sample Type
1	REFERENCE
2	METHOD-BLANK
3	MACHINE BLANK
4	SAMPLE
5	DUPLICATE
6	SAMPLE
7	SAMPLE
8	SAMPLE
9	REFERENCE
10	SAMPLE
11	DUPLICATE
12	SAMPLE
13	SAMPLE
14	SAMPLE

The sequence then repeats itself continuously until all samples are accounted for.

14.2 Control limits consist of upper and lower warning and control limits. If a warning limit is reached, you must ensure that the system is stable and is not going out of control. If a control limit is exceeded, you must abandon the run, troubleshoot the system to find and correct the problem, and carry out the analytical run again.

		UWL	LWL	UCL	LCL
Ag	Dup.	11.5%	0%	15.7%	0%
	Ref.	107%	90.2%	112%	85.8%
	Spike	111%	84.3%	118%	77.7%
Al	Dup.	9.22%	0%	12.5%	0%
	Ref.	112%	89.1%	118%	83.3%
	Spike	125%	77.5%	136%	65.7%
B	Dup.	7.49%	0%	10.0%	0%
	Ref.	107%	87.6%	112%	82.7%
	Spike	112%	79.0%	120%	70.8%
Ba	Dup.	7.54%	0%	10.4%	0%
	Ref.	104%	95.2%	106%	93.0%
	Spike	113%	81.6%	121%	73.6%
Be	Dup.	5.44%	0%	6.93%	0%
	Ref.	106%	94.9%	109%	92.2%
	Spike	110%	81.8%	117%	74.7%
Bi	Dup.	23.2%	0%	30.8%	0%
	Ref.	107%	92.1%	111%	88.3%
	Spike	112%	76.1%	121%	67.0%
Ca	Dup.	5.07%	0%	6.70%	0%
	Ref.	112%	92.6%	117%	87.8%
	Spike	112%	91.3%	117%	86.3%
Cd	Dup.	6.10%	0%	8.29%	0%
	Ref.	117%	86.4%	125%	78.7%
	Spike	117%	78.0%	126%	68.4%

Co	Dup.	42.8%	0%	60.5%	0%
	Ref.	103%	96.4%	105%	94.7%
	Spike	117%	72.3%	128%	61.2%
Cr	Dup.	10.5%	0%	13.8%	0%
	Ref.	107%	88.4%	112%	83.7%
	Spike	114%	79.1%	122%	70.5%
Cu	Dup.	16.4%	0%	22.1%	0%
	Ref.	108%	88.5%	113%	83.5%
	Spike	114%	84.1%	122%	76.6%
Fe	Dup.	11.9%	0%	16.3%	0%
	Ref.	105%	93.1%	108%	90.0%
	Spike	110%	91.5%	114%	86.9%
K	Dup.	13.1%	0%	17.3%	0%
	Ref.	117%	91.1%	124%	84.6%
	Spike	115%	87.4%	122%	80.5%
Li	Dup.	9.90%	0%	13.7%	0%
	Ref	106%	94.5%	109%	91.7%
	Spike	113%	85.6%	120%	78.8%
Mg	Dup.	4.02%	0%	5.26%	0%
	Ref.	105%	93.3%	108%	90.3%
	Spike	111%	86.2%	118%	79.9%
Mn	Dup.	9.45%	9%	12.9%	0%
	Ref.	103%	94.9%	105%	92.9%
	Spike	113%	85.2%	120%	78.3%
Mo	Dup.	9.47%	0%	13.0%	0%
	Ref.	108%	92.5%	111%	88.7%
	Spike	113%	80.1%	121%	71.9%
Na	Dup.	3.80%	0%	5.02%	0%
	Ref.	111%	89.0%	117%	83.4%
	Spike	105%	92.0%	109%	88.7%
Ni	Dup.	11.7%	0%	15.5%	0%
	Ref.	108%	86.8%	114%	81.4%
	Spike	113%	80.2%	122%	71.9%
Pb	Dup.	35.6%	0%	49.0%	0%
	Ref.	107%	87.5%	111%	82.7%
	Spike	113%	81.8%	121%	73.9%
Sb	Dup.	34.2%	0%	47.3%	0%
	Ref.	107%	90.9%	111%	87.0%
	Spike	118%	83.6%	127%	74.9%
Si	Dup.	26.5%	0%	37.5%	0%
	Ref.	115%	85.5%	123%	78.0%
	Spike	140%	58.3%	161%	37.8%
Sn	Dup.	13.0%	0%	17.6%	0%
	Ref.	111%	90.8%	116%	85.7%
	Spike	146%	37.3%	173%	10.2%
Sr	Dup.	8.35%	0%	11.3%	0%
	Ref.	104%	95.8%	106%	93.7%
	Spike	110%	83.6%	116%	77.1%
Ti	Dup.	6.62%	0%	9.08%	0%
	Ref.	107%	94.1%	110%	91.0%
	Spike	114%	80.0%	123%	71.4%

V	Dup.	17.2%	0%	23.4%	0%
	Ref.	106%	95.2%	109%	92.5%
Spike	114%	78.9%	123%	70.1%	
Zn	Dup.	13.6%	0%	18.5%	0%
	Ref.	106%	90.3%	110%	86.4%
	Spike	112%	86.6%	118%	80.3%

PART 2 — SCIENTIFIC BACKGROUND FOR A METHOD

1. DEFINITIONS

Dissolved metals — those constituents (metals) of an unacidified sample that pass through a 0.45-mm membrane filter.

Suspended metals — those constituents (metals) of an unacidified sample that are retained by a 0.45-mm membrane filter.

Total metals — the concentration of metals determined on an unfiltered sample after vigorous digestion, or the sum of the concentrations of metals in both dissolved and suspended fractions.

Acid-extractable metals — the concentration of metals in solution after treatment of an unfiltered sample with hot dilute mineral acid.

2. PARAMETERS MEASURED

2.1 The following elements can be analyzed by ICP analysis and this method. Results are reported in µg/mL.

Silver (Ag)	Magnesium (Mg)
Aluminum (Al)	Manganese (Mn)
Boron (B)	Molybdenum (Mo)
Barium (Ba)	Sodium (Na)
Beryllium (Be)	Nickel (Ni)
Bismuth (Bi)	Lead (Pb)
Calcium (Ca)	Antimony (Sb)
Cadmium (Cd)	Silicon (Si)
Cobalt (Co)	Tin (Sn)
Chromium (Cr)	Strontium (Sr)
Copper (Cu)	Titanium (Ti)
Iron (Fe)	Vanadium (V)
Potassium (K)	Tungsten (W)
Lithium (Li)	Zinc (Zn)

3. EQUIPMENT, REAGENTS, SUPPLIES SPECIFICATIONS

3.1 See Part 1 — Section 7 and 8

3.2 Suggested Suppliers, ICP:

Thermo Instruments Inc.
5716 Coopers Ave., Unit 1
Mississauga, ON L4Z 2E8
Tel: (905) 890-1034
Fax: (905) 890-5775

4. POLLUTION PREVENTION PROCEDURES

There are no harmful chemicals from this analysis.

5. METHOD PERFORMANCE SPECIFICATIONS

5.1 The MDL for metals are as follows:

	mg/mL		mg/mL
Ag	0.0276	Mg	0.0205
Al	0.0138	Mn	0.0064
B	0.0059	Mo	0.0187
Ba	0.0064	Na	0.1435
Be	0.0173	Ni	0.0119
Bi	0.0682	Pb	0.0438
Ca	0.0129	Sb	0.0720
Cd	0.0111	Si	0.0365
Co	0.0114	Sn	0.0380
Cr	0.0243	Sr	0.0119
Cu	0.0183	Ti	0.0353
Fe	0.0130	V	0.0185
K	1.1928	W	0.0329
Li	0.0172	Zn	0.0095

5.2 Other performance specifications include:

		Average (%)	Standard Deviation (%)
Ag	Dup.	2.95	4.26
	Ref.	98.8	4.33
	Spike	97.7	6.67
Al	Dup.	2.63	3.30
	Ref.	101	5.83
	Spike	101	11.8
B	Dup.	2.43	2.53
	Ref.	97.2	4.84
	Spike	95.3	8.17
Ba	Dup.	1.76	2.89
	Ref.	99.2	2.24
	Spike	97.5	7.97
Be	Dup.	2.46	1.49
	Ref	100	2.78
	Spike	95.9	7.06
Bi	Dup.	8.01	7.59
	Ref.	99.7	3.79
	Spike	94.1	9.01
Ca	Dup.	1.79	1.64
	Ref.	102	4.80
	Spike	102	5.08
Cd	Dup.	1.73	2.19
	Ref.	102	7.65
	Spike	97.3	9.62

Co	Dup.	7.32	17.7
	Ref.	99.9	1.74
	Spike	94.5	11.1
Cr	Dup.	3.84	3.33
	Ref.	97.9	4.76
	Spike	96.4	8.63
Cu	Dup.	5.07	5.67
	Ref.	98.3	4.93
	Spike	99.2	7.51
Fe	Dup.	3.11	4.38
	Ref.	99.2	3.05
	Spike	101	4.52
K	Dup.	4.59	4.24
	Ref.	104	6.48
	Spike	101	6.93
Li	Dp.	2.33	3.78
	Ref.	100	2.81
	Spike	99.2	6.83
Mg	Dup.	1.55	1.24
	Ref.	99.3	3.00
	Spike	98.8	6.30
Mn	Dup.	2.57	3.44
	Ref.	99.1	2.09
	Spike	99.0	6.90
Mo	Dup.	2.32	3.57
	Ref.	100	3.77
	Spike	96.7	8.27
Na	Dup.	1.36	1.22
	Ref	100	5.60
	Spike	98.7	3.36
Ni	Dup.	4.30	3.72
	Ref.	97.6	5.39
	Spike	96.7	8.26
Pb	Dup.	8.65	13.5
	Ref.	97.1	4.78
	Spike	97.6	7.92
Sb	Dup.	7.86	13.2
	Ref.	98.8	3.95
	Spike	101	8.66
Si	Dup.	4.51	11.0
	Ref.	100	7.48
	Spike	99.2	20.5
Sn	Dup.	3.82	4.58
	Ref.	101	5.11
	Spike	91.6	27.1
Sr	Dup.	2.36	3.00
	Ref.	100	2.08
	Spike	96.6	6.48
Ti	Dup.	1.70	2.46
	Ref.	100	3.15
	Spike	97.0	8.53

V	Dup.	4.96	6.14
	Ref.	101	2.72
	Spike	96.6	8.84
Zn	Dup.	3.98	4.83
	Ref.	98.0	3.87
	Spike	99.1	6.24

5.3 We have been CAEAL certified for metals since 1994.

6. SAMPLE COLLECTION, PRESERVATION, STORAGE, AND HANDLING

6.1 Preserve samples immediately after sampling by acidifying with concentrated nitric acid (HNO_3) to pH < 2. Filter samples for dissolved metals before preserving.

7. SAMPLE PRETREATMENTS

7.1 Sample pretreatment consists of agitating the sample to disperse any suspended solids.

8. CALCULATIONS/FLAGS/DIGITS

8.1 Prepare standard curves using the instrument software by plotting peak areas of standards processed through the manifold against metal concentration in standards. Compute sample metal concentration by comparing sample peak with that of standard curve.

8.2 Data are generally reported to three significant digits, but with a maximum of two digits after the decimal point. Typical reported data would look like 0.33, 1.33, 5.33, 15.3, 103, 1030, etc.

8.3 Data below the MDL are reported using the "t and w" system where "t" is a measurable result below the calculated MDL and "w" represents no instrument response to the sample. Typical data reported below the MDL would look like 0.05t, or w if the MDL is 0.10 mg/L.

8.4 The reported MDL on data reports must be corrected for dilution of the sample and sample size. The MDL in LIMS is for a typical sample, not necessarily for each actual sample analyzed.

9. QUALITY VALIDATION

9.1 Quality validation requires that the QC_TYPE samples (Section 13.1) are completed and that the results of those samples be reasonable and fall within acceptable range.

9.2 All calibration data must be valid and properly documented.

9.3 The values must be reasonable compared to other data such as Cr^{6+} be lower than total Chem. Rev.

9.4 The MDL must be reasonable compared to sample size and dilutions and data below MDL should have appropriate flags.

9.5 Data should be examined for consistency with historical customer data.

9.6 Method blank should be lower than MDL.

10. CORRECTIVE ACTIONS/INTERVENTIONS

10.1 Upon sample arrival, it is the responsibility of the analyst to ensure that any special instructions given by the customer regarding sample preparation and/or analysis are clearly understood and performed. If there is any doubt as to sample handling or the method of sample preparation the analyst is to seek advise from his/her supervisor or contact the client personally for clarification.

10.2 If duplicate samples submitted by the client do not appear similar (i.e., color or suspended solid content appear different), the analyst performing the analysis shall make a record in his/her lab notebook and the observation is to be noted in the sample comment section of the final report.

10.3 If the method blank contains a concentration above the method detection limit for the analyte but is less than 5% of the average concentrations of that analyte in the samples, the method blank shall be reported as analyzed. If the blank concentrations contribute to more than 5% of the concentration of the analyte, the blank must be reanalyzed. If the blank level remains high and it contributes to a significant amount (> 5%) of the sample concentration, another blank should be digested and analyzed to determine if the original blank was indicative of an ongoing problem in the laboratory. If the high blank reoccurs, it will be necessary to investigate further the source of the high blank by checking the glassware cleaning and rinsing procedure, reagent chemical purity, etc. If the source of the contamination can be found and rectified within a relatively short period of time, it may be necessary to obtain more sample and reanalyze the entire batch of samples affected. If additional sample is unavailable the original set of data and its corresponding high blank shall be recorded and flagged in the sample comment section of the data report.

10.4 Upon analysis, if the duplicate results of any batch fall outside of the acceptable control limits, as defined by previous analyses of similar or like samples, the digested sample shall be remixed and rerun on the instrument. If the precision falls within acceptable limits the second time, these data shall be reported to the client in the final report. If the duplicate precision remains out of control after rerunning the digested samples, additional sample must be obtained and the sample redigested and reanalyzed. If the redigested sample and one of the first two digested samples fall within acceptable precision limits, they shall be reported as duplicates in the final report. If the duplicate precision remains out of the control limits, both of the original sets of duplicate data shall be reported to the client and the data flagged in the sample comment section of the final report. When there is insufficient sample for redigestion and it is feasible to do so, contact the client for additional sample. If there is no sample left, the original data and its duplicate shall be reported and the data flagged in the sample comment section of the final report. If the duplicate precision is poor but the absolute concentration of the analytes fall close to the method detection limit of the analyte in question, the results may be reported and an explanation given in the sample comment section of the final report.

10.5 Upon analysis, if the spike recoveries falls outside of acceptable control limits as defined by previous analyses of similar or like samples, the digested sample shall be remixed and rerun on the instrument. If the spike recoveries fall within acceptable limits for all compounds the second time, these data shall be reported to the client in the final report. If the spike recoveries remain out of control after rerunning the digested samples, additional sample must be obtained and the sample and its spike redigested and reanalyzed. If the second set of sample and spiked sample fall within spike recovery limits, they shall be reported in the final report. If the spike recovery remains out of the control limits, while the duplicate precision of the original unspiked sample and the redigested unspiked sample is acceptable, the original sample shall be reported and the source of the poor spike recovery shall be investigated. When there is insufficient sample for redigestion and it is feasible to do so, contact the client for additional sample. If there is no sample left, the original data and its spiked sample recoveries shall be reported and the data flagged in the sample comment section of the final report.

10.6 If an external reference material falls outside of acceptable control limits as defined by previous analyses of the same material, the digested sample shall be remixed and rerun on the instrument. If the recovery of the reference standard falls within acceptable limits the second time, these data shall be recorded. If the reference recovery remains out of control after rerunning of the digested samples, additional sample must be obtained and the sample redigested and reanalyzed. If the redigested sample falls within acceptable accuracy limits, it shall be reported. If the reference recovery remains out of the control limits, the entire data set must be rejected.

11. PROCEDURE FOR LEGAL SAMPLES

11.1 Legal samples have a Chain of Custody form which must be signed by the person who has the samples at any time. Legal samples must be in a locked fridge or fume hood at all times when they are not attended by the person with custody of the samples. After analysis the samples should be signed back to the customer with the chain of custody form. It is important that all lab notes, books, output, QC samples, and anything else related to the legal case be legible, defensible, and available for the customers when they ask for it.

12. DISPOSAL OF SAMPLES/REAGENTS

12.1 Liquid samples are disposed of down the drain with copious amounts of water.
12.2 Unused portions of samples are returned to the submitter for proper disposal.

13. REFERENCES

13.1 *Standard Methods for the Examination of Water and Wastewater*, 19th ed., Arnold E. Greenberg, Lenore S. Clesceri and Andrew D. Eaton, Eds., 1995.
13.2 AtomScan 25 Spectrometer Operator's Manual, Thermo Jarrel Ash Corp.

Table A3.1 Recommended Wavelengths for Analysis of Metals by Thermo Jarrel Ash - AtomScan 16 ICP

Analyte	Wavelength (nm)	Analyte	Wavelength (nm)
Ag	328.068	Mg	279.079
Al	308.215	Mn	257.610
B	249.678	Mo	203.844
Ba	493.409	Na	588.955
Be	313.042	Ni	231.604
Bi	223.061	Pb	220.353
Ca	315.887	Sb	206.833
Cd	226.502	Si	251.611
Co	228.616	Sn	189.980
Cr	205.552	Sr	407.77
Cu	324.754	Ti	334.941
Fe	259.940	V	292.402
K	766.491	W	207.911
Li	670.784	Zn	213.856

INTERTEK TESTING SERVICES, SAN JOSE, CALIFORNIA (FORMERLY INCHCAPE TESTING SERVICES)

STATE OF CALIFORNIA—HEALTH AND WELFARE AGENCY PETE WILSON, *Governor*

DEPARTMENT OF HEALTH SERVICES

(510)540-2800 March 13, 1997

Steve O'Neil
ITS/Environmental Laboratories
1961 Concourse Drive, Suite E
San Jose, CA 95131

Certificate No.: 2221

Dear Mr. O'Neil:

This is to advise you that the laboratory named above has been certified as an environmental testing laboratory pursuant to the provisions of the California Environmental Laboratory Improvement Act of 1988 (Health and Safety Code, Division 1, Part 2, Chapter 7.5, commencing with Section 100825).

The fields of testing for which this laboratory has been certified under this Act are indicated in the enclosed "List of Approved Fields of Testing and Analytes." Certification shall remain in effect until March 31, 1999 unless revoked. This certificate is subject to an **annual fee** as prescribed by Section 100860(a), Health and Safety Code, **on the anniversary date of the certificate**. Your application for renewal must be received 90 days before the expiration of your certificate to remain in force according to the California Code of Regulations, Title 22, Division 4, Chapter 19, Sections 64801 through 64827.

Please note that your laboratory is required to notify the Environmental Laboratory Accreditation Program of any major changes in the laboratory such as the transfer of ownership, change of laboratory director, change in location, or structural alterations which may affect adversely the quality of analyses (Section 100845(b)(d), California Health & Safety Code).

Your continued cooperation is essential in order to establish a reputation for the high quality of the data produced by environmental laboratories certified by the State of California.

If you have additional questions, please contact Amanda J. Vidal at (510) 540-2800.

Sincerely,

George C. Kulasingam

George C. Kulasingam, Ph.D., Manager
Environmental Laboratory
Accreditation Program

Enclosure

CALIFORNIA DEPARTMENT OF HEALTH SERVICES
ENVIRONMENTAL LABORATORY ACCREDITATION PROGRAM
List of Approved Fields of Testing and Analytes

ITS/Environmental Laboratories
1961 Concourse Drive, Suite E
San Jose, CA

TELEPHONE No: (408) 432-8192
CALIFORNIA COUNTY: Santa Clara

CERTIFICATE NUMBER: 2221
EXPIRATION DATE: 03/31/99

2 Inorganic Chemistry and Physical Properties of Drinking Water excluding Toxic Chemical Elements

2.1 Alkalinity
2.2 Calcium
2.3 Chloride
2.5 Fluoride
2.6 Hardness
2.7 Magnesium
2.8 MBAS
2.9 Nitrate
2.10 Nitrite
2.11 Sodium
2.12 Sulfate
2.13 Total Filterable Residue and Conductivity
2.16 Phosphate, ortho
2.18 Cyanide
2.19 Potassium

3 Analysis of Toxic Chemical Elements in Drinking Water

3.1 Arsenic
3.2 Barium
3.3 Cadmium
3.4 Chromium, total
3.5 Copper
3.6 Iron
3.7 Lead
3.8 Manganese
3.9 Mercury
3.10 Selenium
3.11 Silver
3.12 Zinc
3.13 Aluminum
3.15 Antimony
3.16 Beryllium
3.17 Nickel

9 Physical Properties Testing of Hazardous Waste

9.1 Ignitability by Flashpoint determination (Title 22, CCR, 66261.21)
9.2 Corrosivity - pH determination (Title 22, CCR, 66261.22)

10 Inorganic Chemistry and Toxic Chemical Elements of Hazardous Waste

10.1 Antimony
10.2 Arsenic
10.3 Barium
10.4 Beryllium
10.5 Cadmium
10.6 Chromium, total

Certificate No.: 2221
Expiration Date: 03/31/99
Page 2

10 **Inorganic Chemistry and Toxic Chemical Elements of Hazardous Waste** (continued)

10.7 Cobalt
10.8 Copper
10.9 Lead
10.10 Mercury
10.11 Molybdenum
10.12 Nickel
10.13 Selenium
10.14 Silver
10.15 Thallium
10.16 Vanadium
10.17 Zinc
10.18 Chromium (VI)
10.19 Cyanide
10.20 Fluoride
10.99 Organic Lead

11 **Extraction Tests of Hazardous Waste**

11.1 California Waste Extraction Test (WET) (Title 22, CCR, 66261.100, Appendix II)
11.3 Toxicity Characteristic Leaching Procedure (TCLP) All Classes

12 **Organic Chemistry of Hazardous Waste (measurement by GC/MS combination)**

12.1 EPA Method 8240B
12.3 EPA method 8270B
12.6 EPA Method 8260A

13 **Organic Chemistry of Hazardous Waste (excluding measurements by GC/MS combination)**

13.1 EPA Method 8010B
13.2 EPA Method 8015A
13.3 EPA Method 8020A
13.5 EPA Method 8040A
13.7A EPA Method 8080A
13.7B EPA Method 8081
13.13 EPA Method 8310
13.15 Total Petroleum Hydrocarbons - Gasoline
13.16 Total Petroleum Hydrocarbons - Diesel
13.17 TRPH - Screening by IR
13.18 EPA Method 8011
13.19 EPA Method 8021A

16 **Wastewater Inorganic Chemistry, Nutrients and Demand**

16.2 Akalinity
16.5 Boron
16.7 Calcium
16.10 Chloride
16.11 Chlorine Residual, total
16.12 Cyanide
16.13 Cyanide amenable to chlorination

Certificate No.: 2221
Expiration Date: 03/31/99
Page 3

16 **Wastewater Inorganic Chemistry, Nutrients and Demand** (continued)

16.14 Fluoride
16.15 Hardness
16.17 Magnesium
16.18 Nitrate
16.19 Nitrite
16.20 Oil and Grease
16.21 Organic Carbon
16.23 pH
16.25 Phosphate, ortho
16.26 Phosphorous, Total
16.27 Potassium
16.28 Residue, Total
16.29 Residue, Filterable (Total Dissolved Solids)
16.30 Residue, Nonfilterable (Total Suspended Solids)
16.31 Residue, Settleable (Settleable Solids)
16.33 Silica
16.34 Sodium
16.35 Specific Conductance
16.36 Sulfate
16.37 Sulfide (includes total and soluble)
16.39 Surfactants (MBAS)
16.41 Turbidity
16.44 Total Recoverable PHCs by IR

17 **Toxic Chemical Elements in Wastewater**

17.1 Aluminum
17.2 Antimony
17.3 Arsenic
17.5 Beryllium
17.6 Cadmium
17.8 Chromium, Total
17.9 Cobalt
17.10 Copper
17.13 Iron
17.14 Lead
17.15 Manganese
17.16 Mercury
17.17 Molybdenum
17.18 Nickel
17.24 Selenium
17.25 Silver
17.27 Thallium
17.28 Tin
17.30 Vanadium
17.31 Zinc

Certificate No.: 2221
Expiration Date: 03/31/99
Page 4

18 <u>Organic Chemistry of Wastewater (measurements by GC/MS combination)</u>

18.1 EPA Method 624
18.2 EPA Method 625

19 <u>Organic Chemistry of Wastewater (excluding measurements by GC/MS combination)</u>

19.1 EPA Method 601
19.2 EPA Method 602
19.15 EPA Method 608 PCBs only
19.16 EPA Method 608 Chlorinated Pesticides only

****
(031397)

APPENDIX 4

Groundwater Sampling and Field Analysis Methods (Borden)

A4.1 MONITORING PROGRAMS

Groundwater monitoring programs were divided into the following categories:

1. Background sampling (Work Plan Monitoring Phase Zero — WP MP0)
2. Tracer test (WP MP2)
3. Plume development monitoring (WP MP 3 and 4)
4. Major sampling (WP MP 3, 4 and 5)
5. ORC™ performance and aerobic degradation assessment (Gate 1)
6. Anaerobic zone monitoring in Gate 3
7. Biosparge performance monitoring (Gate 3)
8. External laboratory sampling (WP MP 6)

Monitoring requirements established in the Work Plan (WP) were achieved by these programs, although minor changes to the timing of sampling events were made, as the need arose, to fulfill the project goals. Additional monitoring programs to address issues that were not originally expected (e.g., problems with the ORC™ in Gate 1; the need to closely monitor nutrient pulses and the establishment of anaerobic conditions in Gate 3; etc.) were also completed. The following paragraphs provide a brief description of the purpose of each monitoring phase, the parameters analyzed, and the approximate timeline for sampling events.

Background sampling was completed in May 1996 shortly after the flow system was started (i.e., extraction pumping began). This sampling event provided a snapshot of background conditions at the site. Samples were analyzed for organics, inorganics, and field parameters to provide a comprehensive survey of background geochemical conditions.

The tracer test provided insight into groundwater flow velocities and the existence of preferential pathways. Bromide (added as potassium bromide; KBr) was used as the conservative tracer. Samples were collected from May 1996 to June 1997 and were analyzed for bromide ion (Br^-) concentrations and conductivity.

Plume development monitoring was conducted intermittently, particularly during the early stages of the project. Sampling was completed every 3 to 4 weeks as deemed necessary. Analyses were completed for organics and field parameters.

Major sampling events were intended to provide detailed snapshots of the plume and to permit the comparison of plume development and geochemical characteristics between the three gates. Analyses of organics, inorganics, and field parameters were conducted approximately every 6 weeks once a quasi-steady-state condition was established.

ORC™ performance and aerobic degradation assessment (Gate 1) were conducted to assess the delivery of oxygen to the subsurface and, subsequently, the aerobic degradation of the target organics once adequate oxygen supply was achieved. Installations of the ORC™ socks in cassette C4 and later in the alternate oxygen addition wells (SW-12 to 14) were monitored for DO and pH to observe the release of oxygen from the socks and the impact of the socks on the pH of the water. Effective oxygen release was only observed with the ORC™ in SW-12 to 14. VOC samples were collected to quantify losses of organics within the imposed aerobic environment.

Anaerobic zone monitoring in Gate 3 was conducted to assess the progress of nutrient pulses through the gate and to identify changes in groundwater chemistry resulting from remedial activities. Nutrient pulses were monitored twice weekly for inorganic tracers, conductivity, and benzoate following the addition of chloride or bromide to the metered nutrient concentrate. Sampling for dissolved hydrocarbon gases, anions, and organic acids, in addition to regular sampling events, commenced in June 1997. This program was designed to detect a possible shift in microbial activity and to assess any degradation of added nutrients and contaminants.

Biosparge performance monitoring (Gate 3) commenced in July 1997 to provide an assessment of the oxygen delivery system. DO measurements were conducted in the field before and after daily sparge events to establish the field operating requirements. Monitoring decreased to a minimum of once a week after adequate oxygen levels were observed and could be maintained. Periodic sampling for VOCs and nutrient continued until October 23, 1997.

External laboratory sampling was conducted in accordance with U.S. EPA protocols and provided a performance evaluation with respect to those guidelines. Samples were collected on a regular basis, approximately every 6 weeks, under the guidance of AATDF project manager Dr. S. Fiorenza, and submitted to an independent laboratory for organics analyses.

Table A4.1 provides a summary of the types of analyses completed for each of the monitoring programs.

Table A4.1 Summary of Groundwater Monitoring Programs

Program Number	Monitoring Program Description	Type of Analyses Conducted
1	Background Sampling	VOCs, DHGs, Inorganics
2	Tracer Test	Bromide, Conductivity
3	Plume Development Monitoring	VOCs, DHGs, Field Parameters
4	Major Sampling	VOCs, DHGs, Inorganics, Field Parameters
5	ORC™ Performance and Aerobic Degradation Assessment (Gate 1)	DO, pH; VOCs, DHGs (after aerobic conditions established)
6	Anaerobic Zone Monitoring (Gate 3)	DHGs, Organic Acids, Conductivity, Bromide/Chloride, Nutrient, Anions, Alkalinity
7	Biosparge Performance Monitoring (Gate 3)	DO, nutrient; VOCs (after aerobic conditions established)
8	External Laboratory Sampling (QA/QC)	VOCs

Notes: VOCs: Volatile organic compounds: PCE, TCE, DCE, VC, CT, CF, DCM, CM, CS_2, TOL. DHGs: Dissolved hydrocarbon gases: ethane, ethene, propane, propene, methane. Organic Acids: Formic, acetic, propionic, butyric. Inorganics: Major/ions [Cations (Ca, Fe, K, Mg, Mn, Na); Anions (Cl, Br, NO_3, NO_2, PO_4, SO_4)]; Alkalinity (as calcium carbonate). Field Parameters: DO/pH/Eh/conductivity Nutrients: Benzoate.

A4.2 TYPES OF ANALYSES

A variety of parameter types were monitored over the duration of the project. A description of each is provided here under the following categories:

1. Volatile organic compounds (VOCs)
2. Dissolved hydrocarbon gases (DHGs)
3. Cations
4. Anions and alkalinity
5. Inorganic tracers
6. Field parameters
7. Organic acids
8. Nutrients (benzoate)
9. Water level/head measurements

Table A4.2 provides a summary of these analyses, indicating the specific parameters analyzed, sample handling requirements, and whether samples were analyzed in the field or the laboratory. Field analytical procedures are described in Section A4.4, while laboratory analytical methods are provided in Appendix 3.

A4.2.1 VOCs

Volatile organic compound (VOC) sampling was completed in all three gates during background, plume development, major, and external laboratory sampling events, and provided concentrations of dissolved organic compounds in water. The parameters of interest for this experiment included tetrachloroethene (PCE), trichloroethene (TCE), various isomers of dichloroethene (DCE), vinyl chloride (VC), carbon tetrachloride (CT), chloroform (CF), dichloromethane (DCM), chloromethane (CM), carbon disulfide (CS_2), and toluene (TOL).

VOC samples were collected using the UW sampling apparatus (see Section A4.3.1) to prevent exposure to air and subsequent volatilization of contaminants. VOC samples are typically preserved with hydrochloric acid (HCl) using the method (SW846 Method 8260) described by the U.S. EPA (1990) and have a holding time of 14 days. The holding time for samples without HCl is 7 days. However, only external laboratory sampling was conducted according to U.S. EPA protocol and the laboratory selected to complete these analyses, Barringer Laboratories (formerly MDS) in Mississauga, Ontario, did not require the use of preservative due to the rapid turnaround time (<5 days) they provided. VOC samples from all other monitoring programs were submitted to the Organic Geochemistry Laboratory (OGL) at the University of Waterloo (UW) and used sodium azide as a preservative, rather than HCl, to limit aerobic microbial activity. Holding times for VOCs are listed in Table A4.2.

Table A4.2 Summary of Sampling Specifications

Type of Analysis	Sample Volume (mL)	Sample Bottle	Preservation	Holding Time (days)	Analyst: Field or Lab
VOCs	40	a	b	14	lab
DHGs	40	a	4°C	14	lab
Cations	60	plastic	c	28	lab
Anions and Alkalinity	100	plastic	4°C	anions-28 alkalinity-4	anions-lab alkalinity-both
Inorganic Tracers	20	plastic	4°C	28	field
Field Parameters	100	N/A	N/A	none	field
Organic Acids	40	a	4°C	<7	lab
Nutrients	<20	glass	d	indefinite	lab

Notes: a: glass hypovial, sealed with Teflon-faced septum. b: 10% sodium azide (0.4 mL per 40-mL sample). c: filtered (0.45 μm) and acidified to pH <2 with concentrated HNO_3. d: Formic acid (approx. 2 drops); frozen for indefinite storage.

A4.2.2 DHGs

Dissolved hydrocarbon gas (DHG) sampling was completed in all three gates during background, plume development, major, and external laboratory sampling events, and as an additional assessment tool in Gate 3 during the assessment of the anaerobic zone. The analyses provided concentrations of ethane, ethene, propane, propene, and methane. DHG analysis was completed by the UW OGL.

Like VOC samples, DHG samples contained volatile substances and required collection in a manner similar to VOC samples. To prevent exposure to air and subsequent volatilization of contaminants, DHG samples were collected using the UW sampling apparatus (description provided in Section A4.3.1). DHG samples were not preserved, but were stored at 4°C until analyses were completed. Holding times for samples were 14 days.

A4.2.3 Cations

Samples for cation analyses were collected as part of background and major sampling events. Cation analyses reported concentrations of calcium, iron, potassium, magnesium, manganese, and sodium. Ammonia samples were also collected for major sampling events in March and June 1997 only. For all major sampling rounds, after the sources were started, Water Technology Inc. (WTI) completed inorganic analyses. Holding time limits were 28 days for inorganic samples.

Sampling methodology is provided in Section A4.3.2. Cation samples were filtered using 0.45-μm syringe filters to remove any suspended particles and acidified using nitric acid (pH <2) to prevent precipitation of metals or carbonates. Through the use of this preservation technique, the results provided dissolved, not total, concentrations.

Ammonia samples were not filtered, but were preserved with concentrated sulfuric acid to pH <2.

A4.2.4 Anions and Alkalinity

Anion and alkalinity samples were collected from all gates during background and major sampling events. Alkalinity samples were also collected as an additional tool for the assessment of anaerobic microbial activity in Gate 3 (MP 6). Chloride, nitrite, bromide, nitrate, phosphate, and sulfate concentrations were reported from anion analyses. Alkalinity samples were analyzed in the laboratory and reported in terms of milligrams per liter calcium carbonate ($CaCO_3$). In addition, several samples were analyzed in the field (also as milligrams per liter $CaCO_3$) to permit a comparison between the results of the different methods and to assess whether increased holding time caused a bias. Where comparisons were made, reported concentrations for field and laboratory analyses were similar. Alkalinity measurements were subsequently made in the lab. For all major sampling rounds, after the sources were started, Water Technology Inc. (WTI) completed inorganic analyses. Holding time limits were 28 days for inorganic samples.

The sampling method used was similar to that used for cations and can be found in Section A4.3.2. Anion and alkalinity samples did not require filtering or acidification but were refrigerated as a means of preservation.

A4.2.5 Inorganic Tracers

Inorganic tracer samples were collected from all gates during the tracer test. Tracer samples were also collected as part of tracer dilution testing in Gate 1 and the nutrient pulse tracer test in Gate 3. Samples were collected for bromide or chloride, and for conductivity analysis using a peristaltic pump (method in Section A4.3.3). No preservation of these samples was required. Analyses were typically completed in the field using the method described in Section A4.4.3.

A4.2.6 Field Parameters

The field parameters pH, E_h, conductivity, and dissolved oxygen (DO) were measured in all gates as part of background, plume development, and major sampling programs. DO was monitored in the aerobic portions of Gates 1 and 3 during monitoring programs designed to assess oxygen delivery to the aquifer. Also, sampling for pH was included in the aerobic assessment of Gate 1. Probe measurements were completed in the field using a flow-through cell as described in Section A4.4.1; alternate DO methods are described in Section A4.4.2.

A4.2.7 Organic Acids

Samples for organic acids were collected from Gate 3, and from Gate 2 for comparison, during the anaerobic zone assessment. The analyses provided concentrations of formic, acetic, propionic, and butyric acids.

Like VOCs, organic acids are volatile and samples were collected using the UW sampling apparatus, designed to minimize sample exposure to air (description provided in Section A4.3.1). Samples were stored at 4°C until transported to the Barringer. Organic acids have a 7-day holding time.

A4.2.8 Nutrients

Nutrient samples were collected in Gate 3 to assess pulse mixing in the anaerobic zone, and also as part of the major sampling and aerobic zone monitoring programs. An organic nutrient was used to promote microbial activity in the subsurface to create an anaerobic environment that would enhance the degradation of target chlorinated species (PCE and CT). Sodium benzoate was the organic substrate used in this experiment and was thus the target analyte.

Samples were collected using a peristaltic pump, as described in Section A4.3.3. Samples were preserved with formic acid (2 drops per 20-mL sample) to bring the pH to <2. Samples that were not going to be analyzed within 14 days were frozen. Analysis was completed by the UW OGL. Holding times varied, but was not considered problematic as most samples were analyzed within 14 days or frozen after sample collection.

A4.2.9 Water Level/Head Measurements

Water level measurements were taken regularly to demonstrate that the required gradients had been achieved, to observe fluctuations in the water table, and to permit mass balance calculations for contaminants (calculations based on saturated thickness). Details of the equipment used and the method are provided in Section A4.4.5.

A4.3 SAMPLE COLLECTION METHODOLOGY

A detailed description of the methodology used for sample collection and handling is provided in the following sections.

A4.3.1 UW Sampling Apparatus Methodology

The following provides a detailed description of the field collection method using the UW sampling apparatus:

- Equipment: peristaltic pump UW sampling apparatus (Figure A4.1), prelabeled 40-mL vials with septa, purge volume measure, waste container.

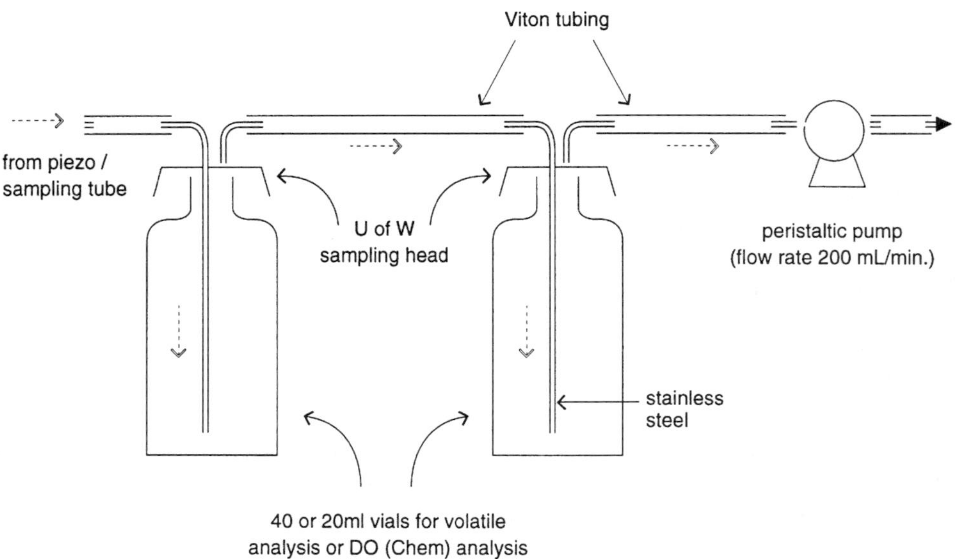

Figure A4.1 UW sampling head setup (schematic).

- The sample train was constructed in the following order (see Figure A4.1): tubing from well (Viton), sample vial(s) for VOCs, DHGs, organic acids, DO (Chemetrics) as required, tubing to pump, peristaltic pump, purge volume measure/waste container. This arrangement prevented agitation of the sample by the pump and also provided a flush of the required purge volume (see Section A4.5.1 for calculations) through the apparatus to rinse between samples.
- The pump was operated at a rate of approximately 200 mL/min. Pumping was stopped when all vials were filled and the appropriate purge volume collected.
- The lead Viton tubing was detached from the well point and held at the same elevation as the UW sampling head while the first vial was carefully unscrewed from the cap. Preservative was added with a syringe (as required). The vial was then filled until brimming with water remaining in the lead Viton tube (the open end of the tubing was lifted above the vial top and the water dripping from the head extension captured). The cap was carefully replaced and tightened. To check for the presence of any air bubbles, the vial was inverted and, if present, the "topping-up" procedure was repeated. This procedure was repeated for each vial in the sampling train.
- Vial labels were checked and Chain of Custody/sample submission forms completed.
- Samples were stored and shipped in dedicated sample coolers equipped with ice packs. Typically, VOC and/or DHG samples were submitted to UW's OGL. In some cases, as required in the WP, samples were shipped to the external laboratory (Barringer), as were organic acid samples. Samples collected for DO (Chemetrics) were analyzed in the field immediately following their collection (method provided in Section A4.4.2).
- Detailed descriptions of laboratory analytical methods are provided in Appendix 3.

A4.3.2 Sampling for Inorganics

Sampling for inorganics included the collection of samples for cation, anion, and alkalinity analyses. The need to filter cation samples made it more convenient to sample using a syringe rather than a peristaltic pump.

- Equipment: 60 mL polypropylene syringe, tubing connectors, syringe filters (0.45 μm pore size), sample containers (analysis specific), waste container, deionized water.
- The tubing and syringe were connected to the desired sampling point and the appropriate purge volume collected (see Section A4.5.1). The water was discarded to the waste container.

- Anion, ammonia, and alkalinity samples were collected first. The tubing and syringe were reconnected to the sampling point and a full syringe volume of water was collected. The syringe was disconnected from the well and discharged into the sample bottle. The procedure was repeated as necessary to fill bottles (100-mL plastic bottle each for anions and field alkalinity; 20-mL plastic scintillation vial for ammonia). Preservative was added, as required.
- Samples for cations were collected next using the same syringe. After a full syringe volume was collected, a 0.45-µm syringe filter was attached to the tip of the syringe. The sample was filtered by carefully pressing on the barrel of the syringe and collecting the filtrate in a 60-mL plastic bottle. Filters were discarded after use to prevent cross-contamination between samples, while syringes were rinsed with deionized water. To the filtered sample, approximately 10 drops of concentrated nitric acid were added to reduce the pH of the sample to below pH 2 (pH checked periodically with pH paper).
- Vial labels were checked and chain of custody/sample submission forms completed.
- Cation and anion samples were stored and shipped in dedicated sample coolers equipped with ice packs. Field alkalinity samples were analyzed immediately in the field (method provided in Section A4.4.4). Background samples were submitted to UW's Water Quality Lab (WQL), while Water Technology International (WTI) Corp. in Burlington, Ontario received all other inorganic samples.
- Detailed descriptions of laboratory analytical methods are provided in Appendix 3.

A4.3.3 Sampling Using Peristaltic Pump

This quick and simple procedure was used for the collection of samples for analyses of nonvolatile parameters:

- Equipment: peristaltic pump, sample containers (analysis specific), purge volume measure, waste container.
- The sampling equipment was arranged as follows: tubing from well to pump tubing, peristaltic pump, purge volume measure/sample container
- The pump was operated at a rate of 300 to 400 mL/min since sample degassing/aeration was not a concern. Following the collection and subsequent disposal of the purge volume, the sample was collected in the appropriate sample container (20-mL plastic scintillation vials for inorganic tracers; 20-mL glass vials for nutrients). Preservative was added, as required.
- Vial labels were checked and Chain of Custody/sample submission forms completed.
- Samples were stored and shipped in dedicated sample coolers equipped with ice packs. Ion-specific electrode analyses for bromide/chloride and conductivity analyses were completed by field personnel as soon as possible after the collection of the samples. In the case of nutrient pulse monitoring, conductivity samples were analyzed in the field within 2 hours of their collection. Nutrient samples were submitted to UW's OGL for analyses. Beginning in 1997, all nutrient samples were frozen upon submission to the laboratory as a long-term preservation measure.
- Detailed descriptions of laboratory analytical methods are provided in Appendix 3.

A4.4 FIELD ANALYSIS METHODS

A detailed description of field analytical techniques is provided in the following sections.

A4.4.1 Probe Measurements of Field Parameters

Probe measurements of pH, E_h, conductivity, and DO provided a fast and effective means of determining these parameters at relatively low cost, but with reasonable accuracy.

- Equipment: peristaltic pump, flow-through cells, waste container.
- Probes and meters:
  - pH — Orion model 260 meter, with Orion model 9107 triode
  - E_h — Markson model 95 meter, with Baxter S500C-ORP Ag/AgCl electrode

- DO — Orion model 835 meter, with Orion model 083010 probe
- Conductivity — Oakton model WD-35607-20

- All probes were calibrated at the beginning of each sampling day and rechecked at the end of the day to ensure the calibration remained valid. Calibration data were recorded in the Field Parameters Log Book. Operation manuals were available at the site in case a problem with calibration arose.
- The sample train was constructed in the following order: tubing from well to pump, peristaltic pump, flow-through cells (analysis specific), purge volume measure/waste container.
- The pump was operated at a rate of 100 mL/min to minimize total volume of water pumped. The well was purged of an appropriate volume (Section A4.5.1) prior to recording values.
- DO and E_h measurements were taken first using the appropriate flow-through cell (Figure A4.2a). The DO probe was seated on the swab in the first cell, while the E_h probe was suspended in the second cell. Once the DO reading stabilized and was recorded, the DO probe was removed and the E_h probe was moved from the second cell to the first cell (minimizing sample exposure to air prior to taking of E_h reading). The E_h probe was allowed to stabilize and the value recorded in the Field Parameters Log Book.
- The pump was disconnected from the DO/Eh flow-through cell and connected to the pH/conductivity flow-through cell (Figure A4.2b). The pump was operated at a rate of approximately 200 mL/min until the cell was filled with water and then slowed to about 100 mL/min while the probes stabilized. Meter readings were recorded in the Field Parameters Log Book.
- Prior to moving to the next sampling point, both flow-through cells were emptied of their contents.

A4.4.2 DO Measurements

During the early stages of the project, a comparison of three methods for measuring DO, including Winkler titration (Greenberg et al., 1985), Chemetrics, and DO probe, was conducted in the field. It was decided that the probe method was preferred for concentrations above 0.5 mg/L, on the basis of accuracy and ease of use. The Chemetrics method was applied to concentrations below 0.5 mg/L.

Methodology for the use of the DO probe is included with the methods for probe measurements of field parameters in Section A4.4.1. The Chemetrics method follows:

- Approximately 20 mL of water was collected in a septa-capped vial using the UW sampling apparatus (Section A4.3.1). Analyses were completed immediately after sample collection.
- The photometer was prepared for analyses according to the product manual. Judgment as to the appropriate filter and ampules to use for a particular sample was made based on previous sampling events. Possible ranges are (1) 1 to 15 mg/L; (2) 0 to 2 mg/L; and (3) 0.0 to 0.8 mg/L. If an error in judgment was made, the analysis was repeated using a different filter and ampule. The photometer was calibrated using a manufacturer-supplied zeroing ampule.
- The sample vial was opened and the appropriate self-filling Chemetrics ampule was broken into the vial immediately. For DO concentration ranges 1 and 2, ampules required 2 minutes stabilization time prior to analysis, while for range 3, samples required analysis within 30 seconds of ampule breaking. Photometric readings were recorded in the Field Parameters Log Book.

A4.4.3 Field Analyses for Inorganic Tracers

Samples for inorganic tracers were collected in 20-mL plastic scintillation vials. Ion selective bromide or chloride probes and conductivity probes could typically be used directly in these sample containers. On some occasions probe measurements were completed in the laboratory at UW.

- Equipment: samples (collected using method in Section A4.3.3), deionized water, waste container.
- Probes and meters:
    - Br^- — Orion model 280 meter, with combination ion-selective probe (Cole Palmer # E-27502-05); also, Corning Br^- electrode (model 476128), in conjunction with Corning Double Junction reference electrode (model 476370)
    - Cl^- — Markson model 95, with Orion model 96-17B ion-selective probe
    - Conductivity — Oakton model WD-35607-20

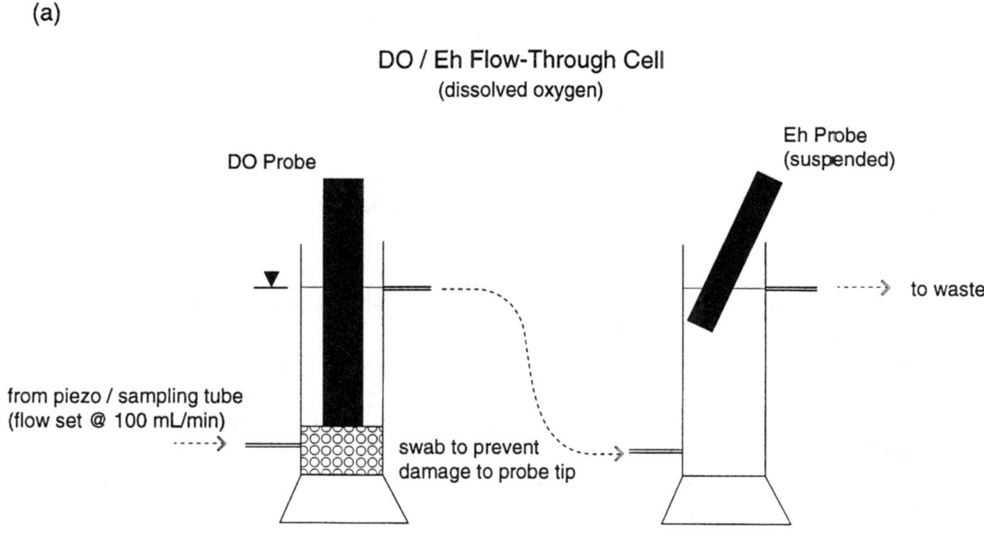

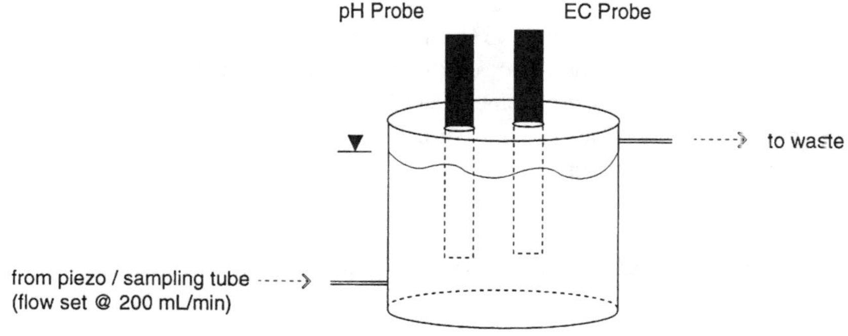

Figure A4.2 Flow-through cells.

- Probes were calibrated (6 standards prepared using deionized water, ranging in concentration from 5 to 1000 mg/L) at the beginning of each sampling day and rechecked at the end of the day to ensure the calibration remained valid. Calibration data were recorded in the Field Parameters Log Book. Operation manuals were available at the site in case a problem with calibration arose.
- Ion-selective probes tend to release ions during analysis that may cause errors in conductivity readings, thus conductivity measurements were completed first. The probe was inserted into the sample and gently stirred. Once the reading stabilized, the value was recorded in the appropriate field book. The probe was removed from the sample and rinsed using deionized water.

- Bromide and/or chloride measurements were then completed using appropriate ion-selective probe. The probe was inserted into the sample and gently stirred. Once the reading stabilized, the value was recorded in the field book. The probe was removed from the sample and rinsed using deionized water.

A4.4.4 Field Alkalinity Measurements

Field alkalinity testing was completed to compare to the results provided by the laboratory. This was considered necessary since acceptable storage times for alkalinity samples are variable and may sometimes be quite short.

- Equipment: samples (collected using method in Section A4.3.3), field alkalinity kit (HACH Test kit, model AL-DT), deionized water, waste container.
- The method used followed precisely that outlined in the manual from the field kit.
- For low alkalinity values, typical of the site, sample volumes of 100 mL were required. The sample was carefully poured into a clean Ehrlenmeyer flask and the colorimetric indicator phenolphthalein was added. Sulfuric acid titrant was added until the desired color change occurred (from pink to clear, transition occurs at about pH 8). This quantity of titrant was recorded in the Field Parameters Log Book. If no color change, i.e., water was a pH <8, was observed upon addition of the indicator (as was typically the case), the analyst proceeded to the next step (i.e., no titrant was added).
- A second colormetric indicator (bromocresol green) was added to the flask and stirred. Titrant was again added until the desired color change (blue-green to violet-grey, pH ~4.5) occurred and this quantity of titrant recorded in the book. Color changes were more noticeable with a piece of white paper placed under the flask.
- The sample was discarded and the Ehrlenmeyer flask rinsed with deionized water.
- Calculations accounting for sample volume, titrant concentration, and volume were completed.

A4.4.5 Water Level/Head Measurements

Water level measurements were completed on a regular basis throughout the duration of the experiment. A water level sounder (model 51453 from Slope Indicator Co., Seattle, WA) was used to measure the depth to water below the top of casing in selected wells. Tops of casing elevations for each well were surveyed with respect to mean sea level on June 3, 1996 using a total station theodolite (Leica model TC600). "Stick-up" measurements for each well were measured on May 22, 1997 to permit the correction of surveyed data-to-ground surface elevations.

A4.5 MISCELLANEOUS SAMPLING INFORMATION

A4.5.1 Purge Volume

The purge volume was calculated as three times the standing well volume. For simplicity, this volume was estimated for the deepest sampling point (i.e., the worst case) and maintained the same for all depths on the multilevel piezometers. The purge volume for the performance wells was estimated based on an average water table elevation and maintained constant over the project duration. Purge volumes were calculated as follows:

1/8 in. stainless steel piezometers:
diameter = 0.318 cm
max. depth = 11 ft = 333 cm bgs
water table depth = 3 ft = 90 cm bgs
standing well volume = $\pi r^2 * h = \pi(0.318/2)^2 * (333 - 90) = 19.3\ cm^3 = 19\ mL$
purge volume = 3*(well volume) = 3*(19 mL) = 57 mL

1.25 in. performance wells:
diameter = 3.18 cm
depth = 12 ft = 363 cm
water table depth = 3 ft = 90 cm
standing well volume = $\pi r^2 * h = \pi\ (3.18/2)^2 * (363 - 90) = 2168\ cm^3 = 2.2\ L$
purge volume = 3*(well volume) = 3*(2.2 L) = 6.6 L

In the case of the performance wells, dedicated low-density polyethylene tubes (LDPE) were installed in each well for purging and sample collection. In addition to the required purging of the well prior to sampling, a 60-mL flush of the sampling equipment between samples was completed to minimize cross-contamination.

A4.5.2 Wastewater Handling

Any wastewater resulting from sampling activities was appropriately disposed in the waste tank located near the extraction wells for Gates 2 and 3. This water was subsequently treated in an aboveground system (Appendix 10).

A4.6 REFERENCES

U.S. EPA, 1990. *Test Methods for Evaluating Solid Waste, Physical/Chemical Methods*. SW-846, 3rd ed., November 1990.

Greenberg, A.E., R.R. Trussell, L.S. Clesceri, and M.A.H. Franson, eds., 1985. *Standard Methods for the Examination of Water and Wastewater*, 16th ed.. American Public Health Association, American Water Works Association, Water Pollution Control Federation, Washington, D.C.

APPENDIX 5

Quality Assurance/Quality Control Procedures and Results

A5.1 INTRODUCTION

In order to maintain the quality objectives for the study, and in accordance with Section 8.2 of the Work Plan, an assessment of the precision and accuracy of the chemical analyses performed was undertaken. In the case of inorganic constituents, laboratory data quality was assessed by several means. Calibration standards, blanks (including trip blanks and equipment rinses), and laboratory duplicates were used to ensure that data quality objectives were met. Reports describing these measures were obtained from each of the laboratories used in this project and are attached as addenda to this appendix. Field data quality was monitored by the collection of field duplicates to assess precision. The results of this program are discussed in the following sections.

Particular emphasis was placed on the assessment of volatile organic data, since these represented the largest block of data collected in the project as well as being the focus of the investigation. The accuracy of these data was assessed on the basis of a batch-by-batch statistical analysis of the calibration curves used in the GC analysis. The method has recently been described by Devlin (1996). This approach was adopted since no reliable way to collect identical samples of water containing volatile organic compounds (VOCs) in the field currently exists. Splitting samples, collecting ground-water samples in immediate succession, or spiking samples are all subject to biases that reflect precision difficulties and hinder the assessment of accuracy determinations. As part of the calibration curve analysis, limits of quantification were determined. These numbers correspond to the lowest concentrations that could be interpreted quantitatively. Lab and equipment blanks were also analyzed to assess analytical carryover between samples and carryover due to equipment contamination, respectively. Precision was assessed by the statistical treatment of the calibration curves as well as by the collection and analysis of field duplicates. The calibration curve analysis afforded the opportunity to assign each analysis a corresponding error estimate at the 95% confidence level, while the field duplicates were assessed on the basis of relative percent differences (RPD) as described below.

A5.2 INORGANIC CONSTITUENTS

A5.2.1 Accuracy

In all cases of inorganic analytes, the detection limits reported by the laboratories were near or below those stated in the Work Plan. It may be concluded from this that the analytical instrumentation used to collect chemical data was sufficiently sensitive to achieve the objectives of the project.

A5.2.2 Precision

On three occasions — December 16, 1996, January 27, 1997, and June 2, 1997 — duplicate field samples were collected for major ions and metals. In total, 27 duplicates were collected for major ion analysis (Table A5.1) and 4 were analyzed for metals (Table A5.1). RPDs were calculated according to the following equation for each duplicate pair, and the average values are tabulated in Table A5.1.

Table A5.1 Summary of QA/QC for Inorganic Constituents

Contents	Units	Avg RPD	Reported MDL
	Major Ions		
Alkalinity as HCO3–	mg/L		3–7
Bromide	mg/L	5.2	0.5
Calcium	mg/L	12.56	0.0645–0.129
Chloride	mg/L	30.91	0.5
Dissolved oxygen probe	mg/L	nd	1
DO, Chemetrics	mg/L	83.41	0.05
Potassium	mg/L	11.78	1.00–1.19
Magnesium	mg/L	4.58	0.0205–0.1
Manganese	mg/L	13.44	0.0064–0.1
Sodium	mg/L	7.86	0.143–1.0
Nitrite	mg/L	nd	0.5
Nitrate	mg/L	nd	0.50–1.50
Phosphate	mg/L	nd	1.00–2.50
Sulfate	mg/L	24.89	0.50–2.00
	Metals		
Silver	mg/L	nd	0.03
Aluminum	mg/L	27.99	0.01
Barium	mg/L	0.98	0.01
Beryllium	mg/L	nd	0.02
Bismuth	mg/L	108.78	0.07
Boron	mg/L	66.67	0.01
Cadmium	mg/L	nd	0.01
Cobalt	mg/L	nd	0.01
Chromium	mg/L	nd	0.02
Copper	mg/L	nd	0.02
Iron	mg/L	69.64	0.10–0.13
Lanthanum	mg/L	80.95	0.03
Lithium	mg/L	nd	0.02
Molybdenum	mg/L	nd	0.02
Nickel	mg/L	133.03	0.01
Lead	mg/L	nd	0.04
Sulfur	mg/L	7.48	
Antimony	mg/L	27.47	0.07
Selenium	mg/L	nd	
Silicon	mg/L	0.65	0.04
Tin	mg/L	77.92	0.04
Strontium	mg/L	1.44	0.01
Titanium	mg/L	nd	0.03
Thalium	mg/L	nd	0.09
Ts	mg/L	nd	0.02
Vanadium	mg/L	31.38	0.03
Tungsten	mg/L	66.67	0.01
Zinc	mg/L	49.21	0.03
Zirconium	mg/L	nd	

$$RPD = \frac{(C_1 - C_2) \times 100\%}{(C_1 + C_2)/2}$$

RPD, as defined above, is equal to the total range of concentrations normalized to the average concentration. Thus, it may be considered, in the first approximation, to be equivalent to twice the uncertainty of a concentration at 95% confidence.

$$RPD = 2(uncertainty\ as \pm \%)$$

The data quality objectives for the inorganic constituents were to obtain precision to within ±10 to 15%, corresponding to a maximum RPD of 30%. This was achieved for all the major ions with measurable concentrations present in the duplicate samples. Ions for which the calculations could not be performed due to low concentrations in the duplicates included nitrate, nitrite, phosphate, and sulfate.

The data quality objectives were not achieved for dissolved oxygen (DO) analyses by the Chemetrics method (±10% vs. ±40%), but this result is not considered compromising since the samples considered in the RPD calculation almost all exhibited DO levels below 0.5 mg/L. The purpose of the DO analyses was to characterize the aquifer as aerobic or anaerobic; this objective was achieved with analyses of 0.5 ± 40%, as indicated by the RPD calculations.

The data quality objectives for metals were not explicitly stated in the Work Plan. By default, a RPD of 30% might be considered the highest acceptable value, as was the case for the major ions. In general, the metals were not analyzed with precision within the desired range. However, the reasons for this were related to the low levels of most metals in the groundwater; concentrations were routinely at or near the method detection limits where reduced precision is to be expected. This result in no way compromised the objectives of the project.

A5.3 ORGANIC CONSTITUENTS

A5.3.1 Accuracy

Although the Work Plan set out instrument detection limits to be achieved, in practice the limits of quantification (LOQ) for the VOCs tested were the real limitations. These were usually one to five times higher than the reported method detection limits (Table A5.2). However, the concentrations for which uncertainty objectives were specified in the Work Plan generally exceeded the achieved LOQs. The compounds found to be the most problematic were carbon tetrachloride (CT) and chloroform (CF), which could not be quantified below a concentration of about 0.025 mg/L, on average (the objective stated in the Work Plan was 0.08 mg/L). Fortunately, with the continuous injection concentrations at or near 1 mg/L, more than a full order of magnitude in concentration could be quantified before CT and CF signals were lost to noise. This range is considered to be sufficient to confidently determine the disappearance of the two compounds due to transformations.

A number of organic substances were not subject to the calibration curve analysis. These include ethene, ethane, propene, propane, methane (analyzed by GC), and benzoate (analyzed by HPLC). In all cases the desired detection limits were achieved.

Blanks, prepared in the laboratory, were analyzed with every analytical batch, while trip and equipment blanks, to assess carryover of analytes in the analytical and field equipment, were analyzed periodically (Table A5.3). In most cases, blanks returned concentrations below detection limits. The exceptions were CT, methane, tetrachloroethene (PCE), trichloroethene (TCE), and toluene (TOL). In all of these cases the concentrations returned in the blanks were more than an

Table A5.2 Summary of QA/QC for Organic Constituents (for Samples Analyzed at the UW Organic Geochemistry Laboratory)

Contents	Units	Avg RPD	Avg ±%	Reported MDL	Average LOQ	Blanks max	Blanks min	Blanks avg
Organics	μg/L							
Chloroform	μg/L	42.31	16.86	9	24.73	nd	nd	nd
Carbon tetrachloride	μg/L	42.88	74.96	5	23.41	6.82	nd	0.35
Carbon disulfide	μg/L	nd	nd	10	15.96	nd	nd	nd
Dichloromethane	μg/L	16.05	24.47	18	22.56	nd	nd	nd
Dichloroethene, *cis*	μg/L			3	7.07	nd	nd	nd
Ethane	μg/L	8.52	nd	1.2		nd	nd	nd
Ethene	μg/L	nd	nd	1.2		nd	nd	nd
Methane	μg/L	25.65	nd	0.7		0.5	nd	0.17
Perchloroethene	μg/L	63.84	31.36	2	7.22	3.68	nd	0.16
Propane	μg/L	9.22	nd	1.8		nd	nd	nd
Propene	μg/L	8.66	nd	1.7		nd	nd	nd
Trichloroethene	μg/L	nd	nd	3	7.34	1.07	nd	0.13
Toluene	μg/L	28.37	11.68	4	16.44	11.28	nd	0.47
Vinyl chloride	μg/L	nd	nd	10	22.3			
Benzoate	μg/L	1.75		5				

Table A5.3 Summary of Trip Blanks and Equipment Rinses (All Samples Analyzed at Barringer Laboratories)

Date	Type	PCE **0.2**	TCE **0.2**	cDCE **0.4**	CT **0.2**	CF **0.2**	DCM **0.6**	CS_2 **0.2**	TOL **0.2**
06-Jan-97	Trip	nd	nd	nd	nd	nd	nd	nd	0.5
06-Jan-97	Equip. Rns	nd	nd	nd	nd	1.7	nd	nd	6
03-Feb-97	Trip	nd	nd	nd	nd	nd	2.4	nd	nd
03-Feb-97	Equip. Rns	nd	nd	nd	nd	1.7	nd	nd	nd
03-Feb-97	Equip. Rns	nd	nd	nd	nd	1.7	nd	nd	nd
03-Mar-97	Equip. Rns	0.6	nd	nd	0.6	2	nd	nd	14.3
03-Mar-97	Trip	nd	nd	nd	nd	nd	nd	nd	nd
03-Mar-97	Equip. Rns	nd	nd	nd	nd	1.7	nd	nd	0.4
31-Mar-97	Trip	nd	nd	nd	nd	nd	2.7	nd	nd
31-Mar-97	Equip. Rns	0.2	nd	nd	nd	2	nd	nd	1.4
31-Mar-97	Equip. Rns	tr	nd	nd	nd	1.7	nd	nd	2.6
04-Apr-97	Equip. Rns	nd	nd	nd	nd	2.1	nd	nd	1.9
04-Apr-97	Trip	nd	nd	nd	nd	nd	nd	nd	nd
04-Apr-97	Equip. Rns	0.3	nd	nd	nd	2.3	nd	nd	2.2
23-Jun-97	Trip	nd	nd	nd	nd	nd	nd	nd	nd

Note: Numbers in bold directly below compound names represent LOQs, in micrograms per liter, reported by Barringer Laboratories.

order of magnitude below the LOQs (determined from the UW geochemistry laboratory), posing no danger of being misidentified as true signals.

Accuracy was also assessed in a test involving the analysis of calibration solutions as part of a normal analytical batch. Two standards containing TOL, PCE, CT, CF, and DCM at concentrations one order of magnitude apart were prepared in sampling vials, transported to the field site, and returned with the other samples. The combined group was analyzed as usual and the measured concentrations of the standards compared with the true concentrations. Small negative biases (–3 to –10%) were noted, with the order of increasing bias being CF < TOL < CT < DCM < PCE. This order does not correlate with either the order of volatility (based on Henry's law constants) or hydrophobicity (based on log K_{ow}s) (refer to Table A5.4), making the cause of the bias uncertain.

Since the biases are below the percent error attributable to analytical uncertainty, the lack of correlation is not surprising and not considered a serious problem.

Table A5.4 Physical and Chemical Properties of Selected VOCs

Compound	Log K_{ow}[a]	H (atm m^3) /mol[b]
Carbon tetrachloride	2.83	0.0298
Chloroform	1.9	0.00358
Methylene chloride	1.25	0.00212
Tetrachloroethene	2.60	0.0174
Toluene	2.65	0.00661[a]

a Data from the Organics Properties Database, University of Waterloo.

b Data from *Dense Chlorinated Solvents and Other DNAPLS in Groundwater*, J.F. Pankow and J.A. Charry, Eds., Waterloo Press, Waterloo, Canada, 1996.

A5.3.2 Precision

RPDs were calculated for about 51 duplicate pairs. The number of detectable concentration sample pairs varied from compound to compound but was in the range of 7 to 27 (except TOL, n = 51) for those compounds found to be present. These included chloroform, CT, methylene chloride (DCM), PCE, and TOL. As discussed above, the RPDs can be considered to be representative of twice the percent uncertainty on either side of the average (taken to be the true) value. Thus, a comparison of the calculated RPDs and the ± % errors determined from the calibration curve analysis should aid in assessing the contribution of analytical error to the observed RPDs from field duplicates. When this is done, it can be seen that the RPDs for all compounds except PCE and CT are less than or equal to 2.5 times the analytical errors. This demonstrates that the majority of the uncertainty in the VOC analyses can be explained as analytical uncertainty.

The data quality objectives for VOCs were about ±25% for all compounds except CT, for which the objective was ±40%. In practice, these objectives were met for all VOCs identified in the field except CT (±75%) and PCE (±30%). In the latter case, the achieved uncertainty level was only slightly in excess of the desired level and posed no significant risk to the data interpretation. In the case of CT, the undesirably high uncertainty estimate is a reflection of the low concentration levels in the aquifer rather than of the analytical performance, in general. CT transformed intrinsically so the measured concentrations were all in the neighborhood of the LOQ. Higher concentrations could have been measured with higher precision, but were not encountered at locations in the treatment gates where the sampling plan called for duplicate analyses to be collected. The apparently low level of CT precision did not compromise the project because CT was attenuated relatively rapidly and did not play a major role in the interpretation of data from within the treatment gates.

A5.4 REFERENCES

Devlin, J.F. 1996. A method to assess analytical uncertainties over large concentration ranges with reference to volatile organics in water, *Ground Water Monitoring and Remediation,* 16(3): 179-185.

A5.5 ALAMEDA EXTERNAL LAB QA/QC SUMMARY

Data Validation checklists were completed once sampling was complete for the samples submitted to Intertek Testing Services (ITS) (see attachments for completed checklists). Common errors in the validation included the following:

- Lab did not specify batch IDs to all data sheets.
- A definition list of short forms and acronyms were not provided with each report.
- Lab work order numbers were not provided.
- Case narratives were not issued.

Most of these errors were corrected when requested — with the exceptions noted on the validation checklists. ITS was contacted several times and requested to provide the information. Unfortunately, most of the information was not retrievable, for two reasons:

1. The laboratory had changed offices (from San Jose, CA to Richardson, TX) and retrieving material from San Jose from Richardson was not feasible.
2. The case narratives were no longer available.

A summary of the VOC concentrations found in the samples submitted to ITS is presented in Table A5.5. The MDL is listed for each compound given the dilution reported by ITS. The missing data files are not considered substantial enough to compromise any conclusions drawn in the report.

Table A5.5 Summary of VOC Concentrations in Groundwater Samples: Alameda

p. 1 of 12

Well ID	R1PA	R1PA Dup	R1PA	R1PA	R1PA	R1PA Dup	R1PA	R1PA Dup
Sample ID	R1PA	R10A	R1PA	R1PA	R1PA	R10	R1PA	R11
Dilution Factor	250.0	1.0	500.0	500.0	1,000.0	500.0	1	1
Date Sampled (m/d/y)	03/04/97	03/04/97	03/19/97	06/17/97	08/04/97	08/04/97	09/30/97	09/30/97
Date Analyzed (m/d/y)	03/10/97	03/12&14/97	03/26/97	06/25&26/97	08/13/97	08/15&16/97	10/08&09/97	10/08&13/97
ITS Report No.	9703042	9703042	9703169	NR	NR	NR	D97-12123	D97-12123
ITS Report Date	03/21/97	03/21/97	04/03/97	07/15/97	08/20/97	08/20/97	10/14/97	10/14/97
Acetone	5,000U	4U	10,000U	10,000U	20,000U	10,000U	20U	20U
Acrylonitrile	NA	NA	NA	NA	NA	NA	5.00U	5.00U
Benzene	1,200U	**79E**	2,500U	2,500U	5,000U	2,500U	**146**	**142**
Bromobenzene	NA	NA	NA	NA	NA	NA	5.00U	5.00U
Bromochloromethane	NA	NA	NA	NA	NA	NA	5.00U	5.00U
Bromodichloromethane	1,200U	0.5U	2,500U	2,500U	5,000U	2,500U	5.00U	5.00U
Bromoform	1,200U	0.5U	2,500U	2,500U	5,000U	2,500U	5.00U	5.00U
Bromomethane	2,500U	0.5U	5,000U	5,000U	10,000U	5,000U	5.00U	5.00U
2-Butanone (MEK)	5,000U	4U	10,000U	10,000U	20,000U	10,000U	100U	100U
Carbon Disulfide	1,200U	0.5U	2,500U	2,500U	5,000U	2,500U	5.00U	5.00U
Carbon Tetrachloride	1,200U	0.5U	2,500U	2,500U	5,000U	2,500U	5.00U	5.00U
Chlorobenzene	1,200U	**10**	2,500U	2,500U	5,000U	2,500U	**22.4**	**19.8**
Chloroethane	2,500U	0.5U	5,000U	5,000U	10,000U	5,000U	5.00U	5.00U
2-Chloroethylvinyl ether	NA	NA	NA	NA	NA	NA	10.0U	10.0U
Chloroform	1,200U	1U	2,500U	2,500U	5,000U	2,500U	5.00U	5.00U
Chloromethane	2,500U	0.5U	5,000U	5,000U	10,000U	5,000U	5.00U	5.00U
2-Chlorotoluene	NA	NA	NA	NA	NA	NA	5.00U	5.00U
4-Chlorotoluene	NA	NA	NA	NA	NA	NA	5.00U	5.00U
Dibromochloromethane	1,200U	0.5U	2,500U	2,500U	5,000U	2,500U	5.00U	5.00U
1,2-Dibromo-3-chloropropane	NA	NA	NA	NA	NA	NA	25.0U	25.0U
1,2-Dibromoethane	NA	NA	NA	NA	NA	NA	5.00U	5.00U
Dibromomethane	NA	NA	NA	NA	NA	NA	5.00U	5.00U
1,2-Dichlorobenzene	1,200U	**51E**	2,500U	2,500U	5,000U	2,500U	**192**	**174**
1,3-Dichlorobenzene	1,200U	0.5U	2,500U	2,500U	5,000U	2,500U	5.00U	5.00U
1,4-Dichlorobenzene	1,200U	**6**	2,500U	2,500U	5,000U	2,500U	**14.2**	**12.3**
trans-1,4-Dichloro-2-butene	NA	NA	NA	NA	NA	NA	100U	100U
1,1-Dichloroethane	1,200U	**56E**	2,500U	2,500U	5,000U	2,500U	**81.4**	**81.1**
1,2-Dichloroethane	1,200U	0.5U	2,500U	2,500U	5,000U	2,500U	5.00U	5.00U
1,1-Dichloroethene	1,200U	**130E**	2,500U	5,000U	10,000U	5,000U	**372**	**396**
cis-1,2-Dichloroethene	**41,000**	**27,000**	**52,000**	**110,000**	**120,000**	**150,000**	**99,600**	**65,300**
trans-1,2-Dichloroethene	1,200U	**170E**	2,500U	5,000U	10,000U	5,000U	**536**	**565**
1,2-Dichloropropane	1,200U	**6**	2,500U	2,500U	5,000U	2,500U	5.00U	5.00U
2,2-Dichloropropane	NA	NA	NA	NA	NA	NA	5.00U	5.00U
1,3-Dichloropropane	NA	NA	NA	NA	NA	NA	5.00U	5.00U
1,1-Dichloropropene	NA	NA	NA	NA	NA	NA	5.00U	5.00U
cis-1,3-Dichloropropene	1,200U	0.5U	2,500U	2,500U	5,000U	2,500U	5.00U	5.00U
trans-1,3-Dichloropropene	1,200U	0.5U	2,500U	2,500U	5,000U	2,500U	5.00U	5.00U
Ethylbenzene	1,200U	**36**	2,500U	5,000U	10,000U	5,000U	**111**	**104**
2-Hexanone	2,500U	4U	5,000U	5,000U	10,000U	5,000U	50.0U	50.0U
Iodomethane	NA	NA	NA	NA	NA	NA	5.00U	5.00U
Methylene Chloride (DCM)	1,200U	1U	2,500U	2,500U	5,000U	2,500U	5.00U	5.00U
4-Methyl-2-Pentanone (MIBK)	2,500U	**58E**	5,000U	5,000U	10,000U	5,000U	**166**	**186**
Styrene	1,200U	0.5U	2,500U	2,500U	5,000U	2,500U	5.00U	5.00U
1,1,1,2-Tetrachloroethane	NA	NA	NA	NA	NA	NA	5.00U	5.00U
1,1,2,2-Tetrachloroethane	1,200U	0.5U	2,500U	2,500U	5,000U	2,500U	5.00U	5.00U
Tetrachloroethene	1,200U	0.5U	2,500U	5,000U	10,000U	5,000U	5.00U	5.00U
1,2,3-Trichlorobenzene	NA	NA	NA	NA	NA	NA	5.00U	5.00U
1,2,4-Trichlorobenzene	NA	NA	NA	NA	NA	NA	5.00U	5.00U
1,1,1-Trichloroethane	1,200U	0.5U	2,500U	2,500U	5,000U	2,500U	5.00U	5.00U
1,1,2-Trichloroethane	1,200U	0.5U	2,500U	2,500U	5,000U	2,500U	5.00U	5.00U
Trichloroethene	1,200U	**5**	2,500U	**5,500**	10,000U	5,000U	**36.3**	**37.3**
Trichlorofluoromethane (Freon 11)	1,200U	0.5U	2,500U	2,500U	5,000U	2,500U	5.00U	5.00U
1,2,3-Trichloropropane	NA	NA	NA	NA	NA	NA	5.00U	5.00U
Trichlorotrifluoroethane (Freon 113)	1,200U	0.5U	2,500U	2,500U	5,000U	2,500U	NA	NA
1,2,4-Trimethylbenzene	NA	NA	NA	NA	NA	NA	**240**	**202**
1,3,5-Trimethylbenzene	NA	NA	NA	NA	NA	NA	**54.9**	**46.4**
Toluene	**1,500**	**570E**	2,500U	**3,800J**	10,000U	**4,200J**	**4,080**	**6,720**
Vinyl Acetate	1,200U	0.5U	2,500U	2,500U	5,000U	2,500U	50.0U	50.0U
Vinyl Chloride	**12,000**	**6,100**	**9,600**	**21,000**	**19,000**	**23,000**	**15,700**	**6,880**
m,p-Xylene	NA	NA	NA	NA	NA	NA	**435**	**400**
o-Xylene	NA	NA	NA	NA	NA	NA	**201**	**192**
Xylenes (total)	1,200U	**160E**	2,500U	2,500U	5,000U	2,500U	NA	NA

Table A5.5 (continued) Summary of VOC Concentrations in Groundwater Samples: Alameda

p. 2 of 12

Well ID	R1PB	R1PB	R1PB	R1PB Dup	R1PB	R1PB	R1PC	R1PC
Sample ID	RIPB	RIPB	R1PB	R2PA	R1PB	RIPB	RIPC	RIPC
Dilution Factor	250.0	1,000.0	1,000.0	200.0	1,000.0	1	100.0	250.0
Date Sampled (m/d/y)	03/04/97	03/19/97	06/17/97	06/17/97	08/04/97	09/30/97	03/04/97	03/19/97
Date Analyzed (m/d/y)	03/07&10/97	03/26/97	06/25/97	06/25&26/97	08/13/97	10/08&09/97	03/10&11/97	03/26/97
ITS Report No.	9703042	9703169	NR	NR	NR	D97-12123	9703042	9703169
ITS Report Date	03/21/97	04/03/97	07/15/97	07/15/97	08/20/97	10/14/97	03/21/97	04/03/97
Acetone	5,000U	20,000U	20,000U	4,000U	20,000U	20U	2,000U	5,000U
Acrylonitrile	NA	NA	NA	NA	NA	5.00U	NA	NA
Benzene	1,200U	5,000U	5,000U	1,000U	5,000U	**156**	500U	1,200U
Bromobenzene	NA	NA	NA	NA	NA	5.00U	NA	NA
Bromochloromethane	NA	NA	NA	NA	NA	5.00U	NA	NA
Bromodichloromethane	1,200U	5,000U	5,000U	1,000U	5,000U	5.00U	500U	1,200U
Bromoform	1,200U	5,000U	5,000U	1,000U	5,000U	5.00U	500U	1,200U
Bromomethane	2,500U	10,000U	10,000U	2,000U	10,000U	5.00U	1,000U	2,500U
2-Butanone (MEK)	5,000U	20,000U	20,000U	4,000U	20,000U	100U	2,000U	5,000U
Carbon Disulfide	1,200U	5,000U	5,000U	1,000U	5,000U	5.00U	500U	1,200U
Carbon Tetrachloride	1,200U	5,000U	5,000U	1,000U	5,000U	5.00U	500U	1,200U
Chlorobenzene	1,200U	5,000U	5,000U	1,000U	5,000U	**28.6**	500U	1,200U
Chloroethane	2,500U	10,000U	10,000U	2,000U	10,000U	5.00U	1,000U	2,500U
2-Chloroethylvinyl ether	NA	NA	NA	NA	NA	10.0U	NA	NA
Chloroform	1,200U	5,000U	5,000U	1,000U	5,000U	5.00U	500U	1,200U
Chloromethane	2,500U	10,000U	10,000U	2,000U	10,000U	5.00U	1,000U	2,500U
2-Chlorotoluene	NA	NA	NA	NA	NA	5.00U	NA	NA
4-Chlorotoluene	NA	NA	NA	NA	NA	5.00U	NA	NA
Dibromochloromethane	1,200U	5,000U	5,000U	1,000U	5,000U	5.00U	500U	1,200U
1,2-Dibromo-3-chloropropane	NA	NA	NA	NA	NA	25.0U	NA	NA
1,2-Dibromoethane	NA	NA	NA	NA	NA	5.00U	NA	NA
Dibromomethane	NA	NA	NA	NA	NA	5.00U	NA	NA
1,2-Dichlorobenzene	1,200U	5,000U	5,000U	1,000U	5,000U	**296**	500U	1,200U
1,3-Dichlorobenzene	1,200U	5,000U	5,000U	1,000U	5,000U	5.00U	500U	1,200U
1,4-Dichlorobenzene	1,200U	5,000U	5,000U	1,000U	5,000U	**26.3**	500U	1,200U
trans-1,4-Dichloro-2-butene	NA	NA	NA	NA	NA	100U	NA	NA
1,1-Dichloroethane	1,200U	5,000U	5,000U	1,000U	5,000U	**105**	500U	1,200U
1,2-Dichloroethane	1,200U	5,000U	5,000U	1,000U	5,000U	5.00U	500U	1,200U
1,1-Dichloroethene	1,200U	5,000U	10,000U	2,000U	10,000U	**516**	500U	1,200U
cis-1,2-Dichloroethene	**74,000**	**62,000**	**76,000**	**71,000**	**110,000**	**117,000**	**23,000**	**11,000**
trans-1,2-Dichloroethene	1,200U	5,000U	10,000U	2,000U	10,000U	**796**	500U	1,200U
1,2-Dichloropropane	1,200U	5,000U	5,000U	1,000U	5,000U	5.00U	500U	1,200U
2,2-Dichloropropane	NA	NA	NA	NA	NA	5.00U	NA	NA
1,3-Dichloropropane	NA	NA	NA	NA	NA	5.00U	NA	NA
1,1-Dichloropropene	NA	NA	NA	NA	NA	5.00U	NA	NA
cis-1,3-Dichloropropene	1,200U	5,000U	5,000U	1,000U	5,000U	5.00U	500U	1,200U
trans-1,3-Dichloropropene	1,200U	5,000U	5,000U	1,000U	5,000U	5.00U	500U	1,200U
Ethylbenzene	1,200U	5,000U	10,000U	2,000U	10,000U	**124**	500U	1,200U
2-Hexanone	2,500U	10,000U	10,000U	2,000U	10,000U	50.0U	1,000U	2,500U
Iodomethane	NA	NA	NA	NA	NA	5.00U	NA	NA
Methylene Chloride (DCM)	1,200U	5,000U	5,000U	1,000U	5,000U	5.00U	500U	1,200U
4-Methyl-2-Pentanone (MIBK)	2,500U	10,000U	10,000U	2,000U	10,000U	**179**	1,000U	2,500U
Styrene	1,200U	5,000U	5,000U	1,000U	5,000U	5.00U	500U	1,200U
1,1,1,2-Tetrachloroethane	NA	NA	NA	NA	NA	5.00U	NA	NA
1,1,2,2-Tetrachloroethane	1,200U	5,000U	5,000U	1,000U	5,000U	5.00U	500U	1,200U
Tetrachloroethene	1,200U	5,000U	10,000U	2,000U	10,000U	5.00U	500U	1,200U
1,2,3-Trichlorobenzene	NA	NA	NA	NA	NA	5.00U	NA	NA
1,2,4-Trichlorobenzene	NA	NA	NA	NA	NA	5.00U	NA	NA
1,1,1-Trichloroethane	1,200U	5,000U	5,000U	1,000U	5,000U	5.00U	500U	1,200U
1,1,2-Trichloroethane	1,200U	5,000U	5,000U	1,000U	5,000U	5.00U	500U	1,200U
Trichloroethene	1,200U	5,000U	10,000U	2,000U	10,000U	**24.7**	500U	1,200U
Trichlorofluoromethane (Freon 11)	1,200U	5,000U	5,000U	1,000U	5,000U	5.00U	500U	1,200U
1,2,3-Trichloropropane	NA	NA	NA	NA	NA	5.00U	NA	NA
Trichlorotrifluoroethane (Freon 113)	1,200U	5,000U	5,000U	1,000U	5,000U	NA	500U	1,200U
1,2,4-Trimethylbenzene	NA	NA	NA	NA	NA	**275**	NA	NA
1,3,5-Trimethylbenzene	NA	NA	NA	NA	NA	**64.4**	NA	NA
Toluene	**2,800**	5,000U	10,000U	**3,900**	10,000U	**4,050**	**2,500**	1,200U
Vinyl Acetate	1,200U	5,000U	5,000U	1,000U	5,000U	50.0U	500U	1,200U
Vinyl Chloride	**16,000**	10,000U	**19,000**	**23,000**	**25,000**	**14,500**	**19,000**	**4,700**
m,p-Xylene	NA	NA	NA	NA	NA	**491**	NA	NA
o-Xylene	NA	NA	NA	NA	NA	**224**	NA	NA
Xylenes (total)	1,200U	5,000U	5,000U	1,000U	5,000U	NA	500U	1,200U

Table A5.5 (continued) Summary of VOC Concentrations in Groundwater Samples: Alameda

p. 3 of 12

Well ID	R1PC	R1PC	R1PC	R1PD	R1PD	R1PE	R1PE
Sample ID	R1PC	R1PC	RIPC	R1PD	RIPD	R1PE	RIPE
Dilution Factor	250.0	100.0	1	500.0	1	200.0	1
Date Sampled (m/d/y)	06/17/97	08/04/97	09/30/97	08/04/97	09/30/97	08/04/97	09/30/97
Date Analyzed (m/d/y)	06/25/97	08/15/97	10/08&09/97	08/15&16/97	10/08&09/97	08/15&16/97	10/08&09/97
ITS Report No.	NR	NR	D97-12123	NR	D97-12123	NR	D97-12123
ITS Report Date	07/15/97	08/20/97	10/14/97	08/20/97	10/14/97	08/20/97	10/14/97
Acetone	5,000U	2,000U	20U	10,000U	20U	4,000U	20U
Acrylonitrile	NA	NA	5.00U	NA	5.00U	NA	5.00U
Benzene	1,200U	500U	**112**	2,500U	**184**	1,000U	**156**
Bromobenzene	NA	NA	5.00U	NA	5.00U	NA	5.00U
Bromochloromethane	NA	NA	5.00U	NA	5.00U	NA	5.00U
Bromodichloromethane	1,200U	500U	5.00U	2,500U	5.00U	1,000U	5.00U
Bromoform	1,200U	500U	5.00U	2,500U	5.00U	1,000U	5.00U
Bromomethane	2,500U	1,000U	5.00U	5,000U	5.00U	2,000U	5.00U
2-Butanone (MEK)	5,000U	2,000U	100U	10,000U	100U	4,000U	100U
Carbon Disulfide	1,200U	500U	5.00U	2,500U	5.00U	1,000U	5.00U
Carbon Tetrachloride	1,200U	500U	5.00U	2,500U	5.00U	1,000U	5.00U
Chlorobenzene	1,200U	500U	**96.2**	2,500U	**13.0**	1,000U	**16.2**
Chloroethane	2,500U	1,000U	5.00U	5,000U	5.00U	2,000U	5.00U
2-Chloroethylvinyl ether	NA	NA	10.0U	NA	10.0U	NA	10.0U
Chloroform	1,200U	500U	5.00U	2,500U	**5.15**	1,000U	5.00U
Chloromethane	2,500U	1,000U	5.00U	5,000U	5.00U	2,000U	5.00U
2-Chlorotoluene	NA	NA	5.00U	NA	5.00U	NA	5.00U
4-Chlorotoluene	NA	NA	5.00U	NA	5.00U	NA	5.00U
Dibromochloromethane	1,200U	500U	5.00U	2,500U	5.00U	1,000U	5.00U
1,2-Dibromo-3-chloropropane	NA	NA	25.0U	NA	25.0U	NA	25.0U
1,2-Dibromoethane	NA	NA	5.00U	NA	5.00U	NA	5.00U
Dibromomethane	NA	NA	5.00U	NA	5.00U	NA	5.00U
1,2-Dichlorobenzene	1,200U	500U	**111**	2,500U	**162**	1,000U	**106**
1,3-Dichlorobenzene	1,200U	500U	5.00U	2,500U	**20.5**	1,000U	5.00U
1,4-Dichlorobenzene	1,200U	500U	**16.0**	2,500U	**15.5**	1,000U	**8.40**
trans-1,4-Dichloro-2-butene	NA	NA	100U	NA	100U	NA	100U
1,1-Dichloroethane	1,200U	500U	**51.3**	2,500U	**103**	1,000U	**74.0**
1,2-Dichloroethane	1,200U	500U	5.00U	2,500U	5.00U	1,000U	5.00U
1,1-Dichloroethene	2,500U	1,000U	**68.3**	5,000U	**499**	2,000U	**298**
cis-1,2-Dichloroethene	**3,200**	**2,900**	**15,300**	**130,000**	**111,000**	**40,000**	**64,800**
trans-1,2-Dichloroethene	2,500U	1,000U	**88.8**	5,000U	**735**	2,000U	**313**
1,2-Dichloropropane	1,200U	500U	5.00U	2,500U	5.00U	1,000U	5.00U
2,2-Dichloropropane	NA	NA	5.00U	NA	5.00U	NA	5.00U
1,3-Dichloropropane	NA	NA	5.00U	NA	5.00U	NA	5.00U
1,1-Dichloropropene	NA	NA	5.00U	NA	5.00U	NA	5.00U
cis-1,3-Dichloropropene	1,200U	500U	5.00U	2,500U	5.00U	1,000U	5.00U
trans-1,3-Dichloropropene	1,200U	500U	5.00U	2,500U	5.00U	1,000U	5.00U
Ethylbenzene	2,500U	1,000U	**70.2**	5,000U	**111**	2,000U	**79.6**
2-Hexanone	2,500U	1,000U	50.0U	5,000U	50.0U	2,000U	50.0U
Iodomethane	NA	NA	5.00U	NA	5.00U	NA	5.00U
Methylene Chloride (DCM)	1,200U	500U	5.00U	2,500U	5.00U	1,000U	5.00U
4-Methyl-2-Pentanone (MIBK)	2,500U	1,000U	100U	5,000U	**129**	2,000U	**117**
Styrene	1,200U	500U	5.00U	2,500U	5.00U	1,000U	5.00U
1,1,1,2-Tetrachloroethane	NA	NA	5.00U	NA	5.00U	NA	5.00U
1,1,2,2-Tetrachloroethane	1,200U	500U	5.00U	2,500U	5.00U	1,000U	5.00U
Tetrachloroethene	2,500U	1,000U	5.00U	5,000U	5.00U	2,000U	5.00U
1,2,3-Trichlorobenzene	NA	NA	5.00U	NA	5.00U	NA	5.00U
1,2,4-Trichlorobenzene	NA	NA	5.00U	NA	5.00U	NA	5.00U
1,1,1-Trichloroethane	1,200U	500U	5.00U	2,500U	5.00U	1,000U	5.00U
1,1,2-Trichloroethane	1,200U	500U	5.00U	2,500U	5.00U	1,000U	5.00U
Trichloroethene	2,500U	1,000U	5.00U	5,000U	**42.6**	2,000U	**33.7**
Trichlorofluoromethane (Freon 11)	1,200U	500U	5.00U	2,500U	5.00U	1,000U	5.00U
1,2,3-Trichloropropane	NA	NA	5.00U	NA	5.00U	NA	5.00U
Trichlorotrifluoroethane (Freon 113)	1,200U	500U	NA	2,500U	NA	1,000U	NA
1,2,4-Trimethylbenzene	NA	NA	**192**	NA	**222**	NA	**148**
1,3,5-Trimethylbenzene	NA	NA	**57.7**	NA	**44.3**	NA	**41.4**
Toluene	**1,500J**	1,000U	**2,160**	5,000U	**4,240**	2,000U	**3,320**
Vinyl Acetate	1,200U	500U	50.0U	2,500U	50.0U	1,000U	50.0U
Vinyl Chloride	**13,000**	**17,000**	**16,300**	**36,000**	**22,100**	**28,000**	**18,600**
m,p-Xylene	NA	NA	**235**	NA	**415**	NA	**270**
o-Xylene	NA	NA	**102**	NA	**189**	NA	**116**
Xylenes (total)	1,200U	500U	NA	2,500U	NA	1,000U	NA

Table A5.5 (continued) Summary of VOC Concentrations in Groundwater Samples: Alameda

p. 4 of 12

Well ID	R5PA	R5PA	R5PA Dup	R5PA	R5PA	R5PA	R5PB
Sample ID	R5PA	R5PA	R10	R5PA	R5PA	R5PA	R5PB
Dilution Factor	1.0	1.0	1.0	10.0	20.0	1	1.0
Date Sampled (m/d/y)	03/04/97	03/19/97	03/19/97	06/17/97	08/04/97	09/30/97	03/04/97
Date Analyzed (m/d/y)	03/12&13/97	03/23&29/97	03/26&29/97	06/25&26/97	08/15/97	10/08/97	03/12&13/97
ITS Report No.	9703042	9703169	9703169	NR	NR	D97-12123	9703042
ITS Report Date	03/21/97	04/03/97	04/03/97	07/15/97	08/20/97	10/14/97	03/21/97
Acetone	4U	4U	4U	200U	400U	20U	4U
Acrylonitrile	NA	NA	NA	NA	NA	5.00U	NA
Benzene	0.5U	0.9	0.9	50U	100U	5.00U	0.5U
Bromobenzene	NA	NA	NA	NA	NA	5.00U	NA
Bromochloromethane	NA	NA	NA	NA	NA	5.00U	NA
Bromodichloromethane	0.5U	0.5U	0.5U	50U	100U	5.00U	0.5U
Bromoform	0.5U	0.5U	0.5U	50U	100U	5.00U	0.5U
Bromomethane	0.5U	0.5U	0.5U	100U	200U	5.00U	0.5U
2-Butanone (MEK)	4U	4U	4U	200U	400U	100U	4U
Carbon Disulfide	0.5U	0.6	0.5	50U	100U	5.00U	0.5U
Carbon Tetrachloride	0.5U	0.5U	0.5U	50U	100U	5.00U	0.5U
Chlorobenzene	0.5U	0.5U	0.5U	50U	100U	5.00U	0.5U
Chloroethane	0.5U	4	0.5U	100U	200U	5.00U	0.5U
2-Chloroethylvinyl ether	NA	NA	NA	NA	NA	10.0U	NA
Chloroform	0.5U	0.5U	0.5U	50U	100U	5.00U	0.5U
Chloromethane	0.5U	0.5U	0.5U	100U	200U	5.00U	0.5U
2-Chlorotoluene	NA	NA	NA	NA	NA	5.00U	NA
4-Chlorotoluene	NA	NA	NA	NA	NA	5.00U	NA
Dibromochloromethane	0.5U	0.5U	0.5U	50U	100U	5.00U	0.5U
1,2-Dibromo-3-chloropropane	NA	NA	NA	NA	NA	25.0U	NA
1,2-Dibromoethane	NA	NA	NA	NA	NA	5.00U	NA
Dibromomethane	NA	NA	NA	NA	NA	5.00U	NA
1,2-Dichlorobenzene	0.5U	0.5U	0.5U	50U	100U	5.00U	0.5U
1,3-Dichlorobenzene	0.5U	0.5U	0.5U	50U	100U	5.00U	0.5U
1,4-Dichlorobenzene	0.5U	0.5U	0.5U	50U	100U	5.00U	0.5U
trans-1,4-Dichloro-2-butene	NA	NA	NA	NA	NA	100U	NA
1,1-Dichloroethane	0.5U	13	12	50U	100U	5.00U	0.5U
1,2-Dichloroethane	0.5U	0.5U	2	50U	100U	5.00U	0.5U
1,1-Dichloroethene	1	0.8	0.8	100U	200U	5.00U	0.5U
cis-1,2-Dichloroethene	310	1,300	1,400	3,200	2,200	55.2	270
trans-1,2-Dichloroethene	2	3	3	100U	200U	5.00U	0.6
1,2-Dichloropropane	0.5U	0.5U	0.5U	50U	100U	5.00U	0.5U
2,2-Dichloropropane	NA	NA	NA	NA	NA	5.00U	NA
1,3-Dichloropropane	NA	NA	NA	NA	NA	5.00U	NA
1,1-Dichloropropene	NA	NA	NA	NA	NA	5.00U	NA
cis-1,3-Dichloropropene	0.5U	0.5U	0.5U	50U	100U	5.00U	0.5U
trans-1,3-Dichloropropene	0.5U	0.5U	0.5U	50U	100U	5.00U	0.5U
Ethylbenzene	0.5U	0.5U	0.5U	100U	200U	5.00U	0.5U
2-Hexanone	4U	4U	4U	100U	200U	50.0U	4U
Iodomethane	NA	NA	NA	NA	NA	5.00U	NA
Methylene Chloride (DCM)	1U	1U	1U	50U	100U	5.00U	1U
4-Methyl-2-Pentanone (MIBK)	4U	4U	4U	100U	200U	100U	4U
Styrene	0.5U	0.5U	0.5U	50U	100U	5.00U	0.5U
1,1,1,2-Tetrachloroethane	NA	NA	NA	NA	NA	5.00U	NA
1,1,2,2-Tetrachloroethane	0.5U	0.5U	0.5U	50U	100U	5.00U	0.5U
Tetrachloroethene	0.5U	0.5U	0.5U	100U	200U	5.00U	0.5U
1,2,3-Trichlorobenzene	NA	NA	NA	NA	NA	5.00U	NA
1,2,4-Trichlorobenzene	NA	NA	NA	NA	NA	5.00U	NA
1,1,1-Trichloroethane	0.5U	0.5U	0.5U	50U	100U	5.00U	0.5U
1,1,2-Trichloroethane	0.5U	0.5U	0.5U	50U	100U	5.00U	0.5U
Trichloroethene	0.8	0.5U	0.5U	100U	200U	5.00U	0.8
Trichlorofluoromethane (Freon 11)	0.5U	0.5U	0.5U	50U	100U	5.00U	0.5U
1,2,3-Trichloropropane	NA	NA	NA	NA	NA	5.00U	NA
Trichlorotrifluoroethane (Freon 113)	0.5U	0.5U	0.5U	50U	100U	NA	0.5U
1,2,4-Trimethylbenzene	NA	NA	NA	NA	NA	5.00U	NA
1,3,5-Trimethylbenzene	NA	NA	NA	NA	NA	5.00U	NA
Toluene	7	10	10	100U	410	143	5
Vinyl Acetate	0.5U	0.5U	0.5U	50U	100U	50.0U	0.5U
Vinyl Chloride	56	1,100	1,200	840	2,000	138	73
m,p-Xylene	NA	NA	NA	NA	NA	5.00U	NA
o-Xylene	NA	NA	NA	NA	NA	5.00U	NA
Xylenes (total)	0.7	0.9	1	50U	100U	NA	0.5U

Table A5.5 (continued) Summary of VOC Concentrations in Groundwater Samples: Alameda

p. 5 of 12

Well ID	R5PB	R5PB	R5PB	R5PB	R5PC	R5PC	R5PC	R5PC
Sample ID	R5PB	R5PB	R5PB	R5PB	R5PC	R5PC	R5PC	R5PC
Dilution Factor	1.0	5.0	10.0	1	1.0	1.0	10.0	5.0
Date Sampled (m/d/y)	03/19/97	06/17/97	08/04/97	09/30/97	03/04/97	03/19/97	06/17/97	08/04/97
Date Analyzed (m/d/y)	03/23&26/97	06/25&26/97	08/15/97	10/08/97	03/12&13/97	03/23&26/97	06/26/97	08/15/97
ITS Report No.	9703169	NR	NR	D97-12123	9703042	9703169	NR	NR
ITS Report Date	04/03/97	07/15/97	08/20/97	10/14/97	03/21/97	04/03/97	07/15/97	08/20/97
Acetone	4U	100U	200U	20U	4U	4U	200U	100U
Acrylonitrile	NA	NA	NA	5.00U	NA	NA	NA	NA
Benzene	0.5U	25U	50U	5.00U	0.5U	0.5U	50U	25U
Bromobenzene	NA	NA	NA	5.00U	NA	NA	NA	NA
Bromochloromethane	NA	NA	NA	5.00U	NA	NA	NA	NA
Bromodichloromethane	0.5U	25U	50U	5.00U	0.5U	0.5U	50U	25U
Bromoform	0.5U	25U	50U	5.00U	0.5U	0.5U	50U	25U
Bromomethane	0.5U	50U	100U	5.00U	0.5U	0.5U	100U	50U
2-Butanone (MEK)	4U	100U	200U	100U	4U	4U	200U	100U
Carbon Disulfide	0.5U	25U	50U	5.00U	0.5U	**0.8**	50U	25U
Carbon Tetrachloride	0.5U	25U	50U	5.00U	0.5U	0.5U	50U	25U
Chlorobenzene	0.5U	25U	50U	5.00U	0.5U	0.5U	50U	25U
Chloroethane	0.5U	50U	100U	5.00U	0.5U	0.5U	100U	50U
2-Chloroethylvinyl ether	NA	NA	NA	10.0U	NA	NA	NA	NA
Chloroform	0.5U	25U	50U	5.00U	0.5U	0.5U	50U	25U
Chloromethane	0.5U	50U	100U	5.00U	0.5U	0.5U	100U	50U
2-Chlorotoluene	NA	NA	NA	5.00U	NA	NA	NA	NA
4-Chlorotoluene	NA	NA	NA	5.00U	NA	NA	NA	NA
Dibromochloromethane	0.5U	25U	50U	5.00U	0.5U	0.5U	50U	25U
1,2-Dibromo-3-chloropropane	NA	NA	NA	25.0U	NA	NA	NA	NA
1,2-Dibromoethane	NA	NA	NA	5.00U	NA	NA	NA	NA
Dibromomethane	NA	NA	NA	5.00U	NA	NA	NA	NA
1,2-Dichlorobenzene	0.5U	25U	50U	5.00U	0.5U	0.5U	50U	25U
1,3-Dichlorobenzene	0.5U	25U	50U	5.00U	0.5U	0.5U	50U	25U
1,4-Dichlorobenzene	0.5U	25U	50U	5.00U	0.5U	0.5U	50U	25U
trans-1,4-Dichloro-2-butene	NA	NA	NA	100U	NA	NA	NA	NA
1,1-Dichloroethane	**4**	25U	50U	5.00U	**1**	**5**	50U	25U
1,2-Dichloroethane	0.5U	25U	50U	5.00U	0.5U	0.5U	50U	25U
1,1-Dichloroethene	0.5U	50U	100U	5.00U	0.5U	0.5U	100U	50U
cis-1,2-Dichloroethene	**620**	**2,900**	**740**	**55.5**	**200**	**480**	**2,700**	**710**
trans-1,2-Dichloroethene	**0.8**	50U	100U	5.00U	0.5U	0.5U	100U	50U
1,2-Dichloropropane	0.5U	25U	50U	5.00U	0.5U	0.5U	50U	25U
2,2-Dichloropropane	NA	NA	NA	5.00U	NA	NA	NA	NA
1,3-Dichloropropane	NA	NA	NA	5.00U	NA	NA	NA	NA
1,1-Dichloropropene	NA	NA	NA	5.00U	NA	NA	NA	NA
cis-1,3-Dichloropropene	0.5U	25U	50U	5.00U	0.5U	0.5U	50U	25U
trans-1,3-Dichloropropene	0.5U	25U	50U	5.00U	0.5U	0.5U	50U	25U
Ethylbenzene	0.5U	50U	100U	5.00U	0.5U	0.5U	100U	50U
2-Hexanone	4U	50U	100U	50.0U	4U	4U	100U	50U
Iodomethane	NA	NA	NA	5.00U	NA	NA	NA	NA
Methylene Chloride (DCM)	1U	25U	50U	5.00U	1U	1U	50U	25U
4-Methyl-2-Pentanone (MIBK)	4U	50U	100U	100U	4U	4U	100U	50U
Styrene	0.5U	25U	50U	5.00U	0.5U	0.5U	50U	25U
1,1,1,2-Tetrachloroethane	NA	NA	NA	5.00U	NA	NA	NA	NA
1,1,2,2-Tetrachloroethane	0.5U	25U	50U	5.00U	0.5U	0.5U	50U	25U
Tetrachloroethene	0.5U	50U	100U	5.00U	0.5U	0.5U	100U	50U
1,2,3-Trichlorobenzene	NA	NA	NA	5.00U	NA	NA	NA	NA
1,2,4-Trichlorobenzene	NA	NA	NA	5.00U	NA	NA	NA	NA
1,1,1-Trichloroethane	0.5U	25U	50U	5.00U	0.5U	0.5U	50U	25U
1,1,2-Trichloroethane	0.5U	25U	50U	5.00U	0.5U	0.5U	50U	25U
Trichloroethene	**0.5**	50U	100U	5.00U	**0.5**	0.5U	100U	50U
Trichlorofluoromethane (Freon 11)	0.5U	25U	50U	5.00U	0.5U	0.5U	50U	25U
1,2,3-Trichloropropane	NA	NA	NA	5.00U	NA	NA	NA	NA
Trichlorotrifluoroethane (Freon 113)	0.5U	25U	50U	NA	0.5U	0.5U	50U	25U
1,2,4-Trimethylbenzene	NA	NA	NA	5.00U	NA	NA	NA	NA
1,3,5-Trimethylbenzene	NA	NA	NA	5.00U	NA	NA	NA	NA
Toluene	**5**	50U	100U	**6.85**	**4**	**4**	100U	50U
Vinyl Acetate	0.5U	25U	50U	50.0U	0.5U	0.5U	50U	25U
Vinyl Chloride	**310**	**600**	**200**	**62.2**	**250**	**360**	**520**	**150**
m,p-Xylene	NA	NA	NA	5.00U	NA	NA	NA	NA
o-Xylene	NA	NA	NA	5.00U	NA	NA	NA	NA
Xylenes (total)	0.5U	25U	50U	NA	0.5U	0.5U	50U	25U

Table A5.5 (continued) Summary of VOC Concentrations in Groundwater Samples: Alameda

p. 6 of 12

Well ID	R5PC	R7PA	R7PA Dup	R7PA	R7PA	R7PA	R7PA	R7PB
Sample ID	R5PC	R7PA	R9A	R7PA	R7PA	R7PA	R7PA	R7PB
Dilution Factor	1	1.0	1.0	1.0	5.0	5.0	1	1.0
Date Sampled (m/d/y)	09/30/97	03/04/97	03/04/97	03/19/97	06/16/97	08/04/97	09/30/97	03/04/97
Date Analyzed (m/d/y)	10/08/97	03/13/97	03/12&13/97	03/23&26/97	06/25&26/97	08/15&16/97	10/07/97	03/13/97
ITS Report No.	D97-12123	9703042	9703042	9703169	NR	NR	D97-12123	9703042
ITS Report Date	10/14/97	03/21/97	03/21/97	04/03/97	07/15/97	08/20/97	10/14/97	03/21/97
Acetone	20U	**39**	**38**	4U	100U	**32J**	20U	**37**
Acrylonitrile	5.00U	NA	NA	NA	NA	NA	5.00U	NA
Benzene	5.00U	0.5U	0.5U	0.5U	25U	25U	5.00U	0.5U
Bromobenzene	5.00U	NA	NA	NA	NA	NA	5.00U	NA
Bromochloromethane	5.00U	NA	NA	NA	NA	NA	5.00U	NA
Bromodichloromethane	5.00U	0.5U	0.5U	0.5U	25U	25U	5.00U	0.5U
Bromoform	5.00U	0.5U	0.5U	0.5U	25U	25U	5.00U	0.5U
Bromomethane	5.00U	0.5U	0.5U	0.5U	50U	50U	5.00U	0.5U
2-Butanone (MEK)	100U	4U	4U	4U	100U	100U	100U	4U
Carbon Disulfide	5.00U	0.5U	0.5U	0.5U	25U	25U	5.00U	0.5U
Carbon Tetrachloride	5.00U	0.5U	0.5U	0.5U	25U	25U	5.00U	0.5U
Chlorobenzene	5.00U	0.5U	0.5U	0.5U	25U	25U	5.00U	0.5U
Chloroethane	5.00U	0.5U	0.5U	0.5U	50U	50U	5.00U	0.5U
2-Chloroethylvinyl ether	10.0U	NA	NA	NA	NA	NA	10.0U	NA
Chloroform	5.00U	0.5U	0.5U	0.5U	25U	25U	5.00U	0.5U
Chloromethane	5.00U	0.5U	0.5U	0.5U	50U	50U	5.00U	0.5U
2-Chlorotoluene	5.00U	NA	NA	NA	NA	NA	5.00U	NA
4-Chlorotoluene	5.00U	NA	NA	NA	NA	NA	5.00U	NA
Dibromochloromethane	5.00U	0.5U	0.5U	0.5U	25U	25U	5.00U	0.5U
1,2-Dibromo-3-chloropropane	25.0U	NA	NA	NA	NA	NA	25.0U	NA
1,2-Dibromoethane	5.00U	NA	NA	NA	NA	NA	5.00U	NA
Dibromomethane	5.00U	NA	NA	NA	NA	NA	5.00U	NA
1,2-Dichlorobenzene	5.00U	0.5U	0.5U	0.5U	25U	25U	5.00U	0.5U
1,3-Dichlorobenzene	5.00U	0.5U	0.5U	0.5U	25U	25U	5.00U	0.5U
1,4-Dichlorobenzene	5.00U	0.5U	0.5U	0.5U	25U	25U	5.00U	0.5U
trans-1,4-Dichloro-2-butene	100U	NA	NA	NA	NA	NA	100U	NA
1,1-Dichloroethane	5.00U	0.5U	0.5U	**4**	25U	25U	5.00U	0.5U
1,2-Dichloroethane	5.00U	0.5U	0.5U	0.5U	25U	25U	5.00U	0.5U
1,1-Dichloroethene	5.00U	0.5U	0.5U	0.5U	50U	50U	5.00U	0.5U
cis-1,2-Dichloroethene	**91.5**	**200**	**240**	**720**	**2,400**	**990**	**81.1**	**260**
trans-1,2-Dichloroethene	5.00U	**1**	**0.9**	**2**	50U	50U	5.00U	**0.8**
1,2-Dichloropropane	5.00U	0.5U	0.5U	0.5U	25U	25U	5.00U	0.5U
2,2-Dichloropropane	5.00U	NA	NA	NA	NA	NA	5.00U	NA
1,3-Dichloropropane	5.00U	NA	NA	NA	NA	NA	5.00U	NA
1,1-Dichloropropene	5.00U	NA	NA	NA	NA	NA	5.00U	NA
cis-1,3-Dichloropropene	5.00U	0.5U	0.5U	0.5U	25U	25U	5.00U	0.5U
trans-1,3-Dichloropropene	5.00U	0.5U	0.5U	0.5U	25U	25U	5.00U	0.5U
Ethylbenzene	5.00U	0.5U	0.5U	0.5U	50U	50U	5.00U	0.5U
2-Hexanone	50.0U	4U	4U	4U	50U	50U	50.0U	4U
Iodomethane	5.00U	NA	NA	NA	NA	NA	5.00U	NA
Methylene Chloride (DCM)	5.00U	1U	1U	1U	25U	25U	5.00U	1U
4-Methyl-2-Pentanone (MIBK)	100U	4U	4U	4U	50U	50U	100U	4U
Styrene	5.00U	0.5U	0.5U	0.5U	25U	25U	5.00U	0.5U
1,1,1,2-Tetrachloroethane	5.00U	NA	NA	NA	NA	NA	5.00U	NA
1,1,2,2-Tetrachloroethane	5.00U	0.5U	0.5U	0.5U	25U	25U	5.00U	0.5U
Tetrachloroethene	5.00U	0.5U	0.5U	0.5U	50U	50U	5.00U	0.5U
1,2,3-Trichlorobenzene	5.00U	NA	NA	NA	NA	NA	5.00U	NA
1,2,4-Trichlorobenzene	5.00U	NA	NA	NA	NA	NA	5.00U	NA
1,1,1-Trichloroethane	5.00U	0.5U	0.5U	0.5U	25U	25U	5.00U	0.5U
1,1,2-Trichloroethane	5.00U	0.5U	0.5U	0.5U	25U	25U	5.00U	0.5U
Trichloroethene	5.00U	**0.9**	**0.9**	**0.6**	50U	50U	5.00U	**0.9**
Trichlorofluoromethane (Freon 11)	5.00U	0.5U	0.5U	0.5U	25U	25U	5.00U	0.5U
1,2,3-Trichloropropane	5.00U	NA	NA	NA	NA	NA	5.00U	NA
Trichlorotrifluoroethane (Freon 113)	NA	0.5U	0.5U	0.5U	25U	25U	NA	0.5U
1,2,4-Trimethylbenzene	5.00U	NA	NA	NA	NA	NA	5.00U	NA
1,3,5-Trimethylbenzene	5.00U	NA	NA	NA	NA	NA	5.00U	NA
Toluene	5.00U	**4**	**4**	**5**	50U	**34J**	**25.9**	**4**
Vinyl Acetate	50.0U	0.5U	0.5U	0.5U	25U	25U	50.0U	0.5U
Vinyl Chloride	**46.7**	**35**	**50**	**350**	**540**	**430**	**91.5**	**54**
m,p-Xylene	5.00U	NA	NA	NA	NA	NA	5.00U	NA
o-Xylene	5.00U	NA	NA	NA	NA	NA	5.00U	NA
Xylenes (total)	NA	0.5U	0.5U	0.5U	25U	25U	NA	0.5U

Table A5.5 (continued) Summary of VOC Concentrations in Groundwater Samples: Alameda

p. 7 of 12

Well ID	R7PB	R7PB	R7PB	R7PB	R7PC	R7PC	R7PC	R7PC
Sample ID	R7PB	R7PB	R7PB	R7PB	R7PC	R7PC	R7PC	R7PC
Dilution Factor	1.0	5.0	5.0	1	1.0	1.0	5.0	5.0
Date Sampled (m/d/y)	03/19/97	06/16/97	08/04/97	09/30/97	03/04/97	03/19/97	06/16/97	08/04/97
Date Analyzed (m/d/y)	03/23&26/97	06/25&26/97	08/15/97	10/08/97	03/13&14/97	03/23&26/97	06/26/97	08/15/97
ITS Report No.	9703169	NR	NR	D97-12123	9703042	9703169	NR	NR
ITS Report Date	04/03/97	07/15/97	08/20/97	10/14/97	03/21/97	04/03/97	07/15/97	08/20/97
Acetone	**54E**	100U	100U	20U	**39**	**45E**	100U	100U
Acrylonitrile	NA	NA	NA	5.00U	NA	NA	NA	NA
Benzene	0.5U	25U	25U	5.00U	0.5U	0.5U	25U	25U
Bromobenzene	NA	NA	NA	5.00U	NA	NA	NA	NA
Bromochloromethane	NA	NA	NA	5.00U	NA	NA	NA	NA
Bromodichloromethane	0.5U	25U	25U	5.00U	0.5U	0.5U	25U	25U
Bromoform	0.5U	25U	25U	5.00U	0.5U	0.5U	25U	25U
Bromomethane	0.5U	50U	50U	5.00U	0.5U	0.5U	50U	50U
2-Butanone (MEK)	4U	100U	100U	100U	4U	4U	100U	100U
Carbon Disulfide	0.5U	25U	25U	5.00U	0.5U	0.5U	25U	25U
Carbon Tetrachloride	0.5U	25U	25U	5.00U	0.5U	0.5U	25U	25U
Chlorobenzene	0.5U	25U	25U	5.00U	0.5U	0.5U	25U	25U
Chloroethane	0.5U	50U	50U	5.00U	0.5U	0.5U	50U	50U
2-Chloroethylvinyl ether	NA	NA	NA	10.0U	NA	NA	NA	NA
Chloroform	0.5U	25U	25U	5.00U	0.5U	0.5U	25U	25U
Chloromethane	0.5U	50U	50U	5.00U	0.5U	0.5U	50U	50U
2-Chlorotoluene	NA	NA	NA	5.00U	NA	NA	NA	NA
4-Chlorotoluene	NA	NA	NA	5.00U	NA	NA	NA	NA
Dibromochloromethane	0.5U	25U	25U	5.00U	0.5U	0.5U	25U	25U
1,2-Dibromo-3-chloropropane	NA	NA	NA	25.0U	NA	NA	NA	NA
1,2-Dibromoethane	NA	NA	NA	5.00U	NA	NA	NA	NA
Dibromomethane	NA	NA	NA	5.00U	NA	NA	NA	NA
1,2-Dichlorobenzene	0.5U	25U	25U	5.00U	0.5U	0.5U	25U	25U
1,3-Dichlorobenzene	0.5U	25U	25U	5.00U	0.5U	0.5U	25U	25U
1,4-Dichlorobenzene	0.5U	25U	25U	5.00U	0.5U	0.5U	25U	25U
trans-1,4-Dichloro-2-butene	NA	NA	NA	100U	NA	NA	NA	NA
1,1-Dichloroethane	**3**	25U	25U	5.00U	0.5U	**3**	25U	25U
1,2-Dichloroethane	0.5U	25U	25U	5.00U	0.5U	0.5U	25U	25U
1,1-Dichloroethene	0.5U	50U	50U	5.00U	0.5U	0.5U	50U	50U
cis-1,2-Dichloroethene	**490**	**3,100**	**860**	**117**	**250**	**450**	**3,000**	**820**
trans-1,2-Dichloroethene	**0.8**	50U	50U	5.00U	**0.8**	**0.7**	50U	50U
1,2-Dichloropropane	0.5U	25U	25U	5.00U	0.5U	0.5U	25U	25U
2,2-Dichloropropane	NA	NA	NA	5.00U	NA	NA	NA	NA
1,3-Dichloropropane	NA	NA	NA	5.00U	NA	NA	NA	NA
1,1-Dichloropropene	NA	NA	NA	5.00U	NA	NA	NA	NA
cis-1,3-Dichloropropene	0.5U	25U	25U	5.00U	0.5U	0.5U	25U	25U
trans-1,3-Dichloropropene	0.5U	25U	25U	5.00U	0.5U	0.5U	25U	25U
Ethylbenzene	0.5U	50U	50U	5.00U	0.5U	0.5U	50U	50U
2-Hexanone	4U	50U	50U	50.0U	4U	4U	50U	50U
Iodomethane	NA	NA	NA	5.00U	NA	NA	NA	NA
Methylene Chloride (DCM)	1U	25U	25U	5.00U	1U	1U	25U	25U
4-Methyl-2-Pentanone (MIBK)	4U	50U	50U	100U	4U	4U	50U	50U
Styrene	0.5U	25U	25U	5.00U	0.5U	0.5U	25U	25U
1,1,1,2-Tetrachloroethane	NA	NA	NA	5.00U	NA	NA	NA	NA
1,1,2,2-Tetrachloroethane	0.5U	25U	25U	5.00U	0.5U	0.5U	25U	25U
Tetrachloroethene	0.5U	50U	50U	5.00U	0.5U	0.5U	50U	50U
1,2,3-Trichlorobenzene	NA	NA	NA	5.00U	NA	NA	NA	NA
1,2,4-Trichlorobenzene	NA	NA	NA	5.00U	NA	NA	NA	NA
1,1,1-Trichloroethane	0.5U	25U	25U	5.00U	0.5U	0.5U	25U	25U
1,1,2-Trichloroethane	0.5U	25U	25U	5.00U	0.5U	0.5U	25U	25U
Trichloroethene	**0.6**	50U	50U	5.00U	**0.9**	**0.6**	50U	50U
Trichlorofluoromethane (Freon 11)	0.5U	25U	25U	5.00U	0.5U	0.5U	25U	25U
1,2,3-Trichloropropane	NA	NA	NA	5.00U	NA	NA	NA	NA
Trichlorotrifluoroethane (Freon 113)	0.5U	25U	25U	NA	0.5U	0.5U	25U	25U
1,2,4-Trimethylbenzene	NA	NA	NA	5.00U	NA	NA	NA	NA
1,3,5-Trimethylbenzene	NA	NA	NA	5.00U	NA	NA	NA	NA
Toluene	**4**	50U	50U	**15.5**	**4**	**4**	50U	50U
Vinyl Acetate	0.5U	25U	25U	50.0U	0.5U	0.5U	25U	25U
Vinyl Chloride	**290**	**550**	**260**	**164**	**51**	**260**	**560**	**260**
m,p-Xylene	NA	NA	NA	5.00U	NA	NA	NA	NA
o-Xylene	NA	NA	NA	5.00U	NA	NA	NA	NA
Xylenes (total)	0.5U	25U	25U	NA	0.5U	0.5U	25U	25U

Table A5.5 (continued) Summary of VOC Concentrations in Groundwater Samples: Alameda

p. 8 of 12

Well ID	R7PC	R8PA	R8PA	R8PA	R8PA	R8PA	R8PB	R8PB
Sample ID	R7PC	R8PA	R8PA	R8PA	R8PA	R8PA	R8PB	R8PB
Dilution Factor	1	1.0	1.0	10.0	10.0	1	1.0	1.0
Date Sampled (m/d/y)	09/30/97	03/04/97	03/19/97	06/16/97	08/04/97	09/30/97	03/04/97	03/19/97
Date Analyzed (m/d/y)	10/08/97	03/11&14/97	03/25&26/97	06/26/97	08/15/97	10/08/97	03/11&14/97	03/25&26/97
ITS Report No.	D97-12123	9703042	9703169	NR	NR	D97-12123	9703042	9703169
ITS Report Date	10/14/97	03/21/97	04/03/97	07/15/97	08/20/97	10/14/97	03/21/97	04/03/97
Acetone	20U	4U	4U	200U	200U	20U	4U	**43E**
Acrylonitrile	5.00U	NA	NA	NA	NA	5.00U	NA	NA
Benzene	5.00U	0.5U	**0.5**	50U	50U	5.00U	0.5U	0.5U
Bromobenzene	5.00U	NA	NA	NA	NA	5.00U	NA	NA
Bromochloromethane	5.00U	NA	NA	NA	NA	5.00U	NA	NA
Bromodichloromethane	5.00U	0.5U	0.5U	50U	50U	5.00U	0.5U	0.5U
Bromoform	5.00U	0.5U	0.5U	50U	50U	5.00U	0.5U	0.5U
Bromomethane	5.00U	0.5U	0.5U	100U	100U	5.00U	0.5U	0.5U
2-Butanone (MEK)	100U	4U	4U	200U	200U	100U	4U	4U
Carbon Disulfide	5.00U	0.5U	0.5U	50U	50U	5.00U	0.5U	0.5U
Carbon Tetrachloride	5.00U	0.5U	0.5U	50U	50U	5.00U	0.5U	0.5U
Chlorobenzene	5.00U	0.5U	0.5U	50U	50U	5.00U	0.5U	0.5U
Chloroethane	5.00U	0.5U	**0.7**	100U	100U	5.00U	0.5U	0.5U
2-Chloroethylvinyl ether	10.0U	NA	NA	NA	NA	10.0U	NA	NA
Chloroform	5.00U	0.5U	0.5U	50U	50U	5.00U	0.5U	0.5U
Chloromethane	5.00U	0.5U	0.5U	100U	100U	5.00U	0.5U	0.5U
2-Chlorotoluene	5.00U	NA	NA	NA	NA	5.00U	NA	NA
4-Chlorotoluene	5.00U	NA	NA	NA	NA	5.00U	NA	NA
Dibromochloromethane	5.00U	0.5U	0.5U	50U	50U	5.00U	0.5U	0.5U
1,2-Dibromo-3-chloropropane	25.0U	NA	NA	NA	NA	25.0U	NA	NA
1,2-Dibromoethane	5.00U	NA	NA	NA	NA	5.00U	NA	NA
Dibromomethane	5.00U	NA	NA	NA	NA	5.00U	NA	NA
1,2-Dichlorobenzene	5.00U	0.5U	**0.5**	50U	50U	5.00U	**0.7**	0.5U
1,3-Dichlorobenzene	5.00U	0.5U	0.5U	50U	50U	5.00U	0.5U	0.5U
1,4-Dichlorobenzene	5.00U	0.5U	0.5U	50U	50U	5.00U	0.5U	0.5U
trans-1,4-Dichloro-2-butene	100U	NA	NA	NA	NA	100U	NA	NA
1,1-Dichloroethane	5.00U	0.5U	**3**	50U	50U	5.00U	0.5U	**0.9**
1,2-Dichloroethane	5.00U	0.5U	0.5U	50U	50U	5.00U	0.5U	0.5U
1,1-Dichloroethene	5.00U	**0.6**	0.5U	100U	100U	5.00U	**0.5**	0.5U
cis-1,2-Dichloroethene	**75.5**	**340**	**420**	**3,500**	**1,100**	**136**	**330**	**210**
trans-1,2-Dichloroethene	5.00U	**1**	**1**	100U	100U	5.00U	**1**	**0.7**
1,2-Dichloropropane	5.00U	0.5U	0.5U	50U	50U	5.00U	0.5U	0.5U
2,2-Dichloropropane	5.00U	NA	NA	NA	NA	5.00U	NA	NA
1,3-Dichloropropane	5.00U	NA	NA	NA	NA	5.00U	NA	NA
1,1-Dichloropropene	5.00U	NA	NA	NA	NA	5.00U	NA	NA
cis-1,3-Dichloropropene	5.00U	0.5U	0.5U	50U	50U	5.00U	0.5U	0.5U
trans-1,3-Dichloropropene	5.00U	0.5U	0.5U	50U	50U	5.00U	0.5U	0.5U
Ethylbenzene	5.00U	0.5U	0.5U	100U	100U	5.00U	0.5U	0.5U
2-Hexanone	50.0U	4U	4U	100U	100U	50.0U	4U	4U
Iodomethane	5.00U	NA	NA	NA	NA	5.00U	NA	NA
Methylene Chloride (DCM)	5.00U	1U	1U	50U	50U	5.00U	1U	1U
4-Methyl-2-Pentanone (MIBK)	100U	4U	4U	100U	100U	100U	4U	4U
Styrene	5.00U	0.5U	0.5U	50U	50U	5.00U	0.5U	0.5U
1,1,1,2-Tetrachloroethane	5.00U	NA	NA	NA	NA	5.00U	NA	NA
1,1,2,2-Tetrachloroethane	5.00U	0.5U	0.5U	50U	50U	5.00U	0.5U	0.5U
Tetrachloroethene	5.00U	0.5U	0.5U	100U	100U	5.00U	0.5U	0.5U
1,2,3-Trichlorobenzene	5.00U	NA	NA	NA	NA	5.00U	NA	NA
1,2,4-Trichlorobenzene	5.00U	NA	NA	NA	NA	5.00U	NA	NA
1,1,1-Trichloroethane	5.00U	0.5U	0.5U	50U	50U	5.00U	0.5U	0.5U
1,1,2-Trichloroethane	5.00U	0.5U	0.5U	50U	50U	5.00U	0.5U	0.5U
Trichloroethene	5.00U	**1**	**0.9**	100U	100U	5.00U	**1**	**0.7**
Trichlorofluoromethane (Freon 11)	5.00U	0.5U	0.5U	50U	50U	5.00U	0.5U	0.5U
1,2,3-Trichloropropane	5.00U	NA	NA	NA	NA	5.00U	NA	NA
Trichlorotrifluoroethane (Freon 113)	NA	0.5U	0.5U	50U	50U	NA	0.5U	0.5U
1,2,4-Trimethylbenzene	5.00U	NA	NA	NA	NA	5.00U	NA	NA
1,3,5-Trimethylbenzene	5.00U	NA	NA	NA	NA	5.00U	NA	NA
Toluene	5.00U	**8**	**8**	100U	100U	**21.8**	**7**	0.5U
Vinyl Acetate	50.0U	0.5U	0.5U	50U	50U	50.0U	0.5U	0.5U
Vinyl Chloride	**54.9**	**63**	**230**	**750**	**460**	**217**	**54**	**57**
m,p-Xylene	5.00U	NA	NA	NA	NA	5.00U	NA	NA
o-Xylene	5.00U	NA	NA	NA	NA	5.00U	NA	NA
Xylenes (total)	NA	**1**	**0.9**	50U	50U	NA	**1**	0.5U

Table A5.5 (continued) Summary of VOC Concentrations in Groundwater Samples: Alameda

p. 9 of 12

Well ID	R8PB	R8PB	R8PB	C-R1P	C-R1P	C-R1P	C-R1P	C-R1P
Sample ID	R8PB	R8PB	R8PB	C-RIP	C-RIP	C-R1P	C-R1P	C-R1P
Dilution Factor	10.0	10.0	1	1.0	1.0	100.0	20.0	1
Date Sampled (m/d/y)	06/16/97	08/04/97	09/30/97	03/04/97	03/19/97	06/16/97	08/04/97	09/30/97
Date Analyzed (m/d/y)	06/26/97	08/15/97	10/08/97	03/11&14/97	03/24&26/97	06/26/97	08/15/97	10/08/97
ITS Report No.	NR	NR	D97-12123	9703042	9703169	NR	NR	D97-12123
ITS Report Date	07/15/97	08/20/97	10/14/97	03/21/97	04/03/97	07/15/97	08/20/97	10/14/97
Acetone	200U	200U	20U	4U	4U	2,000U	400U	20U
Acrylonitrile	NA	NA	5.00U	NA	NA	NA	NA	5.00U
Benzene	50U	50U	5.00U	**64E**	**53E**	500U	100U	5.00U
Bromobenzene	NA	NA	5.00U	NA	NA	NA	NA	5.00U
Bromochloromethane	NA	NA	5.00U	NA	NA	NA	NA	5.00U
Bromodichloromethane	50U	50U	5.00U	0.5U	0.5U	500U	100U	5.00U
Bromoform	50U	50U	5.00U	0.5U	0.5U	500U	100U	5.00U
Bromomethane	100U	100U	5.00U	0.5U	0.5U	1,000U	200U	5.00U
2-Butanone (MEK)	200U	200U	100U	4U	4U	2,000U	400U	100U
Carbon Disulfide	50U	50U	5.00U	0.5U	0.5U	500U	100U	5.00U
Carbon Tetrachloride	50U	50U	5.00U	0.5U	0.5U	500U	100U	5.00U
Chlorobenzene	50U	50U	5.00U	**75E**	**83E**	500U	**100**	**46.1**
Chloroethane	100U	100U	5.00U	0.5U	0.5U	1,000U	200U	5.00U
2-Chloroethylvinyl ether	NA	NA	10.0U	NA	NA	NA	NA	10.0U
Chloroform	50U	50U	5.00U	2U	0.5U	500U	100U	5.00U
Chloromethane	100U	100U	5.00U	0.5U	0.5U	1,000U	200U	5.00U
2-Chlorotoluene	NA	NA	5.00U	NA	NA	NA	NA	5.00U
4-Chlorotoluene	NA	NA	5.00U	NA	NA	NA	NA	5.00U
Dibromochloromethane	50U	50U	5.00U	0.5U	0.5U	500U	100U	5.00U
1,2-Dibromo-3-chloropropane	NA	NA	25.0U	NA	NA	NA	NA	25.0U
1,2-Dibromoethane	NA	NA	5.00U	NA	NA	NA	NA	5.00U
Dibromomethane	NA	NA	5.00U	NA	NA	NA	NA	5.00U
1,2-Dichlorobenzene	50U	50U	5.00U	**30**	**31**	500U	100U	5.00U
1,3-Dichlorobenzene	50U	50U	5.00U	0.5U	0.5U	500U	100U	5.00U
1,4-Dichlorobenzene	50U	50U	5.00U	**10**	**12**	500U	100U	5.00U
trans-1,4-Dichloro-2-butene	NA	NA	100U	NA	NA	NA	NA	100U
1,1-Dichloroethane	50U	50U	5.00U	**50E**	**52E**	500U	100U	5.00U
1,2-Dichloroethane	50U	50U	5.00U	0.5U	0.5U	500U	100U	5.00U
1,1-Dichloroethene	100U	100U	5.00U	**210E**	**200E**	1,000U	200U	5.00U
cis-1,2-Dichloroethene	**2,300**	**1,200**	**131**	**27,000**	**25,000**	**24,000**	**960**	**247**
trans-1,2-Dichloroethene	100U	100U	5.00U	**390E**	**290E**	1,000U	200U	5.00U
1,2-Dichloropropane	50U	50U	5.00U	**10**	0.5U	500U	100U	5.00U
2,2-Dichloropropane	NA	NA	5.00U	NA	NA	NA	NA	5.00U
1,3-Dichloropropane	NA	NA	5.00U	NA	NA	NA	NA	5.00U
1,1-Dichloropropene	NA	NA	5.00U	NA	NA	NA	NA	5.00U
cis-1,3-Dichloropropene	50U	50U	5.00U	0.5U	0.5U	500U	100U	5.00U
trans-1,3-Dichloropropene	50U	50U	5.00U	0.5U	0.5U	500U	100U	5.00U
Ethylbenzene	100U	100U	5.00U	**88E**	**76E**	1,000U	200U	**36.3**
2-Hexanone	100U	100U	50.0U	4U	4U	1,000U	200U	50.0U
Iodomethane	NA	NA	5.00U	NA	NA	NA	NA	5.00U
Methylene Chloride (DCM)	50U	50U	5.00U	1U	1U	500U	100U	5.00U
4-Methyl-2-Pentanone (MIBK)	100U	100U	100U	**60E**	4U	1,000U	200U	100U
Styrene	50U	50U	5.00U	0.5U	0.5U	500U	100U	5.00U
1,1,1,2-Tetrachloroethane	NA	NA	5.00U	NA	NA	NA	NA	5.00U
1,1,2,2-Tetrachloroethane	50U	50U	5.00U	0.5U	0.5U	500U	100U	5.00U
Tetrachloroethene	100U	100U	5.00U	0.5U	0.5U	1,000U	200U	5.00U
1,2,3-Trichlorobenzene	NA	NA	5.00U	NA	NA	NA	NA	5.00U
1,2,4-Trichlorobenzene	NA	NA	5.00U	NA	NA	NA	NA	5.00U
1,1,1-Trichloroethane	50U	50U	5.00U	0.5U	0.5U	500U	100U	5.00U
1,1,2-Trichloroethane	50U	50U	5.00U	0.5U	0.5U	500U	100U	5.00U
Trichloroethene	100U	100U	5.00U	**550**	**1,400**	1,000U	200U	5.00U
Trichlorofluoromethane (Freon 11)	50U	50U	5.00U	0.5U	0.5U	500U	100U	5.00U
1,2,3-Trichloropropane	NA	NA	5.00U	NA	NA	NA	NA	5.00U
Trichlorotrifluoroethane (Freon 113)	50U	50U	NA	0.5U	0.5U	500U	100U	NA
1,2,4-Trimethylbenzene	NA	NA	5.00U	NA	NA	NA	NA	**49.8**
1,3,5-Trimethylbenzene	NA	NA	5.00U	NA	NA	NA	NA	**7.13**
Toluene	100U	100U	5.00U	**1,200**	**1,300**	**1,700**	**140J**	**108**
Vinyl Acetate	50U	50U	50.0U	0.5U	0.5U	500U	100U	50.0U
Vinyl Chloride	**600**	**510**	**121**	**5,500**	**2,600**	**7,300**	**1,900**	**1,250**
m,p-Xylene	NA	NA	5.00U	NA	NA	NA	NA	**55.0**
o-Xylene	NA	NA	5.00U	NA	NA	NA	NA	**19.6**
Xylenes (total)	50U	50U	NA	**340E**	**280E**	500U	100U	NA

Table A5.5 (continued) Summary of VOC Concentrations in Groundwater Samples: Alameda

p. 10 of 12

Well ID	C-R5P	C-R5P	C-R5P	C-R5P	C-R5P	C-R7P	C-R7P	C-R7P
Sample ID	C-R5P	C-R5P	C-R5P	C-R5P	C-R5P	C-R7P	C-R7P	C-R7P
Dilution Factor	1.0	1.0	5.0	10.0	1	1.0	5.0	20.0
Date Sampled (m/d/y)	03/04/97	03/19/97	06/16/97	08/04/97	09/30/97	03/04/97	06/16/97	08/04/97
Date Analyzed (m/d/y)	03/11&13/97	03/24&26/97	06/26/97	08/15/97	10/08/97	03/11&14/97	06/26/97	08/15/97
ITS Report No.	9703042	9703169	NR	NR	D97-12123	9703042	NR	NR
ITS Report Date	03/21/97	04/03/97	07/15/97	08/20/97	10/14/97	03/21/97	07/15/97	08/20/97
Acetone	4U	4U	100U	200U	20U	4U	100U	400U
Acrylonitrile	NA	NA	NA	NA	5.00U	NA	NA	NA
Benzene	**66E**	**46E**	**39**	50U	**19.7**	**8**	**32**	100U
Bromobenzene	NA	NA	NA	NA	5.00U	NA	NA	NA
Bromochloromethane	NA	NA	NA	NA	5.00U	NA	NA	NA
Bromodichloromethane	0.5U	0.5U	25U	50U	5.00U	0.5U	25U	100U
Bromoform	0.5U	0.5U	25U	50U	5.00U	0.5U	25U	100U
Bromomethane	0.5U	0.5U	50U	100U	5.00U	0.5U	50U	200U
2-Butanone (MEK)	4U	4U	100U	200U	100U	4U	100U	400U
Carbon Disulfide	**1**	0.5U	25U	50U	5.00U	0.5U	25U	100U
Carbon Tetrachloride	0.5U	0.5U	25U	50U	5.00U	0.5U	25U	100U
Chlorobenzene	**60E**	**56E**	**56**	50U	**18.6**	**34**	**36**	100U
Chloroethane	0.5U	0.5U	50U	100U	5.00U	0.5U	50U	200U
2-Chloroethylvinyl ether	NA	NA	NA	NA	10.0U	NA	NA	NA
Chloroform	0.5U	0.5U	25U	50U	5.00U	0.5U	25U	100U
Chloromethane	0.5U	0.5U	50U	100U	5.00U	0.5U	50U	200U
2-Chlorotoluene	NA	NA	NA	NA	5.00U	NA	NA	NA
4-Chlorotoluene	NA	NA	NA	NA	5.00U	NA	NA	NA
Dibromochloromethane	0.5U	0.5U	25U	50U	5.00U	0.5U	25U	100U
1,2-Dibromo-3-chloropropane	NA	NA	NA	NA	25.0U	NA	NA	NA
1,2-Dibromoethane	NA	NA	NA	NA	5.00U	NA	NA	NA
Dibromomethane	NA	NA	NA	NA	5.00U	NA	NA	NA
1,2-Dichlorobenzene	**16**	**16**	25U	50U	5.00U	**7**	25U	100U
1,3-Dichlorobenzene	0.5U	0.5U	25U	50U	5.00U	0.5U	25U	100U
1,4-Dichlorobenzene	**10**	**10**	25U	50U	5.00U	**6**	25U	100U
trans-1,4-Dichloro-2-butene	NA	NA	NA	NA	100U	NA	NA	NA
1,1-Dichloroethane	**26**	**24**	25U	50U	5.00U	**8**	25U	100U
1,2-Dichloroethane	0.5U	0.5U	25U	50U	5.00U	0.5U	25U	100U
1,1-Dichloroethene	**26**	**41E**	50U	100U	5.00U	**2**	50U	200U
cis-1,2-Dichloroethene	**6,900**	**11,000**	**9,100**	**1,200**	**201**	**390**	**4,100**	**3,200**
trans-1,2-Dichloroethene	**73E**	**98E**	**28J**	100U	5.00U	**8**	**28J**	200U
1,2-Dichloropropane	0.5U	0.5U	25U	50U	5.00U	0.5U	25U	100U
2,2-Dichloropropane	NA	NA	NA	NA	5.00U	NA	NA	NA
1,3-Dichloropropane	NA	NA	NA	NA	5.00U	NA	NA	NA
1,1-Dichloropropene	NA	NA	NA	NA	5.00U	NA	NA	NA
cis-1,3-Dichloropropene	0.5U	0.5U	25U	50U	5.00U	0.5U	25U	100U
trans-1,3-Dichloropropene	0.5U	0.5U	25U	50U	5.00U	0.5U	25U	100U
Ethylbenzene	**79E**	**75E**	**48J**	100U	**26.6**	**39**	**36J**	200U
2-Hexanone	4U	4U	50U	100U	50.0U	4U	50U	200U
Iodomethane	NA	NA	NA	NA	5.00U	NA	NA	NA
Methylene Chloride (DCM)	1U	1U	25U	50U	5.00U	1U	25U	100U
4-Methyl-2-Pentanone (MIBK)	4U	4U	50U	100U	100U	4U	50U	200U
Styrene	0.5U	0.5U	25U	50U	5.00U	0.5U	25U	100U
1,1,1,2-Tetrachloroethane	NA	NA	NA	NA	5.00U	NA	NA	NA
1,1,2,2-Tetrachloroethane	0.5U	0.5U	25U	50U	5.00U	0.5U	25U	100U
Tetrachloroethene	0.5U	0.5U	50U	100U	5.00U	0.5U	50U	200U
1,2,3-Trichlorobenzene	NA	NA	NA	NA	5.00U	NA	NA	NA
1,2,4-Trichlorobenzene	NA	NA	NA	NA	5.00U	NA	NA	NA
1,1,1-Trichloroethane	0.5U	0.5U	25U	50U	5.00U	0.5U	25U	100U
1,1,2-Trichloroethane	0.5U	0.5U	25U	50U	5.00U	0.5U	25U	100U
Trichloroethene	**0.8**	**1**	50U	100U	5.00U	**0.6**	50U	200U
Trichlorofluoromethane (Freon 11)	0.5U	0.5U	25U	50U	5.00U	0.5U	25U	100U
1,2,3-Trichloropropane	NA	NA	NA	NA	5.00U	NA	NA	NA
Trichlorotrifluoroethane (Freon 113)	0.5U	0.5U	25U	50U	NA	0.5U	25U	100U
1,2,4-Trimethylbenzene	NA	NA	NA	NA	**25.1**	NA	NA	NA
1,3,5-Trimethylbenzene	NA	NA	NA	NA	5.00U	NA	NA	NA
Toluene	**990**	**940**	**890**	100U	**20.9**	**150**	**250**	**140J**
Vinyl Acetate	0.5U	0.5U	25U	50U	50.0U	0.5U	25U	100U
Vinyl Chloride	**6,300**	**2,700**	**3,800**	**1,100**	**232**	**840**	**1,800**	**1,800**
m,p-Xylene	NA	NA	NA	NA	**10.3**	NA	NA	NA
o-Xylene	NA	NA	NA	NA	**5.96**	NA	NA	NA
Xylenes (total)	**140**	**220E**	**130**	50U	NA	**100**	**38**	100U

Table A5.5 (continued) Summary of VOC Concentrations in Groundwater Samples: Alameda

p. 11 of 12

Well ID	C-R8P	C-R8P	Field Blank	Field Blank	Trip Blank	Trip Blank	Trip Blank
Sample ID	C-R8P	C-R8P	FIELDBLK	FLDBLNK	TRIPBLK	TRPBLNK	TRIPBLANK
Dilution Factor	1.0	1	1.0	1.0	1.0	1.0	1.0
Date Sampled (m/d/y)	03/19/97	09/30/97	03/04/97	03/19/97	03/04/97	03/19/97	06/16/97
Date Analyzed (m/d/y)	03/24&26/97	10/08/97	03/12/97	03/22/97	03/12/97	03/22/97	06/30/97
ITS Report No.	9703169	D97-12123	9703042	9703169	9703042	9703169	NR
ITS Report Date	04/03/97	10/14/97	03/21/97	04/03/97	03/21/97	04/03/97	07/15/97
Acetone	4U	20U	4U	4U	4U	4U	4U
Acrylonitrile	NA	5.00U	NA	NA	NA	NA	NA
Benzene	**45E**	5.00U	0.5U	0.5U	0.5U	0.5U	0.5U
Bromobenzene	NA	5.00U	NA	NA	NA	NA	NA
Bromochloromethane	NA	5.00U	NA	NA	NA	NA	NA
Bromodichloromethane	0.5U	5.00U	**0.5**	**0.5**	0.5U	0.5U	0.5U
Bromoform	0.5U	5.00U	0.5U	0.5U	0.5U	0.5U	0.5U
Bromomethane	0.5U	5.00U	0.5U	0.5U	0.5U	0.5U	1U
2-Butanone (MEK)	4U	100U	4U	4U	4U	4U	4U
Carbon Disulfide	0.5U	5,00U	0.5U	0.5U	0.5U	0.5U	0.5U
Carbon Tetrachloride	0.5U	5.00U	0.5U	0.5U	0.5U	0.5U	0.5U
Chlorobenzene	**42E**	**16.3**	0.5U	0.5U	0.5U	0.5U	0.5U
Chloroethane	0.5U	5.00U	0.5U	0.5U	0.5U	0.5U	1U
2-Chloroethylvinyl ether	NA	10.0U	NA	NA	NA	NA	NA
Chloroform	0.5U	5.00U	**19**	**18**	0.5U	0.5U	0.5U
Chloromethane	0.5U	5.00U	0.5U	0.5U	0.5U	0.5U	1U
2-Chlorotoluene	NA	5.00U	NA	NA	NA	NA	NA
4-Chlorotoluene	NA	5.00U	NA	NA	NA	NA	NA
Dibromochloromethane	0.5U	5.00U	0.5U	0.5U	0.5U	0.5U	0.5U
1,2-Dibromo-3-chloropropane	NA	25.0U	NA	NA	NA	NA	NA
1,2-Dibromoethane	NA	5.00U	NA	NA	NA	NA	NA
Dibromomethane	NA	5.00U	NA	NA	NA	NA	NA
1,2-Dichlorobenzene	**7**	5.00U	0.5U	0.5U	0.5U	0.5U	0.5U
1,3-Dichlorobenzene	0.5U	5.00U	0.5U	0.5U	0.5U	0.5U	0.5U
1,4-Dichlorobenzene	**6**	5.00U	0.5U	0.5U	0.5U	0.5U	0.5U
trans-1,4-Dichloro-2-butene	NA	100U	NA	NA	NA	NA	NA
1,1-Dichloroethane	**11**	5.00U	0.5U	0.5U	0.5U	0.5U	0.5U
1,2-Dichloroethane	0.5U	5.00U	0.5U	0.5U	0.5U	0.5U	0.5U
1,1-Dichloroethene	**5**	5.00U	0.5U	0.5U	0.5U	0.5U	2U
cis-1,2-Dichloroethene	**3,400**	**157**	0.5U	0.5U	0.5U	0.5U	2U
trans-1,2-Dichloroethene	**20**	5.00U	0.5U	0.5U	0.5U	0.5U	2U
1,2-Dichloropropane	0.5U	5.00U	0.5U	0.5U	0.5U	0.5U	0.5U
2,2-Dichloropropane	NA	5.00U	NA	NA	NA	NA	NA
1,3-Dichloropropane	NA	5.00U	NA	NA	NA	NA	NA
1,1-Dichloropropene	NA	5.00U	NA	NA	NA	NA	NA
cis-1,3-Dichloropropene	0.5U	5.00U	0.5U	0.5U	0.5U	0.5U	0.5U
trans-1,3-Dichloropropene	0.5U	5.00U	0.5U	0.5U	0.5U	0.5U	0.5U
Ethylbenzene	**54E**	**9.81**	0.5U	0.5U	0.5U	0.5U	2U
2-Hexanone	4U	50.0U	4U	4U	4U	4U	4U
Iodomethane	NA	5.00U	NA	NA	NA	NA	NA
Methylene Chloride (DCM)	1U	5.00U	1U	1U	1U	1U	4U
4-Methyl-2-Pentanone (MIBK)	4U	100U	4U	4U	4U	4U	4U
Styrene	0.5U	5.00U	0.5U	0.5U	0.5U	0.5U	0.5U
1,1,1,2-Tetrachloroethane	NA	5.00U	NA	NA	NA	NA	NA
1,1,2,2-Tetrachloroethane	0.5U	5.00U	0.5U	0.5U	0.5U	0.5U	0.5U
Tetrachloroethene	0.5U	5.00U	0.5U	0.5U	0.5U	0.5U	2U
1,2,3-Trichlorobenzene	NA	5.00U	NA	NA	NA	NA	NA
1,2,4-Trichlorobenzene	NA	5.00U	NA	NA	NA	NA	NA
1,1,1-Trichloroethane	0.5U	5.00U	0.5U	0.5U	0.5U	0.5U	0.5U
1,1,2-Trichloroethane	0.5U	5.00U	0.5U	0.5U	0.5U	0.5U	0.5U
Trichloroethene	**0.7**	5.00U	0.5U	0.5U	0.5U	0.5U	2U
Trichlorofluoromethane (Freon 11)	0.5U	5.00U	0.5U	0.5U	0.5U	0.5U	0.5U
1,2,3-Trichloropropane	NA	5.00U	NA	NA	NA	NA	NA
Trichlorotrifluoroethane (Freon 113)	0.5U	NA	0.5U	0.5U	0.5U	0.5U	0.5U
1,2,4-Trimethylbenzene	NA	**19.0**	NA	NA	NA	NA	NA
1,3,5-Trimethylbenzene	NA	5.00U	NA	NA	NA	NA	NA
Toluene	**330**	**12.8**	0.5U	0.5U	0.5U	0.5U	2U
Vinyl Acetate	0.5U	50.0U	0.5U	0.5U	0.5U	0.5U	4U
Vinyl Chloride	**1,700**	**239**	0.5U	0.5U	0.5U	0.5U	0.5U
m,p-Xylene	NA	**10.6**	NA	NA	NA	NA	NA
o-Xylene	NA	5.00U	NA	NA	NA	NA	NA
Xylenes (total)	**120**	NA	0.5U	0.5U	0.5U	0.5U	0.5U

Table A5.5 (continued) Summary of VOC Concentrations in Groundwater Samples: Alameda

p. 12 of 12

Notes:

All concentrations are in units of micrograms per liter (μg/L).

74.000 - concentration confirmed by re-analysis at a higher dilution as the original amount exceeded the linear range of the instrument calibration.

Dup - field duplicate.

U - analyte was not detected; the associated value is quantitation limit.

J - estimated value; the compound was detected at an amount below the quantitation limit.

E - estimated value; the compound was detected at an amount exceeding the linear range of the instrument calibration, however, the re-analysis was at a too high dilution to detect the compound.

NA - not analyzed.

NR - not reported.

VOC - volatile organic compound.

ITS - Inchcape Testing Services.

Appendix 6

Background Sampling Results (Borden)

Groundwater sampling to provide a comprehensive survey of background geochemical conditions was completed on May 22, 23, and 27, 1996, approximately 1 month following flow system start-up. Samples were collected for volatile organic compounds (VOC), dissolved hydrocarbon gases (DHG), cations, anions, alkalinity, and field parameters (DO, E_h, pH, and conductivity) analyses from all depths at multilevel piezometers G102L, G112L, G205M, G224M, G308, and G321M (shown in Figure A6.1). 321M-1 was dry and no samples were collected from this point. Samples were collected and field analyses completed according to the methods provided in Appendix 4.

Results of this sampling are shown in Tables A6.1, A6.2, and A6.3 and indicate that the shallow aquifer, at the time of sampling, was moderately aerobic, with near-neutral pH and variable, but low, electrical conductivity values. Inorganic results indicated measurable concentrations of calcium, iron, potassium, magnesium, sodium, silica, chloride, and sulfate. Alkalinity values ranged from 121 to 634 mg/L as $CaCO_3$ and were typically higher at shallow sampling points and decreased with depth. VOC analyses detected PCE in low concentrations, 5.7 and 9.7 μg/L at G224M-1 and G321M-2, respectively. This contamination was believed to be the result of a previous University of Waterloo PCE release experiment at an adjacent field site (Brewster et al., 1995). No other organic contaminants were identified. DHG analyses indicated that ethene, ethane, propene, and propane concentrations were low, mostly below detection limits, but that methane levels were quite variable, ranging from trace levels (0.2 μg/L) to values of 78.5 μg/L. Highest concentrations were observed at the influent end of Gates 1 and 2, and at shallower depths. Several sampling locations were resampled for DHGs on June 10, 1996 to confirm the May results.

Geochemical conditions at the site vary from those observed during previous experiments conducted at the Borden site. DO concentrations in the sand pit have been shown to be variable, but typically aerobic conditions are observed (King, 1997). Elevated methane concentrations encountered at shallow sampling locations were likely due to the presence of a wetland area located immediately southeast of the site. Inorganic data were typical of Borden groundwater.

Low chloride concentrations measured in samples indicated that site groundwater was not infuenced by the landfill plume, as is sometimes observed in experiments conducted in the sand pit.

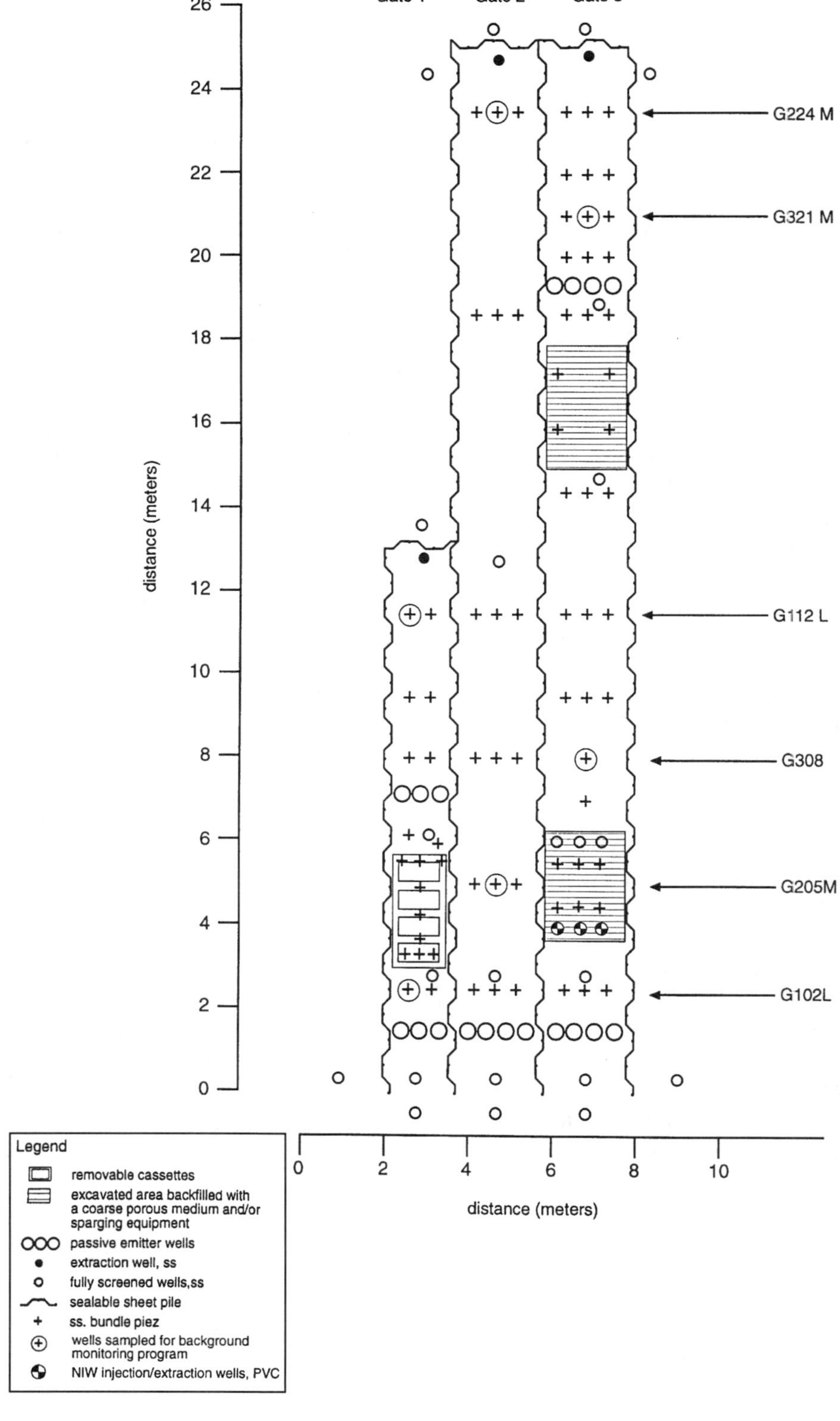

Figure A6.1 Location of wells sampled for background monitoring.

Table A6.1 Background Sampling at UW-AATDF Site, CFB Borden: Volatile Organic Compounds and Dissolved Hydrocarbon Gases

Well No.	PCE (μg/L)	CT (μg/L)	TOL (μg/L)	Ethene (μg/L)	Ethane (μg/L)	Methane (μg/L)	Ethene (μg/L)	Ethane (μg/L)	Methane (μg/L)
	Sampling Conducted May 22, 23, and 27, 1996						DHG Re-sampling on June 10, 1996		
G102L-1	<MDL	<MDL	<MDL	<MDL	<MDL	71.0	ns	ns	ns
G102L-2	<MDL	<MDL	<MDL	<MDL	<MDL	7.1	<MDL	<MDL	7.3
G102L-3	<MDL	<MDL	<MDL	<MDL	<MDL	16.5	ns	ns	ns
G102L-4	<MDL	<MDL	<MDL	<MDL	<MDL	31.5	<MDL	<MDL	15.7
G112L-1	<MDL	<MDL	<MDL	**0.1**	1.8	1.8	ns	ns	ns
G112L-2	<MDL	<MDL	<MDL	**0.4**	3.8	8.4	ns	ns	ns
G112L-3	<MDL	<MDL	<MDL	**0.2**	1.9	13.9	ns	ns	ns
G112L-4	<MDL	<MDL	<MDL	<MDL	1.0	13.4	ns	ns	ns
G205M-1	<MDL	<MDL	<MDL	<MDL	<MDL	17.1	ns	ns	ns
G205M-2	<MDL	<MDL	<MDL	<MDL	0.3	25.9	ns	ns	ns
G205M-3	<MDL	<MDL	<MDL	<MDL	<MDL	22.7	ns	ns	ns
G205M-4	<MDL	<MDL	<MDL	<MDL	<MDL	36.1	ns	ns	ns
G224M-1	**5.7**	<MDL	<MDL	<MDL	<MDL	<MDL	<MDL	<MDL	<MDL
G224M-2	<MDL	<MDL	<MDL	<MDL	0.2	**0.2**	ns	ns	ns
G224M-3	<MDL	<MDL	<MDL	<MDL	<MDL	2.1	<MDL	<MDL	10.0
G224M-4	<MDL	<MDL	<MDL	<MDL	<MDL	30.4	ns	ns	ns
G308-1	<MDL	<MDL	<MDL	<MDL	<MDL	1.1	ns	ns	ns
G308-2	<MDL	<MDL	<MDL	<MDL	<MDL	<MDL	<MDL	<MDL	**0.5**
G308-3	<MDL	<MDL	<MDL	<MDL	<MDL	1.7	ns	ns	ns
G308-4	<MDL	<MDL	<MDL	<MDL	<MDL	1.7	<MDL	<MDL	5.1
G321M-1	ns	ns	ns	ns	ns	ns	ns	ns	ns
G321M-2	**9.7**	<MDL	<MDL	**0.4**	0.7	6.7	ns	ns	ns
G321M-3	<MDL	<MDL	<MDL	<MDL	0.7	9.3	ns	ns	ns
G321M-4	<MDL	<MDL	<MDL	0.7	1.3	17.4	ns	ns	ns

Notes: MDL - method detection limit (MDLs are given in Appendix 6).

ns - not sampled (point was dry). TCE, isomers of DCE, VC, CF, CM were not detected.

Bolded values are >MDL and < LOQ, where LOQ is 2XMDL.

Table A6.2 Background Sampling at UW-AATDF Site, CFB Borden: Inorganics Data

Well No.	Fe (mg/L)	Ca (mg/L)	K (mg/L)	Mg (mg/L)	Mn (mg/L)	Na (mg/L)	Cl (mg/L)	PO_4 (mg/L)	SO_4 (mg/L)	NO_3 (mg/L)	Br (mg/L)
G102L-1	0.52	164	1.28	11.90	<0.05	43.9	17.9	<0.05	3.31	nd	<0.05
G102L-2	0.26	97.5	1.08	8.58	<0.05	2.20	3.82	<0.05	12.1	nd	<0.05
G102L-3	<0.05	62.2	<0.05	5.50	<0.05	4.77	1.75	<0.05	16.8	nd	<0.05
G102L-4	0.14	49.4	<0.05	6.16	<0.05	4.21	1.45	<0.05	18.8	nd	<0.05
G112L-1	0.15	83.7	4.22	7.12	<0.05	2.84	2.66	<0.05	17.9	<0.05	<0.05
G112L-2	<0.05	69.4	6.39	9.10	<0.05	2.02	1.37	<0.05	11.2	<0.05	<0.05
G112L-3	<0.05	51.8	3.62	8.37	<0.05	4.68	1.22	<0.05	5.15	<0.05	<0.05
G112L-4	<0.05	62.3	1.57	4.52	0.29	1.49	1.37	<0.05	10.9	<0.05	<0.05
G205M-1	<0.05	109	<0.05	5.97	<0.05	3.63	4.59	<0.05	16.1	<0.05	<0.05
G205M-2	<0.05	78.3	<0.05	5.85	<0.05	2.76	2.60	<0.05	16.4	<0.05	<0.05
G205M-3	<0.05	70.0	<0.05	5.09	<0.05	1.56	1.95	<0.05	15.8	<0.05	<0.05
G205M-4	0.10	53.1	1.10	4.60	<0.05	5.04	1.87	<0.05	16.6	<0.05	<0.05
G224M-1	<0.05	95.4	<0.05	6.08	0.11	3.28	3.15	<0.05	12.6	<0.05	<0.05
G224M-2	<0.05	65.8	3.42	6.65	<0.05	3.42	4.76	<0.05	9.15	<0.05	<0.05
G224M-3	0.42	82.1	2.57	7.37	<0.05	1.84	1.52	<0.05	22.1	<0.05	<0.05
G224M-4	<0.05	72.7	<0.05	4.96	<0.05	1.67	1.31	<0.05	18.2	<0.05	<0.05
G308-1	<0.05	65.2	1.61	3.31	<0.05	21.2	49.5	<0.05	12.2	<0.05	<0.05
G308-2	<0.05	66.7	1.01	5.05	<0.05	22.9	31.2	<0.05	19.0	<0.05	<0.05
G308-3	<0.05	78.1	1.10	6.72	<0.05	14.1	25	<0.05	21.1	<0.05	<0.05
G308-4	<0.05	95.4	1.20	7.02	<0.05	13.3	19.9	<0.05	18.2	<0.05	<0.05
G321M-1	ns	ns	ns	ns	ns	ns	ns	ns	ns	ns	ns
G321M-2	<0.05	82.7	1.37	5.52	<0.05	5.05	19.3	<0.05	18.6	<0.05	<0.05
G321M-3	<0.05	75.9	2.40	5.39	<0.05	12.3	19.7	<0.05	17.3	<0.05	<0.05
G321M-4	<0.05	76.8	1.79	6.48	<0.05	11.8	11.2	<0.05	13.6	<0.05	<0.05

Notes: Sampling conducted May 22, 23, and 27, 1996.

ns - not sampled (point was dry).

Samples analyzed by UW Water Quality Lab.

Concentrations of Al, As, Ba, Cd, Co, Cr, Cu, Hg, Mo, Ni, Pb, Se, Sn, Ti, Vi, and Zn were all nondetectable.

Analyses for B, Si, and Sr were completed; however, data are not relevant and therefore not provided.

Table A6.3 Background Sampling at UW-AATDF Site, CFB Borden: Field Parameter Data

Well No.	pH	*E*(field) (mV)	E_h (mV)	Cond. (μS)	DO (mg/L)	Alk (mg/L)
G102L-1	7.08	-56	161	1009	1.51	634
G102L-2	7.35	-135	82	462	1.85	231
G102L-3	7.67	-123	94	308	2.89	151
G102L-4	7.75	-90	127	289	3.81	121
G112L-1	7.27	-27	190	469	2.56	260
G112L-2	7.53	-138	79	333	3.48	169
G112L-3	7.82	-147	70	368	3.89	182
G112L-4	7.74	-122	95	420	3.47	195
G205M-1	7.42	102	319	378	2.35	250
G205M-2	7.65	-143	74	372	2.59	183
G205M-3	7.64	-127	90	281	2.92	156
G205M-4	7.71	-128	89	282	2.73	130
G224M-1	7.71	-4	213	319	6.55	251
G224M-2	7.62	-166	51	300	2.10	163
G224M-3	7.4	-150	67	334	6.24	189
G224M-4	7.51	-87	130	357	3.69	164
G308-1	7.62	64	281	470	3.33	136
G308-2	7.89	-35	182	432	1.75	163
G308-3	7.75	-137	80	288	1.68	149
G308-4	7.96	-60	157	360	9.05	177
G321M-1	ns	ns	ns	ns	ns	ns
G321M-2	7.75	-91	126	420	1.89	176
G321M-3	7.57	-139	78	405	3.82	167
G321M-4	7.48	-154	63	411	3.15	179

Notes: Sampling conducted May 22, 23, and 27, 1996.

E(field) - uncorrected E_h values.

E_h - corrected E_h values.

ns - not sampled (point was dry).

DO values are considered estimates.

REFERENCES

Brewster, M.L., A.P. Annan, J.P. Greenhouse, B.H. Kueper, G.R. Olhoeft, J.D. Redman, and K.A. Sander. 1995. Observed Migration of a Controlled DNAPL Release by Geophysical Methods. *Ground Water* 33(6): 977-987.

King, M., 1997. Migration and Natural Fate of a Coal Tar Creosote Plume. Ph.D. thesis, University of Waterloo, Waterloo, Ontario.

APPENDIX 7

Main Tracer Test — Borden

A7.1 BACKGROUND

A tracer test, hereafter referred to as the Main Tracer Test (MTT), was conducted between May 15, 1996 and June 27, 1997 to assess hydraulic parameters in all three gates of the UW-AATDF Borden site. Subsequently, a second smaller tracer test was performed in Gate 3 from September 6, 1996 to October 9, 1996 to assess the effectiveness of source well redevelopment in late August 1996. Table A7.1 shows the time intervals for tracer injection and sample collection phases of the work.

Table A7.1 Tracer Injection and Sample Collection Time Intervals for Tracer Tests Conducted at UW-AATDF Borden Site

	Tracer Injection		Sample Collection		
Gate	Start	Finish	Start	Finish	Description
1	June 10/96	July 15/96	June 10/96	February 1/97	Main tracer test
2	May 15/96	June 15/96	May 15/96	May 30/97	Main tracer test
3	May 15/96	June 15/96	May 15/96	June 27/97	Main tracer test
3	Sept. 6/96	Sept. 9/96	Sept. 6/96	Oct. 9/96	Second tracer test

In both tests, bromide was selected as the tracer due to its conservative behavior and low background concentrations in the Borden aquifer (see Appendix 6 for background sampling results). Figure A7.1 shows the location of source wells used to release tracer into the gates. The installation of these wells is outlined in UW (1997). The tracer was injected into all source wells as a solution of potassium bromide (KBr). The delivery system for each source well comprised a 0.3-cm-diameter stainless steel line extending from a large, common, source tank into 0.6-cm polyethlyene tubes that were perforated at 30-cm intervals and suspended below the water table in the source wells. The perforated lines were utilized to promote a uniform delivery of tracer with depth. The pumping rate was maintained at approximately 10 mL/min using a peristaltic pump. The common tank contained KBr at a concentration of about 90 g/(kg of solution). Based on the injection rate given above and a total extraction rate of 130 mL/min (in Gates 2 and 3; only 100 mL/min in Gate 1, which had a narrower cross-sectional area), it was estimated that a concentration of approximately 700 mg/L would result in the aquifer. This input concentration was chosen to ensure that detectable tracer concentrations would exist at sampling points throughout the entire length of the gates while minimizing the likelihood of pronounced density flow effects.

The approximate masses of tracer introduced to each gate over the duration of the main tracer test are summarized in Table A7.2.

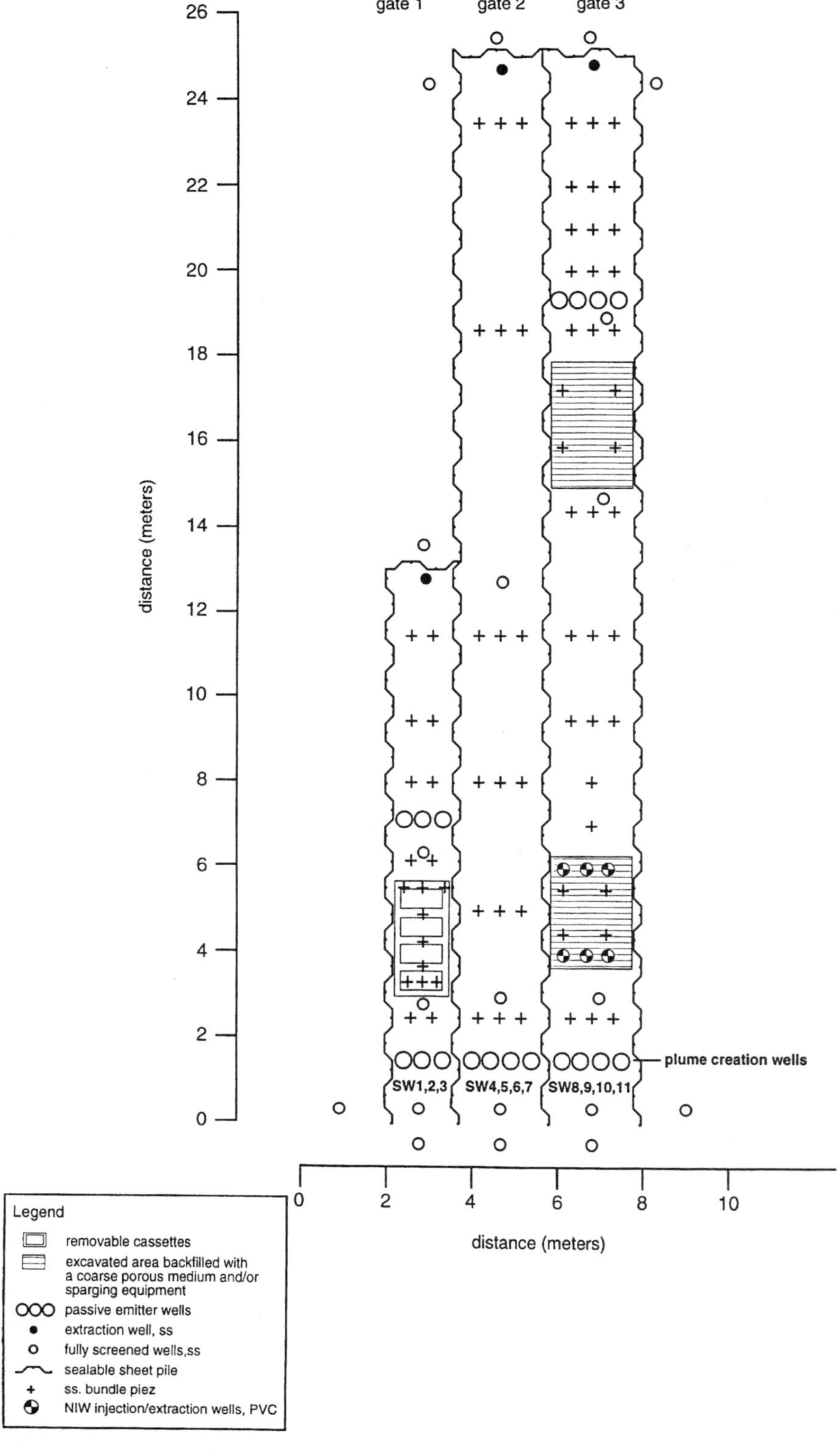

Figure A7.1 Locations of source wells for tracer injection.

Table A7.2 Approximate Mass of Tracer Injected into Each Gate

Gate	Mass KBr Injected (kg)
1	2.83
2	4.585
3	4.585

A7.2 SAMPLING

All multilevel piezometers were sampled for conductivity and potassium bromide (refer to Appendix 4 for analytical details). Analysis of tracer parameters were completed as soon as possible following sample collection (either in the field or at UW). Sample collection was discontinued when concentrations at a sampling point declined to near background levels.

A7.3 DATA HANDLING AND ANALYSIS

The objective of the tracer tests was to characterize the flow sytems in the gates, with particular emphasis on identifying preferential flow paths and estimating velocities (which could then be compared with the design objective of about 10 cm/d). It was also hoped that dispersivities could be estimated for the purposes of interpreting pulse mixing behavior in Gate 3. However, the introduction of the tracer to the aquifer in all three gates, as gauged by concentrations at the first monitoring fences, was too variable, both spatially and over time, for reasonable dispersivity estimates to be calculated from the breakthrough curves. The dispersion in Gate 3 was instead addressed in separate tracer tests conducted from the NIW (Appendix 13). As discussed in Appendix 9, insufficient well development was thought be part of the cause and natural aquifer heterogeneity another part. Nevertheless, fence-averaged velocities were calculated for selected fences in all three gates, and point-specific velocities and breakthrough masses were determined for selected fences in Gates 2 and 3 (Gate 1 was not analyzed in this way due to the complex flow system through the cassettes (refer to Chapter 2), and through the leaky seal(s); see Appendix 14). To assist in the assessment of flow patterns in the gates, a qualitative index based on breakthrough curve areas was also developed.

As alluded to above, two types of average linear groundwater velocity were calculated: fence-averaged and point-specific velocities. In calculating the fence-averaged velocities, two methods were used: the Levenspiel method (Levenspiel, 1993) and the center of mass method (Appendix 24). The Levenspiel method calculates the arrival time of the center of mass of a tracer pulse from

$$t_a = \frac{\Sigma t \times C_i \times \Delta t}{\Sigma C_i \times \Delta t}$$

where t = time, C_i = average concentration at time i, Δt = time interval between measurements, and t_a = travel time of the center of mass of a tracer pulse (to a particular fence). The concentrations used were first fence-averaged as described in Appendix 24, and then normalized to the highest single concentration observed in the first fence (in keeping with the methodology of Mackay et al., 1994). Tracer concentrations observed in the source wells were not considered representative of those entering the aquifer because of incomplete dilution, so the first fence in each gate was used as the input function. Appropriate corrections were made for the time offset from t_o, the tracer test start-up time. The velocity, v, was calculated from

$$v = \frac{d}{t_a}$$

where d = distance from source wells.

In calculating the point-specific velocities, individual breakthrough curves were integrated using the program FENCE (Appendix 24), which determines centers of mass and uses them for the calculation of velocities and total masses breaking through.

A7.4 FENCE-AVERAGED GROUDWATER VELOCITY SUMMARY FOR GATES 1, 2, AND 3

Fence-averaged groundwater velocities for all gates are summarized in Table A7.3. The calculation of velocities beyond fence G317 in Gate 3 was not performed because the breakthrough curves exhibited a bimodal quality that was initially thought to be due to interference from the second tracer test initiated in Gate 3 in early September 1996 (Table A7.1). Subsequent analysis (in the form of mass balances completed in the point-specific velocity determinations) indicated that the peculiar breakthrough curve shapes were probably not interferences of the two tracer tests, but may have been the result of complex flow patterns through the biosparge zone. Sheet piles were installed across the gate, partially penetrating the saturated zone both up- and downgradient of the pea gravel wall. These sheet piles could have caused constriction and subsequently expansion of flow lines as water passed through the biosparge zone and out the other side. The partially penetrating barriers could also have created low-flow zones that might have served as a temporary sink for solute mass from passing pulses. In addition, the pea gravel was not installed uniformly to the base of the aquifer (see UW installation report, 1997). Thus, a dual porosity region might be considered to exist at the base of the biosparge zone. These factors could have combined to produce the complicated breakthrough curves observed.

Examination of Table A7.3 reveals that, on average, the design objectives of the system were achieved. Velocities calculated by both methods were generally on the order of 10 to 15 cm/day in Gates 2 and 3, and in Gate 1 downgradient of the cassettes. Higher velocities were registered within the cassette system in Gate 1, probably because of the narrower cross-sectional area available for flow.

A7.5 QUALITATIVE ASSESSMENT OF FLOW PATTERNS IN GATES 1, 2, AND 3

Based on the breakthrough curves a qualitative index was established for each gate to compare the relative mass of tracer passing each multilevel monitoring point. The index ranked the area under each curve (i.e., mass) into one of four categories, ranging from (-) to (+++). This was done on a fence-specific basis, such that the curves with the greatest area were rated as (+++), and all others with less area on a relative scale from (++) to (-).

The index was established to illustrate the spatial variability of flow between fences in each gate. This assumes that the greatest breakthrough of mass is correlated to regions of active flow as opposed to spatial variability in mass release from source wells. The distribution of mass in the first fences of all gates may have been more strongly influenced by the latter, although this effect would have declined further downgradient as groundwater flow (hence, tracer mass) would have begun to concentrate along preferential flow paths.

Results of the ranking (Tables A7.4 to A7.6) showed that for Gate 1, flow was concentrated along the deeper monitoring depths (i.e., 3 and 4) within the cassette system and immediately downgradient (i.e., fences G106 and G108). Within the cassettes themselves, tracer breakthrough

Table A7.3 Summary of Average Groundwater Velocities (in cm/d) by Method in All Gates

Location	Distance from Source Wells (m)	Velocity by Levenspiel Method (cm/day)	Velocity by Center of Mass Method (determined using FENCE) (cm/day)
G102	1.1	N/A	N/A
G1C1	2.9	18.7	N/A
G103	3.2	20.2	N/A
G104	3.8	20.0	N/A
G105	4.4	18.4	N/A
G106	5.6	10.3	N/A
G108	7.3	9.9	N/A
G110	9.4	10.0	N/A
G112	11.4	9.8	N/A
G202	0.9	N/A	7.9
G205	3.8	8.9	9.5
G208	7.3	12.8	11.3
G212	11.3	12.1	10.4
G218	16.3	10.3	11.0
G224	21.7	10.1	9.9
G302	0.8	N/A	9.3
G304	2.8	14.7	9.2
G306	4.9	20.2	N/A
G307	6.2	17.0	N/A
G308	7.3	16.8	N/A
G310	9.3	16.5	12.3
G312	11.4	12.4	10.7
G315	13.1	16.5	12.3
G316	14.9	12.2	N/A
G317	16.4	12.2	N/A
G318	17.3		9.9
G320	18.7		9.5
G324	22.6		10.2

appeared to be horizontally uniform (both transverse and in the direction of the gradient); relative tracer breakthrough at any given fence closely correlated with those immediately downgradient. Beyond the cassettes, tracer breakthrough became more evenly distributed among the lower three depths, with the distribution at fence G112 being nearly uniform among all four monitoring depths.

For the entire length of Gate 2, tracer breakthrough was dominant in the second and third depths, with the most shallow depth showing the greatest variability between fences (Table A7.5). Up to fence G208, the first depth displayed a relatively high proportion of tracer breakthrough, which shifted to moderate dominance at G212, followed by a low dominance of breakthrough for the remainder of the gate. The fourth depth consistently displayed the lowest proportion of breakthrough.

Tracer breakthrough in the first fence of Gate 3 (G302) showed the greatest degree of spatial variability of any of the three gates, probably because of proximity to the source wells. However, the variable distribution evolved into distinct vertical distributions from fences G304 to G316 (Table A7.6). The greatest proportion of tracer breakthrough occurred at the fourth and third monitoring depths, with a moderate amount at the second depth. The most shallow point consistently displayed either the smallest proportion of breakthrough, or was nearly equivalent to the second depth. Similar trends were shown at fence G317, although the fourth depths could not be ranked due to incomplete breakthrough profiles. Breakthrough curves downgradient of G317 were not ranked for the same reasons that fence-averaged velocities were not determined (see above).

Table A7.4 Ranking of Relative Tracer Mass Breakthrough for Multilevel Points in Gate 1

Location	Relative Rank for Indicated Multilevel Point (fence-specific basis)			
	(-)	(+)	(++)	(+++)
G102L	1,4		3	2
G102R	4	2	1	3
G1C1L		1,2	3	4
G1C1M		1,2	3	4
G1C1R		1,2	3	4
G103L		1,2	3	4
G103M		1,2	3	4
G103R		1,2	3	4
G104L		1,2		3,4
G104M		1,2		3,4
G104R		1,2		3,4
G105L		1,2		3,4
G105M		1,2		3,4
G105R		1,2		3,4
G106L		1	2	3,4
G106R				
G108L			1,2	3,4
G108R				
G110L		1	2,3	4
G110R		1	2,3,4	
G112L			1	2,3,4
G112R			1,4	2,3

Table A7.5 Ranking of Relative Tracer Mass Breakthrough for Multilevel Points in Gate 2

Location	Relative Rank for Indicated Multilevel Point (fence-specific basis)			
	(-)	(+)	(++)	(+++)
G202L	4		1	2,3
G202M	4		2,3	1
G202R	4	2		1,3
G205L	4		3	1,2
G205M	4		2	1,3
G205R	4		3	1,2
G208L	4		2,3	1
G208M	4		2,3	1
G208R	4		1,3	2
G212L	4		1,3	2
G212M	4		1,2,3	
G212R	1,4	2	3	
G218L	4		1,3	2
G218M	4	1	2,3	
G218R	1,4		2,3	
G224L		1,4	3	2
G224M		4	1,2,3	
G224R		1,4	2,3	

Table A7.6 Ranking of Relative Tracer Mass Breakthrough for Multilevel Points in Gate 3

Location	Relative Rank for Indicated Multilevel Point (fence-specific basis)			
	(-)	(+)	(++)	(+++)
G302L	1	2,4	3	
G302M	3,4	1,2		
G302R	4		1	2,3
G304L	1,4	3	2	
G304M	1	3	2,4	
G304R	1	2	3	4
G306L		1	2,4	3
G306M	4	1	2	3
G306R	1	2	3	4
G307	1	2,3		4
G308		1,2	3	4
G310L	1		2,3	4
G310M	1	2	3	4
G310R	1	2	3	4
G312L		1,2	3	4
G312M		1,2	3	4
G312R		1,2	3	4
G315L		1	2,3	4
G315M		1,2	3	4
G315R		1,2	3	4
G316L		1,2	3	4
G316R		1,2	3	4
G317L			1.2	3
G317R			1,2	3

A7.6 POINT-SPECIFIC VELOCITY AND MASS DETERMINATIONS IN GATES 2 AND 3

Breakthrough curves for each point on fences G202, G205, G208, G212, G218, and G224 in Gate 2, as well as G302, G304, G310, G312, G315, G318, G320, and G324 in Gate 3 were integrated using the program FENCE (Appendix 24). Groundwater velocities and total tracer masses crossing the points were calculated. The calculated data were subsequently plotted and contoured on cascading cross sections to provide a three-dimensional view of the parameter distributions through the two gates (Figures A7.1 to A7.4). Mass balances for each fence were also calculated and are summarized in Table A7.7.

The velocity distributions in Figures A7.2 and A7.4 indicate that zones of high velocity existed near the fences located between 8 and 15 m from the source wells. Such variability was expected due to the natural heterogeneity of the aquifer material; however, the velocity variations can also be partially attributed to decreased cross-sectional areas for flow (at constant extraction rates) due to a fluctuating water table. For example, at the time the tracer test was initiated, May 15, 1996, the water level in G213P was about 98.767 masl (see May 9 water levels, Appendix 8). This level rose slightly in June to 98.816 masl contributing to relatively low velocities being measured at G202 and G205. However, from July to September, water levels typically decline in response to dry weather. This can be seen in the 1997 water level data collected at the study site (Appendix 8). Unfortunately, 1996 water levels were not collected between July and August, the times when the tracer pulse was transported through the apparently highest velocity zones of Gates 2 and 3. Nevertheless, the water level at G213P in September 1996 was 98.270 masl, a decline of over 0.5 m from June. A decline in saturated thickness of 3 to 2.5 m could cause a 20% increase in flow velocity, corresponding to 2-cm/day increase at a location with a normal

Table A7.7 Mass Balances for KBr Crossing Fences in Gates 2 and 3

Fence	Mass (kg of KBr)	% of Mass Injected
Input to Gate 2 = 4.6 kg KBr		
G202	5.3	115
G205	4.5	98
G208	5.9	128
G212	4.0	87
G218	4.2	91
G224	6.7	146
Input to Gate 3 not certain. Target = 4.6 kg KBr, best estimate = 3.7 kg KBr used in% calculation		
G302	3.4	92
G304	2.9	78
G310	4.8	130
G312	3.9	105
G315	3.6	97
G318	2.5	68
G320	2.4	65
G324	2.8	76

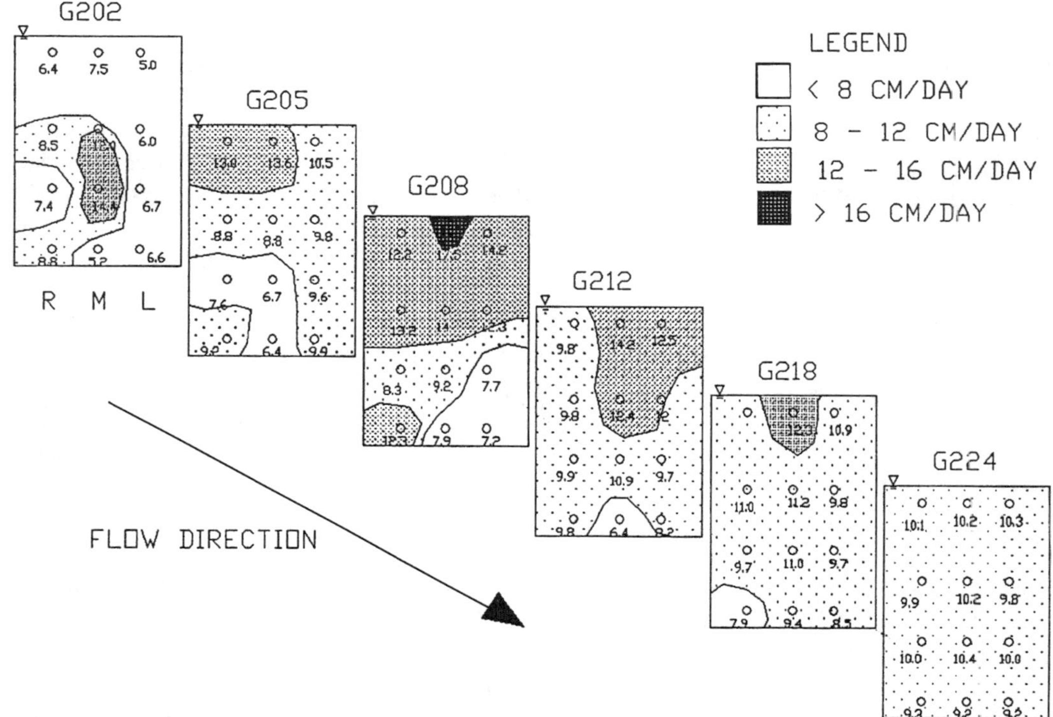

Figure A7.2 Velocity distributions through Gate 2.

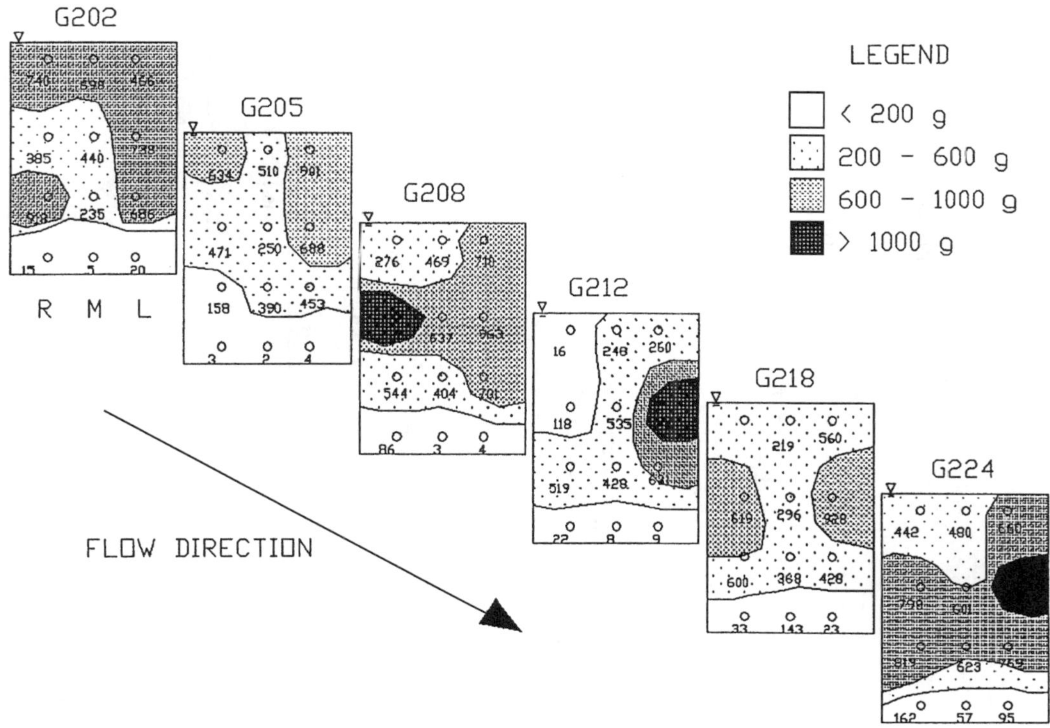

Figure A7.3 Tracer mass breakthrough patterns in Gate 2.

velocity of 10 cm/day. Later in the year, of the water levels recovered until November 21, the level at G213P was 98.767 masl. This does not account for the total range of velocities observed (e.g., 6 to 18 cm/day in Gate 2), but it does help to explain why fences G208 to G212 and G310 to G315 exhibited overall higher velocities than the other fences, both up- and downgradient of them, in the same gates.

The general trend in Gate 2 is for the upper half of the aquifer to exhibit the highest velocities (Figure A7.1). However, this stratification appears to diminish with distance along the gate, until at fence G224, the observed velocities are very uniform. Interestingly, the travel paths followed by the tracer do not appear to correlate very strongly with the high velocity zones (Figure A7.2). They do appear to correspond with the flow patterns described above, based on the qualitative indices. Worthwhile noting here is the lack of appreciable mass that passed through the lowest points in the monitoring network anywhere in Gate 2, despite apparent velocities reasonably consistent with those observed at the shallower monitoring points. These data suggest that the tracer transport was not governed by density flow in Gate 2. This finding is useful in the interpretation of Gate 3 tracer data.

The trend in Gate 3 contrasts somewhat with what was observed in Gate 2. Fence G302 is affected by proximity to the source wells and G304 is located in the NIW, so neither of these fences is considered representative of the gate, in general. Between fences G310 and G315, the higher velocities appear to be in the lower half of the aquifer. Downgradient of G318, the velocity field is remarkably uniform, as it was in Gate 2. The mass distributions do appear to coincide with the high velocity zones more in Gate 3 than in Gate 2, although this correlation is still weak. Nevertheless, in the absence of density effects dominating the tracer flow (cf. the results in Gate 2 above), it explains why the largest portion of the tracer mass was transported along the base of the aquifer, up to G315. Beyond G318, located downgradient of the biosparge zone, tracer mass appears to concentrate along the sides of the gate. However, the mass balance

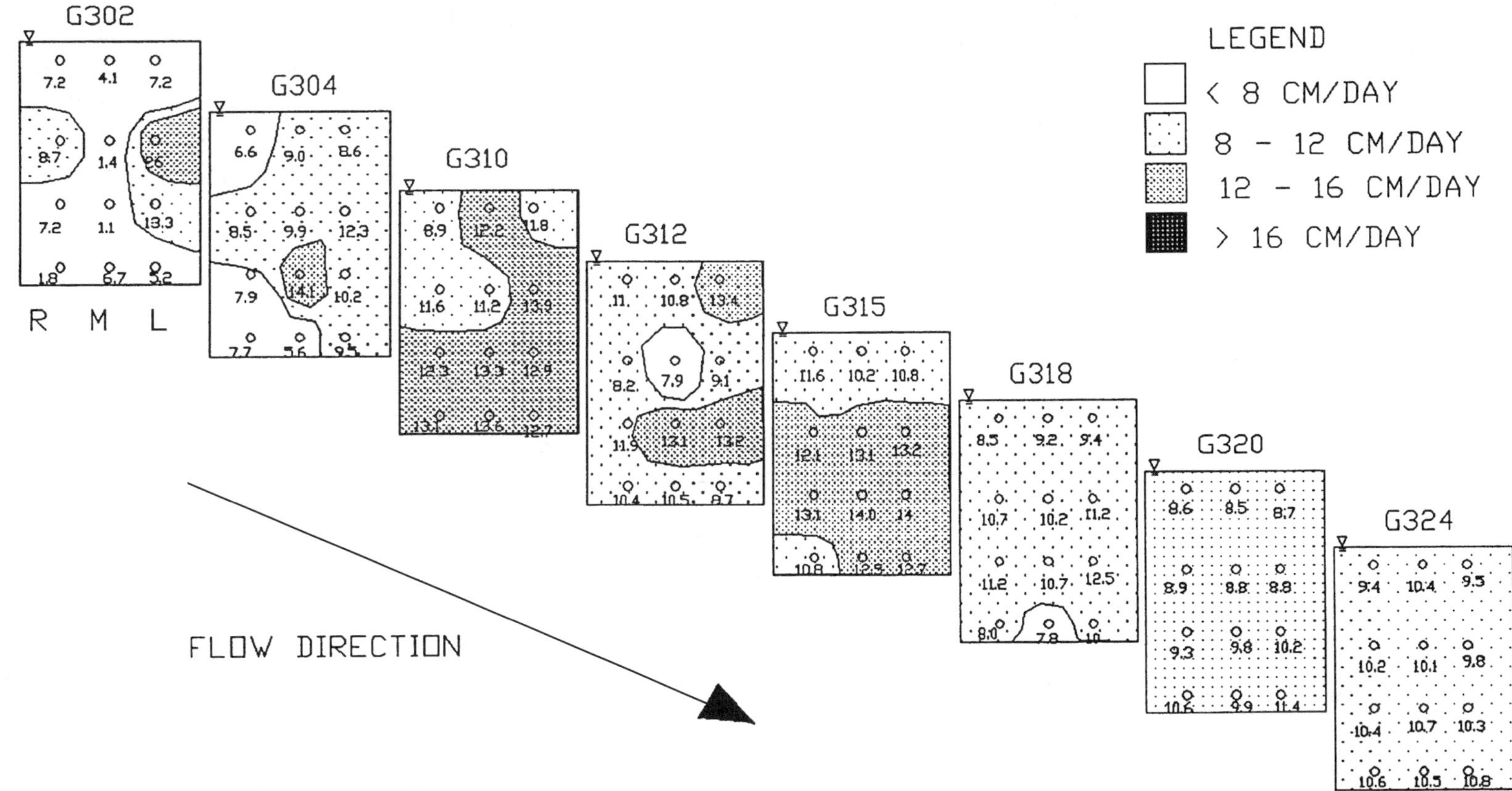

Figure A7.4 Velocity distributions through Gate 3.

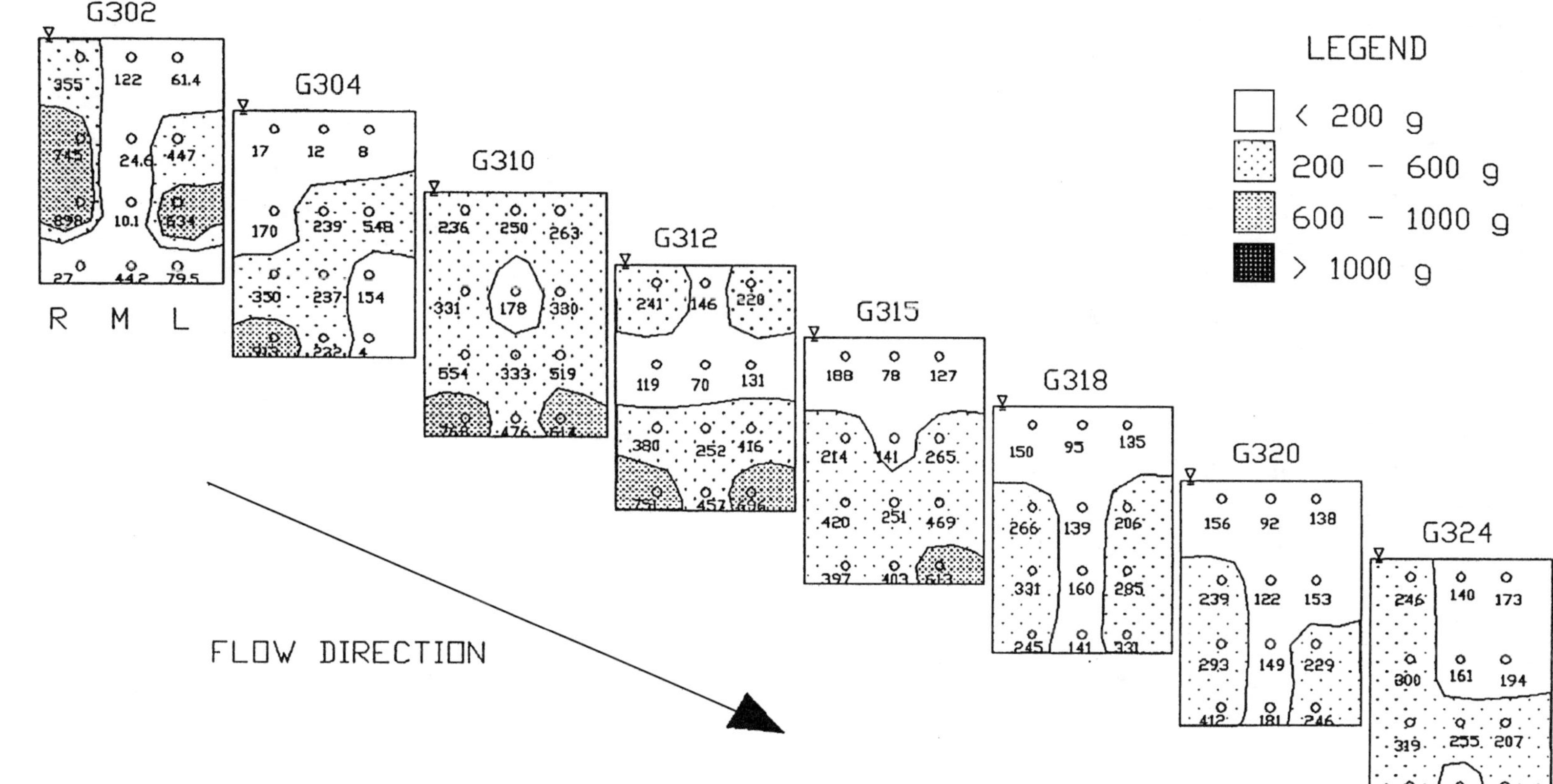

Figure A7.5 Tracer mass breakthrough patterns in Gate 3.

calculations indicate that as much as 30% of the tracer mass is unaccounted for in this region of the gate. This may be due to the hydraulic complexities of the biosparge zone, discussed previously, but further work is needed to confirm this. The mass loss noted for the tracer is not considered to be a long-term problem for solutes released continuously into the gate, because the low-flow zones and dual porosity regions are limited in their capacity for mass storage. Over time, with continuous solute releases, such as occurred throughout the reactive solute testing phase of this work, the storage capacity of these sinks would likely be exceeded and a pseudo-steady-state transport scenario would be expected to develop.

A7.7 REFERENCES

Levenspiel, O., 1993. *The Chemical Reactor Omnibook*. OSU Bookstores, Corvallis, OR.

Mackay, D.M., G. Bianchi-Mosquera, A.A. Kopania, H. Kianjah, and K.W. Thorbjarnarson, 1994. A Forced Gradient Experiment on Solute Transport in the Borden Aquifer. 1. Experimental Methods and Moment Analyses of Results. *Water Resources Research*, 30(2): 369-383.

University of Waterloo (UW), 1997. Passive and Semi-Passive Techniques for Groundwater Remediation: Phase 1 Installation Report. Prepared for AATDF, for Environmental Technology, U.S. Department of Defense DOD-AATDF Administration, Rice University.

Appendix 8

Water Level/Head Measurements and Potentiometric Surface Plots

The UW-AATDF site was instrumented with a variety of monitoring wells, as described in the Phase 1 Installation Report (UW, 1997). The groundwater extraction pumps were started in late April 1996. Water level data were collected from selected piezometers on a weekly basis for most of the operational phase of the experiment. These readings were completed in fully screened wells. Three times over the duration of the experiment, water level readings in all accessible monitoring locations were recorded. Water level reading in the multilevel (0.3 mm ss points) monitoring points could not be recorded, but measurements from the center stalk were taken. It should be noted that these center stalks were wrapped in nytex screen but not slotted, and these readings should be considered more inaccurate than readings from any proximal, fully screened well.

Water level data, collected from selected piezometers on a weekly basis to monitor the flow regime through each gate over the course of the experiment, are presented in Table A8.1. The selected points included those wells surrounding the gates to provide information on the sheet pile seals and the natural water table conditions and those in the gates.

The potentiometric surface through the length of each gate, illustrated in Figures A8.1 to A8.3, was determined through the collection of measurements from all accessible piezometers and fully screened wells on June 29 and November 13, 1996 and on October 23, 1997. This set of data was used to demonstrate whether the expected hydraulic gradient was achieved along the length of each gate throughout the course of the experiment. The water levels were measured with a water level meter. This measurement was then subtracted from the known elevation of the piezometer and the resulting water table elevation recorded in Table A8.1.

Figures A8.1 to A8.3 indicate that the actual hydraulic gradient over the length of each individual gate followed the expected pattern. In general, the highest water levels were observed at the "source well" end of each gate and the lowest water levels at the extraction end of the gates. During the early stages of the experiment a complex potentiometric surface is observed (Figure A8.1); this, however, changes over time and the contours become more evenly spaced and regular in pattern. From Table A8.1 it can be seen that water levels rise through the fall and winter and subside during the summer months. It can also be noted that the water table fluctuated by about 0.5 m over the course of the study.

REFERENCE

University of Waterloo (UW). 1997. Passive and Semipassive Techniques for Groundwater Remediation: Phase I Installation Report.

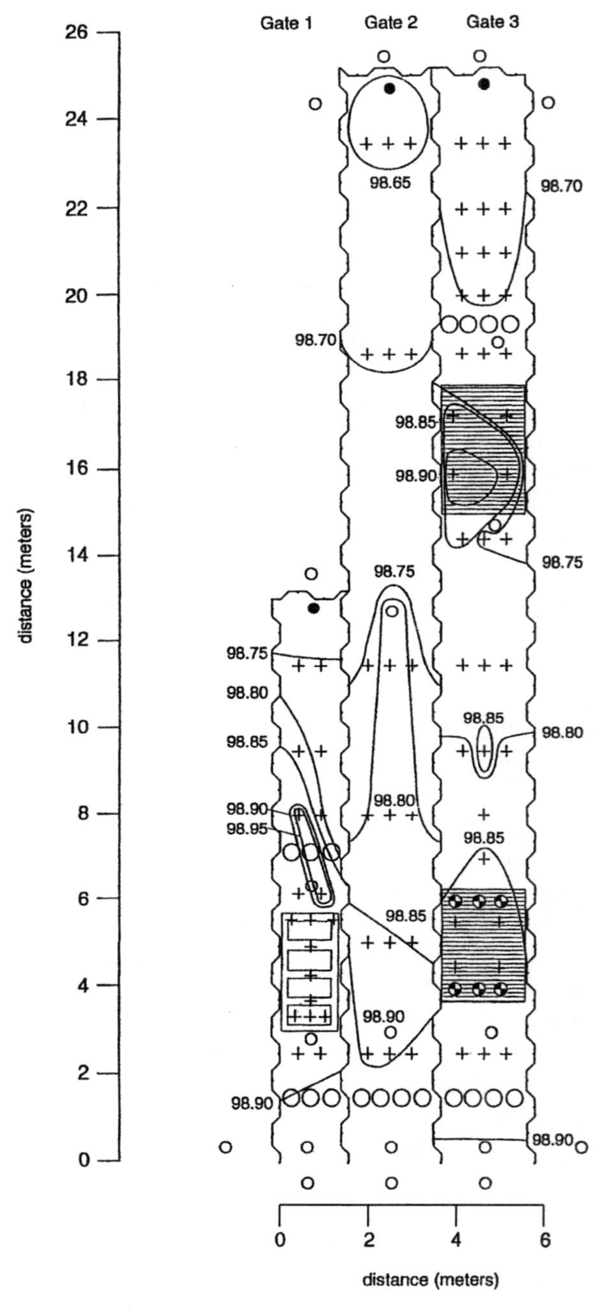

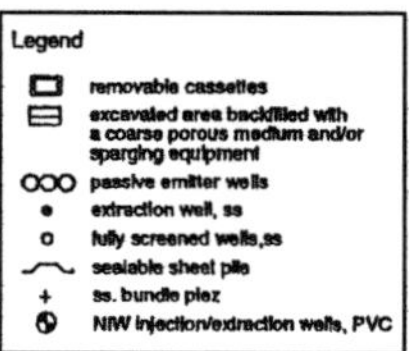

Figure A8.1 Contoured water elevations, June 29, 1996.

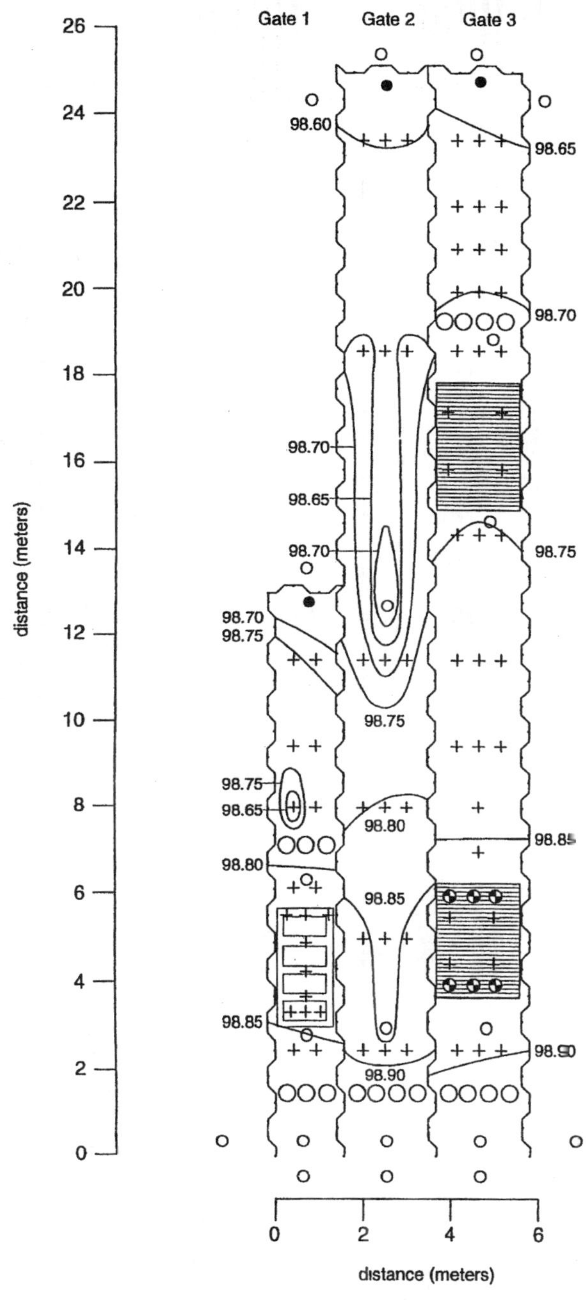

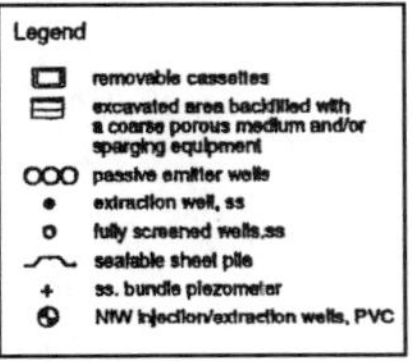

Figure A8.2 Contoured water elevations, November 13, 1996.

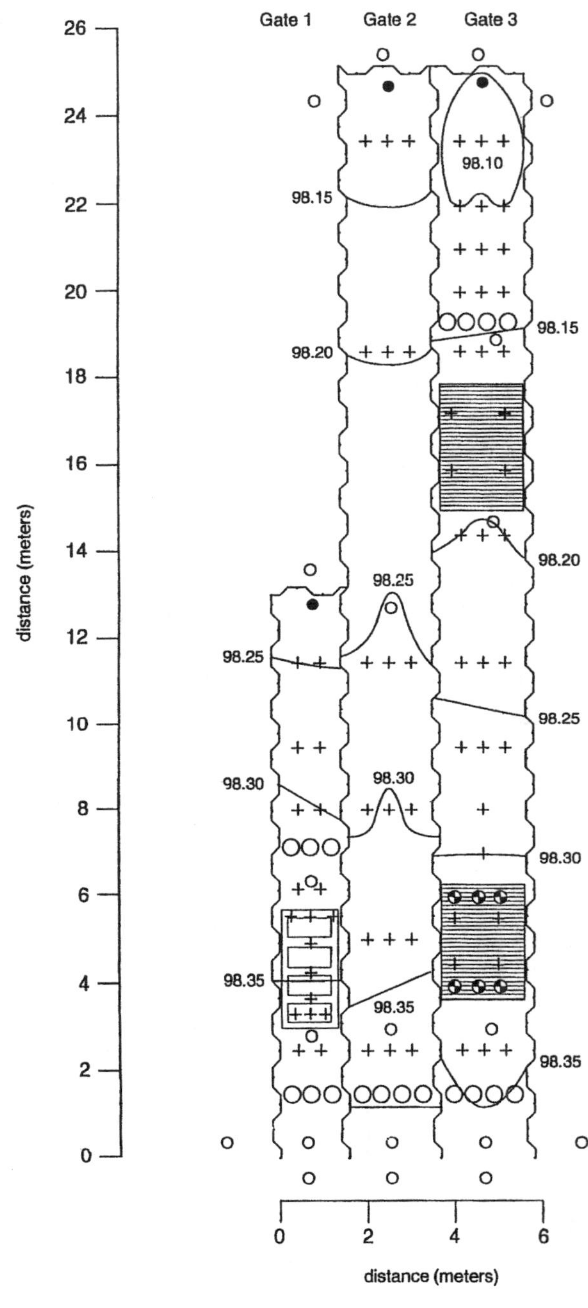

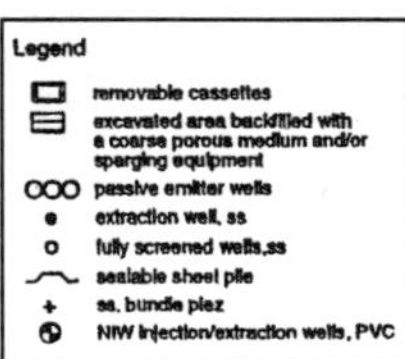

Figure A8.3 Contoured water elevations, October 23, 1997.

Table A8.1 Borden Water Level Elevations

Well	Location		Well Elevation (masl)	Date							
	Northing	Easting		May 9/96	June 29/96	July 11/96	July 23/96	Aug 7/96	Aug 15/96	Aug 27/96	Sept 2/96
					Outside						
G114S	997.779	954.641	99.795	98.713	98.667	98.597	98.701		98.326	98.417	98.317
G224S	1005.860	947.989	99.711			98.474	98.602	98.324	98.214	98.315	98.208
G226S	1007.721	948.342	99.797		98.532	98.441	98.593	98.310	98.172	98.221	98.206
G324S	1009.057	951.620	99.805	98.814	98.586	98.485	98.622	98.315	98.190	98.196	98.168
G326S	1009.118	949.710	99.807		98.582	98.466	98.563	98.292	98.173	98.326	98.195
					Gate 1						
G101P	987.604	963.621	99.884		98.945	98.811	98.994	98.680	98.619	98.613	98.476
G102P	989.571	962.244	99.834	98.974	98.920						
G102R	989.123	962.623	99.955	98.964	98.888	98.800	98.952	98.681	98.501	98.455	98.461
SW13	992.910	958.943	99.833			98.742	98.882	98.641	98.452	98.361	98.410
G112R	996.921	955.840	99.905	98.899	98.762	98.692	98.838	98.604	98.436	98.296	98.381
G113P	997.155	955.186	99.928	98.892	98.718	98.648	98.837	98.529	98.413	98.276	98.395
					Gate 2						
G201P	988.757	964.939	99.915		98.940	98.845	99.016	98.681	98.501	98.543	98.473
G202P	990.540	963.421	99.896	98.921	98.878						
G202M	990.135	962.856	99.954		98.893	98.796	98.969	98.680	98.485	98.491	98.457
G213P	998.386	950.504	99.816	98.767	90.010	98.688	98.847		98.438	98.386	98.332
G218M	1002.309	953.035	99.837	98.782	98.678	00.600	98.731	98.532	00.360	98.261	98.261
G224M	1006.276	949.626	99.867	bent	98.623	98.550	98.687	98.495	98.303	08.227	98.206
G225P	1007.166	948.852	99.694	98.615	98.642	98.530	98.655	98.450	98.271	98.210	98.185

continued

Table A8.1 (continued) Borden Water Level Elevations

Well	Location Northing	Easting	Well Elevation (masl)	Date May 9/96	June 29/96	July 11/96	July 23/96	Aug 7/96	Aug 15/96	Aug 27/96	Sept 2/96
Gate 3											
G301P	990.034	966.403	99.863		98.939	98.796	99.028	98.668	98.494	98.513	98.449
G302P	991.797	965.032	99.866	98.647	98.891						
G302M	991.286	965.385	99.966	98.789	98.887	98.802	98.991	98.689	98.494	98.463	98.442
NUT6	994.346	962.237	100.291								
G315P	1000.906	957.002	99.884	98.640	98.787	98.674	98.842		98.403	98.272	98.284
G324M	1007.577	951.031	99.889	98.578	98.676	98.591	98.737	98.511	98.325	98.197	98.191
G325P	1008.573	950.247	99.894	98.522	98.675	98.599	98.711	98.464	98.294	98.199	98.166
SW16	1003.961	953.936	99.928		99.928	98.614	98.773	98.541	98.370	98.215	98.230
		Sept 18/96	**Sept 25/96**	**Oct 2/96**	**Oct 16/96**	**Oct 23/96**	**Oct 30/96**	**Nov 21/96**	**Nov 29/96**	**Dec 2/96**	**Dec 13/96**
Outside											
G114S	99.795	98.372	98.417	98.469	98.417	98.615	98.670	98.704	98.631	98.655	98.731
G224S	99.711	98.269	98.266	98.355	98.349	98.538	98.580	98.611	98.568	98.598	98.662
G226S	99.797	98.264	98.224	98.325	98.328	98.517	98.578	98.727	98.553	98.614	98.684
G324S	99.805	98.220	98.193	98.333	98.302	98.528	98.638	98.528	98.564	98.686	98.839
G326S	99.807	98.219	98.198	98.323	98.292	98.512	98.585	98.600	98.551	98.615	98.707
Gate 1											
G101P	99.884	98.528	98.488	98.650	98.601	98.790	98.781	98.899	98.839	98.933	99.018
G102P	99.834									98.865	98.941
G102R	99.955	98.516	98.477	98.629	98.596	98.891	98.788	98.876	98.821	98.910	98.974
SW13	99.833	98.458	98.410	98.559	98.541	98.702	98.714	98.818	98.748	98.833	98.876
G112R	99.905	98.402	98.344	98.509	98.491	98.649	98.664	98.802	98.701	98.762	98.820
G113P	99.928	98.352	98.358	98.468	98.416	98.608	98.620	98.742	98.642	98.724	98.764

		Dec 16/96	Dec 27/96	Dec 30/96	Jan 6/97	Jan 13/97	Jan 20/97	Jan 27/97	Feb 3/97	Feb 10/97	Feb 20/97
Gate 2											
G201P	99.915	98.513	98.495	98.638	98.592	98.848	98.748	98.943	99.915	98.964	99.052
G202P	99.896									98.896	99.000
G202M	99.954	98.515	98.479	98.631	98.598	98.787	98.610	98.893	98.832	98.924	99.034
G213P	99.816	98.368	98.371	98.493	98.469	98.643	98.636	98.767	98.676	98.771	98.889
G218M	99.837	98.313	98.270	98.420	98.401	98.557	98.612	98.664	98.621	98.718	98.758
G224M	99.867	98.255	98.230	98.367	98.346	98.511	98.535	98.611	98.566	98.648	98.727
G225P	99.694	98.228	98.219	98.328	98.304	98.481	98.508	98.585	98.545	98.597	98.706
Gate 3											
G301P	99.863	98.519	98.455	98.632	98.586	98.796	98.799	98.903	98.845	98.906	99.101
G302P	99.866									98.915	99.866
G302M	99.966	98.500	98.448	98.622	98.588	98.777	98.820	98.893	98.823	98.954	99.049
NUT6	100.291										
G315P	99.884	98.342	98.287	98.464	98.445	98.610	98.659	98.738	98.680	98.781	98.805
G324M	99.889	98.249	98.249	98.365	98.353	98.527	98.545	98.661	98.575	98.664	98.716
G325P	99.894	98.215	98.147	98.321	98.321	98.492	98.516	98.751	98.538	98.650	98.675
SW16	99.928	98.291	98.294	98.401	98.404	98.550	98.581	98.672	98.617	98.721	98.739
Outside											
G114S	99.795	98.713	98.704	98.707	98.759	98.710	98.661	98.682	98.661	98.667	98.750
G224S	99.711	98.617	98.614	98.662	98.705	98.617	98.614	98.620	98.614	98.620	98.644
G226S	99.797	98.627	98.630	98.648	98.718	98.599	98.611	98.590	98.566	98.599	98.663
G324S	99.805	98.723	98.708	98.708	98.863	98.665	98.616	98.625	98.635	98.595	98.763
G326S	99.807	98.643	98.612	98.637	98.755	98.606	98.576	98.594	98.588	98.563	98.670
Gate 1											
G101P	99.884	98.976	99.037	99.049	99.148	99.015	98.930	98.921	90.070	98.887	98.988
G102P	99.834	98.926	99.011	99.051	99.096	98.996	98.929	98.907	98.849	90.056	98.941
G102R	99.955	98.955	99.031	99.022	99.102	99.294	98.910	98.906	98.864	98.873	98.943
SW13	99.833	98.876	98.958	98.943	99.022	98.919	98.833	98.836	98.794	98.803	98.876
G112R	99.905	98.838	98.875	98.869	98.939	98.838	98.759	98.777	98.716	98.728	98.802
G113P	99.928	98.755	98.828	98.825	98.889	98.785	98.697	98.727	98.678	98.736	98.767

continued

Table A8.1 (continued) Borden Water Level Elevations

Well		Dec 16/96	Dec 27/96	Dec 30/96	Jan 6/97	Jan 13/97	Jan 20/97	Jan 27/97	Feb 3/97	Feb 10/97	Feb 20/97
						Gate 2					
G201P	99.915	98.985	99.062	99.080	99.205	99.052	98.930	98.943	98.927	98.915	99.016
G202P	99.896	98.957	99.040	99.076	99.140	99.067	98.945	98.911	98.866	98.866	98.979
G202M	99.954	98.985	99.052	99.055	99.155	99.018	98.924	98.927	98.869	98.884	98.994
G213P	99.816	98.819	98.844	98.892	99.139	98.831	98.786	98.767	98.728	98.740	98.835
G218M	99.837	98.746	98.795	98.798	98.868	98.761	98.688	98.688	98.664	98.673	98.764
G224M	99.867	98.690	98.733	98.721	98.797	98.678	98.623	98.623	98.596	98.614	98.694
G225P	99.694	98.655	98.688	98.688	98.752	98.652	98.588	98.594	98.566	98.575	98.682
						Gate 3					
G301P	99.863	99.000	99.077	99.089	99.199	99.055	98.943	98.955	98.891	98.903	99.034
G302P	99.866	98.970	99.031	99.064	99.141	99.031	98.888	98.885	98.845	98.857	98.982
G302M	99.966	99.003	99.079	99.076	99.189	99.033	98.917	98.924	98.881	98.881	99.012
NUT6	100.291							99.297	99.246	99.255	99.383
G315P	99.884	98.832	98.899	98.896	98.979	98.851	98.781	98.775	98.726	98.765	98.842
G324M	99.889	98.725	98.798	98.789	98.859	98.746	98.658	98.676	98.612	98.645	98.731
G325P	99.894	98.696	98.745	98.745	98.833	98.736	98.626	98.644	98.589	98.614	98.699
SW16	99.928	98.773	98.834	98.837	98.913	98.806	98.730	98.724	98.678	98.700	98.770
Well		**Feb 24/97**	**Mar 3/97**	**Mar 10/97**	**Mar 17/97**	**Mar 24/97**	**Mar 31/97**	**Apr 7/97**	**Apr 15/97**	**Apr 21/97**	**Apr 30/97**
						Outside					
G114S	99.795	98.777	98.774	98.722	98.716	98.716	98.798	98.798	98.756	98.722	98.728
G224S	99.711	98.705	98.702	98.666	98.602	98.617	98.748	98.745	98.669	98.659	98.653
G226S	99.797	98.730	98.736	98.645	98.624	98.614	98.791	98.800	98.697	98.715	98.663
G324S	99.805	98.878	98.915	98.659	98.696	98.665	98.955	98.991	98.820	98.784	98.760
G326S	99.807	98.771	98.810	98.746	98.630	98.606	98.835	98.859	98.725	98.694	98.679

Well		May 5/97	May 13/97	May 22/97	May 27/97	June 2/97	June 3/97	June 10/97	June 16/97	June 23/97	July 4/97
Gate 1											
G101P	99.884	99.128	99.186	99.168	99.034	98.997	99.159	99.168	99.110	99.098	99.092
G102P	99.834	99.081	99.130	99.038	98.993	99.020	99.124	99.163	99.054	99.048	99.051
G102R	99.955	99.111	99.126	99.056	99.025	98.998	99.144	99.184	99.074	99.065	99.077
SW13	99.833	99.034	99.047	98.952	98.958	98.919	99.080	99.108	98.992	98.998	99.833
G112R	99.905	98.960	98.978	98.890	98.869	98.869	99.015	99.036	98.933	98.899	98.930
G113P	99.928	98.913	98.937	98.849	98.819	98.800	98.934	98.962	98.876	98.883	98.873
Gate 2											
G201P	99.915	99.168	99.208	99.135	99.068	99.040	99.199	99.220	99.150	99.126	99.129
G202P	99.896	99.134	99.171	99.070	99.043	99.006	99.192	99.186	99.107	99.110	99.088
G202M	99.954	99.149	99.162	99.082	99.046	99.006	99.183	99.216	99.140	99.128	99.070
G213P	99.816	98.932	98.956	98.917	98.856	98.835	99.002	99.008	98.926	98.902	98.911
G218M	99.837	98.874	98.904	98.819	98.801	98.752	98.947	98.968	98.856	98.825	98.856
G224M	99.867	98.812	98.825	98.745	98.742	98.675	98.846	98.876	98.782	98.767	98.773
G225P	99.694	98.770	98.804	98.706	98.694	98.658	98.981	98.844	98.774	98.740	98.740
Gate 3											
G301P	99.863	99.199	99.214	99.168	99.089	99.040	99.232	99.253	99.186	99.150	99.147
G302P	99.866	99.144	99.174	99.070	99.049	99.003	99.195	99.223	99.147	99.101	99.104
G302M	99.966	99.192	99.204	99.110	99.085	99.030	99.238	99.286	99.164	99.143	99.131
NUT6	100.291	99.550	99.581	99.495	99.154	99.401	99.605	99.627	99.544	99.517	100.291
G315P	99.884	98.994	99.028	98.933	98.954	98.863	99.067	99.110	98.985	98.957	98.970
G324M	99.889	98.892	98.923	98.844	98.840	98.761	98.956	98.984	98.874	98.844	98.862
G325P	99.894	98.864	98.897	98.800	98.812	98.797	98.928	98.983	98.845	98.818	98.839
SW16	99.928	98.943	98.977	98.895	98.907	98.809	99.008	99.047	98.916	98.889	99.928
Outside											
G114S	99.795	98.801	98.725	98.746	98.725	98.734	98.676	98.618	98.530	98.887	98.582
G224S	99.711	98.705	98.647	98.626	98.598	98.574	98.580	98.510	98.410	98.769	98.452
G226S	99.797	98.803	98.663	98.627	98.624	98.660	98.578	98.483	98.374	98.800	98.416
G324S	99.805	98.891	98.760	98.705	98.659	98.607	98.616	98.488	98.479	98.973	98.424
G326S	99.807	98.774	98.676	98.637	98.597	98.539	98.579	98.493	98.411	98.844	98.435

continued

Table A8.1 (continued) Borden Water Level Elevations

Well		May 5/97	May 13/97	May 22/97	May 27/97	June 2/97	June 3/97	June 10/97	June 16/97	June 23/97	July 4/97
						Gate 1					
G101P	99.884	99.140	99.095	99.028	99.018	98.964	98.933	98.826	98.732	99.162	98.808
G102P	99.834	99.112	99.060	98.865	99.005	98.923	98.929	98.804	98.752	99.045	98.810
G102R	99.955	99.123	99.077	98.970	99.041	98.940	98.922	98.821	98.745	99.074	98.806
SW13	99.833	99.041	98.998	98.675	98.925	98.852	98.833	98.745	98.675	98.967	98.763
G112R	99.905	98.991	98.924	98.841	98.890	98.814	98.796	98.710	98.640	98.896	98.701
G113P	99.928	98.950	98.876	98.593	98.870	98.745	98.739	98.678	98.562	98.864	98.654
						Gate 2					
G201P	99.915	99.193	99.126	99.077	99.080	98.998	98.949	98.827	98.732	99.202	98.812
G202P	99.896	99.149	99.091	99.000	99.036	98.948	98.924	98.814	98.713	99.158	98.790
G202M	99.954	99.158	99.110	99.101	99.058	98.951	98.936	98.823	98.735	99.177	98.802
G213P	99.816	98.963	98.902	98.856	98.841	98.789	98.774	98.682	98.636	98.932	98.676
G218M	99.837	98.923	98.840	98.788	98.807	98.724	98.715	98.630	98.551	98.773	98.673
G224M	99.867	98.825	98.767	98.709	98.706	98.654	98.648	98.559	98.480	98.764	98.572
G225P	99.694	98.804	98.731	98.682	98.676	98.621	98.615	98.530	98.466	98.743	98.557
						Gate 3					
G301P	99.863	99.217	99.153	99.101	99.077	99.037	98.958	98.821	98.760	99.244	98.796
G302P	99.866	99.171	99.101	99.049	99.028	98.924	98.909	98.790	98.769	99.134	98.772
G302M	99.966	99.210	99.143	99.091	99.064	98.969	98.951	98.823	98.741	99.183	98.805
NUT6	100.291	99.575	99.502	99.453	99.428	99.319	99.316	99.185	99.096	99.517	99.185
G315P	99.884	99.043	98.970	98.915	98.903	98.811	98.796	98.689	98.613	98.869	98.701
G324M	99.889	98.932	98.856	98.816	98.789	98.706	98.691	98.594	98.517	98.749	98.624
G325P	99.894	98.903	98.815	98.781	98.757	98.696	98.672	98.559	98.525	98.717	98.599
SW16	99.928	98.980	98.907	98.864	98.846	98.755	98.739	98.642	98.578	98.794	98.675

Well		July 8/97	July 14/97	July 15/97	July 23/97	July 28/97	Aug. 06/97	Aug. 18/97	Aug. 26/97	Sept. 04/97	Sept. 08/97
Outside											
G114S	99.795	98.463	98.335	98.335	98.237	98.170	98.109456	98.20	98.30148	98.216136	98.203944
G224S	99.711	98.339	98.221	98.217	98.123	98.056	97.991928	98.13	98.187	98.08032	98.089464
G226S	99.797	98.307	98.191	98.188	98.087	98.023	97.95296	98.05	98.148032	98.038304	98.032208
G324S	99.805	98.351	98.205	98.202	98.080	98.016	97.924384	98.03	98.1286	98.028016	98.281
G326S	99.807	98.295	98.164	98.176	98.070	98.003	97.929432	98.03	98.136696	98.017824	98.014776
Gate 1											
G101P	99.884	98.650	98.540	98.522	98.390	98.330	98.274656	98.37524	98.448392	98.326472	98.323424
G102P	99.834	98.645	98.533	98.508	98.371	98.356	98.221608	98.352672	98.419728	98.31	98.337432
G102R	99.955	98.660	98.553	98.529	98.382	98.324	98.23288	98.32432	98.434048	98.336512	98.315176
SW13	99.833	98.599	98.477	98.461	98.318	98.266	98.165744	98.257184	98.363864	98.260232	98.26328
G112R	99.905	98.570	98.448	98.424	98.271	98.232	98.033528	98.219456	98.316992	98.25908	98.201168
G113P	99.928	98.486	98.374	98.389	98.215	98.191	98.14492	98.1754	98.263792	98.157112	98.147968
Gate 2											
G201P	99.915	98.656	98.549	98.531	98.391	98.327	98.247744	98.354424	98.448912	98.320896	98.323944
G202P	99.896	98.585	98.530	98.521	98.378	98.314	98.222648	98.344568	98.423816	98.31104	98.301896
G202M	99.954	98.692	98.531	98.531	98.384	98.323	98.23188	98.317224	98.43	98.32332	98.314176
G213P	99.816	98.524	98.423	98.414	98.277	98.210	98.12436	98.218848	98.292	98.212752	98.227992
G218M	99.837	98.493	98.407	98.389	98.237	98.170	98.072208	98.13012	98.230704	98.172792	98.139264
G224M	99.867	98.404	98.319	98.319	98.163	98.124	98.010768	98.08392	98.17536	98.102208	98.08392
G225P	99.694	98.380	98.289	98.280	98.143	98.076	97.981024	98.054176	98.145616	98.072464	98.090752
Gate 3											
G301P	99.863	98.635	98.537	98.522	98.373	98.315	98.226224	98.30852	98.427392	98.317664	98.299376
G302P	99.866	98.610	98.516	98.497	98.345	98.287	98.192648	98.2658	98.378576	98.274944	98.25056
G302M	99.966	98.640	98.552	98.527	98.369	98.311	98.216448	98.295696	98.417616	98.30484	98.292648
NUT6	100.291	99.014	98.929	98.907	98.749	98.691	98.59936	98.669464	98.78224	98.687752	98.669464
G315P	99.884	98.531	98.454	98.421	98.275	98.202	98.10092	98.16188	98.247224	98.189312	98.164928
G324M	99.889	98.441	98.368	98.338	98.194	98.115	98.01448	98.0602	98.160784	98.398528	98.0602
G325P	99.894	98.413	98.330	98.309	98.163	98.080	97.979856	98.031672	98.129208	98.37	98.028624
SW16	99.928	98.465	98.407	98.386	98.218	98.148	98.044336	98.111392	98.20588	98.16016	97.834024

continued

Table A8.1 (continued) Borden Water Level Elevations

Well		Sept. 16/97	Sept. 22/97	Sept. 29/97	Oct. 7/97	Oct. 16/97	Oct. 21/97	Oct. 23/97
				Outside				
G114S	99.795	98.14908	98.267952	98.271	98.289288	98.289288	98.298432	98.289288
G224S	99.711	98.058984	98.153472	98.251008	98.177856	98.187	98.177856	98.208336
G226S	99.797	97.992584	98.111456	98.154128	98.138888	98.251664	98.129744	98.144984
G324S	99.805	97.997536	98.079832	98.277952	98.110312	98.085928	98.165176	98.107264
G326S	99.807	97.966008	98.078784	98.08488	98.109264	98.087928	98.090976	98.118408
				Gate 1				
G101P	99.884	98.271608	98.384384	98.427056	98.45144	98.448392	98.381336	98.393528
G102P	99.834	98.245992	98.349624	98.441064	98.383152	98.364864	98.352672	98.35572
G102R	99.955	98.26336	98.366992	98.41576	98.394424	98.37004	98.351752	98.373088
SW13	99.833	98.199272	98.296808	98.333384	98.327288	98.309	98.299856	98.309
G112R	99.905	98.158496	98.246888	98.338328	98.277368	98.268224	98.252984	98.25908
G113P	99.928	98.184544	98.19064	98.410096	98.227216	98.245504	98.181496	98.20588
				Gate 2				
G201P	99.915	98.287368	98.077056	98.445864	98.415384	98.424528	98.369664	98.424528
G202P	99.896	98.289704	98.381144	98.414672	98.396384	98.35676	98.344568	98.362856
G202M	99.954	98.265408	98.372088	98.396472	98.396472	98.36904	98.356848	98.372088
G213P	99.816	98.182272	98.234088	98.304192	98.27676	98.249328	98.252376	98.252376
G218M	99.837	98.102688	98.157552	98.215464	98.20632	98.197176	98.178888	98.188032
G224M	99.867	98.032104	98.117448	98.202792	98.150976	98.132688	98.120496	98.132688
G225P	99.694	98.00236	98.087704	98.173048	98.12428	98.130376	98.099896	98.127328
				Gate 3				
G301P	99.863	98.24756	98.357288	98.38472	98.387768	98.424344	98.351192	98.360336
G302P	99.866	98.210936	98.31152	98.399912	98.335904	98.317616	98.305424	98.314568
G302M	99.966	98.246928	98.323128	98.371896	98.368848	98.344464	98.326176	98.332272
NUT6	100.291	98.623744	98.715184	98.770048	98.75176	98.724328	98.715184	98.724328
G315P	99.884	98.122256	98.189312	98.292944	98.228936	98.274656	98.201504	98.198456
G324M	99.889	98.017528	98.081536	98.19736	98.139448	98.112016	98.102872	98.096776
G325P	99.894	97.979856	98.043864	98.184072	98.11092	98.113968	98.077392	98.04996
SW16	99.928	98.120536	98.12968	98.239408	98.166256	98.14492	98.141872	98.117488

Note: All water levels given in meters above sea level.

Appendix 9

Source Well Evaluation

A9.1 BACKGROUND

The CFB Borden test site was uncontaminated at the outset of the experiment. Therefore, a plume with the desired contaminants (CT, TOL, and PCE) had to be generated for the experiment. Several different methods to create a plume of contaminated groundwater were considered, including direct injection, source emplacement, and a controlled NAPL spill. The disadvantages of these methods are the need for continuous pumping during operation of a direct injection, disturbance of the aquifer to emplace the source, and difficult cleanup afterwards in the case of either a controlled NAPL spill or an emplaced source. An alternative method of delivering contaminants (or nutrients) into a relatively homogeneous aquifer was developed and described by Wilson and Mackay (1995), and was chosen for use in this experiment. The premise of this system was to circulate solutions of high-concentration contaminants inside polymer tubing and allow the diffusional gradient to deliver the desired compounds (contaminants) to the groundwater in the well. Subsequent migration into the aquifer occurs by diffusion and advection of groundwater through the well screen. The final concentration in the aquifer is controlled by varying (a) the concentrations of compounds being circulated in the tubing, (b) the length of the tubing, or (c) the tubing wall thickness. The potential advantages of this semipassive source generation are the controlled nature of the system and the ease of ending the source input without expensive cleanup costs.

Laboratory tests were completed in early 1996 and a field prototype test of the final source well configuration was field tested from July to September 1996 to determine the quantity of LDPE tubing required to deliver appropriate target organic concentrations (UW, 1997). From these tests, tubing lengths of 10 m for CT and 12 m each for PCE and TOL were chosen for use in each well. Results of the field prototype test demonstrated that the selected lengths of tubing were satisfactory, but that delivery of constant levels of contaminant and the development of a homogeneous plume across the saturated thickness of the aquifer were unlikely. The arrival of contaminant to the first fence was not uniform and the time to steady state was much longer than anticipated.

A9.2 INSTALLATION

For the UW-AATDF Borden project, large-diameter PVC wells (25-cm OD) were used to deliver the contaminants to the aquifer. A total of 11 wells were installed in close proximity to each other across the inlet ends of the gates (Figure A9.1). Wells were screened across the saturated thickness, from 0.9 to 3.1 m below ground surface. These wells were termed source wells. The well installation process was described in detail in the Phase I Installation Report (UW, 1997).

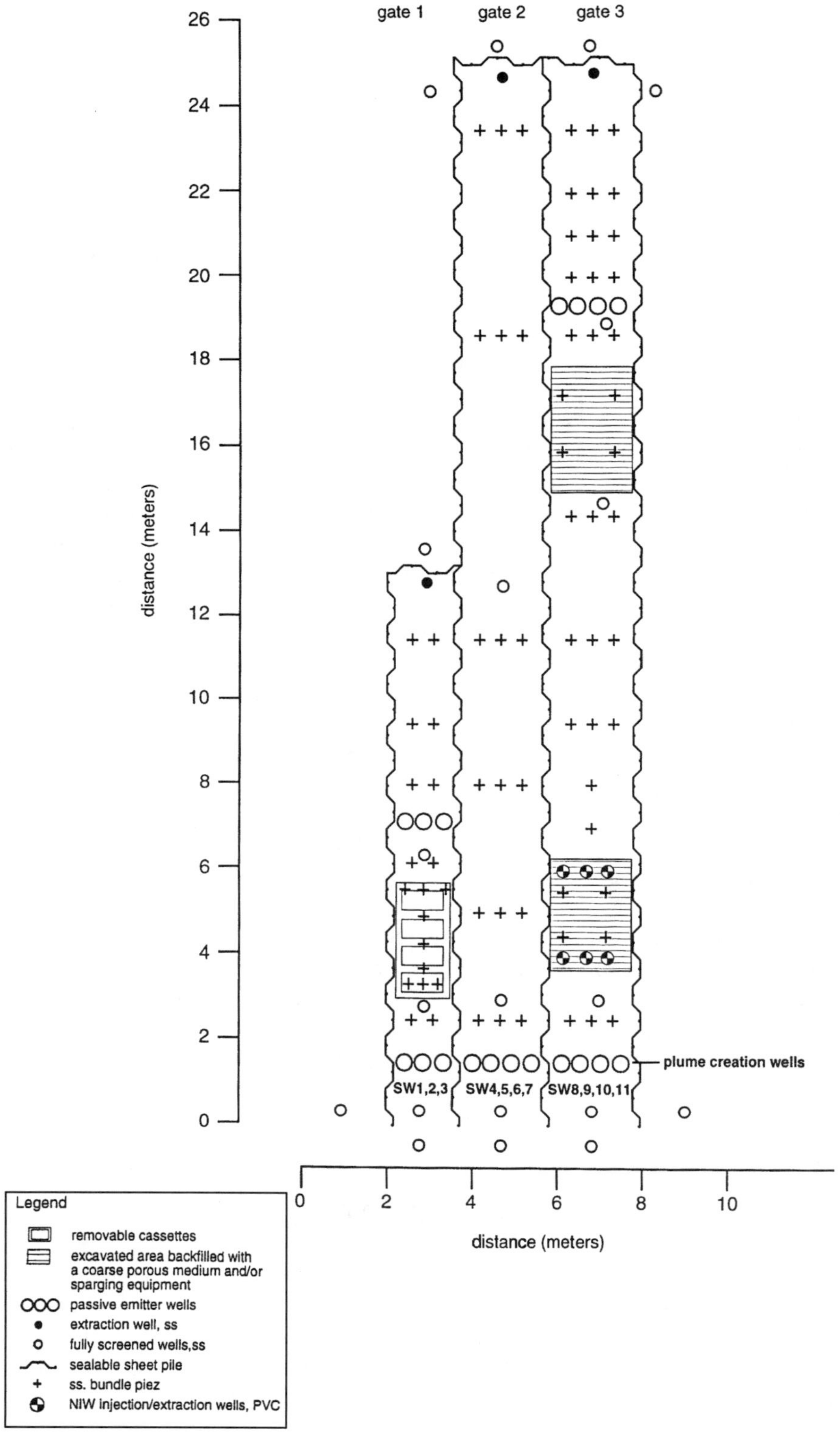

Figure A9.1 Location of plume creation wells, all gates.

A source diffuser apparatus is shown schematically in Figure A9.2. Source wells were equipped with a PVC frame to support coils of LDPE (with nominal wall thickness of 0.3 mm and an OD of 0.4 cm), which were subsequently connected to source drums using 3.2-mm stainless steel tubing and Swagelok fittings. Dedicated, sealed source drums (114-L steel drums, coated to prevent corrosion from solvents) for each contaminant were filled with water and a small quantity (approximately 1 L) of pure phase TOL, PCE, or CT was added. In no cases were the contaminants mixed between the drums. The drums were stirred gently with a drill-adapted mixing rod and then sealed. Additional solvent was added at various times throughout the duration of the project, as required.

Peristaltic pumps recirculated the near-saturated solutions through the tubing in the wells and back to the drums. Precautions were taken to avoid the accidental transport of NAPL through the down hole tubing. Bubbles of pure product in the tubing would be expected to cause nonuniform, rapid diffusion of the compound(s) across the tubing and result in undesired fluctuations in the delivery concentration. For this reason, the intake line was located at least 20 cm from the bottom of the drum above the level of the DNAPL phases (PCE, CT) and below the level of the LNAPL phase (TOL). The recycle return from the well discharged above the intake line to promote mixing within the drums.

Each source drum was placed inside an overpack drum (205-L drum) for secondary containment in case of leaks. The bottom third of the overpack drums were filled with vermiculite. The drums were fitted with drum heaters over the winter months to keep the temperature of the mixture at a constant 16°C.

One sampling point, consisting of 0.32-mm ss tube, was installed in each source well at about 2.3 m below ground surface.

A9.3 OPERATION AND RESULTS

The diffusive source release was started on October 17, 1996. However, after 4 days of operation, the system was shut off due to a PCE leak. Approximately 40 L of PCE solution from the source drum leaked into SW-1. One of the fittings had been overtightened and had caused a fracture in the tubing. The diffusion apparatus was removed from the well and tested on the surface. The leak was confirmed visually and the fitting replaced. SW-1 was then pumped to recover the PCE which had been leaked. About 100 L of water was pumped from the well and treated in the aboveground treatment system (Appendix 10). The diffusive source apparatus was then returned to SW-1. For the duration of the experiment each source drum was retrofitted with a float level switch to shut the recirculation pump off when the level decreased by about 500 mL.

The source generation system was restarted on November 18, 1996 and operated until October 28, 1997. Pumps were operated continuously from November 1996 until June 1997 to prevent possible freezing during the winter months. Subsequently, the system was operated in intermittent mode (3 hours circulating and 9 hours stagnant, controlled by a timer). The contaminant delivery system functioned well over the duration of the project. No major leaks occurred, but the float sensors were actuated several times. These occurrences were apparently due to declining volumes in the source drums, resulting from sampling and volatilization. The CT drum was the most prone to volumetric declines, with losses of about 500 mL every 3 weeks. All CT lines were tested on two different occasions without detecting any tubing leaks. Two possible causes of the lost volume are (1) slow leaks in several wells which are sufficiently small that no detection during leak tests was observed; and (2) volatilization through the drum seal and fittings.

In the first 12 weeks of the experiment, source wells were monitored on a regular basis (every 2 weeks) to determine the delivery of the contaminants to the wells. After this time the source well monitoring was less frequent (about every 10 weeks). The concentrations of each contaminant, as measured by sampling the groundwater in the source wells, is shown in Figure A9.3. Individual source wells are not identified, for clarity. The data used for these graphs are presented in Table

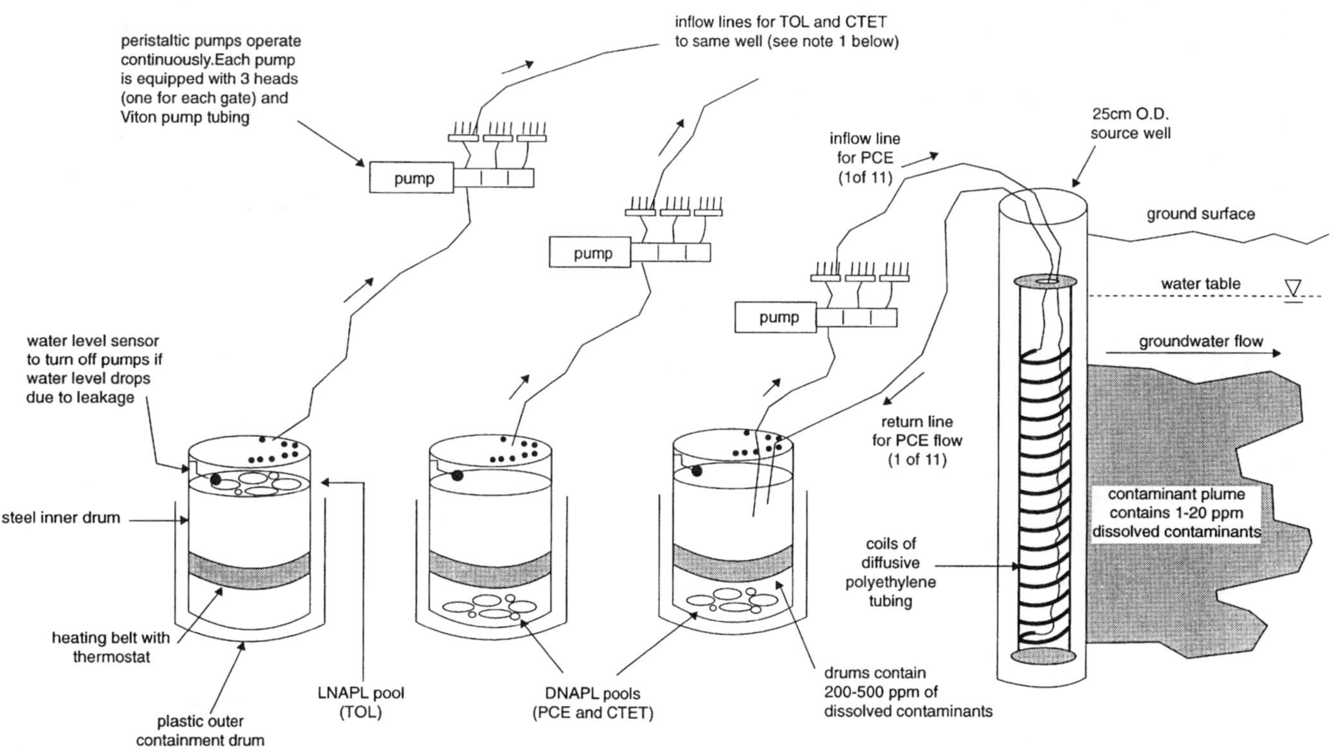

Note 1: One peristaltic pump is used for each contaminant solution, with flow to each of the 11 source wells subdivided through stainless steel manifolds. Most of the 0.3 cm O.D. ss tubing is not shown above for simplicity.

Note 2: Each drum is equipped with a mechanical mixer (not shown above)

Figure A9.2 Diffusive VOC source emitter equipment, all gates.

A9.1. The analytical uncertainties are also given and are sufficiently small that they are not plotted on Figure A9.3, for clarity. As Table A9.1 shows, the concentrations reported steadily increase over time, implying that steady-state conditions were not reached. The behavior of TOL over the project suggests that a shorter length (<12 m) of tubing in the diffusion apparatus would have been sufficient. The variability between wells can be attributed to numerous factors, including preferential flow paths which allowed contaminants to advect and diffuse out of some wells more rapidly than others, or allowed uneven circulation of compounds within the diffusion tubing, such that some wells received more contaminant mass than others.

On February 10, 1997, day 83, the concentrations of TOL in the source wells ranged from 40 to 90 mg/L. At the first fence, 0.7 m downgradient (G102, 202, or 302), TOL concentrations in selected points exceeded 10 mg/L. The TOL circulation system was temporarily disengaged and the wells were allowed to flush through, producing input concentrations within the target range (1–10 mg/L). The TOL graph in Figure A9.3 shows this event — concentrations of TOL level off or decline. The TOL source was restarted on April 7, 1997 and continued to operate until October 1997. On June 11, 1997, day 204, the entire system was set on a timer (3 hours on and 9 hours off) with no noticeable decline in source well concentrations being observed (see Figure A9.3).

More NAPL was added to each drum on July 4, 1997 (day 227), resulting in an increase in the drum and in-well concentrations (Figure A9.3 and Table A9.2). The concentrations of each contaminant in the source drums is presented in Table A9.2. The analytical uncertainties given in Table A9.2 are most pronounced at the first two time points due to insufficient laboratory dilution.

A9.4 REMOVAL OF SOURCES

Source injection was terminated on October 28, 1997 by stopping the pumps and draining all of the lines in the source apparati. The diffusive source apparati were removed from the source wells and dismantled. The tubing was checked for integrity, and no cracks or leaks were found after almost 1 year of use. The recovered water was treated using the aboveground treatment system (Appendix 10). The source dismantling is the focus of a co-op work term report (F. Lyle, UW, 1997).

A9.5 ASSESSMENT OF DELIVERY OF TARGET ORGANICS TO AQUIFER

The evaluation of the tracer test that was completed in all gates (Appendix 7) concluded that preferential flow paths in the aquifer within each gate existed. Normally, the response to such a conclusion would be to redevelop the source wells, complete aquifer response tests (i.e., slug tests), and then conduct another tracer test. Due to time constraints, this was not possible in this experiment. It was decided that well development by pumping would nullify the tracer results (determination of average groundwater velocities at each fence and in each gate), as the conditions would be substantially altered. Instead, the source wells were surge-blocked, to attempt to remove fines from the well screens with a plungerlike device. The surging action becomes increasingly difficult as the well diameter increases because the methodology requires that the column of water above the plunger device be lifted quickly and repeatedly. Nevertheless, all 11 source wells were surged prior to the VOC emitter source emplacement.

In order to gain confidence in the functioning of the diffusive emitters, it was decided that the water in each source well should be initially sampled at 2-week intervals as described above. The first fence (102, 202, and 302) in each gate would also be sampled more frequently than originally intended to observe the target organic delivery to the first fence in each gate.

a)

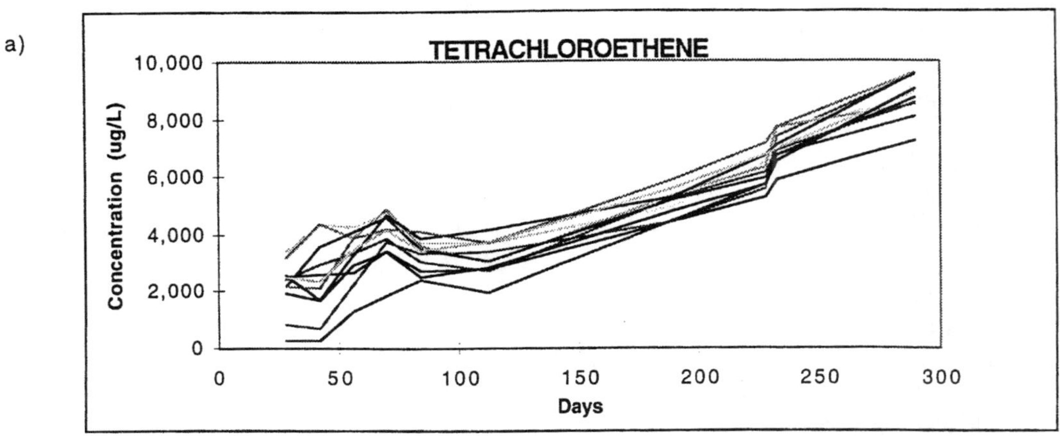

b)

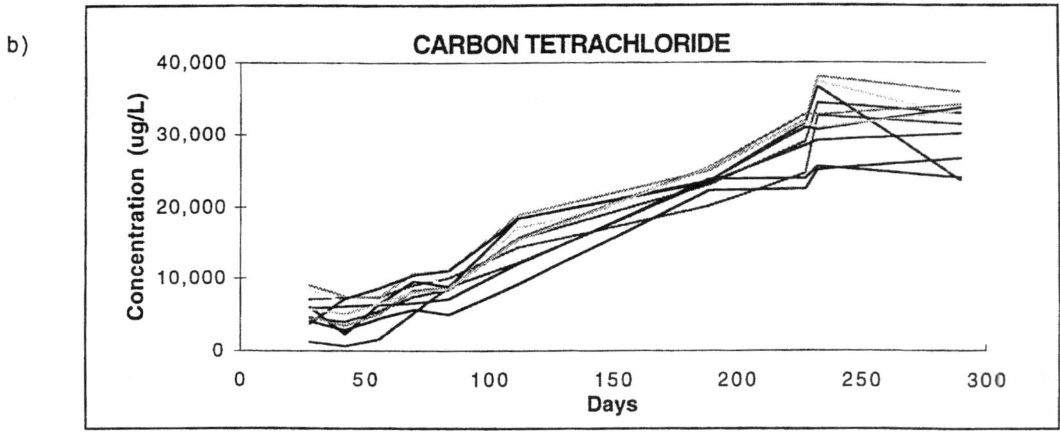

c)

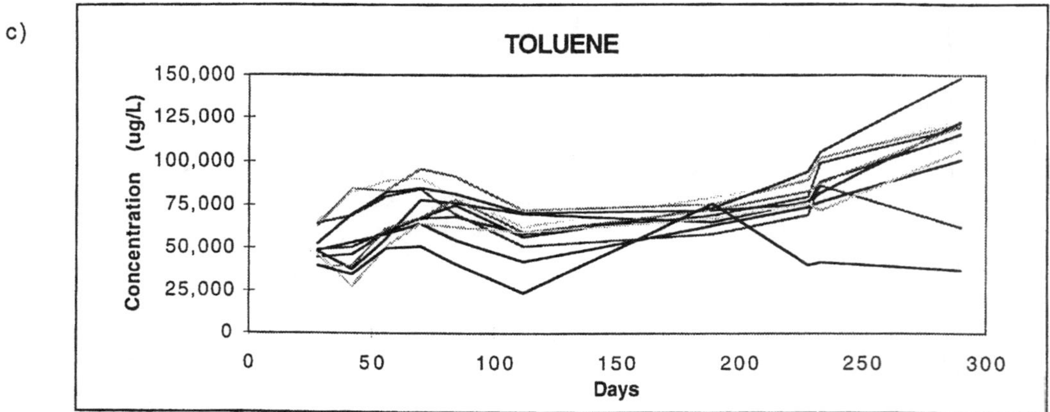

Figure A9.3 Concentrations of target organics in source wells.

Individual Source Well Evaluation

Contaminant concentrations in the individual source wells (see Table A9.1) are provided as Figure A9.4a through k. Note that in these figures the axes have been fixed to a maximum value to allow comparisons between wells. PCE concentrations in the wells are typically less than 10 mg/L, CT

Table A9.1 Source Well Concentrations of Target Organics, Including Analytical Errors (units are in μg/L)

						TOLUENE						
DATE	**DAY**	**SW-1**	**SW-2**	**SW-3**	**SW-4**	**SW-5**	**SW-6**	**SW-7**	**SW-8**	**SW-9**	**SW-10**	**SW-11**
12/16/96	28	48,450	64,330	65,280	62,870	44,550	52,460	49,280	48,040	39,250	46,670	39,740
12/30/96	42	#N/A	68,880	82,120	84,350	46,230	69,540	50,500	37,490	39,250	27,030	34,470
1/13/97	56	57,960	79,760	88,700	83,330	59,410	82,320	59,540	55,390	61,370	50,780	49,810
1/27/97	70	64,100	84,230	90,490	95,690	#N/A	84,700	67,100	78,100	67,400	64,100	51,200
2/10/97	84	54,830	69,670	80,350	91,330	75,050	81,400	68,230	76,260	78,300	62,100	40,750
3/10/97	112	41,910	51,160	62,210	72,200	56,390	70,360	58,230	69,820	59,310	59,290	23,440
5/26/97	189	#N/A	58,340	79,460	75,790	73,520	72,070	69,120	65,300	72,330	67,040	75,850
7/4/97	228	74,150	69,410	88,860	89,580	94,160	75,930	80,060	77,370	83,330	76,300	39,950
7/9/97	233	77,090	86,390	101,900	102,200	105,400	88,390	99,290	82,290	87,870	72,150	41,520
9/4/97	290	10,1100	62,100	124,400	121,900	148,500	116,100	120,000	122,700	120,400	106,500	37,000
						CT						
DATE	**DAY**	**SW-1**	**SW-2**	**SW-3**	**SW-4**	**SW-5**	**SW-6**	**SW-7**	**SW-8**	**SW-9**	**SW-10**	**SW-11**
12/16/96	28	5,828	7,091	8,421	8,891	1,217	3,752	4,518	5,972	4,670	5,790	4,093
12/30/96	42	#N/A	7,329	6,677	7,459	645.9	7,139	4,015	2,336	3,440	5,071	2,818
1/13/97	56	#N/A	7,371	8,973	7,284	1,638	8,541	5,451	6,398	5,052	6,441	4,455
1/27/97	70	6,560	9,036	10,370	10,640	#N/A	10,410	7,480	9,535	8,275	8,110	5,694
2/10/97	84	7,158	9,974	9,080	11,100	8,796	11,000	8,421	8,748	8,707	8,468	4,962
3/10/97	112	12,000	14,380	17,190	18,930	12,180	18,440	15,610	18,470	15,750	15,490	9,172
5/26/97	189	23,860	20,140	22,720	24,970	23,520	23,420	23,100	23,600	25,420	25,310	22,280
7/4/97	228	24,020	24,700	31,710	31,990	31,620	28,490	29,020	31,000	32,810	31,770	22,520
7/9/97	233	25,620	32,620	37,340	37,990	36,650	29,180	34,420	30,710	32,800	30,950	25,130
9/4/97	290	24,090	31,440	32,520	35,010	23,700	30,140	33,000	33,800	34,120	33,350	26,670

continued

Table A9.1 (continued) Source Well Concentrations of Target Organics, Including Analytical Errors

						PCE						
DATE	**DAY**	**SW-1**	**SW-2**	**SW-3**	**SW-4**	**SW-5**	**SW-6**	**SW-7**	**SW-8**	**SW-9**	**SW-10**	**SW-11**
12/16/96	28	2,488	811.8	3,372	3,175	259.6	2,136	2,378	2,521	2,127	2,475	1,908
12/30/96	42	#N/A	674.1	4,343	4,366	248.4	3,558	2,923	1,670	2,068	2,302	1,642
1/13/97	56	2,627	2,204	4,247	3,872	1,268	4,092	3,323	3,276	3,878	3,458	2,888
1/27/97	70	3,404	3,735	4,739	4,876	#N/A	4,593	3,860	4,668	4,192	4,167	3,382
2/10/97	84	2,681	3,303	3,691	3,577	2,360	3,472	3,011	3,858	4,081	3,427	2,462
3/10/97	112	2,744	3,373	3,685	2,987	1,924	3,039	2,688	4,147	3,699	3,716	2,803
5/26/97	189	#N/A	4,395	4,905	5,320	4,443	5,172	5,218	5,260	5,844	5,565	4,778
7/4/97	228	5,256	5,553	6,452	6,292	5,712	5,925	6,126	6,713	7,145	6,732	5,702
7/9/97	233	5,859	7,377	7,707	7,684	6,734	6,707	6,891	7,087	7,701	6,956	6,508
9/4/97	290	7,244	9,528	8,638	8,501	8,078	8,734	8,561	9,572	9,587	8,933	9,039
						TOLUENE - 99% Confidence Range for Reported Value						
DATE	**DAY**	**SW-1**	**SW-2**	**SW-3**	**SW-4**	**SW-5**	**SW-6**	**SW-7**	**SW-8**	**SW-9**	**SW-10**	**SW-11**
12/16/96	28	2,374	2,433	2,437	2,427	2,361	2,125	2,113	2,108	2,079	2,103	2,081
12/30/96	42	#N/A	3,951	4,029	4,043	3,847	3,954	3,864	3,817	3,823	2,381	2,413
1/13/97	56	4,099	4,256	4,332	4,285	4,108	3,290	4,109	4,083	4,012	4,057	4,052
1/27/97	70	3,100	3,263	3,322	3,373	#N/A	3,150	3,120	3,210	3,120	3,200	3,010
2/10/97	84	3,482	3,589	3,679	3,783	3,633	3,689	3,577	3,643	3,661	3,531	3,401
3/10/97	112	1,858	1,902	1,969	2039	1,933	2,026	1,944	2,022	1,951	1,951	859
5/26/97	189	#N/A	610	703	686	675	669	655.5	639	670	646	686
7/4/97	228	1,470	1,417	1,643	1,652	1,708	1,490	1,538	1,507	1,577	1,495	1,134
7/9/97	233	1,395	1,477	1,621	1,624	1,655	1,495	1,596	1,440	1,490	1,354	1,140
9/4/97	290	2,622	1,684	3,195	3,133	3,791	2,989	3,086	3,152	3,095	2,754	1,122

CT - 99% Confidence Range for Reported Value

DATE	DAY	SW-1	SW-2	SW-3	SW-4	SW-5	SW-6	SW-7	SW-8	SW-9	SW-10	SW-11
12/16/96	28	592		1,418	1,420	272	514	518	528	519	526	515
12/30/96	42			694	702	292	699	673	665	670	680	666
1/13/97	56			1,627	926	441	1,625	900	912	895	913	888
1/27/97	70	7,305		1,670	1,680	132	1,680	1,650	1,670	1,660	1,660	717
2/10/97	84	1,308		1,318	1,332		1,331	1,315	1,316	1,316	1,315	950
3/10/97	112	862		1,092	1,182	864	1,179	1,075	1,180	1,077	1,074	833
5/26/97	189	1,053		1,041	1,064	1,049	1,048	1,045	1,050	1,069	1,068	1,037
7/4/97	228	796	802	874	877	872	840	845	866	886	875	783
7/9/97	233	925	990	793	803	878	881	877	835	875	887	786
9/4/97	290	343	414	421	455	352	415	435	444	441	424	433

PCE - 99% Confidence Range for Reported Value

DATE	DAY	SW-1	SW-2	SW-3	SW-4	SW-5	SW-6	SW-7	SW-8	SW-9	SW-10	SW-11
12/16/96	28	258	88	262	261	18	228	228	229	228	229	227
12/30/96	42	#N/A	117	221	221	35	217	215	212	213	213	212
1/13/97	56	298	296	308	305	293	607	301	301	305	302	299
1/27/97	70	1,870	188	195	196		194	190	195	192	192	187
2/10/97	84	592	602	609	607	588	605	597	612	616	604	590
3/10/97	112	73	78	80	75	69	75	73	82	79	80	74
5/26/97	189	#N/A	125	128	195	125	194	194	194	196	195	127
7/4/97	228	365	366	371	371	367	368	369	372	375	373	367
7/9/97	233	262	268	269	270	265	266	270	266	269	265	264
9/4/97	290	365	9,528	8,638	8,501	8,078	8,734	8,561	9,572	9,587	8,933	9,039

Table A9.2 Concentrations of Target Organics in Source Drums

Date	Day	Toluene		Carbon Tetrachloride		Perchloroethene	
			Percent		Percent		Percent
11/18/96	**0**	**Concentration**	**Error**	**Concentration**	**Error**	**Concentration**	**Error**
12/2/96	14	237,167	17.53	*167,750*	17.30	15,813	26.41
12/16/96	28	239,700	5.70	*112,500*	48.61	17,570	12.94
12/30/96	42	272,567	8.74	81,590	27.07	*18,230*	11.64
1/13/97	56	265,767	8.96	174,667	10.49	20,343	14.52
1/27/97	70	279,700	6.08	171,667	12.87	14,150	12.69
2/10/97	84	237,567	10.25	123,050	10.92	10,394	26.23
6/11/97	205	Recirculation switched from constant to timed (3 hr On, 9 Off)					
7/4/97	228	500,600	6.99	151,010	9.86	10,627	16.61
7/4/97	228	500 mL of neat PCE, TOL, and CT added to appropriate source drums					
7/9/97	233	52,967	10.43	166,833	5.17	29,911	8.17
9/4/97	290	*266,850*	2.78	*115,400*	1.45	*54,120*	4.05

Notes: Values are the average of three determinations unless noted. Italicized values are the average of two determinations. Concentrations are given in µg/L.

values reach as high as 38 mg/L, and TOL concentrations range from 37 mg/L to almost 150 mg/L. Ideally a well that is in "good hydraulic connection" with the surrounding aquifer should not show a buildup of concentrations within the well, as the water is continually replenished by flushing. However, the concentrations of contaminants in the source wells was not representative of what was actually delivered to the aquifer and, as such, the first fence in each gate was considered the "input" concentration for each gate. Insufficient development of the source wells, leading to a buildup of contaminant concentrations within each well, was apparently the cause of this nonideality.

Examination of Figure A9.4 reveals that similar inputs of contaminants were introduced to each gate although some variability is evident. For example, at SW-5 (Figure A9.4e), the highest in-well concentrations of TOL were reported, while SW-11 (Figure A9.4i) showed the lowest in-well concentration of TOL. This range of in-well concentrations suggests one of two things: either wells such as SW-11 delivered the most contaminant to the aquifer (well flushing was most pronounced) or the delivery in wells was significantly different than wells such as SW-5 which received higher TOL concentrations.

Target Organic Arrival at the First Fence

The first fence in each gate was sampled several times over the experiment. The concentrations of the target organics and CF are presented in Table A9.3. The first section of the table reports the values at each point in a multilevel, and the second section of the table reports the fence average and standard deviation of the average. In most cases the standard deviation is greater than the average, indicating extreme variability in the delivery of the target organics to the first fence points. This indicates that selected points received most of the mass, while others received very little. These results are similar to the tracer results (Appendix 7) and therefore suggest that the surge block efforts to develop the wells were not entirely successful. The FENCE program (see Appendix 24) was written to assist in the interpretation of nonuniformly distributed plumes. It integrates breakthrough on a point-by-point basis and calculates the mass crossing the fence. Thus, fence-wide as well as point-specific breakthrough masses are determined.

It is unclear why the October 21, 1997 sampling results for the first fence were significantly lower in concentration than previous sampling events. The source wells were not sampled but the trend in concentrations in the wells suggests ever-increasing solute levels. The seasonally low water table may have had some effect. In general, the arrival of the target organics at the first fence was both laterally and vertically varied. Brief summaries of the arrival of the target organics to the first fence in each gate are given below.

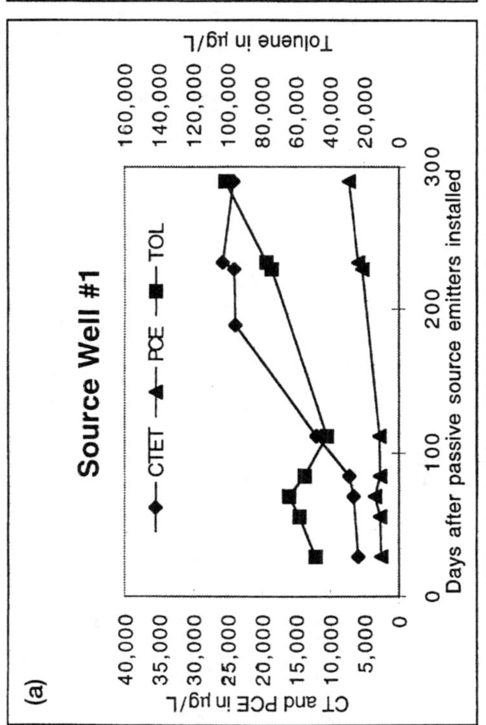

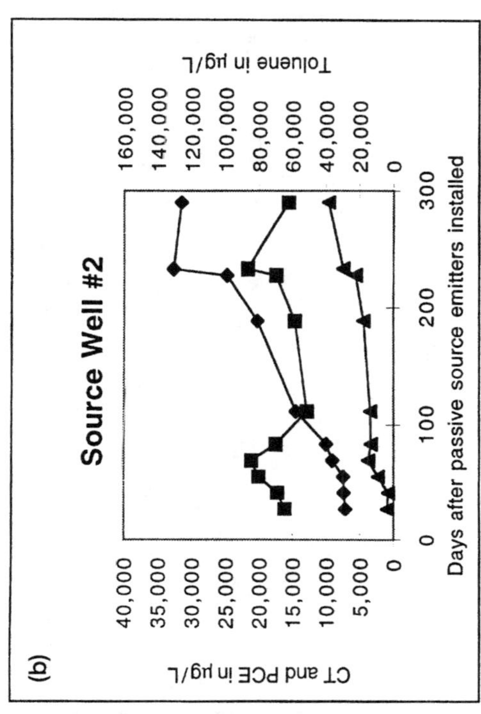

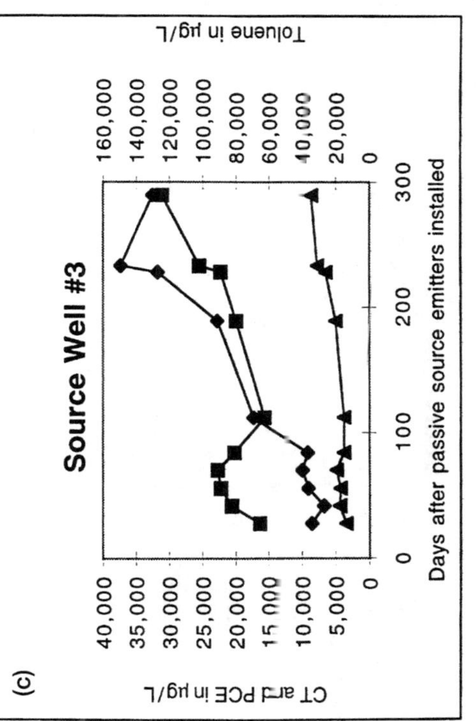

Figure A9.4 Individual source well target organic concentrations over time.

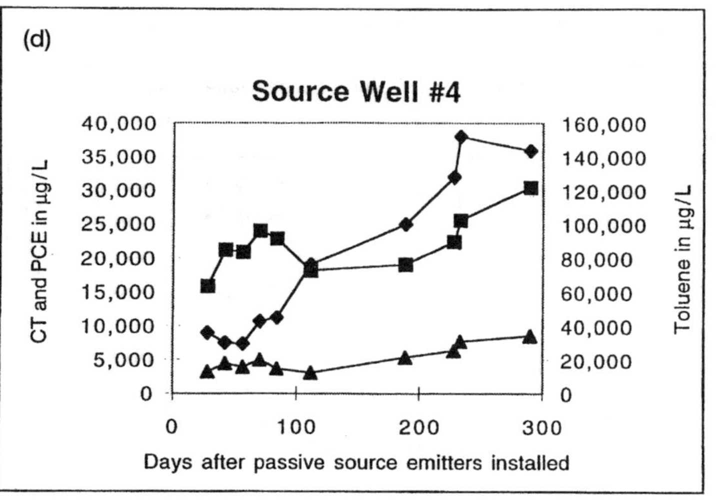

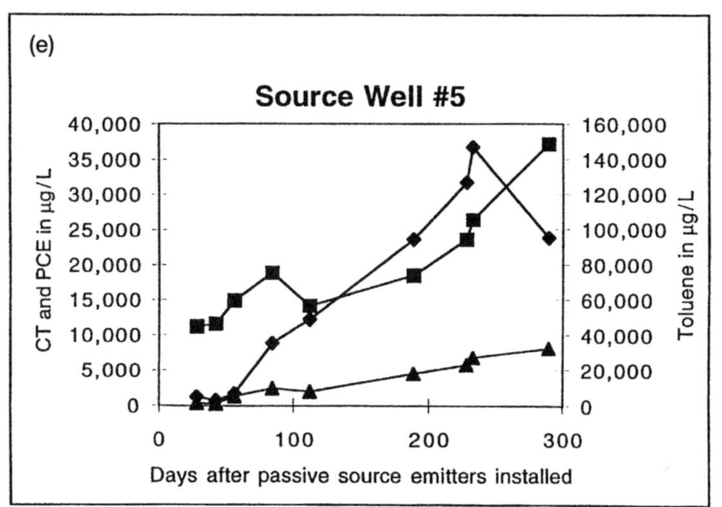

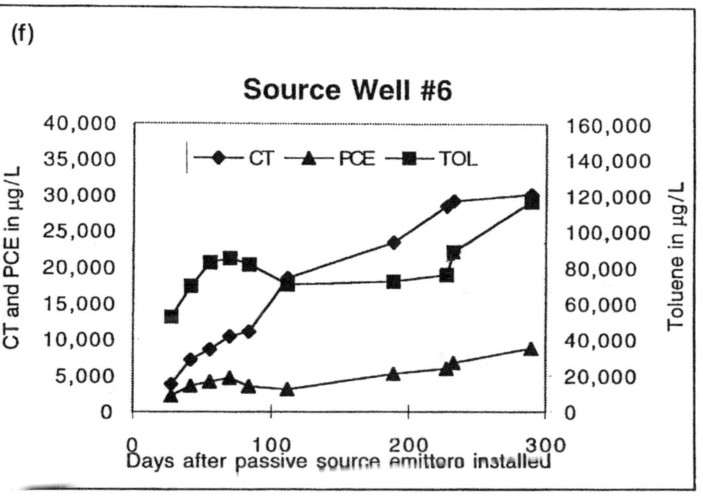

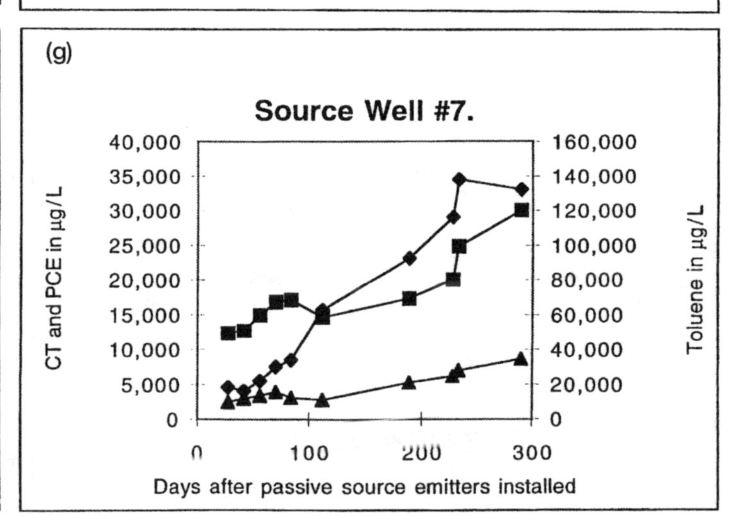

Figure A9.4 (continued) Individual source well target organic concentrations over time.

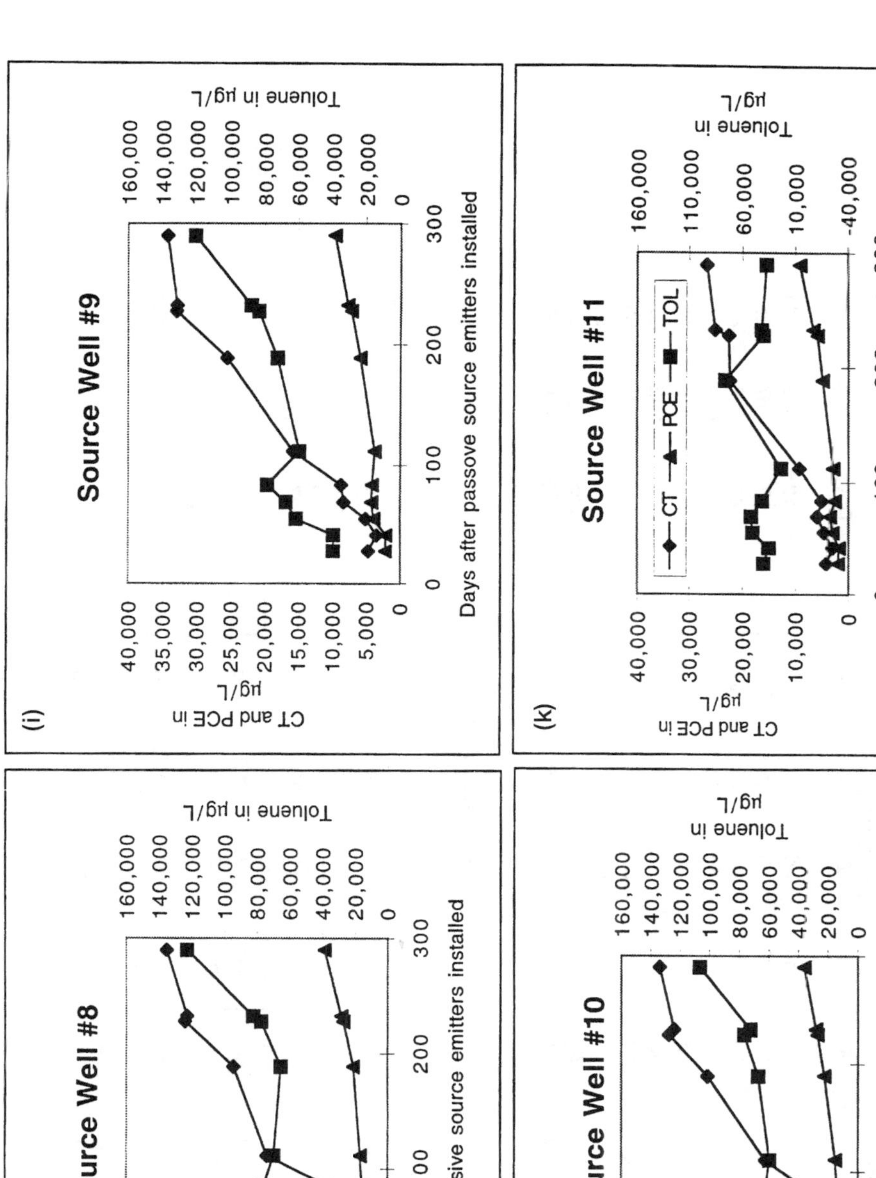

Figure A9.4 (continued) Individual source well target organic concentrations over time.

Table A9.3 Concentrations of TOL, PCE, CT, and CF at the First Fence of Each Gate

PCE Concentrations (μg/L)

Sampling Date	Days	G102L 1	G102L 2	G102L 3	G102L 4	G102R 1	G102R 2	G102R 3	G102R 4	Days	Avg.	Std. Dev.
2-Dec-96	13	3.5	712	1.5	2.9	nd	nd	458	3.2	13	148	311
16-Dec-96	27	53	643	21	16	42	10	533	20.5	27	167	262
30-Dec-96	41	21	1079	3.5	2.0	4	nd	707	6.3	41	228	445
13-Jan-97	55	8.1	2339	4.5	2.9	nd	nd	801	5.7	55	395	943
27-Jan-97	69	29	1733	4.7	nd	15	nd	779	nd	69	320	758
10-Feb-97	83	ns	1199	nd	ns	ns	nd	817	ns	83	252	270
10-Mar-97	111	5.0	1388	3.6	0.9	0.9	nd	952	2.8	111	294	583
14-Jul-97	237	1.5	440	1.5	nd	2.5	nd	1316	2.5	237	221	530
4-Sep-97	288	24	172	nd	23	21	4.4	816	nd	288	132	319
21-Oct-97	335	N/A	31	nd	nd	42	13	535	2.1	335	89	230

CT Concentrations (μg/L)

Sampling Date	Days	G102L 1	G102L 2	G102L 3	G102L 4	G102R 1	G102R 2	G102R 3	G102R 4	Days	Avg.	Std. Dev.
2-Dec-96	13	14	205	nd	nd	44	nd	367	nd	13	79	163
16-Dec-96	27	11	654	21	16	8.0	11	884	38	27	205	353
30-Dec-96	41	794	602	nd	nd	1006	nd	1034	nd	41	430	202
13-Jan-97	55	20	1417	nd	nd	nd	nd	822	nd	55	282	701
27-Jan-97	69	1351	nd	nd	nd	nd	903	nd	nd	69	282	317
10-Feb-97	83	ns	920	nd	ns	ns	1119	1004	ns	83	380	100
10-Mar-97	111	6.3	2506	nd	nd	nd	nd	2987	nd	111	687	1600
14-Jul-97	237	nd	804	nd	nd	nd	nd	4717	3.6	237	691	2522
4-Sep-97	288	nd	193	nd	nd	4.4	17	1523	16	288	219	660
21-Oct-97	335	N/A	9.5	nd	nd	8.6	10	1247	nd	335	182	619

CF Concentrations (μg/L)

Sampling Date	Days	G102L 1	G102L 2	G102L 3	G102L 4	G102R 1	G102R 2	G102R 3	G102R 4	Days	Avg.	Std. Dev.
2-Dec-96	13	37	208	17	nd	nd	36	170	nd	13	59	88
16-Dec-96	27	185	448	nd	nd	32	nd	219	16	27	112	174
30-Dec-96	41	nd	848	nd	nd	nd	nd	222	nd	41	134	442
13-Jan-97	55	nd	1313	1562	nd	nd	nd	191	nd	55	383	730
27-Jan-97	69	33	656	nd	nd	29	nd	216	nd	69	117	295
10-Feb-97	83	ns	437	nd	ns	ns	nd	171	ns	83	76	189
10-Mar-97	111	nd	655	nd	nd	nd	nd	315	nd	111	121	241
14-Jul-97	237	nd	312	nd	nd	nd	nd	1011	nd	237	165	494
4-Sep-97	288	12	nd	nd	nd	40	nd	361	nd	288	52	194
21-Oct-97	335	N/A	nd	nd	nd	32	nd	148	nd	335	26	82

TOL Concentrations (μg/L)

Sampling Date	Days	G102L 1	G102L 2	G102L 3	G102L 4	G102R 1	G102R 2	G102R 3	G102R 4	Days	Avg.	Std. Dev.
2-Dec-96	13	111	3915	9.4	6.2	258	nd	3966	7.8	13	1034	1887
16-Dec-96	27	1102	18110	627	399	896	216	13610	636	27	4449	7150
30-Dec-96	41	241	27060	48	12	25	5.8	16810	126	41	5541	10483
13-Jan-97	55	81	33780	71	17	14	8.9	17750	93	55	6477	12652
27-Jan-97	69	752	30010	36	15	578	7.2	16020	13	69	5929	11193
10-Feb-97	83	ns	19660	nd	ns	ns	3.9	15650	ns	83	4414	10386
10-Mar-97	111	58	17840	36	10	6	3.8	16330	42	111	4291	7907
14-Jul-97	237	24	4856	14	nd	072	nd	17500	38	237	2850	6976
4-Sep-97	288	791	1659	46	23	291	21	9089	8.9	288	1491	0123
21-Oct-97	335	N/A	126	nd	nd	490	80	5888	12	335	942	2561

continued

Table A9.3 (continued) Concentrations of TOL, PCE, CT, and CF at the First Fence of Each Gate

Gate 2 Organics Input (Concentrations at Fence G202)

PCE Concentrations (μg/L)

Sampling		G202L				G202M				G202R						Std.
Date	Days	1	2	3	4	1	2	3	4	1	2	3	4	Days	Avg.	Dev.
2-Dec-96	13	114	335	413	<mdl	58	47	22	<mdl	379	562	959	<mdl	13	241	374
16-Dec-96	27	196	417	374	<mdl	126	95	24	<mdl	227	380	739	<mdl	27	215	263
30-Dec-96	41	208	441	257	<mdl	151	177	45	<mdl	76	226	840	<mdl	41	202	295
13-Jan-97	55	198	485	260	<mdl	175	174	98	<mdl	85	316	899	<mdl	55	224	309
27-Jan-97	69	243	552	480	<mdl	281	128	29	<mdl	229	503	962	<mdl	69	284	337
10-Mar-97	111	190	392	185	<mdl	184	465	292	<mdl	24	52	691	<mdl	111	206	257
14-Jul-97	237	366	na	1318	<mdl	318	76	30	<mdl	246	422	2207	<mdl	237	415	825
21-Oct-97	335	8.0	34	157	<mdl	<mdl	8.0	5.0	<mdl	348	347	287	<mdl	335	100	177

CT Concentrations (μg/L)

Sampling		G202L				G202M				G202R						Std.
Date	Days	1	2	3	4	1	2	3	4	1	2	3	4	Days	Avg.	Dev.
2-Dec-96	13	4	58	1	<mdl	62	39	<mdl	<mdl	115	199	773	3	13	105	290
16-Dec-96	27	53	38	21	<mdl	90	32	<mdl	<mdl	55	175	n/a	20	27	40	62
30-Dec-96	41	21	58	4	<mdl	102	162	33	<mdl	18	64	1424	6	41	158	517
13-Jan-97	55	8	95	80	<mdl	105	116	93	<mdl	37	102	676	6	55	110	230
27-Jan-97	69	29	163	179	<mdl	419	95	30	<mdl	131	250	743	nd	69	170	266
10-Mar-97	111	5	133	194	<mdl	592	1164	1200	<mdl	27	57	2719	3	111	508	983
14-Jul-97	237	2	N/A	3591	<mdl	592	65	21	<mdl	232	829	8015	3	237	1112	2936
21-Oct-97	335	N/A	<mdl	223	<mdl	<mdl	<mdl	<mdl	<mdl	409	203	576	2	335	118	249

CF Concentrations (μg/L)

Sampling Date	Days	G202L 1	G202L 2	G202L 3	G202L 4	G202M 1	G202M 2	G202M 3	G202M 4	G202R 1	G202R 2	G202R 3	G202R 4	Days	Avg.	Std. Dev.
2-Dec-96	13	47	328	337	<mdl	<mdl	22	<mdl	<mdl	206	364	172	<mdl	13	123	140
16-Dec-96	27	109	377	329	<mdl	67	67	<mdl	<mdl	67	295	251	<mdl	27	130	114
30-Dec-96	41	163	526	261	<mdl	87	119	48	<mdl	60	164	342	<mdl	41	148	109
13-Jan-97	55	172	413	261	<mdl	68	59	32	<mdl	24	201	250	<mdl	55	123	96
27-Jan-97	69	213	550	401	<mdl	144	66	<mdl	<mdl	64	365	247	<mdl	69	171	129
10-Mar-97	111	250	621	202	<mdl	151	518	167	<mdl	23	58	349	<mdl	111	195	188
14-Jul-97	237	646	671	1070	<mdl	369	90	<mdl	<mdl	345	792	761	<mdl	237	395	299
21-Oct-97	335	<mdl	<mdl	123	<mdl	<mdl	<mdl	<mdl	<mdl	306	444	101	<mdl	335	81	173

TOL Concentrations (μg/L)

Sampling Date	Days	G202L 1	G202L 2	G202L 3	G202L 4	G202M 1	G202M 2	G202M 3	G202M 4	G202R 1	G202R 2	G202R 3	G202R 4	Days	Avg.	Std. Dev.
2-Dec-96	13	815	3068	2583	9	106	265	65	5	1043	4100	10690	125	13	1906	3754
16-Dec-96	27	3096	6356	6630	255	3672	4148	557	161	4613	7772	21770	961	27	4999	7065
30-Dec-96	41	4712	11130	6076	189	4375	6198	845	5	1382	3887	22290	88	41	5098	7381
13-Jan-97	55	3934	9827	5442	55	2992	3509	2662	12	1371	5683	17970	199	55	4471	5826
27-Jan-97	69	5119	11820	11270	43	6697	2008	350	6	4327	7931	14180	59	69	5318	4981
10-Mar-97	111	5634	9816	3878	15	4812	11240	7842	46	555	1083	17650	40	111	5218	6436
14-Jul-97	237	6216	6274	15880	41	4722	868	173	5	3610	6703	2390	115	237	3916	2495
21-Oct-97	335	dry	28	2126	5	dry	16	44	<mdl	4300	3864	3051	11	335	1120	2074

continued

Table A9.3 (continued) Concentrations of TOL, PCE, CT, and CF at the First Fence of Each Gate

Gate 3 Organics Input (Concentrations at Fence G302)

PCE Concentrations (μg/L)

Sampling Date	Days	G302L 1	G302L 2	G302L 3	G302L 4	G302M 1	G302M 2	G302M 3	G302M 4	G302R 1	G302R 2	G302R 3	G302R 4	Days	Avg.	Std. Dev.
2-Dec-96	13	<mdl	<mdl	22	<mdl	70	27	<mdl	<mdl	908	523	201	<mdl	13	146	370
16-Dec-96	27	12	21	138	219	17	<mdl	<mdl	<mdl	767	587	339	16	27	176	336
30-Dec-96	41	<mdl	82	320	10	17	<mdl	<mdl	<mdl	918	838	575	<mdl	41	230	407
13-Jan-97	55	<mdl	103	506	14	39	<mdl	<mdl	<mdl	1119	986	665	13	55	287	519
27-Jan-97	69	<mdl	24	152	<mdl	35	<mdl	<mdl	<mdl	970	1047	414	21	69	222	493
10-Mar-97	111	90	225	430	<mdl	30	<mdl	<mdl	<mdl	663	796	497	15	111	229	361
21-Oct-97	335	ns	<mdl	<mdl	<mdl	ns	38	5	<mdl	4	987	1121	103	335	188	528

CT Concentrations (μg/L)

Sampling Date	Days	G302L 1	G302L 2	G302L 3	G302L 4	G302M 1	G302M 2	G302M 3	G302M 4	G302R 1	G302R 2	G302R 3	G302R 4	Days	Avg.	Std. Dev.
2-Dec-96	13	25	<mdl	13	<mdl	21	22	<mdl	<mdl	955	20	140	25	13	102	374
16-Dec-96	27	<mdl	30	269	<mdl	20	<mdl	<mdl	<mdl	1750	300	464	30	27	239	717
30-Dec-96	41	<mdl	124	409	<mdl	<mdl	<mdl	<mdl	<mdl	1313	432	558	<mdl	41	236	476
13-Jan-97	55	<mdl	82	615	<mdl	<mdl	<mdl	<mdl	<mdl	1264	386	464	17	55	236	525
27-Jan-97	69	<mdl	18	177	<mdl	45	21	<mdl	<mdl	1238	604	373	<mdl	69	206	500
10-Mar-97	111	183	471	1185	<mdl	<mdl	<mdl	<mdl	<mdl	1536	1365	904	<mdl	111	470	327
21-Oct-97	335	ns	<mdl	<mdl	<mdl	ns	26	11	<mdl	<mdl	2256	2671	24	335	416	1346

CF Concentrations (μg/L)

Sampling Date	Days	G302L 1	G302L 2	G302L 3	G302L 4	G302M 1	G302M 2	G302M 3	G302M 4	G302R 1	G302R 2	G302R 3	G302R 4	Days	Avg.	Std. Dev.
2-Dec-96	13	<mdl	<mdl	14	<mdl	53	22	23	22	133	325	61	<mdl	13	54	110
16-Dec-96	27	<mdl	28	103	<mdl	<mdl	<mdl	<mdl	<mdl	156	700	202	<mdl	27	99	302
30-Dec-96	41	<mdl	97	202	<mdl	13	<mdl	<mdl	<mdl	114	571	212	<mdl	41	101	243
13-Jan-97	55	<mdl	36	124	33	47	<mdl	<mdl	<mdl	97	493	198	12	55	87	194
27-Jan-97	69	<mdl	<mdl	95	<mdl	<mdl	<mdl	<mdl	<mdl	135	495	101	23	69	71	210
10-Mar-97	111	29	180	160	<mdl	<mdl	<mdl	<mdl	<mdl	47	502	179	33	111	94	218
21-Oct-97	335	ns	<mdl	<mdl	<mdl	ns	390	449	<mdl	<mdl	37	<mdl	201	335	90	188

TOL Concentrations (μg/L)

Sampling Date	Days	G302L 1	G302L 2	G302L 3	G302L 4	G302M 1	G302M 2	G302M 3	G302M 4	G302R 1	G302R 2	G302R 3	G302R 4	Days	Avg.	Std. Dev.
2-Dec-96	13	<mdl	11	184	11	813	421	137	<mdl	10120	3506	1535	209	13	1412	3604
16-Dec-96	27	403	664	3533	na	358	205	197	172	17240	12740	7161	572	27	3604	6811
30-Dec-96	41	48	2141	5794	162	237	5	8	<mdl	19090	13560	9554	111	41	4226	7967
13-Jan-97	55	113	2023	5164	11	775	29	4	<mdl	16560	14610	9658	324	55	4106	7418
27-Jan-97	69	87	469	2804	55	786	183	120	<mdl	15300	13980	5856	587	69	3352	6724
10-Mar-97	111	1890	5410	8193	160	567	109	28	17	5596	6621	4151	646	111	2782	2773
21-Oct-97	335	ns	6	46	<mdl	ns	204	39	48	98	10390	12880	979	335	2058	5600

Notes:

ns = not sampled; n/a = not available; nd = not detected.

<mdl = less than method detection limit.

G102 fence results suggest that G102L-2 and G102R3 received most of the mass from the three source wells in Gate 1. The first fence in Gate 2 (G202L, M, and R) shows the most laterally uniform delivery of target organics of the three gates. The deepest point (4) received the lowest percentage of mass of all four depths. G202R-3 received the largest percentage of mass. Breakthrough of TOL at G302 is dominated at the R location, supporting the contention that the best hydraulic connection to the aquifer exists at the right side of Gate 3 (with SW-11). The nutrient flush zone ensures that once every 28 days the plume is homogenized across the saturated thickness of the aquifer and downgradient the heterogeneities are not as apparent.

A9.6 CONCLUSIONS

Implementation of the source generation system generally proceeded according to plan. In spite of a number of development-related difficulties encountered initially, once installed, it performed satisfactorily. Target delivery concentrations, as reported from the first fence, were not ideal: TOL concentrations exceeded target levels, while PCE and CT were slightly below target levels. The nonuniform vertical distributions of contaminants suggest natural aquifer heterogeneities intersected the source wells and that the contaminants were delivered preferentially to some of these layers. Alternately, or in addition, insufficient source well development may have contributed to the observed "patchy" plume creation.

Although in need of further testing, the diffusive emitters successfully met their primary objective: to provide a controlled delivery of contaminants through the use of variable tubing lengths and recirculation systems.

A9.7 REFERENCES

University of Waterloo (UW). 1995. Passive and Semipassive Groundwater Remediation Techniques. Revised Work Plan. October 1995.

University of Waterloo (UW). 1997. Passive and Semipassive Groundwater Remediation Techniques. Phase 1 Installation Report. April 1997.

Wilson, R.D. and D.M. Mackay. 1995. A method for passive release of solutes from an unpumped well. *Ground Water* 33(6): 936-945.

APPENDIX 10

Aboveground Treatment System Design — Borden

Groundwater extracted from all gates, as well as any wastewater resulting from sampling/purging activities, was treated prior to its disposal onto the ground surface at the north end of the site. A storage tank, a peristaltic pump (flow rate set at approximately 350 mL/min to balance combined inflow from all gates with outflow through the treatment system), a treatment cylinder containing approximately 225 kg of commercial-grade granular iron, and a prefabricated granular activated carbon (GAC) cannister (containing approximately 75 kg GAC; Calgon Carbon Corporation's Floursorb Cannister) comprised the treatment system. A schematic of the system is provided in Figure A10.1.

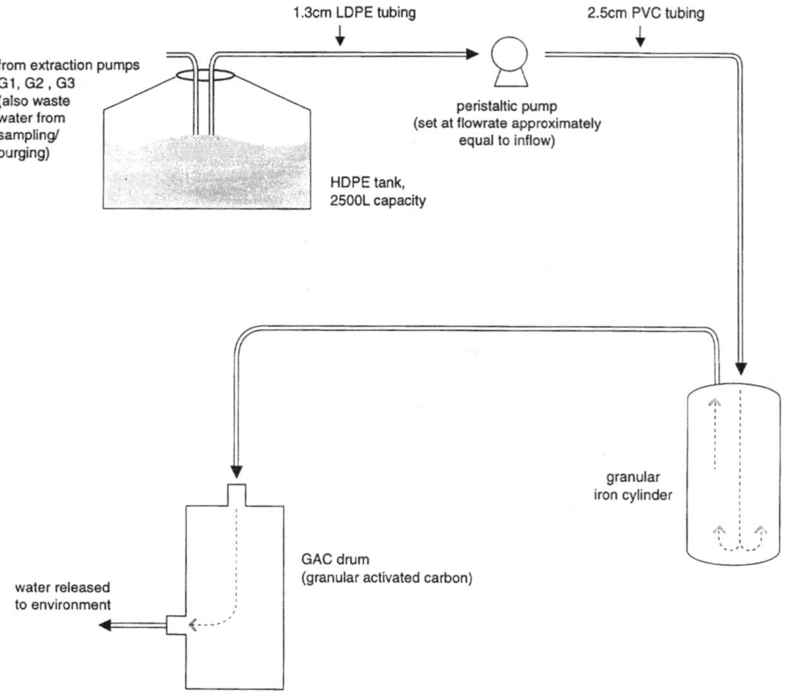

Figure A10.1 Aboveground treatment system (schematic).

Sampling of individual gate extraction lines in June 1996 indicated a need to bring the treatment system online prior to the originally anticipated time (i.e., when injected contaminants were expected to have traversed the length of the gates). Table A10.1 provides data from this sampling

Table A10.1 VOC Sampling Results for Surface Discharged Water

Parameter	Date Sampled											
Analyzed	96/06/19	96/06/19	96/06/19	96/11/01 after treatment	97/05/26 after treatment	97/06/18 after treatment	97/09/04 after treatment	97/10/21	97/10/21	97/10/21	97/10/28 after treatment	Guideline Value (MOEE)
Sample Point	G113P	G225P	G325P					G113P	G225P	G325P		
Laboratory	OGL	OGL	OGL	OGL	OGL	OGL	OGL	BL	BL	BL	OGL	
PCE	<MDL	**3.15**	10.95	<MDL	<MDL	<MDL	<MDL	12.1	15.6	<MDL	<MDL	5
CT	<MDL	<MDL	<MDL	<MDL	**5.6**	<MDL	<MDL	<MDL	<MDL	<MDL	<MDL	5
CF	<MDL	<MDL	<MDL	<MDL	<MDL	<MDL	<MDL	<MDL	<MDL	<MDL	<MDL	5
Toluene	<MDL	<MDL	<MDL	<MDL	<MDL	<MDL	<MDL	<MDL	182	65.6	<MDL	24

Notes: Aboveground treatment initiated in July 1996. All concentrations given in µg/L. OGL - Analyses conducted at Organic Geochemistry Laboratory, UW. BL - Analyses conducted at Barringer Laboratories, Mississauga, Ontario. Bolded values are >MDL but <LOQ. Guideline values from MOEE (1996), Guidelines for Use at Contaminated Sites in Ontario, Table A, Potable Water Criteria.

event. Low concentrations of PCE were detected in the extraction lines of Gates 2 and 3, likely due to the previous UW release experiment (Brewster et al., 1995). No detectable contaminant concentrations were measured in extracted water from Gate 1.

The treatment system was originally activated in July 1996, but at that time did not include the granular iron cylinder. Low concentrations of PCE from Gates 2 and 3 were anticipated and as such were thought to be better handled by the GAC alone. Also, it was felt that a less complicated treatment system would be easier to maintain over the winter months. In April 1997, when increased contaminant concentrations were expected to reach the extraction wells, the iron cylinder was added to the system to eliminate the chlorinated constituents and reduce the total contaminant loading on the GAC.

Intermittent monitoring of effluent organic concentrations ensured that released water (combined outflow, after treatment) complied with appropriate guidelines. A summary of the data and the guideline values for comparison is provided in Table A10.1. Concentrations of target organics from the extraction wells are provided for pretreatment concentrations. The results indicate that suitable treatment of site wastewater has been provided by this system throughout the duration of the project.

REFERENCE

Brewster, M.L., A.P. Annan, J.P. Greenhouse, B.H. Kueper, G.R. Olhoeft, J.D. Redman, and K.A. Sander. 1995. Observed migration of a controlled DNAPL release by geophysical methods. *Ground Water*, 33(6): 977-987.

Appendix 11

Laboratory Microcosm Experiments

Laboratory experiments were conducted, either to obtain preliminary data with respect to aquifer response to manipulation, to obtain data supportive of field-derived hypotheses, or to further examine certain points of interest that arose during the field experiments. The major advantage of field studies is that observed behavior is "real," and not subject to questions of "lab artifacts." However, one disadvantage is that field experiments, once initiated, can be monitored but often not substantially altered. Typically, collected field data are used to construct hypotheses concerning events occurring in the experimental area. Supplementary laboratory studies provide a means by which these hypotheses may be tested. For example, a hypothesized biotransformation may be confirmed in the lab, where biological and abiological transformation may be unequivocally separated through the use of sterile controls, an option not available in the field study. Explanation of unforseen field occurrences may also become possible if potential scenarios can be examined in the relatively simple, easily manipulated lab microcosm environment.

Five laboratory studies were conducted. Section A11.1 describes an investigation of the dichloromethane-degrading potential in the Borden aquifer. Sections A11.2 and A11.3 describe experiments investigating the potential for biodegradation of aromatic hydrocarbons and chlorinated solvents in microcosms simulating the Borden aquifer and the Alameda aquifer, respectively. In Section A11.4, the effect of benzoate concentration on toluene biodegradation is investigated in microcosms simulating the aerobic zone of Gate 3 at the Borden site. Section A11.5 reports an experiment conducted to confirm an aerobic biodegradation potential within the biosparge zone at the Alameda site.

A11.1 DICHLOROMETHANE DEGRADATION UNDER AEROBIC CONDITIONS

Small amounts of dichloromethane (DCM) may arise as a degradation product during the abiotic reduction of carbon tetrachloride (CT) by granular iron, and will persist through the treatment gate because DCM is unaffected by the metallic catalyst (Gillham and O'Hannesin, 1994). Therefore, DCM attenuation must occur downgradient of the iron cassette in Gate 1, if it occurs at all. A laboratory experiment was devised to investigate the likelihood of biologically mediated DCM transformation within the native aquifer. It was anticipated that aerobic conditions would predominate in the region downgradient of the iron cassette, with the assistance of the aerobic treatment cassette and/or the passive dissolved oxygen (DO) release wells included in the Gate 1 design.

DCM is also a potential daughter product in biologically mediated dechlorination of CT, and so could accumulate (along with chloroform (CF)) in Gate 3, although perhaps only transiently. Some CT-transforming pure cultures do accumulate DCM. However, CT metabolism typically involves multiple, competing pathways, so that a number of products may form and differing product ratios may result depending on the environmental conditions and microorganisms involved (Egli et al., 1988; Criddle et al., 1990; Criddle and McCarty, 1991). DCM is potentially further degradable under anaerobic conditions. A mixed anaerobic culture consumed DCM under methanogenic conditions, producing principally CO_2, but also CH_4 (when methanogenesis was uninhibited) and acetate (when methanogenesis was partially or completely prevented with metabolic inhibitors), probably as a result of acetogenic DCM transformation to CO_2 and acetate, and further methanogenic metabolism of these DCM degradation products to CH_4 (Freedman and Gossett, 1991). In view of the possibility of DCM accumulation in Gate 3, the lab experiment also tested the potential for aerobic, benzoate-driven DCM biotransformation in site aquifer material, an activity that might conceivably occur in or after the biosparge zone of Gate 3, should accumulated DCM and excess benzoate travel through that area.

Most aerobic, DCM-degrading bacteria are at least facultative C1-degraders, including *Methylobacterium* strain DM4, *Hyphomicrobium* strain DM2, and some methylotrophic *Pseudomonas* sp. (Gälli and Leisinger, 1985; Stucki et al., 1981; Brunner et al., 1980). The DCM is converted to formaldehyde and inorganic chloride. Isolates able to grow on DCM possess glutathione-dependent reductive dehalogenases with stringent substrate specificities (dihalomethanes only); the dehalogenase is an inducible enzyme in *Hyphomicrobium* at least (Fetzner and Lingens, 1994). These microorganisms are believed relatively rare, and most likely arise in natural populations exposed to DCM for extended periods (Stucki et al., 1981). This belief results from testing 35 obligate and facultative methylotrophs from 12 different genera and finding that none would grow on DCM, indicating that the potential for DCM metabolism is not a capability inherent to certain methylotrophs. Nonetheless, one DCM degrader (strain DM1) was fortuitously obtained "from thin air" during chemostat culture of a facultatively methylotrophic *Pseudomonas* fed 0.04% methanol (utilizable C source) + 0.2% DCM (nonutilizable C source). As Brunner et al. (1980) relate, their intent was to select for spontaneously arising DCM-utilizing mutants; however, the DCM degrader eventually obtained differed from the pseudomonad in a number of properties and so was not likely a mutant but an airborne contaminant.

Cometabolism of DCM under aerobic conditions has also been reported. For example, this reaction occurs via methane monooxygenase activity in methanotrophs, and ammonia monooxygenase catalyzes a similar reaction in nitrifying bacteria (Henson et al., 1988; Rasche et al., 1991).

In this experiment, static microcosms were prepared to examine the microbially mediated fate of DCM under aerobic conditions (1) simulating the undisturbed aquifer (active microcosm); (2) at alkaline pH (active, pH 9 microcosm) to simulate the high pH environment resulting from the granular iron wall activity; (3) in the presence of benzoate, the organic substrate injected into Gate 3 (active, benzoate microcosm); and (4) in the presence of methanol, added at substrate levels to encourage growth of any C1-degrading microorganisms present in the native aquifer material (active, methanol microcosm). The methanol added was sufficient to support significant microbial growth initially, but was intended to be depleted mid-experiment so that competitive inhibition of DCM biotransformation by methanol, if it occurred, would not be an issue over the entire experimental course. Levels of DCM in the active microcosms were compared to sterile control microcosms at ambient pH or at pH 9. To ensure that microbial activity was carbon-limited and not restricted by inorganic nutrient limitations, modified Bushnell–Haas (MBH) medium (Mueller et al., 1991), a mineral medium typical of those used in microbiological studies, was added to all microcosms.

A11.1.1 Methods and Materials

A11.1.1.1 Site Materials

Aquifer material was collected from the CFB Borden site in November 1995, prior to installation of the experimental gates. The material was obtained from a depth of 1.5 to 3 m below ground surface by the method of Starr and Ingleton (1992). Briefly, a hand auger was used to remove material between the surface and the water table, then a drive point/piston sampler was vibrated into position in the subsurface using a gasoline-powered jackhammer, and the sample collected over the desired interval in a 1.5 m × 5 cm aluminum core barrel. The core barrel assembly was then removed from the ground using an electric winch, the sample-containing barrel cut free, and both ends sealed with plastic core caps and duct tape. Uncontaminated groundwater was obtained from a 5-cm well (designated PFAT 8800SW) located upgradient of the sand pit experimental area at CFB Borden in June 1995. Several liters were pumped from the well, then water was collected into sterile 20-L glass carboys using a peristaltic pump. Site materials were returned to the University of Waterloo (UW) on the day of collection and stored at 4°C until required.

In the laboratory, core barrels were cut into 1-ft sections, then the sand removed using alcohol-sterilized tools. The processing was conducted within a sterile containment hood to avoid the introduction of airborne microbial contaminants into the aquifer material. About 1 cm of material was pared away from the ends of each section and discarded. In addition, only the center portion of each sand cylinder was collected, leaving behind about 0.5 cm of material adjacent to the aluminum barrel. The collected sand was mixed in a sterilized pan, then allocated into microcosms.

A11.1.1.2 Microcosms

Aseptic techniques were used during microcosm preparation and equipment was sterilized before use. Microcosm vessels consisted of 1.1-L glass bottles fitted with o-ring stopcocks and screw-cap mininert™ valves (Dynatech, Precision Sampling Corp., Baton Rouge, LA) on the sidearm used for sampling (Figure A11.1). The "double-seal" design of the bottles is intended to allow repeated sampling of the microcosm and minimize the loss of volatile components over the incubation period. Six microcosms were prepared; each contained 200 g (wet weight) of core material, 500 mL of groundwater, and 9.1 mL of MBH. Added inorganic salt concentrations in the microcosms were 18 mg K_2HPO_4, 18 mg KH_2PO_4, 18 mg NH_4NO_3, 3.6 mg $MgSO_4{\cdot}7H_2O$, 0.36 mg $CaCl_2{\cdot}2H_2O$, and 0.17 mg $FeCl_3{\cdot}6H_2O$ per liter. The control microcosms were sterilized by autoclaving the sand-containing vessels for 1 h on three successive days. Groundwater, either at ambient pH or after adjustment to pH 9 (by addition of NaOH), was then added, as was 5 mL of a 10% sodium azide solution. Azide is a metabolic inhibitor, added to restrict activity of microorganisms introduced into the control microcosms in the groundwater or during sampling of the bottles. The active-adjusted active microcosms and the pH 9 were identical except that the groundwater added to the latter was adjusted with NaOH. Sodium benzoate was added to the active benzoate microcosm from a sterile stock solution, to a target concentration of 100 mg benzoate/L groundwater. The active methanol microcosm was amended to 20 mg/L methanol using neat methanol sterilized by passage through a sterile Teflon filter (0.2 μm pore size). DCM was added to all microcosms by injection of neat solvent through the sidearm, to obtain an initial concentration of about 1 mg/L. All microcosms were incubated in the dark at 10°C to simulate the natural CFB Borden groundwater environment.

A11.1.1.3 Sampling and Analysis

For sampling, the o-ring taps were loosened, and the microcosms placed on their sides within the sterile containment hood so that water filled the sidearm. A 20-mL glass syringe was used to remove water (14.3 mL) from each microcosm through the mininert™ valve, for volatile halocarbon analysis. The water was transferred to a 22-mL headspace autosampler vial that was then sealed with a Teflon-faced septum and aluminum crimp seal. Additional water was removed in a similar fashion for benzoate analysis (6 mL into a scintillation vial) or methanol analysis (2 mL into a 2-mL crimp seal vial) as required. Benzoate samples were frozen until analyzed, and samples for methanol analysis were preserved with 10% Na azide (0.1 mL). DCM samples were not preserved because analyses were conducted on the day of sample collection. Headspace

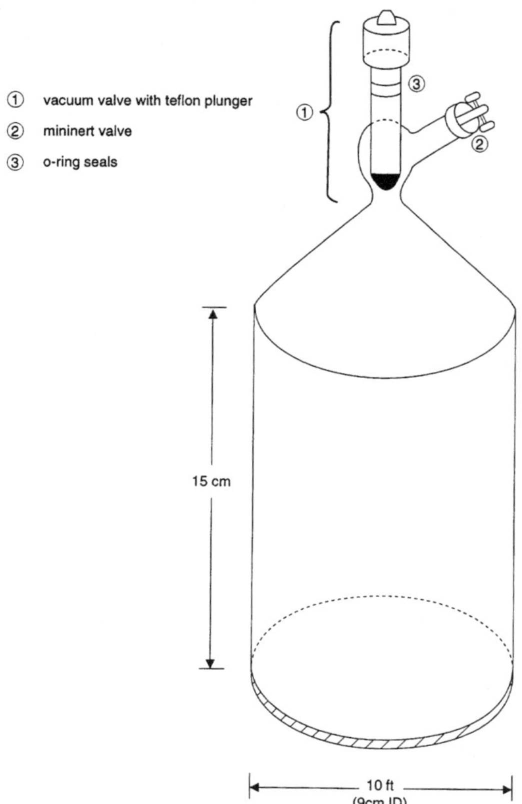

Figure A11.1 Schematic of double-seal microcosms.

gas samples (3 mL) were also periodically obtained via syringe through the microcosm sidearm to conduct O_2/CO_2 analyses. DCM, methanol, and O_2/CO_2 were analyzed by gas chromatography as described in Appendix 3.

A11.1.2 Results And Discussion

The microcosms were periodically sampled over 250 days, but no evidence of DCM biotransformation was observed under any experimental condition (Figure A11.2). Both benzoate and methanol were rapidly depleted in their respective microcosms, indicating biologically active aquifer materials. Benzoate decreased from 67.1 mg/L on day 1 to 3.2 mg/L on day 6, and was no longer detectable by day 15 in the active, benzoate microcosm. Methanol (initially 19.0 mg/L) was undetectable by the next sampling time (day 6) in the active, methanol microcosm. Metabolism of these substrates was also reflected in detectable accumulations of CO_2 in the gas phase, from 0.16–0.18% on day 1 to 3.2% (active, benzoate) and 0.5% (active, methanol) by day 15. In comparison, headspace CO_2 hovered in the 0.05–0.32% range in the other four microcosms. Despite this biological activity, DCM metabolism was not stimulated.

A11.1.3 Conclusions

The rapid metabolism of the organic additions indicates that the tested DCM level was not inhibitory to the indigenous aquifer microorganisms. However, the microcosm microbial populations were incapable of metabolizing DCM, suggesting that this compound may persist *in situ* if it is formed from CT.

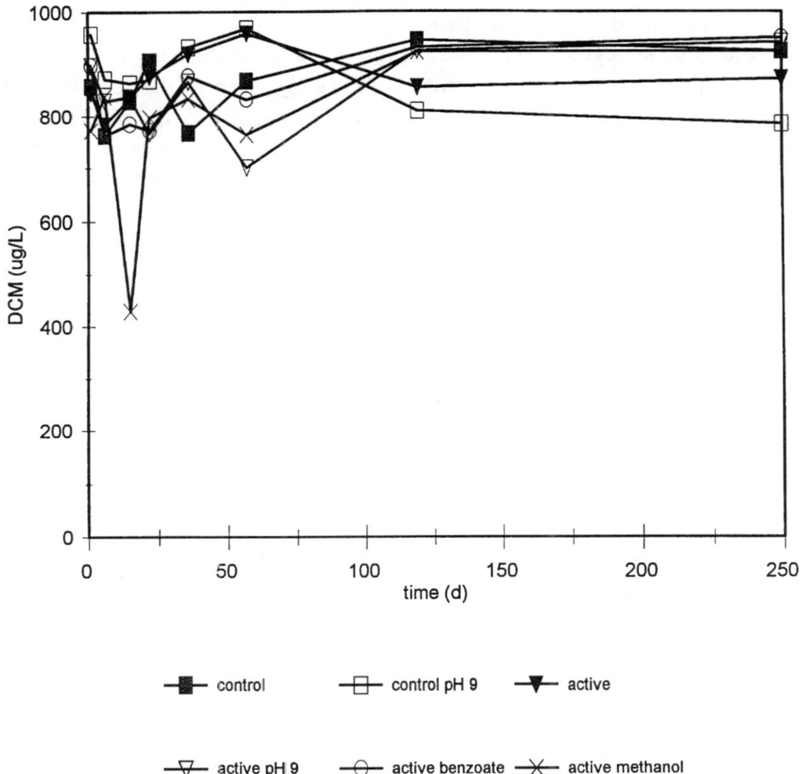

Figure A11.2 Dichloromethane in aerobic Borden microcosms.

A11.2 MICROCOSMS SIMULATING CONDITIONS IN THE BORDEN AQUIFER

A series of microcosms was constructed to examine the inherent transformation potential of the microbial communities initially present at the CFB Borden site, under conditions that simulated the perturbations to be conducted in various zones within Gate 3. This information, plus data collected on microbial biotypes inhabiting the aquifer over the course of the experiment, will be used in assessing changes in microbial communities that have resulted from contaminant influx and/or the purposeful manipulation of the *in situ* environment (in progress). The intent was a short-term microcosm experiment of 6–8 weeks duration, because the focus was initial degradative potential, not whether aquifer populations would eventually evolve; however, the experiment was later adapted to a longer term and steps taken to probe observed restrictions on microbial activity. Two additional microcosm experiments (short term) will be conducted to assess biodegradative potential in certain segments of Gates 2 and 3, during and at the end of the CFB Borden experiment. Material for the second experiment was recently collected.

Both aerobic and anaerobic microcosms were constructed using three composites of core material collected from the CFB Borden experimental site in October 1996. The microcosms were amended with the contaminant mix used in the field experiment, i.e., tetrachloroethene (PCE), toluene, and CT, at concentrations similar to the proposed field concentrations. One set of microcosms was also amended with benzoate at a level similar to the intended field concentration. After it was apparent that microbial activity was lacking in some instances, certain microcosms were further amended with inorganic nutrients, to determine if this would alleviate the observed inhibition.

Abundant evidence in the literature plus previous experience at CFB Borden (e.g., Barker et al., 1987) led to the expectation of ready toluene degradation in aerobically incubated microcosms. Whether anaerobic toluene degradation would occur was more uncertain; such activities have been reported (e.g., Hutchins et al., 1991; Edwards et al., 1992), even in the landfill leachate-affected zone of the Borden aquifer (Barbaro et al., 1992), but our recent experience with toluene degradation under denitrifying conditions is that this activity cannot always be established within the Borden environment (J. R. Barbaro; Butler, unpublished). Previous field studies at CFB Borden showed that CT could be dechlorinated to CF *in situ* (Devlin, 1994), and this reaction does not require highly reducing conditions to be favorable (Vogel et al., 1987), so it was thought it might occur in the anaerobic microcosms, even though the aquifer material was not previously exposed to CT. Although a report of aerobic PCE biotransformation has appeared (Deckhard et al., 1994), reducing conditions and a developed anaerobic microbial community are generally recognized as prerequisites for biologically mediated PCE transformation (Vogel and McCarty, 1985; Freedman and Gossett, 1989), conditions not likely to be met by the sampled aquifer material. Therefore, PCE biotransformation was not expected even in anaerobic microcosms, although benzoate-amended microcosms could produce conditions suited for PCE dechlorination.

A11.2.1 Methods and Materials

A11.2.1.1 Site Materials

Aquifer core was collected on October 1, 1996 as described in Section A11.1.1.1. Six 1.5-m-long cores (designated MC-1 to MC-6) were obtained from 1.5 to 3 m below ground surface, in locations that were actually adjacent to the experimental area, about 0.5 m outside of Gate 3. Outside-gate samples were more representative of "initial" aquifer conditions than inside-gate samples would have been at the time of coring, because pretesting of various Gate 3 components (e.g., the O_2 sparge) had been ongoing, as had the tracer tests in the three gates. Two cores were collected about 15–20 cm apart at each of three locations along the Gate 3 length. Cores MC-1 and MC-2 were obtained between the G307 and G310 fences, i.e., adjacent to the anaerobic treatment zone, MC-3 and MC-4 between G322 and G324, i.e., adjacent to the aerobic treatment zone, and MC-5 and MC-6 between the source wells and G302, i.e., near source.

Material was removed from the core barrels as in Section A11.1.1.1, except that this processing was conducted under flowing N_2 within an Atmosbag glovebag (Aldrich) to avoid exposure to the air. Three composites were ultimately prepared from (1) MC-1 and MC-2, (2) MC-3 and MC-4, and (3) MC-5 and MC-6 material, by intermixing the sand obtained from each of these core pairs. After mixing, the sand was allocated into sterile Whirlpak bags and placed inside canning jars previously gassed with argon. After the jars were tightly sealed, they were removed from the glovebag, then stored at 4°C until required. Groundwater was obtained from the PFAT 8800SW well.

A11.2.1.2 Microcosms

Microcosms consisted of 200 g (wet weight) aquifer material and 500 mL of groundwater plus amendments (described below) in the 1.1-L double-seal bottles. Aerobic microcosms were prepared in the sterile containment hood, while anaerobic microcosms were prepared in an anaerobic glovebox (2.5% H_2, 1% CO_2, balance N_2 atmosphere, equipped with palladium catalyst O_2 scrubbing). Groundwater used in anaerobic microcosm preparation was first purged with sterile N_2, then placed into the glovebox. Groundwater for use in the aerobic microcosms was similarly gassed, but with air. The dissolved oxygen content of these two waters was 0.3 and 9.3 mg/L, respectively, as determined by the azide modification of the Winkler method (Eaton et al., 1995). Sterile controls were prepared as in Section A11.1.1.2, except that 2.5 mL of a 4% $HgCl_2$ solution replaced azide in the anaerobic microcosms (Trevors, 1996). Microcosms were incubated in the dark at room temperature, either in the glovebox (anaerobic) or under normal atmospheric conditions (aerobic), and were sampled as described previously. Anaerobic microcosms were sampled within the glovebox except when headspace gas was removed for O_2/CO_2 analyses. This latter operation was conducted under the normal atmosphere, but with a stream of N_2 gas directed over the sidearm during sampling.

In total, 22 microcosms were prepared. Single aerobic and anaerobic, sterile controls were prepared from each composite sample (MC-1,2, MC-3,4, MC-5,6), providing a total of three aerobic and three anaerobic controls. Active aerobic and anaerobic microcosms, prepared in duplicate for each composite sample, were amended to concentrations of 5 mg/L toluene, and 1 mg/L each of CT and PCE. Two additional pairs of active, benzoate-amended microcosms were also prepared (aerobic and anaerobic for MC-1,2 only), containing the contaminant mix plus 300 mg/L benzoate (as sodium benzoate). Microcosm pH (measured on day 40) ranged from pH 7 to 7.9. After 64 days of incubation, all four benzoate-containing microcosms were further amended with MBH, to concentrations of 24 mg K_2HPO_4, 24 mg KH_2PO_4, 24 mg NH_4NO_3, 0.48 mg $MgSO_4 \cdot 7H_2O$, 0.48 mg $CaCl_2 \cdot 2H_2O$, and 0.22 mg $FeCl_3 \cdot 6H_2O$ per liter.

Microcosm water was analyzed for PCE and potential products TCE, DCE isomers, VC, CT, and potential products CF, DCM, CM, and TOL as in Appendix 4. A product peak identified as DCM was later recognized to be carbon disulfide (CS_2). Thereafter (from day 40 onward) this compound was included in the analytical suite. Estimates of previous CS_2 concentrations (on day 21) were made by applying the day 40 CS_2 calibration to the appropriate GC peak areas obtained on day 21. Benzoate analyses were conducted as described in Appendix 4.

A11.2.2 Results and Discussion

A11.2.2.1 Aerobic Microcosms

Results from the aerobic sterile controls and the aerobic active microcosms prepared with MC-1,2, MC-3,4, and MC-5,6 core material are shown in Figure A11.3 and Figures A11.4, A11.5, and A11.6, respectively. The values are plotted as means of the three sterile controls (up to day 21) or duplicate active microcosms, as behavior of replicate microcosms was similar. Biological activity was apparent in the MC-3,4 sterile control on day 40, thus the bottle was excluded thereafter. A comparison of Figures A11.4, A11.5, and A11.6 indicates that response of the core material to the contaminants was similar, no matter where the material was obtained along the length of the experimental gates. This suggests the initial aerobic biological potential at the site was fairly uniform.

Toluene was biodegraded readily in the active microcosms, at a rate on the order of 100 µg/L/d, reaching nondetectable levels after 40–64 days. Some decline in the levels of CT and PCE was observed over the first 21 days, but this occurred in the sterile controls as well and therefore cannot be attributed to biological activity. The pattern was first thought to be related to alteration in the microcosm headspace to liquid ratio due to water sampling, but this seems unlikely as the decline ceased after day 21. Slow sorption of the chlorinated compounds is a possible explanation, but the cause of the decline remains unexplained. No biotransformation of CT or PCE was evident in any of the active aerobic microcosms. Daughter products were never detected, with the exception of a single recording of 19.5 µg/L CF in one MC-5,6 microcosm on day 40.

The presence of ca. 300 mg/L benzoate substantially modified biological activity in the aerobic microcosms (Figure A11.7). Aerobic metabolism of toluene was restricted in the presence of benzoate, and little degradation of benzoate itself was detectable over the first 64 days. Provision of inorganic nutrients after the day 64 sampling event (arrow in Fig. A11.7) led to a burst of microbial activity resulting in rapid degradation of both aromatic compounds (toluene ~85 µg/L/d, benzoate ~5.7-7 mg/L/d over the day 64–98 interval). Toluene was below detection by day 98, and benzoate decreased to about 1.5 mg/L by day 182. This aromatic hydrocarbon degradation was paralleled by CT dechlo-

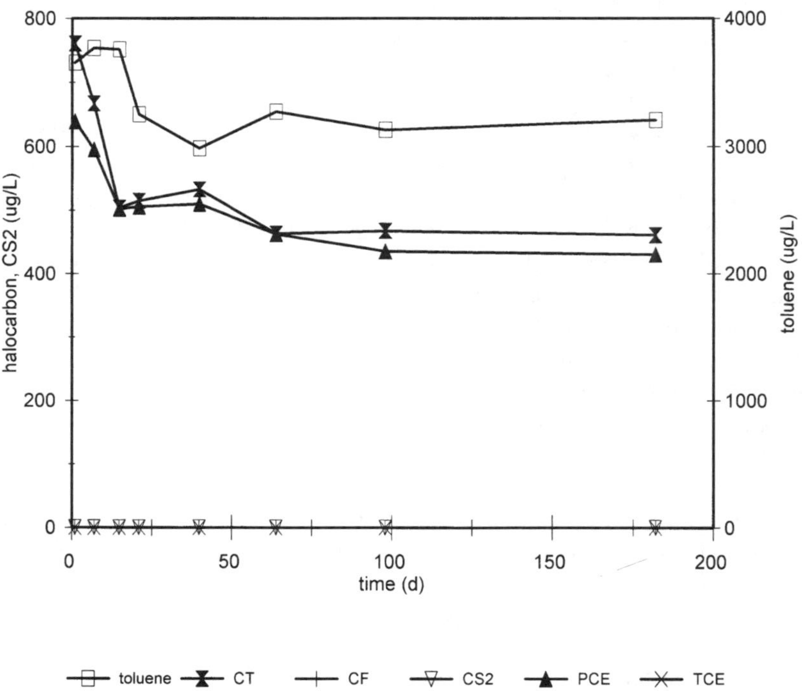

Figure A11.3 Aerobic, sterile toluene-, CT-, and PCE-amended Borden microcosms.

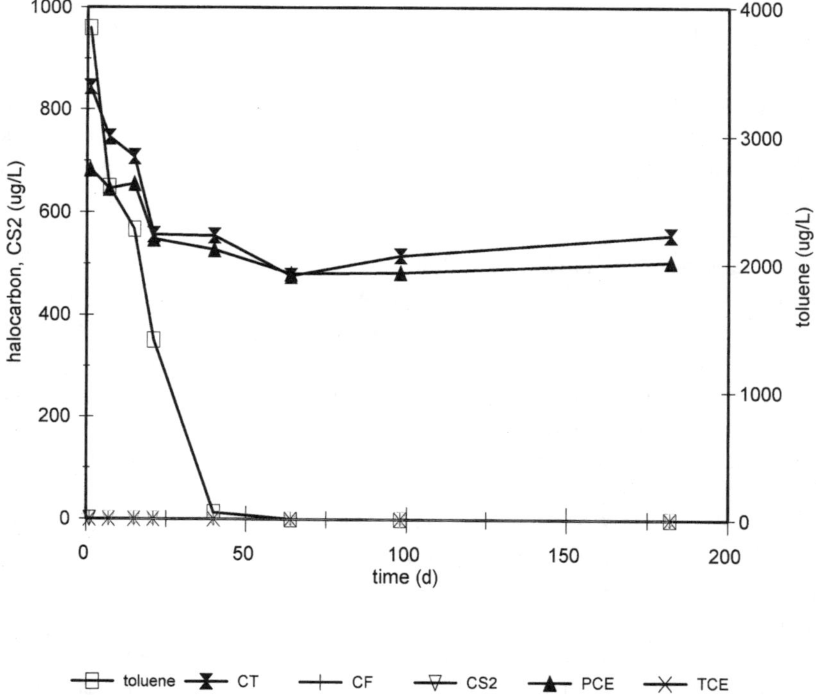

Figure A11.4 Aerobic toluene-, CT-, and PCE-amended MC-1,2 microcosms.

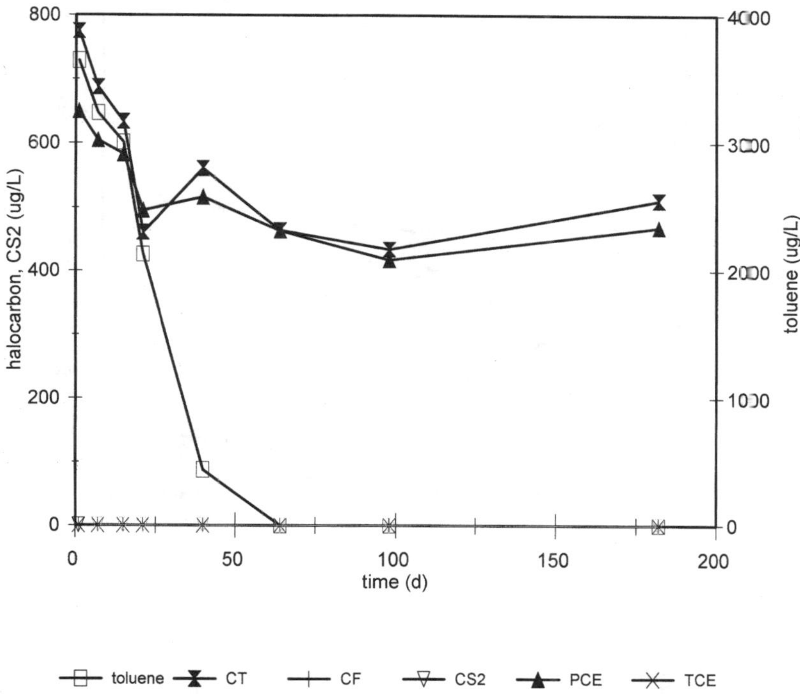

Figure A11.5 Aerobic toluene-, CT-, and PCE-amended MC-3,4 microcosms.

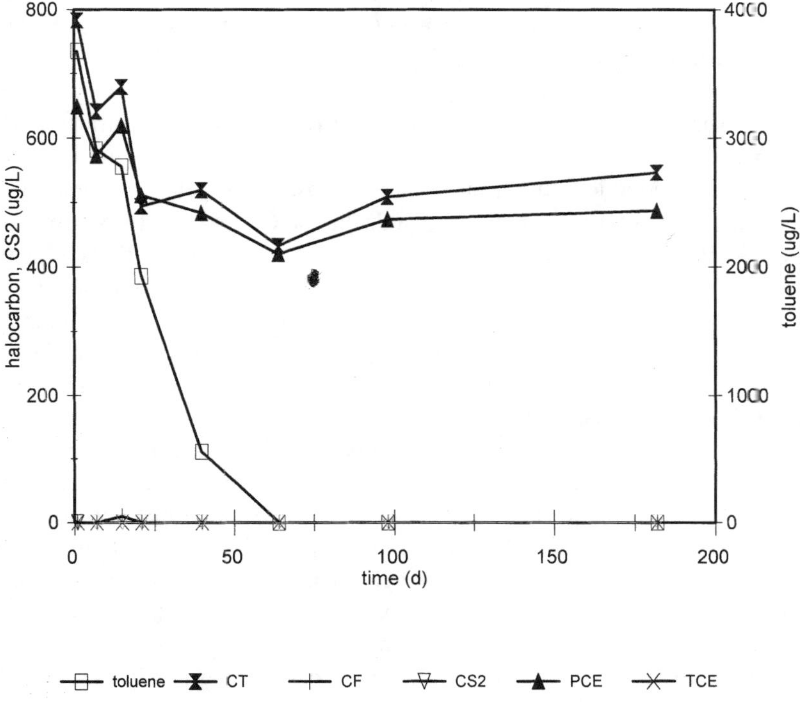

Figure A11.6 Aerobic toluene-, CT-, and PCE-amended MC-5,6 microcosms.

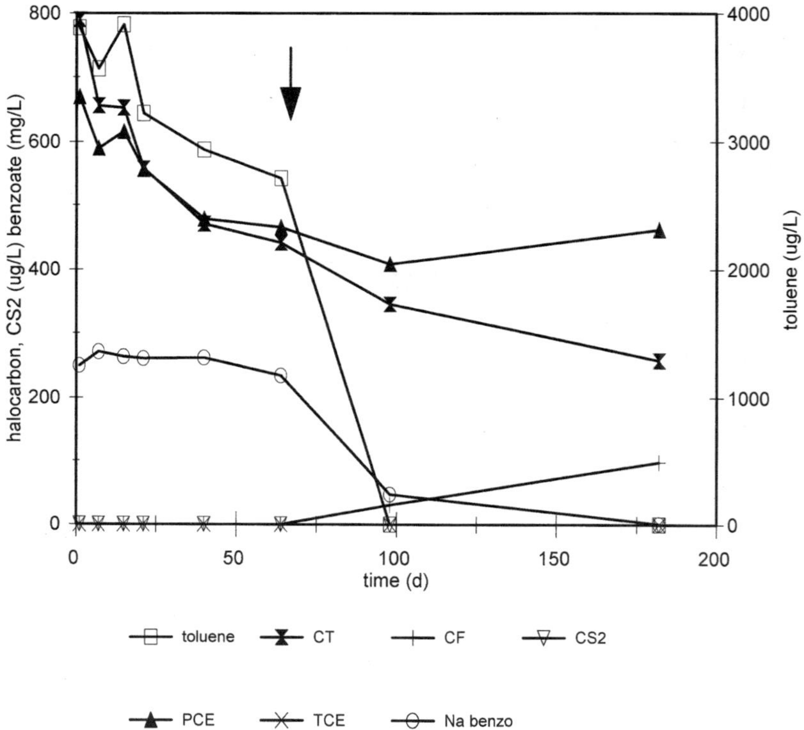

Figure A11.7 Aerobic toluene-, CT-, PCE-, and benzoate-amended MC-1,2 microcosms.

rination to CF. CT levels declined to 258 μg/L by day 182. CF was detected on day 98 and increased from 31 to 98 μg/L by day 182. No other products of CT transformation were detected. PCE remained completely recalcitrant in the benzoate-containing microcosms.

The limitation on biodegradation resulting from a paucity of inorganic nutrients was not unexpected, as earlier experimentation with Borden material revealed similar restrictions (Butler et al., 1992 and unpublished; Allen-King et al., 1994; Barbaro et al., 1994), but the intent of the initial microcosm setup was to mimic natural aquifer conditions. Oxygen was present in excess in all microcosms throughout the experiment, although a decline in O_2 was clearly evident in the benzoate microcosms after MBH addition. On day 182, about 9% O_2 was recorded in the headspace of benzoate-containing microcosms, vs. levels of about 19% in the other aerobic microcosms and 20% in the sterile controls. Nevertheless, the CT dechlorination detected in the benzoate-containing microcosms after relief of the inorganic nutrient limitation suggests that oxygen-depleted microsites developed in the sand–water matrix as a result of aromatic hydrocarbon metabolism. CT transformation may have been directly biologically mediated and/or may have been abiotic but followed from biologically driven changes in the sand–water matrix and/or redox potential within the matrix. The continued recalcitrance of PCE indicated that the altered environment was not sufficiently reduced and/or did not contain microorganisms able to support reductive dechlorination of that compound.

Clearly, the presence of a high level of benzoate blocked toluene degradation in the benzoate-amended microcosms, and the substrate metabolism that did occur under the inorganic nutrient-limited condition was apparently mainly of benzoate and probably undetectable by the methods used, because the relative benzoate concentration change would be small. The benzoate level was not inhibitory to the microbiota, as aromatic hydrocarbon degradation rapidly ensued upon addition of MBH. It is evident that toluene and benzoate biotransformations are competing reactions in the Borden microcosms, although we cannot distinguish whether benzoate is actually the preferred substrate or simply able to outcompete toluene for enzyme-active site(s) because it is present in excess (preliminary results of an experiment in progress suggest the latter). Given the commonality of aerobic aromatic hydrocarbon degradation pathways, this result is not surprising. Certain enzymes and intermediates are common to both toluene and benzoate biodegradation, and in some pathways of toluene degradation (e.g., the TOL plasmid-mediated route) benzoate is a direct pathway intermediate.

A11.2.2.2 Anaerobic Microcosms

In contrast to the aerobic microcosms, some differences in compound loss were observed in duplicate active, anaerobic microcosms. Mean results from the three anaerobic sterile controls and the duplicate anaerobic active microcosms prepared with MC-3,4 and MC-5,6 material are shown in Figure A11.8, and Figures A11.10 and A11.11. The contaminant behavior in these sets of similarly prepared microcosms was similar. However, individual microcosm data are plotted in Figures A11.9a and A11.9b for the MC-1,2 material, as toluene fate in these microcosms differed substantially. One microcosm of this pair (Figure A11.9a) followed the trend of the other anaerobic active microcosms wherein little change in toluene levels occurred. However, toluene was eventually completely depleted in the second MC-1,2 microcosm (Figure A11.9b). The extended plateau at lower toluene concentration has been observed previously with toluene biotransformation under denitrifying conditions, both in the field (Barbaro et al., 1992) and the laboratory (Barbaro et al., 1992; J. R. Barbaro, Butler unpublished). These results suggest that distribution of anaerobic microbial potential for aromatic hydrocarbon metabolism across the experimental site is nonuniform.

Dechlorination of CT occurred in all active anaerobic microcosms (Figures A11.9 to A11.11) including the MC-1,2 microcosms amended with benzoate (Figure A11.12), but not in the sterile controls (Figure A11.8). CT declined over time, while CF appeared and subsequently increased to

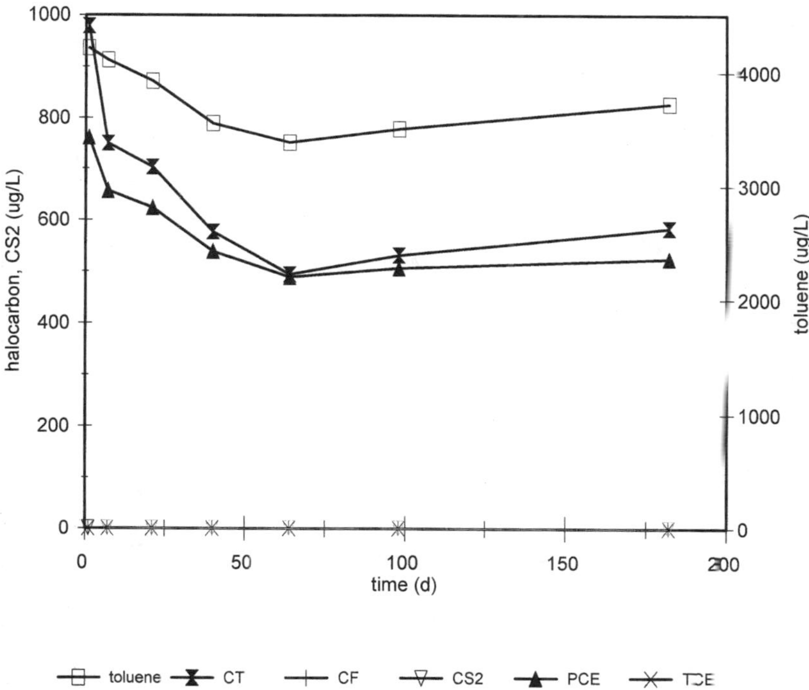

Figure A11.8 Anaerobic, sterile toluene-, CT-, and PCE-amended Borden microcosms.

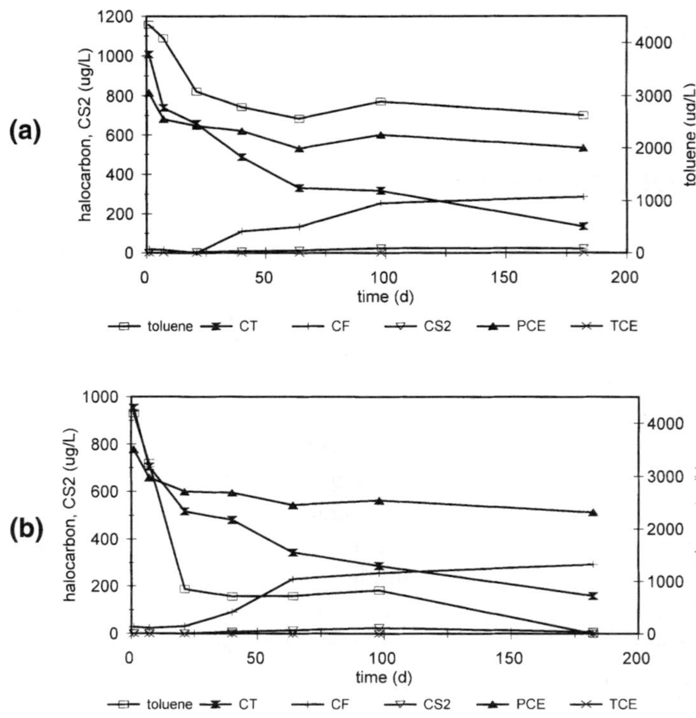

Figure A11.9 (a) Anaerobic toluene-, CT-, and PCE-amended MC-1,2 microcosms replicate a. (b) Anaerobic toluene-, CT-, and PCE-amended MC-1,2 microcosms replicate b.

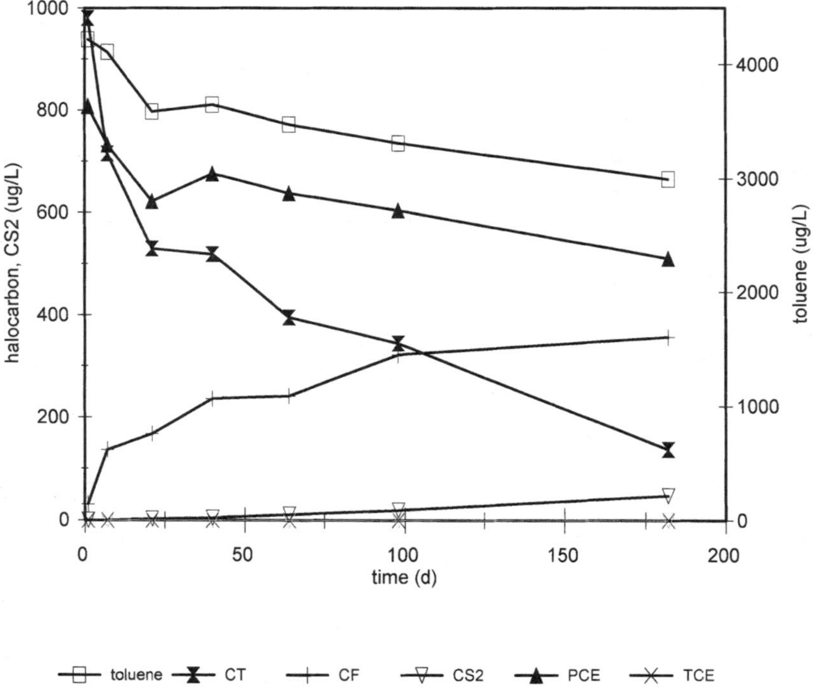

Figure A11.10 Anaerobic toluene-, CT-, and PCE-amended MC-3,4 microcosms.

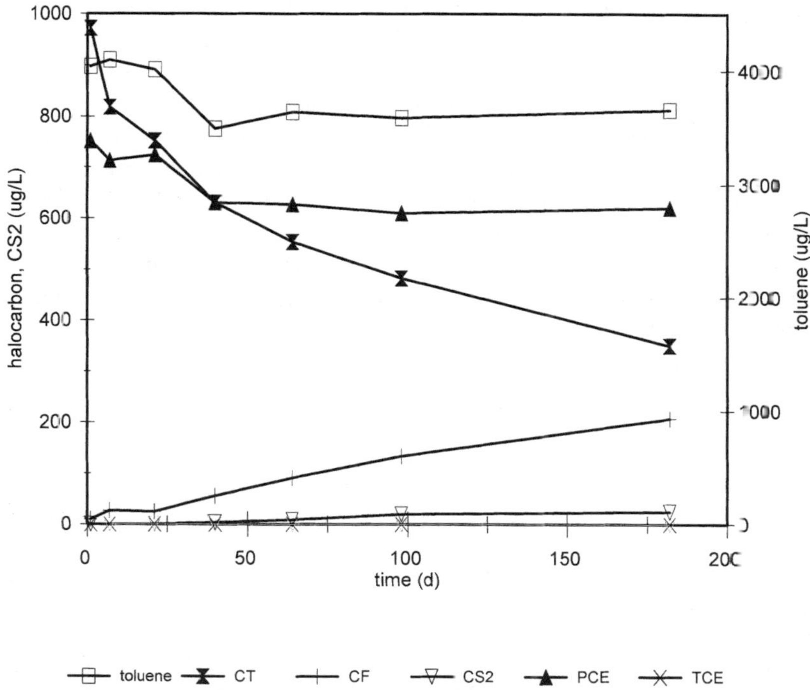

Figure A11.11 Anaerobic toluene-, CT-, and PCE-amended MC-5,6 microcosms.

levels ranging from 127 to 370 μg/L by day 182. Neither DCM nor CM was detected but carbon disulfide was, at levels as high as 49 μg/L in one microcosm. There was a general trend upward in CS_2 in most active microcosms, but this was not as pronounced as CF accumulation. Typically, CS_2 levels were 10 to 25 μg/L over the day 64 to 182 period (Figures A11.9 to A11.12). The results suggest reductive dechlorination was the dominant route of CT biotransformation, although the evidence is not absolutely conclusive. CS_2 may have been hydrolyzed to CO_2 via abiotic reaction, but CO_2, CO, hexachloroethane, and formate, other potential products of CT transformation (Criddle and McCarty, 1991; Kriegman-King and Reinhard, 1992), were not monitored. Mass balances for the CT-related compounds, based on the aqueous phase distribution of compounds monitored (Table A11.1), were poor because of the unexplained decline in contaminant concentrations early in the experiment. However, sums of recovered mass were similar for active and sterile microcosms at every sampling time, so that a major missing product of CT biotransformation seems unlikely. As Table A11.1 indicates, 55 to 68% of the initial contaminant was recovered as either CT or detected product on day 182.

Table A11.1 Distribution of CT-Related Compounds in the Aqueous Phase of Anaerobic Borden Microcosms on Day 182

Condition	CT (μmol/L)	CF (μmol/L)	CS_2 (μmol/L)	Sum (μmol/L)	%of C_1[a]
Sterile	3.80	0	0	3.80	60
Active MC-1,2	0.96	2.42	0.20	3.58	55
Active MC-3,4	0.90	2.98	0.64	4.52	68
Active MC-5,6	2.28	1.74	0.32	4.34	68
Active + benzoate (MC-1,2)	0.21	3.08	0.34	3.63	57

[a] % of day 1 sum.

PCE remained completely recalcitrant in the anaerobic microcosms and no daughter products expected in the reductive dechlorination of PCE were detected during the experiment.

No detectable benzoate loss was recorded in the benzoate-amended microcosms over the first 64 days of the experiment (Figure A11.12), and unlike the aerobic microcosms, addition of MBH (arrow in Figure A11.12) did not lead to benzoate consumption. Thus, although inorganic nutrient addition is required for metabolism of the benzoate concentration present in the microcosms, this alone was insufficient for significant anaerobic benzoate biotransformation. Some other condition(s) remains unsatisfied — either a lack of the requisite microorganisms or some other critical environmental parameter has not been suitably adjusted. Lack of electron acceptor is a possibility if either CT or PCE will not serve. MBH addition supplied some nitrate and sulfate, but not a great deal relative to the amount of benzoate present. If Figure A11.12, and Figures A11.10 and A11.11 are compared, there is indication that benzoate presence "drives" CT dechlorination as the CT level was lowest in the benzoate-amended case on day 182 (Table A11.1).

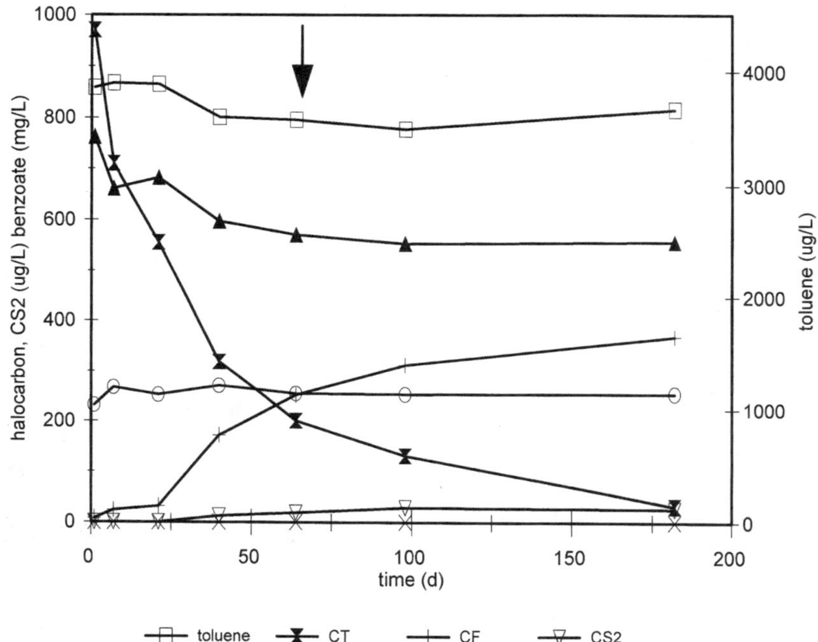

Figure A11.12 Anaerobic toluene-, CT-, PCE-, and benzoate-amended MC-1,2 microcosms.

A11.2.3 Conclusions

The microcosm experiment indicated that ~5 mg/L toluene was readily biodegradable if O_2 was available, but the presence of excess benzoate or a lack of O_2 could impede this process. The implication for the field experiment is that if benzoate remains in excess downgradient of the injection zone and moves into the biosparge zone in Gate 3, toluene biodegradation may be poor, even in the presence of oxygen, because benzoate levels are significantly greater. Excess downgradient benzoate will occur *in situ*, because the natural aquifer supplies inorganic nutrient levels insufficient to support biodegradation of all the added primary carbon and energy substrate (i.e., benzoate and toluene), and because the native aquifer microbiota is not readily capable of anaerobic benzoate metabolism even when inorganic nutrients are available.

The readily initiated biotransformation of CT in the microcosms suggests this activity will be observed in the field, but PCE dechlorination seems considerably less likely, as that activity would require a very low redox environment and the development of an extensive, complex anaerobic community (1) to reduce the E_h and (2) to increase chances of a population of suitable dechlorinating strains appearing. If the supplied growth substrate (i.e., benzoate) is not readily utilized by that portion of the native microbiota able to grow in anoxic conditions, then this sequence of events may not occur within the time frame of the field experiment. The chance occurrence of conditions suited to anaerobic benzoate metabolism is more favored in the field situation, since a vast array of microsites and their microbial inhabitants may be encountered by the contaminant plume, rather than in the laboratory, where only a few hundred grams of aquifer matrix were tested.

A11.3 MICROCOSMS SIMULATING CONDITIONS IN THE ALAMEDA AQUIFER

The intrinsic ability of microorganisms inhabiting the Alameda subsurface to degrade halogenated alkenes and aromatic hydrocarbons was investigated. The activity in microcosms intended to simulate *in situ* conditions, plus that in additional sets of microcosms amended with inorganic nutrients and/or additional electron donor were compared with sterile control microcosms.

MBH was added to some microcosms to test whether provision of inorganic nutrients would affect microbial activity in the Alameda soil. Methanol was added to some microcosms to test the effect of an exogenous electron donor on the microbial activity. Methanol is known to stimulate reductive dechlorination of halogenated alkenes (Freedman and Gossett, 1989; Major et al., 1991; Maymó-Gatell et al., 1995) and occurs in Alameda groundwater (M. Morkin, unpublished), although inherent methanol in the groundwater used in this experiment was low or absent. The site groundwater obtained for the experiment was not specifically analyzed for methanol, but water samples from microcosms that were not CH_3OH-amended contained no detectable methanol.

TCE was present only at low levels (~30 µg/L) in the groundwater obtained from the site, although this compound (and/or PCE) was presumably the original, parent halogenated ethene contaminating the Alameda site. *c*DCE, the chlorinated contaminant currently present at highest levels, would rarely be used industrially, but it is a major daughter product of solvent biotransformation. To test whether the indigenous microbiota biotransformed TCE, this compound was added to microcosm water.

Toluene was present in the groundwater, but levels of this contaminant were also adjusted upward to ensure that its fate could be monitored over the experimental period. Toluene is the monoaromatic hydrocarbon mostly readily and widely degradable under anaerobic conditions, as well as the most prominent BTEX contaminant at Alameda, and so seemed the most reasonable "model" aromatic compound to focus upon in this study.

A11.3.1 Methods and Materials

A11.3.1.1 Site Materials

During excavation for the Funnel-and-Gate at Alameda NAS in December 1996, approximately 5 kg of subsurface soil was collected in small canning jars, which were filled to capacity. This wet, dark-colored, sandy material was shipped on ice to UW, then stored at 4°C until required. Groundwater was collected on March 3, 1997 from upgradient of the treatment gate (row 1 PB; M. Morkin, personal communication), and shipped to UW where it was stored at 4°C. The loss of volatiles and introduction of oxygen into the water were minimized by filling 4-L glass bottles to overflowing, then tightly sealing each bottle before shipment.

A11.3.1.2 Microcosms

The sacrificial microcosms used in this experiment were prepared and incubated in the anaerobic glovebox. Each consisted of 20 g (wet weight) soil and groundwater (amended as described below) to fill a 60-mL hypovial. After preparation, the hypovials were sealed with Teflon-faced silicon septa and aluminum crimp seals. To retain both target contaminants and any other unanalyzed components present, the groundwater was not purged to specifically strip O_2. Reducing agents present in the soil were used to scavenge any O_2 that may have been introduced into the groundwater during handling, shipping, and storage. Starting groundwater had a DO content of 0.46 to 0.42 mg/L, but DO in the microcosms typically ranged from 0.25 mg/L to undetectable.

Some groundwater was purged with N_2 to remove volatile compounds. This was then mixed with nonpurged groundwater to dilute the target contaminants to desired levels. Additional TCE was added to the water as a spike of neat compound. Toluene levels (~2400 µg/L in the obtained site water) were adjusted upward in a similar manner. Three batches of groundwater were required for the entire microcosm setup. The target compound content of these batches is given in Table A11.2. It is apparent that the GW-3 batch of water inadvertently received twice as much TCE as intended, and no additional toluene. This is reflected in the TCE and toluene data for two of the microcosm sets, but did not unduly influence the pattern of activity in those microcosms in comparison to microcosm sets prepared with the GW-1 and GW-2 batches of water.

Eight sets of microcosms were prepared for the experiment, with quadruplicate hypovials for each experimental condition (sterile, unamended, MBH-amended, CH_3OH-amended, MBH + CH_3OH-amended), for a total of 160 hypovials. Sterile controls were prepared with $HgCl_2$ as described in Section A11.2.1.2. This procedure was not wholly successful, as failure of some controls was visibly evident by blackening of the soil, beginning on day 62. Blackened controls were excluded from the analyses, and the experiment was terminated on day 167 when no uncompromised controls remained. MBH was added to provide 18 mg K_2HPO_4, 18 mg KH_2PO_4, 18 mg NH_4NO_3, 3.6 mg $MgSO_4 \cdot 7H_2O$, 0.36 mg $CaCl_2 \cdot 2H_2O$, and 0.17 mg $FeCl_3 \cdot 6H_2O$ per liter in some microcosms. Methanol was added to some active microcosms and to the sterile controls to a calculated concentration of 50 mg/L. GC analyses conducted on day 1, however, recorded concentrations on the order of 150 mg/L in all methanol-amended microcosms. Similar levels were obtained in the sterile controls on day 27 (methanol was completely depleted in the active microcosms), but analyses on day 136 recorded 47 mg/L methanol in the controls. The MBH + CH_3OH-amended microcosms were re-amended with methanol (calculated concentration of 50 mg/L) on day 95, and the GC analyses that were conducted recorded values close to the expected level (60.5 and 61.8 mg/L). It is therefore unclear whether too much methanol was added initially, or whether there was some unrecognized analytical problem (an interfering peak?) on days 1 and 27.

Three microcosms from each set of four replicates were sampled for volatile halocarbons and aromatic hydrocarbons by decapping the hypovials and quickly transferring water directly into 22-mL autosampler vials, using a 30-mL glass syringe. The autosampler vials were preloaded with 0.2 mL of 10% Na azide to restrict further microbial activity. After crimp-sealing, the sample vials were analyzed as described in Appendix 3. Water was removed from the fourth microcosm in a similar manner to fill two 18-mL hypovials, which were used for methanol analysis, and for methane, ethene, and ethane analysis (Appendix 3). A second, composited sample for methanol analysis was also prepared, using water pooled from the first three microcosms. Data are plotted as the mean of three values for TCE, the DCE isomers, VC, and BTEX, as a single datum for methane, ethene, ethane, and as a mean of two values for methanol.

Normally, microcosms were removed from the glovebox and sacrificed on the lab bench. However, on days 95 and 136, sampling was conducted within the glovebox, so that the solids and water remaining in the sacrificed vials could be retained, potentially for future work. This may have had a deleterious effect on the recovery of gaseous compounds in particular, because sampling manipulations were considerably more cumbersome to complete.

Table A11.2 Initial Groundwater Composition

Compound	GW-1	GW-2	GW-3
VC	6402	8068	6271
1,1-DCE	57.55	68.27	50.57
*c*DCE	40000	35320	34440
*t*DCE	31.96	37.96	26.86
TCE	2000	2686	6153
Benzene	51.29	59.48	46.48
Toluene	4047	4646	2577
Ethylbenzene	23.83	28.29	21.12
p- and *m*-Xylene	91.07	107.2	84.02
o-Xylene	33.96	40.18	30.5

Note: All data are in µg/L. GW-1 was used in the sterile and MBH + CH_3OH-amended microcosms, GW-2 was used in the CH_3OH-amended microcosms, and GW-3 was used in the unamended and MBH-amended microcosms.

A11.3.2 Results and Discussion

A11.3.2.1 Chlorinated Contaminants

Site groundwater was purposely diluted, because it was feared that high contaminant levels might inhibit microbial activity. Such fears quickly proved groundless. All the nonsterile microcosms were actively dechlorinating, showing the typical sequential reductive dechlorination of a halogenated ethene (deBruin et al., 1992; Major et al., 1991). The results for TCE and compounds related to its degradation are shown in Figures A11.13 to A11.17.

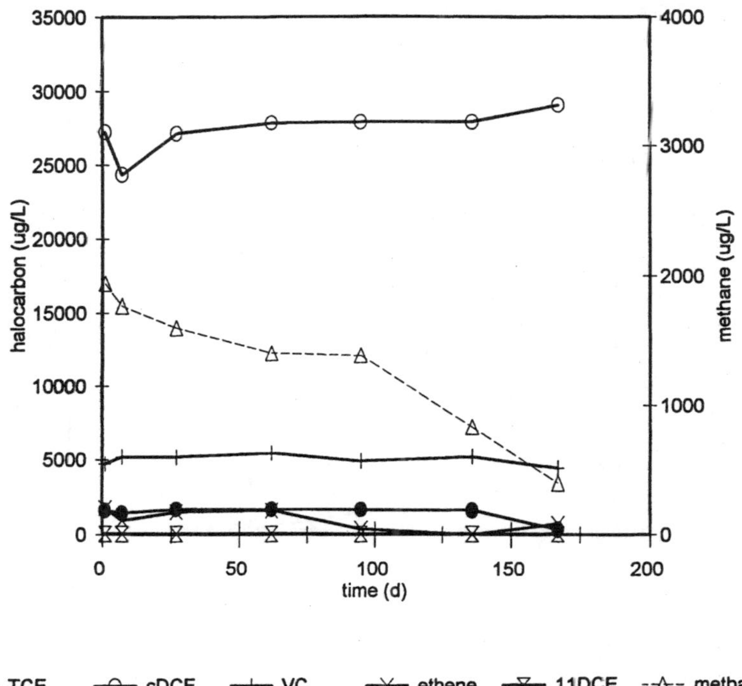

Figure A11.13 Sterile Alameda microcosms — chloroethenes.

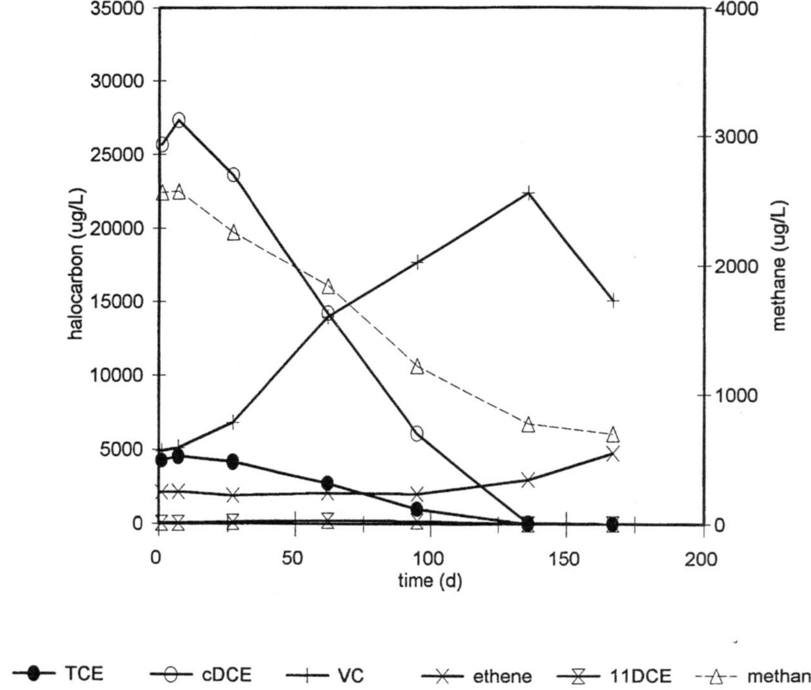

Figure A11.14 Unamended Alameda microcosms — chloroethenes.

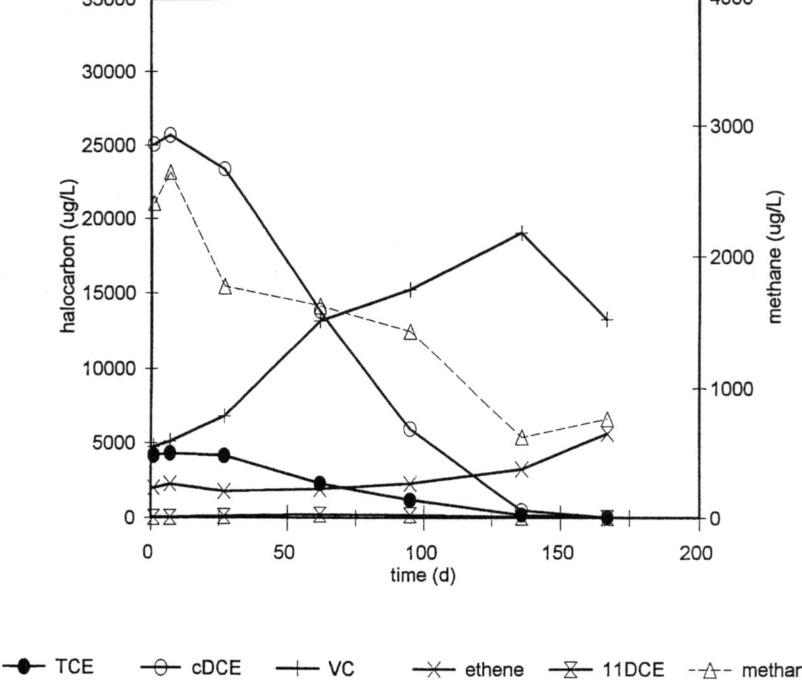

Figure A11.15 MBH-amended Alameda microcosms — chloroethenes.

Comparison of the unamended (Figure A11.14) and MBH-amended (Figure A11.15) microcosms indicates that addition of inorganic nutrients did not improve dechlorination over that obtained with the unamended materials. In contrast, addition of methanol increased the speed of the reaction sequence considerably (Figures A11.16 and A11.17). Methanol initially added was undetectable by day 27, although clearly the "reducing power" added into the system exerted a detectable effect over the course of the experiment. An additional aliquot of methanol (ca. 60 mg/L) was added to all the remaining MBH + CH_3OH-amended microcosms after the day 95 sampling event, in an attempt to drive the dechlorination reaction as far toward completion as possible. By day 136, this second methanol addition was completely depleted as well, but there was ultimately little difference between contaminant levels in these microcosms and the CH_3OH-amended ones, so that the additional amendment was perhaps unnecessary. TCE and *c*DCE were below detection in both microcosm types by day 136 (Figures A11.16 and A11.17). At the end of the experiment, slightly higher levels of VC were present in the CH_3OH-amended microcosms (39–44 μg/L vs. 22–33 μg/L), as was also true for 1,1-DCE and *t*DCE (2–9 μg/L vs. undetectable–5 μg/L), but indications are that complete dechlorination to ethene will occur in both types of methanol-amended microcosms.

Little 1,1-DCE or *t*DCE was present in the microcosms at any time, a finding common in systems where the reductive dechlorination is biologically mediated. *t*DCE levels remained constant, ranging from 23–42 μg/L under all experimental conditions until day 62, when they had noticeably dropped in microcosms that had received methanol. 1,1-DCE was considerably more responsive, its behavior paralleling that of *c*DCE. Initial 1,1-DCE levels of 41–52 μg/L rose to >200 μg/L over days 62–95 in the active unamended and MBH-amended microcosms before declining again. Peak concentrations in CH_3OH- and MBH + CH_3OH-amended microcosms were earlier (day 27) and smaller (121–138 μg/L).

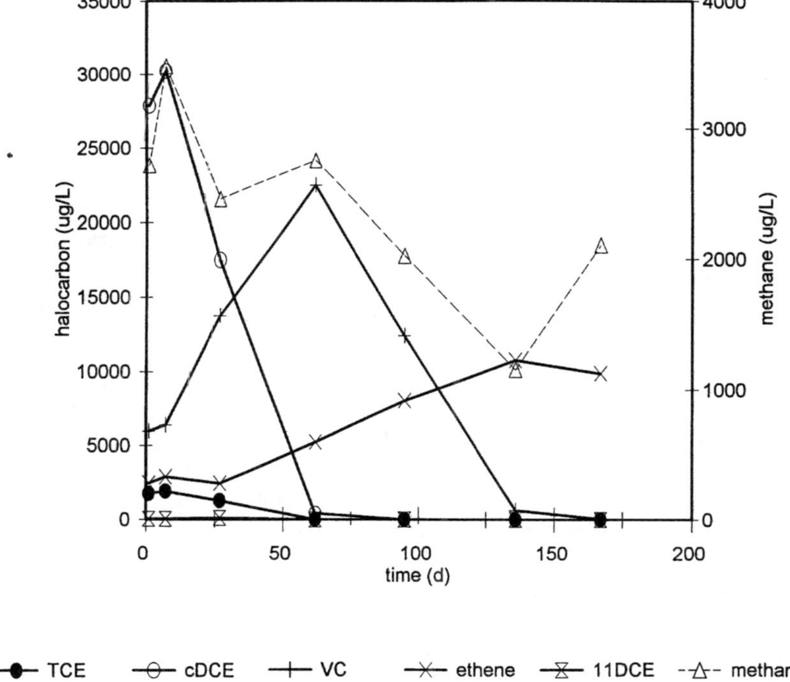

Figure A11.16 Methanol-amended Alameda microcosms — chloroethenes.

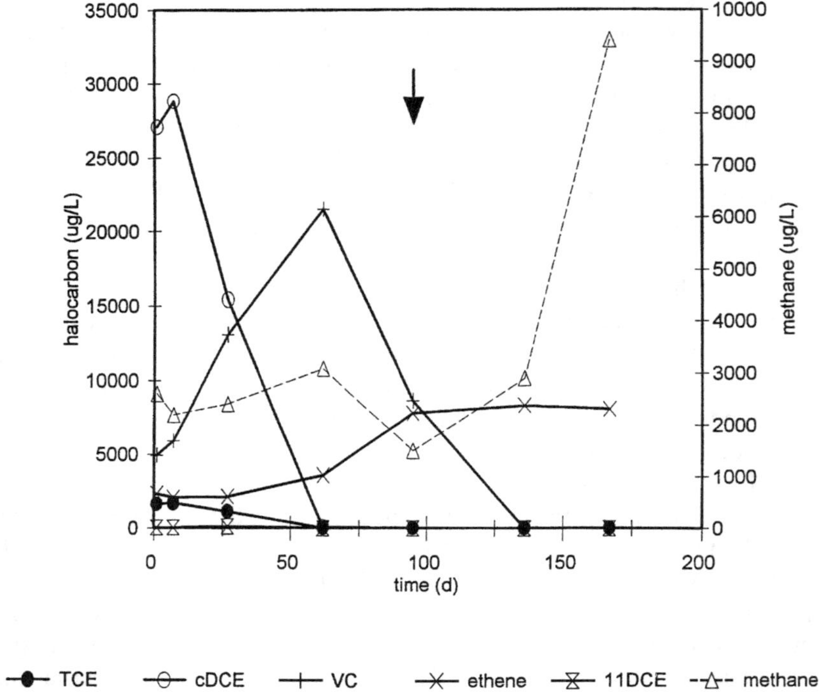

Figure A11.17 MBH + methanol-amended microcosms — chloroethenes. *Note:* After the day 95 sampling event, methanol was added to all remaining microcosms (arrow).

Ethene accumulation was apparent in all the active microcosms (Figures A11.14 to A11.17), although it was far more extensive when methanol was provided. Ethene was initially present in the sterile controls, but did not increase further (Figure A11.13). Ethane was never detected in active microcosms, although ethane peaks (and concurrently lower ethene) were recorded in the sterile controls on days 7 and 95. The reason for this latter observation is unclear.

The sum of aqueous phase mass recovered for TCE and associated products ranged from 87 to 113% over the experiment, with the exception of three points. On day 167 there was 73% recovery in the CH_3OH-amended microcosms, and days 136 and 167 recoveries for the MBH + CH_3OH-amended microcosms were 76 and 52%, respectively. Figures A11.18, A11.19, and A11.20 depict three examples of the data, for the sterile control, unamended, and CH_3OH-amended microcosms, respectively. The falloff of recovery in the CH_3OH-amended and the MBH + CH_3OH-amended microcosms could reflect slow diffusional loss of gaseous products from the aqueous phase through the microcosm septa, or could be an indicator of significant CO_2 production from VC (Vogel and McCarty, 1985). CO_2 was not monitored in the microcosms.

Interestingly, the dechlorinating microcosms were not methanogenic, which differs from microcosms derived from another naturally dechlorinating subsurface system we have examined where haloalkene levels were substantially lower (Major et al., 1991). Methane levels tended to remain roughly the same or to decline over the course of the experiment, except in the microcosms that received two methanol amendments. Since methane is present in significant quantities in the Alameda site water, there must be methanogenic zones *in situ*, but either the area from which the soil was obtained was not methanogenic, or microcosm conditions did not favor methanogenesis. The obvious blackening of the soil in active microcosms (and in the failed sterile controls) indicated that sulfidogenic activity occurred in the microcosms, although this does not imply that sulfidogenic microorganisms were responsible for dechlorination. Analyses conducted on day 167

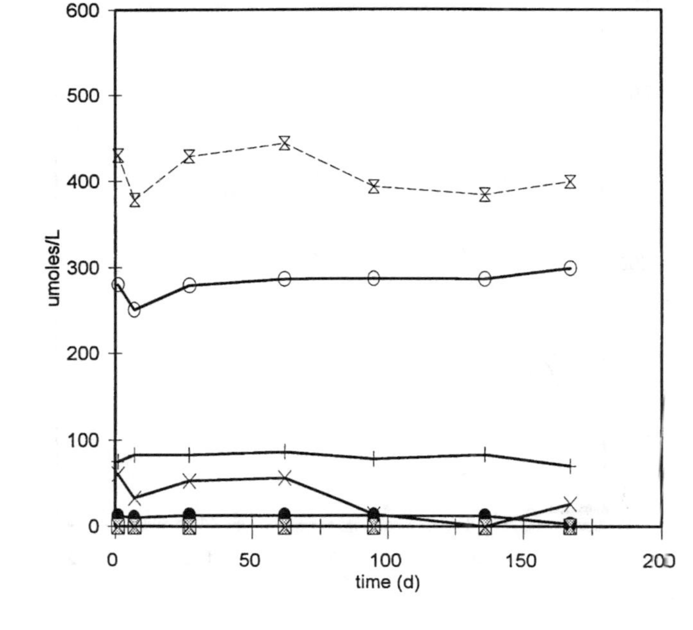

TCE cDCE tDCE 11DCE VC ethene sum

Figure A11.18 Aqueous phase mass distribution in sterile microcosms.

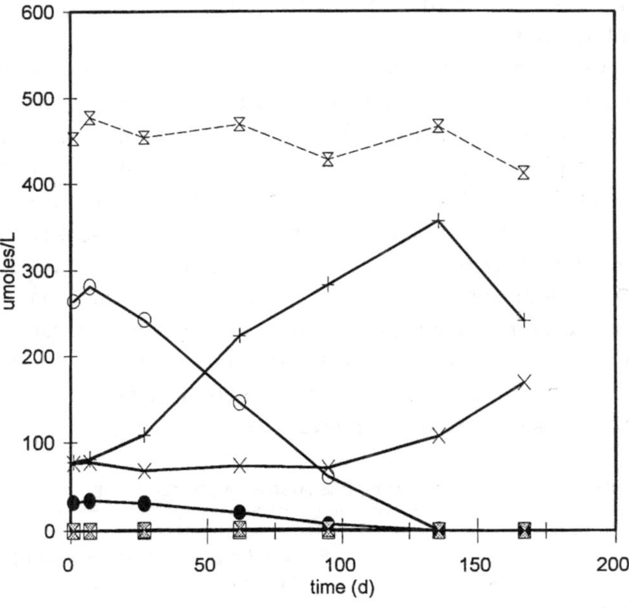

TCE cDCE tDCE 11DCE VC ethene sum

Figure A11.19 Aqueous phase mass distribution in unamended microcosms.

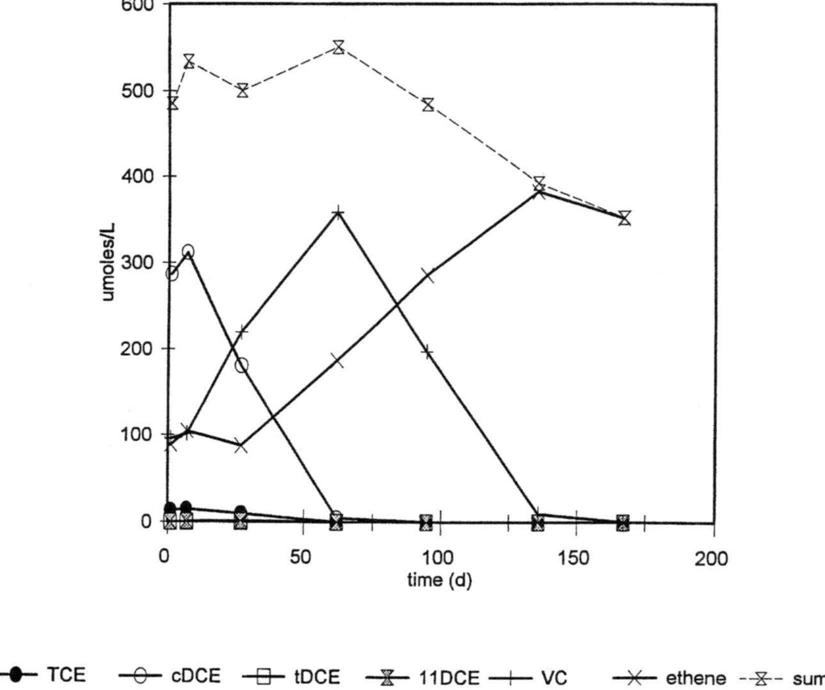

Figure A11.20 Aqueous phase mass distribution in CH_3OH-amended microcosms.

microcosms confirmed complete sulfate depletion in those highly reactive microcosms that had received methanol (Table A11.3). Partial sulfate loss was evident in active microcosms that had not received methanol addition.

Work conducted by Gossett's group with a CH_3OH- and PCE-fed dechlorinating enrichment culture revealed that adding PCE at high levels (550 μmol/L) inhibited methanogenesis, and the electron flow in the culture was eventually directed away from methanogenesis to acetogenesis (69%) and dechlorination (31%) (DiStefano et al., 1991). Eventually, a derivative, strictly dechlorinating culture was obtained, which metabolized H_2 (the direct electron donor in the CH_3OH-fed culture) and PCE to produce VC and ethene, but no methane or acetate (DiStefano et al., 1992; Maymó-Gatell et al., 1995). It is plausible that similar selective pressures have acted *in situ* at Alameda, resulting in development of a "natural" microbial community that grows by employing haloalkene as electron acceptor. Evidence in support of this suggestion can be seen especially in Figure A11.17, but also in Figure A11.16. A substantial increase in methane occurred between days 136 and 167, especially in the microcosms that receive two methanol amendments, indicating that methanogenesis was initiated once haloalkene levels were very low.

Table A11.3 Sulfate in Day 162 Alameda Microcosms

Condition	Sulfate (mg/L) Mean	Sulfate (mg/L) SD
Sterile	60.0	2.4
Unamended	29.4	14.5
MBH-amended	39.9	7.9
CH_3OH-amended	<0.05	—
MBH + CH_3OH-amended	<0.05	—

Methanol was the ultimate source of electron donor in those microcosms amended with the alcohol. The *in situ* electron donor(s) — which presumably also serve that function in the unamended and MBH-amended microcosms — have not been identified, although it has been established that BTEX do not serve this function (see Section A11.3.2.2).

A11.3.2.2 Monoaromatic Contaminants

Information on the fate of BTEX in the Alameda microcosms is presented in Figures A11.21 to A11.25. Through 167 days of incubation there was relatively little degradation of BTEX

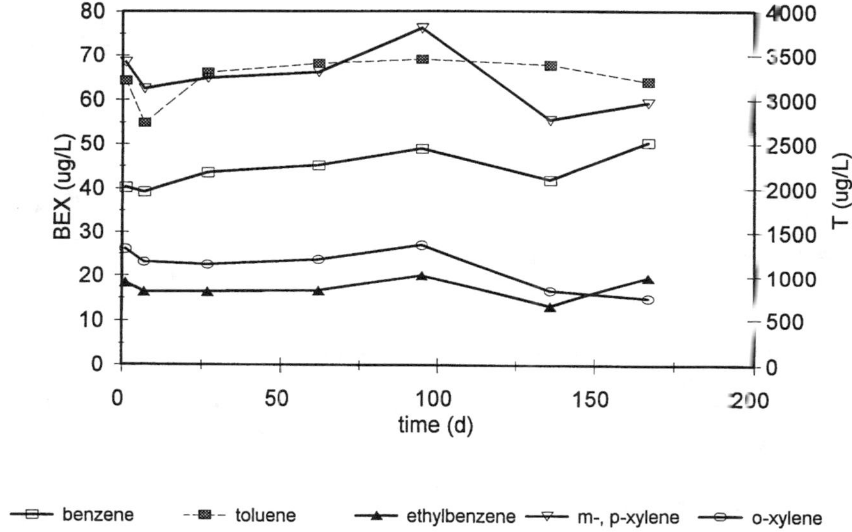

Figure A11.21 Sterile Alameda microcosms — BTEX.

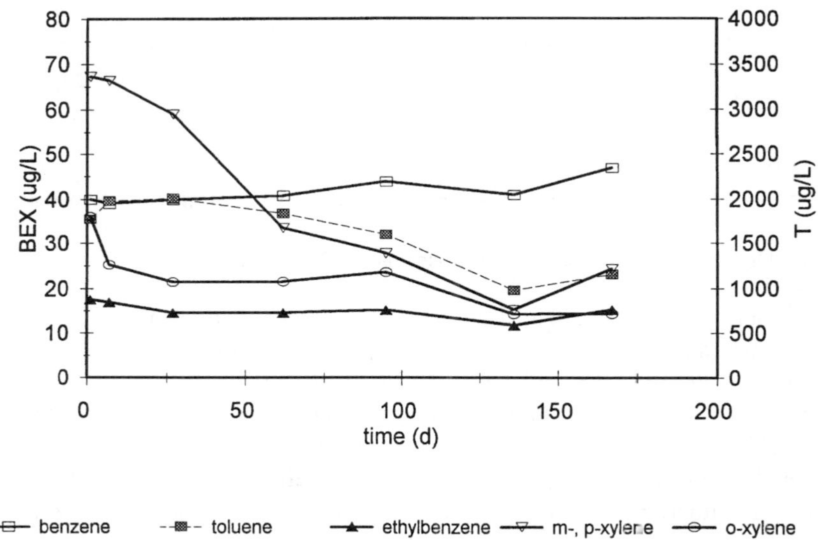

Figure A11.22 Unamended Alameda microcosms — BTEX.

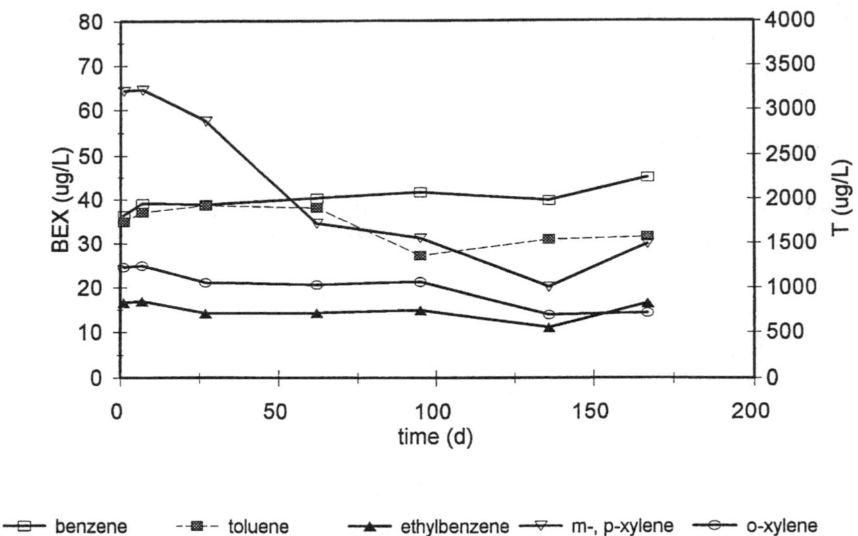

Figure A11.23 MBH-amended Alameda microcosms — BTEX.

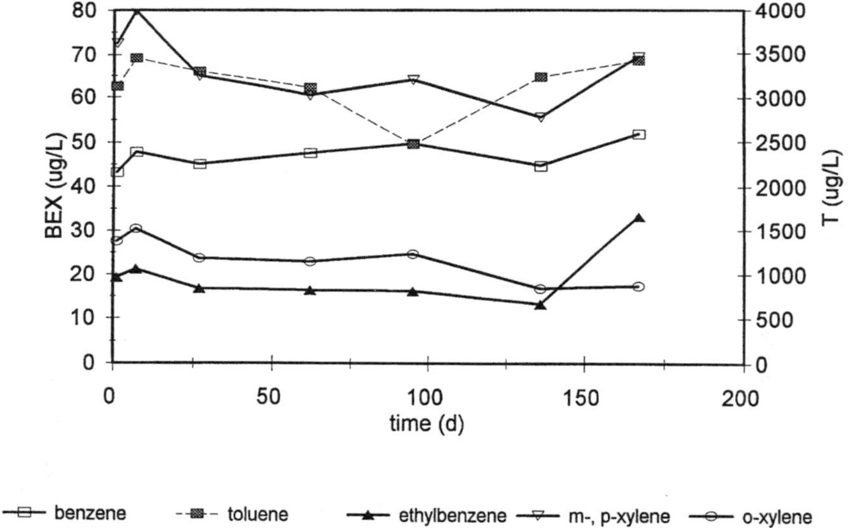

Figure A11.24 Methanol-amended Alameda microcosms — BTEX.

compounds in any of the active microcosms, clearly demonstrating that BTEX compounds are not the electron donors supporting reductive dechlorination. Initial levels of *m-/p*-xylene (the analytical protocol does not distinguish the isomers) were low in the microcosms, but slow *m-/p*-xylene biotransformation occurred in those active microcosms not supplied with methanol (Figures A11.22 and A11.23). It appears that some toluene depletion also occurred in these microcosms toward the end of the incubation period. In contrast, all BTEX compounds persisted in the methanol-supplied microcosms. The evidence suggests that conditions favoring accelerated biologically mediated reductive dechlorination are not conducive to coincident anaerobic aromatic hydrocarbon biotransformation at the Alameda site.

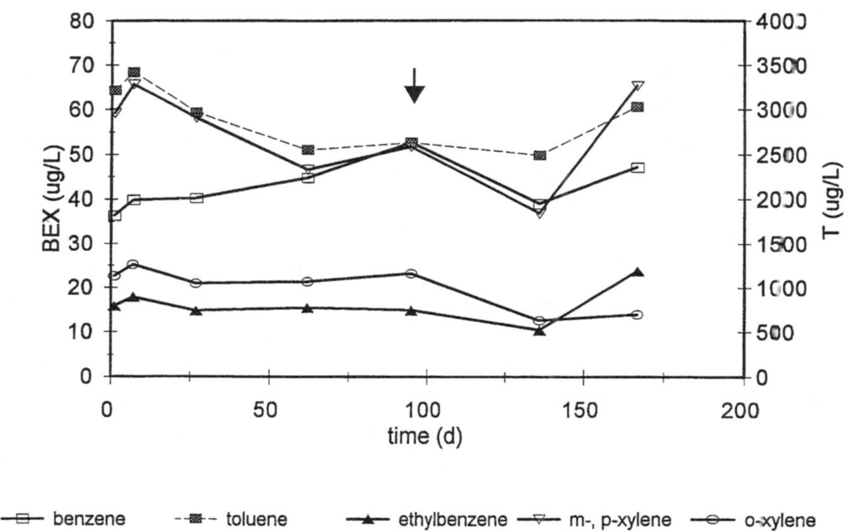

Figure A11.25 MBH + methanol-amended Alameda microcosms — BTEX. *Note:* After the day 95 sampling event, methanol was added to all remaining microcosms (arrow).

A11.3.3 Conclusions

Microorganisms within the Alameda soil actively dechlorinated chloroalkenes to produce less-chlorinated products and ethene. Presumably this activity has contributed significantly to the current makeup of the contaminant plume under study at the site. This natural contaminant attenuation was supported by unidentified electron donor(s) present in the soil and/or groundwater, but could be significantly increased by the addition of methanol. Dechlorination of ca. 500 µmol/L total haloalkene was close to completion within 167 days under methanol-amended conditions. BTEX degradation was limited and did not serve as a source of electrons for the reductive dechlorination observed.

A11.4 EFFECT OF BENZOATE ON AEROBIC TOLUENE DEGRADATION IN AEROBIC BORDEN MICROCOSMS

The initial Borden microcosm experiment (Section A11.2 above), plus data collected from Gate 3, suggested that significant benzoate presence could impede aerobic toluene biodegradation. The finding is not unexpected, since both compounds are metabolized via similar pathways; in fact benzoate may be an intermediary in toluene degradation (e.g., in the TOL plasmid-directed route of toluene degradation). This experiment investigated this benzoate-toluene interaction. A number of interactive events are possible at the cellular level. In the presence of benzoate the microbial population may preferably utilize this compound rather than toluene. A second alternative is that the compounds are metabolized in parallel, but the vastly greater number of benzoate molecules effectively outcompete toluene molecules for some enzyme in a shared metabolic pathway, or possibly some common transport system. Benzene is believed to enter cells by diffusing through the cell membrane, as it is highly lipid-soluble (Atlas, 1997). Toluene probably exhibits similar behavior. The more hydrophilic benzoate likely moves into the cell via a transport system, which could well be a more efficient mechanism than simple diffusion. A combination of effects is also possible, both because the alternatives noted above could affect any given organism in the community simultaneously, and because a variety of different organisms within the mixed, indigenous microbial community are likely active in compound biodegradation.

The experiment was conducted under aerobic conditions to mimic the biosparge zone of Gate 3, using a single concentration of toluene, in the presence of varied levels of benzoate. Inorganic nutrients were also added to avoid a restriction on microbial activity due to a lack of these. At the time of the experiment, it had been decided to lower the field target level for benzoate from 300 to 180 mg/L. Hence, the "standard" ratio of benzoate to toluene was considered to be 180 to 5 mg/L, or 1470 nmol benzoate per liter to 54.3 nmol toluene per liter. The microbial population would thus be exposed to 27 molecules of benzoate for every molecule of toluene. In addition to this "standard" benzoate-to-toluene ratio, we also examined equimolar concentrations (54.3 nmol/L each, or 5 mg toluene per liter and 6.627 mg benzoate per liter), and a 2:1 ratio of benzoate to toluene (108.6 nmol benzoate to 54.3 nmol/L toluene, or 5 mg toluene per liter and 13.254 mg benzoate per liter) and a 10:1 ratio (543 nmol/L (66.27 mg/L) benzoate to 54.3 nmol/L (5 mg/L) toluene). The conditions are designated 27X, 1X, 2X, and 10X below. Rates of compound degradation were compared to the condition where only toluene (5 mg/L) was provided (designated T only), since the primary question to be addressed was the effect of benzoate presence on the biodegradation of toluene. However, two sets of microcosms supplied with only benzoate (6.627 and 180 mg/L, designated 1X Benzo only and 27X Benzo only) were also included in the experiment.

A11.4.1 Methods and Materials

Aerobic microcosms were constructed in 250-mL screw-cap bottles, each containing 50 g Borden aquifer material (Core MC-3,4) and 125 mL Borden groundwater, amended with toluene and/or sodium benzoate, as necessary. Sterile controls were sterilized by autoclaving and poisoned with sodium azide (1% w/v). The microcosms were sealed with steam-sterilized mininert™ caps (Dynatech Precision Sampling Corp., Baton Rouge, LA) to permit repeated sampling. Six milliliters of the aqueous phase were periodically removed for analysis (2 mL for benzoate, 4 mL for toluene analysis). Benzoate analyses were conducted by HPLC as described in Appendix 3. A microsolvent extraction procedure followed by gas chromatography, described by Patrick et al. (1985), was used to determine toluene levels.

A11.4.2 Results and Discussion

Results are presented in Figures A11.26a (benzoate) and A11.26b (toluene) on a compound mass-per-volume basis. No benzoate data appear for the 1X condition in Figure A11.26a because all the benzoate had degraded by the first sampling time. Benzoate analyses were not conducted after day 6 because benzoate levels were below the MDL (about 2 mg/L) in all active microcosms (including the 1X benzoate only and 27X benzoate only conditions, which are not depicted in the figure).

Focusing on Figure A11.26b, it is apparent that benzoate level does indeed influence toluene biodegradation, which was delayed considerably under "standard" (i.e., 27X) conditions, compared to the rate when toluene only is provided. Interestingly, toluene degradation is actually stimulated in the presence of 1X benzoate and 2X benzoate. One can hypothesize that this occurs because benzoate-derived biomass contributes to toluene degradation. In the presence of 10X benzoate, toluene degradation was not similarly stimulated, but roughly equivalent to what occurred under the toluene-only condition, indicating that at this ratio of the compounds, benzoate was more likely to be utilized. The pattern of results suggests that the "preference" for benzoate is most strongly related to the amount of each substrate available to the degrader population. It is probably not due to a catabolite repression type of mechanism, where competing substrates are used sequentially because one metabolic pathway is suppressed in the presence of a more easily degraded substrate. Indications here are that the two substrates are used simultaneously.

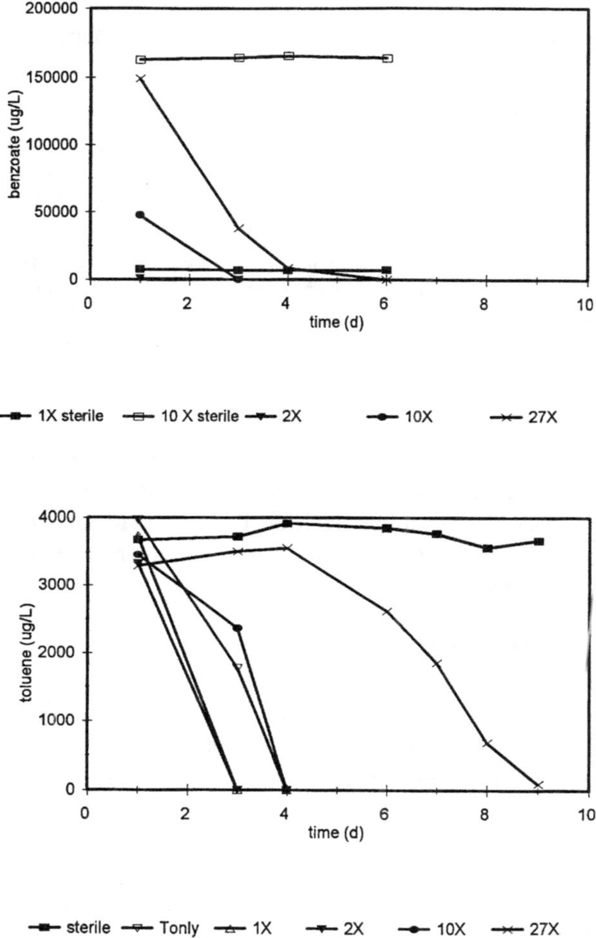

Figure A11.26a Benzoate- and toluene-amended microcosms — (top) benzoate, (bottom) toluene.

It is also clear that benzoate was metabolized far more rapidly than toluene. Estimates of degradation rates under each of the conditions examined (expressed as a maximal rate obtained from the linear portion of each degradation curve) are summarized in Table A11.4.

The results of the experiment suggest that toluene persistence may occur in the aerobic zone of Gate 3, in areas where benzoate is present in excess. On the plus side, however, degradation of the two compounds is not mutually exclusive. When present in near-equimolar amounts, benzoate may actually enhance toluene degradation by rapidly increasing the size of the degrader population, assuming that inorganic nutrient limitations do not become a controlling factor.

A11.5 CONTAMINANT DEGRADATION IN AEROBIC ALAMEDA GROUNDWATER MICROCOSMS

A simple microcosm experiment was conducted to confirm the potential for contaminant biodegradation in the oxygenated portion of the Alameda treatment gate, by proving the existence

Table A11.4 Degradation of Toluene and Benzoate

Condition	Toluene	Benzoate[a]
T only	1,260 μg/L/d	
1X benzoate only		>7,700 μg/L/d[b]
27X benzoate only		44,300 μg/L/d
1X	>1,870 μg/L/d[b]	>7,700 μg/L/d[b]
2X	1,670 μg/L/d	14,400 μg/L/d
10X	1,070 μg/L/d	25,400 μg/L/d
27X	723 μg/L/d	48,200 μg/L/d

[a] Where necessary, a "time 0" value of 7700, 15400, or 77000 μg benzoate per liter was assumed for the 1X, 2X, and 10X conditions (based on benzoate content in the sterile 1X control at time 0).

[b] Rate cannot be estimated because compound was totally degraded before the first sampling event (benzoate), or between the first and second sampling event (toluene).

of biologically mediated contaminant loss under controlled laboratory conditions. This work supports field-data-based calculations that indicate contaminant loss across the biosparge zone cannot be due to volatilization alone, but must in part be attributed to bioattenuation (Chapter 5). Toluene was included in the experiment as a representative aromatic hydrocarbon, since it may eventually break through into the biosparge zone.

A11.5.1 Methods and Materials

Sacrificial microcosms were prepared in triplicate for each sampling event, in 60-mL glass hypovials. Each vial contained 50 mL of contaminant-containing groundwater plus 1 mL of inoculum, and was sealed with a teflon-faced septum and crimp seal. The headspace within each vial ensured oxygen in excess, resembling biosparge zone conditions. Sterile controls contained autoclaved inoculum and were also poisoned with Na azide (0.1% w/v).

The groundwater used in microcosm preparation was collected from the row 6 fully screened wells on October 1, 1997 and shipped to UW where it was stored at 4°C until required. Levels of target organics (*c*DCE, VC) in the groundwater proved very low, so that it was necessary to amend the water with these, as well as with toluene. An initial target concentration of 500 μg/L of each compound was chosen — a level that was similar to "high-end" breakthrough data recorded in row 5 for *c*DCE and VC (Chapter 5), and also analytically convenient. Groundwater was amended with the pure chemicals, mixed, and then distributed among the hypovials, so that some losses, particularly of VC, were expected during the setup procedure.

At the time of groundwater collection, a number of "bioballs" were also collected from the biosparge zone. These were placed in Ziploc plastic bags along with enough groundwater to keep the material moist, and shipped on ice to UW. In the laboratory, the bags were shaken to dislodge the solids adhering to the plastic balls into the groundwater. The solids–groundwater mixture was then aseptically collected and pooled. This thick, brown-colored slurry constituted the inoculum used in the experiment. Brief microscopic examination confirmed the presence of living microorganisms in the inoculum — bacterial cells and protozoan cells were observed by phase contrast microscopy.

A11.5.2 Results and Discussion

The mean values of data from triplicate active and control microcosms are tabulated in Table A11.5. Toluene biodegradation was very rapid, as no toluene remained in the active microcosms by the time of initial sampling. VC was also rapidly biotransformed to levels below detection. However, *c*DCE was only partially biotransformed, with little concentration change evident after day 15. The latter compound loss may have been a cometabolic reaction supported by toluene. In the absence of further available growth substrate (i.e., more toluene), such cometabolism would cease. Cometabolic biotransformation of *c*DCE by toluene-oxidizing bacteria has been documented (Wackett and Gibson, 1988). Under aerobic conditions VC may serve as a growth substrate (Hartmans and de Bont, 1992). It is also potentially subject to cometabolism. Although the toluene-degrading strain studied by Wackett and Gibson did not cometabolize VC, this reaction is carried out by other oxygenase-possessing bacteria (e.g., the methanotrophs; Fogel et al., 1986), so it is possible that toluene-oxidizing strains able to cometabolize VC also exist. The experiment conducted does not allow distinction between these two routes of VC biotransformation. *t*DCE, apparently present in the groundwater at background levels, did not undergo detectable biotransformation during the experiment.

Table A11.5 Biotransformation in Aerobic Alameda Groundwater Microcosms

Time (d)	Toluene (μg/L)	*c*DCE (μg/L)	*t*DCE (μg/L)	VC (μg/L)
		Sterile control		
1	489	482	12	372
8	497	492	7	350
15	501	501	8	350
27	482	484	4	335
		Active		
1	<mdl	411	11	292
8	<mdl	316	10	6
15	<mdl	270	7	<mdl
27	<mdl	250	9	<mdl

Note: Values are the mean from triplicate microcosms.

A11.6 ACKNOWLEDGMENTS

Excellent technical assistance in the field was provided by Paul Johnson. The technical assistance of Shirley Chatten, Kimberley Hamilton, John Tiffin, and Marianne VanderGriendt in the laboratory, and Michaye McMaster's assistance in sample collection and her handling of the logistics, are also greatly appreciated.

A11.7 REFERENCES

Allen-King, R.M., J.F. Barker, R.W. Gillham, and B.K. Jensen. 1994. Substrate- and nutrient-limited toluene biotransformation in sandy soil. *Environ. Toxicol. Chem.* 13: 693-705.

Atlas, R.M. 1997. *Principles of Microbiology*, 2nd ed., Wm. C. Brown Publishers, Dubuque, IA.

Barbaro, J.R., J.F. Barker, L.A. Lemon, and C.I. Mayfield. 1992. Biotransformation of BTEX under anaerobic denitrifying conditions: field and laboratory observations. *J. Contam. Hydrol.* 1: 245-272.

Barbaro, S.E., H.-J. Albrechtsen, B.K. Jensen, C.I. Mayfield, and J.F. Barker. 1994. Relationships between aquifer properties and microbial populations in the Borden aquifer. *Geomicrobiol. J.* 12: 203-219.

Barker, J.F., G.C. Patrick, and D. Major. 1987. Natural attenuation of aromatic hydrocarbons in a shallow sand aquifer. *Ground Water Monit. Rev.* 7: 64-71.

Brunner, W., D. Staub, and T. Leisinger. 1980. Bacterial degradation of dichloromethane. *Appl. Environ. Microbiol.* 40: 950-958.

Butler, B.J., M. VanderGriendt, and J.F. Barker. 1992. Impact of methanol on the biodegradative activity of aquifer microcosms. SETAC 13th Ann. Meeting, November 8-12, 1992, Cincinnati, OH.

Criddle, C.S. and P.L. McCarty. 1991. Electrolytic model system for reductive dehalogenation in aqueous environments. *Environ. Sci. Technol.* 25: 973-978.

Criddle, C.S., J.T. DeWitt, and P.L. McCarty. 1990. Reductive dechlorination of carbon tetrachloride by *Escherichia coli* K-12. *Appl. Environ. Microbiol.* 56: 3247-3254.

deBruin, W.P., J.J. Kotterman, M.A. Posthumus, G. Schraa, and A.H.B. Zehnder. 1992. Complete biological reductive transformation of tetrachloroethene to ethane. *Appl. Environ. Microbiol.* 58: 1996-2000.

Deckhard, L.A., J.C. Willis, and D.B. Rivers. 1994. Evidence for the aerobic degradation of tetrachloroethylene by a bacterial isolate. *Biotechnol. Lett.* 16: 1221-1224.

Devlin, J.F. 1994. Enhanced *In Situ* Biodegradation of Carbon Tetrachloride and Trichloroethylene Using a Permeable Wall Injection System. Ph.D. thesis, Department of Earth Sciences, University of Waterloo.

DiStefano, T.D., J.M. Gossett, and S.H. Zinder. 1991. Reductive dechlorination of high concentrations of tetrachloroethene to ethene by an anaerobic enrichment culture in the absence of methanogenesis. *Appl. Environ. Microbiol.* 57: 2287-2292.

DiStefano, T.D., J.M. Gossett, and S.H. Zinder. 1992. Hydrogen as an electron donor for dechlorination of tetrachloroethene by an anaerobic mixed culture. *Appl. Environ. Microbiol.* 58: 3622-3629.

Eaton, A.D., L.S. Clesceri, and Greenberg, A.E. (Eds.). 1995. *Standard Methods for the Examination of Water and Wastewater*, 19th ed., American Public Health Association, American Water Works Association, Water Environment Federation, Washington, D.C.

Edwards, E.A., L.E. Wills, M. Reinhard, and D. Grbic-Galic. 1992. Anaerobic degradation of toluene and xylene by aquifer microorganisms under sulfate-reducing conditions. *Appl. Environ. Microbiol.* 58: 794-800.

Egli, C., T. Tschan, R. Scholtz, A.M. Cook, and T. Leisinger. 1988. Transformation of tetrachloromethane to dichloromethane and carbon dioxide by Acetobacterium woodii. *Appl. Environ. Microbiol.* 54: 2819-2824.

Fetzner, S. and F. Lingens. 1994. Bacterial dehalogenases: biochemistry, genetics, and biotechnological applications. *Microbiol. Rev.* 58: 641-685.

Fogel, M.M., A.R. Taddeo, and S. Fogel. 1986. Biodegradation of chlorinated ethenes by a methane-utilizing mixed culture. *Appl. Environ. Microbiol.* 51: 720-724.

Freedman, D.L. and J.M. Gossett. 1989. Biological reductive dechlorination of tetrachloroethylene and trichloroethylene to ethylene under methanogenic conditions. *Appl. Environ. Microbiol.* 55: 2144-2151.

Freedman, D.L. and J.M. Gossett. 1991. Biodegradation of dichloromethane and its utilization as a growth substrate under methanogenic conditions. *Appl. Environ. Microbiol.* 57: 2847-2857.

Gälli, R. and T. Leisinger. 1985. Specialized bacterial strains for the removal of dichloromethane from industrial waste. *Conserv. Recycl.* 8: 91-100.

Gillham, R.W. and S.F. O'Hannesin. 1994. Enhanced degradation of halogenated aliphatics by zero-valent iron. *Ground Water.* 32: 958-967.

Hartmans, S. and J.A.M. de Bont. 1992. Aerobic vinyl chloride metabolism in *Mycobacterium aurum* L1. *Appl. Environ. Microbiol.* 58: 1220-1226.

Henson, J.M., M.V. Yates, J.W. Cochran, and D.L. Shackleford. 1988. Microbial removal of halogenated methanes, ethanes, and ethylenes in an aerobic soil exposed to methane. *FEMS Microbiol. Ecol.* 53: 1193-201.

Hutchins, S.R., G.W. Sewell, D.A. Kovacs, and G.A. Smith. 1991. Biodegradation of aromatic hydrocarbons by aquifer microorganisms under denitrifying conditions. *Environ. Sci. Technol.* 25: 68-76.

Kriegman-King, M.R. and M. Reinhard. 1992. Transformation of carbon tetrachloride in the presence of sulfide, biotite, and vermiculate. *Environ. Sci. Technol.* 26: 2198-2206.

Major, D.W., E.W. Hodgins, and B.J. Butler. 1991. Field and laboratory evidence of *in situ* biotransformation of tetrachloroethene to ethene at a chemical transfer facility in North Toronto. In Hinchee, R.E. and Olfenbuttel, R.F. (Eds.), *On-site Bioreclamation Processes for Xenobiotic and Hydrocarbon Treatments.* Butterworth-Heinemann, Boston, 147-171.

Maymó-Gatell, X., V. Tandoi, J.M. Gossett, and S.H. Zinder. 1995. Characterization of an H_2-utilizing enrichment culture that reductively dechlorinates tetrachloroethene to vinyl chloride and ethene in the absence of methanogenesis and acetogenesis. *Appl. Environ. Microbiol.* 61: 3928-3933.

Mueller, J.G., S.E. Lantz, B.O. Blattmann, and P.J. Chapman. 1991. Bench-scale evaluation of alternative biological treatment processes for the remediation of pentachlorophenol- and creosote-contaminated materials: solid-phase bioremediation. *Environ. Sci. Technol.* 25: 1045-1055.

Patrick, G.C., Ptacek, C.J., Gillham, R.W., Barker, J.F., Cherry, J.A., Major, D., Mayfield, C.I., and Dickout, R.D. 1985. The behavior of soluble petroleum product derived from hydrocarbons in groundwater. Phase I. PACE Report No. 83-3. Petroleum Association for Conservation of the Canadian Environment, Ottawa, ON.

Rasche, M.E., M.R. Hyman, and D.J. Arp. 1991. Factors limiting aliphatic chlorocarbon degradation by *Nitrosomonas europaea*: cometabolic inactivation of ammonia monooxygenase and substrate specificity. *Appl. Environ. Microbiol.* 57: 2986-2994.

Starr, R.C. and R.A. Ingleton. 1992. A new method for collecting core samples without a drilling rig. *Ground Water Monit. Rev.* 12: 91-95.

Stucki, G., R. Gälli, H. Ebersold, and T. Leisinger. 1981. Dehalogenation of dichloromethane by cell extracts of *Hyphomicrobium* DM2. *Arch. Microbiol.* 130: 366-371.

Trevors, J.T. 1996. Sterilization and inhibition of microbial activity in soil. *J. Microbiol. Methods*. 26: 53-59.

Vogel, T.M. and P.L. McCarty. 1985. Biotransformation of tetrachloroethylene to trichloroethylene, dichloroethylene, vinyl chloride, and carbon dioxide under methanogenic conditions. *Appl. Environ. Microbiol.* 49: 1080-1083.

Vogel, T.M., C.S. Criddle, and P.L McCarty. 1987. Transformation of halogenated aliphatic compounds. *Environ. Sci. Technol.* 21: 722-736.

Wackett, L.P. and D.T. Gibson. 1988. Degradation of trichloroethylene by toluene dioxygenase in whole-cell studies with *Pseudomonas putida* F1. *Appl. Environ. Microbiol.* 54: 1703-1708.

Appendix 12

Granular Iron Performance Analyses

A12.1 MONITORING PROGRAM

Detailed sampling in the vicinity of the iron cassettes was completed on March 24, June 2, July 14, and September 4, 1997 to assess the rates of chlorinated solvent degradation due to contact with granular iron. Samples were collected according to methods described in Appendix 4 and analyzed for volatile organic compounds (VOCs) and dissolved hydrocarbon gases (DHGs) using analytical methods described in Appendix 3.

A12.2 DEGRADATION RATE CALCULATIONS

A12.2.1 Theory

The reductive dehalogenation of target chlorinated compounds in contact with granular iron has been previously shown to follow an approximately first-order kinetic model (Gillham and O'Hannesin, 1994):

$$C = C_o\, e^{-kt} \qquad \text{(A12.1)}$$

where C = concentration in solution at time t
C_o = initial concentration of the organic
k = first-order rate constant
t = time

which, by rearranging and taking the natural log, becomes

$$\ln(C/C_o) = -kt \qquad \text{(A12.2)}$$

The half-life, determined as the time at which the initial concentration has declined by one half ($C/C_o = 0.5$), is calculated using

$$t_{1/2} = 0.693/k \qquad \text{(A12.3)}$$

The first-order rate constant (k) is determined as the slope of a fitted line through a semilog plot of C/C_o versus time. Values of r^2 provided an indication of how closely the first-order model fit the experimental data. In the case of two-point calculations, a perfect fit was forced ($r^2 = 1$).

A12.2.2 Assumptions Used in Calculations

At the outset of the project, abiotic reductive dechlorination using granular iron for the remediation of groundwater containing chlorinated solvents (particularly PCE and CT) was a recognized and well-documented commercial technology. The purpose of the iron performance assessment and half-life calculations included here was to document the overall effectiveness of the granular iron in a removable cassette system and to permit comparison of the calculated values with those presented in the literature. These types of calculations are typically performed for one-dimensional laboratory columns where flow systems are simple and controllable. In a field application, the situation is very different; flow is three-dimensional with the possibility of preferential flow paths that complicate the analysis.

A12.2.2.1 Groundwater Velocity and Retention Time

The main tracer test (Appendix 7) established that a groundwater velocity of approximately 20 cm/day existed in the iron cassettes. However, a subsequent tracer test completed in October 1997 (Appendix 14) indicated that flow through the cassettes had decreased significantly from that previously observed. By October 1997, flow through the cassettes was minimal, while flow through the leak between the sheet-piling wall and the outer cassette box was believed to have increased significantly. This change was possibly due to a decrease in hydraulic conductivity of the reactive media within the cassettes over time, as a result of precipitate formation on the influent screen. Another reason for this change might be that the leak worsened over time and hence became the preferred flow path over the cassettes. If so, some time between the completion of the main tracer test (completed in the cassette portion of Gate 1 in December 1996) and the cassette tracer test (October 1997), a substantial change to the flow system occurred. Several lines of evidence suggest that this change did not occur before mid-July 1997:

- Concentrations of PCE and CF at fence G106 increased slightly between January and June 1997, but between early June and mid-July 1997 these compounds were measured at relatively constant concentrations. Subsequently, PCE, which undergoes complete abiotic degradation in the granular iron, but which should experience minimal intrinsic biodegradation within this aquifer (based on Gate 2 behavior) and thus act as a nondegrading tracer through the leak, showed rapid concentration increases. This suggests that at some time following the July 14, 1997 sampling event, the flow system changed and the groundwater at fence G106 had a greater contribution from the leak rather than from flow through the iron cassettes.
- Although average concentrations of PCE and CF entering the iron cassettes (fence G1C0) were relatively constant with time, the average concentration measured at fence G1C1 (at the mid-point of C1) decreased with time. This behavior suggested an increased rate of reaction within the iron zone, or, alternatively, a decreased rate of flow through the system and consequently a longer residence time.

For the purposes of the half-life calculations presented here, the groundwater velocity of 20 cm/day, based on the main tracer test results, was used and assumed to be valid for the March, June, and July 1997 sampling events. Data from the September 4, 1997 sampling event, as they related to the performance of the iron, were disregarded.

Based on a velocity of 20 cm/day, the retention time in each of the iron cassettes was 48 hours. A pore volume through the iron portion of the gate represents approximately 6 days, making the March 24, June 2, and July 14, 1997 sampling events represent approximately 16, 28, and 35 pore volumes, respectively (accounting for plume development and transport to the cassettes). No retardation factors are considered in these estimates.

A12.2.2.2 Data Handling and Interpretation

Because of the complexity in the flow pattern, fence-averaged concentrations were considered to give the best estimates of contaminant half-lives. Each concentration value representing a monitoring fence was calculated as an arithmetic average of all measured concentrations along that fence (nondetectable concentrations taken as zero in calculations). Tables of average concentration data are included. A full description of the method used in these calculations is provided in Appendix 24.

Using average data for fences G1C0, G1C1, and G103, three concentration values along the flow path were available for C1. For C2, only influent and effluent concentrations (from fences G103 and G104) were available. Upon examination of the data to be included in the performance analysis, the following problems were identified:

- Contaminant concentrations measured at fence G1C0 were typically much lower than those measured at fence G102 (typically an order of magnitude for PCE). An open water zone exists immediately upgradient of C1 and it is within this zone that the monitoring points of fence G1C0 were located. The porosity change between the natural aquifer and open water would represent a decrease in average concentration of approximately a factor of three. Further, tracer data (Appendix 7) showed that the presence of this open water caused mixing of the highly variable plume identified at fence G102 into a more uniform plume, particularly in the transverse direction, at fence G1C1. No tracer data were collected at fence G1C0. The increase in cross-sectional area that the plume covered at fence G1C1, as compared to its coverage at fence G102, would also contribute to the decrease in concentrations observed. The PCE plume apparently behaved in a similar manner to the tracer plume.
- CT concentrations at fence G1C0 were usually at or below the limit of quantification (LOQ) due to the reasons suggested above, and also due to intrinsic degradation within the aquifer between fence G102 and the cassette system. Therefore, a half-life for CT could not be calculated. The degradation of CT produced CF at measurable concentrations at fence G102 and at concentrations of approximately 100 μg/L at fence G1C0. Half-life calculations for the chlorinated methane component of the plume were performed for CF rather than for CT.
- The low concentrations entering the first iron cassette, C1, containing 25% iron, produced effluent concentrations of chlorinated compounds typically at or below the LOQ, and often below the method detection limit (MDL) for the analyses. The two-cassette design for the granular iron technology was undertaken to ensure that measurable concentrations of the parent compounds would be observed at the midpoint of the first cassette (fence G1C1) and that breakthrough of these compounds at the downgradient side of the cassette (fence G103) would also occur. Thus, calculation of a contaminant half-life based on three sampling points could be performed. The second cassette was installed to duplicate typical field installations and to ensure complete degradation of all chlorinated compounds. Half-life calculations for C2 were not possible since influent concentrations were so low and effluent concentrations were typically below the MDL.
- In cases where average concentrations were below the LOQ (and in some cases the MDL, i.e., if a few quantifiable sample concentrations occurred at a particular fence, but once averaged with several nondetect samples, which are added as zeroes, the average then becomes less than the LOQ), the calculated average value was used in the half-life calculations. In instances where all sampling points considered in the fence average showed nondetectable concentrations, the average MDL value (Appendix 5) was used in the calculations. This complication only arose for CF data from July 14, 1997. An average MDL of 9 μg/L was used in the calculations, providing a conservative estimate of the actual half-life. The half-life reported was reported as a maximum value (shown as <value hr).
- Concentrations of PCE and CF at fence G1C1 (at the mid-point of C1) were often lower than at fence G103. However, no chemical reason was determined for this behavior. It is possible that a leak developed around the cassette due to a problem with the inflatable seal, although adequate

pressure was believed to have been maintained throughout the project and this possibility is considered unlikely. It is also possible that a zone of undetected high concentration organics passed through the iron/sand mixture and contributed to the apparent increase in concentrations at fence G103. Whatever the reason, the occurrence of the non-first-order degradation profile through the cassette brings uncertainty to the half-life calculation (slope determination gives very low r^2 values). Thus, both three-point and two-point half-lives were calculated for each sampling date. Only two-point values were normalized for comparison to literature values.

The effectiveness of the iron at degrading target compounds (assessed using calculated half-lives) varies depending upon the iron surface-area to solution-volume ratio (Gillham and O'Hannesin, 1994). To permit a comparison of calculated half-lives with literature values, half-life values were normalized to 1 m^2 iron surface area/mL solution. The area-to-volume ratio for C1 was calculated to be 0.55 m^2/mL based on a calculated porosity of 0.44 and a specific surface area for the iron of 0.7 m^2/g (measured by BET analyses), and considering that granular iron comprised only 25% of the total media weight in C1. Calculations are provided as an attachment to this appendix.

A12.2.3 Calculated Half-Life Values

Table A12.1 provides a summary of the calculated half-life values, determined using both three- and two-point methods.

Table A12.1 Degradation Rates in C1 (25% Iron)

		Concentration (μg/L)						
Sampling Date	Compound	Initial (G1C0)	Middle (G1C1)	Final (G103)	Method of Calc'n	Calc'd r^2	Calc'd Half-Life (hr)	Normalized Half-Life (hr)
24-Mar-97	PCE	71	22	26	3 pt	0.626	33.1	
					2 pt		14.2	8
	CF	87	45	59	3 pt	0.343	85.7	
					2 pt		25.2	14
2-Jun-97	PCE	47	15	15	3 pt	0.75	29.1	
					2 pt		14.6	8
	CF	86	36	48	3 pt	0.432	57.0	
					2 pt		19.1	10
14-Jul-97	PCE	38	**7**	**5**	3 pt	1	17.9	
					2 pt		9.8	5
	CF	69	<MDL (9)	17	3 pt	0.452	23.7	
					2 pt		<8.2	<5

Notes: Average velocity in C1 = 20 cm/day. Length of flow path in C1 = 40 cm. Retention time of groundwater in C1 = 48 hr. Concentrations <LOQ are not measured with certainty. The average MDL for the project (9 μg/L) used for the purpose of these calculations (only required for CF on July 14, 1997). Concentrations shown in bold are between MDL and LOQ. NA - data not appropriate for half-life calculations, since concentrations are below LOQ. Normalization factor - calculated to normalize half-life values to 1m^2 (iron surface area)/mL (solution); factor used in calculations: C1: 0.55.

Calculated half-lives for PCE degradation were very similar for the March 24 and June 2, 1997 sampling events. Half-lives calculated for July 14, 1997 data were lower, possibly because of a slight reduction in flow through the cassettes, and may not be representative of the true conditions. September 4, 1997 sampling data were discarded for this reason. CF half-lives varied considerably between sampling events; analytical uncertainty could be the cause, although a varying CT input might also have contributed to this behavior. If CT degradation produced CF in the cassettes, the calculated half-life would have been larger than the true value, although in general CT concentrations were low.

Three-point calculations were completed for C1, but resulted in low r^2 values (0.95 to 0.28) because the middle concentration point (at fence G1C1) typically did not follow the trend of declining organic concentration with distance traveled through the iron. Two-point determinations force a perfect fit to the first-order kinetic model.

Half-life estimates determined using only two sampling fences were significantly lower than the values calculated using all three sampling fences in the vicinity of C1. Higher concentrations at the downgradient fence are the reason for this result. The discussion in Section A12.2.2 provides some possible causes of this discrepancy, but no chemical reason for this behavior was determined. The two-point half-life values that were calculated using only sampling fences G1C0 and G1C1 were accepted as more realistic. The higher concentrations at fence G103 could have been the result of physical flow or heterogeneous plume effects, as described earlier. If half-life values were calculated using only data from fences G1C0 and G103, the results would be the same as for the three-point determinations, although the r^2 value would equal one, since a linear fit to the suggested kinetic model would be assumed.

A12.2.4 Normalized Half-Lives

Calculated half-life values were normalized to 1 m^2 of iron surface per milliliter of solution (Table A12.1). Only half-lives determined using two sampling fences (G1C0 and G1C1) were converted. Normalized half-lives of approximately 8 hours for PCE, and between 10 and 14 hours for CF were determined at the March and June 1997 sampling events. The July 14, 1997 half-life is not considered in the comparison with literature values that follows.

Normalized half-life values reported in the literature vary over a considerable range. Laboratory studies using pure iron, laboratory grade chemicals, and deionized water suggest that half-lives typically range from minutes to several hours for CT, CF, and PCE, with the fastest values being reported for CT and the slowest for PCE (Gillham and O'Hannesin, 1994; Matheson and Tratnyek, 1994; and others). Laboratory studies using commercial-grade and simulated groundwater report normalized half-life values approximately an order of magnitude higher than electrolytic iron studies (Gillham, 1995).

Data for field-scale installations of the granular iron technology are limited at this time, likely due to a lack of monitoring data, and the limited number of field installations. Typically, the design of permeable reactive barriers includes a factor of safety to ensure complete elimination of contaminants at the downgradient side of the wall. As mentioned above, this does not permit the calculation of a true half-life, but rather an estimate of the maximum since it is unknown at what distance the concentration was actually reduced to below detection. Obviously, however, the ability to accurately complete this calculation depends upon the monitoring network at a particular site. At full-scale field sites, the groundwater monitoring network is often widely spaced and representative data cannot be collected. Where contaminant half-lives are reported, the values are typically higher than laboratory studies, possibly due to temperature effects (it was suggested by Sensaki and Kumagai (1988) that a temperature change from 20 to 10°C would cause a 10% increase in half-life), and the presence of inorganic compounds which lead to precipitate formation (Schuhmacher, 1995). Also, complex flow systems and nonuniform contaminant distribution may cause high loading at some points and, if considered representative of the overall behavior, may lead to the calculation of erroneously high half-life values.

The normalized half-life for PCE of 8 hours calculated in this study compares well to the upper end of the range suggested by Gillham (1995) for laboratory studies using commercial-grade iron. The range quoted was 2.1 to 10.8 hours. A field study at Borden by O'Hannesin (1993) reported half-lives for PCE and TCE of 117 and 147 hours, respectively. Once normalized, these values become approximately 55 and 75 hours, respectively, much higher than the values estimated for the AATDF study. The flow environment at the O'Hannesin site was less controlled and likely caused some of the problems in interpretation and the subsequently high half-lives. Also, the commercial-grade iron used in the O'Hannesin study was very different than that used at the AATDF site, and was less reactive.

Gillham (1995) reports a half-life for CF of 4.8 hours for laboratory studies using commercial-grade iron, which is considerably faster than the values determined for this study (10 and 14 hours on the first two sampling dates). Production of CF through the iron cassette due to the degradation of CT could have caused some of this difference, although other reasons suggested above could also have contributed. No field data for CF degradation in granular iron were available at the time of writing.

A12.3 MASS BALANCE CALCULATIONS

The carbon mass balance for sampling points within the iron was calculated for March 24, June 2, and July 14, 1997 data. These calculations attempt to account for the fate of influent contaminants. Organic sampling results for VOCs and DHGs were reported by the laboratory in micrograms per liter. To permit a stoichiometric comparison of these data, all concentrations were converted to micromoles per liter. The original concentration data and the conversions are provided as an attachment to this appendix. For each sampling fence, the sum of the

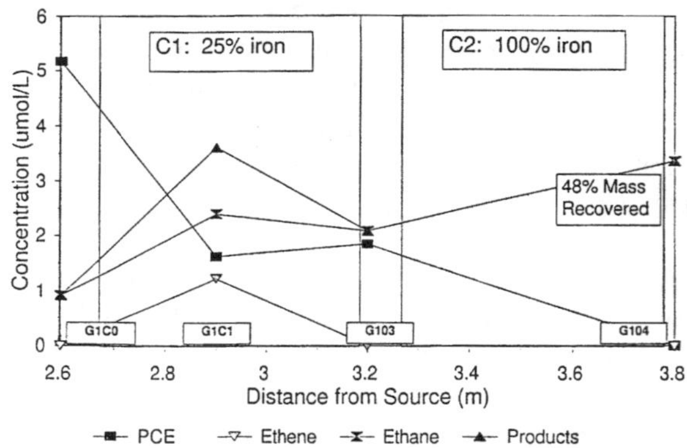

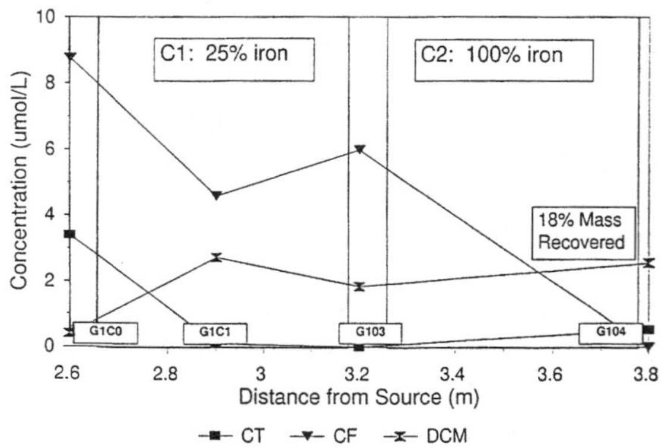

Figure A12.1a PCE and CT mass balances for March 24, 1997 data.

concentrations of each compound was determined (i.e., micromolarities for all multilevel sampling points were summed). Since each fence considered had the same number of sampling points, a direct comparison of these values with subsequent fences was possible. Plots of concentration of PCE and CT, with their respective breakdown products, versus distance traveled through the iron were prepared and are provided in Figure A12.1a, b, and c. In the case of PCE, concentrations of TCE and all isomers of DCE were below the MDL. For CT, methane was excluded from the plots since the influent concentrations were high. Also, the role in the degradation pathway was expected to be minimal in this case since DCM was likely the end product of the CT dechlorination process, given that DCM concentration profiles through the iron were relatively stable and CM was not observed.

Included in Figure A12.1 are the calculated carbon mass balances (MB), determined as

$$MB = \frac{[\text{breakdown products gained}]}{[\text{parent compounds lost}]} \times 100\%$$

where:

breakdown products gained	=	sum of the concentrations of breakdown products gained through the iron, accounting for non-zero influent concentrations (i.e., $\Sigma(C_f - C_i)$)
parent compounds lost	=	sum of the concentrations of parent compounds lost through the iron, accounting for non-zero effluent concentrations (i.e., $\Sigma(C_i - C_f)$)
I	=	initial concentration at fence G1C0
f	=	final concentration at fence G103

Carbon mass balances for PCE ranged from –5 to 48%. The maximum value was obtained for March 24, 1997 data and compares poorly with that observed in a previous study of TCE by Orth and Gillham (1996), where a maximum of 73% of the injected mass was accounted for in the identified products and only 3 to 3.5% was observed as chlorinated products. For CT, the mass balance was complicated by the intrinsic degradation of CT and the production of CF upgradient of the iron cassettes. The mass balances ranged from 18 to 28%. Losses may be attributed to unobserved CM, carbon dioxide, or other products not analyzed, or production of methane that was unquantifiable due to the elevated influent concentrations.

A combination of factors was expected to contribute to the poor mass balances even before they were attempted. To begin with, laboratory analyses were completed only for the expected breakdown products from the stepwise dechlorination of PCE and CT. However, other degradation pathways are known to exist (Roberts et al., 1996). Specific analyses for these compounds (e.g., acetylene, acetate, etc.) were not conducted, and thus a perfect mass balance could not be expected. Sorption of chlorinated solvents in the granular iron has been demonstrated in the laboratory (Burris et al., 1995) and could be acting in the cassettes. Volatile losses could be another cause of the poor mass balances. Ethane and ethene are generally reactive compounds within a groundwater system; however, the presence of the open water zones and the mixing that was observed within them would be expected to enhance these losses. Maximum concentrations of these compounds were typically not measured at fence G103 (considered as the effluent concentration), suggesting that losses either due to volatilization or breakdown to carbon dioxide were occurring along the length of the groundwater flow path. Uncertainty with respect to the changing flow regime within the cassettes could also cause some problems, allowing longer residence times within the open water zones.

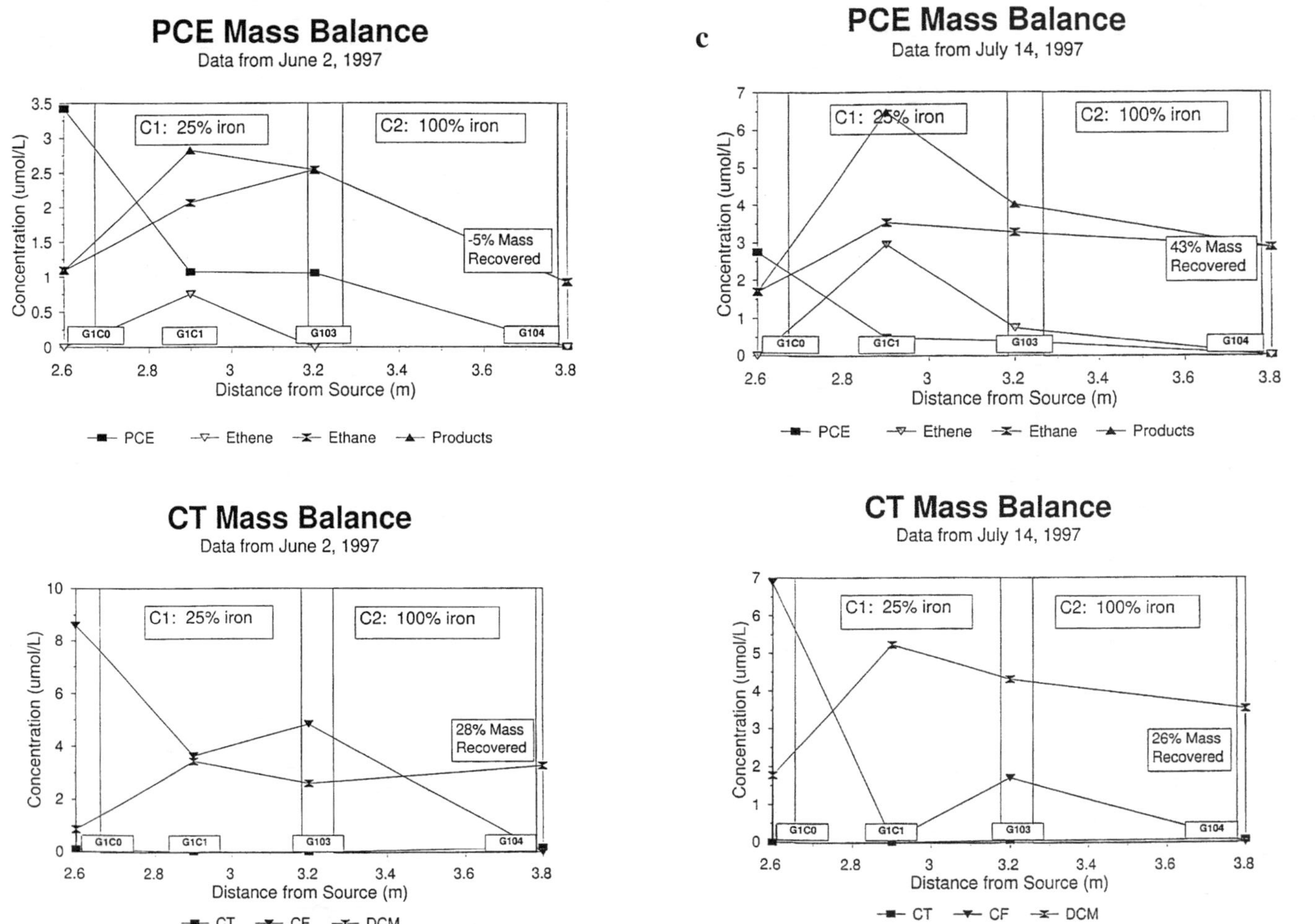

Figure A12.1 (b-c) (b) PCE and CT mass balances for June 2, 1997 data; (c) PCE and CT mass balances for July 14, 1997 data.

A12.4 REFERENCES

Burris, D.R., T.J. Campbell, and V.S. Manoranjan, 1995. Sorption of trichloroethylene and tetrachloroethylene in a batch reactive metallic iron–water system. *Environmental Science and Technology*. 29(11): 2850.

Gillham, R.W. 1995. In-situ treatment of groundwater: metal-enhanced degradation of chlorinated organic contaminants. Published in the *Proceedings of the Recent Advances in Ground-Water Pollution Control and Remediation Conference*, A NATO Advances Study Institute, Kemer, Antalya, Turkey, May 20-June 1, 1995.

Gillham, R.W. and S.F. O'Hannesin. 1994. Enhanced degradation of halogenated aliphatics by zero-valent iron. *Ground Water*. 32(6): 958.

Matheson, L.J. and P.G. Tratnyek. 1994. Reductive dehalogenation of chlorinated methanes by iron metal. *Environmental Science and Technology*. 28: 2045.

O'Hannesin, S.F. 1993. A Field Demonstration of a Permeable Reactive Wall for the In-Situ Abiotic Degradation of Halogenated Aliphatic Organic Compounds. M.Sc. Thesis, University of Waterloo, Waterloo, Ontario, Canada.

Orth, W.S. and R.W. Gillham, 1996. "Dechlorination of Trichloroethene in Aqueous Solution using Fe^0." *Environmental Science and Technology*. 30(1): 66.

Roberts, A.L., L.A. Totten, W.A. Arnold, D.R. Burris, and T.J. Campbell. 1996. Reductive elimination of chlorinated ethylenes by zero-valent metals. *Environmental Science and Technology*. 30(8): 2654.

Schuhmacher, T.T. 1995. Identification of Precipitates Formed on Zero-Valent Iron in Anaerobic Aqueous Solutions. M.Sc. Thesis, University of Waterloo, Waterloo, Ontario, Canada.

Sensaki, T. and Y. Kumagai. 1988. Removal of chlorinated organics from wastewater by reduction process: I. Treatment of 1,1,2,2-tetrachloroethane with iron powder. *Kogyo Yosui*, 357: 2.

ATTACHMENT

FENCE-AVERAGED CONCENTRATION DATA

At a particular fence, there are either 8 or 12 sampling points (4 depths, 2 or 3 piezometers).

$$\text{Average} = (C1 + C2 + C3 + \ldots + Cn)/n$$

where n = # of sampling points.

USE OF THESE AVERAGES IN HALF-LIFE CALCULATIONS

- In the average calculation, $c = 0$ for any reported concentration that is "<MDL." However, $c = 0$ cannot be used in half-life calculations (division by 0 error); so use c = "average project MDL," as specified in Appendix 4.
- If one or more concentration values were detectable (i.e., $c > 0$), the average did not equal "<MDL," but rather a nonzero value. Therefore, this value was used in calculations (even though it might be <"average project MDL" or <"average reported LOQ."

TWO-POINT HALF-LIFE CALCULATION

$$\begin{aligned}\text{Retention Time} &= \text{Flow Distance/Velocity}\\ &= 40\text{ cm/20cm/day}\\ 2\text{ day} &= 48\text{ hr}\end{aligned}$$

Calculate slope of line

$$\begin{aligned}\text{1st order rate constant, } k &= \ln \text{slope}\\ &= \log \text{slope} \times 2.30258\\ &= \frac{\log(C_1/C_0) \times 2.30258}{t_2 - t_1}\end{aligned}$$

Rearranging gives:

$$\begin{aligned}\ln(C/C0) &= -kt\\ t_{1/2} &= \ln 2/k = 0.693/k\end{aligned}$$

THREE-POINT HALF-LIFE DETERMINATION

Log slope of line calculated using regression analysis (Quattro Pro). From spreadsheet:

$$\begin{aligned}\text{"x coeff"} &= \text{log slope}\\ \text{"slope conv."} &= \text{log slope} \times 2.30258 = \ln \text{slope} = k\\ \text{"half life"} &= 0.693/\text{"slope conv."} = \ln 2/k\end{aligned}$$

CALCULATING SURFACE AREA TO VOLUME RATIO

Volume of reactive media in cassette C1, iron/sand mixture: $V = l * w * h$

w	= 0.97 m (transverse width)
l	= 0.40 m (flow length)
depth to iron	= 0.71 m, inside depth = 3.66 m
h	= 3.66 – 0.71 = 2.95 m
V	= 1.145 m^3

Saturated Volume = $V * h_{sat}/h$, where h_{sat} = saturated thickness of 2.44 m.

Therefore, saturated volume = 0.947 m^3

Mass of reactive media in C1: M = M iron + M sand
= 870 lb + 2558 lb = 1555 kg

Bulk Density: $\rho_b = M/V = 1555/1.145 = 1.358$ g/cm^3

Porosity: $\rho = 1 - \rho_b/\rho_s$ where ρ_s = particle density of 2.65 g/cm^3 for silica sand
$\rho = 1 - 1.358/2.65 = 0.44$

$$\text{Area to Volume Ratio} = \frac{\text{saturated surface area (iron)}}{\text{volume (solution)}}$$

$$\begin{aligned}\text{Saturated SA (iron)} &= \text{specific SA} * \text{saturated mass of iron} \\ &= \text{specific SA} * \text{total mass} * (h_{sat}/h)\end{aligned}$$

$$\begin{aligned}\text{Volume (solution)} &= \text{saturated volume} * \text{porosity} \\ &= \text{total volume} * (h_{sat}/h) * \text{porosity}\end{aligned}$$

$$\begin{aligned}\text{Area to Volume Ratio} &= \frac{\text{specific SA} * \text{total mass (iron)} * (h_{sat}/h)}{\text{total volume} * (h_{sat}/h) * \text{porisity}} \\ &= \frac{\text{specific SA} * \text{total mass (iron)}}{\text{total volume} * \text{porosity}}\end{aligned}$$

In this case, specific SA = 0.5 m^2/g (determined by BET)
mass iron = 395 kg
total volume = 1.145 m^2
porosity = 0.44

Area to Volume Ratio = 0.549 m^2/mL

To normalize a calculated half-life to 1 m^2/mL:

$$t_N \text{ (normalized)} = \text{area to volume ratio} * t_{1/2}$$

Average Concentration Data for March 24, 1997

Location	Distance	PCE	TCE	DCE	VC	Ethene	Ethane	Propene	Propane
G102	1.1	NS	NS	NS	NS	NS	NS	NS	NS
G1CO	2.6	71.4	<MDL	<MDL	<MDL	<MDL	2.3	**0.8**	1.0
G1C1	2.9	22.3	<MDL	<MDL	<MDL	2.9	6.0	1.6	1.0
G103	3.2	25.5	<MDL	<MDL	<MDL	<MDL	5.2	1.3	1.1
G104	3.8	nd	<MDL	<MDL	<MDL	<MDL	8.4	1.8	1.2

Location	Distance	CT	CF	DCM	Methane	Toluene
G102	1.1	NS	NS	NS	NS	NS
G1C0	2.6	43.5	87.2	**2.9**	510.8	1711.9
G1C1	2.9	**1.2**	45.4	19.2	542.1	1960.2
G103	3.2	<MDL	59.4	12.9	464.8	1839.3
G104	3.8	6.8	<MDL	18.0	359.4	1257.2

Average Concentration Data for June 2, 1997

Location	Distance	PCE	TCE	DCE	VC	Ethene	Ethane	Propene	Propane
G102	1.1	NS	NS	NS	NS	NS	NS	NS	NS
G1CO	2.6	47.3	<MDL	<MDL	<MDL	<MDL	2.7	**0.8**	**0.9**
G1C1	2.9	14.9	<MDL	<MDL	<MDL	1.8	5.2	1.7	1.2
G103	3.2	14.5	<MDL	<MDL	<MDL	<MDL	6.4	1.7	1.4
G104	3.8	<MDL	<MDL	<MDL	<MDL	<MDL	8.0	1.9	1.4

Location	Distance	CT	CF	DCM	Methane	Toluene
G102	1.1	NS	NS	NS	NS	NS
G1C0	2.6	**1.4**	85.6	**6.1**	600.0	1010.1
G1C1	2.9	<MDL	36.2	24.2	576.1	1096.5
G103	3.2	nd	48.1	18.5	528.3	939.1
G104	3.8	**2.3**	<MDL	23.1	376.9	851.7

Notes: All concentrations measured in μg/L. Distances measured in m from organic source.
<MDL — approx. 5 μg/L for VOCs; 0.2 μg/L for DHGs.
LOQ — approx. 15 μg/L for VOCs; 1 μg/L for DHGs.
— Values between MDL and LOQ shown in bold.
NS — not sampled.

Average Concentration Data for July 14, 1997

Location	Distance	PCE	TCE	DCE	VC	Ethene	Ethane	Propene	Propane
G102	1.1	220.5	<MDL	<MDL	<MDL	0.2	0.2	0.5	0.3
G1CO	2.6	38.1	<MDL	<MDL	<MDL	<MDL	4.3	1.4	1.1
G1C1	2.9	**6.7**	<MDL	<MDL	<MDL	6.9	8.9	2.8	1.7
G103	3.2	**5.5**	<MDL	<MDL	<MDL	1.9	8.9	2.5	1.5
G104	3.8	nd	<MDL	<MDL	<MDL	<MDL	7.2	1.8	1.2

Location	Distance	CT	CF	DCM	Methane	Toluene
G102	1.1	690.6	165.4	nd	2521.9	2249.5
G1C0	2.6	<MDL	68.5	12.6	672.0	1136.0
G1C1	2.9	<MDL	**0.6**	37.0	688.7	1294.5
G103	3.2	**0.7**	17.0	30.5	601.5	1037.2
G104	3.8	**0.9**	<MDL	25.0	421.3	674.0

Average Concentration Data for September 4, 1997

Location	Distance	PCE	TCE	DCE	VC	Ethene	Ethane	Propene	Propane
G102	1.1	132.5	<MDL	<MDL	<MDL	<MDL	<MDL	<MDL	<MDL
G1CO	2.6	57.9	<MDL	<MDL	<MDL	<MDL	0.2	0.6	0.2
G1C1	2.9	**3.0**	<MDL	<MDL	<MDL	<MDL	6.1	2.5	1.0
G103	3.2	**1.9**	<MDL	<MDL	<MDL	<MDL	6.4	2.3	1.1
G104	3.8	0.4	<MDL	<MDL	<MDL	<MDL	5.5	1.6	0.7

Location	Distance	CT	CF	DCM	Methane	Toluene
G102	1.1	219.2	51.7	0.5	986.3	1491.1
G1C0	2.6	2.2	65.9	20.4	983.4	990.7
G1C1	2.9	5.2	25.9	43.0	811.3	1312.0
G103	3.2	**1.0**	2.7	33.6	554.0	939.5
G104	3.8	**4.2**	3.8	32.4	307.9	686.9

Notes: All concentrations measured in µg/L. Distances measured in m from organic source.

<MDL — approx. 5 µg/L for VOCs; 0.2 µg/L for DHGs.

LOQ — approx. 15 µg/L for VOCs; 1 µg/L for DHGs.

— Values between MDL and LOQ shown in bold.

NS — not sampled.

Appendix 13

Assessment of Bioremediation Efforts in Gate 1

A13.1 ORC™ PERFORMANCE MONITORING

Monitoring was conducted to assess the release of oxygen from the ORC™ socks and its transport into the downgradient aquifer. Also, elevated pH, commonly associated with ORC™ usage (Chapman et al., 1997), was of particular concern since microbial populations may be adversely affected by a highly alkaline environment.

A13.1.1 Monitoring of the ORC™ in the Cassette System (C4)

The performance of the ORC™ installed in C4 was based on comparative DO and pH measurements taken at upgradient (G105) and downgradient (G106) piezometers immediately prior to the installation (October 8 and November 12, 1996), and frequently after the installation. Points of interest are identified in Figure A13.1. DO and pH probe measurements were completed according to methods provided in Appendix 4. Data are provided in the attached tables. Figure A13.2 shows fence-averaged DO and pH measurements at fences G105 and G106.

Measurements taken prior to the installation of the ORC™ indicated that DO concentrations were low (DO <1 mg/L) and that elevated pH (approximately 9.5), due to the granular iron located immediately upgradient in C1 and C2, existed. Over time, the ORC™ released a minimal amount of oxygen, but pH values increased slightly from about 9.5 to 10.5. The high influent pH may have prevented the oxygen release. This has been supported by preliminary laboratory experiments (details in Appendix 15). Attempts to neutralize the water through acid addition were successful in that the pH of the water entering C4 could be lowered. However, DO levels in C4 remained low, indicating that a second unknown parameter controlling oxygen release from the ORC™ may exist. Details of the acid addition experiments are also provided in Appendix 15.

A13.1.2 Monitoring of ORC™ in SW-12 to 14

On June 19, 1997, a new batch of 90 ORC™ socks was installed in the alternate oxygen addition wells (SW-12 to 14) located in the natural aquifer, downgradient of the cassette system. Monitoring of DO and pH was again completed at upgradient (G106) and downgradient (G108) piezometers and in the wells themselves immediately prior to the installation of the socks. Monitoring continued in SW-12 to 14 and in the downgradient fence on a regular basis from the time of installation until late October 1997. Figure A13.1 shows the locations of all points of interest. Monitoring data are provided in the attached tables. Figures A13.3a and A13.3b show fence-averaged DO and pH measurements, respectively, in SW-12 to 14. Similar plots for fence G108 are provided in Figures A13.4a and A13.4b.

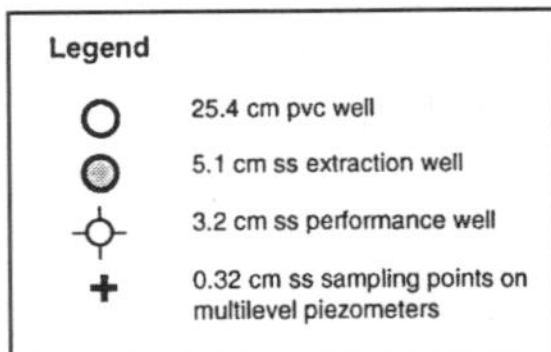

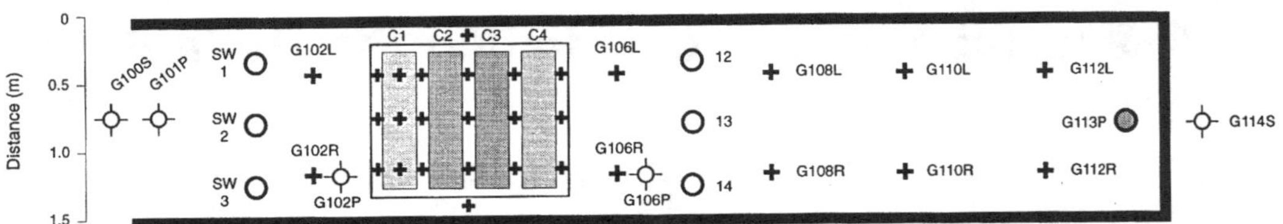

Figure A13.1 Plan view of Gate 1 showing points of interest for ORC™ monitoring.

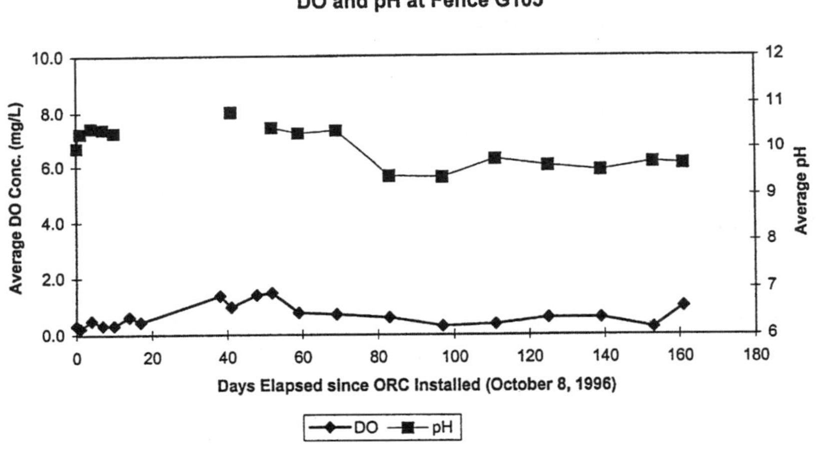

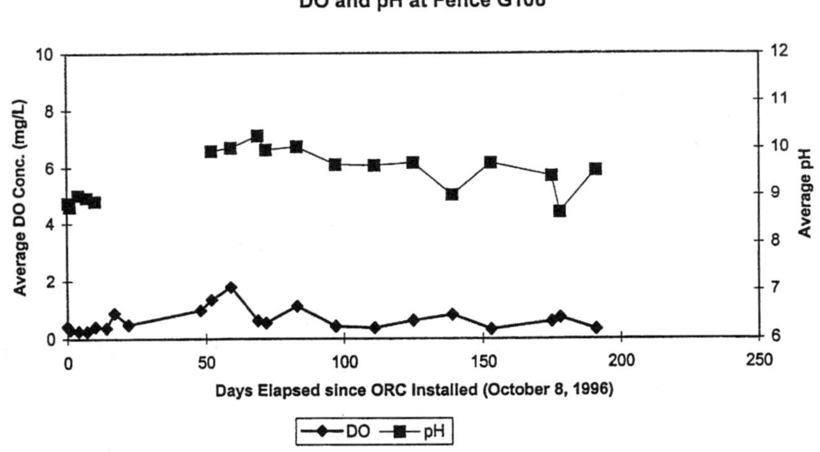

Figure A13.2 DO and pH monitoring results at fences G105 and G106 with ORC in C4.

ORC PERFORMANCE MONITORING
DO CONCENTRATIONS (mg/L)

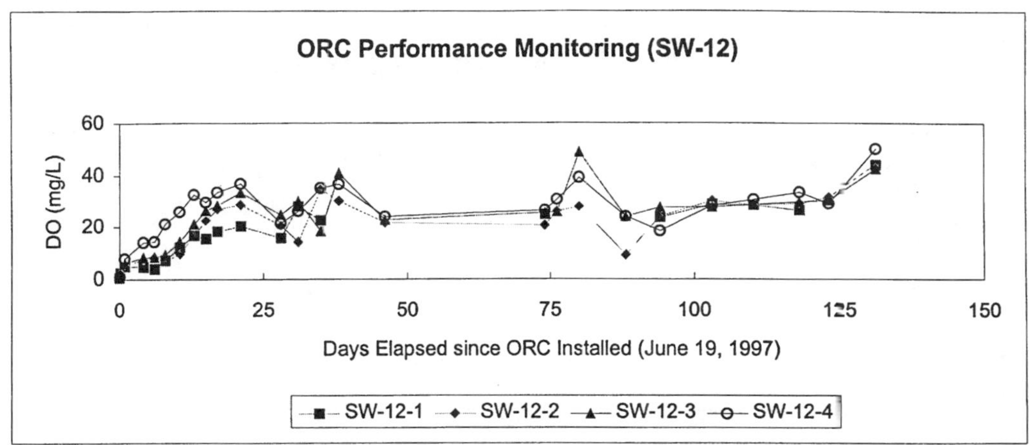

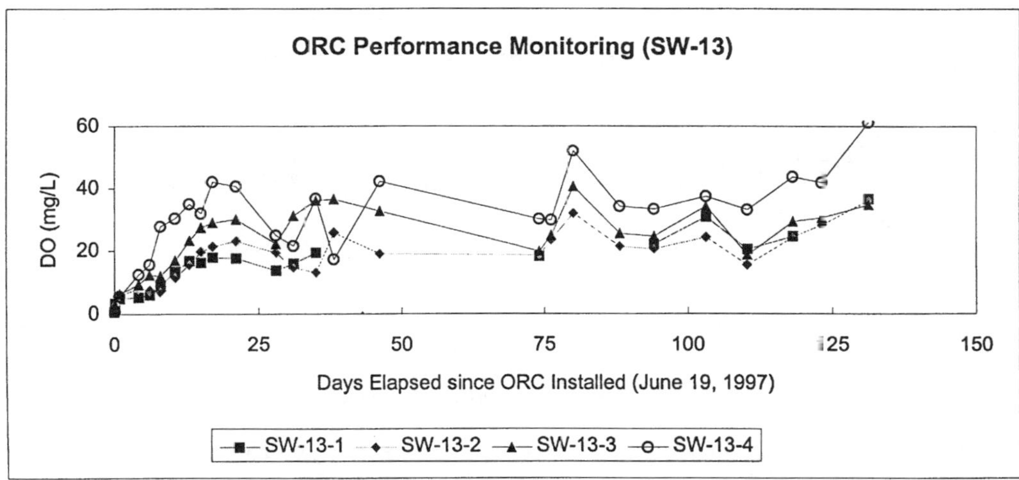

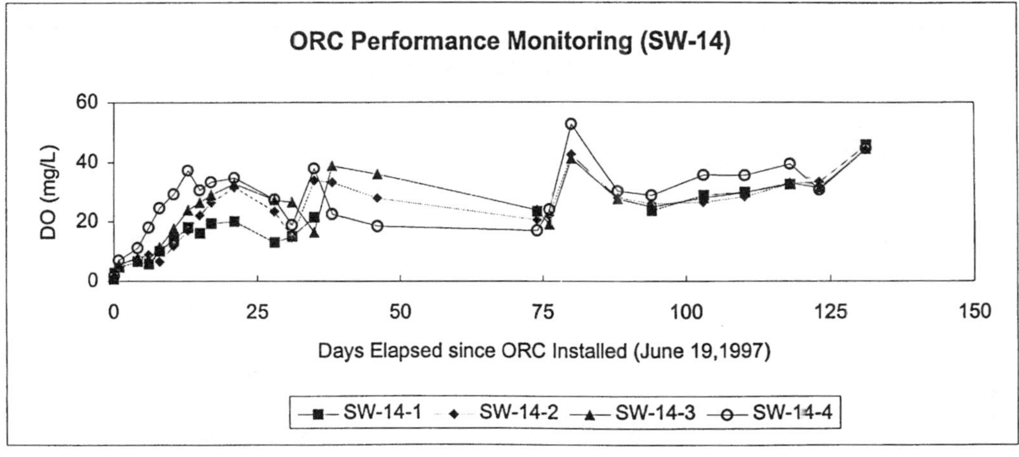

Figure A13.3a DO monitoring results in SW-12, SW-13, and SW-14 with ORC in the alternate oxygen addition wells (SWs).

ORC PERFORMANCE MONITORING

pH MEASUREMENTS

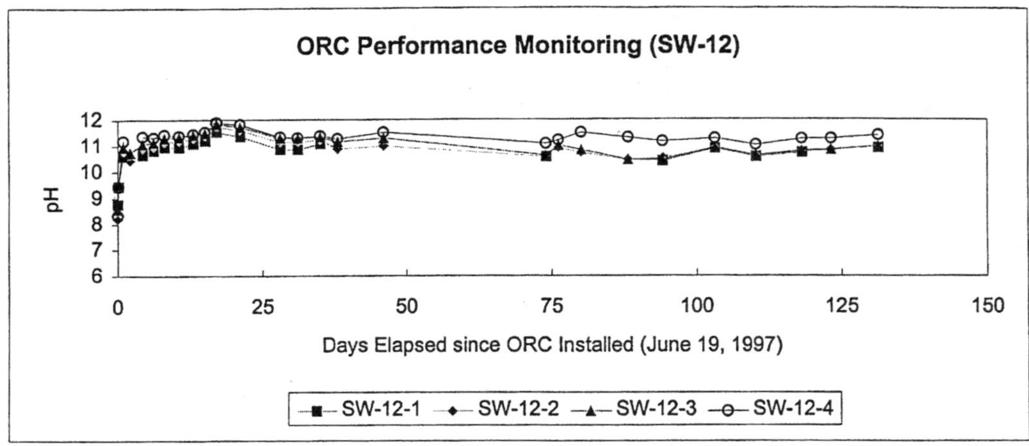

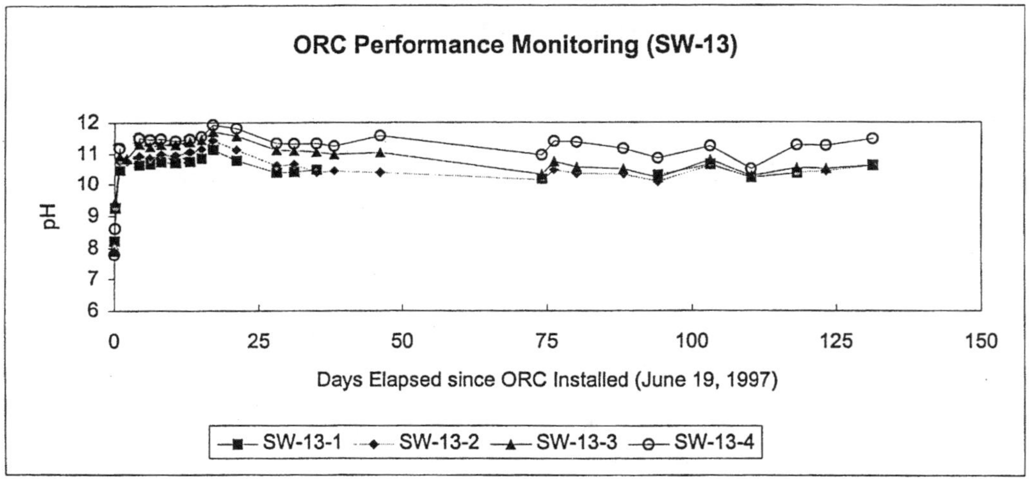

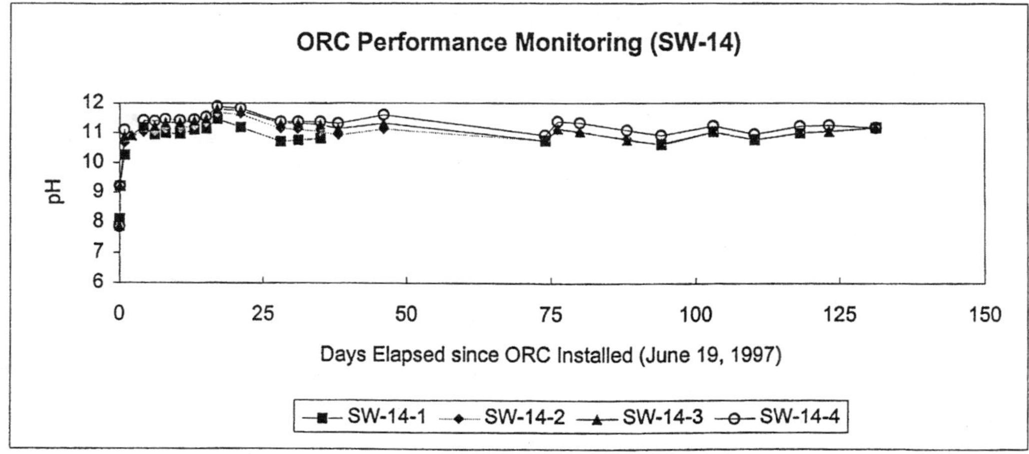

Figure A13.3b pH monitoring results in SW-12, SW-13, and SW-14 with ORC in the SWs.

ORC PERFORMANCE MONITORING
DO CONCENTRATIONS (mg/L)

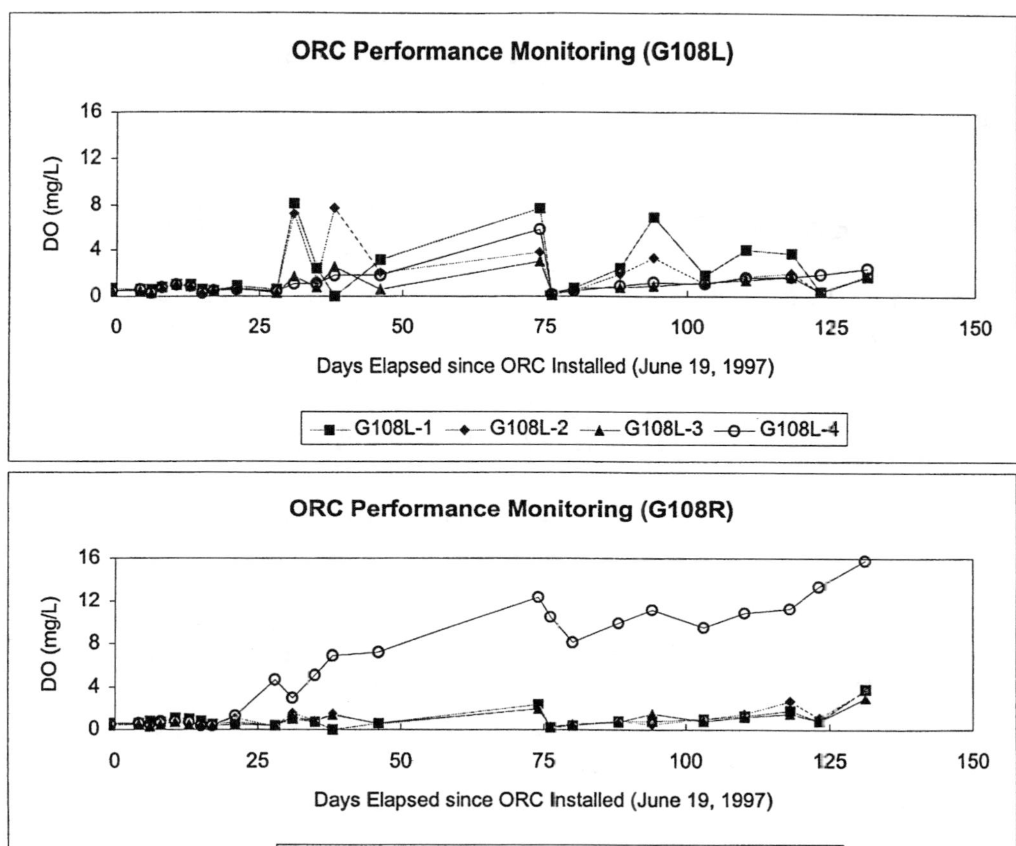

Figure A13.4a DO monitoring results at G108L and G108R with ORC in the SWs.

Prior to the installation of the socks, DO readings were less than 1 mg/L at all monitoring points and pH values ranged from 7 to 8.5. Following installation, significant increases in DO were observed in the alternate oxygen addition wells (SW-12 to 14), bringing DO concentrations from background levels to greater than 10 mg/L within 10 days. Concentrations in excess of 40 mg/L were measured, although a more moderate average concentration of 30 mg/L was sustained over the entire installation period (132 days from June 19 to October 29, 1997). pH measurements in the wells themselves were high (pH >11), but the release of oxygen from the ORC™ did not appear to decrease (high pH intolerance of the ORC™ is discussed in Appendix 15). Upgradient water at fence G106 maintained a near-neutral pH, and it was this water that contacted and activated the release of oxygen from the socks. Monitoring of the downgradient fence (G108) showed increases in pH at some points (as high as 10) only 7 days after installation, but elevated DO concentrations were not observed until after an additional 25 days. This behavior suggests that some oxygen consuming reaction, either biotic or abiotic, was at work.

DO concentrations at fence G108 (Figure A13.4a) were quite variable, indicating the existence of preferential flow paths, typical of those observed in the tracer test. At sampling point G108R-4, DO concentrations up to 15 mg/L were observed in October 1997, while at other sampling

ORC PERFORMANCE MONITORING

pH MEASUREMENTS

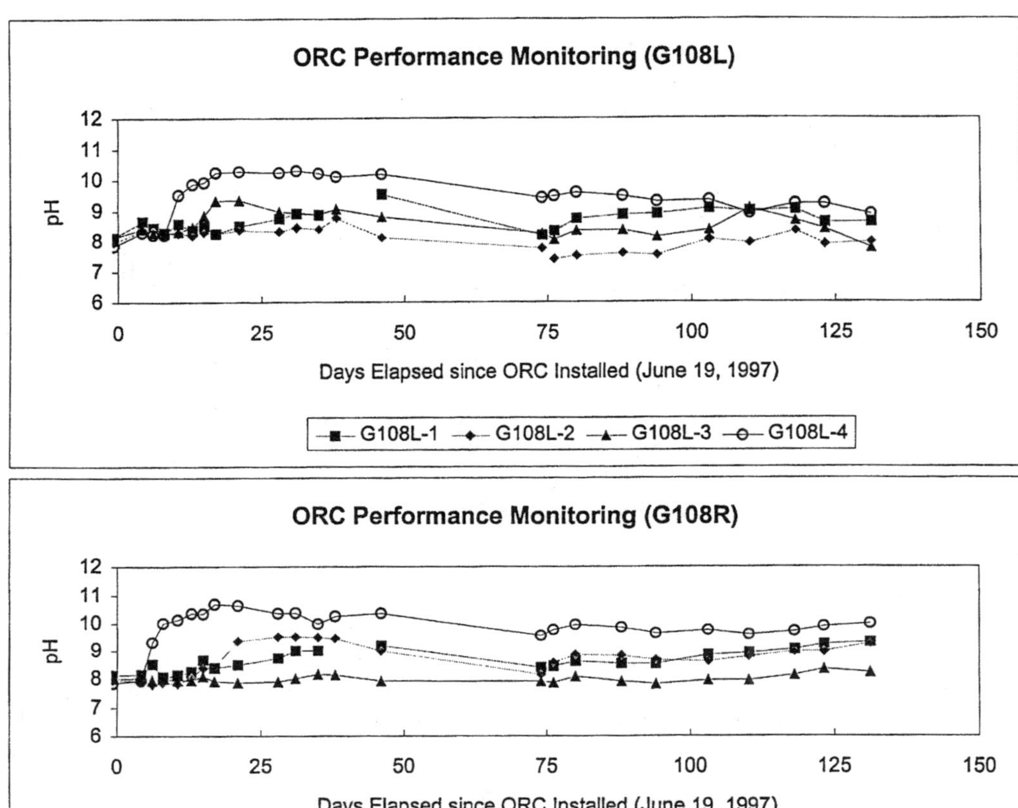

Figure A13.4b pH monitoring results at G108L and G108R with ORC in the SWs.

points, concentrations rarely exceeded 4 mg/L. These DO values are considerably lower than observed within the oxygen addition wells themselves (average between September 4 and October 29, 1997 of approximately 3 mg/L at fence G108 vs. 30 mg/L in SW-12 to 14). Several possible reasons for the observed behavior exist:

- Oxygen delivery from the alternate oxygen addition wells to the aquifer was minimal due to poor hydraulic connectivity between the wells and the groundwater flow field;
- Oxygen consumption due to the existence of inorganic sinks including reduced metal species (iron or manganese utilizes oxygen during transformation into oxidized forms), hydrogen sulfide (uses oxygen to form sulfate), etc.;
- Oxygen uptake by microbial populations for the degradation of natural organic matter; and
- Biodegradation of anthropogenic organic contaminants (aerobic toluene degradation and/or cometabolic degradation of chlorinated compounds) consumed the oxygen produced.

The lack of oxygen delivery to the aquifer from the alternate oxygen addition wells (SW-12 to 14) due to poor hydraulic connection can be ruled out on the basis of the tracer test conducted

TRACER DILUTION TESTS (SW-12 to 14)
AFTER 1st DEVELOPING

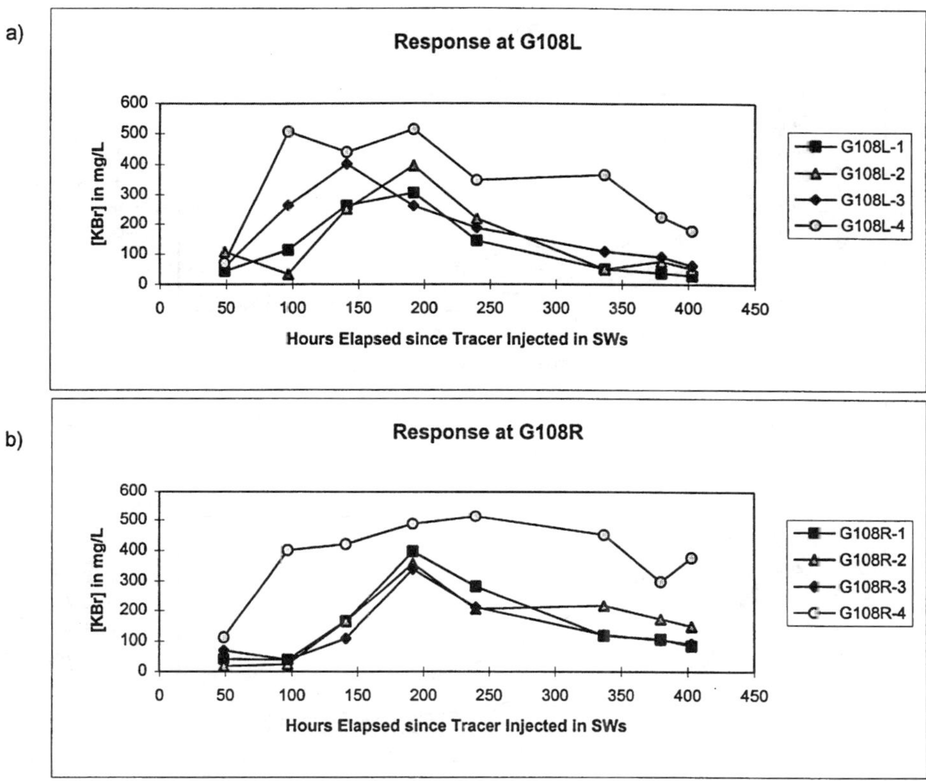

Figure A13.5 Tracer dilution test results – breakthrough curves for G108L and G108R.

as part of the well development work. The test showed that good hydraulic connection between the wells and the monitoring fence existed. Figure A13.5 provides the tracer dilution test results for comparison to DO profiles presented in Figure A13.4a. Tracer test results indicated that the standing water within SW-12 to 14 was effectively flushed within approximately 150 hours (Figure A13.6). The DO profiles for these wells (Figure A13.3a) do not show evidence of flushing. Such profiles might be expected if the ORC™ had stopped functioning and the well screens had clogged, or if replenishment of oxygen by the ORC™ occurs at the same rate as the well volume was replaced due to groundwater flow. If the first suggestion were true, a decrease in DO concentration within the oxygen addition wells themselves would likely result since equilibration with atmospheric DO would cause levels to stabilize at approximately 10 mg/L. Thus, this scenario is unlikely and some oxygen production was probably continuous. There is no way of knowing, however, if the ORC™ continued to release oxygen at the same rate as during the early stages of its life, or if slightly reduced oxygen production over time was balanced by a similar reduction in the flushing of the wells. A reduced release rate was observed by Chapman et al. (1997); however, this behavior cannot be automatically assumed for this site since the arrangement of the ORC™ socks within the wells, the geochemical conditions (particularly related to inorganic compounds), and the biological demand, both natural and anthropogenic, were different. Nonetheless, DO levels within the oxygen addition wells remained at near-saturated levels over the long term (approximately 30 mg/L), as desired.

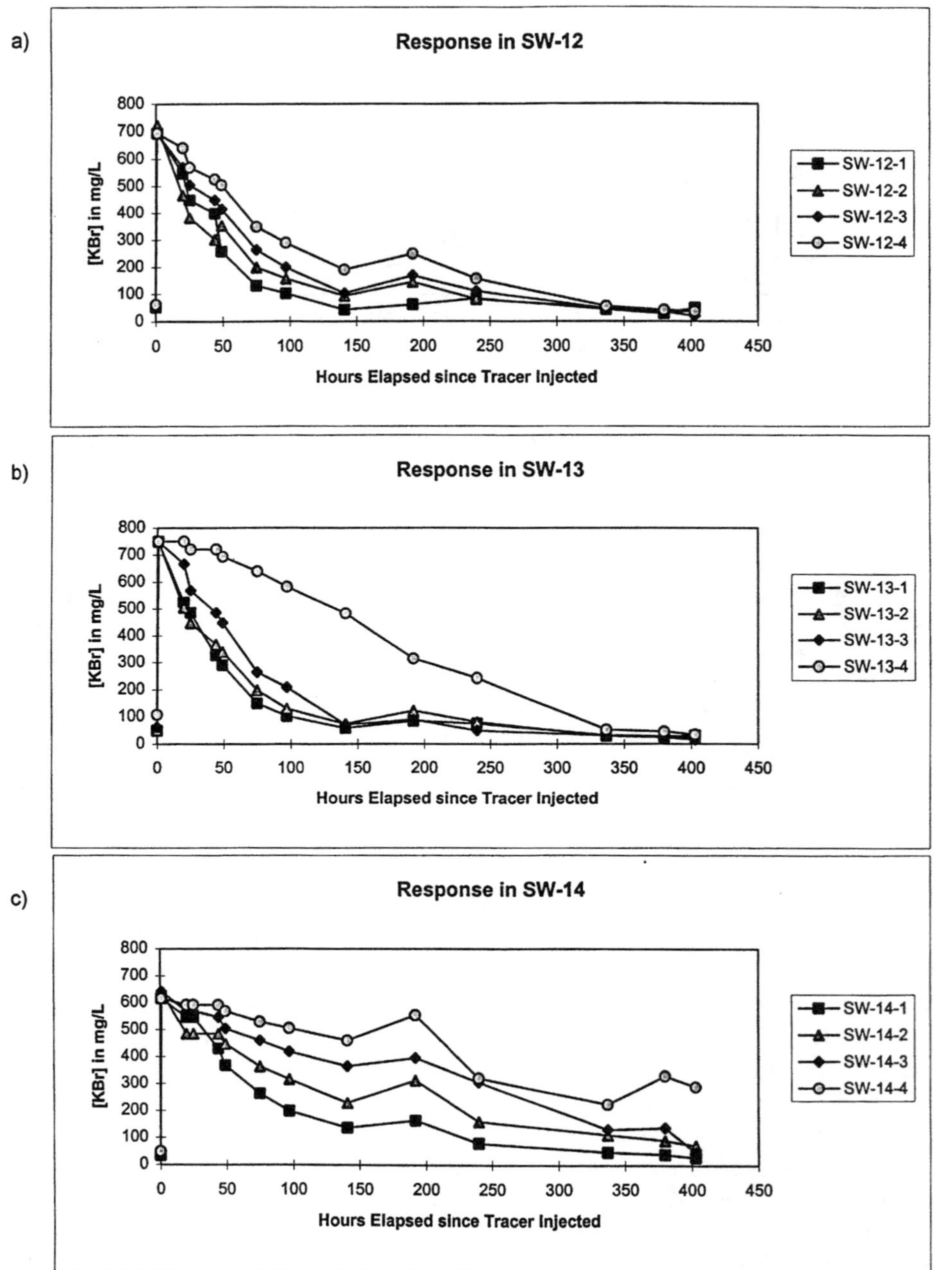

Figure A13.6 Tracer dilution testing of SW-12, SW-13, and SW-14.

Inorganic oxygen sinks would be significant if reduced metal species existed in the groundwater at significant concentrations, or if dissolved sulfide species or sulfide solids were present. Inorganic data for this site suggested that total dissolved metal concentrations were low (at fence G106, average iron concentrations were typically below 1 mg/L, except on July 14, 1997 when 8.9 mg/L was measured, and average manganese concentrations were below 0.2 mg/L). However, in a previous laboratory study using Borden sand, abiotic oxygen consumption was significant (up to 60% losses) without the presence of high dissolved metal concentrations (Reardon, personal communication). A rapid increase in redox conditions, caused by the addition of an oxygen-saturated mineral salts medium, resulted in a ferro-magnesium substitution within the biotite present in the sand and consequently the release of reduced iron species. Oxidation and precipitation of these species followed. A similar process might be expected to occur in this case. Sulfide species were not quantified, although their presence in the groundwater was indicated by the existence of the characteristic sulfide odor in water samples. Sulfate concentrations (from approximately 7 to 20 mg/L) through the aerobic portion of the gate increased, indicating that sulfate production and consequently the utilization of oxygen were likely occurring (probably a biologically driven process). If sulfide levels remain constant, the potential oxygen sink from this source should also remain relatively constant.

Consumption of oxygen for the degradation of natural organic material would be significant if the fraction of organic carbon in the aquifer materials is high. Typically, Borden sand has a very low organic content (f_{oc} = 0.021%; Ball et al., 1990). The presence of the swampy area immediately upgradient of the site could imply a slightly higher organic content in the solids, although no measurements have been made for this portion of the Borden site. Also, groundwater might have a higher dissolved organic carbon content for the same reason. The presence of methane at concentrations in the hundreds of micrograms per liter range, as observed at fence G106, could also significantly increase the overall oxygen demand.

Microbial uptake of the released oxygen within the alternate oxygen addition wells themselves or in the aquifer between these wells and fence G108 was a likely cause for the unexpectedly low DO concentrations observed. Toluene is known to degrade very rapidly at Borden under aerobic conditions (e.g., Barker et al., 1987; Allen-King et al., 1994), and as Section A13.2.1 discusses, the significant reduction in toluene concentration between fence G106 and G108 could be responsible for the low DO concentrations observed. Cometabolic degradation of the chlorinated compounds within the aerobic zone, particularly the chloromethanes, could also be a factor in these observations, as discussed in Section A13.2.2.

Occasional DO monitoring downgradient of fence G108 showed very little evidence of an increase in oxygen in the subsurface (DO concentrations typically <1 mg/L). The reasons suggested above for low DO levels also apply to this portion of the aquifer, but based on the degradation observed and subsequently discussed, some oxygen must make its way into the aquifer. It is perhaps not detected because it is rapidly consumed. Anaerobic zones may still persist within the aquifer and anaerobic microbes may also contribute to the degradation of the target organics. Nonetheless, for the purposes of the following discussion, the region in Gate 1 between fences G106 and G112, where marginally aerobic conditions were expected due to the use of the ORC™, will be referred to as the *aerobic zone*.

A13.2 BIODEGRADATION OF TARGET ORGANICS

The existence of a leak between the sheet-piling wall and the outer cassette box (Appendix 14) resulted in the presence of chlorinated compounds at concentrations above background levels within the aerobic portion of the gate. Thus, not only was toluene present, which was the original target compound for aerobic biodegradation, but also the behavior of chlorinated organics, particularly PCE and CF, within the enhanced aerobic environment could be examined (Section A13.2.2). The intrinsic anaerobic degradation of the chlorinated components, as it relates to the behavior of these compounds in the aerobic zone, was of importance and is discussed in Section A13.2.3.

A13.2.1 Aerobic Toluene Degradation

Figure A13.7 shows the fence-averaged concentrations of toluene through the aerobic zone on several sampling dates. Profiles from June 2 and June 18, 1997 show the toluene concentrations in the groundwater prior to the installation of the ORC™ within the alternate oxygen addition wells (SW-12 to 14). The relatively constant concentration profiles for these two sampling events suggest that a quasi-steady-state condition had been achieved. The profile for July 14, 1997 (25 days after ORC™ installed) also appears relatively constant, although a slight decline at fence G108 might exist. This profile was expected due to the delayed oxygen release from the ORC™ and the slow delivery of oxygen to the aquifer (increases in DO at fence G108 were not evident until 30 days after installation).

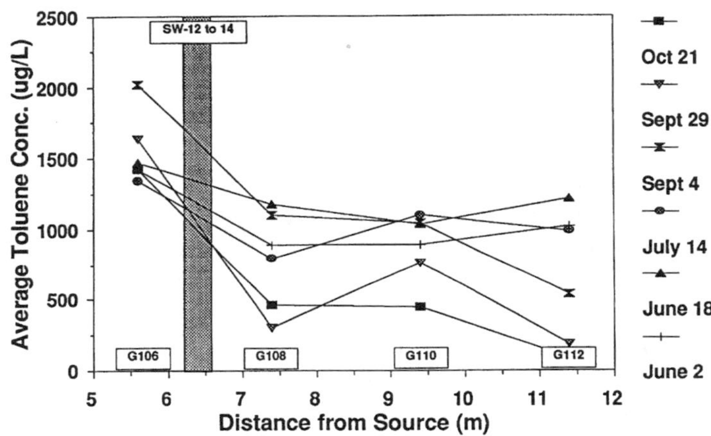

Figure A13.7 Toluene behavior in aerobic zone (sampling snapshots on days noted).

The toluene profile from September 4, 1997 (77 days after ORC™ installed) provided the first evidence of a concentration decline through the aerobic portion of the gate. In this snapshot, the fence-averaged concentration decreased from approximately 2000 to 500 μg/L in a distance of 5.8 m (between fences G106 and G112). Given that the groundwater flow velocity in the natural aquifer material is approximately 10 cm/day, the water leaving the system (at fence G112) on this sampling date is representative of what existed at the upgradient end of the aerobic zone (at fence G106) on July 14, 1997. Thus, a significant reduction in concentration was evident (from approximately 1500 μg/L at fence G106 on July 14, 1997 to approximately 500 μg/L at fence G112 on September 4, 1997).

Further evidence of toluene concentration decreases were observed for sampling events on September 29 and October 21, 1997 (102 and 124 days after ORC™ installation, respectively). Fence-averaged concentrations declined from approximately 1500 μg/L at fence G106 to less than 200 μg/L at fence G112 in both cases.

Breakthrough curves at fences G106, G108, G110, and G112 are provided in Figure A13.8 for toluene. These curves confirm the results discussed above for the various sampling events and show that although toluene input at fence G106 was relatively constant over the sampling interval, concentrations within the aerobic zone declined significantly following the installation of the ORC™ in the alternate oxygen addition wells (SW-12 to 14). The greatest decline was observed at fence G112, where average concentrations decreased from a maximum of 1200 μg/L on June

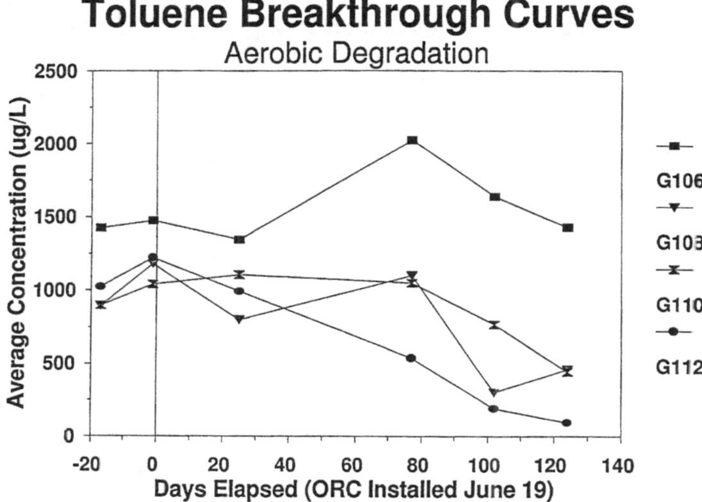

Figure A13.8 Toluene breakthrough curves for piezometers in aerobic zone of Gate 1.

18, 1997 to a minimum of 95 μg/L on October 21, 1997. Degradation at fence G108 could be limited by microbial inhibition, unsuitable geochemical conditions possibly related to the high pH of the water, or the distribution of oxygen. Given sufficient pH buffering, acclimatization time for the microbial populations and dispersion, which will be dependent upon the length of the gate, complete degradation might occur.

Complete degradation of toluene could also be limited by a lack of oxygen within the subsurface. Measurable increases in DO concentration were only observed at a few points at fence G108. For complete degradation to occur, oxygen must be delivered at rate approximately three times (stoichiometrically, assuming complete mineralization and no cell growth) that at which toluene is present (Norris et al., 1994). For an average toluene input concentration of 1500 μg/L, an average DO concentration of 4 to 5 mg/L would be required. Although the average concentration between September 2 and October 29, 1997 was approximately 3 mg/L at fence G108 (0.75 m downgradient of SW-12 to 14), the concentration that was actually delivered to the aquifer could not be measured (i.e., immediately downgradient of SW-12 to 14). Nonetheless, the DO profile was quite variable (Section A17.1.2) and complete toluene degradation across the full width of the gate did not occur. The variable nature of the toluene and DO concentration data is evident in the October 21, 1997 data (Table A13.1). A correlation between monitoring points with high DO concentration can be made. At G108R-4, where the highest DO concentrations were observed (up to 15 mg/L on October 29, 1997), the toluene concentration was below the method detection limit. At G108L-4, where DO concentrations up to 2.5 mg/L were observed, a toluene concentration of approximately 400 μg/L was measured compared to other sampling points at this piezometer where toluene levels ranged from 760 to 1260 μg/L. In general, toluene concentrations measured in piezometers along the right side of the gate were considerably lower than those on the left side. Again, this suggests some preferential flow paths, and possibly a more thorough delivery of oxygen to the right portion of the gate as compared to the left.

The influence of seasonal fluctuations of the water table on the behavior of organic contaminants is twofold. First, contaminants delivered to the aerobic zone during high water table may have been trapped within the vadose zone as the water table lowered. These contaminants could be remobilized if the water table rises again or could undergo aerobic biodegradation within the vadose zone istelf. Nonetheless, losses above the water table either temporal or permanent are insignificant when compared to acutal losses observed (typically >80%).

The second influence of a fluctuating water table involves the delivery of oxygen to the groundwater itself. If water table fluctuations are rapid, air trapped within the capillary fringe can enter the groundwater and degradation within the shallow portion of the plume might be enhanced. The assessment of the aerobic degradation of contaminants conducted at this site occurred during the time of low water table, when significant water level fluctuations were not observed. In addition, the most significant toluene (and CF) losses occurred at the deepest sampling points, while at the shallowest sampling points degradation rates appeared slower (Table A13.1 shows high contaminant concentrations at depth 1 at fences G108 and G110). As discussed above, the preferential delivery of oxygenated water from the ORC™ at depth showed good correlation to points with more rapid losses. Fluctuations in the water table did not appear to influence oxygen delivery and subsequent contaminant degradation.

Table A13.1 Toluene and DO Concentrations in Aerobic Zone on October 21, 1997 (Gate 1)

		Left Piezometer		Right Piezometer	
Sampling Fence	Depth	Toluene (μg/L)	DO (mg/L)	Toluene (μg/L)	DO (mg/L)
G106	1	1130	0.4	1230	0.4
	2	1400	0.6	1520	0.4
	3	1610	0.5	1650	0.6
	4	1550	0.8	1310	0.5
G108	1	832	0.4	248	0.8
	2	1260	0.4	12.1	1.1
	3	761	0.4	149	0.8
	4	429	1.9	<MDL	13.4
G110	1	1150	0.4	1020	0.5
	2	182	0.8	195	0.4
	3	771	0.6	15.7	0.4
	4	28.3	0.3	208	0.9
G112	1	25.2	0.4	<MDL	0.5
	2	<MDL	0.4	<MDL	0.4
	3	504	0.5	<MDL	0.5
	4	139	0.8	N/A	0.4

Notes: LOQ for toluene was 13.5–17.4 on this date. Concentrations noted as <MDL were not detected. "N/A" indicates that a sample was not analyzed.

A13.2.2 Degradation Of Chlorinated Compounds in the Aerobic Environment

The persistence of DCM through the granular iron was expected based on previous work (Gillham and O'Hannesin, 1994). Aerobic biodegradation of this compound has been suggested (Brunner et al., 1980) and the sequential treatment system implemented here was designed to study this possibility. Due to low chloromethane concentrations entering the iron, and consequently low concentrations of DCM exiting the cassettes (18 to 25 μg/L) and making its way into the aquifer, an assessment of DCM degradation was not possible.

PCE and chloroform persisted downgradient of the iron cassettes due to the leak between the sheet-piling wall and the outer cassette box. Figure A13.9 shows PCE and chloroform behavior through the length of the aerobic zone on several sampling occasions. In both plots, input concentrations (i.e., at fence G106) are relatively constant for the first three sampling events (June 2, June 18, and July 14, 1997). For PCE, the average influent concentration was approximately 80 μg/L, while for chloroform it was approximately 150 μg/L. This stable input might hint at the time when the leak around the cassettes was minimal. At later times, the greater contaminant contributions from the leak resulted in increased concentrations at fence G106 (average PCE concentrations at later time range from 127 to 177 μg/L; chloroform concentrations

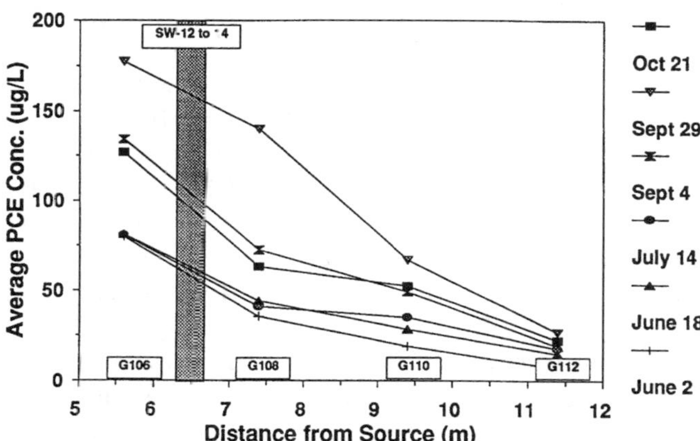

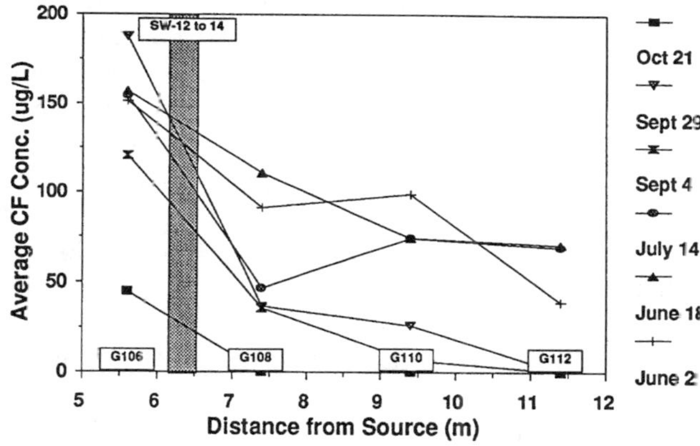

Figure A13.9 PCE and CF behavior in the aerobic zone (sampling snapshots on days noted).

were variable, but tended to decrease with time; discussed in Section A13.2.3). The constant contaminant input over the early part of the sampling interval allows for an evaluation of the degradation of these compounds, at least up to the time when the pore volume of known input exits the aerobic zone. Without accounting for retardation (thus, giving the most conservative estimate of travel time) a pore volume travels the length of the aerobic zone in approximately 60 days. Thus, groundwater arriving at fence G106 on July 14, 1997 should pass fence G112 on approximately September 14, 1997. All data within this timeframe can be considered in the degradation assessment. However, for slow-moving compounds like PCE (retardation factor, R = 2.7 to 3.9; Roberts et al., 1986), this timeframe would be extended.

In Figure A13.9, PCE concentrations appear to decline with distance; however, this observation is more likely related to slow plume development rather than the result of degradation. The relatively low input concentrations, which were due to the effectiveness of the granular iron (while the leak was minimal, seemingly until mid-July 1997), did not allow for the creation of a steady-state condition. The PCE breakthrough curves (Figure A13.10) support this hypothesis. At fence G112, average PCE concentrations remained constant through the monitoring interval, suggesting that no degradation was occurring. Additional time might bring the development of a steady-state condition or provide concrete proof of biodegradation, but further monitoring is beyond the scope of this project.

Chloroform behavior in the aerobic portion of the gate suggests some possibility for degradation. Slow plume development is unlikely in this case since retardation of chloroform was minimal in other experiments at CFB Borden (R = 1.0 to 1.2; Devlin, 1994). Concentration snapshots for chloroform (Figure A13.9) show that after the introduction of the ORC™ on June 19, 1997, the rate of contaminant loss between fence G106 and G108 increased. The losses that existed prior to this introduction were likely due to intrinsic anaerobic biodegradation (also seen in Gate 2). However, the increased rate of reaction could be due to the imposed aerobic conditions and consequent cometabolism of chloroform by toluene-degrading bacteria, as observed by McClay et al. (1996), or by methanotrophic activity (Little et al., 1988).

Looking at the concentrations of chloroform, methane, and toluene on several sampling dates through the aerobic zone (Table A13.2), trends suggest that cometabolic degradation may have occurred. In March 1997, concentrations of chloroform and toluene at fence G106 were variable, but generally high, and methane concentrations were quite constant in the 150- to 300-μg/L range. By June 1997, all compounds showed relatively stable concentrations at fence G106, with chloroform, toluene, and methane averaging approximately 150, 1500, and 800 μg/L, respectively. At downgradient fences, slight decreases in all compounds were observed due to dispersive effects, sorption, and possibly some intrinsic anaerobic losses. By September 1997, significant chloroform losses were seen at fences G108, G110, and G112, particularly at sampling points where toluene concentrations also decreased. Methane data were not collected for this sampling event. Reduced concentrations of all compounds were evident in the October 1997 data. Beyond fence G106 chloroform was not detected, and in several instances, both toluene and methane concentrations also decreased to less than 100 μg/L. Reduced methane levels were not observed downgradient of fence G108, perhaps because oxygen levels were low and toluene was the preferred substrate. These data are difficult to assess since this water reflects a changing input function (Figure A13.10; concentrations of chloroform at fence G106 decreased considerably between July 14 and September 4, 1997).

A13.2.3 Intrinsic Anaerobic Biodegradation of Target Organics

October 21, 1997 (day 335) snapshot data showed detectable concentrations of TCE and *c*DCE at several sampling points on fences G106, G108, and G110. Similar data were obtained for Gate 2 during this sampling event at fences G202 through G212. No vinyl chloride was detected. The presence of these breakdown products is likely due to intrinsic anaerobic biodegradation of PCE since TCE and *c*DCE are typically observed as a result of this process. The granular iron was unlikely involved in the partial degradation of PCE discussed here since concentrations of TCE and *c*DCE were below detection limits at fence G104, as they were throughout the entire project. Also, a suspected lack of flow through the cassettes (Appendix 14) would minimize the persistence of any breakdown products in water reentering the aquifer, since time for complete degradation within the granular iron medium would surely exist.

Intrinsic anaerobic degradation was likely responsible for the reduction in chloroform concentrations observed at fence G106 for the October 1997 sampling (Figure A13.9). Concentrations at fence G102 decreased with time, although CT supplied to the system through the source wells was believed to have increased, given that concentrations within the wells themselves were higher (Appendix 9). The increase in methane concentrations (Table A13.2) over time at fence G102 hints at the production of methane within the natural aquifer due to methanogenic conditions upgradient. Chloroform degradation would be expected in this environment. Similar behavior was observed for Gate 2, and a full discussion related to this progression is provided in Chapter 3.

Table A13.2 CF, Methane, and Toluene Behavior (Gate 1)

DATE:	10-Mar-97			02-Jun-97			04-Sep-97			21-Oct-97		
Well No.	CF (μg/L)	Methane (μg/L)	Toluene (μg/L)	CF (μg/L)	Methane (μg/L)	Toluene (μg/L)	CF (μg/L)	Methane (μg/L)	Toluene (μg/L)	CF (μg/L)	Methane (μg/L)	Toluene (μg/L)
LOQ	14.4-19.2	0.2	5.7-19.1	18.5-32.9	0.2	11.0-17.1	28.4-37.8	1.4	24.3-33.5	7.87-21.9	1.4	13.5-17.4
G102L-1	<MDL	4630	58.3	N/A	N/A	N/A	12.4	105	791	N/A	N/A	N/A
G102L-2	655	120	17840	N/A	N/A	N/A	<MDL	32.3	1660	<MDL	210	126
G102L-3	<MDL	6.5	35.6	N/A	N/A	N/A	<MDL	17.2	45.5	<MDL	8.8	<MDL
G102L-4	<MDL	10.4	9.6	N/A	N/A	N/A	<MDL	20.4	23.4	<MDL	13.4	<MDL
G102R-1	<MDL	1980	5.8	N/A	N/A	N/A	40.0	7670	291	32.3	535	490
G102R-2	<MDL	4.6	3.8	N/A	N/A	N/A	<MDL	10.3	21.2	<MDL	84	79.5
G102R-3	315	37.7	16330	N/A	N/A	N/A	361	22.0	9090	148	12.0	5890
G102R-4	<MDL	6.8	41.8	N/A	N/A	N/A	<MDL	10.0	8.9	<MDL	10.0	11.9
G106L-1	<MDL	243	489	100	370	1130	35.6	794	1140	<MDL	9790	1130
G106L-2	56.0	281	1570	183	944	1710	53.7	1450	2270	<MDL	1120	1400
G106L-3	21.8	258	803	167	756	1460	221	1250	2170	120	1060	1610
G106L-4	51.6	226	1440	196	787	1680	195	1260	2420	21.5	1420	1550
G106R-1	<MDL	263	610	110	658	1200	<MDL	851	1330	<MDL	1060	1230
G106R-2	121	311	2500	155	1000	1530	169	1450	2230	69.6	1120	1520
G106R-3	47.2	256	1040	164	712	1610	290	1130	2250	151	924	1650
G106R-4	41.1	408	841	136	800	1070	<MDL	1310	2380	<MDL	1910	1310
G108L-1	<MDL	254	351	47.6	398	750	<MDL	N/A	930	<MDL	607	832
G108L-2	28.9	259	638	104	466	933	<MDL	N/A	2410	<MDL	986	1260
G108L-3	63.8	302	1640	85.8	495	913	36.1	N/A	2020	<MDL	260	761
G108L-4	40.8	250	845	63.9	518	882	231	N/A	1120	<MDL	108	429
G108R-1	<MDL	249	308	84.8	489	860	<MDL	N/A	796	<MDL	143	248
G108R-2	43.7	263	892	131	557	977	<MDL	N/A	565	<MDL	22.3	12.1
G108R-3	26.3	164	167	114	519	973	<MDL	N/A	886	<MDL	680	149
G108R-4	36.0	289	624	99.6	524	856	18.9	N/A	101	<MDL	26.9	<MDL
G110L-1	<MDL	230	63.7	29.7	304	592	<MDL	N/A	682	<MDL	735	1150
G110L-2	32.4	222	347	83.4	433	957	<MDL	N/A	1150	<MDL	557	182
G110L-3	32.2	153	230	124	438	1040	<MDL	N/A	1360	<MDL	812	771
G110L-4	<MDL	236	13.2	120	557	953	<MDL	N/A	1210	<MDL	997	28.3
G110R-1	<MDL	296	90.7	29.2	574	469	<MDL	N/A	952	<MDL	862	1020
G110R-2	34.8	235	667	125	589	1030	<MDL	N/A	1080	<MDL	893	195
G110R-3	70.5	160	2060	167	767	1160	13.7	N/A	843	<MDL	661	15.7
G110R-4	43.4	210	506	113	513	980	38.4	N/A	1120	<MDL	103	208
G112L-1	N/A	N/A	N/A	<MDL	200	577	<MDL	N/A	187	<MDL	596	25.2
G112L-2	N/A	N/A	N/A	94.1	378	1420	<MDL	N/A	851	<MDL	756	<MDL
G112L-3	N/A	N/A	N/A	38.4	256	815	<MDL	N/A	1420	<MDL	1030	504
G112L-4	N/A	N/A	N/A	47.7	326	934	<MDL	N/A	934	<MDL	868	139
G112R-1	N/A	N/A	N/A	8.8	321	513	<MDL	N/A	125	<MDL	397	<MDL
G112R-2	58.2	174	<MDL	<MDL	387	1890	<MDL	N/A	3.2	<MDL	823	<MDL
G112R-3	25.6	217	<MDL	59.4	260	1210	<MDL	N/A	57.1	<MDL	139	<MDL
G112R-4	N/A	N/A	N/A	61.3	292	851	<MDL	N/A	726	N/A	495	N/A

Notes: - LOQ REFERS TO THE LIMIT OF QUANTIFICATION FOR THE LABORATORY ANALYSES; RANGE FOR ANALYTICAL EVENT GIVEN.
- MDL REFERS TO THE METHOD DETECTION LIMIT; CONCENTRATIONS NOTED AS "<MDL" WERE NOT DETECTED.
- "N/A" INDICATES THAT A SAMPLE WAS NOT ANALYZED.

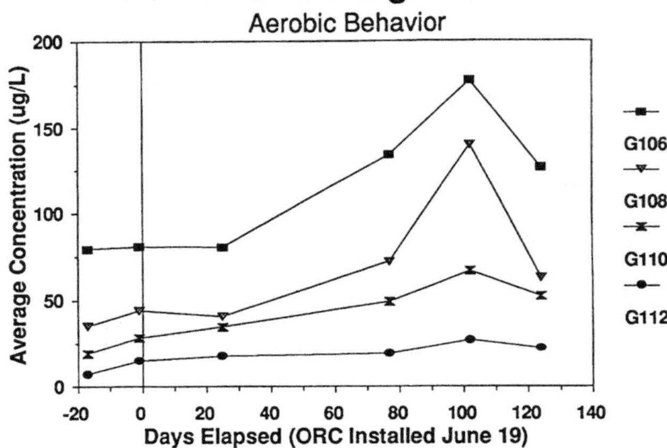

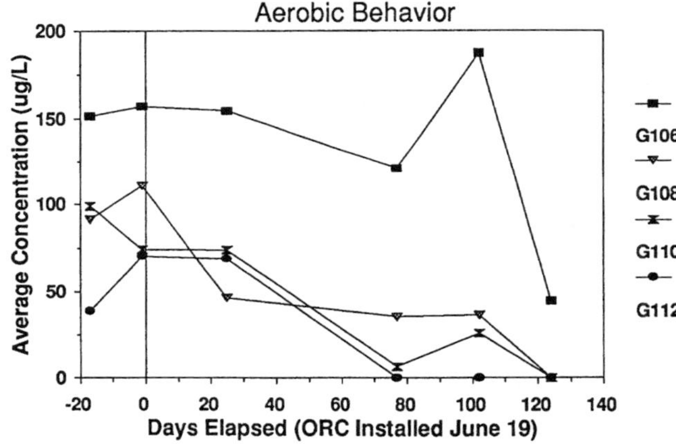

Figure A13.10 PCE and CF breakthrough curves for piezometers in the aerobic zone of Gate 1.

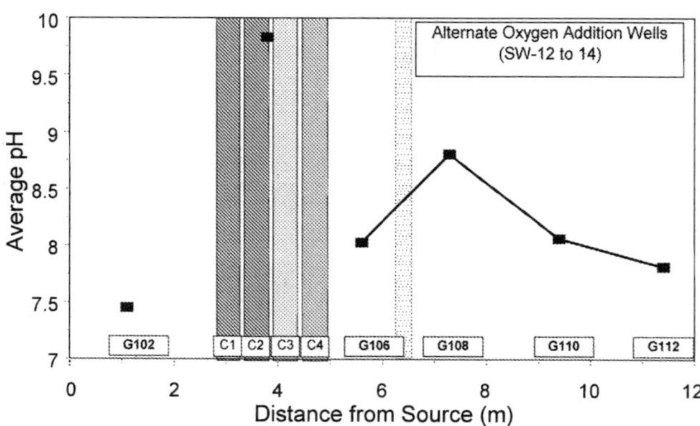

Figure A13.11 Natural pH buffering in the aerobic zone.

A13.2.4 Influence of pH on Biodegradation

The significant pH increase observed at fence G108 following ORC™ installation (Figure A13.4b) was expected based on previous experience (Smyth et al., 1995); however, its implications with respect to the behavior of microbial populations and their degradation efficiencies is unknown. Natural pH buffering by aquifer materials was expected to reduce the pH to near neutrality within a short distance from the ORC™, and thus prevent adverse impacts to the existing microbial population. Figure A13.11 shows the pH profile downgradient of the ORC™ on October 21, 1997, and demonstrates that, in fact, elevated pH levels do not persist. The desired pH range, close to that of the natural aquifer (pH 7 to 8.5), was achieved at fence G110 (within 3 m of the oxygen source).

A13.3 REFERENCES

Allen-King, R.M., J.F. Barker, R.W. Gillham, and B.K. Jensen. 1994. Substrate- and nutrient-limited toluene biotransformation in sandy soil. *Environmental Toxicology and Chemistry*, 13(5): 693.

Ball, W.P., C.H. Buehler, T.C. Harmon, D.M. Mackay, and P.V. Roberts. 1990. Characterization of a sandy aquifer material at the grain scale. *Journal of Contaminant Hydrology*, 5: 253.

Barker, J.F., G.C. Patrick, and D. Major. 1987. Natural attenuation of aromatic hydrocarbons in shallow sand aquifer. *Ground Water Monitoring Review*, Winter: 64.

Brunner, W., D. Staub, and T. Leisinger. 1980. Bacterial degradation of dichloromethane. *Appl. Environ. Microbial.* 40: 950-958.

Chapman, S.W., B.T. Byerley, D.J.A. Smyth, and D.M. Mackay. 1997. A pilot test of passive oxygen release for enhancement of in-situ bioremediation of BTEX-contaminated ground water. *Ground Water Monitoring and Remediation*. Spring, 1997: 93.

Devlin, J.F. 1994. Enhanced in-situ biodegradation of CT and TCE using a permeable wall injection system. Department of Earth Sciences, University of Waterloo, Waterloo, Ontario. pp 633.

Little, C.D., A.V. Palumbo, S.E. Herbes, M.E. Lidstrom, R.L. Tyndall, and P.J. Gilmer. 1988. Trichloroethylene biodegradation by a methane-oxidizing bacterium. *Applied and Environmental Microbiology*. 54: 951.

McClay, K., B.G. Fox, and R.J. Steffan. 1996. Chloroform mineralization by toluene-oxidizing bacteria. *Applied and Environmental Microbiology*. 62(8): 2716.

Norris, R.D. R.E. Hinchee, R. Brown, P.L. McCarty, J.T. Wilson, D.H. Kampbell, M. Reinhard, E.J. Bouwer, R.C. Borden, T.M. Vogel, J.M. Thomas, and C.H. Ward. 1994. *Handbook of Bioremediation*. Lewis Publishers, Boca Raton.

Roberts, R.V., M.N. Gotz, and D.M. Mackay. 1986. A natural gradient experiment on solute transport in a sand aquifer: III. Retardation estimates and mass balances for organic solutes. *Water Resources Research*. 22(13): 2047.

Smyth, D.J.A., B.T. Byerley, S.W. Chapman, R.D. Wilson, and D.M. Mackay. 1995. Oxygen-enhanced in-situ biodegradation of petroleum hydrocarbons in groundwater using a passive interception system. *Proceedings 5th Annual Symposium on Groundwater and Soil Remediation (GASRep)*, Toronto, Ontario, October 26, 1995.

DATA

Table A13.3a DO Monitoring Results at G105 with ORC in C4

Sampling Date	Days Elapsed	G105L 1	2	3	4	G105M 1	2	3	4	G105R 1	2	3	4
08/10/96	0	0.4	0.5	0.6	0.5	0.2	0.2	0.3	0.4	0.3	0.2	0.1	0.1
09/10/96	1	ns	ns	ns	ns	0.3	0.2	0.2	0.2	ns	ns	ns	ns
12/10/96	4	ns	ns	ns	ns	0.4	0.4	0.5	0.7	ns	ns	ns	ns
15/10/96	7	ns	ns	ns	ns	0.3	0.3	0.4	0.3	ns	ns	ns	ns
18/10/96	10	ns	ns	ns	ns	0.3	0.4	0.3	0.3	ns	ns	ns	ns
22/10/96	14	ns	ns	ns	ns	0.7	0.6	0.5	0.7	ns	ns	ns	ns
25/10/96	17	ns	ns	ns	ns	0.5	0.3	0.3	0.7	ns	ns	ns	ns
15/11/96	38	ns	ns	ns	ns	1.3	1.5	1.2	1.5	ns	ns	ns	ns
18/11/96	41	ns	ns	ns	ns	1.1	1	0.9	0.9	ns	ns	ns	ns
25/11/96	48	ns	ns	ns	ns	1.4	1.6	1.4	1.2	ns	ns	ns	ns
29/11/96	52	ns	ns	ns	ns	1.3	1.6	1.6	1.4	ns	ns	ns	ns
06/12/96	59	ns	ns	ns	ns	1.1	0.4	0.5	1.1	ns	ns	ns	ns
16/12/96	69	0.7	0.9	0.6	0.5	1.2	0.7	0.5	0.7	0.7	0.6	0.4	1.1
30/12/96	83	ns	ns	ns	ns	ns	0.6	0.6	ns	ns	ns	ns	ns
13/01/97	97	ns	ns	ns	ns	ns	0.5	0.3	ns	ns	ns	ns	ns
27/01/97	111	0.4	0.4	0.5	0.3	0.5	0.3	0.2	0.4	0.4	0.3	0.6	0.2
10/02/97	125	ns	ns	ns	ns	ns	0.5	0.6	ns	ns	ns	ns	ns
24/02/97	139	ns	ns	ns	ns	ns	0.6	0.6	ns	ns	ns	ns	ns
10/03/97	153	ns	ns	ns	ns	0.2	0.2	0.3	0.3	ns	ns	ns	ns
18/03/97	161	ns	ns	ns	ns	ns	1.2	0.8	ns	ns	ns	ns	ns

Notes: ns = not sampled.

Table A13.3b pH Monitoring Results at G105 with ORC in C4

Sampling	Days	G105L				G105M				G105R			
Date	Elapsed	1	2	3	4	1	2	3	4	1	2	3	4
08/10/96	0	10.15	10.14	10.14	10.16	9.25	9.84	10.07	10.1	10.01	10.08	10.13	10.12
09/10/96	1	ns	ns	ns	ns	10.08	10.32	10.51	10.48	ns	ns	ns	ns
12/10/96	4	ns	ns	ns	ns	10.08	10.29	10.7	10.8	ns	ns	ns	ns
15/10/96	7	ns	ns	ns	ns	10.08	10.24	10.6	10.82	ns	ns	ns	ns
18/10/96	10	ns	ns	ns	ns	10.01	10.2	10.51	10.7	ns	ns	ns	ns
22/10/96	14	ns	ns	ns	ns	ns	ns	ns	ns	ns	ns	ns	ns
25/10/96	17	ns	ns	ns	ns	ns	ns	ns	ns	ns	ns	ns	ns
15/11/96	38	ns	ns	ns	ns	ns	ns	ns	ns	ns	ns	ns	ns
18/11/96	41	ns	ns	ns	ns	10.65	10.83	10.94	10.83	ns	ns	ns	ns
25/11/96	48	ns	ns	ns	ns	ns	ns	ns	ns	ns	ns	ns	ns
29/11/96	52	ns	ns	ns	ns	10.17	10.27	10.33	11.09	ns	ns	ns	ns
06/12/96	59	ns	ns	ns	ns	10.4	9.96	10.14	10.87	ns	ns	ns	ns
16/12/96	69	10.15	10.2	10.03	10.68	10.4	10.29	10.32	10.78	10.52	10.33	10.29	10.76
30/12/96	83	ns	ns	ns	ns	ns	9.33	9.46	ns	ns	ns	ns	ns
13/01/97	97	ns	ns	ns	ns	ns	9.33	9.41	ns	ns	ns	ns	ns
27/01/97	111	9.74	9.75	9.74	9.74	9.76	9.76	9.74	9.74	9.79	9.79	9.76	9.75
10/02/97	125	ns	ns	ns	ns	ns	9.59	9.63	ns	ns	ns	ns	ns
24/02/97	139	ns	ns	ns	ns	ns	9.5	9.51	ns	ns	ns	ns	ns
10/03/97	153	ns	ns	ns	ns	9.84	9.67	9.61	9.62	ns	ns	ns	ns
18/03/97	161	ns	ns	ns	ns	9.62	9.65	9.65	9.65	ns	ns	ns	ns

Notes: ns = not sampled.

Table A13.3c Fence-Averaged DO and pH Monitoring Results at G105 with ORC in C4

Sampling Date	Days Elapsed	Ave. DO (mg/L)	Ave. pH
08/10/96	0	0.3	10.02
09/10/96	1	0.2	10.35
12/10/96	4	0.5	10.47
15/10/96	7	0.3	10.44
18/10/96	10	0.3	10.36
22/10/96	14	0.6	ns
25/10/96	17	0.5	ns
15/11/96	38	1.4	ns
18/11/96	41	1.0	10.81
25/11/96	48	1.4	ns
29/11/96	52	1.5	10.47
06/12/96	59	0.8	10.34
16/12/96	69	0.7	10.40
30/12/96	83	0.6	9.40
13/01/97	97	0.3	9.37
27/01/97	111	0.4	9.76
10/02/97	125	0.6	9.61
24/02/97	139	0.6	9.51
10/03/97	153	0.3	9.69
18/03/97	161	1.0	9.65

Notes: ns = not sampled.

Table A13.4a DO and pH Monitoring Results at G106 with ORC in C4

DO Concentrations (mg/L)

Sampling Date	Days Elapsed	G106L 1	2	3	4	G106R 1	2	3	4
10/8/96	0	0.3	0.4	0.3	0.5	0.3	0.4	0.9	0.2
10/9/96	1	0.1	0.2	0.3	0.5	0.2	0.2	0.5	0.4
10/12/96	4	0	0.4	0.4	0.8	0.4	0	0	0
10/15/96	7	0.2	0.2	0.2	0.4	0.2	0.2	0.2	0.3
10/18/96	10	0.8	0.4	0.4	0.3	0.3	0.3	0.4	0.3
10/22/96	14	0.3	0.3	0.4	0.4	0.2	0.3	0.6	0.4
10/25/96	17	0.5	0.4	0.8	0.7	1.9	0.7	1	0.9
10/30/96	22	ns	ns	ns	ns	0.4	0.5	0.6	0.4
11/25/96	48	0.8	0.6	0.9	1	1.2	1.2	1.2	0.8
11/29/96	52	1.4	1.4	1.4	1.7	0.8	1.1	1.8	1
12/6/96	59	0.4	0.3	0.6	4.3	1.4	2.2	3.3	1.7
12/16/96	69	1	0.4	0.6	0.4	ns	ns	ns	ns
12/19/96	72	1	0.5	0.4	0.4	0.3	0.3	0.6	0.7
12/30/96	83	ns	1	1.1	ns	ns	ns	ns	ns
1/13/97	97	ns	0.4	0.4	ns	ns	ns	ns	ns
1/27/97	111	0.3	0.2	0.5	0.3	0.4	0.2	0.6	0.3
2/10/97	125	ns	0.6	0.6	ns	ns	ns	ns	ns
2/24/97	139	ns	ns	ns	ns	ns	0.7	0.9	ns
3/10/97	153	0.3	0.3	0.4	0.3	0.3	0.3	0.3	0.3
4/1/97	175	ns	ns	ns	ns	0.6	0.3	0.7	0.7
4/4/97	178	ns	ns	ns	ns	ns	1	0.4	ns
4/17/97	191	0.3	0.3	0.4	0.4	0.3	0.3	0.3	0.2

pH Measurements

Sampling Date	Days Elapsed	G106L 1	2	3	4	G106R 1	2	3	4
10/8/96	0	8.91	8.84	9.33	8.81	8.84	8.32	9.33	8.32
10/9/96	1	8.95	8.9	9.36	8.82	7.87	8.49	9.26	8.36
10/12/96	4	9.07	8.99	9.41	8.88	8.74	9.02	9.38	8.49
10/15/96	7	9.04	8.96	9.32	8.9	8.8	8.83	9.26	8.47
10/18/96	10	8.97	8.78	9.18	8.66	8.86	8.94	9.2	8.37
10/22/96	14	ns	ns	ns	ns	ns	ns	ns	ns
10/25/96	17	ns	ns	ns	ns	ns	ns	ns	ns
10/30/96	22	ns	ns	ns	ns	ns	ns	ns	ns
11/25/96	48	ns	ns	ns	ns	ns	ns	ns	ns
11/29/96	52	9.85	9.72	10.25	9.97	10.35	9.66	10.39	9.38
12/6/96	59	10.38	9.93	10.44	10.31	10.03	9.35	10.31	9.35
12/16/96	69	10.14	10.32	10.25	10.35	ns	ns	ns	ns
12/19/96	72	10.14	9.9	10.29	10.35	10.21	9.12	10.32	9.42
12/30/96	83	ns	9.67	10.39	ns	ns	ns	ns	ns
1/13/97	97	ns	9.37	9.93	ns	ns	ns	ns	ns
1/27/97	111	9.89	9.75	9.38	10.1	9.22	9.6	9.18	9.94
2/10/97	125	ns	9.1	10.28	ns	ns	ns	ns	ns
2/24/97	139	ns	ns	ns	ns	ns	8.04	9.95	ns
3/10/97	153	9.91	9.4	10.2	10.06	10	8.62	9.67	9.64
4/1/97	175	ns	ns	ns	ns	10.26	8.1	9.98	9.28
4/4/97	178	ns	ns	ns	ns	ns	7.47	9.78	ns
4/17/97	191	10.13	8.79	10.2	9.67	10.33	7.95	9.88	9.24

Notes: ns = not sampled.

Table A13.4b Fence-Averaged DO and pH Monitoring Results at G106 with ORC in C4

Sampling Date	Days Elapsed	Ave. DO (mg/L)	Ave. pH
10/8/96	0	0.4	8.84
10/9/96	1	0.3	8.75
10/12/96	4	0.3	9.00
10/15/96	7	0.2	8.95
10/18/96	10	0.4	8.87
10/22/96	14	0.4	ns
10/25/96	17	0.9	ns
10/30/96	22	0.5	ns
11/25/96	48	1.0	ns
11/29/96	52	1.3	9.95
12/6/96	59	1.8	10.01
12/16/96	69	0.6	10.27
12/19/96	72	0.5	9.97
12/30/96	83	1.1	10.03
1/13/97	97	0.4	9.65
1/27/97	111	0.4	9.63
2/10/97	125	0.6	9.69
2/24/97	139	0.8	9.00
3/10/97	153	0.3	9.69
4/1/97	175	0.6	9.41
4/4/97	178	0.7	8.63
4/17/97	191	0.3	9.52

Notes: ns = not sampled.

Table A13.5a DO Monitoring Results in SW-12, SW-13, and SW-14 with ORC in the SWs

Date	Time	Days Elapsed	SW-12-1	SW-12-2	SW-12-3	SW-12-4
19-Jun-97	10:00	0.0	0.7	ns	0.9	ns
19-Jun-97	13:30	0.1	2.4	2.6	2.2	1.0
20-Jun-97	08:00	0.9	4.8	5.5	6.1	7.8
23-Jun-97	15:00	4.2	4.6	7.3	8.2	13.9
25-Jun-97	13:00	6.1	4.0	8.1	8.3	14.2
27-Jun-97	10:00	8.0	7.1	6.7	9.3	21.1
29-Jun-97	22:00	10.5	12.1	9.7	14.2	25.9
02-Jul-97	09:00	13.0	16.6	16.6	21.0	32.7
04-Jul-97	10:00	15.0	15.3	22.4	26.4	29.7
07-Jul-97	10:00	17.0	18.1	27.0	28.5	33.6
11-Jul-97	11:00	21.0	20.2	28.7	33.5	36.9
18-Jul-97	11:00	28.0	15.5	20.4	24.7	21.3
21-Jul-97	12:00	31.1	28.0	14.0	30.1	26.1
25-Jul-97	09:00	35.0	22.4	34.7	18.2	35.2
28-Jul-97	12:00	38.1	dry	30.2	40.9	36.7
05-Aug-97	12:00	46.1	dry	21.5	22.6	23.9
02-Sep-97	12:00	74.1	24.8	20.4	25.5	26.5
04-Sep-97	15:00	76.2	dry	26.1	25.8	30.8
08-Sep-97	12:00	80.1	dry	27.8	48.8	39.2
16-Sep-97	14:00	88.2	dry	9.0	23.9	24.0
22-Sep-97	12:00	94.1	23.6	24.0	27.7	18.3
01-Oct-97	13:00	103.1	28.9	30.3	27.8	28.4
08-Oct-97	16:00	110.3	28.5	28.6	28.8	30.8
16-Oct-97	14:00	118.2	26.5	29.2	29.9	33.6
21-Oct-97	16:00	123.3	ns	31.8	30.5	29.2
29-Oct-97	12:00	131.3	43.9	44.3	42.5	50.2

Date	Time	Days Elapsed	SW-13-1	SW-13-2	SW-13-3	SW-13-4
19-Jun-97	10:00	0.0	0.8	ns	0.7	ns
19-Jun-97	13:30	0.1	3.3	3.1	2.9	0.9
20-Jun-97	08:00	0.9	4.9	6.0	6.3	5.6
23-Jun-97	15:00	4.2	5.3	7.9	9.3	12.4
25-Jun-97	13:00	6.1	6.0	7.3	12.2	15.5
27-Jun-97	10:00	8.0	8.9	7.0	11.9	27.8
29-Jun-97	22:00	10.5	13.3	11.4	16.9	30.4
02-Jul-97	09:00	13.0	16.7	15.5	23.3	35.0
04-Jul-97	10:00	15.0	16.2	19.7	27.5	32.0
07-Jul-97	10:00	17.0	17.8	21.4	29.0	42.0
11-Jul-97	11:00	21.0	17.6	23.1	30.1	40.6
18-Jul-97	11:00	28.0	13.6	19.3	22.2	24.9
21-Jul-97	12:00	31.1	15.8	14.4	31.2	21.5
25-Jul-97	09:00	35.0	19.3	12.9	36.1	36.7
28-Jul-97	12:00	38.1	dry	25.8	36.5	17.2
05-Aug-97	12:00	46.1	dry	19.0	32.7	42.2
02-Sep-97	12:00	74.1	18.1	18.5	19.8	30.1
04-Sep-97	15:00	76.2	dry	23.5	24.9	29.8
08-Sep-97	12:00	80.1	dry	31.9	40.5	51.9
16-Sep-97	14:00	88.2	dry	21.3	25.5	34.1
22-Sep-97	12:00	94.1	22.1	20.5	24.6	33.3
01-Oct-97	13:00	103.1	30.8	24.3	34.2	37.4
08-Oct-97	16:00	110.3	20.4	15.4	18.6	33.1
16-Oct-97	14:00	118.2	24.6	24.4	29.3	43.5
21-Oct-97	16:00	123.3	ns	28.2	30.5	41.7
29-Oct-97	12:00	131.3	36.5	36.1	34.7	61.3

Table A13.5a (continued) DO Monitoring Results in SW-12, SW-13, and SW-14 with ORC in the SWs

Date	Time	Days Elapsed	SW-14-1	SW-14-2	SW-14-3	SW-14-4
19-Jun-97	10:00	0.0	0.7	ns	0.7	ns
19-Jun-97	13:30	0.1	2.9	2.2	2.4	1.8
20-Jun-97	08:00	0.9	4.7	5.5	5.8	7.1
23-Jun-97	15:00	4.2	6.7	7.9	8.0	11.3
25-Jun-97	13:00	6.1	5.9	9.0	7.6	18.1
27-Jun-97	10:00	8.0	10.2	6.7	11.5	24.6
29-Jun-97	22:00	10.5	14.2	12.0	17.8	29.2
02-Jul-97	09:00	13.0	18.1	17.2	23.9	37.3
04-Jul-97	10:00	15.0	16.2	22.1	26.4	30.5
07-Jul-97	10:00	17.0	19.4	26.3	29.0	33.3
11-Jul-97	11:00	21.0	20.1	31.5	32.7	34.7
18-Jul-97	11:00	28.0	13.1	23.3	27.2	27.4
21-Jul-97	12:00	31.1	15.1	14.8	26.5	18.8
25-Jul-97	09:00	35.0	21.4	33.8	16.4	37.9
28-Jul-97	12:00	38.1	dry	33.1	38.8	22.5
05-Aug-97	12:00	46.1	dry	27.8	35.9	18.5
02-Sep-97	12:00	74.1	23.3	20.5	23.8	17.0
04-Sep-97	15:00	76.2	dry	22.0	19.1	24.0
08-Sep-97	12:00	80.1	dry	42.6	41.3	52.7
16-Sep-97	14:00	88.2	dry	27.9	27.5	30.1
22-Sep-97	12:00	94.1	23.6	25.9	24.8	28.7
01-Oct-97	13:00	103.1	28.6	26.3	27.8	35.6
08-Oct-97	16:00	110.3	29.8	28.2	29.6	35.5
16-Oct-97	14:00	118.2	32.4	32.7	32.9	39.4
21-Oct-97	16:00	123.3	ns	33.3	31.3	30.6
29-Oct-97	12:00	131.3	45.8	45.8	44.6	44.7

Notes: ns = not sampled.

Table A13.5b pH Monitoring Results in SW-12, SW-13, and SW-14 with ORC in the SWs

Date	Time	Days Elapsed	SW-12-1	SW-12-2	SW-12-3	SW-12-4
19-Jun-97	10:00	0.0	8.77	8.19	8.57	8.33
19-Jun-97	13:30	0.1	9.45	9.47	9.53	9.45
20-Jun-97	08:00	0.9	10.60	10.80	10.95	11.18
21-Jun-97	12:00	2.1	ns	10.43	10.74	ns
23-Jun-97	15:00	4.2	10.66	10.89	11.07	11.36
25-Jun-97	13:00	6.1	10.82	10.96	11.13	11.32
27-Jun-97	10:00	8.0	10.95	11.14	11.29	11.42
29-Jun-97	22:00	10.5	10.94	11.18	11.33	11.37
02-Jul-97	09:00	13.0	11.08	11.29	11.44	11.45
04-Jul-97	10:00	15.0	11.19	11.37	11.48	11.53
07-Jul-97	10:00	17.0	11.53	11.73	11.84	11.89
11-Jul-97	11:00	21.0	11.35	11.60	11.72	11.81
18-Jul-97	11:00	28.0	10.86	11.13	11.30	11.33
21-Jul-97	12:00	31.1	10.85	11.15	11.30	11.30
25-Jul-97	09:00	35.0	11.06	11.21	11.33	11.37
28-Jul-97	12:00	38.1	dry	10.87	11.15	11.26
05-Aug-97	12:00	46.1	dry	10.98	11.29	11.51
02-Sep-97	12:00	74.1	10.56	10.56	10.61	11.07
04-Sep-97	15:00	76.2	dry	10.89	11.00	11.20
08-Sep-97	12:00	80.1	dry	10.71	10.82	11.49
16-Sep-97	14:00	88.2	dry	10.44	10.44	11.30
22-Sep-97	12:00	94.1	10.39	10.50	10.43	11.15
01-Oct-97	13:00	103.1	10.87	10.90	10.89	11.26
08-Oct-97	16:00	110.3	10.56	10.60	10.60	11.01
16-Oct-97	14:00	118.2	10.72	10.74	10.79	11.25
21-Oct-97	16:00	123.3	ns	10.82	10.83	11.26
29-Oct-97	12:00	131.3	10.89	10.95	10.96	11.39

Date	Time	Days Elapsed	SW-13-1	SW-13-2	SW-13-3	SW-13-4
19-Jun-97	10:00	0.0	8.22	7.98	7.90	7.79
19-Jun-97	13:30	0.1	9.26	9.25	9.44	8.61
20-Jun-97	08:00	0.9	10.46	10.74	10.95	11.19
21-Jun-97	12:00	2.1	ns	10.76	10.85	ns
23-Jun-97	15:00	4.2	10.64	10.91	11.32	11.51
25-Jun-97	13:00	6.1	10.67	10.86	11.24	11.46
27-Jun-97	10:00	8.0	10.74	10.98	11.31	11.49
29-Jun-97	22:00	10.5	10.71	10.94	11.30	11.42
02-Jul-97	09:00	13.0	10.75	11.06	11.39	11.47
04-Jul-97	10:00	15.0	10.86	11.15	11.46	11.56
07-Jul-97	10:00	17.0	11.14	11.43	11.72	11.93
11-Jul-97	11:00	21.0	10.78	11.13	11.58	11.82
18-Jul-97	11:00	28.0	10.38	10.61	11.12	11.34
21-Jul-97	12:00	31.1	10.41	10.65	11.10	11.32
25-Jul-97	09:00	35.0	10.48	10.39	11.06	11.33
28-Jul-97	12:00	38.1	dry	10.43	10.98	11.25
05-Aug-97	12:00	46.1	dry	10.38	11.04	11.58
02-Sep-97	12:00	74.1	10.18	10.14	10.32	10.95
04-Sep-97	15:00	76.2	dry	10.47	10.74	11.40
08-Sep-97	12:00	80.1	dry	10.35	10.56	11.37
16-Sep-97	14:00	88.2	dry	10.33	10.50	11.16
22-Sep-97	12:00	94.1	10.30	10.09	10.21	10.84
01-Oct-97	13:00	103.1	10.64	10.61	10.79	11.23
08-Oct-97	16:00	110.3	10.21	10.23	10.27	10.49
16-Oct-97	14:00	118.2	10.36	10.37	10.53	11.26
21-Oct-97	16:00	123.3	ns	10.43	10.51	11.25
29-Oct-97	12:00	131.3	10.61	10.61	10.61	11.46

Table A13.5b (continued) pH Monitoring Results in SW-12, SW-13, and SW-14 with ORC in the SWs

Date	Time	Days Elapsed	SW-14-1	SW-14-2	SW-14-3	SW-14-4
19-Jun-97	10:00	0.0	8.10	7.92	7.88	7.86
19-Jun-97	13:30	0.1	9.22	9.25	9.21	9.22
20-Jun-97	08:00	0.9	10.26	10.66	10.90	11.11
21-Jun-97	12:00	2.1	ns	10.91	10.91	ns
23-Jun-97	15:00	4.2	11.18	11.02	11.24	11.42
25-Jun-97	13:00	6.1	10.94	11.04	11.22	11.40
27-Jun-97	10:00	8.0	10.99	11.14	11.35	11.46
29-Jun-97	22:00	10.5	10.97	11.15	11.35	11.42
02-Jul-97	09:00	13.0	11.10	11.24	11.41	11.45
04-Jul-97	10:00	15.0	11.14	11.35	11.48	11.54
07-Jul-97	10:00	17.0	11.47	11.69	11.81	11.89
11-Jul-97	11:00	21.0	11.18	11.62	11.74	11.83
18-Jul-97	11:00	28.0	10.71	11.14	11.32	11.37
21-Jul-97	12:00	31.1	10.75	11.11	11.32	11.37
25-Jul-97	09:00	35.0	10.80	11.04	11.27	11.37
28-Jul-97	12:00	38.1	dry	10.92	11.13	11.31
05-Aug-97	12:00	46.1	dry	11.12	11.32	11.60
02-Sep-97	12:00	74.1	10.75	10.73	10.73	10.91
04-Sep-97	15:00	76.2	dry	11.11	11.13	11.37
08-Sep-97	12:00	80.1	dry	11.01	11.03	11.33
16-Sep-97	14:00	88.2	dry	10.80	10.76	11.08
22-Sep-97	12:00	94.1	10.63	10.59	10.61	10.91
01-Oct-97	13:00	103.1	11.03	11.02	11.03	11.23
08-Oct-97	16:00	110.3	10.77	10.77	10.77	10.95
16-Oct-97	14:00	118.2	10.99	11.00	11.00	11.22
21-Oct-97	16:00	123.3	ns	11.07	11.05	11.26
29-Oct-97	12:00	131.3	11.17	11.17	11.17	11.17

Notes: ns = not sampled.

Table A13.6a DO Monitoring Results at G108L and G108R with ORC in the SWs

Date	Time	Days Elapsed	G108L-1	G108L-2	G108L-3	G108L-4
18-Jun-97	15:00	-0.8	0.7	0.5	0.5	0.5
23-Jun-97	15:00	4.2	0.6	0.7	0.5	0.6
25-Jun-97	13:00	6.1	0.6	0.4	0.3	0.3
27-Jun-97	10:00	8.0	0.8	0.7	0.8	0.8
29-Jun-97	22:00	10.5	1.0	1.1	1.0	1.0
02-Jul-97	09:00	13.0	1.0	0.9	0.9	0.9
04-Jul-97	10:00	15.0	0.6	0.4	0.4	0.3
07-Jul-97	10:00	17.0	0.5	0.6	0.6	0.5
11-Jul-97	11:00	21.0	0.9	0.8	0.7	0.6
18-Jul-97	11:00	28.0	0.6	0.4	0.3	0.5
21-Jul-97	12:00	31.1	8.1	7.2	1.7	1.1
25-Jul-97	09:00	35.0	2.4	1.3	0.8	1.1
28-Jul-97	12:00	38.1	ns	7.7	2.5	1.8
05-Aug-97	12:00	46.1	3.1	2.0	0.6	1.8
02-Sep-97	12:00	74.1	7.7	3.8	3.0	5.8
04-Sep-97	15:00	76.2	0.1	0.2	0.3	0.2
08-Sep-97	12:00	80.1	0.7	0.5	0.7	0.5
16-Sep-97	14:00	88.2	2.4	1.9	0.8	0.9
22-Sep-97	12:00	94.1	6.9	3.3	0.9	1.2
01-Oct-97	13:00	103.1	1.8	1.1	1.2	1.1
08-Oct-97	16:00	110.3	4.0	1.7	1.4	1.6
16-Oct-97	14:00	118.2	3.7	2.0	1.7	1.7
21-Oct-97	16:00	123.3	0.4	0.4	0.4	1.9
29-Oct-97	12:00	131.3	1.7	1.8	1.7	2.4

Date	Time	Days Elapsed	G108R-1	G108R-2	G108R-3	G108R-4
18-Jun-97	15:00	-0.8	0.6	0.5	0.4	0.5
23-Jun-97	15:00	4.2	0.6	0.6	0.5	0.6
25-Jun-97	13:00	6.1	0.8	0.2	0.3	0.5
27-Jun-97	10:00	8.0	0.7	0.6	0.5	0.8
29-Jun-97	22:00	10.5	1.1	0.8	0.7	0.8
02-Jul-97	09:00	13.0	1.0	0.7	0.5	0.7
04-Jul-97	10:00	15.0	0.8	0.6	0.5	0.4
07-Jul-97	10:00	17.0	0.5	0.5	0.4	0.4
11-Jul-97	11:00	21.0	0.7	1.1	0.5	1.3
18-Jul-97	11:00	28.0	0.4	0.3	0.4	4.7
21-Jul-97	12:00	31.1	1.0	1.6	1.2	3.0
25-Jul-97	09:00	35.0	0.7	0.8	0.8	5.1
28-Jul-97	12:00	38.1	ns	1.5	1.4	6.9
05-Aug-97	12:00	46.1	0.6	0.5	0.6	7.2
02-Sep-97	12:00	74.1	2.4	2.0	2.0	12.4
04-Sep-97	15:00	76.2	0.2	0.2	0.3	10.5
08-Sep-97	12:00	80.1	0.4	0.5	0.5	8.1
16-Sep-97	14:00	88.2	0.8	0.8	0.7	9.9
22-Sep-97	12:00	94.1	0.8	0.5	1.5	11.2
01-Oct-97	13:00	103.1	1.0	1.0	0.8	9.5
08-Oct-97	16:00	110.3	1.3	1.5	1.2	10.9
16-Oct-97	14:00	118.2	1.8	2.7	1.5	11.3
21-Oct-97	16:00	123.3	0.8	1.1	0.8	13.4
29-Oct-97	12:00	131.3	3.8	3.7	3.0	15.8

Note: ns = not sampled. dry = sampling point was above water table on sampling date.

Table A13.6b pH Monitoring Results at G108L and G108R with ORC in the SWs

Date	Time	Days Elapsed	G108L-1	G108L-2	G108L-3	G108L-4
18-Jun-97	15:00	-0.8	8.13	7.96	8.14	7.84
23-Jun-97	15:00	4.2	8.66	8.35	8.41	8.30
25-Jun-97	13:00	6.1	8.45	8.34	8.27	8.22
27-Jun-97	10:00	8.0	8.29	8.17	8.25	8.22
29-Jun-97	22:00	10.5	8.58	8.27	8.32	9.52
02-Jul-97	09:00	13.0	8.37	8.19	8.47	9.86
04-Jul-97	10:00	15.0	8.51	8.30	8.85	9.92
07-Jul-97	10:00	17.0	8.25	8.25	9.33	10.24
11-Jul-97	11:00	21.0	8.51	8.37	9.35	10.27
18-Jul-97	11:00	28.0	8.75	8.32	8.97	10.23
21-Jul-97	12:00	31.1	8.91	8.45	8.94	10.30
25-Jul-97	09:00	35.0	8.86	8.39	8.88	10.21
28-Jul-97	12:00	38.1	ns	8.76	9.06	10.10
05-Aug-97	12:00	46.1	9.53	8.11	8.80	10.18
02-Sep-97	12:00	74.1	8.19	7.75	8.25	9.41
04-Sep-97	15:00	76.2	8.33	7.39	8.04	9.47
08-Sep-97	12:00	80.1	8.74	7.50	8.35	9.58
16-Sep-97	14:00	88.2	8.87	7.58	8.36	9.47
22-Sep-97	12:00	94.1	8.90	7.52	8.14	9.29
01-Oct-97	13:00	103.1	9.07	8.05	8.37	9.33
08-Oct-97	16:00	110.3	8.98	7.93	9.07	8.91
16-Oct-97	14:00	118.2	9.04	8.33	8.68	9.21
21-Oct-97	16:00	123.3	8.61	7.88	8.39	9.21
29-Oct-97	12:00	131.3	8.62	7.95	7.75	8.87

Date	Time	Days Elapsed	G108R-1	G108R-2	G108R-3	G108R-4
18-Jun-97	15:00	-0.8	8.11	7.96	8.00	7.84
23-Jun-97	15:00	4.2	8.14	7.91	8.01	7.94
25-Jun-97	13:00	6.1	8.50	7.82	7.95	9.30
27-Jun-97	10:00	8.0	8.05	7.87	8.00	10.01
29-Jun-97	22:00	10.5	8.11	7.82	7.94	10.13
02-Jul-97	09:00	13.0	8.23	8.04	7.95	10.36
04-Jul-97	10:00	15.0	8.65	8.36	8.08	10.35
07-Jul-97	10:00	17.0	8.37	8.40	7.91	10.69
11-Jul-97	11:00	21.0	8.48	9.34	7.86	10.64
18-Jul-97	11:00	28.0	8.72	9.51	7.90	10.36
21-Jul-97	12:00	31.1	8.99	9.51	8.01	10.38
25-Jul-97	09:00	35.0	8.99	9.48	8.16	9.98
28-Jul-97	12:00	38.1	ns	9.44	8.13	10.26
05-Aug-97	12:00	46.1	9.17	8.97	7.92	10.36
02-Sep-97	12:00	74.1	8.38	8.14	7.91	9.54
04-Sep-97	15:00	76.2	8.43	8.55	7.87	9.76
08-Sep-97	12:00	80.1	8.61	8.83	8.08	9.95
16-Sep-97	14:00	88.2	8.52	8.80	7.91	9.84
22-Sep-97	12:00	94.1	8.51	8.64	7.82	9.62
01-Oct-97	13:00	103.1	8.82	8.60	7.95	9.74
08-Oct-97	16:00	110.3	8.89	8.76	7.95	9.57
16-Oct-97	14:00	118.2	9.02	8.94	8.13	9.70
21-Oct-97	16:00	123.3	9.21	8.94	8.33	9.88
29-Oct-97	12:00	131.3	9.28	9.23	8.22	9.98

Note: ns = not sampled. dry = sampling point was above water table on sampling date.

Appendix 14

Leak Around Cassettes (Gate 1)

A14.1 INSTALLATION OF CASSETTE STRUCTURE

Figure A14.1 shows a detailed view of the cassette system and its location relative to the sheet-piling walls. Items of particular interest include bentonite seals between the outer cassette box and the sheet piling, and the presence of pea gravel at both ends of the structure.

During installation of the cassette system in December 1995, field observations indicated that a potential problem with the sealing of the outer cassette box within the sheet-piling walls might exist. Mixing of the pea gravel placed at the ends of the cassette structure and the bentonite placed in the corners may have inadvertently occurred as the two materials were installed in their respective lifts. Although the importance of creating a good seal was recognized at the time and every effort was made to achieve this, the potential for a leak was evident. To assess this potential leak, a multilevel piezometer was installed on either side of the cassette box between the sheet-piling walls (labeled as G1CKL and G1CKR in Figure A14.1). Installations occurred on June 25, 1996. Monitoring of these wells for tracer compounds, VOCs, and inorganics occurred regularly to document the status of any leakage that might exist.

A14.2 EVIDENCE OF A LEAK

A14.2.1 Tracer Test Results

Piezometers, installed to check the integrity of the seals (G1CKL and G1CKR), were sampled regularly during the main tracer test and measurable concentrations of the bromide tracer were detected at G1CKR, particularly at sampling point 3 (approximately 2.7 m below ground). These results indicated that the bentonite seal on the right side of the cassette (adjacent to Gate 2) was breached. Samples from G1CKL showed no evidence of tracer compounds (except for an anomalous point in the data, likely resulting from mislabeling of sample vials). Figure A14.2 shows the breakthrough curves for the G1CK piezometers.

Despite the chemical evidence of poor seal performance, groundwater velocities through the cassettes, based on tracer data, compared reasonably well with those predicted based on the Darcy equation. Thus, it was believed that the leak was a minor contributor to the mass flux through the gate and no effort to repair the leak was deemed necessary.

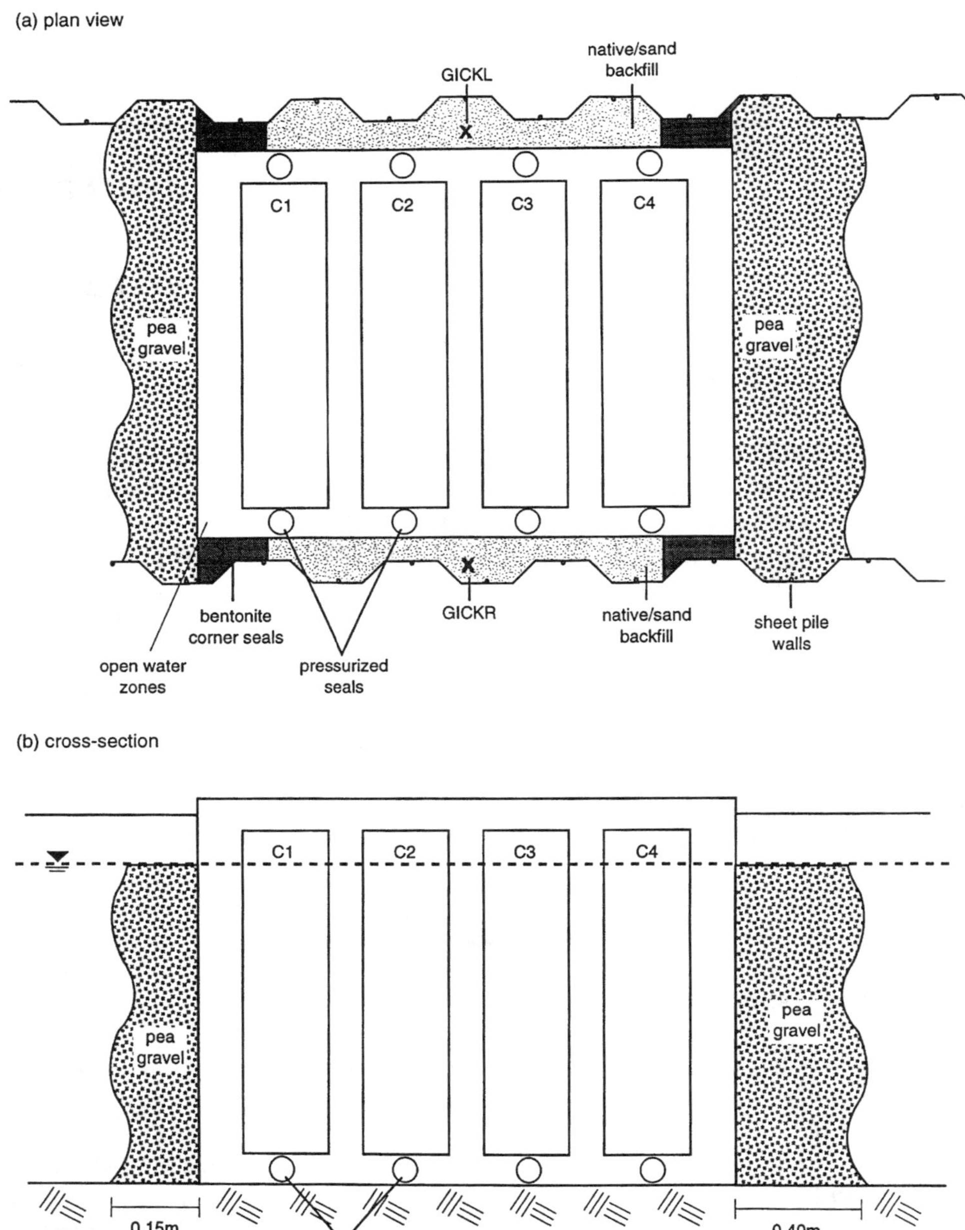

Figure A14.1 Details of cassette installation.

A14.2.2 Behavior of Organic Compounds

The behavior of volatile organics at G1CK piezometers was similar to that observed for the tracer compounds. Table A14.1 provides a summary of the VOC concentrations measured over time. As with the tracer compounds, sampling point G1CKR-3 shows the highest concentrations of contaminants, ranging from 25 to 190 μg/L for PCE, CT, and CF, and as high as 1660 μg/L for toluene. Concentrations at other depths of G1CKR (located adjacent to Gate 2) were typically at

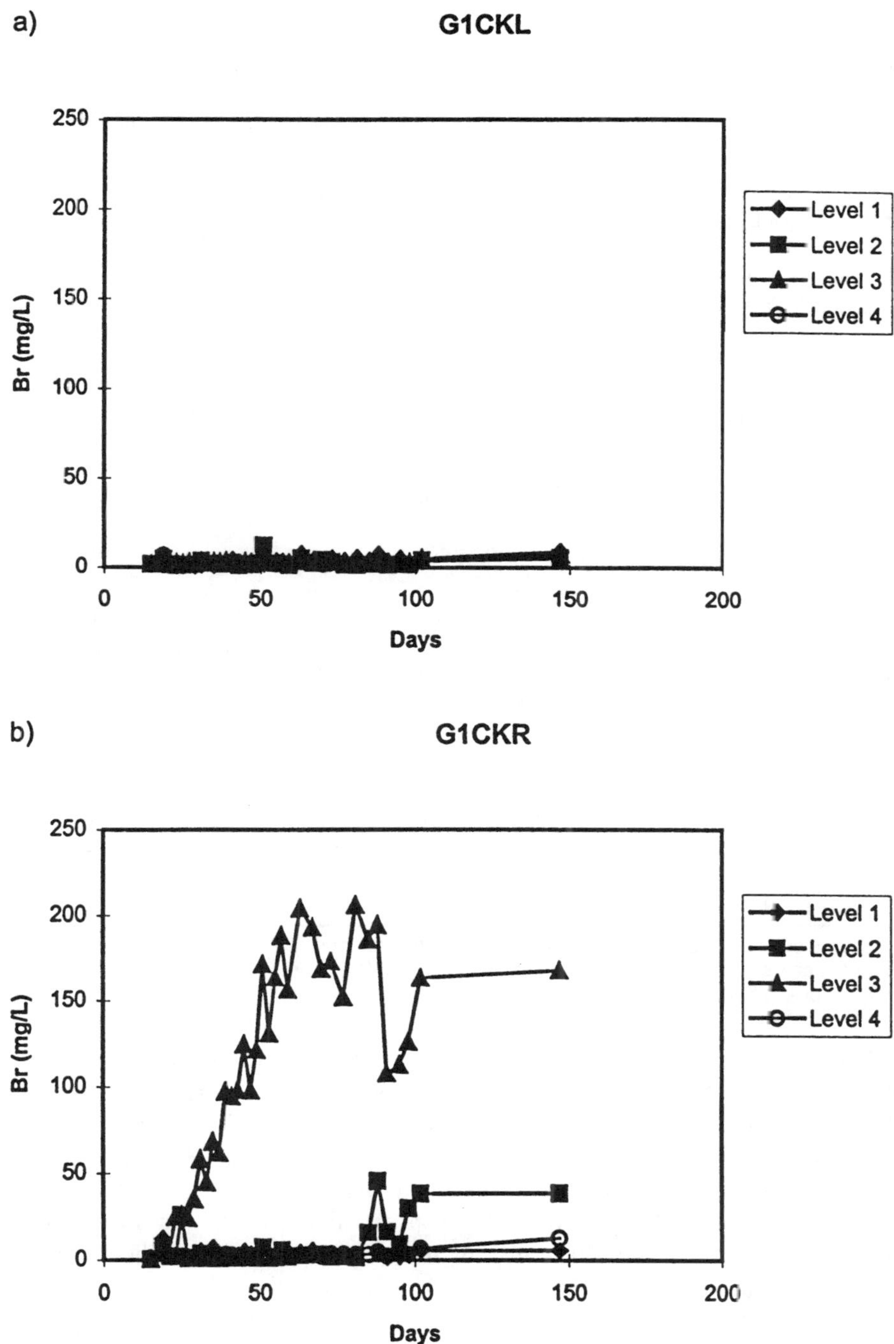

Figure A14.2 Tracer breakthrough curves at G1CK piezometers.

or below the limit of quantification (LOQ; Appendix 5) and no VOC parameters were detected at quantifiable concentrations on the left side (G1CKL).

Some trends in organic data at piezometer G1CKR may be explained through a comparison with data from the upgradient fence (G102), provided in Table A14.2. Only organics data from later sampling times (July, September, and October 1997) are given, since samples were not collected from

Table A14.1 VOC Data for Seal Checking Piezometer (G1CK)

Sampling Date	Concentrations (μg/L)							
	G1CKL-1	G1CKL-2	G1CKL-3	G1CKL-4	G1CKR-1	G1CKR-2	G1CKR-3	G1CKR-4
				PCE				
97/03/24	<MDL	<MDL	<MDL	<MDL	<MDL	7.1	105	<MDL
97/06/02	<MDL	<MDL	<MDL	<MDL	<MDL	**4.7**	91.7	**2.2**
97/07/14	<MDL	<MDL	<MDL	<MDL	<MDL	<MDL	92.0	<MDL
97/09/04	**2.0**	<MDL	<MDL	<MDL	<MDL	**6.2**	143	**1.9**
97/10/21	N/A	<MDL	<MDL	<MDL	<MDL	**3.3**	115	**2.0**
				CT				
97/03/24	<MDL	<MDL	<MDL	<MDL	<MDL	<MDL	91.5	<MDL
97/06/02	<MDL	<MDL	<MDL	<MDL	<MDL	<MDL	29.0	<MDL
97/07/14	<MDL	<MDL	<MDL	<MDL	<MDL	<MDL	**13.1**	<MDL
97/09/04	<MDL	<MDL	**4.8**	<MDL	<MDL	**6.9**	35.9	<MDL
97/10/21	N/A	<MDL	<MDL	<MDL	<MDL	<MDL	25.3	<MDL
				CF				
97/03/24	<MDL	<MDL	<MDL	<MDL	<MDL	<MDL	126.5	<MDL
97/06/02	<MDL	<MDL	<MDL	<MDL	<MDL	<MDL	189.1	<MDL
97/07/14	<MDL	<MDL	<MDL	<MDL	<MDL	<MDL	68.2	<MDL
97/09/04	<MDL	<MDL	<MDL	<MDL	**8.4**	**19.4**	63.4	<MDL
97/10/21	N/A	<MDL	<MDL	<MDL	<MDL	<MDL	40.2	<MDL
				TOL				
97/03/24	<MDL	<MDL	<MDL	<MDL	**2.0**	**5.2**	1660	**2.3**
97/06/02	<MDL	**4.0**	<MDL	<MDL	**2.2**	**3.8**	757	10.3
97/07/14	<MDL	<MDL	<MDL	<MDL	<MDL	<MDL	490	<MDL
97/09/04	<MDL	<MDL	<MDL	<MDL	**3.5**	192	1290	**3.8**
97/10/21	N/A	**3.7**	<MDL	<MDL	**3.8**	41.7	503	<MDL

Notes: <MDL = concentration values are analyte-specific; Appendix 5. <LOQ = concentration varies for different compounds and sampling dates; Appendix 5. Values between MDL and LOQ shown as bold. N/A = not available.

fence G102 on March 24 or June 2, 1997. For the purposes of identifying trends, these data are sufficient. Maximum concentrations at G1CKR range from 5 to 10% of the maximum input concentrations (at fence G102). Interestingly, maximum concentrations at both locations were measured at depth 3 on the right side, suggesting a hydraulic connection between G102R-3 and G1CKR-3.

Although concentrations of chlorinated compounds decreased to below detection immediately downgradient of the iron (at fence G104), VOC concentrations measured at fence G106 were of similar magnitude to those obtained for sampling fence G102, located upgradient of the cassettes. This result suggests that a considerable portion of the mass of contaminant in the system bypassed the remedial materials in the cassettes. However, no measure of the actual mass flux through the leak was possible nor was confirmation of its exact location possible. Nevertheless, observations made during the cassette installation (discussed above) would suggest that the problem was related to the integrity of the corner seals (Figure A14.1).

The presence of elevated concentrations of VOCs across the full width of the gate suggests the existence of a serious and gatewide leak, although no definitive evidence of such a leak exists. The flow environment created by the installation of the cassette system, complete with pea gravel zones, bentonite seals, and relatively conductive reactive media zones, was complex. It was considered that the hydraulics of the system might be sufficient to explain the contaminant distribution

Table A14.2 Fence G102 VOC Data for Comparison

Parameter Analyzed	Concentrations (μg/L) G102L-1	G102L-2	G102L-3	G102L-4	G102R-1	G102R-2	G102R-3	G102R-4
				PCE				
97/07/14	**1.5**	440	**1.5**	<MDL	2.5	<MDL	1320	**2.5**
97/09/04	24.2	172	<MDL	22.8	21.1	**4.4**	876	<MDL
97/10/21	N/A	31.0	<MDL	<MDL	42.1	**13.1**	535	**2.1**
				CT				
97/07/14	<MDL	804	<MDL	<MDL	<MDL	<MDL	4720	**3.6**
97/09/04	<MDL	193	<MDL	<MDL	**4.4**	**16.9**	1520	**16.4**
97/10/21	N/A	**9.5**	<MDL	<MDL	**8.6**	**10.5**	1250	<MDL
				CF				
97/07/14	<MDL	312	<MDL	<MDL	<MDL	<MDL	1011	<MDL
97/09/04	**12.4**	<MDL	<MDL	<MDL	40.0	<MDL	360.8	<MDL
97/10/21	N/A	<MDL	<MDL	<MDL	32.3	<MDL	147.9	<MDL
				TOL				
97/07/14	23.6	48.6	14.2	<MDL	372	<MDL	17500	37.6
97/09/04	820	1660	45.5	**23.4**	291	**21.2**	9090	**8.9**
97/10/21	N/A	126	<MDL	<MDL	490	79.5	5890	**11.9**

Notes: <MDL = concentration values are analyte-specific; Appendix 5. <LOQ = concentration varies for different compounds and sampling dates; Appendix 5. Values between MDL and LOQ shown as bold. N/A = not available.

observed, based on a narrow leak on one side of the cassette only. Groundwater flow modeling to investigate this hypothesis was conducted using Flownet/trans (Guiguer et al., 1995).

Figure A14.3 provides the results of three scenarios that may have existed during this experiment. In the first case, the seals at all corners of the outer cassette box were sound and the ideal flow system was modeled. Hydraulic conductivities of 10^{-5}, 10^{-3}, 10^{-1}, 10^{0}, and 10^{-9} m/s were used for the native sand aquifer, the reactive media and the loosely packed sand between the corner seals, the pea gravel, the open water zones, and for the seals themselves and the steel walls of the outer cassette box, respectively (values were estimated using UW, 1997; Freeze and Cherry, 1979). Porosities of 0.38 for the native sand, 1.0 for the open water zones, and 0.5 for all other material types were also assigned. Boundary conditions included no flow along the sheet-piling walls, a constant head of 2.7 m at the upgradient end of the gate, and a pumping rate of 105 mL/min at the extraction well. Velocities similar to those obtained during the main tracer test were obtained using this parameter set. Figure A14.3b shows the modeling results for a scenario of leaky seals. Bentonite seals on the right side of the gate (shown at the bottom in the figure) were removed and replaced with pea gravel (hydraulic conductivity of 10^{-1} m/s). The hydraulic conductivity of the sand between the corner seals was maintained at 10^{-3} m/s (i.e., the same as for the reactive media). Flow through the leak was observed in this case, although downgradient of the cassettes it was confined to the right portion of the gate. The leak did not provide a preferential pathway for groundwater flow, since it was no more conducive to flow than the cassettes themselves.

The third scenario (Figure A14.3c) shows the worst-case scenario for a leak, and demonstrates that the observations made during this experiment can be duplicated by varying only material hydraulic conductivities. The actual values used in this analysis are not as important as the contrast between them. The hydraulic conductivity of the sand backfill between the outer cassette box and the sheet-piling wall consisted of some native material, but also contained silica sand and pea

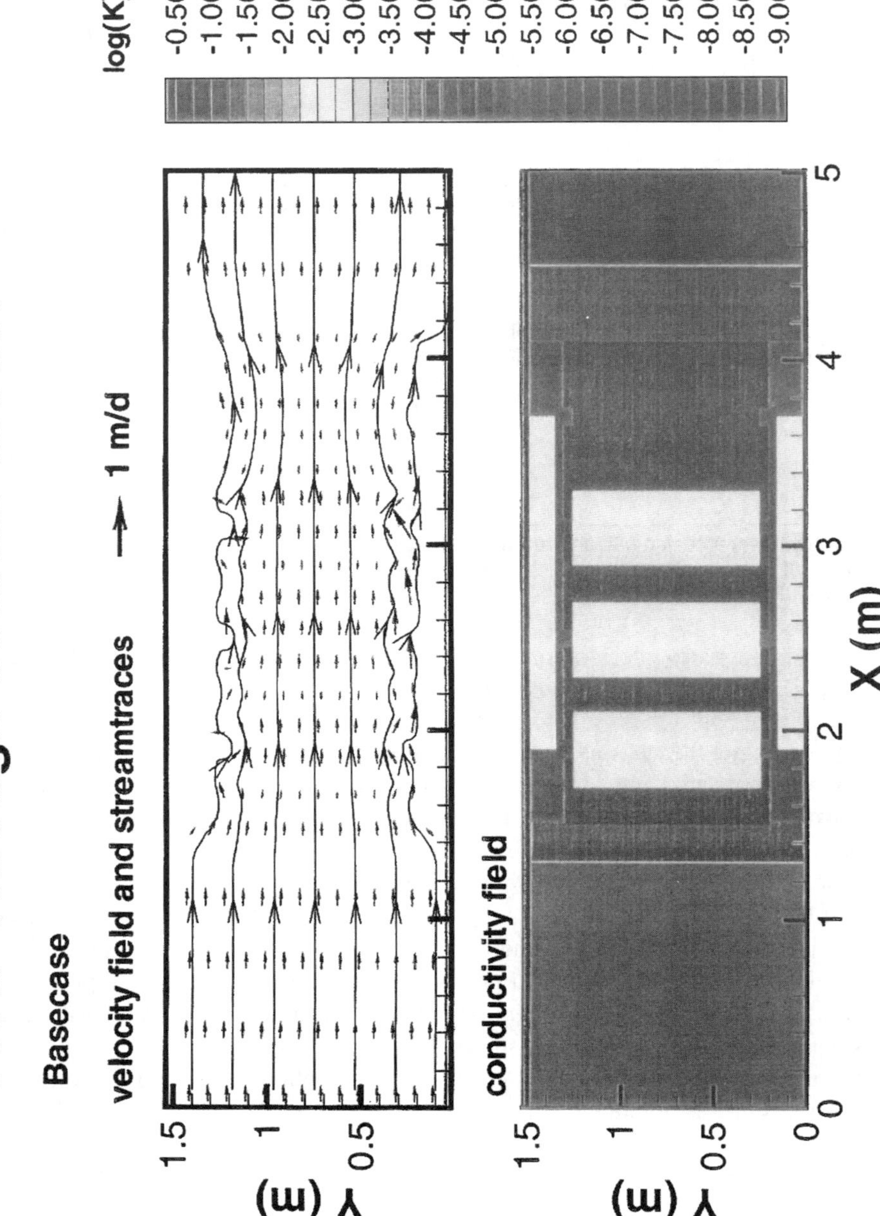

Figure A14.3a Base-case with sound bentonite seals.

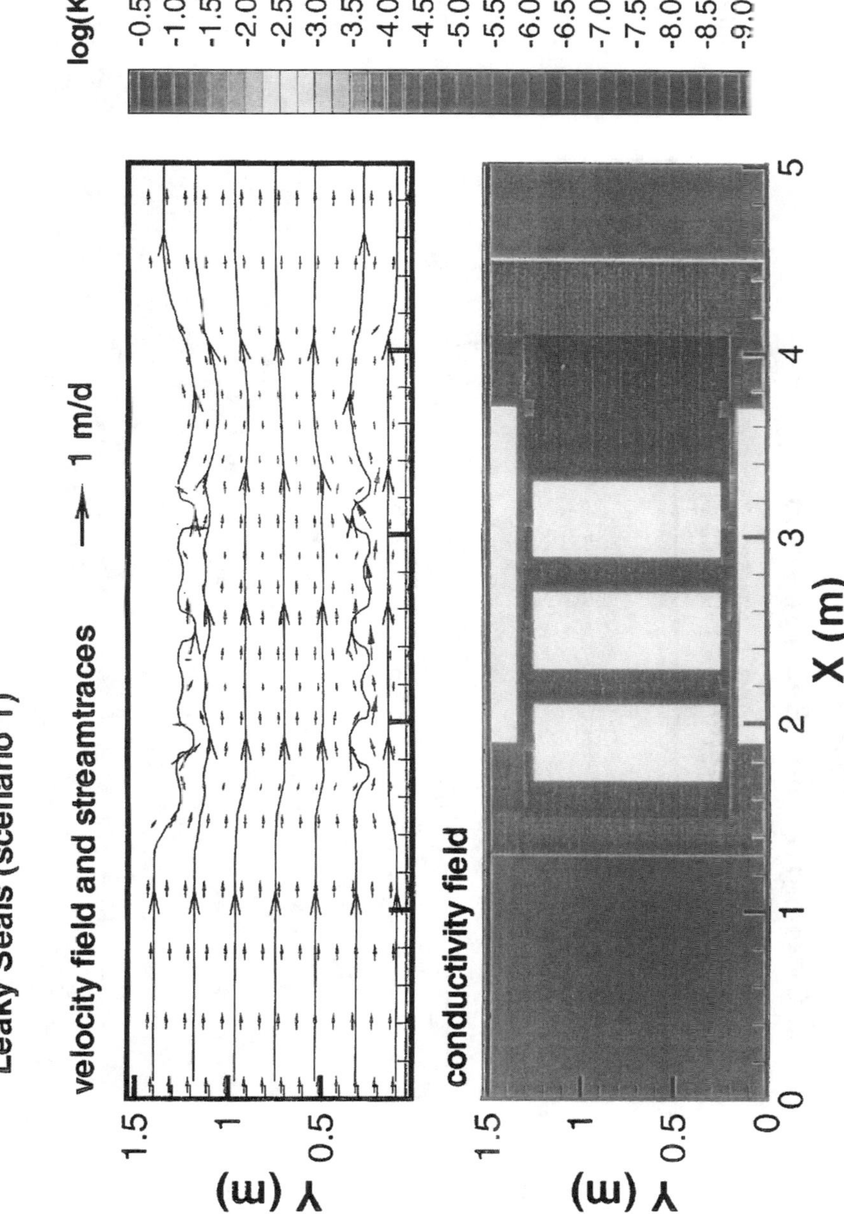

Figure A14.3b Scenario showing leaky seals.

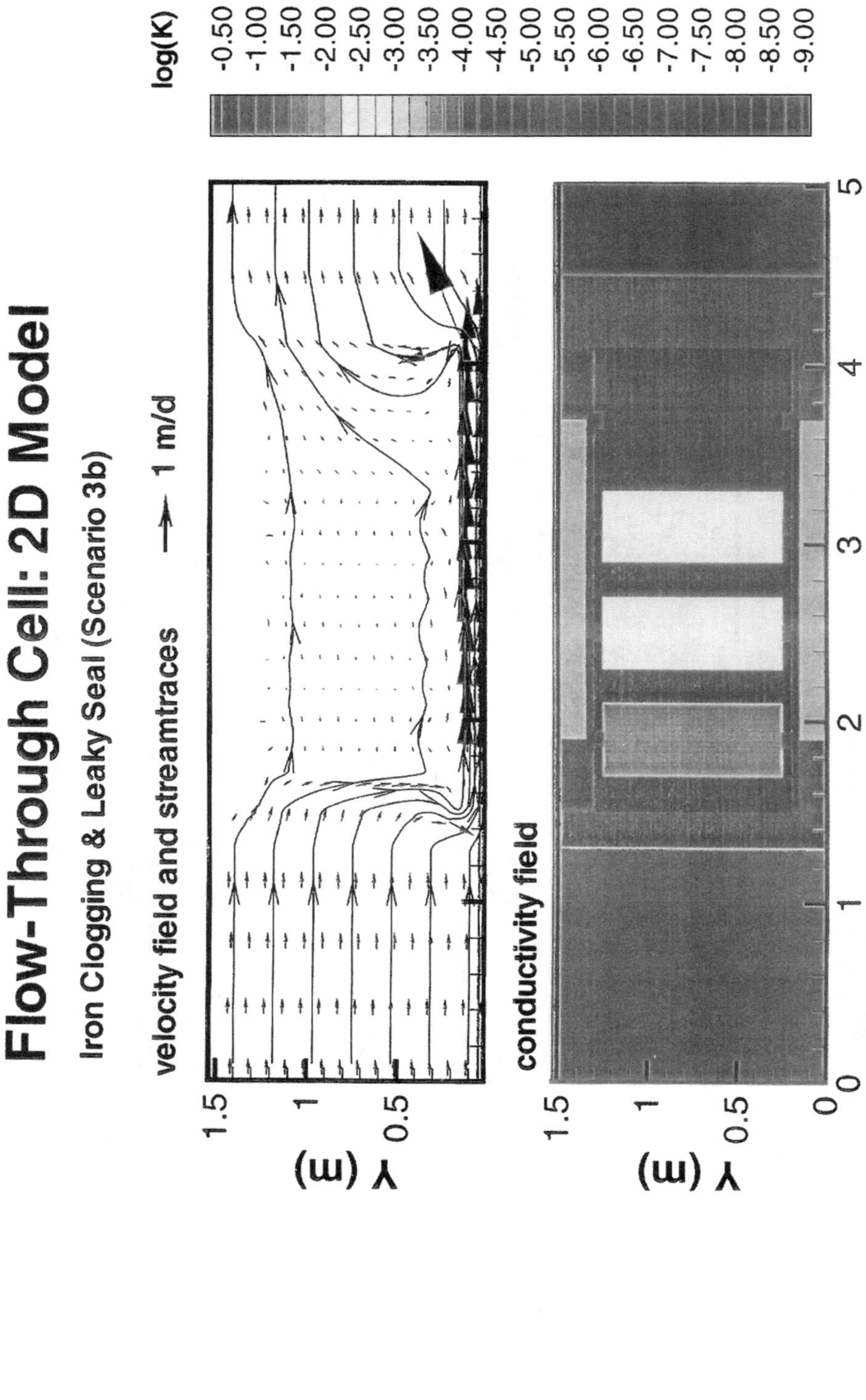

Figure A14.3c Worst-case scenario for leaky seals.

gravel. Further, compaction of this material was not undertaken due to the manner in which backfilling was conducted (i.e., simultaneously with bentonite and pea gravel placement). It is reasonable that this material could be highly conductive, with a hydraulic conductivity almost equal to the pea gravel itself. Alternatively, the hydraulic conductivity of the granular iron could have decreased over time due to settling, or due to the formation of precipitates, particularly iron carbonates or hydroxides, on the iron surfaces or in pore spaces between the grains (Appendix 16). Precipitate formation has been observed at the influent end of laboratory columns, and has been recognized as a possible limitation of the iron technology, although work in this area continues. Nonetheless, a decrease in hydraulic conductivity of at most an order of magnitude might be expected as a result of this phenomenon. However, given the relatively short time frame of this experiment it is more likely that this loss in conductivity would be less, and would be confined to the upgradient portion of the first granular iron cassette. For the purposes of the modeling, the conductivity of the leak was increased by one order of magnitude (from 10^{-3} to 10^{-2} m/s) while the conductivity of the reactive media was decreased by the same (from 10^{-3} to 10^{-4} m/s). A two-order-of-magnitude difference in hydraulic conductivity was expected to be the worst possible configuration for the leak. Results showed that the leak dominated the flow-field downgradient of the cassettes, diverging from the right portion of the gate to encompass almost the entire width of the gate. Very high velocities through the leak were observed, while flow through the cassettes themselves was minimal. Within the open water zones in and around cassette C4, groundwater flowed against the overall hydraulic gradient, which would explain the elevated contaminant concentrations observed at fences G105 and G1C4.

A14.2.3 Behavior of Inorganic Constituents

Samples from all depths of piezometers G1CKL and G1CKR were analyzed for inorganics on September 4, 1997, while on June 2, 1997 only samples from depth 3 were analyzed (Table A14.3). The data from the September sampling will be discussed thoroughly here, since they are more complete, although it is interesting to note that the June results compare well to those obtained for depth 3 in September. Very different water quality was observed at G1CKL as compared to G1CKR (averages for each piezometer are given in Table A14.3b). Further, a difference between results from point G1CKR-3 and all other sampling points was noted. Concentrations of major ions in groundwater at G1CKR-3, which the tracer experiment had shown was influenced by the leak, were similar to those measured at upgradient aquifer locations (G102), indicating a possible connection between these locations. Inorganic data from September 4, 1997 for fence G102 were fence-averaged and are provided in Table A14.3b for comparison to G1CK piezometer data.

Samples from G1CKL showed abnormally high average concentrations of sulfate (456 mg/L), sodium (474 mg/L), and chloride (131 mg/L) compared to values of approximately 15 mg/L measured for each compound at point G1CKR-3. Other sampling points at G1CKR showed concentrations between these extremes, although values generally more closely resembled those measured at G1CKL points than data from G1CKR-3. These data confirm the tracer test results indicating that G1CKL was isolated from the flow system by good bentonite seals. However, points G1CKR-1, -2, and -4 show evidence of limited hydraulic connection. Bromide concentration data in Table A14.3b confirm this hypothesis. During the tracer experiment, significant levels of bromide were measured at point G1CKR-3. Minor traces were observed at other depths at this piezometer, while essentially no tracer was observed at piezometer G1CKL. Months after the tracer test was completed (i.e., September 1997), all bromide tracer had been flushed from point G1CKR-3, while the low concentrations that were measured during the tracer test at other depths remained. The presence of bromide at depths 1, 2, and 4 of piezometers G1CKR during the tracer test could have been due to diffusion from the highly concentrated solution at point G1CKR-3. Hence, bromide was later trapped at the points with little flow. No detectable bromide concentrations at piezometer G1CKL was not surprising, since it was believed that tracer never penetrated this area.

Table A14.3a Inorganic Data for Piezometers G1CK with Average Values from Fence G102 for Comparison (June 2, 1997 Data)

Parameter	Concentrations (mg/L, unless noted)								Average of
Analyzed	G1CKL-1	G1CKL-2	G1CKL-3	G1CKL-4	G1CKR-1	G1CKR-2	G1CKR-3	G1CKR-4	G102
DO	0.1	0.3	0.2	0.2	0.2	0.3	0.5	0.5	0.6
pH (units)	7.1	7.5	7.6	7.9	7.1	7.4	7.3	8.0	7.5
E_h (mV)	53	47	64	61	81	43	76	60	133
Cond. (µS)	1349	1826	2630	576	1846	844	433	486	377
Alk. ($CaCO_3$)	ns	ns	ns	ns	ns	ns	ns	ns	255
Iron	ns	ns	0.695	ns	ns	ns	2.17	ns	7.4
Calcium	ns	ns	89.1	ns	ns	ns	83.9	ns	92.7
Potassium	ns	ns	3.45	ns	ns	ns	2.68	ns	3.1
Magnesium	ns	ns	10.2	ns	ns	ns	7.09	ns	6.8
Manganese	ns	ns	1.32	ns	ns	ns	2.08	ns	0.2
Sodium	ns	ns	690	ns	ns	ns	14.8	ns	3.1
Chloride	ns	ns	151	ns	ns	ns	6.91	ns	3.3
Sulfate	ns	ns	993	ns	ns	ns	13.7	ns	11.7
Phosphate	ns	ns	<MDL	ns	ns	ns	<MDL	ns	<MDL
Nitrate/Nitrite	ns	ns	<MDL	ns	ns	ns	<MDL	ns	<MDL
Bromide	ns	ns	<MDL	ns	ns	ns	<MDL	ns	<MDL

Notes: Data for piezometer G1CK from June 2, 1997; data for fence G102 from March 10, 1997.

<MDL = concentration values are analyte-specific; Appendix 5.

ns = not sampled.

Table A14.3b Inorganic Data for Piezometers G1CK with Average Values from Fence G102 for Comparison (September 4, 1997 Data)

Parameter Analyzed	Concentrations (mg/L, unless noted)										Average of G102
	G1CKL-1	G1CKL-2	G1CKL-3	G1CKL-4	Ave. L	G1CKR-1	G1CKR-2	G1CKR-3	G1CKR-4	Ave. R	
DO	1.4	0.3	0.3	0.4	0.6	0.4	0.3	0.5	0.6	0.5	0.6
pH (units)	7.2	7.6	7.7	7.6	7.5	7.2	7.4	7.4	7.9	7.5	7.3
E_h (mV)	122	92	103	104	105	105	69	92	89	89	115
Cond. (μS)	ns	ns	ns	ns	ns	ns	ns	ns	ns	ns	ns
Alk. ($CaCO_3$)	469	742	523	368	526	640	504	255	244	411	252
Iron	13.4	9.99	0.817	3.58	6.9	22.1	26.0	2.51	0.279	12.7	11.1
Calcium	117	76.8	70.7	129	98.4	114	69.1	80.9	25.5	72.4	96.3
Potassium	1.63	2.92	4.02	6.79	3.8	3.84	2.61	2.61	1.79	2.7	2.5
Magnesium	9.39	7.05	7.96	16.6	10.3	10.4	9.78	6.95	2.57	7.4	33.4
Manganese	3.03	2.76	1.12	0.281	1.8	3.32	3.06	2.19	0.028	2.1	0.4
Sodium	149	486	712	550	474	486	157	18.8	152	203	9.6
Chloride	109	126	155	135	131.3	166	41.8	15.2	28.4	62.9	23.8
Sulfate	17.6	242	787	779	456.4	390	<MDL	13	91.8	123.7	10.3
Phosphate	<MDL	<MDL	<MDL	<MDL	<MDL	<MDL	<MDL	<MDL	<MDL	<MDL	<MDL
Nitrate/Nitrite	<MDL	<MDL	<MDL	<MDL	<MDL	<MDL	<MDL	<MDL	<MDL	<MDL	<MDL
Bromide	<MDL	<MDL	<MDL	<MDL	<MDL	10	3.94	<MDL	2.52	4.1	<MDL

Notes: <MDL = concentration values are analyte-specific; Appendix 5.

ns = not sampled.

All evidence suggests that a leak between the sheet-piling wall and the right side of the cassette exists; however, its vertical extent appears limited. Only sampling point G1CKR-3, located at approximately 2.8 m below ground surface (bgs), shows evidence of good hydraulic connection to the upgradient aquifer. Other depths on G1CKR appear to be influenced only by diffusion, while piezometer G1CKL appears to be isolated from the flow system.

A14.2.4 Individual Cassette Seals

Poor performance of the silicon pressure seals was suggested (Appendix 12) as a possible reason for higher concentrations of chlorinated compounds at fence G103 (i.e., downgradient of the first iron cassette, C1) than at fence G1C1 (i.e., in the middle of cassette C1). However, other possibilities were also considered that were equally as likely. Assessment of the true cause is difficult and has not been undertaken. Nonetheless, nondetectable concentrations observed downgradient of cassette C2 (at fence G104) would suggest that the pressure seal around C2 was secure, and, thus, a possible leak around C1 would not affect the ultimate performance of the system.

A14.3 TRACER TEST IN THE CASSETTES

A flow-through test of the cassette system was completed in the fall of 1997 to assess whether the influence of the leak had worsened with time. On September 29, 1997, an inorganic tracer (KCl) was injected between cassettes C1 and C2 (Figure A14.2) as a pulse. An input concentration of approximately 200 mg/L of KCl within the entire injection zone (between the cassettes) was desired to limit density effects. Injection was completed in a manner similar to that used for the tracer dilution test within the alternate oxygen addition wells to ensure good distribution of tracer through the entire depth of the open water. Over time, tracer samples were collected using the method described in Appendix 4 and were submitted to UW's Water Quality Laboratory for ion-specific analysis (Appendix 3, analytical method similar to that used at WTI).

The results are summarized in Figure A14.4 and suggest that flow through the cassettes was dominated by density effects, rather than advective flow (even though the selected input tracer concentration was believed to be lower than the range affected by density). Within the tracer injection zone (at fence G103), chloride concentrations rapidly declined at depths 2 and 3, but at depth 4, an increase in concentration (up to the injection levels observed in the upper points) was observed. At the up and downgradient monitoring fences, G1C1 and G104, respectively, tracer concentrations at depth 4 ultimately reached the input concentration (approximately 120 mg/L). At fence G1C1, this concentration was achieved very quickly, while at fence G104, approximately 350 hours passed before it was observed. The rapid migration of tracer to the upgradient fence suggested that density flow dominated the system and that advective flow was limited. Depths 2 and 3 showed minimal increases in chloride concentrations over time, which would be expected since it was apparent that injected tracer at these depths at fence G103 moved downward rather than forward. Chloride concentrations above background levels were not observed at fences G1C0 and G105 (further up- and downgradient, respectively) during the duration of the monitoring program (405 hours after tracer injected).

The results of the cassette tracer test imply that flow through the leak between the sheet-piling wall and the outer cassette box must have increased significantly. This change was possibly due to a decrease in hydraulic conductivity of the reactive media or clogging of the screen material on the cassettes ultimately worsening of the leak over time, making it the preferred flow path over the cassettes. In either case, at some time between the completion of the main tracer test (completed in the cassette portion of Gate 1 in December 1996) and the cassette tracer test (October 1997) a substantial change to the flow system occurred. Several lines of evidence suggest that flow through the cassettes remained significant until July 1997:

- Concentrations of PCE and chloroform at fence G106 (immediately downgradient of the cassettes) increased slightly between January and June 1997 (likely as a result of plume development), but between early June and mid-July 1997 these compounds were measured at relatively constant concentrations (indicating that quasi-steady-state conditions had developed). Subsequently, PCE, which undergoes complete abiotic degradation in the granular iron, but which should experience minimal intrinsic biodegradation within this aquifer (based on Gate 2 behavior) and thus act as an organic tracer through the leak, showed rapid concentration increases. This suggests that at some time following the July 14, 1997 sampling event, the flow system changed and the groundwater at fence G106 had a greater contribution from the leak rather than through the iron cassettes.
- Water flowing through the granular iron experienced declines in chlorinated contaminant concentrations (at least 90%; Appendix 12) due to degradation, but also toluene concentrations declined (approximately 10%), likely due to sorption. Concentrations of contaminants passing through the leak were approximately 10% lower than at fence G102. At early time, when the majority of the mass flux was through the cassettes and only a small portion passed through the leak, the water from within the cassettes should have dominated the flow system downgradient of the cassettes. Thus, at fence G106, concentrations in the right portion of the gate would be higher, since flow from the leak would be limited to this zone. At later time, when the flow regime was dominated by water from the leak, the distribution of contaminants would be more evenly spread across the gate. In this case, essentially all water at fence G106 passed through the leak and was distributed across the width of the gate due to the revised hydraulics. Modeling results support this hypothesis (discussed in Section A14.2.2 and shown in Figure A14.3). The distribution of contaminants at fence G106 on March 10 and July 14, 1997 are provided in Figure A14.5. The March data show that typical concentrations are lower on the left side as compared to the right, and indicate that influence of the leak is mainly in the right portion of the gate. By July, concentrations across the width of the gate are similar, indicating that the influence of the leak has increased.
- Average concentrations of PCE and chloroform entering the iron cassettes (at fence G1C0) were relatively constant with time; however, the average concentration measured at fence G1C1 (at the midpoint of C1) decreased with time. This behavior suggested an increased rate of reaction within the iron zone (considered unlikely) or, alternatively, a decreased rate of flow through the system and consequently a longer residence time (considered likely).

For the purposes of evaluating iron performance for the degradation of chlorinated compounds, the groundwater velocity of 20 cm/day, based on the main tracer test results, was assumed to be valid for the March, June, and July 1997 sampling events. Later data, as they related to the performance of the iron, were disregarded.

The influence of the leak on biodegradation efforts in the aerobic portion of the gate was less critical to the performance evaluation of this remedial technology, since the oxygen delivery wells and the treatment zone itself were downgradient of the cassettes. The leak obviously influenced the distribution of contaminants within the aerobic zone. However, the presence of chlorinated compounds in this zone simply permitted some evaluation of their behavior in an aerobic environment. Appendix 13 provides a discussion of these results. The domination of the leak in the groundwater flow system at later time was perhaps beneficial for the delivery of oxygen from the ORC™. Although the natural ability of the aquifer to buffer the high pH water emanating from the iron cassettes was demonstrated at early time (Appendix 15), the shift in flow regime did not permit iron-influenced water after July 1997 to reach fence G106, and near-neutral pH values would be expected (typical of those observed at fence G102). The intolerance of the ORC™ to high pH water (Appendix 15) was sidestepped through this shift; however, the installation of the ORC™ in the alternate oxygen addition wells on June 19, 1997 was early enough to create and observe the effects of sequencing the technologies.

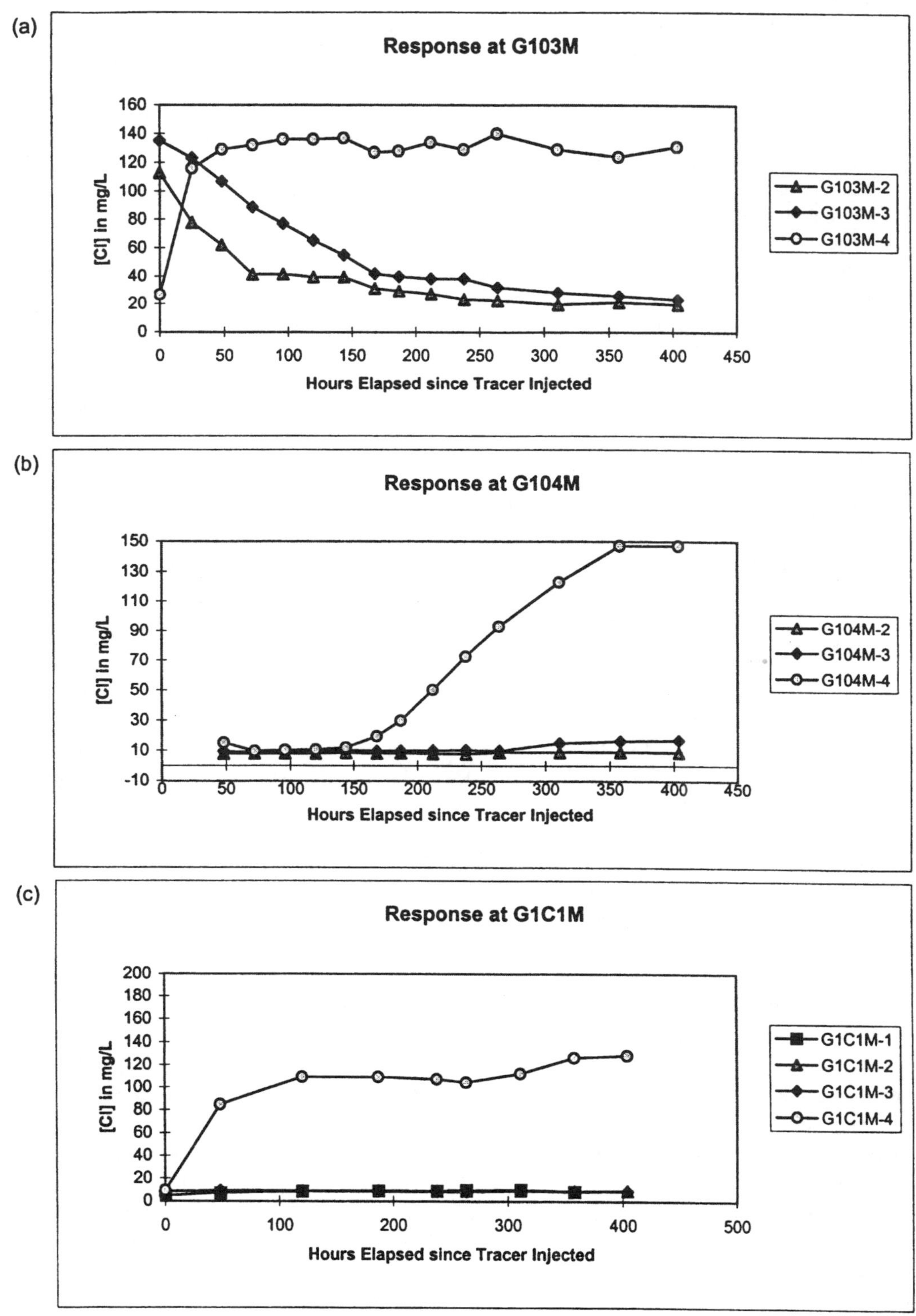

Figure A14.4 Cassette tracer test results. (a) Response in injection zone, (b) response at downgradient wells, (c) response at upgradient wells.

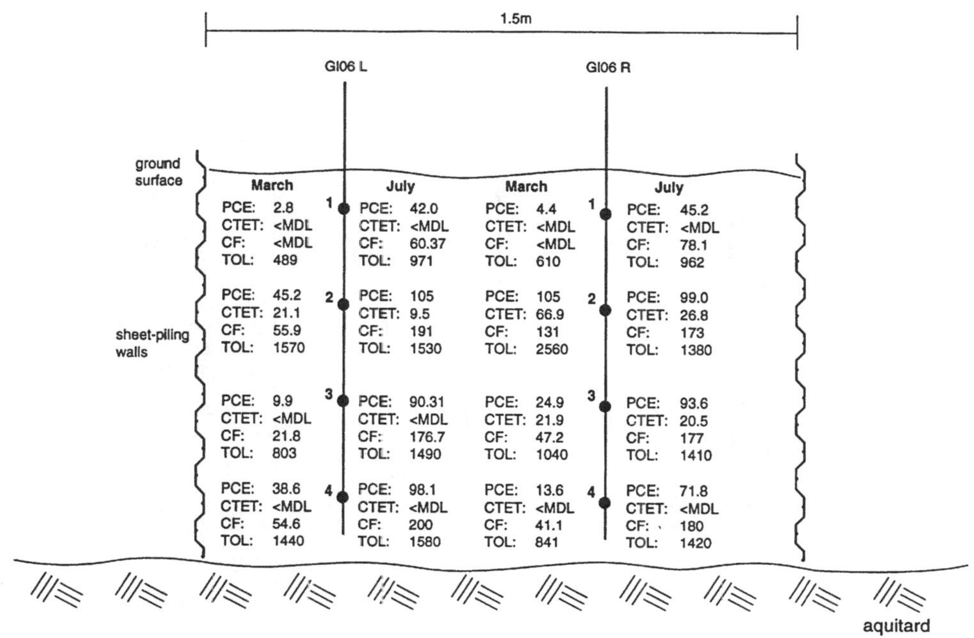

Figure A14.5 Contaminant distribution at fence G106 (March 10 and July 14, 1997 data).

A14.4 SEAL INTEGRITY

Cores were collected to investigate the conditions of the seals on the right side of the cassettes were collected on November 6, 1997, using the methods described by Starr and Ingleton (1992). The locations of the cores are shown in Figure A14.6. A total of four cores were collected, but only two were completed to the depth of the suspected leak (approximately 3 m) due to refusal (coring equipment was not designed to penetrate the hard and/or coarse materials encountered). The suspicion that bentonite and pea gravel layers were mixed during installation (discussed above) was confirmed in all cores. Coarse gravel zones were observed between sound bentonite layers. Small gravel layers (0.1 m long) were observed in core R-3 at depths of 0.3, 1.0, 2.0, and 2.5 m bgs. A larger gravel zone at approximately 2.8 to 3.2 m bgs was also observed and was possibly responsible for the penetration of water through the seal. It could not be confirmed that these layers penetrated through the entire seal at this depth because cores R-1 and R-2 were terminated at approximately 2 m bgs due to refusal. Similar observations were made for core R-4, with small gravel layers (0.1 m long) occurring at 1.5 and 2.4 m bgs, and a thicker gravel layer at a depth of approximately 2.7 m. The presence of this layer ultimately prevented further penetration of this core.

Cores R-1 and R-2 were collected approximately 0.15 m apart within the front seal on the right side of the cassettes (Figure A14.6). Although these cores could only be completed to depths of approximately 2 m bgs, interesting observations could still be made. In core R-1, several gravel layers were observed through the length of the core (final depth 2.0 m), and once completed, the hole quickly filled with water to 0.75 m bgs. Core R-2 generally consisted of sound bentonite through the length of the core (final depth 2.0 m). The bentonite in this core was hard and dry, even below the water table (1.2 m bgs), and the hole remained open and relatively dry (approximately 2 cm of water accumulated in the base of the hole after 4 hours) over time. It is evident that, although these cores were located only 0.15 m apart, their composition is quite different. The gravel layers observed in core R-1 did not penetrate through the entire seal and, thus, should not

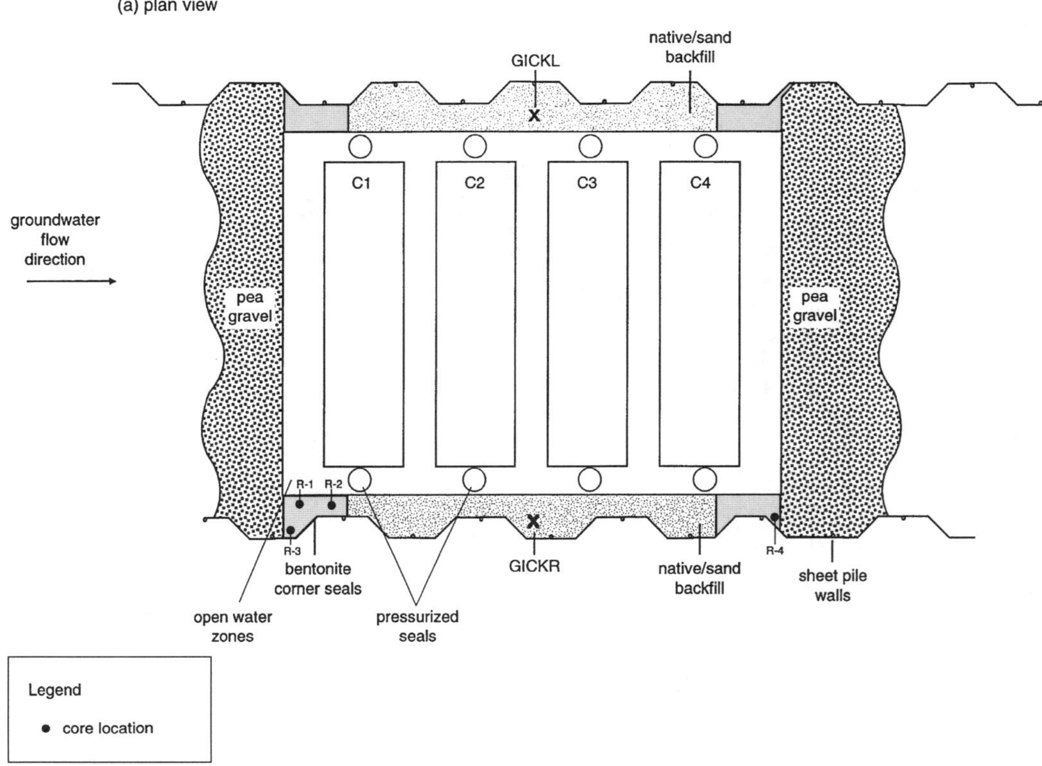

Figure A14.6 Locations of cores to investigate Bentonite seal integrity.

permit flow. Although at shallower depths installation of the seals would have been easier, it is possible that many of the gravel layers observed at depth in cores R-3 and R-4 also did not penetrate through the entire seal. This cannot be investigated through standard coring methods.

Although repairs to the leaky cassette box seals could not be undertaken as part of this project, UW personnel, whose interests lie specifically in the use of Funnel-and-Gate technologies, feel repair efforts are justifiable and are currently considering possible alternatives for the sealing of the identified leak. Jet-grouting appears to be the favored alternative at this time.

A14.5 REFERENCES

Freeze, R.A. and J.A. Cherry. 1979. *Groundwater.* Prentice-Hall, Englewood Cliffs, NJ, p. 29.

Guiguer, N., T. Franz, J. Molson, and E. Frind. 1995. Flownet — User Guide. Waterloo Hydrogeologic and University of Waterloo.

Starr, R.C. and R.A. Ingleton. 1992. A new method for collecting core samples without a drill rig. *Groundwater Monitoring Review.* 12(4): 91.

University of Waterloo (UW). 1997. Passive and Semipassive Techniques for Groundwater Remediation: Phase 1 Installation Report. Prepared for AATDF, Rice University.

APPENDIX 15

Iron and ORC™ Incompatibility

The methodology used to address questions related to the incompatibility between the granular iron and ORC™ systems is presented in the following sections. Laboratory experiments that were conducted to demonstrate the pH dependence of the oxygen-releasing reaction and to determine an appropriate pH range for this release are presented in Section A15.1. Several options to achieve the pH range in which the iron/ORC™ sequence might be successful are presented in Section A15.2. Section A15.3 provides some modifications to the iron technology that might prevent the production of high pH water within the iron itself and consequently eliminate the incompatibility problem all together.

A15.1 LABORATORY EXPERIMENTS

A15.1.1 Materials and Equipment

water:	• low DO, high pH water (water allowed to equilibrate, at room temperature, with granular iron for at least 24 hours)
	• deionized
ORC™:	• same formulation as used in field (50:50 mix of ORC™ powder and silica sand)
chemicals:	• laboratory-grade hydrochloric acid (HCl; 0.1 M)
probes:	• same probes as used at field site:
	pH:Orion Model 260 meter with Orion Model 9107 triode
	DO:Orion Model 835 meter with Orion Model 083010 probe
other:	• 1-L glass beaker
	• laboratory scale accurate to 0.01 g
	• magnetic stirrer and stir bars
	• stop watch

A15.1.2 Method

A15.1.2.1 Experimental Method

A beaker containing a stirring bar was placed on the magnetic stirrer. Approximately 500 mL of low DO water (equilibrated with iron) was added to the beaker and the magnetic stirrer switched "on" (set at midway). Probes were inserted into the beaker and initial measurements recorded. The ORC™ was then added to the beaker and the stop watch started. pH and DO measurements were recorded with time, while stirring continued. A schematic of the laboratory setup is provided in Figure A15.1.

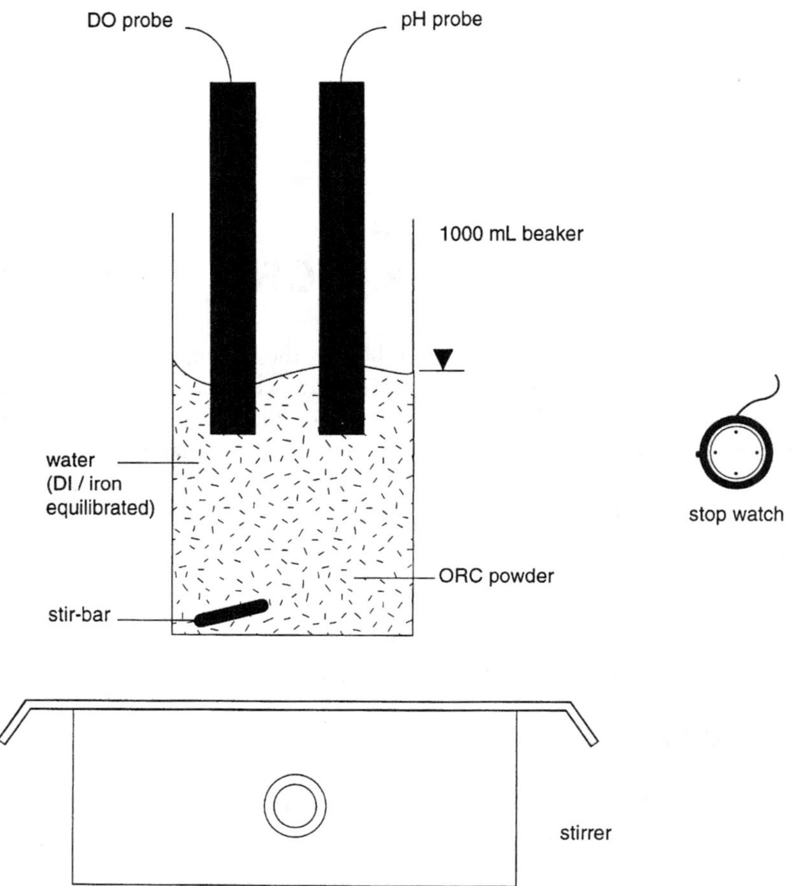

Figure A15.1 Laboratory experimental setup.

To investigate the pH dependence of the ORC™ reaction, the above experimental method was followed, with the following variation: after addition of the ORC™ to the water and the subsequent increase in pH occurred, hydrochloric acid was added dropwise to maintain the pH within several desired ranges (7 to 8; 8 to 9; and 9 to 10). Deionized water was used in most experiments, although for one, iron-equilibrated water was used and a near-neutral pH was maintained. All experiments were repeated to ensure reproducible results.

A15.1.3 Results

pH and DO measurements taken during unconstrained pH experiments are provided in Figure A15.2. The results indicate that

1. A sharp increase in pH (from 8.7 to greater than 10 within 30 seconds) results from the addition of the ORC™ to the water.
2. The marginal increase in DO observed (from 2 to 4 mg/L over 10 minutes) is likely the result of equilibration of the low DO water with atmospheric oxygen, rather than the result of oxygen release from the ORC™. Increases in DO concentration to near saturation (approximately 40 mg/L) observed in previous work using ORC™ as an oxygen source (Chapman et al., 1997; Smyth et al., 1995) were not observed here.

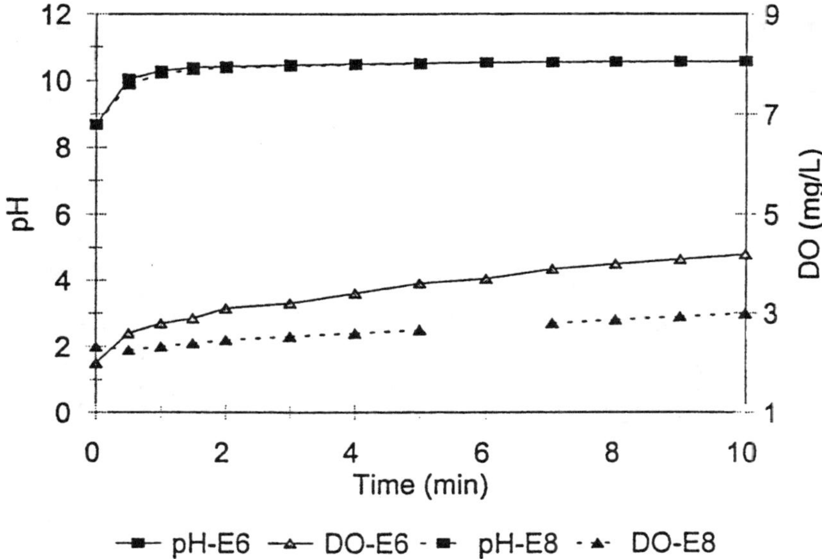

Figure A15.2 Results for "basic" ORC™ laboratory experiments.

The results of pH-constrained experiments conducted in the laboratory (using deionized water) are provided in Figure A15.3 and demonstrate that effective oxygen release from the ORC™ can be achieved when pH values are maintained in the near neutral range (7 to 8). DO concentrations of greater than 15 mg/L were obtained after only 10 minutes of contact between the water and ORC™. At higher pH, increases in DO were considerably less. DO increased from 8 to approximately 11 when pH was maintained between 8 and 9, while for a pH range of 9 to 10, only a very slight increase in DO from 8 to 9 was observed. The use of iron-equilibrated water, rather than deionized water, did not hinder the effective release of oxygen from the ORC™ when the pH was maintained near neutral. Thus, the performance of the ORC™ does not appear to depend upon the reductive potential (E_h) of the system.

A15.2 OPTIONS FOR REDUCING PH OF WATER EXITING IRON WALLS

A15.2.1 Sparging with Carbon Dioxide

Sparging with carbon dioxide was completed in the biosparge portion of the Alameda site. Details of the program are provided in Chapter 5. Results indicate that pH could be successfully maintained below 7.5.

A15.2.2 Acid Addition

The slow addition of hydrochloric acid (HCl) to the ORC™ system during laboratory batch experiments was shown to provide an effective means of activating the oxygen releasing reaction. A modification of this idea was implemented at the field site between March 18 and May 7, 1997.

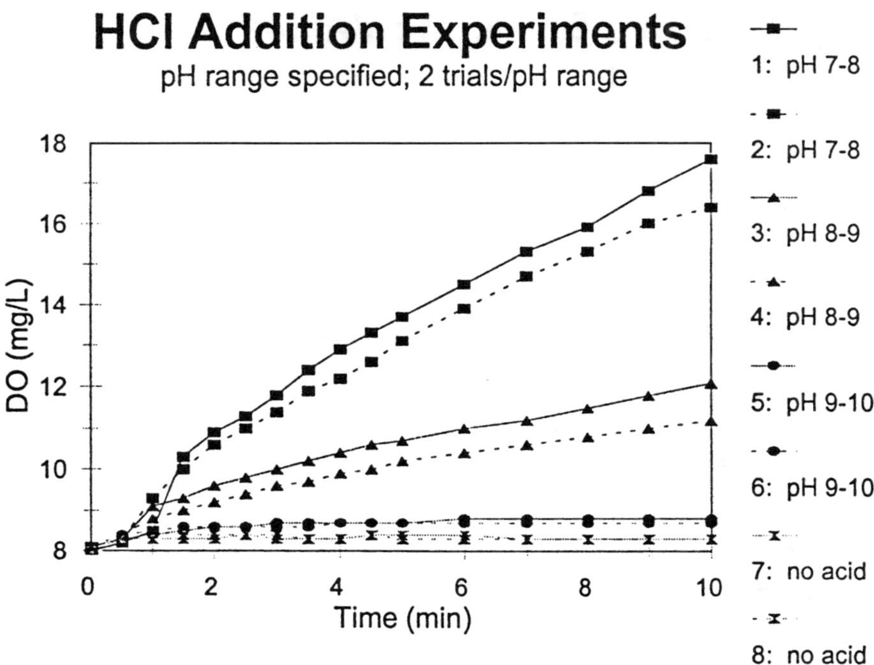

Figure A15.3 Results of laboratory HCl acid addition experiments.

A15.2.2.1 Method

pH monitoring, using the method described in Appendix 4, was completed to assess background conditions in the open water zone between cassettes C3 and C4 (Figure A15.4). Hydrochloric acid (HCl) was injected into this zone at times when monitoring indicated that water pH was greater than 8.5. Monitoring and injection were conducted at least twice per week. The volume of acid (1 M solution) added was determined based on the background conditions measured. The appropriate

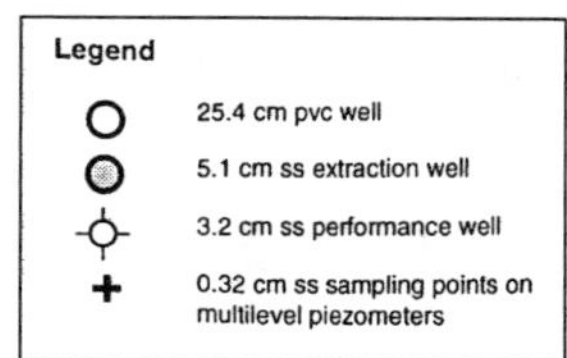

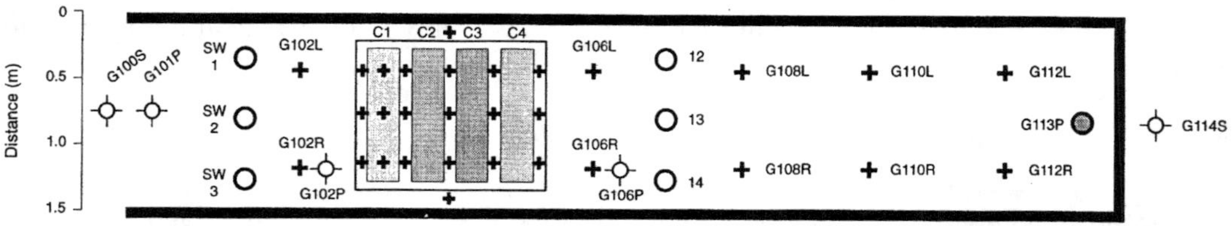

Figure A15.4 Plan view of Gate 1.

volume (typically 200 mL) was diluted using deionized water, to create a total injection volume of 1 L. Low density polyethylene (LDPE) tubing, perforated over its length and weighted at the bottom, was used for the injections. Three sections of tubing were used to provide lateral distribution of the acid across the width of the cassettes. Disruption to the flow system was minimized, although minor efforts to mix the injected acid and the groundwater were undertaken using the LDPE tubing.

A15.2.2.2 Results

Figure A15.5 provides a summary of the results of the field acid addition experiments. Solid lines show data for fence G105 (located between C3 and C4), while data for fence G1C4 (located downgradient of C4) are shown as dotted lines. pH at fence G105 was maintained below 8.5, while at fence G1C4, it measured approximately 10. No increases in DO above those measured at the upgradient fence (G105) were observed at fence G104. Also, monitoring for DO within C4 directly, using a typical multilevel piezometer, was conducted on occasion, and maximum DO concentrations of 5 mg/L were obtained. Values up to 40 mg/L were expected for a properly functioning system. pH measurements taken within C4 were high (pH >10), suggesting that influent pH control was not the only problem here.

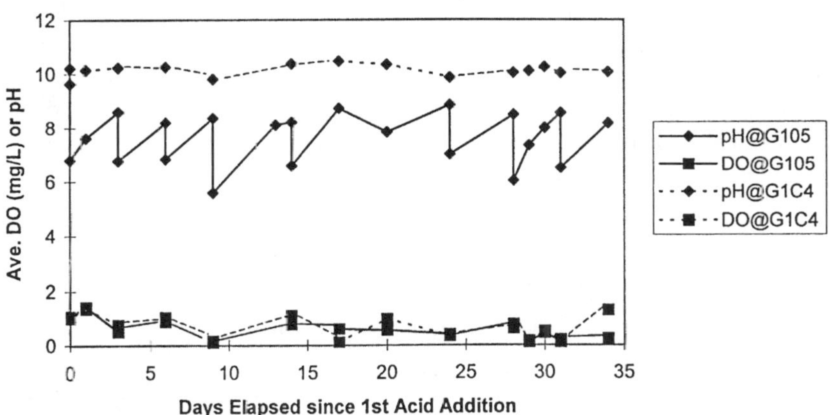

Figure A15.5 Results of field acid addition experiment.

A15.2.3 Sulfide Minerals

In the sulfide ore mining industry, the production of acidic waters from mine tailings is a critical environmental problem. For this project, however, it was suggested that the oxidation of sulfide minerals might solve the pH problem associated with using iron and ORC™ sequentially. The following reaction, for pyrite specifically, was the basis for the proposed idea:

$$FeS_2 + 7/2\ O_2 + H_2O \rightarrow Fe_{2^+} + 2\ SO_4^{2-} + 2\ H^+$$

Under the right conditions, pH decreases should be observed; however, the reaction is limited by the need for oxygen. The highly reducing conditions typical of effluent from iron barriers may not be suitable for this reaction to initiate. Also, control of the reaction to prevent an over-compensation to acidic conditions may also be complicated.

Two methods of utilizing sulfide minerals in conjunction with granular iron were proposed. The first method involved the mixing of the two materials directly within an *in situ* wall to simultaneously degrade the target organics and maintain the pH within a near-neutral range. This concept was demonstrated in the laboratory by Burris et al. (1995) for the treatment of PCE and TCE. Batch systems of granular iron and pyrite were tested and pH was held between 6.4 and 6.7 over a period of 456 hours. The alternate suggestion for the use of sulfide minerals with granular iron followed the overall UW-AATDF project design and involved the concept of sequential treatment. In this design, individual cassettes of each material could be installed to work independently, but ultimately, the same result should be achieved — effluent pH should be near neutral.

Side effects related to increased concentrations of inorganics, particularly iron, that may cause increased precipitation or fouling need to be considered prior to any field implementation. Nonetheless, the conceptual design seems promising.

A15.2.4 Natural Buffering by Aquifer Minerals

Installation of ORC™ in C4 did not provide a significant increase in DO, possibly due, at least in part, to the pH compatibility problem identified above. An alternative to the cassette system for ORC™ installation was the large diameter (25 cm) wells (SW-12 to 14) located in the aquifer approximately 2 m downgradient. The pH at fences G106 and G108 was considerably lower than in C4, typically between 7.5 and 8.5. This decrease in pH was likely the result of natural buffering by aquifer materials, and was evidenced in early pH monitoring data (Figure A15.6). However, the presence of the leak could also have some effect. Water bypassing the iron system through the leak has a pH typical of what was seen at fence G102. If the volume of this leak is significant, the interaction between the iron and ORC™ is eliminated. Nonetheless, successful implementation of this idea demonstrates that the desired reduction in pH causes effective oxygen release from the ORC™.

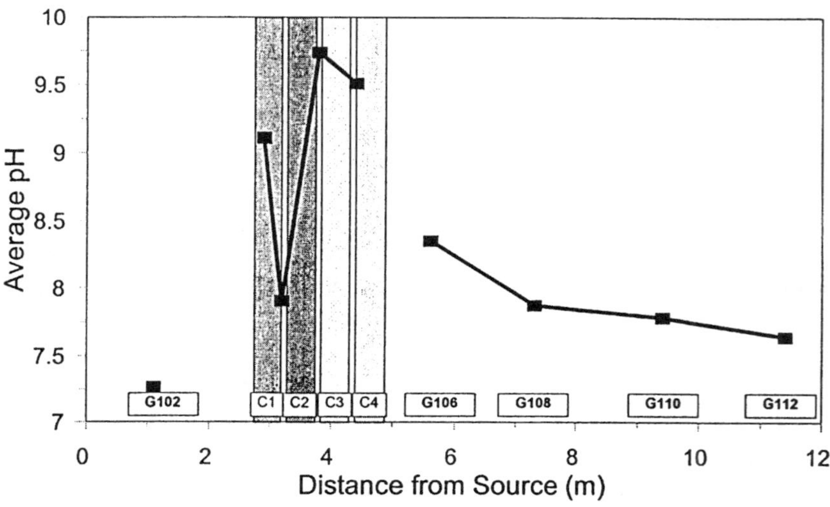

Figure A15.6 Natural pH buffering by aquifer materials.

The ORC™ socks were installed in wells SW-12 to 14 on June 19, 1997, and monitoring has yielded encouraging results. Oxygen levels in the downgradient aquifer have increased slightly, but the impact that this potential DO increase has had on the microbial population is unknown at this time. Results from VOC sampling completed in September and October 1997 indicate that toluene concentrations are declining with time and with distance traveled through the gate. This suggests that the provision of oxygen using ORC™ has successfully stimulated the microbes to degrade the target compounds.

A15.3 MODIFICATIONS OF IRON TO ELIMINATE PH PROBLEM

A brief literature review shows two possible modifications to the granular iron portion of the sequential design that may reduce the pH of effluent water and hence eliminate the incompatibility problem before it arises. Alternatives include the use of nickel-plated iron and the mixing of sulfide minerals with iron (discussed above).

Dehalogenation of chlorinated solvents using nickel-plated iron does not cause a pH increase. This technology may provide a suitable alternative to the commercial-grade granular iron commonly used, particularly when pH incompatibility is an issue. At this time, however, the nickel-enhanced iron technology is still in the early stages of development and is not practical for field applications (Gillham, personal communication). This work concentrated on improving the reactivity of the iron by maintaining a near-neutral pH, but ultimately the outcome is the same — no increase in pH was observed.

Neither of these modifications have been evaluated in this study, but their implementation in the future could be considered.

A15.4 REFERENCES

Burris, D.R., T.J. Campbell, and V.S. Manoranjan. 1995. Sorption of trichloroethylene and tetrachloroethylene in a batch reactive metallic iron — water system. *Environmental Science and Technology*. 29(11): 2850.

Chapman, S.W., B.T. Byerley, D.J.A. Smyth, and D.M. Mackay. 1997. A pilot test of passive oxygen release for enhancement of in-situ bioremediation of BTEX-contaminated ground water. *Ground Water Monitoring and Remediation*. Spring 1997: 93.

Smyth, D.J.A., B.T. Byerley, S.W. Chapman, R.D. Wilson, and D.M. Mackay. 1995. Oxygen-enhanced in-situ biodegradation of petroleum hydrocarbons in groundwater using a passive interception system. Published in the *Proceedings from the 5th Annual Symposium on Groundwater and Soil Remediation (GASRep)*, Toronto, Ontario, October 2-6, 1995.

Appendix 16

Surface Investigations of Granular Iron

This Appendix addresses questions related to changes in surface characteristics that occur when the granular iron is exposed to groundwater. These changes include autoreduction of surface films and the formation of precipitates that can affect the performance of the iron technology. Section A16.1 provides a detailed description of the sample collection methods, while analytical procedures are given in Section A16.2. Results of this work are discussed in Section A16.3.

A16.1 SAMPLE COLLECTION

Solid samples of "as received" materials were retained for analysis. "Exposed" samples were obtained on August 29, 1996, after approximately 4 months of exposure to Borden groundwater. "Final" samples were collected on September 3, 1997, after approximately 16 months of exposure to Borden groundwater and 9 months of exposure to organic contaminants.

During the initial sampling program, three cores of the reactive media in cassettes C1 to C3 (consisting of an iron/silica sand mixture, iron alone, and a limestone/silica sand mixture, respectively) and one from the natural aquifer formation were collected. Cores were designated as C1-1, C2-1, C3-1, and F-1, respectively, and their locations are shown in Figure A16.1.

Samples were collected in aluminum core barrels (typically 5 cm in diameter and 1.5 m in length using the method described by Starr and Ingleton (1992)). A brief description of the method is provided here.

A hand auger was used to remove material between ground surface (or the surface of the reactive material) and the water table (located at a depth of approximately 1.3 m). A drive point/piston sampler was vibrated into position in the subsurface using a gasoline-powered jackhammer, and the sample taken over the desired interval. The core barrel assembly was then removed from the ground using either an electric winch or a manually operated jack and the inner barrel containing the sample removed. The core barrel was then cut to a maximum length of 0.3 m centered about the sample itself and the ends of the sample were then sealed with plastic core caps and duct tape, and the core labeled for identification. The cores were transported to the University of Waterloo (UW) on the day of collection in coolers containing ice packs and were placed in an anaerobic chamber immediately upon arrival at the UW laboratory.

Within the cassettes, the medium was loosely placed and coarse grained. To minimize sample loss out of the core barrel, a 2.5-cm core barrel, rather than the standard 5-cm barrel, was used. The material tended to compact and bind within the smaller barrel resulting in improved recovery. Every effort was made to obtain the sample from approximately the middle of the saturated material thickness (corresponding to a depth of 2.7 m bgs). Depending on the degree of compaction of the media, coring commenced at depths ranging from 2.2 to 2.4 m and extended to depths ranging

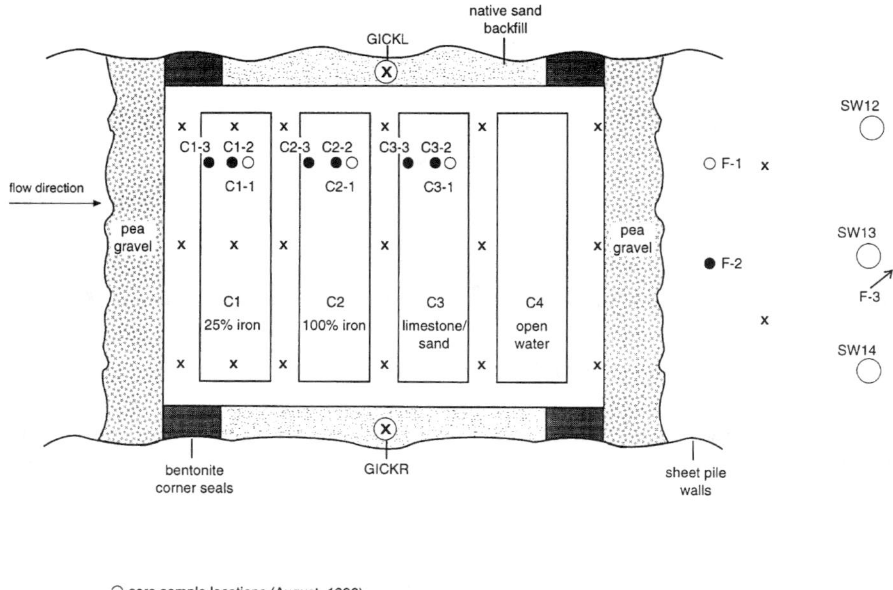

Figure A16.1 Solid sampling locations.

from 3.0 to 3.3 m, centered around the desired sampling depth (2.7 m bgs). During sampling, the material compacted from 2 to 3 times its *in situ* thickness, resulting in a completed core less than 30 cm in length. This maximum length of 30 cm was established to allow easy placement of the sample in an anaerobic glove box for storage. Cores were sealed with duct tape immediately after removal from below the water table to minimize contact with air.

The fourth core (designated as F-1) was collected from the aquifer approximately 0.5 m downgradient of the cassette system. The sample was collected from a depth interval of approximately 2.55 to 2.85 m bgs.

The final solid sampling event was completed in a manner similar to the August 1996 program. To provide the best comparison of surface characteristics, samples from within the cassettes were collected from approximately the same locations as before (designated C1-2, C2-2, and C3-2). Additional samples were also obtained from near the front face of each cassette (designated C1-3, C2-3, and C3-3), since precipitates would most likely form upon initial contact with a new media. Two aquifer samples were collected from depths of approximately 2.7 m. The first core was collected downgradient of C4 (similar location as previous sampling; designated as F-2), and the second, designated as F-3, was collected downgradient of the alternate oxygen addition wells (SW-12 to 14). Figure A16.1 shows the September 1997 sampling locations.

A16.2 ANALYTICAL PROCEDURES

A16.2.1 Sample Preparation

Samples were prepared for analysis in a Lab-Line Instruments, Inc., Programmed Anaerobic Controlled Environment. Each sample tube was cut across its diameter using a pipe cutter, and a subsample from this location was transferred wet to a sealable glass bottle. Each bottle was then filled with site water (collected from fence G1C1) and stored in the anaerobic chamber. This handling procedure ensured the collection of samples that had experienced little or no exposure to air and thus minimized the chances of surface modification.

A16.2.2 Raman Microspectroscopy

Immediately prior to analysis, samples were further prepared in the anaerobic chamber. A few grains of sample material were transferred from the sealed bottle into a specially designed spectroscopic cell (Figure A16.2). The cell was then filled with site water and tightly sealed, allowing removal of the sample from the anaerobic chamber without the opportunity for contact with oxygen.

Raman spectra were obtained using a Renishaw 1000 Raman microscope system consisting of an Olympus microscope, a single spectrograph fitted with holographic notch filters for spectroscopy mode, and a Peltier-cooled CCD detector. The optical sensitivity of the instrument is high and the detection of a very weak signal (1 photon/s) is possible. Excitation was achieved using the 632.8-nm line of a Melles Griot low-power HeNe laser, and measurements were conducted in the backscattering geometry. For spectral studies of *in situ* samples (wet), the long, working-length objective lens with an objective magnification of 50 (OLYMPUS IC50/MSPLAN 50, 4/0, f = 180, NA = 0.55) was used. The laser was operated at its maximum power (30 mW), equivalent to approximately 5 mW at the sample surface; however, at this low light intensity, surface film composition was unlikely altered. With this configuration, at a well-defined, flat sample surface, a spot size of 2 μm could be examined with a confocal depth of field of 3 μm. For the background samples, *ex situ* studies were completed at an objective magnification of 100, giving a spatial resolution of 1 μm.

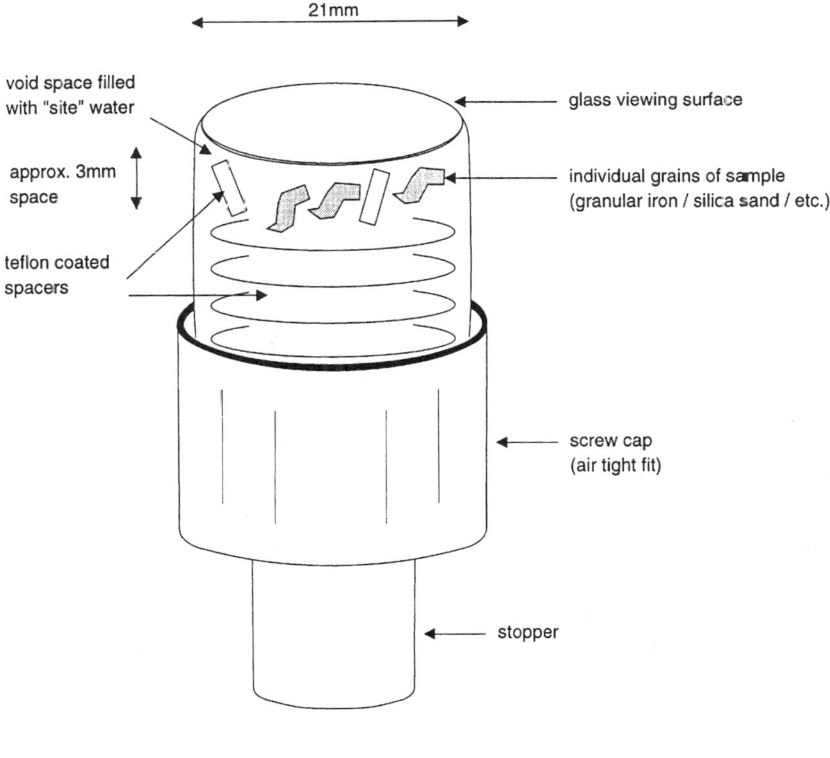

Figure A16.2 Schematic of spectroscopic cell. *Note:* not to scale.

A16.3 RESULTS

A16.3.1 Reference Spectra

Raman spectra of reference iron oxides and selected hydroxides are provided in Figure A16.3. A good correlation between the Raman results from this study and available literature data is evident from Table A16.1, which provides a summary of observed and expected band positions for Raman vibrational modes.

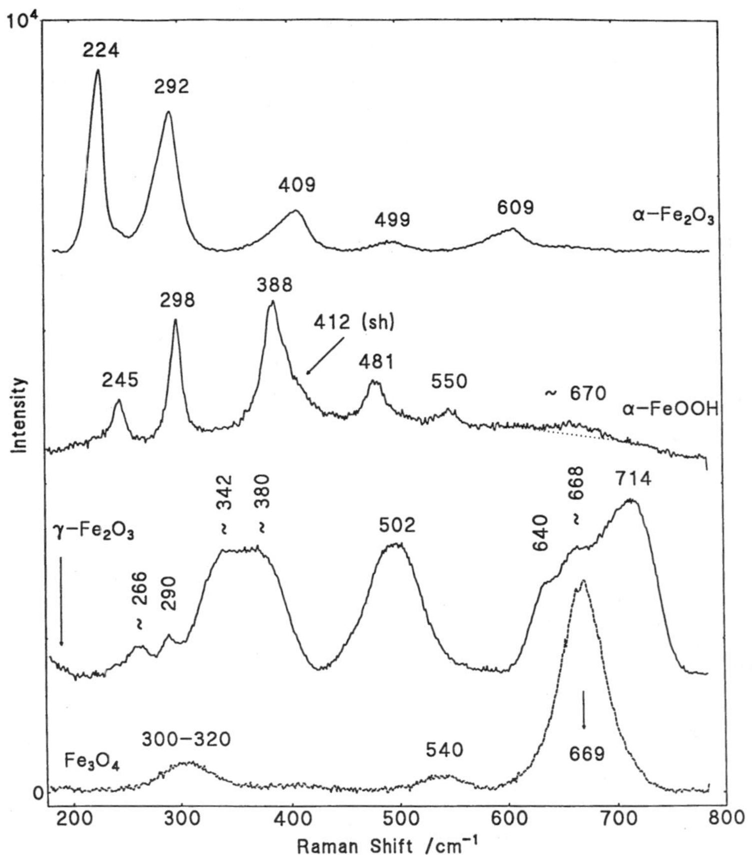

Figure A16.3 Reference spectra.

A16.3.2 Background Samples

Raman spectra of the unused Peerless iron are shown in Figure A16.4. These results clearly indicate both the horizontal and vertical heterogeneity of each iron grain. The vertical heterogeneity was manifested by the existence of outer and inner surface oxide films. Outer oxide films were identified through the examination of the undisturbed surface of the iron grains, while inner oxide films were examined following the mechanical removal of the outer layers (i.e., scraping of the grain's surface using a sharp knife). The outer surface layer consisted of α-Fe_2O_3 and γ-Fe_2O_3 (shown as A and B, respectively, in Figure A16.4) with randomly distributed islands of separated α and γ phases. Considerable variability between individual iron grains and on any particular grain was evident. These oxide films are visible to the naked eye (appear as rusty coating) and result from the manufacturing of the product. The granular iron filings

Table A16.1 Raman Bands of Selected Iron Oxides in 200–800 cm^{-1} Region

Species	Raman Band Position (cm–1)									Ref.
α-Fe_2O_3	225	245	295		415	500	615			(1)
α-Fe_2O_3	227	245	293	298	414	501	612			(2)
α-Fe_2O_3	226	245	293		413	500	612			(3)
α-Fe_2O_3	224	244	291		409	499	609			this
	vs	vw	s		m	vw	w			work
γ-Fe_2O_3	265	300	345		395	515	645	670	715	(1)
γ-Fe_2O_3			350			505		660	710	(4)
γ-Fe_2O_3	263		350		380	505	650		740	(5)
γ-Fe_2O_3	266	290	342		380	502	640	668	714	this
	vw	vw	s, sh		s, sh	s	s, sh	s, sh	s, sh	work
Fe_3O_4		300	320		420	560	680			(6)
Fe_3O_4		294	319		415	540	669			(7)
Fe_3O_4						540	665			(1)
Fe_3O_4			310			540	669			this
			vb			w	vs			work
α - FeOOH	245	300	390	420	480	550	685			(1)
α - FeOOH		299	397		479	550	685			(8)
α - FeOOH	250	300	385		470	560				(5)
α - FeOOH		298	397	414	474	550				(2)
α - FeOOH	245	298	388	412	481	550	570			this
	m	s	s	m, sh	m	w	vb			work

Notes: A_1 modes and other lines of a hight intensity are underlined.

Intensity statements and abreviations:

vs - very strong
s - strong
m - medium
w - weak
vw - very weak
vb - very broad
sh - shoulder

References:

(1) Ohtsuka, Kubo and Sato, 1986
(2) Thibeau, Brown, and Heicersbach, 1978
(3) Beatties and Gilson, 1970
(4) Boucherit, Hugot-Le Goff, and Joiret, 1991
(5) Thierry et al., 1991
(6) Verble, 1974
(7) Odziemkowski, Flis, and Irish, 1994
(8) Nauer et al., 1985

which are produced in metal tooling industries undergo a high temperature treatment to remove grease and oil (i.e., contaminants) from their surfaces. The presence of these Fe(III) oxides hinders the transfer of electrons at the interface (Pourbaix, 1973; Frankenthal and Kruger, 1978) and, therefore, might slow the reductive dechlorination reaction (assuming that charge transfer at the iron surface is the mechanism acting in the reaction). In contrast, the inner layer, shown as C in Figure A16.4, consists of magnetite (Fe_3O_4), which possesses less protective properties than the passive α-Fe_2O_3 and γ-Fe_2O_3 layers and allow charge transfer to occur through it.

A16.3.3 Exposed Samples

Figure A16.5 provides a comparison of Raman spectra data for exposed (A and B) iron samples and for magnetite (C). It is evident from these spectra that, after equilibration, the γ-Fe_2O_3 disappeared from the iron surface, as had most of the α-Fe_2O_3 film (many of the α-Fe_2O_3 bands were no longer visible). This supports the autoreduction theory presented in the literature (Odziemkowski and Gillham, 1997). Without the autoreduction of these passive surface films, one might expect a reduced rate of degradation of chlorinated solvents in contact with this material (again, assuming that the reductive dechlorination process results from surface contact). The mechanisms responsible

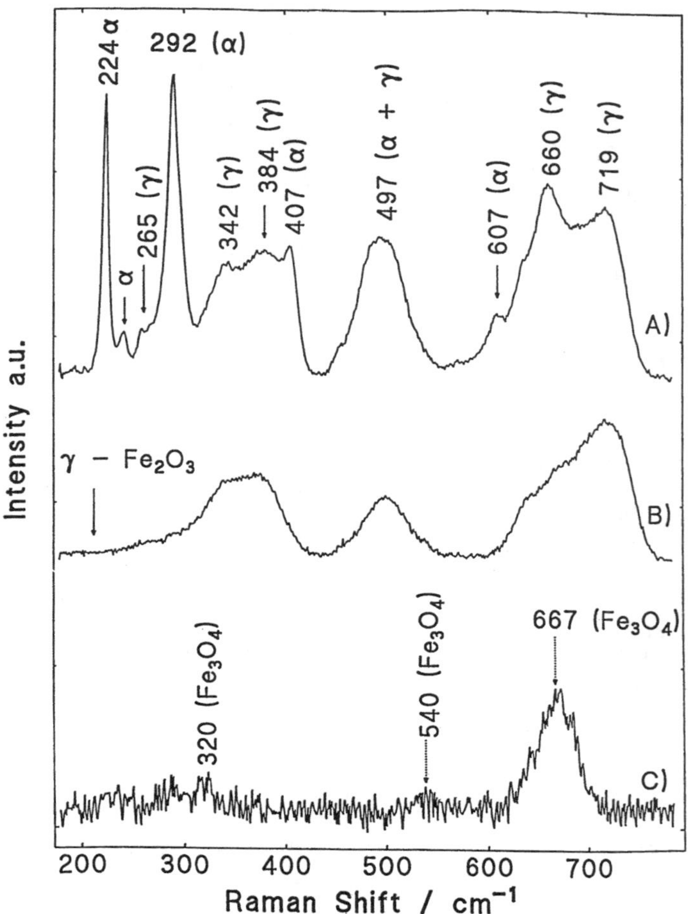

Figure A16.4 Spectra of background samples (Fe(III) oxides in outer surface layer — A and B; magnetite in inner layer — C).

for the degradation are not known; however, charge transfer from the iron surface has been suggested as a likely cause (Matheson and Tratnyek, 1994; Weber, 1996).

The formation of two new broad bands at 420 and 507 cm^{-1} indicate the presence of "green rust" (Boucherit and Hugot-Le Goff, 1992), formed as a result of the autoreduction process. Green rust is a thick unstable colloidal hydroxy salt layer, comprising Fe^{2+}, Fe^{3+}, hydroxyl ions, and interlayer hydroxyl or nonhydroxyl ions, and may include sulfates, halides, nitrates, and carbonates (Simard et al., 1997). Green rust behaves like magnetite, allowing charge transfer.

A16.3.4 Final Samples

The surface characteristics of the final samples were very similar to those observed for the exposed samples. Surface heterogeneity was evident and identified compounds include green rust, magnetite (Fe_3O_4), oxyhydroxides, and in certain spots α-Fe2O3 and γ-Fe2O3 films observed on the background samples. One spot of calcite was also identified on sample C2-3 obtained from the upgradient side of the cassette. Raman spectra for three spots on the iron samples that were typically observed and that demonstrate the variability of the surface characteristics are provided in Figure A16.6. Spectrum A shows evidence of green rust and Fe_3O_4, while Spectrum B indicates the

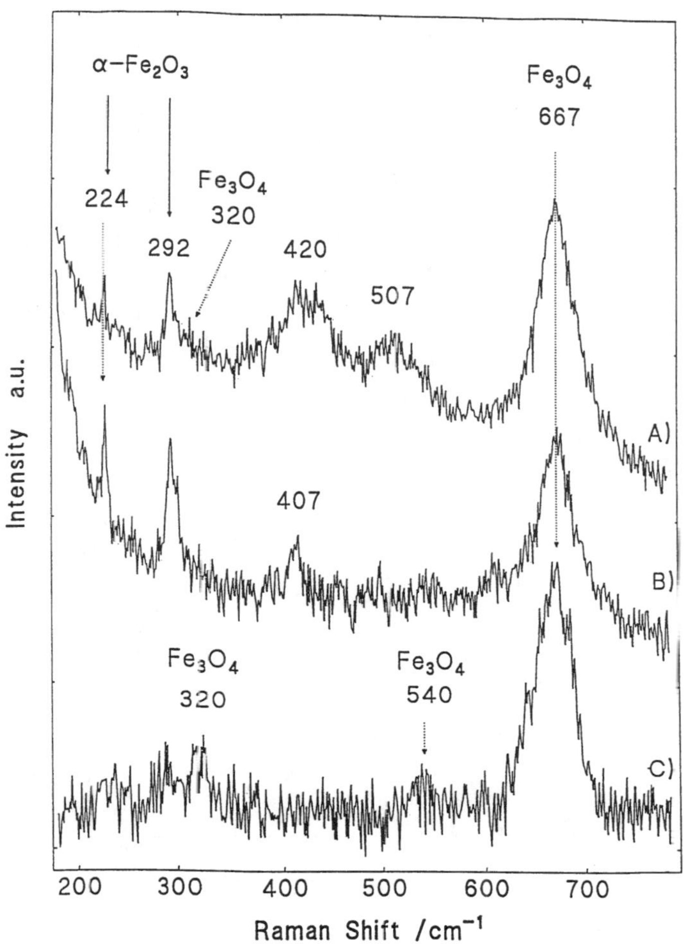

Figure A16.5 Spectra of exposed samples (A and B) and magnetite (C).

presence of α-Fe_2O_3, Fe_3O_4, green rust, and a very small amount of α-FeOOH, and Spectrum C identifies only magnetite. Figure A16.7 shows complete spectra for all wavelengths from a different spot on sample C2-3. The band at approximately 670 cm^{-1} indicates the presence of magnetite, green rust is evident based on bands at 432 and 503 cm^{-1}, and the bands at 156, 282, 714, and 1087 cm^{-1} are indicative of calcite (Herman et al., 1987).

The presence of passive oxide films (consisting of α-Fe_2O_3 and γ-Fe_2O_3) indicates that for this particular commercial-grade granular iron (produced by Peerless), complete autoreduction of these films did not occur on most of the surfaces investigated. Degradation rates for chlorinated compounds within the iron depend upon the surface area of the iron (Gillham and O'Hannesin, 1994). However, it is more likely only the unprotected portion of the surface that acts in the reductive dechlorination reaction and thus, only this percentage of the surface area should be considered. Calculated half-life values are normalized for comparison to other studies using the measured specific surface area (determined by BET analysis), without accounting for removal of the passive films. The ability of the iron to autoreduce these surface films is expected to vary depending upon the nature of the films, which likely depends on the manufacturing/treatment process. The normalized half-life values determined for this study were approximately 8 for PCE and between 10 and 14 hours for CF, while for previous laboratory studies using commercial-

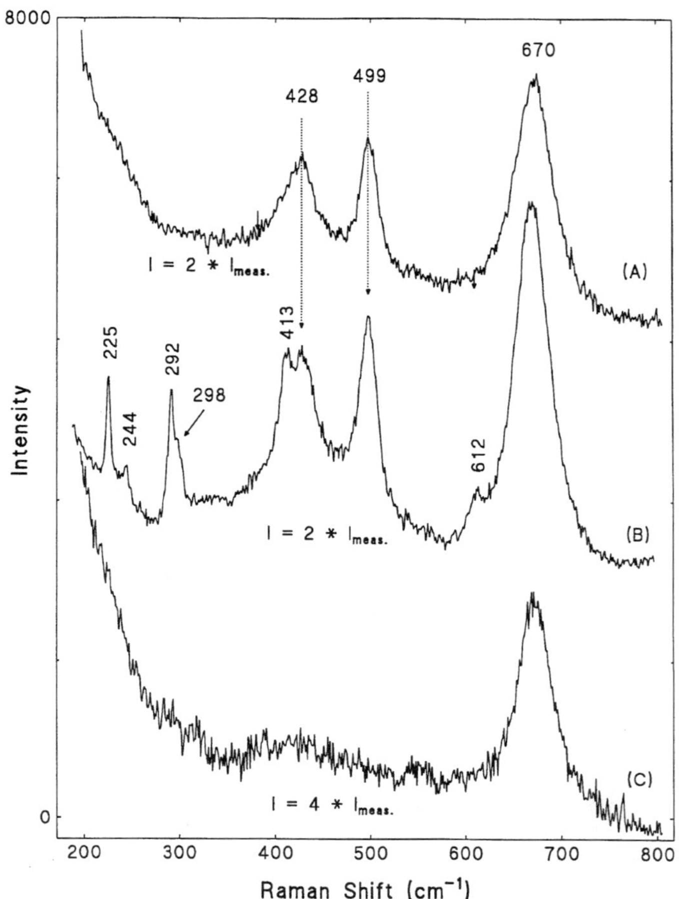

Figure A16.6 Spectra of final samples (showing surface heterogeneity).

grade iron, values ranging from 2 to 10.8 hours for PCE and a value of 4.8 hours for CF were estimated. The slower rates obtained here could be the result of the reduced surface reactivity of the Peerless iron compared to other commercially available granular iron materials (e.g., Master Builder, Fischer Scientific, Connelly).

As discussed above, the presence of magnetite and green rust on the metal surface does not substantially hinder the transfer across the metal/solution interface. The presence of these compounds after long-term exposure to Borden groundwater and contaminants is not surprising since they form as a result of contact with water, and in the case of green rust, water containing a variety of inorganic species.

Very small quantities of iron oxyhydroxides were identified on the final samples. These compounds were not present on background samples, but rather, formed only after contact with water and as a result of the autoreduction process. The formation of large quantities of iron oxyhydroxide films on the iron surface will hinder the degradation process, regardless of the mechanism of the reductive dechlorination reaction. If charge transfer at the surface is responsible for the reaction, passive FeOOH films will hinder charge transfer at the surface, while if charge transfer occurs between dissolved species, the reaction would be slowed since dissolution of the metal would be limited by the presence of the oxyhydroxides. Also, hydrogen production, which has been proposed as the driving force behind hydrogenolysis, cannot occur if extensive FeOOH films exist on the granular iron surfaces.

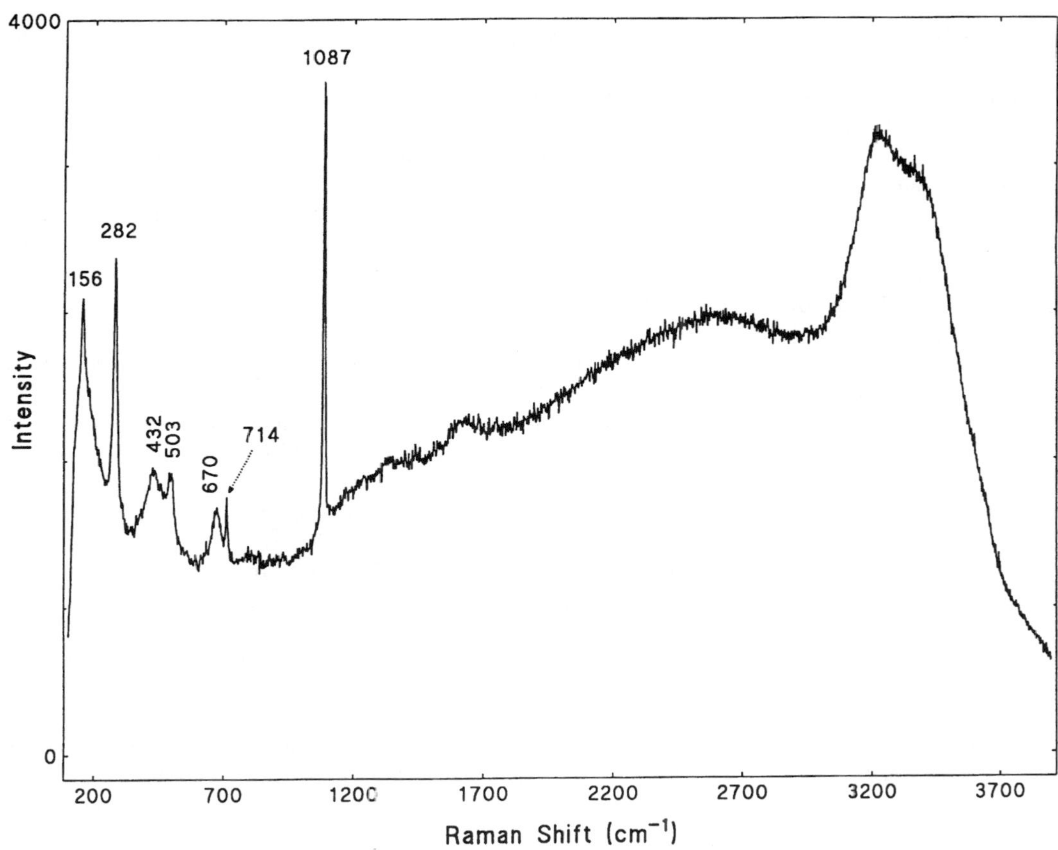

Figure A16.7 Full spectra for sample C2-3 (calcite identified).

Identification of only one spot of calcite on the iron samples was encouraging, since it indicated minimal formation of precipitates on the iron surfaces, and little change in hydraulic conductivity through the reactive wall. It was not surprising to identify this compound on the sample obtained on the upgradient side of the wall. Precipitate formation was typically observed, in the laboratory at least, at the influent end of columns containing iron (Schuhmacher, 1995). Thus, the upgradient portion of a reactive iron wall would be most susceptible to this behavior. Precipitate formation on samples from the upgradient side of cassette C1 would be expected; however, these samples were not analyzed as part of this work.

A16.4 REFERENCES

Beatties, I.R. and T.R. Gilson. 1970. *Journal Chemical Society America*. p. 980.

Bonin, P.M.L., M.S. Odziemkowski, and R.W. Gillham. In preparation.

Boucherit, N., A. Hugot-Le Goff, and S. Joiret. 1991. Raman studies of corrosion films on Fe and Fe-6Mo in pitting conditions. *Corrosion Science*. 32: 497.

Boucherit, N. and A. Hugot-Le Goff. 1992. *Faraday Discuss*. 94: 17.

Dean, K.J., W.F. Sherman, and G.R. Wilkinson. 1982. Temperature and pressure dependence of the raman active modes of vibration of "α-quartz." *Spectrochimica Acta*. 38A(10): 1105.

Frankenthal, R.P. and J. Kruger (Eds.). 1978. *Passivity of Metals*. The Electrochemical Society Inc., Princeton, NJ.

Herman, R.G., C.E. Bogdan, A.J. Sommer, and D.R. Simpson. 1987. Discrimination among carbonate minerals by raman spectroscopy using the laser microprobe. *Applied Spectroscopy*. 41(3): 437.

Gillham, R.W. and S.F. O'Hannesin. 1994. Enhanced degradation of halogenated aliphatics by zero-valent iron. *Ground Water*. 32(6): 958.

Matheson, L.J. and P.G. Tratnyek. 1994. Reductive dehalogenation of chlorinated methanes by iron metal. *Environmental Science and Technology*. 28: 2045.

Nauer, G., P., Strecha, N. Brinda-Konopik, and G. Lipstay. 1985. Spectroscopic and thermoanalytical characterization of standard substances for the identification of reaction products of iron electrodes. *J. Therm. Anal.* p. 813.

Odziemkowski, M.S. and R.W. Gillham. 1997. Surface redox reactions on commercial grade granular iron (steel) and their influence on the reductive dechlorination of solvent: micro raman spectroscopic studies. Presented at 213th ACS National Meeting, San Francisco, CA., American Chemical Society, Division of Environmental Chemistry, Preprints of Papers, 37(1): 177, April 1997.

Odziemkowski, M., J. Fils, and D.E. Irish. 1994. Raman spectral and electrochemical studies of surface film formation on iron and its alloys with carbon in $Na_2CO_3/NaHCO_3$ solution with reference to stress corrosion cracking. *Electrochim. Acta*. 39: 2225.

Ohtsuka, T., K. Kubo, and N. Sata. 1986. Raman spectroscopy of thin corrosion films on iron at 100 to 150°C in air. *Corrosion*. 42: 476.

Pourbaix, M. 1973. *Lectures on Electrochemical Corrosion*. Plenum Press, New York.

Schuhmacher, T.T. 1995. Identification of Precipitates Formed on Zero-Valent Iron in Anaerobic Aqueous Solutions. M.Sc. Thesis, University of Waterloo, Waterloo, Ontario, Canada.

Scott, J.F. and S.P.S. Porto. 1967. Longitudinal and transverse optical lattice vibrations in quartz. *Physical Review*. 161(3): 903.

Simard, S., M. Odziemkowski, D.E. Irish, L. Brossard, and H. Ménard. In-situ micro-raman spectroscopic and electrochemical study of pitting corrosion of 1024 mild steel in phosphate and bicarbonate solutions containing chloride and sulfate. Submitted to *Journal Electrochemical Society*. May 1997.

Starr, R.C. and R.A. Ingleton. 1992. A new method for collecting core samples without a drill rig. *Groundwater Monitoring Review*. 12(4): 91-95.

Thibeau, R.J., C.W. Brown, and R.H. Hediersbach. 1978. Raman spectra of possible corrosion products of iron. *Applied Spectroscopy*. 32: 532.

Thierry, D., D. Persson, C. Leygraf, N. Boucherit, and A. Hugot-Le Goff. 1991. Raman spectroscopy and XPS investigations of anodic corrosion films formed on Fe-Mo alloys in alkaline solutions. *Corrosion Science*. 32: 273.

Verble, J.L. 1974. *Physical Review Bulletin*. 9: 5236.

Weber, E.J. 1996. Iron mediated reductive transformations: investigations of reaction mechanisms. *Environmental Science and Technology*. 30(2): 716.

Appendix 17

Gate 3 Modeling - Design of Nutrient Pulse Delivery

A17.1 DESIGN MODELING

Model simulations were conducted to aid in estimating the frequency of nutrient injections necessary to maintain a constant supply of benzoate-amended water at a selected point downgradient of the nutrient injection wall (NIW) in Gate 3. The procedure was to simulate pulse profiles over time at various distances downgradient of the permeable wall (Figure A17.1) using assumed or, where possible, measured hydraulic parameters.

A fixed nutrient threshold concentration (C_x) was selected to allow for substrate losses due to degradation (since the rate of field utilization was unknown), while ensuring a consistent overlap between pulses. The time interval (dt) between the chosen concentration (C_x) on the rising and falling limbs of the pulse profile determined the nutrient injection frequency necessary to achieve pulse overlap at the selected concentration, C_x. For nearly symmetrical breakthrough curves, dt is approximately equal to the time interval between centers of mass of the pulses, and thus is equivalent to the pulse interval (i.e., dt_2).

The model, PULSEPE, was written by J. F. Devlin (University of Waterloo) and solves a 1-D form of the advection–dispersion equation (Devlin and Barker, 1996). In a predictive capacity, all hydraulic parameters (and initial conditions) are specified for the generation of a set of theoretical breakthrough curves. In addition, the model can be used as a fitting routine for the estimation of hydraulic parameters.

All model simulations used a time interval of 0 to 100 days, considered long enough to encompass the entire period during which a simulated pulse traveled past the selected sampling points. Of the user-defined options, those that remained constant during each simulation or between simulaitons are summarized in Table A17.1.

The diffusion coefficient and dispersivity were based on similar experimental work completed at Borden (Devlin and Barker, 1996). Site-specific dispersivity was not available because the main tracer test did not produce tracer data suitable for accurate dispersivity determinations. The pulse width represents the length of the NIW (i.e., the distance between the extraction and reinjection wells). The retardation factor was taken as 1.0, the expected value for an anionic species such as benzoate. This expectation is supported by the findings of Gibbens and Bowman (1992) for difluorobenzoate analogs used as groundwater tracers. Input values for the injection concentration (in units of mg/L), the distance from source (measured as from the center of the wall to a selected point downgradient), and groundwater velocity were varied according to simulation requirements and data availability.

Calculated concentrations over time were subsequently arranged into plots, similar to Figure A17.1, and interpreted for nutrient flush frequency. The original operational designs were aimed

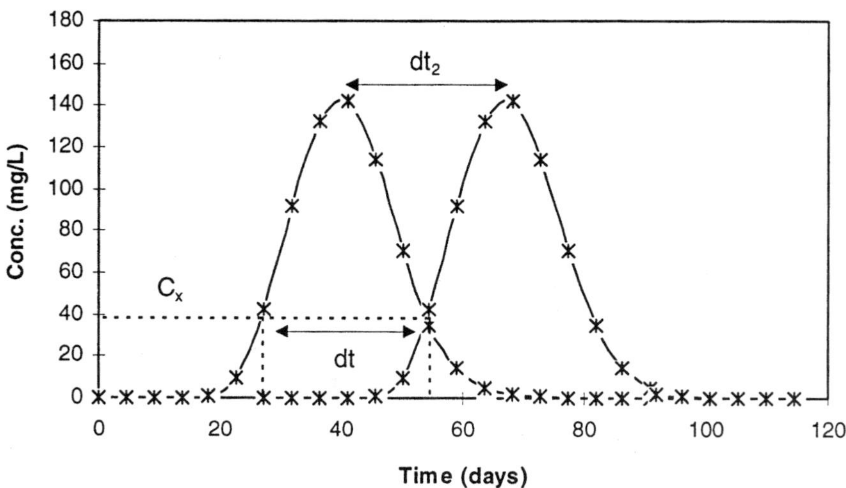

Figure A17.1 Simulated profiles of consecutive nutrient pulses, and determination of injection frequency.

Table A17.1 User-Defined Options Remaining Constant During Each and/or Between Each PULSEPE Simulation

Option	Value
Diffusion coefficient (m^2/sec)	1.00E-10
Dispersivity (m)	9.00E-02
Pulse width (m)	2.17
Retardation factor	1

at achieving pulse mixing sufficient to provide about 50 mg/L of sodium benzoate 5.2 m downgradient of the center of the permeable wall (i.e., fence G310).

Results of the initial model simulation are given in Figure A17.2. Early tracer test data indicated an average linear velocity of 13 cm/d based on measurement along middle (M) piezometers only. A nutrient input concentration of 300 mg/L sodium benzoate was used as per initial trials in the field. A nutrient injection interval of once approximately every 28 days was determined from this simulation. These design parameters formed the basis for operating the nutrient injection system from November 1996 to February 1997.

Subsequent analysis of tracer data up to fence G317 (using arithmetically averaged fence concentrations; see Appendix 7) indicated an average groundwater velocity of approximately 11 cm/d up to that fence, which was used as the input velocity in all subsequent model simulations.

Given the limited residual gas phase content of the pea gravel that filled the biosparge gate (Appendix 22), and the presence of residual sodium benzoate found downgradient of the gate in February 1997, an effort was made to reduce the biochemical oxygen demand (BOD) associated with the influx of sodium benzoate. This was to allow for a greater proportion of the dissolved oxygen within the gate to be available for toluene degradation.

Multiple model simulations were done using varying nutrient concentrations (100 to 250 mg/L) to estimate the lowest input concentrations that could be used to achieve an average overlap of 50 mg/L sodium benzoate at fence G310 while maintaining a 28-day nutrient injection interval. A more frequent injection interval, which would allow lower nutrient concentrations to be introduced into the aquifer, could not have been as low as to be within the time frame that a slug of amended

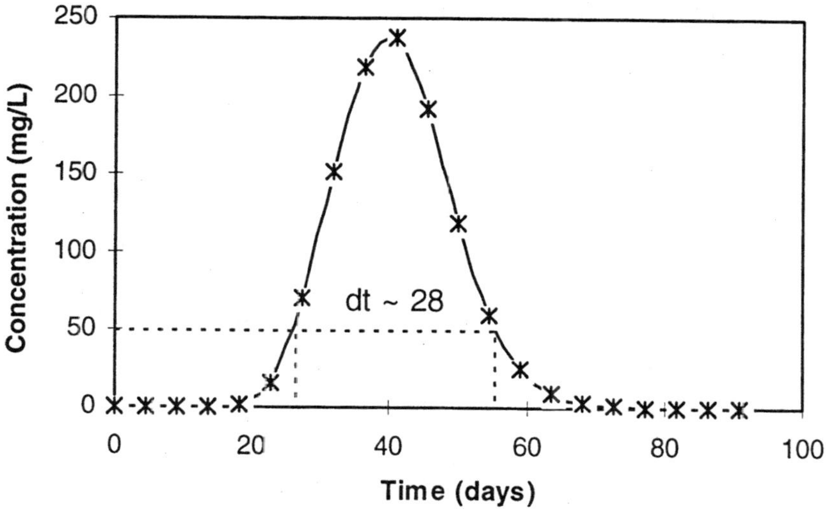

Figure A17.2 PULSEPE simulation result at G310.

groundwater would need to leave the nutrient injection wall. This would have detracted from the goal of utilizing dispersive mixing between pulses to maintain nutrient concentrations downgradient of the wall, as well as the semipassive mode of operation. Less frequent injection intervals would have dictated the use of higher input concentrations, which would subsequently increase the oxygen demand on the biosparge gate.

Input parameters were as given in Table A17.2. Results of the model simulations are summarized in Table A17.3 and Figures A17.3 to A17.8.

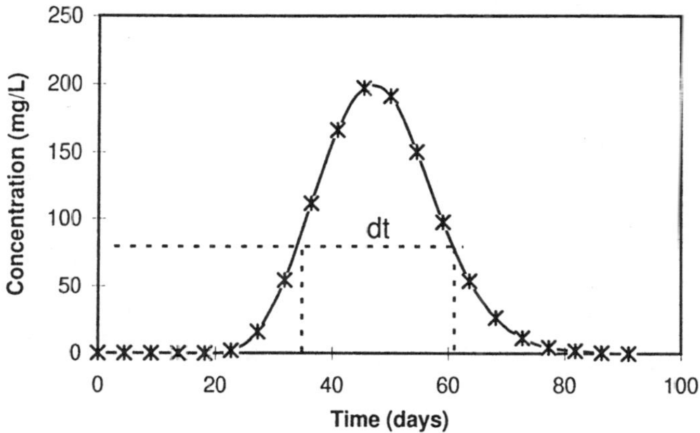

Figure A17.3 PULSEPE simulation results at G310, 250 mg/L nutrient input concentration, 11 cm/d groundwater velocity.

A minimum input concentration of 180 mg/L was selected for injections beginning April 18, 1997 and continued for the duration of the experiment. An average overlap concentration of approximately 54 mg/L (or 30% of the input concentration) was expected at fence G310 given an average groundwater velocity of 11 cm/d (Table A17.2 and Figure A17.9). This overlap concentration was assumed to be a maximum, since the rate and extent of substrate utilization in the field were unknown.

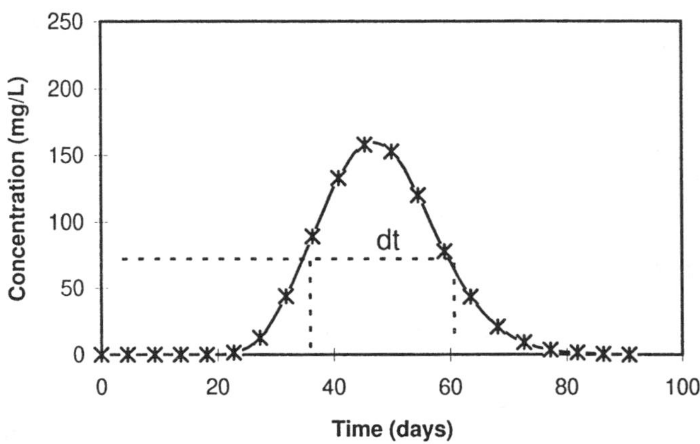

Figure A17.4 PULSEPE simulation results at G310, 200 mg/L nutrient input concentration, 11 cm/d groundwater velocity.

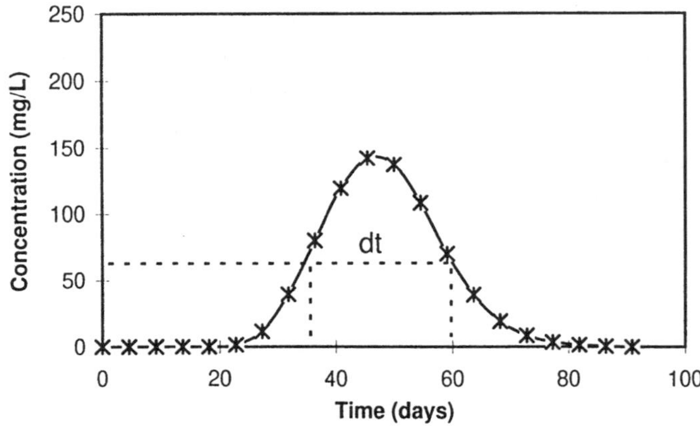

Figure A17.5 PULSEPE simulation results at G310, 180 mg/L nutrient input concentration, 11 cm/d groundwater velocity.

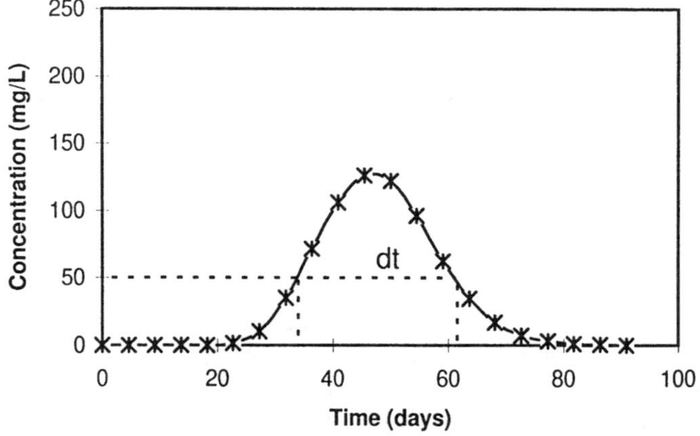

Figure A17.6 PULSEPE simulation results at G310, 160 mg/L nutrient input concentration, 11 cm/d groundwater velocity.

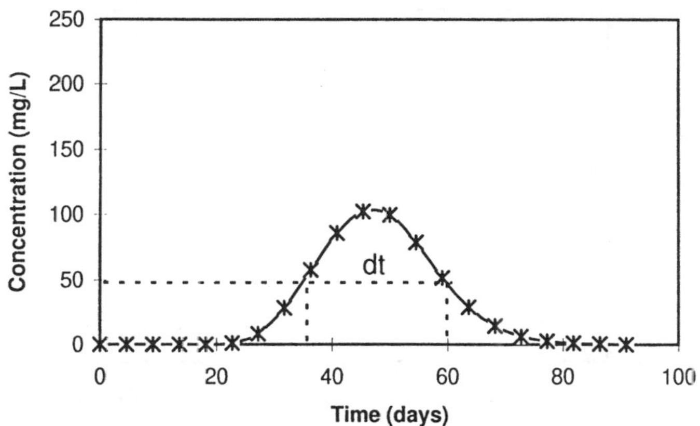

Figure A17.7 PULSEPE simulation results at G310, 130 mg/L nutrient input concentration, 11 cm/d groundwater velocity.

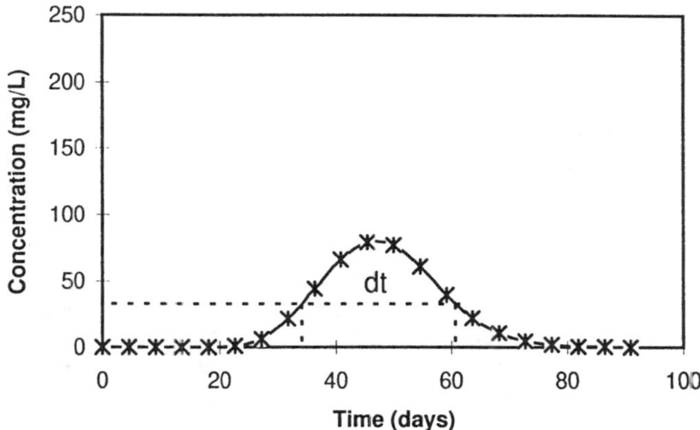

Figure A17.8 PULSEPE simulation results at G310, 100 mg/L nutrient input concentration, 11 cm/d groundwater velocity.

Table A17.2 Average Modeled Overlap Concentrations at G310 for Varying Input Concentrations

Input Concentration (mg/L)	Average Overlap Concentration (mg/L)
100	31
130	40
160	48
180	54
200	61
250	76
300	91

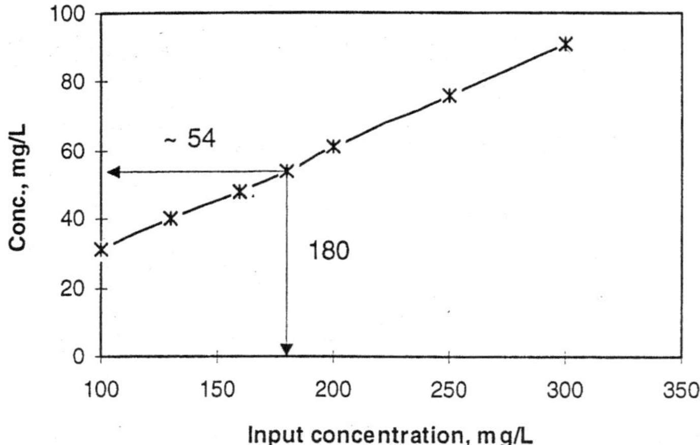

Figure A17.9 Comparison of average pulse overlap concentrations as a function of the input concentration at a fixed injection interval (28 days).

Table A17.3 Summary of Modeled Pulse Data for Varying Nutrient Input Concentrations

Time (days)	**250**	**200**	**180**	**160**	**130**	**100**
Input conc.:			**Calculated Concentrations**			
0	0	0	0	0	0	0
4.545	2.56E-31	2.05E-31	1.84E-31	1.64E-31	1.07E-31	8.20E-32
9.091	2.68E-11	2.14E-11	1.93E-11	1.71E-11	1.25E-11	9.62E-12
13.64	6.07E-05	4.86E-05	4.37E-05	3.89E-05	2.96E-05	2.27E-05
18.18	5.31E-02	4.25E-02	3.83E-02	3.40E-02	2.64E-02	2.03E-02
22.73	2.01	1.61	1.45	1.29	1.01	0.78
27.27	16.15	12.92	11.63	10.34	8.21	6.32
31.82	54.79	43.83	39.45	35.06	28.06	21.58
36.36	111.50	89.17	80.26	71.34	57.38	44.14
40.91	165.60	132.50	119.20	106.00	85.60	65.85
45.45	197.10	157.70	141.90	126.10	102.30	78.71
50	190.70	152.50	137.30	122.00	99.50	76.54
54.55	149.90	119.90	107.90	95.94	78.70	60.53
59.09	97.59	78.08	70.27	62.46	51.55	39.65
63.64	54.21	43.37	39.03	34.69	28.80	22.16
68.18	26.46	21.17	19.05	16.94	14.14	10.88
72.73	11.65	9.32	8.39	7.46	6.26	4.81
77.27	4.72	3.77	3.40	3.02	2.55	1.96
81.82	1.79	1.43	1.29	1.14	0.97	0.75
86.36	0.64	0.51	0.46	0.41	0.35	0.27
90.91	0.22	0.18	0.16	0.14	0.12	0.09

Note: Concentrations were calculated at G310 for varying input concentrations at the nutrient injection wall (mg/L); 11 cm/d groundwater velocity.

A17.2 HYDRAULIC PARAMETER ESTIMATES

Following the Nutrient Pulse Tracer Test (Appendix 18), hydraulic parameter estimates from inorganic tracer pulse profiles at G312 could be estimated using PULSEPE as a fitting routine. The data indicated average velocities were approximately 15 cm/d in Gate 3. Profiles at fence G312 (vs. G310) were examined in detail, since more substantial overlap of the benzoate pulses was observed there (see Appendix 18).

The data-fitting simulations were done for each sampling point on G312, using the input concentrations as well as velocity and dispersivity as fitting parameters, since the distribution of mass vertically and laterally across the fence was observed to be nonuniform. Assuming a uniform concentration within the wall itself would have either over- or underestimated the mass passing by each point, subsequently yielding poor parameter estimates. Results are summarized in Table A17.4, with modeled data shown in Figures A17.10 to A17.21. These results are extended to the benzoate pulses in Appendix 18.

Table A17.4 Summary of Fitted NIW Input Concentrations (C_o) and Hydraulic Parameters at Fence G312 for the April 18, 1997 Nutrient Injection

	Optimized Parameter		
Location	*Co* (mg/L)	Groundwater Velocity (cm/d)	Dispersivity (cm)
G312L - 1	97.2	15.5	11
G312L - 2	58.1	13.9	19
G312L - 3	168.2	14.7	12
G312L - 4	285.6	12.5	26.9
G312M - 1	97.8	17.5	4.8
G312M - 2[a]	11.2	15.9	0.07
G312M - 3	156.7	16.2	9.4
G312M - 4	281.8	14.8	23.3
G312R - 1	145.1	17.9	7.9
G312R - 2[a]	17.3	15.9	0.06
G312R - 3	139.6	13.9	4.8
G312R - 4	234.4	13.9	4.8
Average	166.4 ± 73.9	15.4 ± 1.9	10.2 ± 5.4

Note: Parameter estimates for the May 16, 1997 tracer pulse were not considered reliable due to data handling problems with the model.

[a] Points omitted from average computations due to poor model fit of the field data (Figures A17.15 and A17.19); ± error indicates 1 standard deviation.

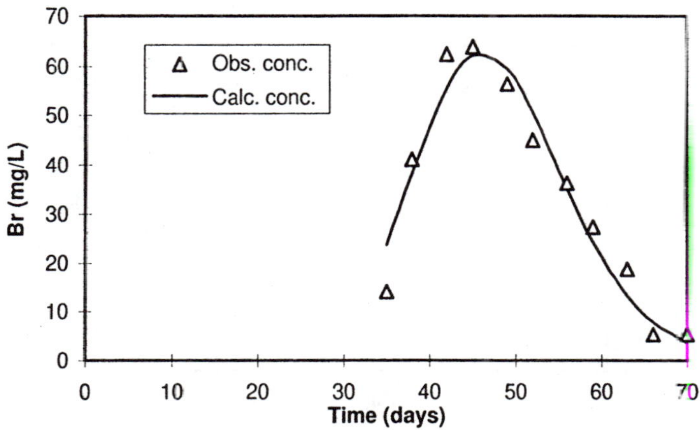

Figure A17.10 PULSEPE fitting of field data at G312L-1.

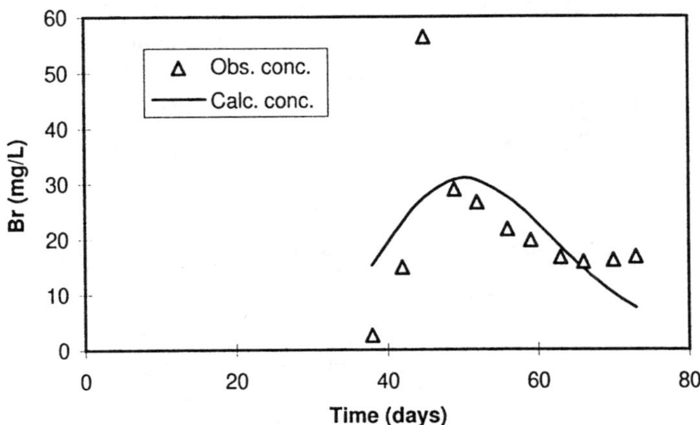

Figure A17.11 PULSEPE fitting of field data at G312L-2.

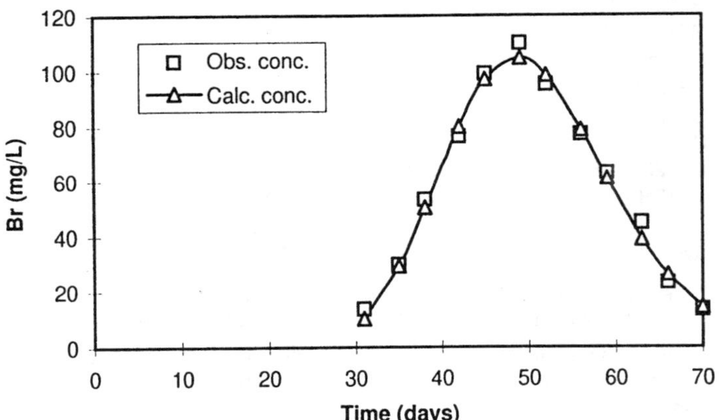

Figure A17.12 PULSEPE fitting of field data at G312L-3.

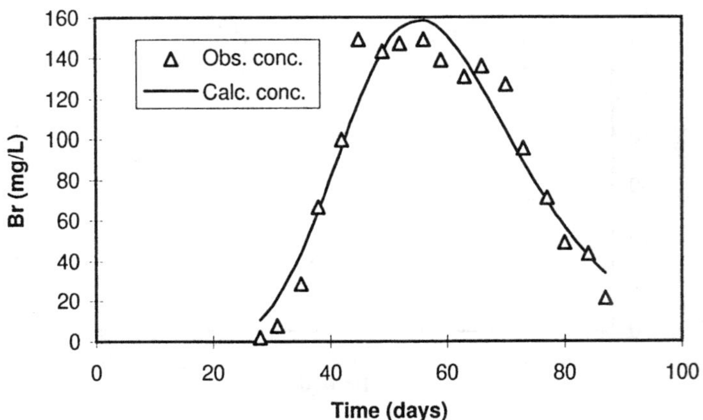

Figure A17.13 PULSEPE fitting of field data at G312L-4.

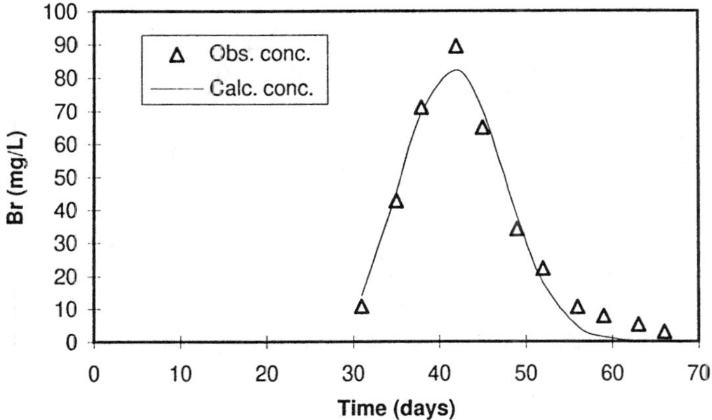

Figure A17.14 PULSEPE fitting of field data at G312M-1.

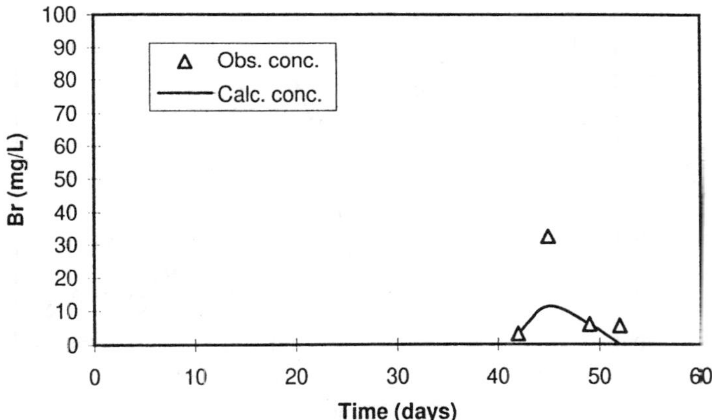

Figure A17.15 PULSEPE fitting of field data at G312M-2.

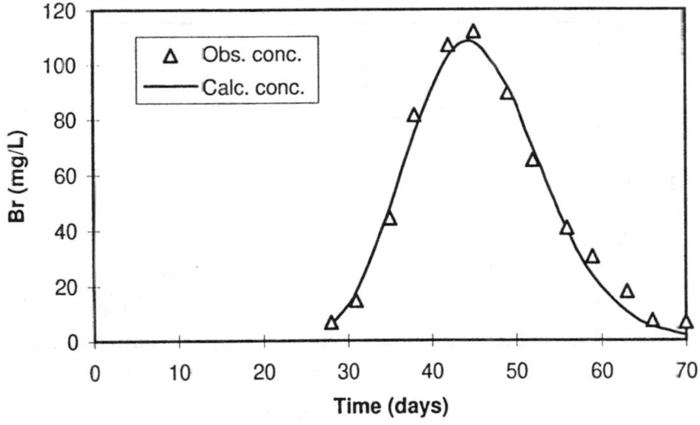

Figure A17.16 PULSEPE fitting of field data at G312M-3.

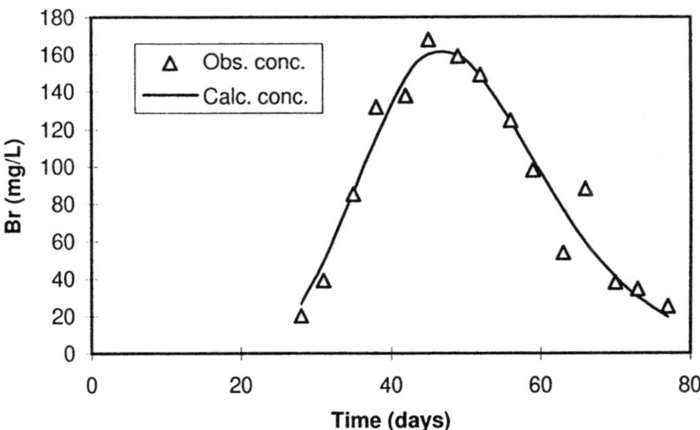

Figure A17.17 PULSEPE fitting of field data at G312M-4.

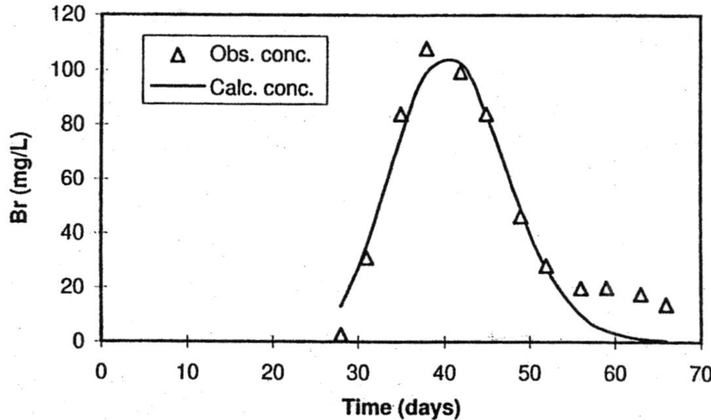

Figure A17.18 PULSEPE fitting of field data at G312R-1.

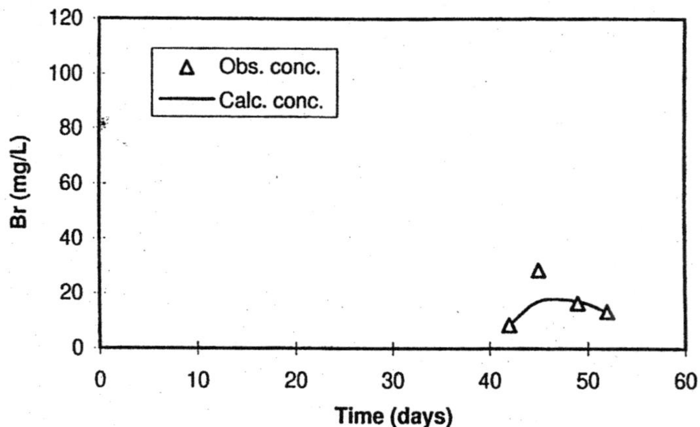

Figure A17.19 PULSEPE fitting of field data at G312R-2.

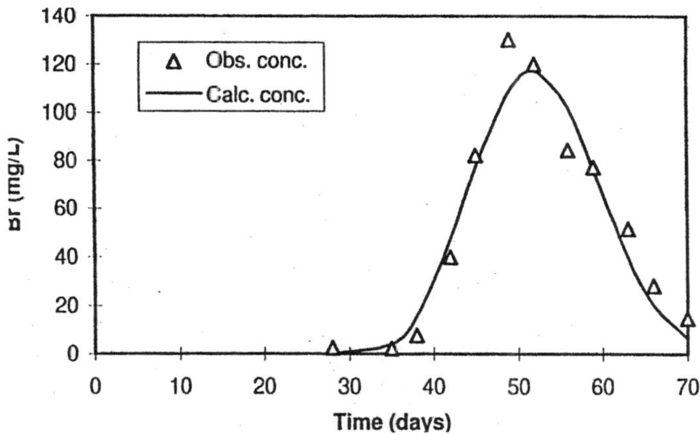

Figure A17.20 PULSEPE fitting of field data at G312R-3.

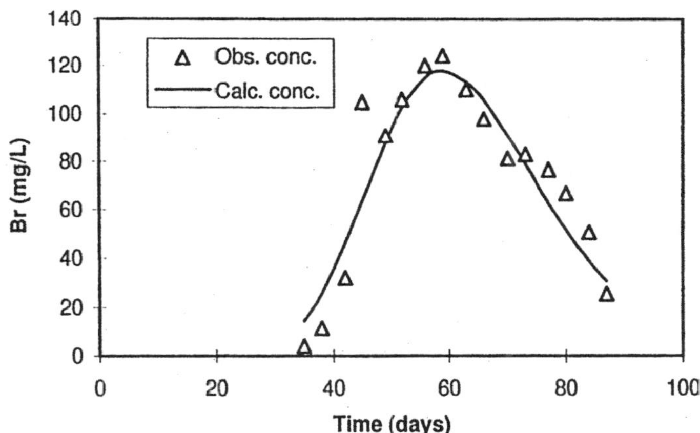

Figure A17.21 PULSEPE fitting of field data at G312R-4.

A17.3 REFERENCES

Devlin, J.F. and Barker, J.F. 1996. Field investigation of nutrient pulse mixing in an *in situ* biostimulation experiment. *Water Resources Research*. 32(9): 2869-2877.

Gibbens, J.F. and Bowman, R.S. 1992. Difluorobenzoates as nonreactive tracers in soil groundwater. *Ground Water*. 30(1) 8-14.

APPENDIX 18

Nutrient Pulse Tracer Test

A18.1 BACKGROUND

A tracer test was initiated in Gate 3, beginning April 18, 1997, to achieve two goals simultaneously: a more accurate assessment of the hydraulic parameters within the gate, and determination of the extent of pulse overlap between sequential nutrient pulses.

Inorganic tracers were added to the regular nutrient concentrate (containing only benzoate at this time) for the April 18, 1997 and May 16, 1997 nutrient injections; target and calculated concentrations are listed in Table A18.1.

Different tracers (bromide and chloride) were utilized in the two successive injections to permit accurate quantification of each tracer downgradient of the permeable wall in the region where pulse overlap was to occur. The appearance of both tracers in samples from the "overlap" region would constitute unambiguous evidence that pulse merging had occurred.

Beginning with the first injection, simultaneous collection of tracer and nutrient samples (see Appendix 4) from all monitoring points downgradient of the permeable wall to the end of the anaerobic zone (i.e., G307 to G315) was done on a semiweekly basis to ensure adequate characterization of pulse progress. Sample collection was initiated at each fence following the arrival of the tracer at the nearest upgradient fence.

A18.2 SAMPLE ANALYSIS

All samples were analyzed for conductivity in the field (as described in Appendices 4 and 7) as a preliminary assessment of pulse breakthrough, and were subsequently analyzed for inorganic tracers by ion-specific electrodes, either in the field or at UW (Appendix 4). Ion-selective probes are subject to interferences between similar species. Such interferences were encountered between chloride and bromide, limiting the usefulness of these analyses in establishing the degree of pulse overlap. To overcome this difficulty, the probe data were used to select a subset of samples, collected at G312, for analysis by ion chromatography at Water Technology International (WTI) Laboratories.

A18.3 DATA ANALYSIS

Ion probe data were expressed as milligrams per liter of potassium bromide or potassium chloride. Results are listed in Section A18.4.

Table A18.1 Average Tracer Concentrations Introduced to the Aquifer

	Average Injected Concentrations (mg/L)		
Date	**Br^-**	**Benzoate**	**Cl^-**
April 18, 1997	117	148	n/a
May 16, 1997	n/a	193	127

Note: n/a: not added.

Overlap profiles (probe data corrected to concentrations of the free ion) for points at fence G310 (Figures A18.1 to A18.12) indicated minimal pulse overlap in 10 of the 12 cases (taking into account the potential for analytical interference between the two ions). Pulse overlap was then studied at fence G312. The extent of merging appeared to be more significant (Figures A18.13 to A18.24) at this fence, with nearly complete pulse mixing occurring within 2 m downgradient of the design objective. As stated above, selected samples were subsequently submitted to WTI for quantification of bromide and/or chloride (Appendix 3). Sampling points where probe analyses indicated negligible overlap (i.e., G312M-2, G312R-2) were omitted from submission to WTI. Combined bromide and chloride profiles are given in Figures A18.25 to A18.34. Although qualitatively similar in the arrival times of peaks and valleys, quantitative differences in concentrations of tracer between probe analyses and those of WTI are evident. The cause of these differences is unclear. Analytical interference from benzoate was determined to be negligible (Appendix 3).

Estimates of groundwater velocity were determined from the WTI profiles at fence G312 by using the model PULSEPE (Appendix 17) and were verified using the program FENCE (Appendix 24). Excellent agreement was obtained between the two types of velocity estimates. An average velocity and dispersivity of 15.2 cm/d and 10 cm, respectively, were determined. The breakthrough curves at each point were noticeably different in terms of their maximum concentrations and durations. It was apparent that the various monitoring points lay along flow paths that received quite different tracer (and benzoate) masses. This is reflected in the variability of the apparent initial concentrations (C_o) fitted for each bromide breakthrough curve.

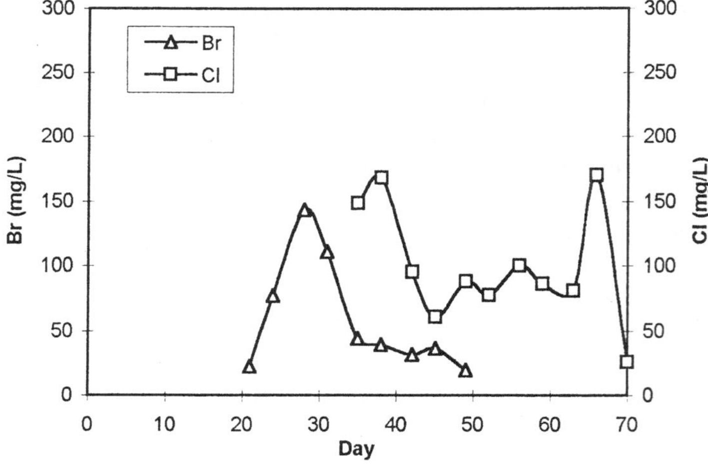

Figure A18.1 Pulse tracer overlap at G310L-1 (ion probe data).

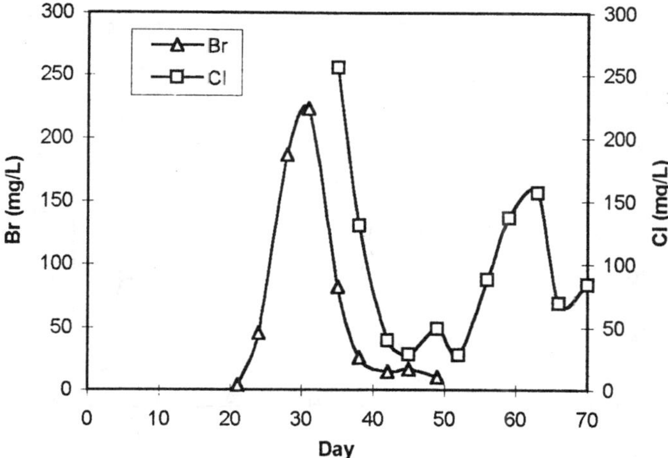

Figure A18.2 Pulse tracer overlap at G310L-2 (ion probe data).

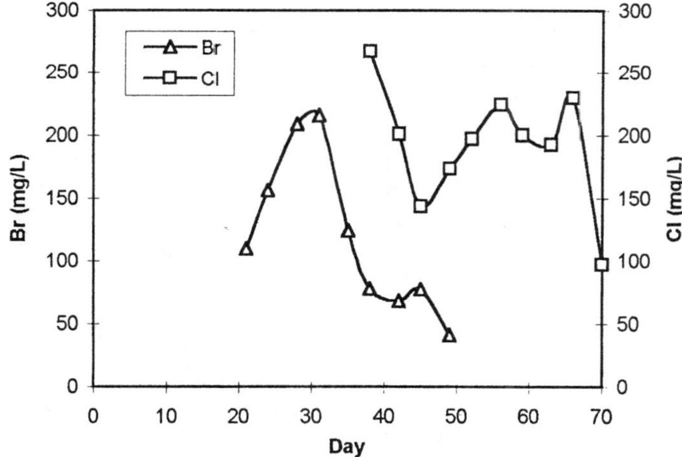

Figure A18.3 Pulse tracer overlap at G310L-3 (ion probe data).

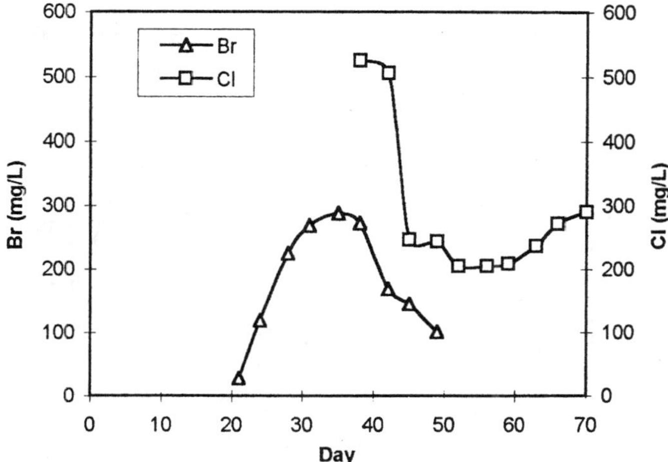

Figure A18.4 Pulse tracer overlap at G310L-4 (ion probe data).

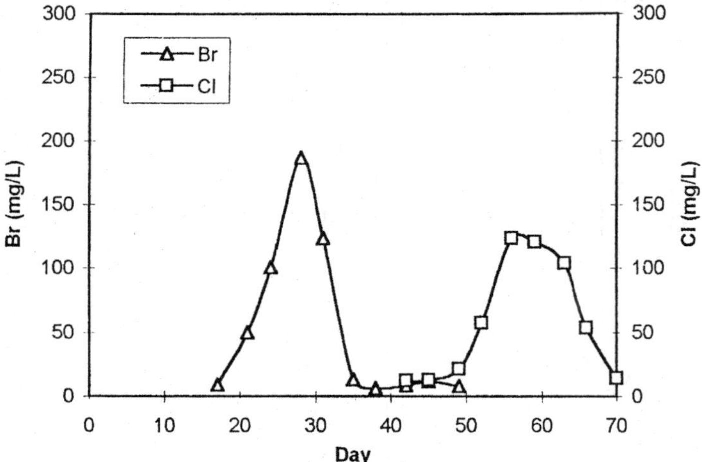

Figure A18.5 Pulse tracer overlap at G310M-1 (ion probe data).

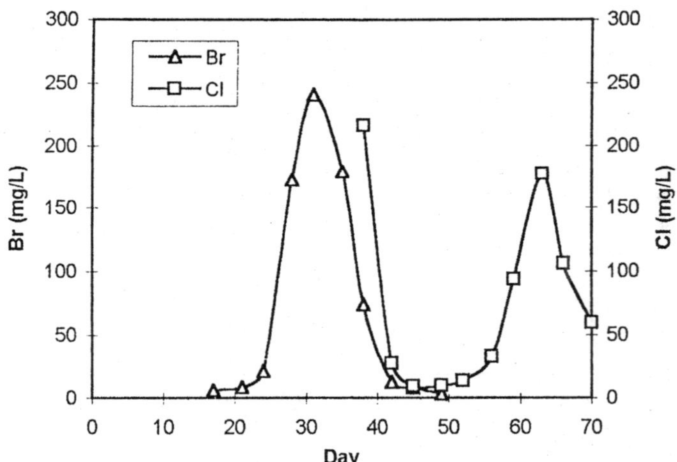

Figure A18.6 Pulse tracer overlap at G310M-2 (ion probe data).

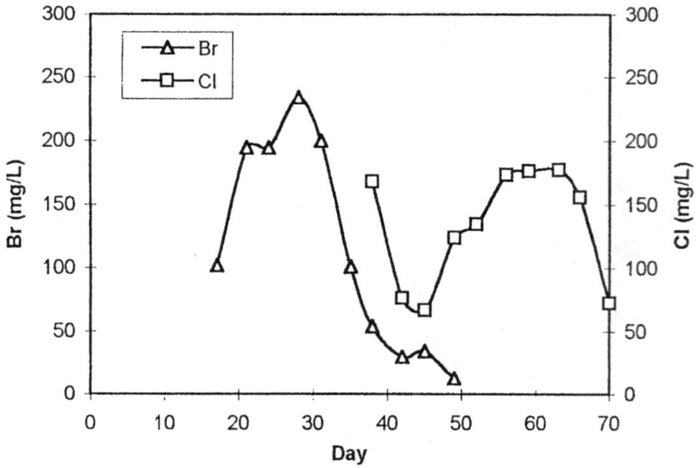

Figure A18.7 Pulse tracer overlap at G310M-3 (ion probe data).

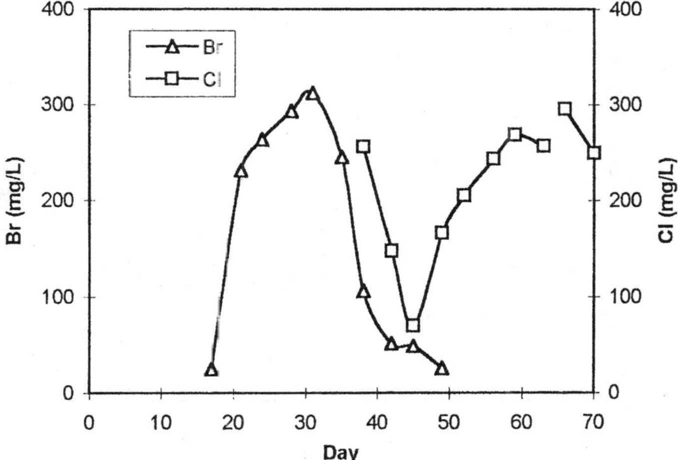

Figure A18.8 Pulse tracer overlap at G310M-4 (ion probe data).

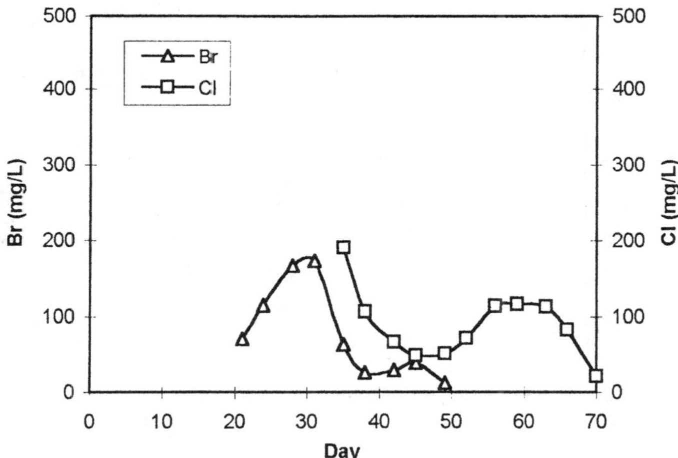

Figure A18.9 Pulse tracer overlap at G310R-1 (ion probe data).

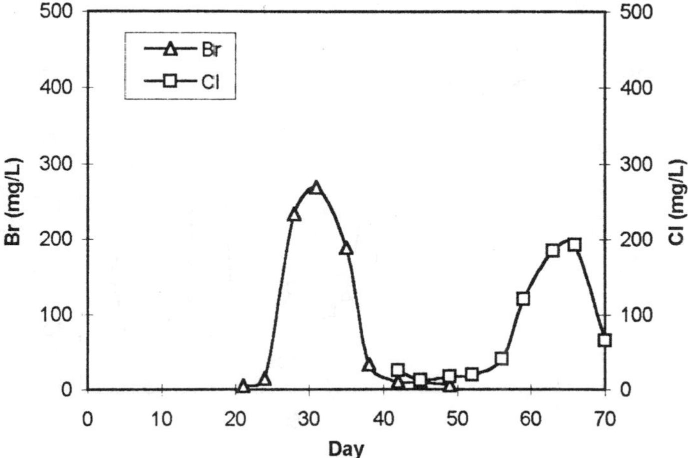

Figure A18.10 Pulse tracer overlap at G310R-2 (ion probe data).

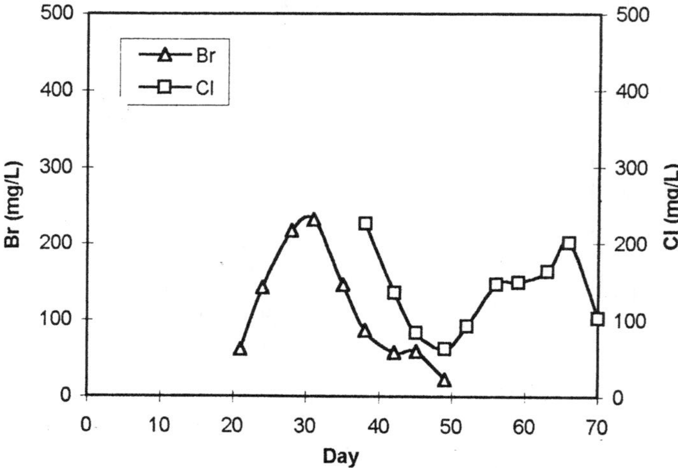

Figure A18.11 Pulse tracer overlap at G310R-3 (ion probe data).

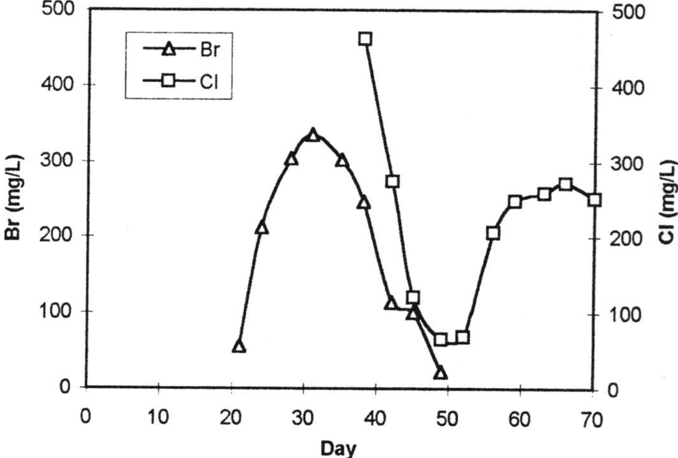

Figure A18.12 Pulse tracer overlap at G310R-4 (ion probe data).

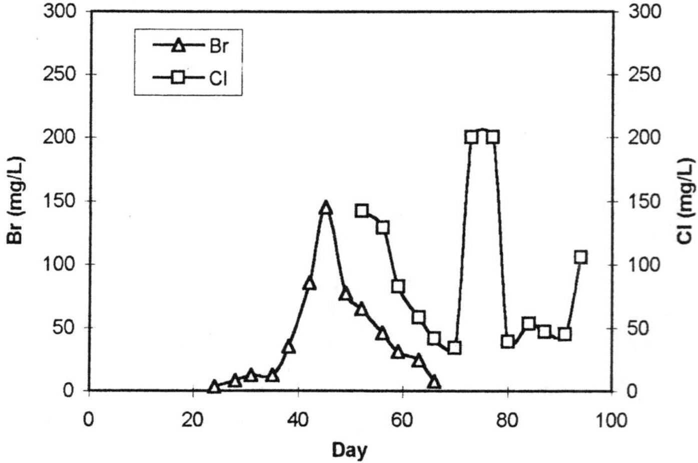

Figure A18.13 Pulse tracer overlap at G312L-1 (ion probe data).

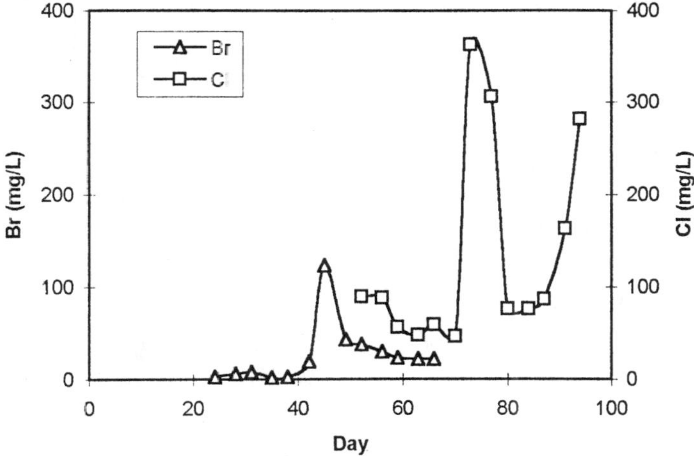

Figure A18.14 Pulse tracer overlap at G312L-2 (ion probe data).

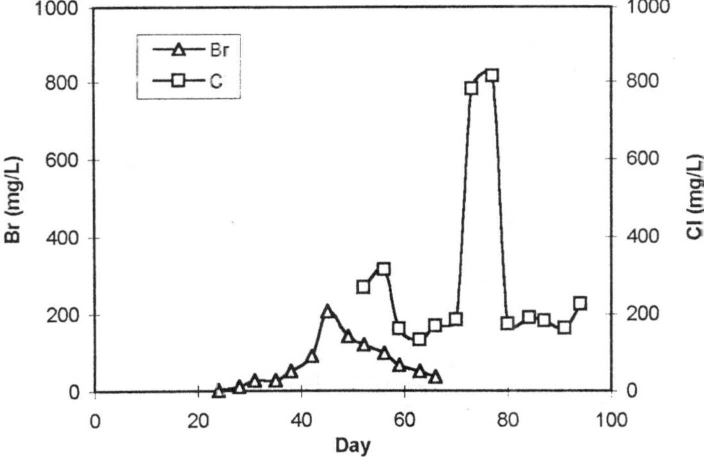

Figure A18.15 Pulse tracer overlap at G312L-3 (ion probe data).

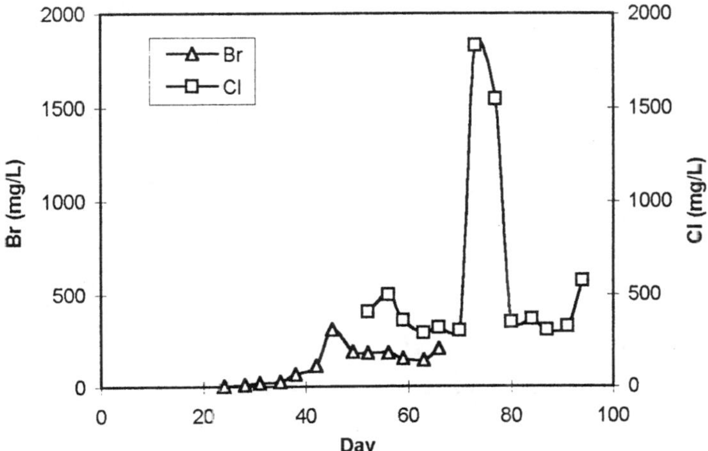

Figure A18.16 Pulse tracer overlap at G312L-4 (ion probe data).

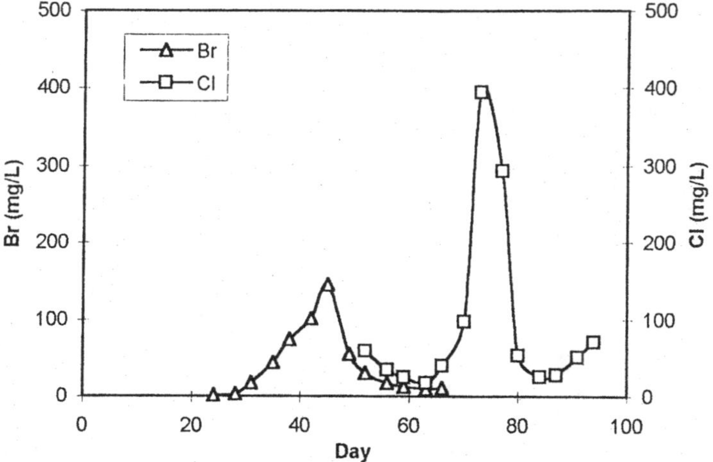

Figure A18.17 Pulse tracer overlap at G312M-1 (ion probe data).

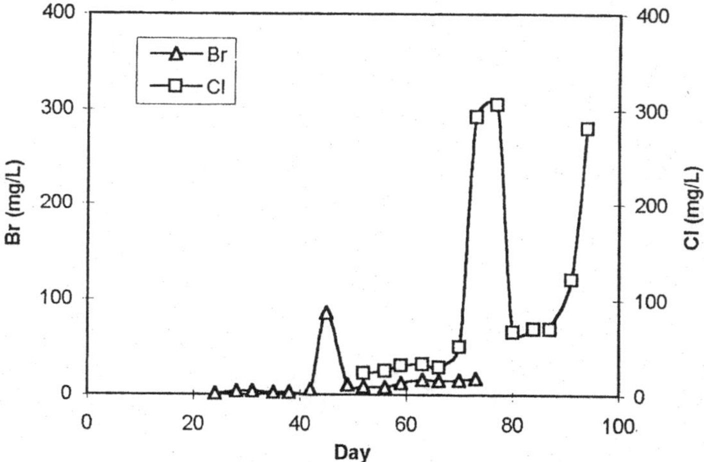

Figure A18.18 Pulse tracer overlap at G312M-2 (ion probe data).

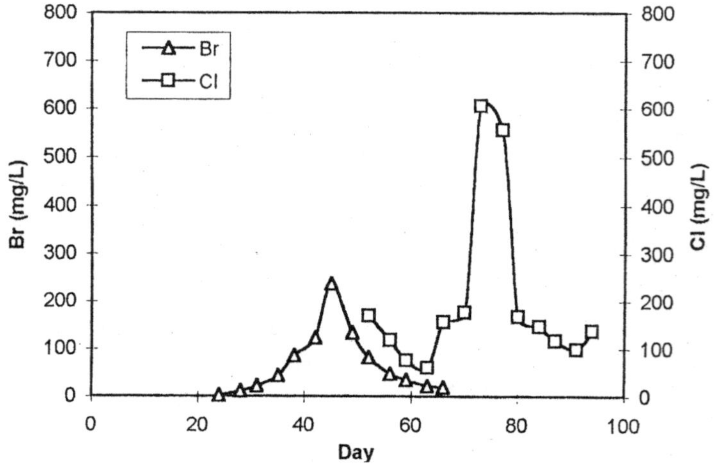

Figure A18.19 Pulse tracer overlap at G312M-3 (ion probe data).

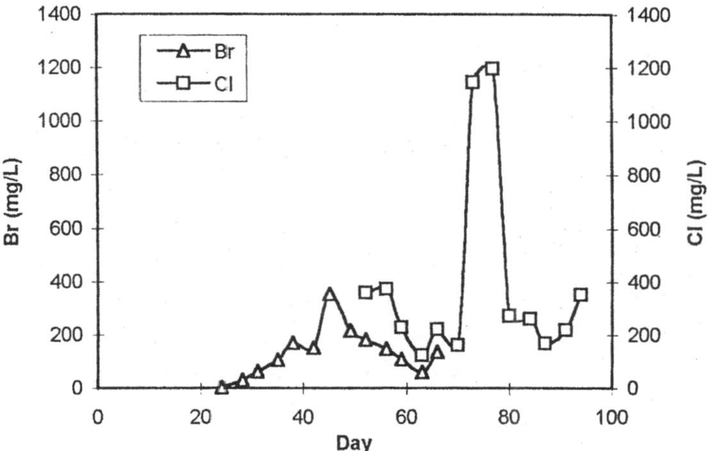

Figure A18.20 Pulse tracer overlap at G312M-4 (ion probe data).

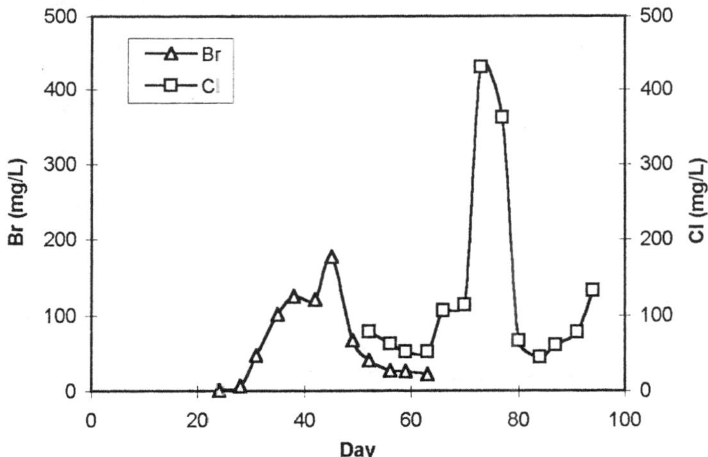

Figure A18.21 Pulse tracer overlap at G312R-1 (ion probe data).

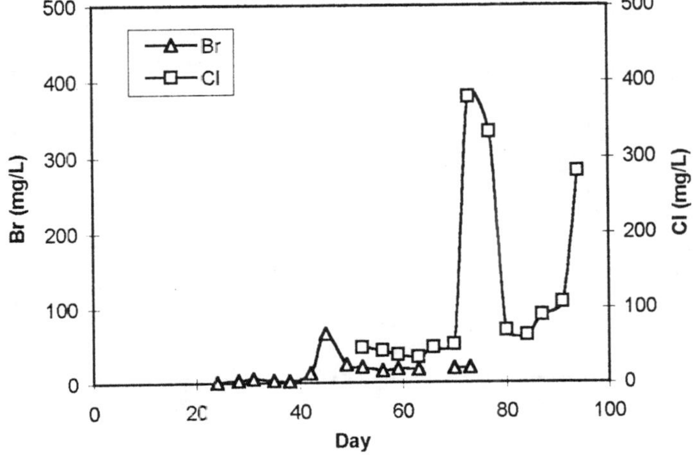

Figure A18.22 Pulse tracer overlap at G312R-2 (ion probe data).

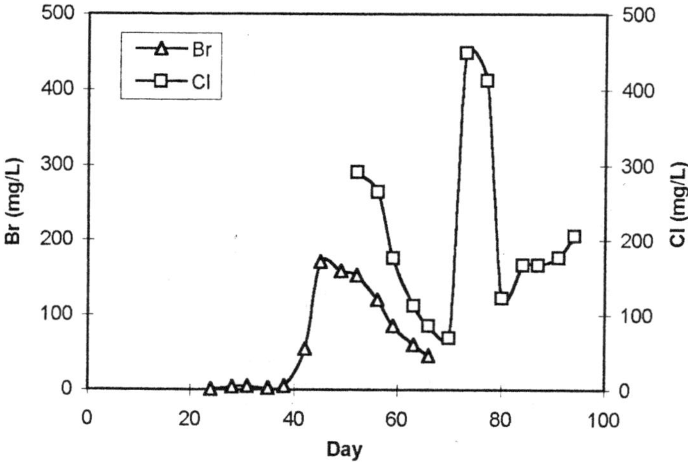

Figure A18.23 Pulse tracer overlap at G312R-3 (ion probe data).

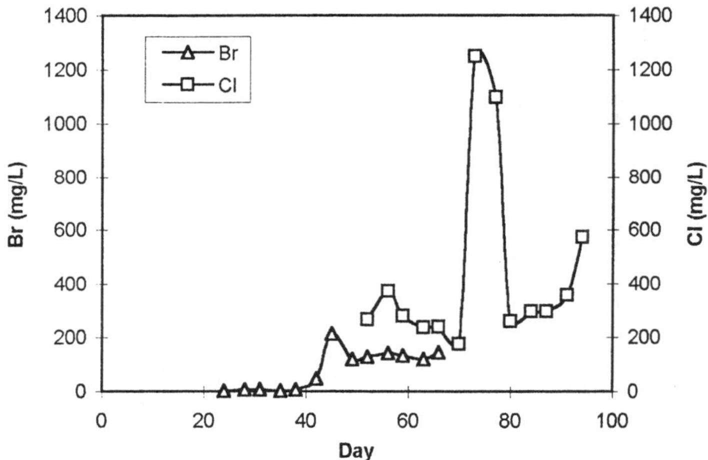

Figure A18.24 Pulse tracer overlap at G312R-4 (ion probe data).

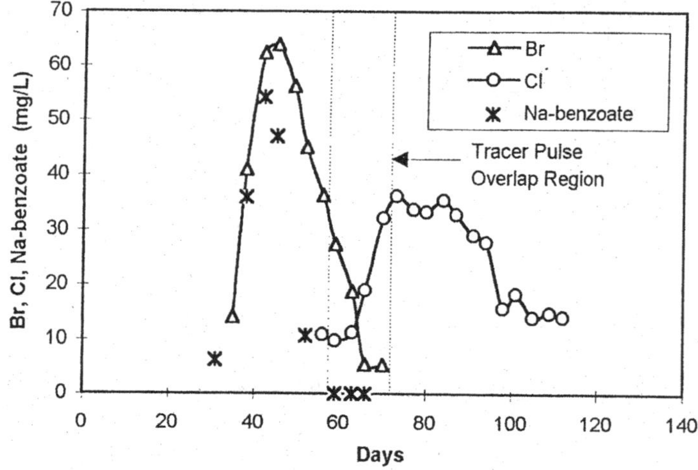

Figure A18.25 Pulse tracer overlap (WTI data) and April 18, 1997 benzoate profile at G312L-1.

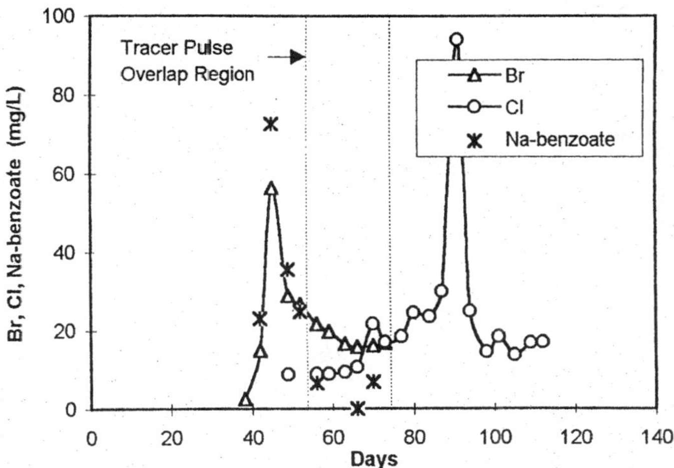

Figure A18.26 Pulse tracer overlap (WTI data) and April 18, 1997 benzoate profile at G312L-2.

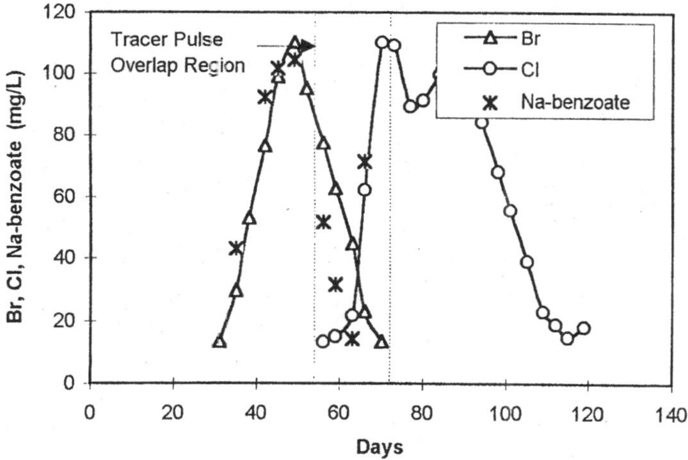

Figure A18.27 Pulse tracer overlap (WTI data) and April 18, 1997 benzoate profile at G312L-3.

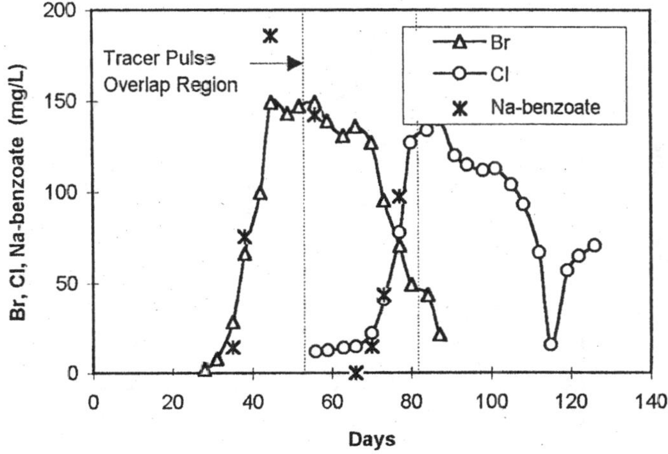

Figure A18.28 Pulse tracer overlap (WTI data) and April 18, 1997 benzoate profile at G312L-4.

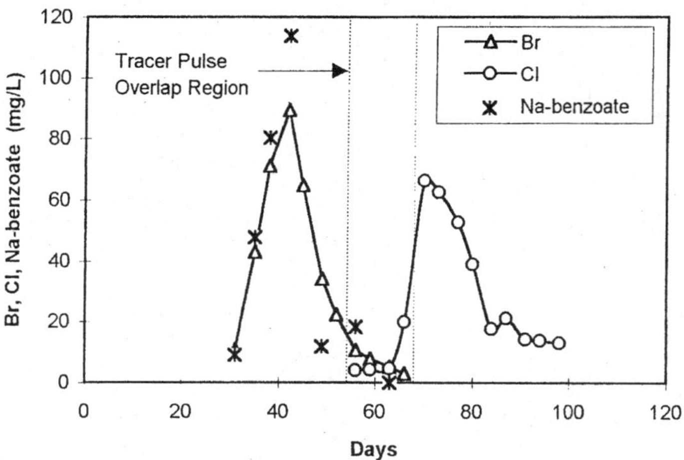

Figure A18.29 Pulse tracer overlap (WTI data) and April 18, 1997 benzoate profile at G312M-1.

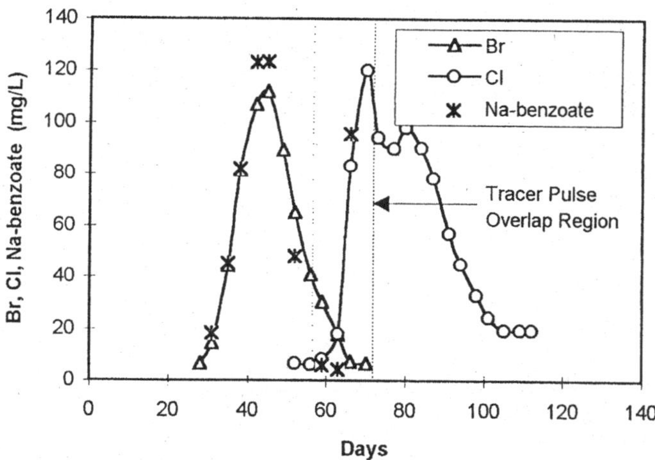

Figure A18.30 Pulse tracer overlap (WTI data) and April 18, 1997 benzoate profile at G312M-3.

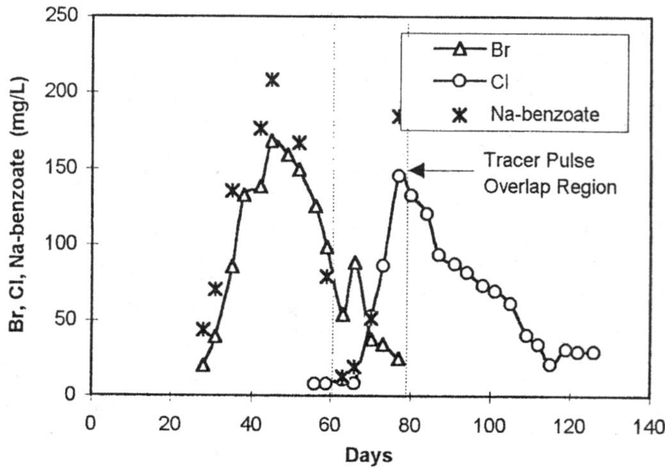

Figure A18.31 Pulse tracer overlap (WTI data) and April 18, 1997 benzoate profile at G312M-4.

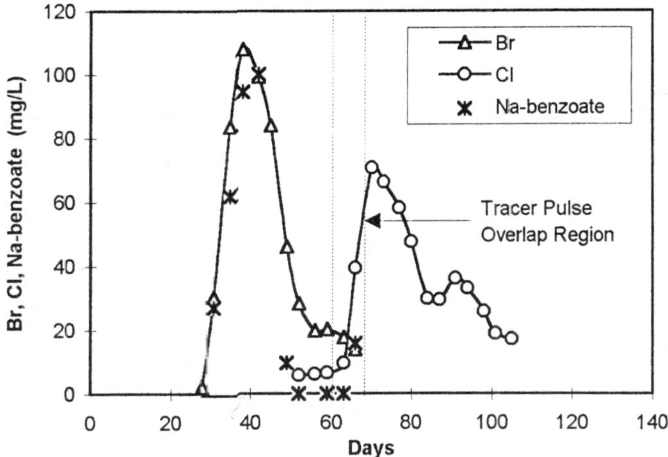

Figure A18.32 Pulse tracer overlap (WTI data) and April 18, 1997 benzoate profile at G312R-1.

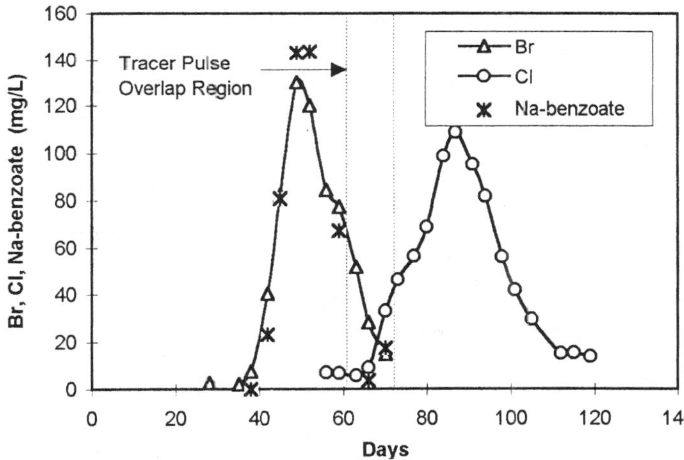

Figure A18.33 Pulse tracer overlap (WTI data) and April 18, 1997 benzoate profile at G312R-3.

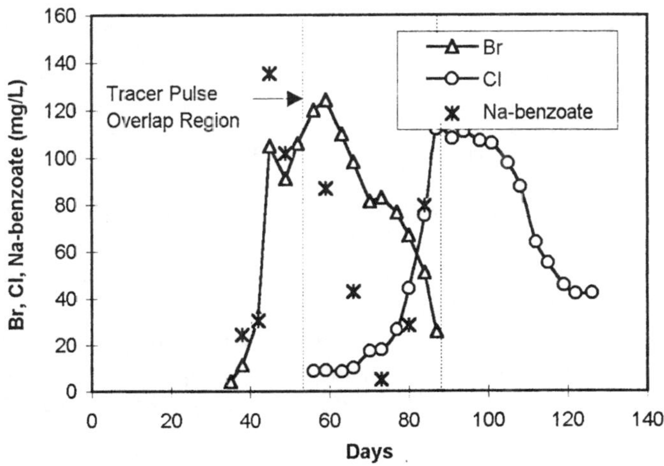

Figure A18.34 Pulse tracer overlap (WTI data) and April 18, 1997 benzoate profile at G312R-4.

Table A18.2 Summary of Estimated Hydraulic and Reaction Parameters at Fence G312

Location	Velocity (cm/d)	Bromide Dispersivity (cm)	Streamtube C_o (mg/L)	Streamtube C_o (mg/L)	Benzoate $k \times 10^3$ (day^{-1})	$t_{1/2}$ (days)
G312L - 1	15.5	11.7	98.2	125	14.7	47
G312L - 2	13.9	21.1	60.3	76.8	4.16	166
G312L - 3	14.7	17.6	203	259	10.4	67
G312L - 4	12.5	28.2	348	443	3.18	218
G312M - 1	17.5	4.9	98.8	125	2.65	261
G312M - 2	15.9	2.6	17.3	22	3.87E-7	>1000[a]
G312M - 3	16.2	9.9	161	205	4.83	143
G312M - 4	14.8	24.2	336.5	428	3.18	218
G312R - 1	17.9	8.8	151	193	11.2	62
G312R - 2	15.9	3.0	21.8	27.7	2.74E-9	>1000[a]
G312R - 3	13.9	5.4	146	186	3.17	219
G312R - 4	13.9	24.3	248	315	10.1	69
Average	15.2 +/- 1.5	10.6 +/- 8.6	158 +/- 105	200 +/- 134	5.63 +/- 4.6	147 +/- 76

[a] Points omitted from calculation of average; ± error indicates 1 standard deviation.

Selected samples were analyzed for benzoate and are plotted with Figures A18.25 to A18.34. Although pronounced overlap of benzoate pulses was achieved at only five points along the fence, definitive pulse overlap was shown by the inorganic tracers at ten of the points. Table A18.3 summarizes where quantitatively significant overlap in fence G312 occurred. Overlap was considered significant when the minimum concentration between pulses was above background values (approximately 3 mg/L for chloride, and zero for both bromide and benzoate).

Table A18.3 Summary of Pulse Overlap at Fence G312

	Pulse Overlap	
Location	Inorganic Tracers	Sodium benzoate
G312L - 1	+	–
G312L - 2	+	–
G312L - 3	+	+
G312L - 4	+	–
G312M - 1	+	–
G312M - 2	–	–
G312M - 3	+	+
G312M - 4	+	+
G312R - 1	+	–
G312R - 2	–	–
G312R - 3	+	+
G312R - 4	+	+

Notes: + = positive overlap, – = no overlap.

The lack of significant benzoate overlap between pulses was likely due to its utilization by aquifer microorganisms. In conjunction with observed trends in sulfate, organic acids, and VOCs (relative to Gate 2), success was achieved in terms of stimulating microbial activity within the anaerobic zone of the gate using benzoate.

An attempt was made to quantify the rate of benzoate utilization along the G312 fence. Since the bromide and benzoate were introduced together, a correction to the initial benzoate concentrations was made based on the calculated (119 mg/L) and observed C_o (Table A18.2) for bromide,

$$C_{o_{benzoate}} = C_{o_{bromide}} \times \frac{152}{119}$$

The average concentrations (152 mg/L benzoate and 119 mg/L bromide) were subsequently recalculated and found to be slightly lower (148 and 117 mg/L, Table A18.1). However, the values used above are reasonably close to the best estimates, and the ratio is only 0.5% different from that calculated with the numbers from Table A18.1. Velocity and dispersivity were assumed identical to those determined for bromide. Breakthrough curves were fit by adjusting the pseudo-first-order decay term, k (Table A18.2).

A comparison between the distributions of velocity, C_o (benzoate), and pseudo-half-lives for benzoate on fence G312 is given in Figure A18.35. There were no clear correlations between benzoate utilization and groundwater velocity, but there was some indication that benzoate utilization depended on a steady supply of benzoate (recall that TOL was still available even when benzoate was not). For example, the R2 and M2 locations, which received relatively small bromide and benzoate loadings, were not very active; estimated pseudo-half-lives were in excess of 1000 days. In all other locations, where benzoate was delivered in abundance, there was some suggestion of benzoate utilization, with pseudo-half-lives varying between 47 and 261 days. Four zones of apparently rapid ($t_{1/2}$ < 100 days) benzoate utilization were identified: two near the water table and two at depth. The shallow zones may have been consuming benzoate aerobically, but further work is needed to confirm this. The deeper active zones may be due to natural heterogeneous population distributions in the aquifer, but, again, further work is required to demonstrate this hypothesis.

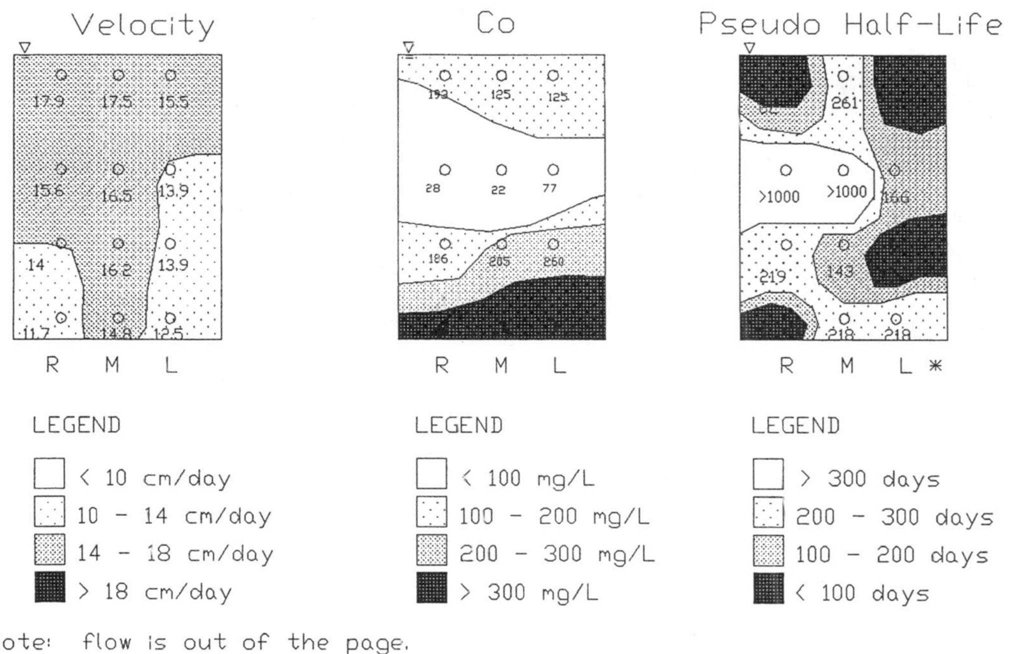

Figure A18.35 Comparison of velocity, Co, and pseudo half-life distributions for benzoate at G312 following the pulse tracer test of April 18, 1997.

A18.4 POTASSIUM BROMIDE, POTASSIUM CHLORIDE, AND CONDUCTIVITY DATA

The following pages contain the data from the Nutrient Pulse Tracer Test. Values listed as "n/s" mean "not sampled."

A18.5 REFERENCES

Devlin, J.F. and Barker, J.F. 1996. Field investigation of nutrient pulse mixing in an *in situ* biostimulation experiment. *Water Resources Research.* 32(9): 2869-2877.

DATA

DATE	LOCATION:	G304L				G304M			
	DEPTH:	-1	-2	-3	-4	-1	-2	-3	-4
	PARAMETER								
04/18/97	Lab Conductivity (us)	791.00	961.00	811.00	289.00	808.00	832.00	809.00	378.00
	KBr (mg/L)	486.41	506.47	456.77	11.09	448.64	488.60	482.06	45.23

DATE	LOCATION:	G304R				G306L			
	DEPTH:	-1	-2	-3	-4	-1	-2	-3	-4
	PARAMETER								
04/18/97	Lab Conductivity (us)	740.00	841.00	808.00	811.00	515.00	497.00	618.00	731.00
	Field Conductivity (us)	n/s	n/s	n/s	n/s	621.00	778.00	781.00	722.00
	KBr (mg/L)	440.66	484.23	467.14	173.18	154.79	323.26	303.56	199.04
05/16/97	KCl (mg/L)	n/s	n/s	n/s	n/s	440.96	574.12	1383.57	655.09

DATE	LOCATION:	G306M				G306R			
	DEPTH:	-1	-2	-3	-4	-1	-2	-3	-4
	PARAMETER								
04/18/97	Lab Conductivity (us)	452.00	674.00	721.00	691.00	574.00	813.00	578.00	705.00
	Field Conductivity (us)	590.00	799.00	692.00	788.00	606.00	815.00	806.00	696.00
	KBr (mg/L)	98.36	323.26	344.23	32.59	117.18	336.59	94.04	318.93
05/05/97	Lab Conductivity (us)	371.00	289.00	373.00	612.00	n/s	n/s	n/s	n/s
	Field Conductivity (us)	423.00	320.00	433.00	816.00	n/s	n/s	n/s	n/s
	KBr (mg/L)	22.56	6.41	6.82	416.63	n/s	n/s	n/s	n/s
05/16/97	KCl (mg/L)	175.11	626.90	655.09	1649.68	284.05	549.41	549.41	1578.70

DATE	LOCATION:	G307				G308			
	DEPTH:	-1	-2	-3	-4	-1	-2	-3	-4
	PARAMETER								
04/15/97	Field Conductivity (us)	n/s	n/s	n/s	n/s	522.00	455.00	557.00	640.00
04/18/97	Lab Conductivity (us)	418.00	289.00	301.00	492.00	n/s	n/s	n/s	n/s
	KBr (mg/L)	7.31	3.36	3.64	4.46	n/s	n/s	n/s	n/s
04/21/97	Lab Conductivity (us)	397.00	383.00	501.00	673.00	n/s	n/s	n/s	n/s
	KBr (mg/L)	3.12	3.39	12.39	6.04	n/s	n/s	n/s	n/s
04/25/97	Lab Conductivity (us)	606.00	742.00	841.00	763.00	444.00	410.00	498.00	555.00
	KBr (mg/L)	179.40	394.51	478.37	222.24	5.98	3.82	10.18	6.22
04/28/97	Lab Conductivity (us)	690.00	804.00	858.00	837.00	446.00	432.00	600.00	659.00
	KBr (mg/L)	292.32	431.64	470.24	377.97	15.29	24.49	155.75	200.53
05/05/97	Lab Conductivity (us)	645.00	561.00	425.00	446.00	542.00	593.00	556.00	647.00
	Field Conductivity (us)	537.00	509.00	726.00	881.00	710.00	771.00	712.00	856.00
	KBr (mg/L)	54.38	68.05	332.92	540.06	332.92	441.14	350.96	528.31
05/09/97	Lab Conductivity (us)	n/s	n/s	n/s	n/s	n/s	n/s	n/s	n/s
	Field Conductivity (us)	425.00	361.00	537.00	745.00	658.00	594.00	786.00	740.00
	KBr (mg/L)	14.54	13.92	62.07	535.55	242.66	302.34	411.34	559.63
05/12/97	Lab Conductivity (us)	n/s	n/s	n/s	n/s	n/s	n/s	n/s	n/s
	Field Conductivity (us)	405.00	378.00	485.00	889.00	507.00	499.00	752.00	883.00
	KBr (mg/L)	9.79	8.21	8.58	512.51	26.92	62.07	302.34	490.46
05/16/97	Lab Conductivity (us)	n/s	n/s	n/s	n/s	n/s	n/s	n/s	n/s
	Field Conductivity (us)	409.00	415.00	473.00	816.00	428.00	412.00	526.00	779.00

	LOCATION:	G307				G308			
	DEPTH:	-1	-2	-3	-4	-1	-2	-3	-4
DATE	PARAMETER								
05/16/98	KBr (mg/L)	6.61	5.07	26.72	390.12	16.35	12.55	151.77	335.43
	KCl (mg/L)	26.42	34.40	107.94	1723.85	n/s	n/s	n/s	n/s
5/19/97	Field Conductivity (us)	412.00	553.00	425.00	824.00	409.00	368.00	606.00	642.00
	KBr (mg/L)	15.34	9.83	16.52	345.38	17.80	10.20	117.87	147.23
	KCl (mg/L)	17.79	46.81	42.86	1383.57	26.42	13.07	369.83	440.96
5/23/97	Field Conductivity (us)	587.00	892.00	1033.00	904.00	400.00	405.00	525.00	584.00
	KBr (mg/L)	6.89	7.65	8.06	110.03	4.54	2.55	21.77	58.76
	KCl (mg/L)	208.79	403.83	503.15	931.32	22.16	20.30	112.80	271.83
05/26/97	Field Conductivity (us)	818.00	964.00	1110.00	1085.00	403.00	443.00	631.00	734.00
	KBr (mg/L)	6.21	6.54	6.89	25.46	1.68	2.69	7.65	9.43
	KCl (mg/L)	338.69	440.96	525.78	684.54	15.59	28.85	182.98	284.05
05/30/97	Field Conductivity (us)	755.00	944.00	1127.00	1187.00	652.00	812.00	786.00	1070.00
	KBr (mg/L)	n/s	n/s	n/s	34.28	n/s	n/s	19.93	28.07
	KCl (mg/L)	248.94	369.83	503.15	626.90	191.21	338.69	284.05	503.15
06/02/97	Field Conductivity (us)	774.00	735.00	968.00	1233.00	784.00	1007.00	933.00	1171.00
	KCl (mg/L)	201.00	133.91	361.41	518.56	288.40	413.81	330.22	473.81
06/06/97	Field Conductivity (us)	489.00	489.00	754.00	1219.00	763.00	915.00	977.00	1171.00
	KCl (mg/L)	12.83	25.27	163.00	556.93	249.00	396.83	396.83	533.83
06/09/97	Field Conductivity (us)	444.00	472.00	661.00	1183.00	720.00	784.00	982.00	1159.00
	KCl (mg/L)	32.58	43.83	163.00	606.18	193.11	282.75	450.61	556.93
06/13/97	Field Conductivity (us)	445.00	451.00	618.00	1178.00	525.00	541.00	903.00	1151.00
	KCl (mg/L)	33.99	35.47	131.89	511.68	33.99	69.86	364.59	556.93
06/16/97	Field Conductivity (us)	394.00	443.00	610.00	1131.00	465.00	469.00	798.00	1059.00
	KCl (mg/L)	n/s	n/s	104.71	477.31	27.19	22.97	264.61	457.61
06/20/97	Field Conductivity (us)	508.00	518.00	702.00	991.00	452.00	440.00	709.00	898.00
	KBr (mg/L)	10.62	33.57	159.26	31.37	5.77	4.26	21.62	22.36
	KCl (mg/L)	n/s	n/s	329.69	405.14	38.66	28.97	218.32	343.56
06/23/97	Field Conductivity (us)	730.00	793.00	849.00	961.00	463.00	445.00	695.00	973.00
	KCl (mg/L)	n/s	n/s	n/s	n/s	n/s	n/s	433.51	541.38
06/27/97	Field Conductivity (us)	724.00	775.00	875.00	880.00	689.00	720.00	698.00	763.00
	KBr (mg/L)	164.74	215.97	264.61	231.10	139.09	182.35	125.66	182.35
	KCl (mg/L)	n/s	n/s	n/s	n/s	n/s	n/s	2053.53	3825.60
06/30/97	Field Conductivity (us)	618.00	713.00	843.00	891.00	796.00	776.00	737.00	819.00
	KBr (mg/L)	58.57	162.55	249.40	284.51	266.38	266.38	204.69	257.75

	LOCATION:	G310L				G310M			
	DEPTH:	-1	-2	-3	-4	-1	-2	-3	-4
DATE	PARAMETER								
04/15/97	Field Conductivity (us)	n/s	n/s	n/s	n/s	582.00	468.00	581.00	616.00
05/05/97	Lab Conductivity (us)	n/s	n/s	n/s	n/s	369.00	379.00	489.00	475.00
	Field Conductivity (us)	n/s	n/s	n/s	n/s	419.00	433.00	593.00	566.00
	KBr (mg/L)	n/s	n/s	n/s	n/s	13.73	9.36	152.18	37.25
05/09/97	Field Conductivity (us)	561.00	441.00	563.00	752.00	481.00	351.00	693.00	633.00
	KBr (mg/L)	33.54	6.03	163.34	41.78	74.01	12.74	289.33	344.98
05/12/97	Field Conductivity (us)	631.00	581.00	733.00	774.00	594.00	420.00	720.00	820.00
	KBr (mg/L)	114.89	67.78	232.22	178.36	149.59	32.09	289.33	393.64
05/16/97	Field Conductivity (us)	714.00	749.00	790.00	843.00	700.00	657.00	788.00	895.00
	KBr (mg/L)	213.21	277.72	311.03	335.43	277.72	257.51	348.34	436.92

	LOCATION:	G310L				G310M			
	DEPTH:	-1	-2	-3	-4	-1	-2	-3	-4
DATE	PARAMETER								
05/19/97	Field Conductivity (us)	676.00	794.00	778.00	886.00	652.00	745.00	716.00	889.00
	KBr (mg/L)	164.55	332.81	320.70	400.59	183.91	358.43	297.78	464.62
05/23/97	Field Conductivity (us)	490.00	641.00	714.00	883.00	490.00	722.00	610.00	829.00
	KBr (mg/L)	65.24	122.15	185.57	428.25	19.61	267.55	150.56	366.10
	KCl (mg/L)	311.60	538.89	n/s	n/s	n/s	n/s	n/s	n/s
05/26/97	Field Conductivity (us)	613.00	581.00	707.00	944.00	452.00	645.00	613.00	722.00
	KBr (mg/L)	58.76	38.68	115.93	406.44	9.43	110.03	80.41	158.64
	KCl (mg/L)	353.59	274.60	562.08	1103.06	n/s	455.30	353.59	538.89
05/30/97	Field Conductivity (us)	582.00	518.00	668.00	912.00	416.00	488.00	551.00	629.00
	KBr (mg/L)	46.92	21.71	101.44	252.93	12.98	19.36	44.32	76.24
	KCl (mg/L)	199.80	82.91	421.99	1062.67	26.42	58.32	160.36	310.17
06/02/97	Field Conductivity (us)	623.00	589.00	711.00	808.00	439.00	461.00	635.00	582.00
	KBr (mg/L)	54.16	24.58	115.50	215.92	17.68	12.31	50.71	72.85
	KCl (mg/L)	128.00	59.43	301.72	518.56	27.59	20.12	140.09	146.55
06/06/97	Field Conductivity (us)	641.00	581.00	729.00	757.00	418.00	407.00	711.00	747.00
	KBr (mg/L)	29.11	15.17	61.92	151.08	12.35	5.06	19.28	38.30
	KCl (mg/L)	185.10	102.28	364.59	511.68	45.73	21.33	259.78	349.46
06/09/97	Field Conductivity (us)	625.00	528.00	784.00	715.00	494.00	422.00	782.00	916.00
	KBr (mg/L)	13.09	2.86	28.70	88.67	n/s	n/s	7.27	14.44
	KCl (mg/L)	163.00	58.97	414.00	431.92	121.17	28.69	282.75	431.92
06/13/97	Field Conductivity (us)	692.00	678.00	897.00	748.00	647.00	475.00	842.00	1020.00
	KBr (mg/L)	n/s	n/s	21.38	57.03	n/s	n/s	6.92	11.87
	KCl (mg/L)	210.18	185.10	470.11	431.92	259.78	69.86	364.59	511.68
06/16/97	Field Conductivity (us)	719.00	827.00	924.00	847.00	713.00	626.00	892.00	1065.00
	KBr (mg/L)	n/s	n/s	32.17	49.49	n/s	n/s	n/s	n/s
	KCl (mg/L)	181.09	287.87	420.63	438.73	253.69	197.02	370.68	564.94
06/20/97	Field Conductivity (us)	684.00	901.00	950.00	960.00	732.00	895.00	926.00	1113.00
	KBr (mg/L)	n/s	n/s	n/s	41.45	n/s	n/s	n/s	n/s
	KCl (mg/L)	170.49	329.69	405.14	497.85	218.32	373.08	373.08	540.63
06/23/97	Field Conductivity (us)	867.00	641.00	976.00	1028.00	617.00	739.00	830.00	1123.00
	KCl (mg/L)	357.11	145.98	481.17	570.56	113.06	223.51	327.94	621.30
06/27/97	Field Conductivity (us)	541.00	755.00	777.00	1157.00	476.00	663.00	685.00	1112.00
	KBr (mg/L)	13.46	15.94	17.06	33.57	5.97	10.62	15.41	24.75
	KCl (mg/L)	53.85	176.69	204.97	609.12	29.73	124.94	152.30	525.05
06/30/97	Field Conductivity (us)	533.00	645.00	766.00	1111.00	468.00	606.00	636.00	958.00
	KBr (mg/L)	15.69	13.31	28.39	48.07	9.27	10.92	42.14	34.59
	KCl (mg/L)	171.79	298.88	904.69	1947.60	164.63	286.42	985.14	1445.44
07/04/97	Field Conductivity (us)	498.00	552.00	715.00	913.00	493.00	468.00	634.00	711.00
	KCl (mg/L)	311.89	212.57	1028.02	1713.96	286.42	94.62	1072.75	1327.39
07/07/97	Field Conductivity (us)	701.00	652.00	823.00	940.00	660.00	546.00	723.00	844.00
	KCl (mg/L)	198.79	55.90	270.02	456.46	182.14	34.55	207.68	498.20
07/11/97	Field Conductivity (us)	679.00	581.00	728.00	890.00	710.00	466.00	616.00	794.00
	KCl (mg/L)	258.46	134.09	366.78	476.87	456.46	90.45	321.66	520.48
07/14/97	KCl (mg/L)	383.18	436.92	520.48	400.31	456.46	294.71	568.07	647.74

	LOCATION:	G310R				G312L			
	DEPTH:	-1	-2	-3	-4	-1	-2	-3	-4
DATE	PARAMETER								
05/09/97	Field Conductivity (us)	445.00	384.00	578.00	507.00	n/s	n/s	n/s	n/s
	KBr (mg/L)	105.22	7.19	92.21	84.45	n/s	n/s	n/s	n/s
05/12/97	Field Conductivity (us)	627.00	482.00	689.00	798.00	625.00	651.00	601.00	719.00
	KBr (mg/L)	170.69	21.60	212.67	315.94	5.29	3.56	3.41	4.84
05/16/97	Field Conductivity (us)	705.00	791.00	758.00	892.00	631.00	654.00	622.00	739.00
	KBr (mg/L)	247.97	348.34	323.00	453.73	12.09	9.28	17.64	12.55
05/19/97	Field Conductivity (us)	692.00	819.00	759.00	886.00	608.00	642.00	641.00	718.00
	KBr (mg/L)	256.74	400.59	345.38	500.38	18.47	12.28	40.23	28.81
05/23/97	Field Conductivity (us)	602.00	764.00	699.00	886.00	585.00	658.00	669.00	738.00
	KBr (mg/L)	94.06	281.90	217.07	451.23	18.61	2.83	40.76	40.76
	KCl (mg/L)	401.24	n/s	n/s	n/s	n/s	n/s	n/s	n/s
05/26/97	Field Conductivity (us)	564.00	613.00	681.00	888.00	638.00	674.00	717.00	805.00
	KBr (mg/L)	38.68	50.23	128.71	366.10	52.93	3.49	76.31	99.11
	KCl (mg/L)	222.43	n/s	474.90	972.08	n/s	n/s	n/s	n/s
05/30/97	Field Conductivity (us)	528.00	517.00	646.00	756.00	716.00	677.00	757.00	888.00
	KBr (mg/L)	43.07	16.32	85.47	169.59	127.47	28.88	134.96	164.82
	KCl (mg/L)	140.54	53.41	284.05	574.12	n/s	n/s	n/s	n/s
06/02/97	Field Conductivity (us)	547.00	504.00	630.00	698.00	611.00	744.00	803.00	935.00
	KBr (mg/L)	57.85	14.04	88.76	150.31	215.92	183.15	310.17	460.47
	KCl (mg/L)	102.14	25.21	175.55	251.88	n/s	n/s	n/s	n/s
06/06/97	Field Conductivity (us)	553.00	434.00	543.00	560.00	662.00	699.00	639.00	859.00
	KBr (mg/L)	18.01	8.76	33.39	33.39	114.82	64.08	212.91	280.16
	KCl (mg/L)	106.71	35.47	131.89	137.59	n/s	n/s	n/s	n/s
06/09/97	Field Conductivity (us)	611.00	454.00	618.00	602.00	638.00	692.00	734.00	921.00
	KBr (mg/L)	9.29	n/s	9.29	12.47	96.72	55.86	179.35	270.71
	KCl (mg/L)	149.76	40.27	193.11	143.55	298.74	187.93	562.08	856.64
06/13/97	Field Conductivity (us)	727.00	562.00	766.00	923.00	610.00	698.00	714.00	942.00
	KBr (mg/L)	10.76	n/s	9.29	10.25	68.63	43.93	145.98	270.71
	KCl (mg/L)	238.67	86.34	307.75	431.92	271.02	185.10	659.78	1051.48
06/16/97	Field Conductivity (us)	754.00	736.00	786.00	1053.00	555.00	638.00	663.00	878.00
	KBr (mg/L)	n/s	n/s	n/s	n/s	47.05	34.55	100.09	228.03
	KCl (mg/L)	243.22	253.69	313.18	519.28	173.62	118.82	340.72	758.75
06/20/97	Field Conductivity (us)	752.00	968.00	884.00	1102.00	537.00	644.00	655.00	908.00
	KBr (mg/L)	n/s	n/s	n/s	n/s	37.01	32.26	76.07	212.91
	KCl (mg/L)	237.08	388.78	343.56	540.63	122.60	99.77	279.58	611.79
06/23/97	Field Conductivity (us)	695.00	960.00	955.00	1172.00	575.00	679.00	801.00	951.00
	KBr (mg/L)	n/s	n/s	n/s	n/s	11.36	32.45	53.91	302.97
	KCl (mg/L)	173.10	405.79	423.45	570.56	87.56	123.11	357.11	676.55
06/27/97	Field Conductivity (us)	526.00	711.00	782.00	1151.00	633.00	694.00	925.00	1001.00
	KBr (mg/L)	7.32	13.46	17.64	27.40	n/s	n/s	n/s	n/s
	KCl (mg/L)	44.18	137.94	215.38	525.05	72.48	97.54	390.12	640.03
06/30/97	Field Conductivity (us)	501.00	650.00	686.00	948.00	618.00	679.00	908.00	931.00
	KBr (mg/L)	10.23	20.42	22.54	29.34	n/s	n/s	n/s	n/s
	KCl (mg/L)	252.06	616.60	700.65	1385.16	420.24	762.96	1642.48	3850.35
07/04/97	Field Conductivity (us)	519.00	589.00	669.00	788.00	601.00	667.00	855.00	979.00
	KCl (mg/L)	498.31	520.00	731.14	1028.02	420.24	643.43	1713.96	3247.13
07/07/97	Field Conductivity (us)	698.00	684.00	779.00	825.00	719.00	787.00	940.00	1102.00
	KCl (mg/L)	198.79	117.60	236.81	236.81	82.87	159.73	366.78	738.58
07/11/97	Field Conductivity (us)	688.00	553.00	682.00	682.00	650.00	698.00	903.00	1146.00
	KCl (mg/L)	383.18	107.75	307.89	207.68	112.56	159.73	400.31	771.61

	LOCATION:	G310R				G312L			
	DEPTH:	-1	-2	-3	-4	-1	-2	-3	-4
DATE	PARAMETER								
07/14/97	KCl (mg/L)	366.78	174.34	294.71	321.66	98.72	182.14	383.18	647.74
07/18/97	Field Conductivity (us)	713.00	709.00	695.00	762.00	628.00	882.00	845.00	1074.00
	KCl (mg/L)	n/s	n/s	n/s	n/s	95.28	341.90	341.90	688.97

	LOCATION:	G312M				G312R			
	DEPTH:	-1	-2	-3	-4	-1	-2	-3	-4
DATE	PARAMETER								
04/15/97	Field Conductivity (us)	509.00	552.00	541.00	568.00	n/s	n/s	n/s	n/s
05/12/97	Field Conductivity (us)	513.00	627.00	522.00	620.00	493.00	577.00	583.00	720.00
	KBr (mg/L)	2.29	2.10	2.99	2.86	1.76	1.69	1.69	3.72
05/16/97	Field Conductivity (us)	472.00	611.00	510.00	601.00	490.00	592.00	555.00	704.00
	KBr (mg/L)	5.07	5.47	16.98	45.33	9.64	5.27	5.47	8.94
05/19/97	Field Conductivity (us)	460.00	626.00	518.00	652.00	558.00	603.00	552.00	717.00
	KBr (mg/L)	25.78	5.43	33.42	90.93	70.15	8.80	7.87	9.83
05/23/97	Field Conductivity (us)	543.00	657.00	565.00	742.00	642.00	614.00	503.00	697.00
	KBr (mg/L)	65.24	3.49	65.24	158.64	150.56	4.78	3.49	5.04
05/26/97	Field Conductivity (us)	633.00	667.00	653.00	837.00	719.00	634.00	507.00	697.00
	KBr (mg/L)	110.03	3.49	128.71	253.92	185.57	4.09	7.26	10.47
05/30/97	Field Conductivity (us)	718.00	705.00	750.00	898.00	733.00	677.00	581.00	766.00
	KBr (mg/L)	151.29	8.46	184.76	225.63	179.56	19.93	80.72	69.98
06/02/97	Field Conductivity (us)	674.00	708.00	773.00	934.00	711.00	714.00	687.00	828.00
	KBr (mg/L)	215.92	127.50	353.83	525.29	263.09	97.97	254.57	320.55
06/06/97	Field Conductivity (us)	526.00	663.00	682.00	885.00	555.00	645.00	701.00	792.00
	KBr (mg/L)	81.47	16.24	198.79	321.37	100.09	35.76	235.99	179.35
06/09/97	Field Conductivity (us)	497.00	670.00	628.00	870.00	519.00	652.00	724.00	811.00
	KBr (mg/L)	45.47	13.68	122.97	270.71	59.83	31.17	228.03	192.09
	KCl (mg/L)	123.31	48.80	353.59	754.92	165.62	99.88	611.50	562.08
06/13/97	Field Conductivity (us)	461.00	668.00	566.00	841.00	488.00	644.00	681.00	571.00
	KBr (mg/L)	26.26	12.78	71.02	220.34	39.64	24.52	179.35	212.91
	KCl (mg/L)	72.88	54.18	249.00	781.63	131.89	90.07	556.93	781.63
06/16/97	Field Conductivity (us)	446.00	640.00	529.00	723.00	471.00	625.00	620.00	805.00
	KBr (mg/L)	18.63	18.63	52.16	161.81	38.30	28.13	127.26	198.79
	KCl (mg/L)	51.16	65.87	159.59	477.31	109.22	77.97	370.68	589.25
06/20/97	Field Conductivity (us)	436.00	592.00	557.00	653.00	503.00	626.00	580.00	830.00
	KBr (mg/L)	13.68	24.52	32.26	87.26	32.26	27.18	90.30	179.35
	KCl (mg/L)	37.10	68.85	127.76	257.45	108.34	71.75	237.08	497.85
06/23/97	Field Conductivity (us)	488.00	691.00	765.00	789.00	661.00	632.00	551.00	824.00
	KBr (mg/L)	16.49	23.13	29.32	201.84	n/s	n/s	68.33	215.97
	KCl (mg/L)	83.91	62.27	327.94	461.11	223.51	99.49	180.63	502.11
06/27/97	Field Conductivity (us)	699.00	719.00	914.00	786.00	798.00	684.00	597.00	826.00
	KBr (mg/L)	n/s	23.65	n/s	n/s	n/s	28.90	n/s	n/s
	KCl (mg/L)	204.97	107.70	371.28	336.28	237.79	107.70	144.94	371.28
06/30/97	Field Conductivity (us)	703.00	n/s	n/s	n/s	n/s	n/s	n/s	n/s
	KBr (mg/L)	n/s	26.67	n/s	n/s	n/s	30.08	n/s	n/s
	KCl (mg/L)	830.81	616.60	1272.04	2409.91	904.69	796.16	944.05	2624.22
07/04/97	Field Conductivity (us)	675.00	671.00	774.00	1046.00	679.00	670.00	635.00	805.00
	KCl (mg/L)	616.60	643.43	1168.15	2514.78	762.96	700.65	866.96	2309.40
07/07/97	Field Conductivity (us)	740.00	786.00	929.00	1063.00	768.00	779.00	790.00	934.00
	KCl (mg/L)	112.56	140.09	351.08	568.07	140.09	146.35	258.46	543.75

	LOCATION:	G312M				G312R			
	DEPTH:	-1	-2	-3	-4	-1	-2	-3	-4
DATE	PARAMETER								
07/11/97	Field Conductivity (us)	540.00	700.00	839.00	1024.00	596.00	684.00	805.00	1082.00
	KCl (mg/L)	53.51	146.35	307.89	543.75	94.49	134.09	351.08	620.01
07/14/97	KCl (mg/L)	58.40	146.35	247.40	351.08	128.35	190.28	351.08	620.01
07/18/97	Field Conductivity (us)	556.00	667.00	715.00	875.00	637.00	679.00	840.00	1022.00
	KCl (mg/L)	107.82	256.21	208.50	456.25	162.82	226.41	371.28	748.17

	LOCATION:	G315L				G315M			
	DEPTH:	-1	-2	-3	-4	-1	-2	-3	-4
DATE	PARAMETER								
04/15/97	Field Conductivity (us)	n/s	n/s	n/s	n/s	551.00	576.00	530.00	582.00
05/19/97	Field Conductivity (us)	642.00	598.00	570.00	574.00	545.00	522.00	524.00	554.00
	KBr (mg/L)	7.04	6.07	6.54	4.68	3.12	2.89	4.51	3.61
05/23/97	Field Conductivity (us)	627.00	606.00	548.00	526.00	505.00	487.00	520.00	538.00
	KBr (mg/L)	2.42	1.68	7.26	3.49	1.44	1.29	3.32	3.49
05/26/97	Field Conductivity (us)	621.00	621.00	609.00	614.00	499.00	464.00	543.00	599.00
	KBr (mg/L)	1.77	1.44	18.61	52.93	5.04	1.44	9.43	33.07
05/30/97	Field Conductivity (us)	645.00	611.00	689.00	800.00	548.00	472.00	580.00	763.00
	KBr (mg/L)	20.50	6.54	80.72	179.56	44.32	6.54	52.60	142.89
06/02/97	Field Conductivity (us)	689.00	609.00	755.00	877.00	581.00	537.00	541.00	822.00
	KBr (mg/L)	91.73	13.59	215.92	445.55	119.37	52.41	47.48	353.83
06/06/97	Field Conductivity (us)	639.00	546.00	733.00	827.00	574.00	589.00	666.00	814.00
	KBr (mg/L)	87.26	22.89	185.61	300.05	110.94	122.97	179.35	261.58
06/09/97	Field Conductivity (us)	635.00	598.00	748.00	848.00	565.00	638.00	717.00	840.00
	KBr (mg/L)	93.45	55.86	212.91	310.53	110.94	161.81	205.73	270.71
06/13/97	Field Conductivity (us)	668.00	641.00	716.00	338.00	598.00	684.00	708.00	822.00
	KBr (mg/L)	96.72	110.94	173.30	270.71	96.72	173.30	185.61	270.71
06/16/97	Field Conductivity (us)	636.00	654.00	620.00	796.00	504.00	632.00	560.00	759.00
	KBr (mg/L)	87.26	136.30	141.06	235.99	93.45	151.08	156.35	228.03
06/20/97	Field Conductivity (us)	657.00	672.00	633.00	733.00	558.00	580.00	593.00	700.00
	KBr (mg/L)	84.31	136.30	96.72	179.35	71.02	100.09	93.45	167.46
	KCl (mg/L)	366.78	738.58	456.46	676.71	294.71	418.22	400.31	738.58
06/23/97	Field Conductivity (us)	629.00	703.00	706.00	837.00	491.00	570.00	623.00	815.00
	KBr (mg/L)	99.15	208.79	134.46	302.97	52.12	134.46	70.68	273.72
	KCl (mg/L)	265.03	502.11	372.65	768.78	108.34	301.16	233.24	621.30
06/27/97	Field Conductivity (us)	603.00	639.00	808.00	776.00	468.00	470.00	712.00	681.00
	KBr (mg/L)	44.00	125.66	47.09	188.63	19.53	33.57	30.32	129.99
	KCl (mg/L)	137.94	289.87	353.35	579.70	46.42	51.25	215.38	289.87
06/30/97	KBr (mg/L)	43.55	86.96	43.55	116.95	21.11	15.69	30.32	66.82
	KCl (mg/L)	700.65	1218.99	1385.16	1947.60	325.46	187.07	1072.75	1168.15
07/04/97	Field Conductivity (us)	557.00	544.00	786.00	844.00	504.00	432.00	711.00	903.00
	KBr (mg/L)	15.93	20.36	13.75	13.75	4.68	3.32	10.25	9.76
	KCl (mg/L)	700.65	731.14	1327.39	1573.98	402.72	203.70	1028.02	1508.34
07/07/97	Field Conductivity (us)	710.00	631.00	923.00	1022.00	632.00	579.00	867.00	1059.00
	KBr (mg/L)	10.25	13.75	10.25	8.02	6.27	5.97	6.59	8.02
	KCl (mg/L)	166.88	146.35	418.22	520.48	98.72	82.87	282.10	476.87
07/11/97	Field Conductivity (us)	648.00	617.00	930.00	1048.00	587.00	697.00	892.00	1026.00
	KBr (mg/L)	17.17	20.15	16.49	12.97	9.41	10.61	11.97	13.50
	KCl (mg/L)	152.90	198.79	436.92	520.48	128.35	247.40	351.08	498.20
07/14/97	KCl (mg/L)	146.35	226.67	383.18	593.47	112.56	294.71	321.66	520.48

	LOCATION:	G315L				G315M			
	DEPTH:	-1	-2	-3	-4	-1	-2	-3	-4
DATE	PARAMETER								
07/18/97	Field Conductivity (us)	630.00	757.00	871.00	1109.00	585.00	814.00	830.00	1074.00
	KCl (mg/L)	208.50	314.85	420.15	608.84	184.25	403.18	386.90	584.25
07/21/97	Field Conductivity (us)	741.00	914.00	1019.00	1269.00	716.00	930.00	991.00	1229.00
	KCl (mg/L)	217.27	328.10	437.82	608.84	184.25	386.90	403.18	560.66
07/25/97	Field Conductivity (us)	n/s	721.00	801.00	973.00	616.00	626.00	811.00	950.00
	KCl (mg/L)	n/s	379.40	517.85	646.70	347.14	332.05	473.81	646.70
07/28/97	Field Conductivity (us)	n/s	723.00	806.00	975.00	679.00	678.00	784.00	976.00
	KCl (mg/L)	n/s	347.14	433.51	541.38	317.61	254.33	362.91	591.70

	LOCATION:	G315R			
	DEPTH:	-1	-2	-3	-4
DATE	PARAMETER				
05/19/97	Field Conductivity (us)	567.00	584.00	588.00	637.00
	KBr (mg/L)	3.23	2.89	3.00	2.59
05/23/97	Field Conductivity (us)	541.00	535.00	558.00	666.00
	KBr (mg/L)	2.18	1.68	2.55	1.23
05/26/97	Field Conductivity (us)	561.00	532.00	569.00	694.00
	KBr (mg/L)	9.43	2.07	5.31	1.36
05/30/97	Field Conductivity (us)	602.00	545.00	583.00	723.00
	KBr (mg/L)	68.01	6.73	35.27	6.36
06/02/97	Field Conductivity (us)	656.00	558.00	592.00	697.00
	KBr (mg/L)	145.45	12.72	85.88	11.91
06/06/97	Field Conductivity (us)	637.00	514.00	590.00	638.00
	KBr (mg/L)	151.08	17.40	93.45	6.66
06/09/97	Field Conductivity (us)	645.00	531.00	634.00	634.00
	KBr (mg/L)	145.98	55.86	136.30	9.38
06/13/97	Field Conductivity (us)	641.00	665.00	702.00	599.00
	KBr (mg/L)	131.70	127.26	185.61	22.89
06/16/97	Field Conductivity (us)	566.00	677.00	626.00	549.00
	KBr (mg/L)	110.94	156.35	192.09	43.93
06/20/97	Field Conductivity (us)	548.00	681.00	695.00	653.00
	KBr (mg/L)	76.07	151.08	167.46	78.72
	KCl (mg/L)	270.02	593.47	647.74	270.02
06/23/97	Field Conductivity (us)	557.00	613.00	640.00	745.00
	KBr (mg/L)	80.93	129.99	129.99	182.35
	KCl (mg/L)	180.63	288.60	276.57	388.87
06/27/97	Field Conductivity (us)	555.00	568.00	868.00	616.00
	KBr (mg/L)	53.91	109.75	273.72	83.71
	KCl (mg/L)	124.94	204.97	706.64	237.79
06/30/97	KBr (mg/L)	43.55	71.37	73.76	314.05
	KCl (mg/L)	762.96	866.96	1218.99	4017.91
07/04/97	Field Conductivity (us)	574.00	458.00	580.00	837.00
	KBr (mg/L)	13.75	21.38	28.70	225.16
	KCl (mg/L)	796.16	566.24	1168.15	3247.13
07/07/97	Field Conductivity (us)	775.00	559.00	739.00	897.00
	KBr (mg/L)	8.02	14.44	11.30	113.32
	KCl (mg/L)	207.68	82.87	247.40	676.71

	LOCATION:	G315R			
	DEPTH:	-1	-2	-3	-4
DATE	PARAMETER				
07/11/97	Field Conductivity (us)	722.00	581.00	793.00	723.00
	KBr (mg/L)	14.63	22.72	17.87	117.46
	KCl (mg/L)	216.97	159.73	321.66	520.48
07/14/97	KCl (mg/L)	207.68	207.68	321.66	383.18
07/18/97	Field Conductivity (us)	706.00	781.00	791.00	728.00
	KCl (mg/L)	266.99	371.28	386.90	437.82
07/21/97	Field Conductivity (us)	855.00	955.00	915.00	879.00
	KCl (mg/L)	278.23	403.18	386.90	608.84
07/25/97	Field Conductivity (us)	678.00	740.00	744.00	759.00
	KCl (mg/L)	495.34	565.98	517.85	1008.56
07/28/97	Field Conductivity (us)	752.00	819.00	857.00	876.00
	KCl (mg/L)	541.38	541.38	495.34	1204.76

Appendix 19

Redox Status of the Anaerobic Zone of Gate 3

A19.1 BACKGROUND

Benzoate and mineral nutrients were injected via the NIW to encourage the development of a microbial community able to reductively dechlorinate PCE, CT, and CF. Such microorganisms typically require highly reduced environments for development and activity. Therefore, geochemical data were collected to detect the evolution of microbial habitats downgradient of the NIW. Disappearance of SO_4^{2-} would indicate sulfate reduction and the production of CH_4 would typify methanogenesis.

Dissolved H_2 data are presented in Table A19.1. Measurement of dissolved H_2 has been proposed as a useful aid in deducing the likely dominant terminal electron acceptor process active in groundwater environments (Lovley and Goodwin, 1988). Several studies (e.g., Chapelle et al., 1995 and 1997; Bjerg et al., 1997) have reported the use of dissolved H_2 measurements in groundwater studies. The method is based on the fact that H_2 is a central intermediate in anaerobic microbial metabolism, being both produced by fermentative metabolism and rapidly consumed by respiratory activities of other microbial groups, a process termed interspecies H_2 transfer. Organisms using more electrochemically positive electron acceptors, like NO_3^-, are more efficient in H_2 usage than sulfate reducers or methanogens. As a result, the latter groups are essentially restricted by lack of H_2 substrate when the former are active, because H_2 levels are maintained by the denitrifiers at a level too low for the sulfate reducers or methanogens to obtain sufficient H_2. Thus, anaerobic systems may develop discrete zones where a single electron-accepting process dominates, and each of these zones should have a characteristic steady-state H_2 concentration (Lovley and Goodwin, 1988).

Chapelle et al. (1997) indicated the following steady-state ranges of H_2 concentration in water (nM or 10^{-6} molar) for each terminal electron accepting process:

methanogenesis	5–30 nM
sulfate reduction	1–4 nM
Fe(III) reduction	0.2–0.8 nM
nitrate reduction	< 0.1 nM

Organic acid anions, including acetate, propionate, butyrate, and formate, are indicators of fermentative activity, an anaerobic process. In addition, acetate is one of the products of benzoate fermentation and its occurrence in groundwaters downgradient of the NIW could indicate that benzoate was being utilized, as designed, by anaerobes. Organic acid data appear in Table A19.2.

Table A19.1 Dissolved H_2 in Selected Multilevel Monitoring Points in Gates 2 and 3 and Inferred Dominant Electron-Accepting Process

Gate 2 Point	H_2 nM (SD)	Process	Gate 3 Point	H_2 nM (SD)	Process
			G300	0.2	F
G202M-1	2.6 (0.9)	S	G302M-1	0.3 (.2)	F
G20M-2	4.4 (0.2)	S-M	G302M-2	5.2 (0.6)	M
G202M-3	2.7 (0.2)	S	G302M-3	6.9 (1.8)	M
G202M-4	4.9 (1.7)	S-M	G302M-4	0.2 (0.1)	F
			G310M-1	0.3	F
			G310M-2	1.7 (0.2)	S
			G310M-3	0.4 (0.1)	F
			G310M-4	2.2 (0.3)	S
G212M-1	0.7	F	G312M-1	2.2 (0.5)	S
G212M-2	1.6 (0.3)	S	G312M-2	3.9 (0.1)	S
G212M-3	0.6	F	G312M-3	5.2 (0.2)	M
G212M-4	0.7 (0.1)	F	G312M-4	1.6 (0.1)	S
			G315-P	4.0 (0.6)	S
			G315M-1	dry	
			G315M-2	3.7 (0.2)	S
			G315M-3	5.3 (0.3)	M
			G315M-4	2.4 (0.2)	S
			G318M-1	0.1	N-F
			G318M-3	0.7 (0.1)	F
G224M-1	0.8 (0.1)	F	G324M-1	dry	
G224M-2	0.3 (0.1)	F	G324M-2	0.5 (0.2)	F
G224M-3	0.1 (0.2)	N-F	G324M-3	0.7 (0.2)	F
G224M-4	1.8 (0.5)	S	G324M-4	0.7 (0.1)	F

Notes: Dominant terminal electron-accepting process: M: methanogenesis; S: sulfate reduction; F: Fe(III) reduction; N: nitrate reduction.

A19.2 MATERIALS AND METHODS

Sampling of all listed analytes/parameters was conducted as outlined in Appendix 4, except for H_2. Groundwater H_2 data were collected October 16–18, 1997. The bubble strip method (Chapelle et al., 1997) was used and triplicate (usually) samples were analyzed for dissolved H_2 by gas chromatography with reduction gas detection (Trace Analytical, Menlo Park, CA) with the assistance of T. Anderson from Dr. Lovley's lab at the University of Massachusetts. Results are presented in Table A19.1.

A19.3 RESULTS OF THE ASSESSMENT

A19.3.1 Dissolved H_2

The standard interpretation of the H_2 data suggests strongly anaerobic conditions had developed at the upgradient end of both Gates 2 and 3, at fences G202 and G302, by October 1997. Sulfate reduction and methanogenesis were indicated to be the dominant terminal electron accepting processes. Background sampling results suggested this area was mildly aerobic before the experiment began (Section 4.2.3.1), so addition of contaminants (TOL, PCE, and CT) induced a significant change in the microbial habitat. Anaerobic toluene degraders may have contributed to the utilization of O_2, NO_3^-, and to the lowering of the redox levels. Reductive dechlorination of CT to CF appeared to be occurring here (discussed in Chapter 4), and it is possible that the

H_2 levels may be influenced by such microbes. Note that no typical H_2 ranges are listed for the situation of a halogenated compound like CT serving as the dominant terminal electron acceptor. Dechlorinating cultures using the chlorinated compound as the terminal electron acceptor have been reported (e.g., Holliger and Schumacher, 1994; Maymo-Gatell et al., 1995) and dechlorinators do appear to be better scavengers of low levels of H_2 than are methanogens (Smatlak et al., 1996; Ballapragada et al., 1997).

In the zone (fences G310–G315) where benzoate addition was to have created more reducing conditions, the H_2 data suggest sulfate reduction is the dominant electron-accepting process. This zone does not appear more reducing than does the upgradient, unamended area at G302, but it does appear to be more reducing than the corresponding zone in the unamended Gate 2. Based on the H_2 data, the segment of Gate 3 from G310 to G315 does not appear to have evolved to strictly methanogenic conditions, desired for the development of dechlorinating microbial populations.

Gates G318 and G324 are downgradient of the aerobic biosparge zone and groundwater from these fences contained significant DO in October 1997. The interpretation of the H_2 data suggests conditions were still anaerobic, with Fe(III) reduction dominant. While the H_2 data certainly indicate this segment of the gate is more oxidizing than the upgradient segment, they do not reasonably reflect the general aerobic conditions suggested by the DO data.

A19.3.2 Measured Redox Potential (E_h) and DO, Fe, SO_4^{2-}, Acetic Acid, and CH_4 Concentrations

Chapelle et al. (1995) suggested a hierarchy of certainty, with respect to redox level determinations, that is based on the number of agreeing lines of evidence. Thus, the interpretation of the H_2 data is supplemented with evidence derived from other redox-sensitive geochemical parameters: DO, SO_4^{2-}, and CH_4 concentrations. Organic acids are typical fermentation products and so are also indicators of anaerobic conditions. Measured E_h is considered of less value in this regard, but it will also be discussed briefly.

Typical E_h profiles down the anaerobic section of Gate 3 are illustrated in Figure A19.1 for day 42 (December 30, 1996) and day 338 (October 23, 1997). As will be seen later in this section, the apparent increase in E_h from December 1996 to October 1997 is not consistent with other geochemical indicators of redox level.

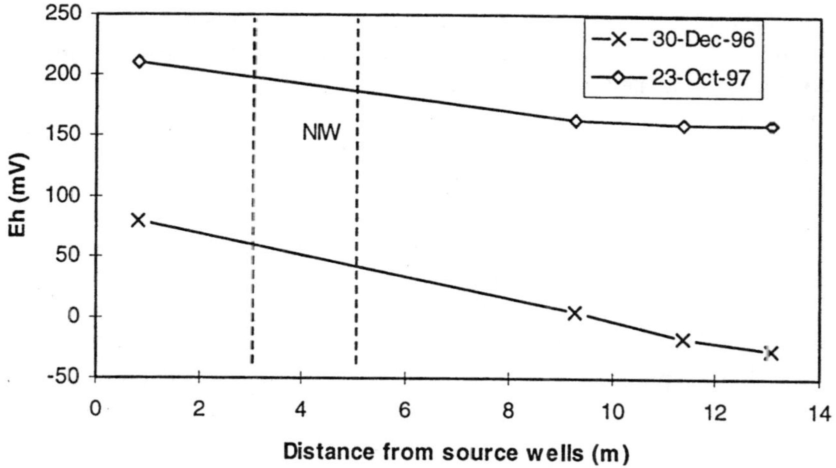

Figure A19.1 Fence-averaged redox potential along the anaerobic zone of Gate 3, December 30, 1996 and October 23, 1997.

DO was typically below 1 mg/L in groundwaters from fences G302 to G315, except for generally higher DO values (>2 mg/L) in October 1997. These DO concentrations >2 mg/L are indicative of aerobic conditions, but all other redox parameters suggest this zone was strongly anaerobic in October 1997. Therefore, these October 1997 DO values have been disregarded.

Iron levels increased slightly through the anaerobic zone (see Figure A19.2), perhaps as reduced groundwaters solubilized Fe(II) from the aquifer sand. The decline in Fe beyond the biosparge zone could reflect the conversion of Fe(II) to the less soluble Fe(III) and its precipitation as oxyhydroxides.

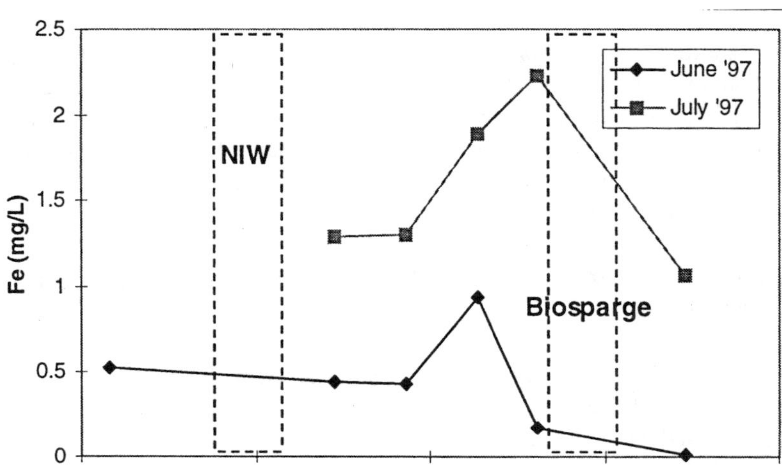

Figure A19.2 Profile of dissolved iron along selected point –3 of middle multilevel piezometers in Gate 3.

The SO_4^{2-} concentrations over the summer and fall of 1997 are indicated in Figure A19.3 for points in piezometers G310M, G312M, and G315M. This is the zone in which benzoate addition should have induced strongly anaerobic conditions. The onset of sulfate reduction is indicated by declining SO_4^{2-} concentrations. At points G310M-4, G312M-2, -3, -4, and G315M-3 and -4, SO_4^{2-} was already below 1 mg/L by June 26, 1997, indicating that sulfate reduction was already active in the deeper anaerobic zone upgradient of piezometer G310M. Upgradient or at all other points except perhaps G310M-2, sulfate reduction appeared to become important after June 30, 1997. The variations in sulfate concentrations in G312M-1 may reflect changing redox conditions with more oxidizing conditions generated as the water table declines, bringing the water table and thus soil O_2 closer to the shallowest monitoring points.

Sulfate reduction in the upper part of the gate was therefore alternatively fostered by the benzoate additions from the NIW, and inhibited by the influx of oxygen from the water table. Certainly sulfate-reducing conditions were present in lower segments of the anaerobic zone by July 1997.

More generally, fence-averaged SO_4^{2-} concentrations declined downgradient of the NIW (Figure A19.4), indicating anaerobic, sulfate-reducing conditions had developed at or upgradient of fences G312 and G315 by June 2, 1997 (day 196) and were developed before fence G310 by July 14, 1997. After July 14 (day 238), little SO_4^{2-} was available beyond fence G310, so active sulfate reduction probably became a less significant terminal electron-accepting process downgradient of fence G310 after day 238. This is consistent with the conclusion that sulfate reduction and/or methanogenesis were the major terminal electron-accepting processes in this segment of Gate 3, based on the H_2 data interpretation (Table A19.1).

Methane concentrations exceeded background values (1 to 2 μg/L found in G308) at fences G310 to G315, particularly at the second, third, and fourth points, after June 1997. This increased

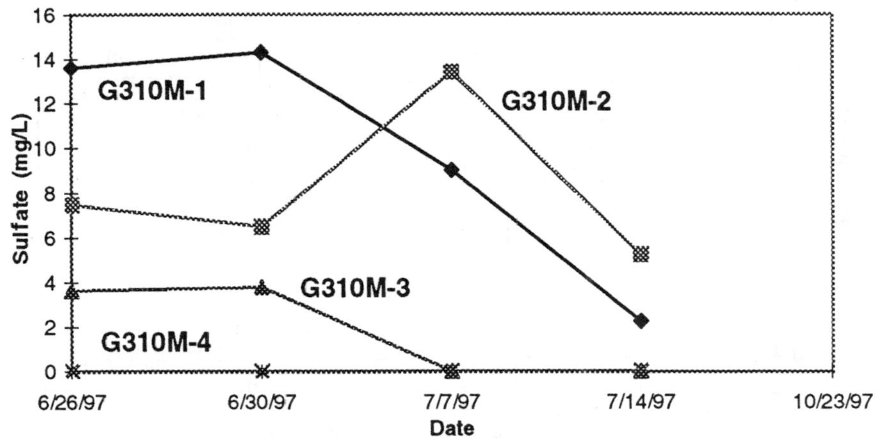

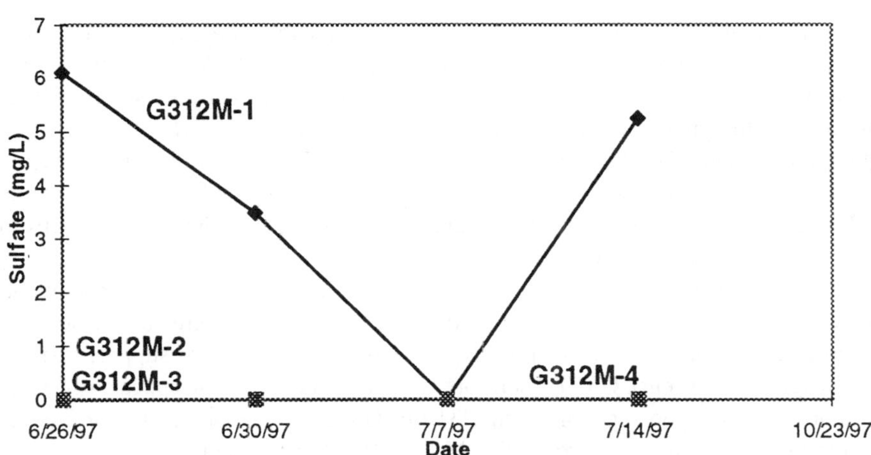

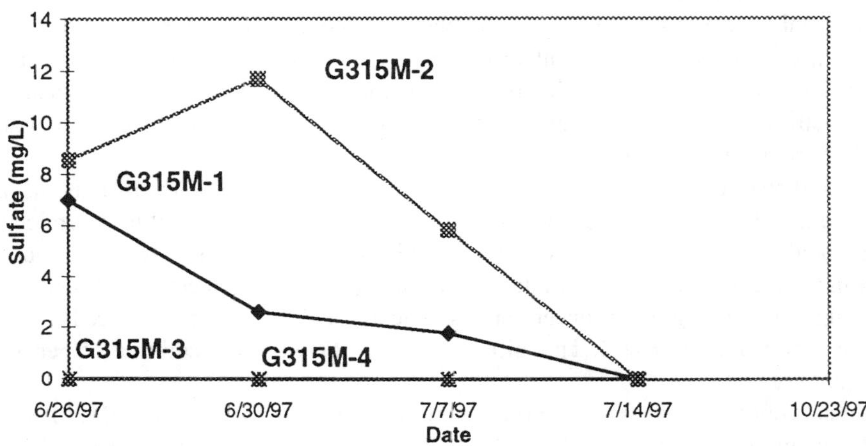

Figure A19.3 Temporal trends in sulfate concentrations in groundwaters at G310M, G312M, and G315M.

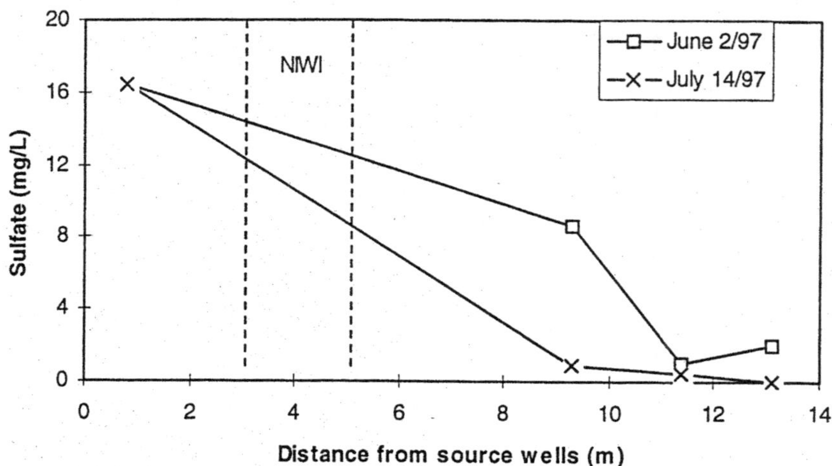

Figure A19.4 Fence-averaged sulfate concentrations along the anaerobic zone, June 2, 1997 and July 14, 1997.

CH_4 is evidence of methanogenic activity and so supports the inference of strongly anaerobic conditions being developed in the anaerobic segment of Gate 3. Again, point 1 appears less anaerobic, with consistently the lowest methane concentrations at that piezometer (Figure A19.5). This likely reflects the proximity of point 1 to the water table and soil O_2.

The generally higher CH_4 concentrations in October 1997 compared to June 1997 indicate that methanogenesis had become more significant. The development of high CH_4 concentrations at points 2, 3, and 4 suggests that a strongly reducing environment conducive to reductive dechlorination had been established by October 1997.

Thus, the development of elevated CH_4 concentrations in the G310 to G315 segment suggests methanogenesis was a significant terminal electron-accepting process. Given that SO_4^{2-} concentrations were already low at fence G310, little SO_4^{2-} would be available to support sulfate-reducing microbes downgradient of fence G310. Therefore, active sulfate reduction appears to be a potentially less important process in the segment from G310 to G315 by October 1997. On the other hand, CH_4 concentrations generally rose from G310M to G315M in October 1997. This suggests methanogenesis was active during this time. Perhaps the H_2 concentrations had not attained a steady-state distribution typical of active methanogenesis by October 1997 but rather were in transition from lower levels typical of sulfate reduction.

Methanogenic bacteria are sensitive to higher redox conditions and their activity can be suppressed if redox levels increase. The enhanced CH_4 production in the anaerobic segment from June to October 1997 suggests that sufficiently low redox conditions had been maintained. Thus, not only were methanogenic redox conditions attained, they were maintained for the 4 months prior to the end of the experiment in October 1997.

Acetic acid was detected at concentrations up to 250 mg/L (G315M-4) in the benzoate-stimulated zone of Gate 3. Acetic acid was not detected in background sampling. Formic, propionic, and butyric acids were generally below MDLs (Table A19.2). Acetic acid was also found at point G212M-4 in Gate 2 on day 226 (see Table A19.2). Acetic acid was detected in all points in G312M in Gate 3, suggesting that greater concentrations of acetic acid were being produced in this segment of Gate 3 as compared to Gate 2. This difference is attributed to the addition of benzoate in the NIW in Gate 3.

Acetic acid concentrations (Figure A19.6) increased with depth at G310M, G312M, and G315M, increased downgradient from G310M to G315M, and generally increased in October 1997 samples compared to June 1997. For example, the average acetic acid concentration at G315M increased from 30 mg/L on day 220 (June 26, 1997) to >120 mg/L on day 228 (October 21, 1997). Note

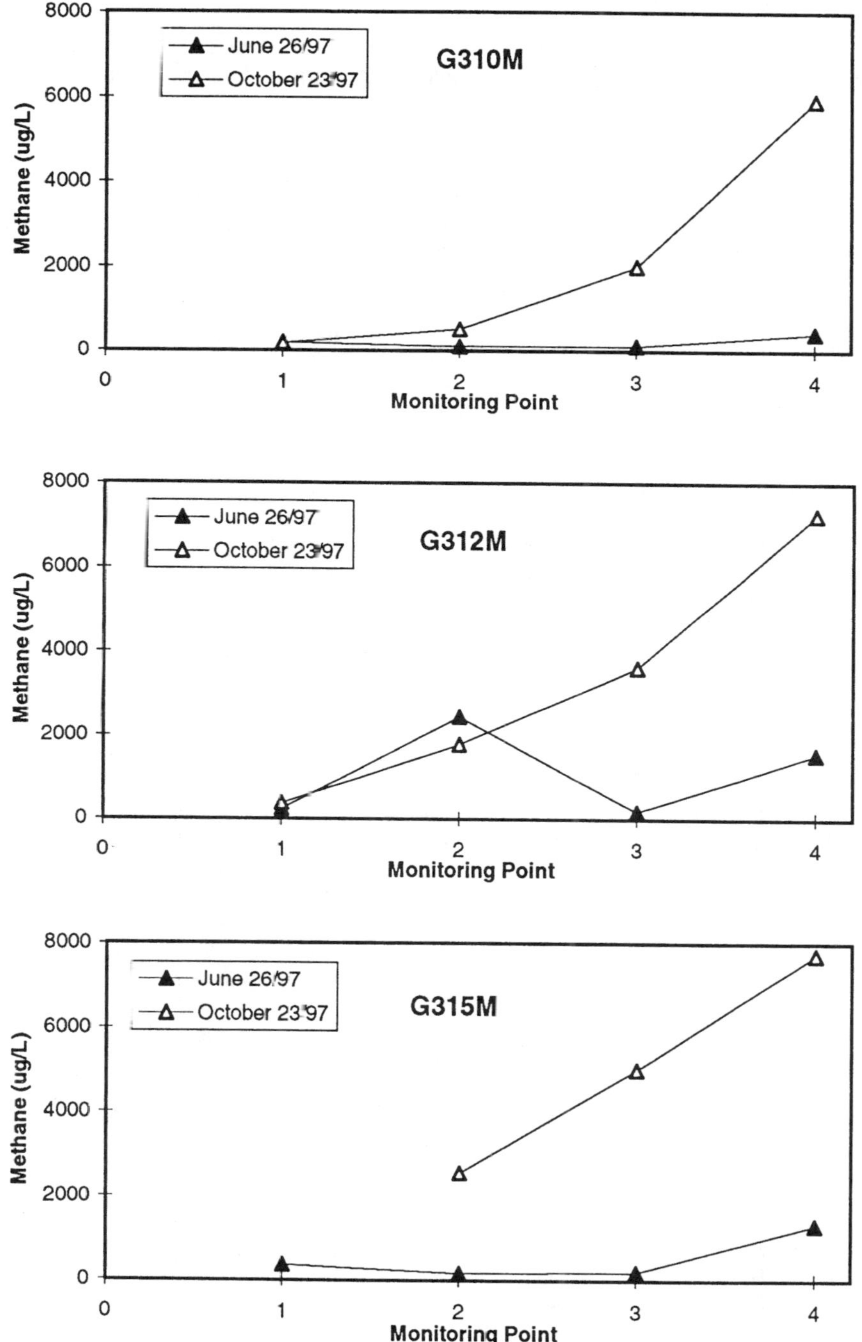

Figure A19.5 Depth profiles for methane at fences G310M, G312M, and G315M in the Gate 3 anaerobic zone at two sampling dates.

that the latter average assumed 0 mg/L in G315M-1, since that point was dry on day 338. These data suggest that fermentation of benzoate was occurring, so the groundwater was strongly reducing in this segment of Gate 3. The very high acetate concentrations suggest conversion of much of the benzoate to acetate occurred in groundwater as it migrated downgradient of the NIW. Thus, the benzoate appeared to be utilized by fermentative microbes, as designed.

Table A19.2 Organic Acids in Groundwater from Gates 2 and 3 (Concentrations in mg/L)

Location	Position	Level	Date					
			6/26/97	6/30/97	7/7/97	7/14/97	7/30/97	10/21/97
			Acetic acid					
212	M	1	< 1.5	0.5	< 1.5	< 1.5	n/s	< 3.0
212	M	2	< 1.5	n/d	< 1.5	< 1.5	n/s	< 3.0
212	M	3	< 1.5	n/d	< 1.5	< 1.5	n/s	< 3.0
212	M	4	< 1.5	71.0	68.2	31.8	n/s	85.0
301	P		n/s	n/s	n/s	n/s	n/d	n/s
302	M	1	n/s	n/s	n/s	n/s	n/s	< 3.0
302	M	2	n/s	n/s	n/s	n/s	n/s	< 3.0
302	M	3	n/s	n/s	n/s	n/s	n/s	< 3.0
302	M	4	n/s	n/s	n/s	n/s	n/s	< 3.0
310	L	1	7.1	n/s	n/s	n/s	n/s	n/s
310	L	2	28	n/s	n/s	n/s	n/s	n/s
310	L	3	34.4	n/s	n/s	n/s	n/s	n/s
310	L	4	47.6	n/s	n/s	n/s	n/s	n/s
310	M	1	3.4	0.9	14.4	27.2	n/s	11
310	M	2	25.6	21.1	3.7	25.9	n/s	12
310	M	3	33.8	30.6	35.5	36.2	n/s	91
310	M	4	54.4	37.4	43.2	49.5	n/s	184
310	R	1	10.2	n/s	n/s	n/s	n/s	n/s
310	R	2	34.1	n/s	n/s	n/s	n/s	n/s
310	R	3	33.3	n/s	n/s	n/s	n/s	n/s
310	R	4	35.4	n/s	n/s	n/s	n/s	n/s
312	L	1	3.7	n/s	n/s	n/s	n/s	n/s
312	L	2	58.3	n/s	n/s	n/s	n/s	n/s
312	L	3	48.6	n/s	n/s	n/s	n/s	n/s
312	L	4	155	n/s	n/s	n/s	n/s	n/s
312	M	1	19.9	22.7	25.7	6.7	n/s	25
312	M	2	87.4	100	95.1	103	n/s	69
312	M	3	37.4	42.8	47.1	55.1	n/s	104
312	M	4	62.7	60.7	66.3	91	n/s	202
312	R	1	24.8	n/s	n/s	n/s	n/s	n/s
312	R	2	74.7	n/s	n/s	n/s	n/s	n/s
312	R	3	30.3	n/s	n/s	n/s	n/s	n/s
312	R	4	87.9	n/s	n/s	n/s	n/s	n/s
315	L	1	19	n/s	n/s	n/s	n/s	n/s
315	L	2	39.8	n/s	n/s	n/s	n/s	n/s
315	L	3	45.3	n/s	n/s	n/s	n/s	n/s
315	L	4	60	n/s	n/s	n/s	n/s	n/s
315	M	1	2.1	12.9	18.3	18.6	n/s	n/a
315	M	2	17.1	7.7	22.6	40.5	n/s	86
315	M	3	36.7	44.5	45.6	51.4	n/s	147
315	M	4	62.8	53.8	56.5	65.3	n/s	244
315	R	1	25.9	n/s	n/s	n/s	n/s	n/s
315	R	2	42.6	n/s	n/s	n/s	n/s	n/s
315	R	3	41.2	n/s	n/s	n/s	n/s	n/s
315	R	4	105	n/s	n/s	n/s	n/s	n/s
324	S		n/s	n/s	n/s	n/s	n/d	n/s
LOQ			0.3	0.3	0.3	0.3	0.3	3
			Butyric acid					
212	M	1	n/d	< 2.5	n/d	< 5.0	n/s	< 10.0
212	M	2	n/d	< 2.5	n/d	< 5.0	n/s	< 10.0
212	M	3	n/d	< 2.5	n/d	< 5.0	n/s	< 10.0
212	M	4	n/d	< 2.5	n/d	< 5.0	n/s	< 10.0

Table A19.2 (continued) Organic Acids in Groundwater from Gates 2 and 3

			Date					
Location	**Position**	**Level**	**6/26/97**	**6/30/97**	**7/7/97**	**7/14/97**	**7/30/97**	**10/21/97**
301	P		n/s	n/s	n/s	n/s	n/d	n/s
302	M	1	n/s	n/s	n/s	n/s	n/s	n/s
302	M	2	n/s	n/s	n/s	n/s	n/s	n/s
302	M	3	n/s	n/s	n/s	n/s	n/s	n/s
302	M	4	n/s	n/s	n/s	n/s	n/s	n/s
310	L	1	n/d	n/s	n/s	n/s	n/s	n/s
310	L	2	n/d	n/s	n/s	n/s	n/s	n/s
310	L	3	n/d	n/s	n/s	n/s	n/s	n/s
310	L	4	n/d	n/s	n/s	n/s	n/s	n/s
310	M	1	n/d	< 2.5	n/d	< 5.0	n/s	< 10.0
310	M	2	n/d	< 2.5	n/d	< 5.0	n/s	< 10.0
310	M	3	n/d	< 2.5	n/d	< 5.0	n/s	< 10.0
310	M	4	n/d	< 2.5	n/d	< 5.0	n/s	< 10.0
310	R	1	n/d	n/s	n/s	n/s	n/s	n/s
310	R	2	n/d	n/s	n/s	n/s	n/s	n/s
310	R	3	n/d	n/s	n/s	n/s	n/s	n/s
310	R	4	n/d	n/s	n/s	n/s	n/s	n/s
312	L	1	n/d	n/s	n/s	n/s	n/s	n/s
312	L	2	n/d	n/s	n/s	n/s	n/s	n/s
312	L	3	n/d	n/s	n/s	n/s	n/s	n/s
312	L	4	n/d	n/s	n/s	n/s	n/s	n/s
312	M	1	n/d	< 2.5	n/d	< 5.0	n/s	< 10.0
312	M	2	n/d	< 2.5	n/d	< 5.0	n/s	< 10.0
312	M	3	n/d	< 2.5	n/d	< 5.0	n/s	< 10.0
312	M	4	n/d	< 2.5	n/d	< 5.0	n/s	< 10.0
312	R	1	n/d	n/s	n/s	n/s	n/s	n/s
312	R	2	n/d	n/s	n/s	n/s	n/s	n/s
312	R	3	n/d	n/s	n/s	n/s	n/s	n/s
312	R	4	n/d	n/s	n/s	n/s	n/s	n/s
315	L	1	n/d	n/s	n/s	n/s	n/s	n/s
315	L	2	n/d	n/s	n/s	n/s	n/s	n/s
315	L	3	n/d	n/s	n/s	n/s	n/s	n/s
315	L	4	n/d	n/s	n/s	n/s	n/s	n/s
315	M	1	n/d	< 2.5	n/d	< 5.0	n/s	n/a
315	M	2	n/d	< 2.5	n/d	< 5.0	n/s	< 10.0
315	M	3	n/d	< 2.5	n/d	< 5.0	n/s	< 10.0
315	M	4	n/d	< 2.5	n/d	< 5.0	n/s	< 10.0
315	R	1	n/d	n/s	n/s	n/s	n/s	n/s
315	R	2	n/d	n/s	n/s	n/s	n/s	n/s
315	R	3	n/d	n/s	n/s	n/s	n/s	n/s
315	R	4	n/d	n/s	n/s	n/s	n/s	n/s
324	S		n/s	n/s	n/s	n/s	n/d	n/s
LOQ			1	1	1	1	1	10
				Formic acid				
212	M	1	n/d	n/d	n/d	< 1.0	n/s	< 2.0
212	M	2	n/d	n/d	n/d	< 1.0	n/s	< 2.0
212	M	3	n/d	n/d	n/d	< 1.0	n/s	< 2.0
212	M	4	n/d	n/d	n/d	< 1.0	n/s	< 2.0
301	P		n/s	n/s	n/s	n/s	n/d	n/s
302	M	1	n/s	n/s	n/s	n/s	n/s	n/s
302	M	2	n/s	n/s	n/s	n/s	n/s	n/s
302	M	3	n/s	n/s	n/s	n/s	n/s	n/s
302	M	4	n/s	n/s	n/s	n/s	n/s	n/s

continued

Table A19.2 (continued) Organic Acids in Groundwater from Gates 2 and 3

			Date					
Location	**Position**	**Level**	**6/26/97**	**6/30/97**	**7/7/97**	**7/14/97**	**7/30/97**	**10/21/97**
310	L	1	n/d	n/s	n/s	n/s	n/s	n/s
310	L	2	n/d	n/s	n/s	n/s	n/s	n/s
310	L	3	n/d	n/s	n/s	n/s	n/s	n/s
310	L	4	n/d	n/s	n/s	n/s	n/s	n/s
310	M	1	n/d	n/d	n/d	< 1.0	n/s	< 2.0
310	M	2	n/d	n/d	n/d	< 1.0	n/s	3
310	M	3	n/d	n/d	n/d	< 1.0	n/s	< 2.0
310	M	4	n/d	n/d	n/d	< 1.0	n/s	3
310	R	1	n/d	n/s	n/s	n/s	n/s	n/s
310	R	2	n/d	n/s	n/s	n/s	n/s	n/s
310	R	3	n/d	n/s	n/s	n/s	n/s	n/s
310	R	4	n/d	n/s	n/s	n/s	n/s	n/s
312	L	1	n/d	n/s	n/s	n/s	n/s	n/s
312	L	2	n/d	n/s	n/s	n/s	n/s	n/s
312	L	3	n/d	n/s	n/s	n/s	n/s	n/s
312	L	4	n/d	n/s	n/s	n/s	n/s	n/s
312	M	1	n/d	n/d	n/d	< 1.0	n/s	2
312	M	2	n/d	n/d	n/d	< 1.0	n/s	< 2.0
312	M	3	n/d	n/d	n/d	< 1.0	n/s	< 2.0
312	M	4	n/d	n/d	n/d	< 1.0	n/s	3
312	R	1	n/d	n/d	n/s	n/s	n/s	n/s
312	R	2	n/d	n/d	n/s	n/s	n/s	n/s
312	R	3	n/d	n/d	n/s	n/s	n/s	n/s
312	R	4	n/d	n/s	n/s	n/s	n/s	n/s
315	L	1	n/d	n/s	n/s	n/s	n/s	n/s
315	L	2	n/d	n/s	n/s	n/s	n/s	n/s
315	L	3	n/d	n/s	n/s	n/s	n/s	n/s
315	L	4	n/d	n/s	n/s	n/s	n/s	n/s
315	M	1	n/d	n/d	n/d	< 1.0	n/s	n/a
315	M	2	n/d	n/d	n/d	< 1.0	n/s	< 2.0
315	M	3	n/d	n/d	n/d	< 1.0	n/s	< 2.0
315	M	4	n/d	n/d	n/d	< 1.0	n/s	< 2.0
315	R	1	n/d	n/s	n/s	n/s	n/s	n/s
315	R	2	n/d	n/s	n/s	n/s	n/s	n/s
315	R	3	n/d	n/s	n/s	n/s	n/s	n/s
315	R	4	n/d	n/s	n/s	n/s	n/s	n/s
324	S		n/s	n/s	n/s	n/s	n/d	n/s
LOQ			0.2	0.2	0.2	0.2	0.2	2
				Propionic acid				
212	M	1	n/d	< 1.8	< 3.5	< 3.5	n/s	< 7.0
212	M	2	n/d	< 1.8	< 3.5	< 3.5	n/s	< 7.0
212	M	3	n/d	< 1.8	< 3.5	< 3.5	n/s	< 7.0
212	M	4	n/d	< 1.8	< 3.5	< 3.5	n/s	< 7.0
301	P		n/s	n/s	n/s	n/s	< 1.8	n/s
302	M	1	n/s	n/s	n/s	n/s	n/s	n/s
302	M	2	n/s	n/s	n/s	n/s	n/s	n/s
302	M	3	n/s	n/s	n/s	n/s	n/s	n/s
302	M	4	n/s	n/s	n/s	n/s	n/s	n/s
310	L	1	n/d	n/s	n/s	n/s	n/s	n/s
310	L	2	n/d	n/s	n/s	n/s	n/s	n/s
310	L	3	n/d	n/s	n/s	n/s	n/s	n/s
310	L	4	n/d	n/s	n/s	n/s	n/s	n/s
310	M	1	n/d	< 1.8	< 3.5	< 3.5	n/s	< 7.0
310	M	2	n/d	< 1.8	< 3.5	< 3.5	n/s	< 7.0

Table A19.2 (continued) Organic Acids in Groundwater from Gates 2 and 3

Location	Position	Level	Date					
			6/26/97	6/30/97	7/7/97	7/14/97	7/30/97	10/21/97
310	M	3	n/d	< 1.8	< 3.5	< 3.5	n/s	< 7.0
310	M	4	n/d	< 1.8	< 3.5	< 3.5	n/s	< 7.0
310	R	1	n/d	n/s	n/s	n/s	n/s	n/s
310	R	2	n/d	n/s	n/s	n/s	n/s	n/s
310	R	3	n/d	n/s	n/s	n/s	n/s	n/s
310	R	4	n/d	n/s	n/s	n/s	n/s	n/s
312	L	1	n/d	n/s	n/s	n/s	n/s	n/s
312	L	2	n/d	n/s	n/s	n/s	n/s	n/s
312	L	3	n/d	n/s	n/s	n/s	n/s	n/s
312	L	4	n/d	n/s	n/s	n/s	n/s	n/s
312	M	1	n/d	< 1.8	< 3.5	< 3.5	n/s	< 7.0
312	M	2	n/d	< 1.8	< 3.5	< 3.5	n/s	< 7.0
312	M	3	n/d	< 1.8	< 3.5	< 3.5	n/s	< 7.0
312	M	4	n/d	< 1.8	< 3.5	< 3.5	n/s	< 7.0
312	R	1	n/d	n/s	n/s	n/s	n/s	n/s
312	R	2	n/d	n/s	n/s	n/s	n/s	n/s
312	R	3	n/d	n/s	n/s	n/s	n/s	n/s
312	R	4	n/d	n/s	n/s	n/s	n/s	n/s
315	L	1	n/d	n/s	n/s	n/s	n/s	n/s
315	L	2	n/d	n/s	n/s	n/s	n/s	n/s
315	L	3	n/d	n/s	n/s	n/s	n/s	n/s
315	L	4	n/d	n/s	n/s	n/s	n/s	n/s
315	M	1	n/d	< 1.8	< 3.5	< 3.5	n/s	n/a
315	M	2	n/d	< 1.8	< 3.5	< 3.5	n/s	< 7.0
315	M	3	n/d	< 1.8	< 3.5	< 3.5	n/s	< 7.0
315	M	4	n/d	< 1.8	< 3.5	< 3.5	n/s	< 7.0
315	R	1	n/d	n/s	n/s	n/s	n/s	n/s
315	R	2	n/d	n/s	n/s	n/s	n/s	n/s
315	R	3	n/d	n/s	n/s	n/s	n/s	n/s
315	R	4	n/d	n/s	n/s	n/s	n/s	n/s
324	S		n/s	n/s	n/s	n/s	< 1.8	n/s
LOQ			0.7	0.7	0.7	0.7	0.7	7

Notes: n/a: not available; n/s: not sampled; n/d: not detected (<MDL).

A19.3.3 Overall Assessment of the Redox Level in the Anaerobic Zone

The geochemical indicators of redox conditions roughly agree and indicate that the zone between fences G310 and G315 became strongly anaerobic by June/July 1997. Interpretation of the H_2 data suggests sulfate reduction and methanogenesis were the dominant electron-accepting processes in this zone. The evidence of depleted SO_4^{2-}, elevated CH_4, lack of significant DO, nitrate, and the presence of significant Fe in groundwaters are consistent with this interpretation. Methanogenesis may have become the dominant terminal electron-accepting process by October 1997.

Benzoate was being utilized by the microbes with acetate being generated. Overall, the addition of benzoate via the NIW appears to have produced a strongly anaerobic habitat between fences G310 and G315 by June/July 1997. This habitat is considered conducive to reductive dechlorinators. Thus, the periodic addition of sodium benzoate and mineral nutrients (included after day 207, June 13, 1997; see Section 4.2.1.1) created and maintained the desired geochemical conditions in groundwater downgradient of the NIW from June/July 1997 until the end of the experiment in October 1997.

Shallow groundwater in this zone was less reduced than deeper groundwater. This could have been due to the proximity of the water table which may have allowed O_2 to diffuse into the shallowest groundwater, raising its redox level.

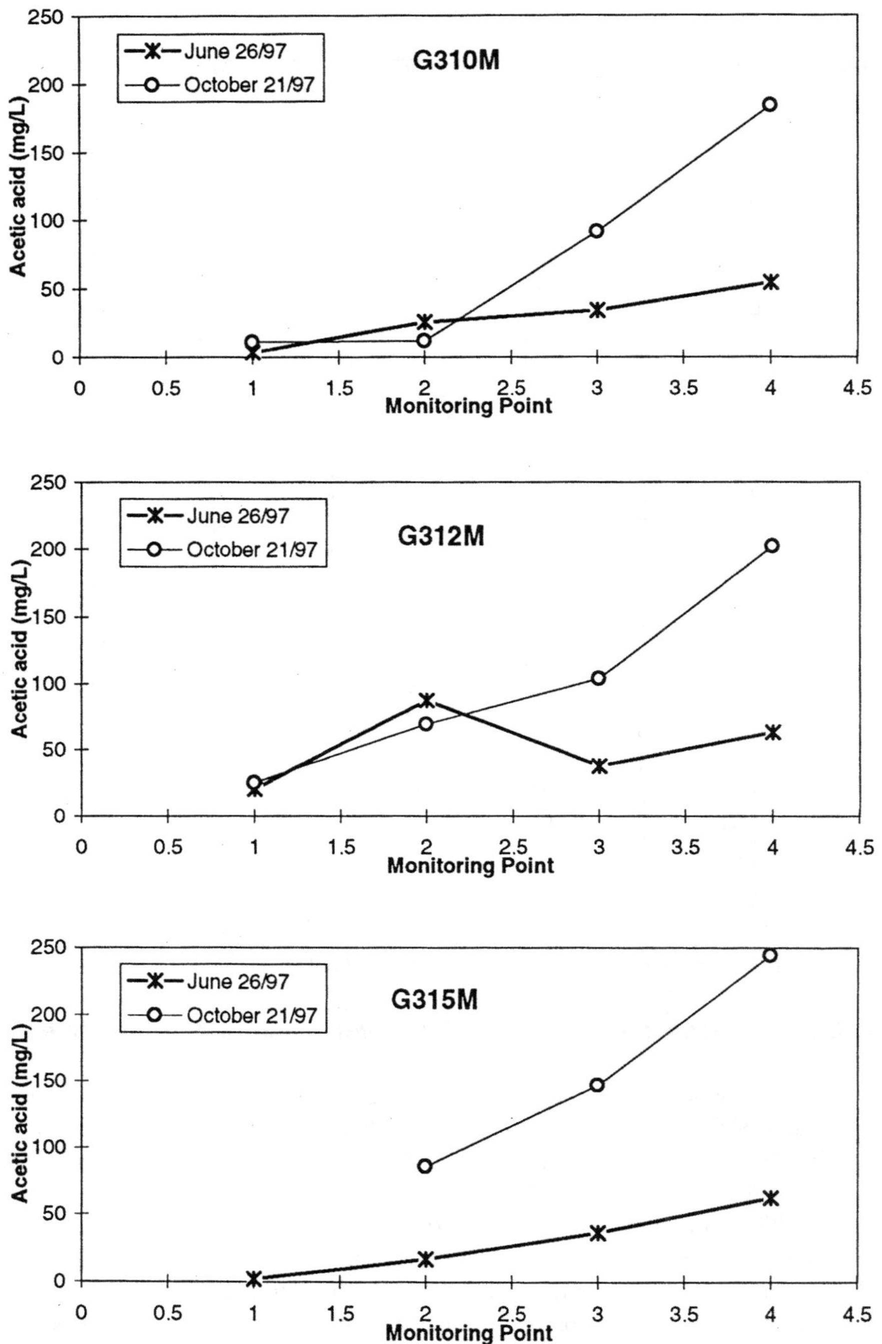

Figure A19.6 Depth profiles for acetic acid at fences G310M, G312M, and G315M in the Gate 3 anaerobic zone at two sampling dates.

While conditions appeared to be strongly reducing in June 1997, they were even more strongly reducing in October 1997, perhaps as a result of the inclusion of mineral nutrients with benzoate. Redox conditions appear to have become very conducive to microbial reductive dechlorination between June and October 1997.

A19.4 REFERENCES

Ballapragada, B.S., H.D. Stensel, J.A. Puhakka, and J.F. Ferguson. 1997. Effect of hydrogen on reductive dechlorination of chlorinated ethenes. *Environ. Sci. Technol.* 31: 1728-1734.

Bjerg, P.L., R. Jakobsen, H. Bay, M. Rasmussen, H.-J. Albrechtsen, and T.H. Christensen. 1997. Effects of sampling well construction on H_2 measurements made for characterization of redox conditions in a contaminated aquifer. *Environ. Sci. Technol.* 31: 3029-3031.

Chapelle, F.H., D.A. Vroblesky, J.C. Woodward, and D.R. Lovley. 1997. Practical considerations for measuring hydrogen concentrations in groundwater. *Environ. Sci. Technol.* 31: 2873-2877.

Chapelle, F.H., P.B. McMahon, N.M. Dubrovsky, R.F. Fujii, E.T. Oaksford, and D.A. Vroblesky. 1995. Deducing the distribution of terminal electron-accepting processes in hydrologically diverse groundwater systems. *Water Resour. Res.* 31: 359-371.

Holliger, C. and W. Schumacher. 1994. Reductive dehalogenation as a respiratory process. *Antonie van Leeuwenhoek* 66: 239-246.

Lovley, D.R. and S. Goodwin. 1988. Hydrogen concentrations as an indicator of the predominant terminal electron-accepting reactions in aquatic systems. *Geochim. Cosmochim. Acta* 52: 2993-3003.

Maymo-Gatell, X., V. Tandoi, J.M. Gossett, and S.H. Zinder. 1995. Characterization of an H_2-utilizing enrichment culture that reductively dechlorinates tetrachloroethene to vinyl chloride in the absence of methanogenesis and acetogenesis. *Appl. Environ. Microbiol.* 61: 3928-3933.

Scholz-Muramatsu, H., R. Szewzyk, U. Szewzyk, and S. Gaiser. 1990. Tetrachloroethylene as electron acceptor for the anaerobic degradation of benzoate. *FEMS Microbiol. Lett.* 66: 81-86.

Smatlak, C.R., J.M. Gossett, and S.H. Zinder. 1996. Competitive hydrogen utilization for reductive dechlorination of tetrachloroethene and methanogenesis in an anaerobic enrichment culture. *Environ. Sci. Technol.* 30: 2850-2858.

APPENDIX 20

Gate 3 Biosparge Zone Assessment

A20.1 INTRODUCTION

As part of the aerobic system performance assessment in Gate 3, detailed sampling rounds of the biosparge gate and selected monitoring points downgradient were completed during July, September, and October 1997. The sampling was done to assess the attenuation of TOL, PCE, and its dechlorination products *c*DCE and VC, as well as benzoate added in the NIW.

In November 1997 a direct measurement was made of the volume of oxygen delivered to the biosparge zone during a one-minute sparge. This measurement was useful in assessing the adequacy of oxygen delivered and the potential for significant off gas to be generated by this sparging.

Sampling of the unsaturated zone headspace gas was also completed to quantify VOC mass loss via volatilization. The concern was that the addition of O_2 gas in a sparge event would force gas already in the headspace out of the biosparge system and into the atmosphere, thus volatilizing VOCs.

A20.2 SAMPLING METHODS AND TIMING

Table A20.1 lists the groundwater sampling points monitored. Their locations are shown in Figure A20.1. Groundwater samples were analyzed for VOCs, benzoate, dissolved oxygen (DO), redox potential, and pH, as per Appendix 4.

Table A20.1 Monitoring of Groundwater in the Aerobic Section of Gate 3

Date and Day of Sampling	Fences Sampled
July 14, 1997 (day 238)	G315 to G318
September 30, 1997 (day 315)	G315 to G318
October 7, 1997 (day 322)	G315 to G320, G324
October 14, 1997 (day 329)	G315 to G320, G324
October 23, 1997 (day 339)	G315 to G320, G324

The oxygen sparge system shown in Figure 4.5 in Chapter 4 was temporarily modified to conduct the oxygen gas flow measurement in November 1997. A flow sensor (McMillan Co. gas flow sensor, model 100-14, calibrated for oxygen) was installed after the pressure gauge and output was provided to a digital flow meter (Cole-Palmer Flow Rate Monitor/Totalizer, model 94787). A solenoid valve (a brass, Cole-Palmer General Purpose Solenoid Valve, two-way direct-lift diaphragm type) was installed after the flow sensor. This solenoid was activated for 1.0 minute by a

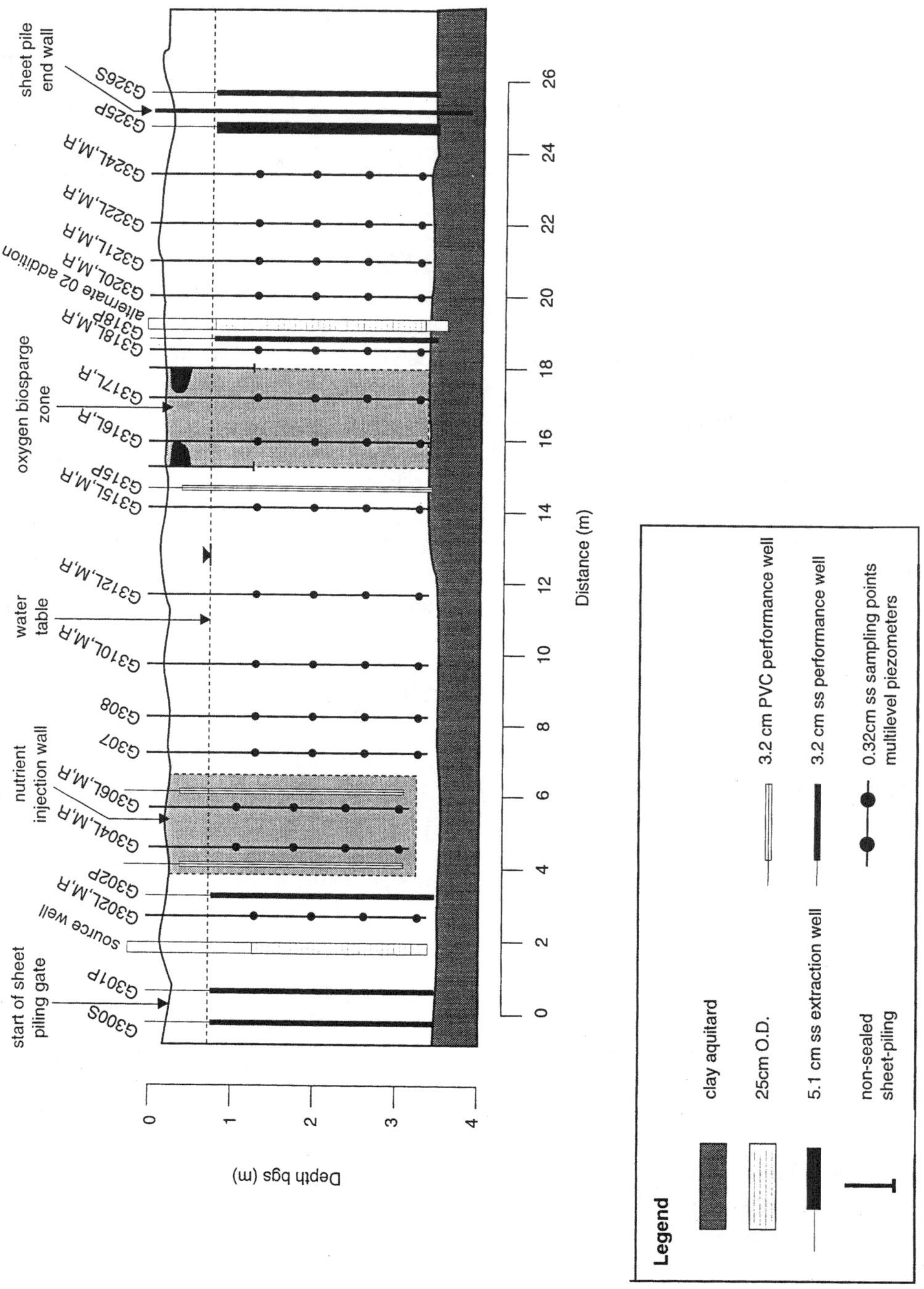

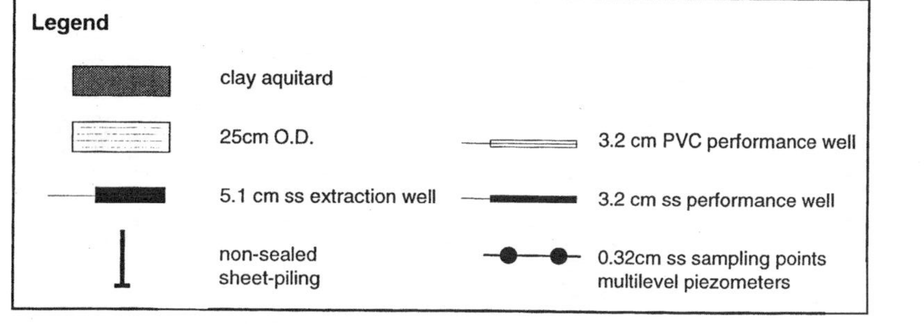

Figure A20.1 Longitudinal cross section of Gate 3, showing the NIW, biosparge zone, and wells.

digital timer (Noma Outdoor Timer, model N1507) to provide oxygen to the biosparge system under a delivery pressure typical of the routine operation. The total flow over the one-minute interval was displayed with the Flow Rate Monitor/Totalizer. This process was repeated with essentially the same oxygen volume delivered in one minute. The ambient temperature and gauge pressure were recorded so that the oxygen gas flow could be converted to STP conditions. In the one-minute sparge, 41 L (at STP) of oxygen was being delivered to the biosparge system.

On days 315, 322, and 329, gas samples were collected both prior to and after the daily oxygen sparge. The former determined the composition of the headspace gas that might potentially be expelled during the next sparge event. Sampling after the sparge event provided a better estimate of the gas recently transferred into the headspace. Twenty-milliliter ground glass syringes (pre-moistened with organic-free water) were used to collect gas samples from stainless steel soil gas probes fitted with Mininert valves (Figure A20.2). The sample inlet port for each probe was located approximately 22 cm below ground surface, well above the water table.

The purge volume for each probe was estimated as approximately 5 mL, and each was purged of 10 mL using a dedicated purging syringe prior to the collection of each sample. Approximately 15 to 20 mL of gas was collected from each SGP using glass syringes. To facilitate later subsampling, the needle tip was replaced with a silicone sleeve stopper immediately after sampling, while a positive pressure was maintained on the syringe barrel to prevent inflow of air. Sample syringes were subsequently placed into a small cooler to maintain similar temperature among the samples prior to analysis.

Samples were analyzed using a PhotoVac model 10S70 field gas chromatograph (GC), with Ultra Zero Air as the carrier gas. Samples were withdrawn from the ground glass syringes using clean 50-μL Hamilton gas tight syringes (model 1705) by piercing the sleeve stopper, flushing the gas tight syringe barrel twice, and withdrawing a third aliquot and injecting 50 μL into the GC for quantitation. Sleeve stoppers were replaced on the ground glass syringes following injection (maintaining positive pressures on the barrel) to keep a reserve volume for additional analyses, if required.

A20.3 RESULTS AND DISCUSSION

A25.3.1 Groundwater

Nonuniform bubble distributions were observed at the water table surface within the biosparge gate in December 1996 (Figure A20.3). Bubble breakout occurred over 60% of the surface area. Bubbles appeared sporadically at locations marked "intermittent."

Sparging was done twice weekly from November 22, 1996 to July 18, 1997 (day 277). On July 14, 1997 (day 273) only one point (G317R-3) had DO > 5 mg/L. Piezometer G317R was immediately downgradient of an area apparently receiving significant O_2 during sparging (see Figure A20.3). Overall, however, the twice-weekly sparging was ineffective in significantly raising the DO throughout the gate. The oxygen flow test conducted in November 1997 subsequently revealed that only 41 L (at STP) of oxygen was being provided by each one-minute sparge event. The twice-weekly sparge schedule had been premised on providing almost 400 L (at STP) of oxygen with each sparge event.

The sparging frequency was increased to once per day beginning July 18, 1997 (day 277). The DO measured in the biosparge zone groundwater on day 339 was significantly higher than the DO measured on day 273. Thirteen of the 24 points had DO > 5 mg/L and 10 points had DO > 10 mg/L. The higher concentrations were along the left side of the gate, in contrast to July, where the only high concentration was on the right side at G317R-3. The high DO concentrations on day 339 did not correlate well with the bubble distribution observed in December 1996 (Figure A20.3). Increased lateral mixing from the more frequent sparging and/or

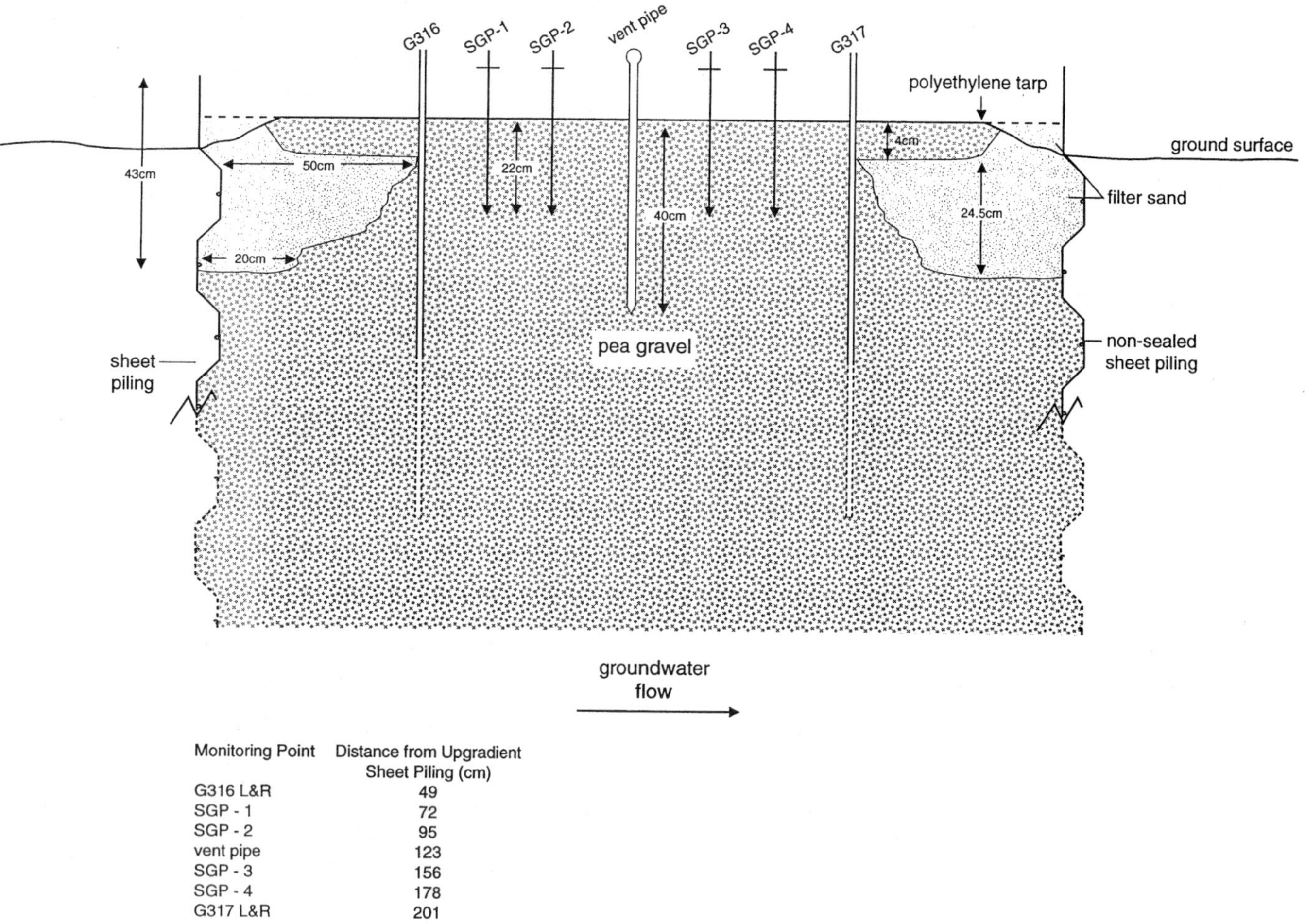

Monitoring Point	Distance from Upgradient Sheet Piling (cm)
G316 L&R	49
SGP - 1	72
SGP - 2	95
vent pipe	123
SGP - 3	156
SGP - 4	178
G317 L&R	201

Figure A20.2 Biosparge zone soil gas monitoring probe locations.

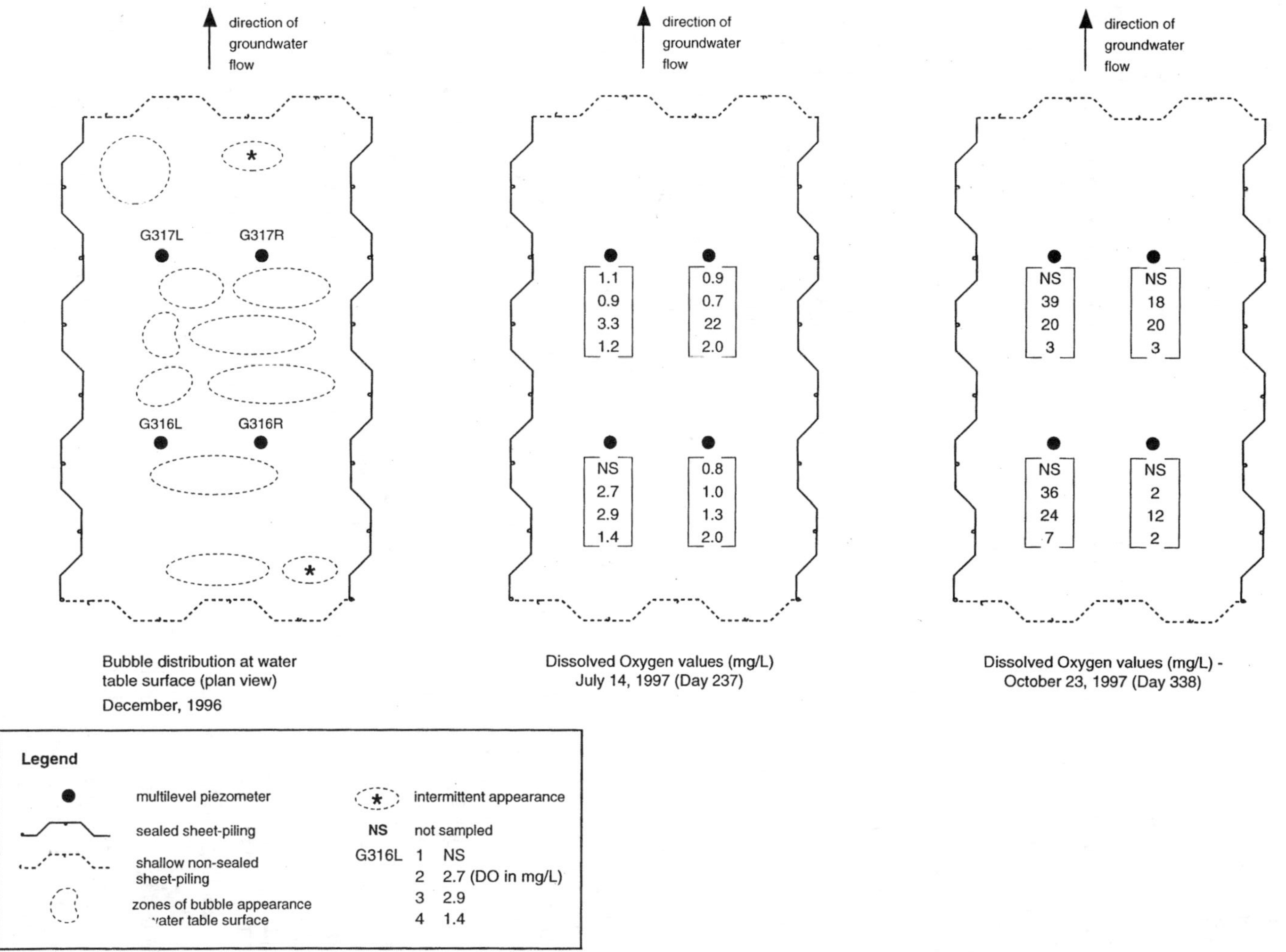

Figure A20.3 Bubble and dissolved oxygen distribution within the biosparge zone.

transport and dispersion of well oxygenated water from upgradient in the zone may have contributed to the better distribution of DO by day 339.

During October 1997, the DO was usually highest at levels 2 and 3 within the biosparge gate (i.e., fences G316 and G317). Level 4 appeared to be the most weakly oxygenated, while the oxygenation in level 1 monitoring points could not be accurately determined because they frequently did not yield water and were likely above the water table. Weak oxygenation observed at level 4 is likely because point 4 was at about the elevation of the O_2 delivery pipes (Figure 4.3, Chapter 4), and so little O_2 gas would spread laterally to these points.

A wider distribution and higher concentration of DO observed at levels 2 and 3 may be partly attributed to the spreading of oxygen bubbles that rose from the vent ports. Also, the groundwater mixing induced by daily sparging could have helped produce a more uniform aeration of groundwater in the biosparge zone.

TOL, sodium benzoate, and DO data for days 273 to 339 for each point in the aerobic segment of Gate 3 are shown in Figure A20.4. Data for CVOCs are briefly discussed here. For nearly all points within the biosparge gate under the revised sparge regime, after day 277, the general trend was increased DO with a concurrent decline in TOL and benzoate. In most cases, benzoate dropped to below detection on day 339. Sufficient DO remained at those points where TOL was found to support the aerobic biodegradation of the remaining TOL (using a ratio of 3:1 DO to TOL by mass).

Only at monitoring points G316R-3, G316R-4, and G317R-4 were both TOL and benzoate still present on day 339. These points also had lower DO values relative to other monitoring points within the gate. Groundwater at the depth of monitoring points 4 may have escaped significant aeration, as discussed above, and so TOL and benzoate could escape the biosparge zone at higher concentrations at this level.

Downgradient of the biosparge zone at fence G318, benzoate concentrations had declined to below detection by day 339, except at points G318M-4 and G318R-4. TOL had declined to below LOQ except at G318M-3, G318M-4, and G318L-4. At the level 4 points, DO was <4 mg/L, suggesting the benzoate and TOL persistence at this level reflected the minimal aeration of the deepest segment of the biosparge zone.

At fences G320 and G324, benzoate was no longer detected (< 0.6 mg/L) on day 338. TOL was >25 μg/L only in points G320L-4, G320M-4, G320R-3 and -4, and G324L-4. DO was >1.8 mg/L at all points. TOL was still found in the lowest points, perhaps related to poor aeration of the deepest groundwater in the biosparge gate. However, sufficient DO was found in all these points to provide for complete aerobic biodegradation of the remaining TOL.

The persistence of benzoate and TOL in the lowest points beyond the biosparge zone points out the benefit of locating the O_2 delivery pipes below the contaminated groundwater to ensure effective delivery of oxygen to all of the contaminated groundwater. However, even with this limitation, by day 338 the biosparge zone appeared to be delivering sufficient oxygen to support complete aerobic biodegradation of essentially all the TOL and sodium benzoate passing through.

CT and CF were usually not detected in and beyond the biosparge zone from day 273 to the final sampling on day 338. Concentrations above LOQs were found only at G318M-3 (33 μg/L CF on day 315). Essentially no significant chloromethanes were transported to the biosparge zone, so their fate in this zone cannot be evaluated.

PCE was found at >10 μg/L only at points G317R-4 on day 329 and G316R-4 and G316L-4 on day 338. This result indicates PCE was rapidly declining from average concentrations of about 60 μg/L upgradient of the biosparge zone in fence G315. While volatilization could have contributed to this rapid decline, no PCE was noted in the headspace gas (Section A20.3.2). Because PCE is not readily degraded aerobically, some of the decrease in PCE concentration between fences G315 and G316 may be attributed to anaerobic biodegradation immediately downgradient of fence G315 and before the onset of aerobic conditions at fence G316. *c*DCE, a dechlorination product of PCE, was found in the biosparge zone, presumably generated in the

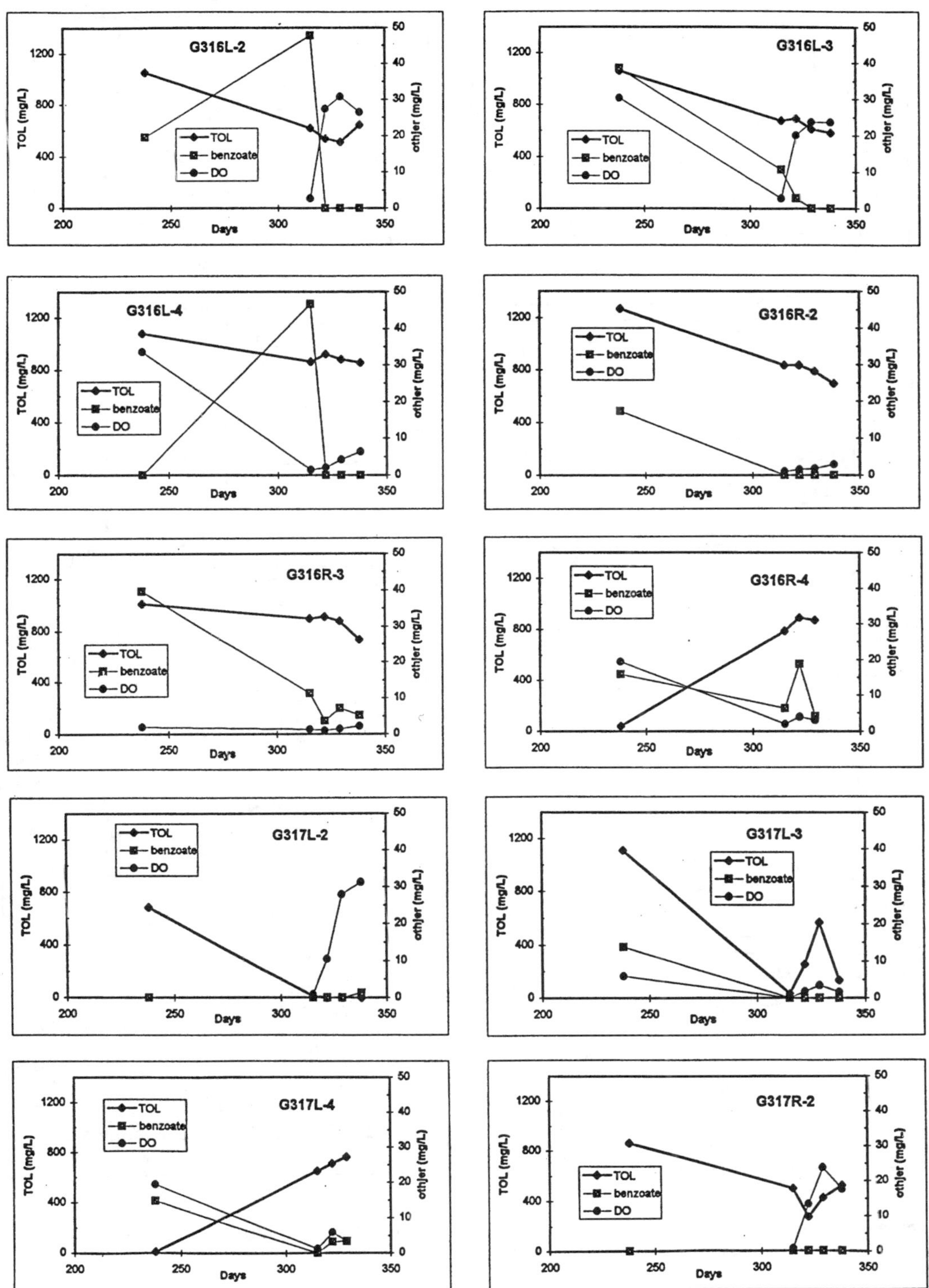

Figure A20.4 Toluene, sodium benzoate, and dissolved oxygen concentrations in groundwater from fences G316 to G324 from day 238 to 338.

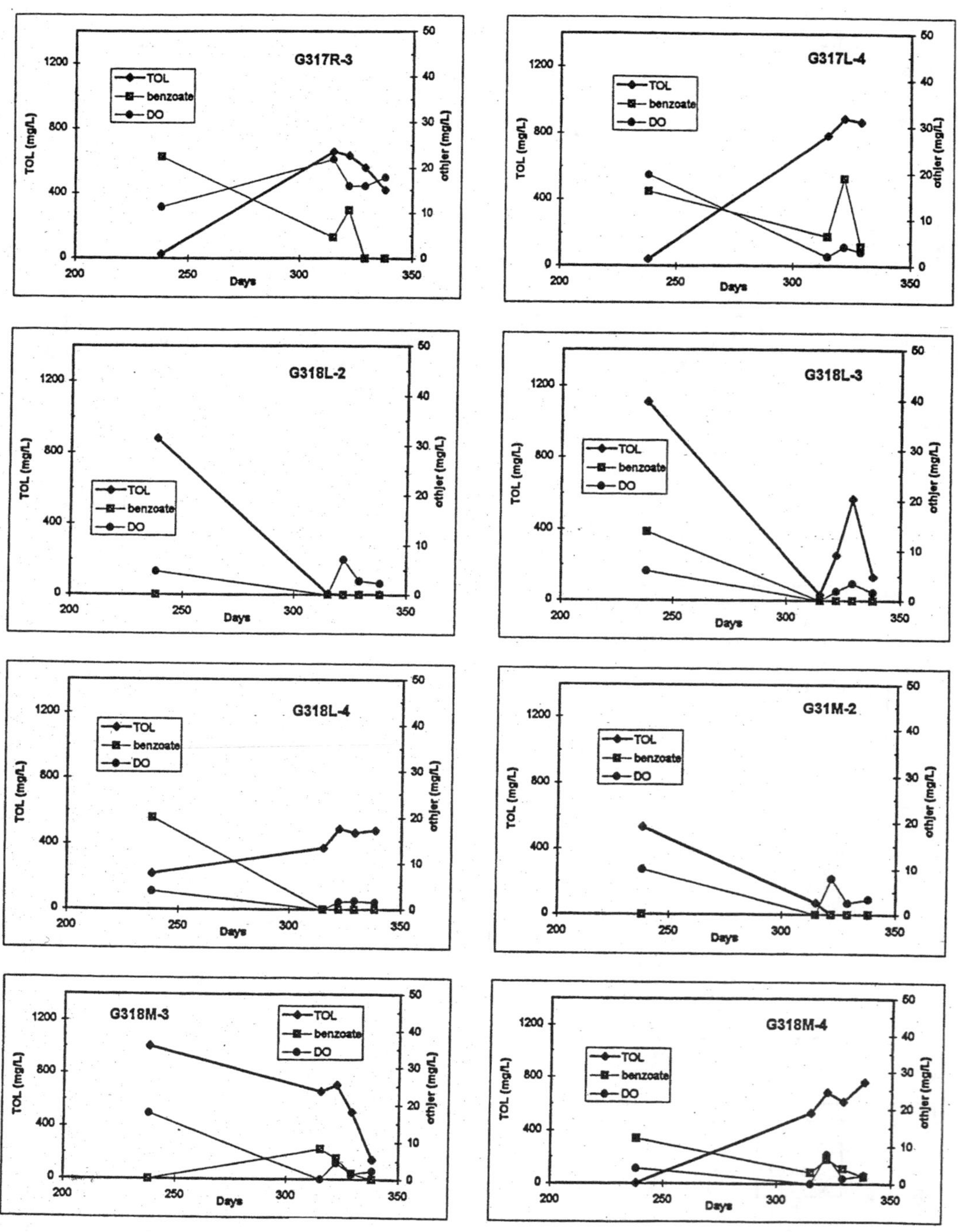

Figure A20.4 (continued) Toluene, sodium benzoate, and dissolved oxygen concentrations in groundwater from fences G316 to G324 from day 238 to 338.

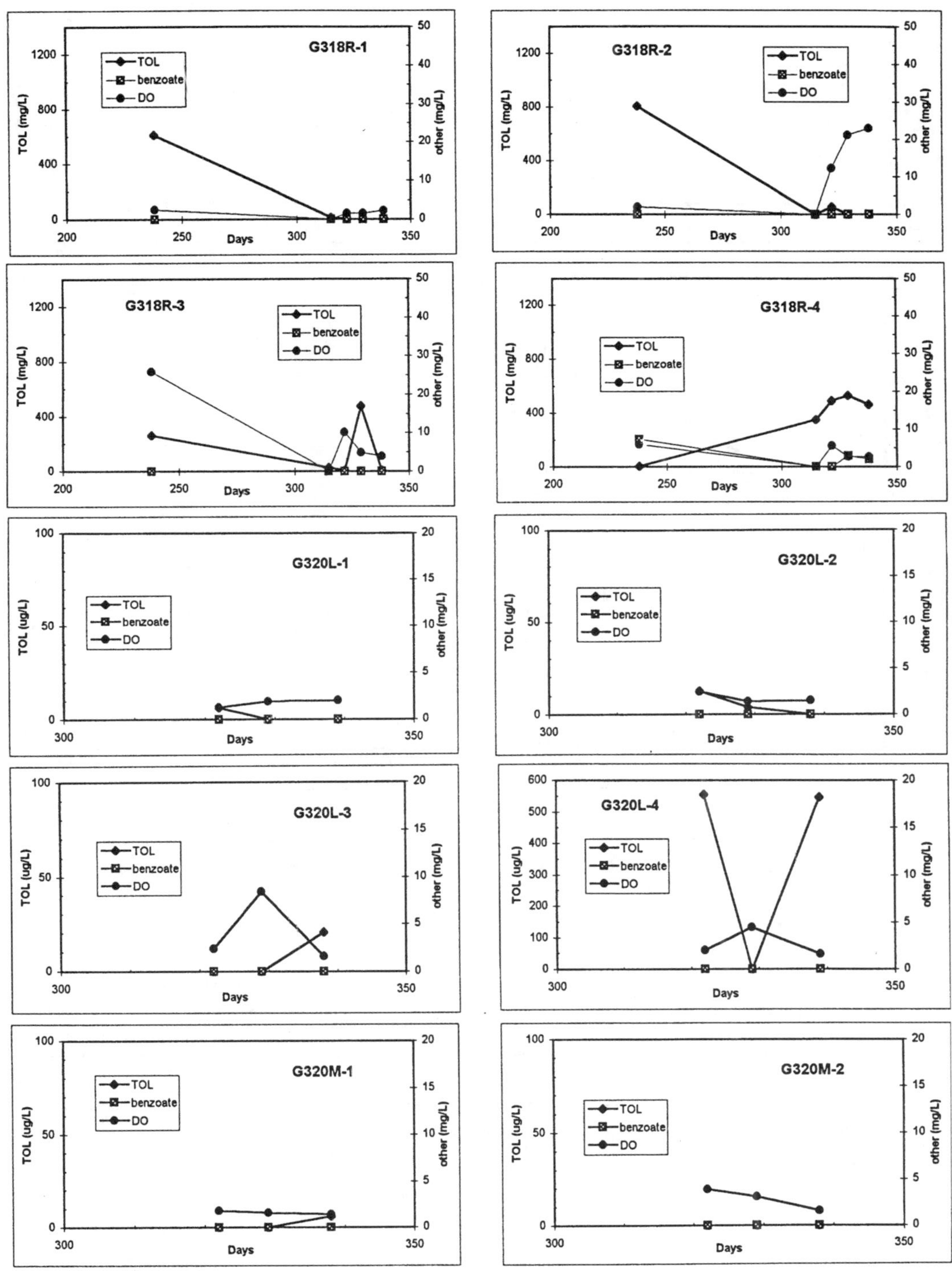

Figure A20.4 (continued) Toluene, sodium benzoate, and dissolved oxygen concentrations in groundwater from fences G316 to G324 from day 238 to 338.

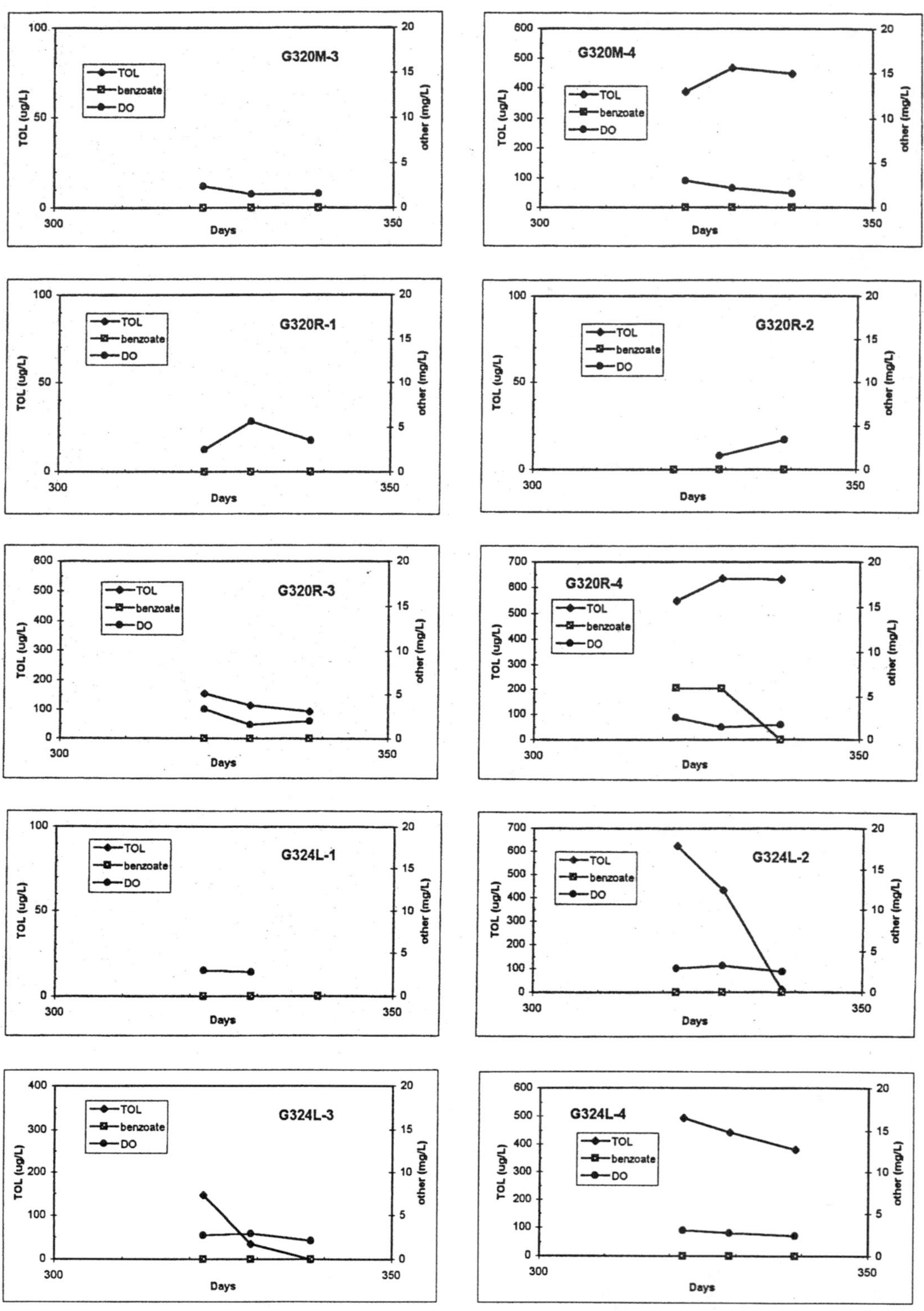

Figure A20.4 (continued) Toluene, sodium benzoate, and dissolved oxygen concentrations in groundwater from fences G316 to G324 from day 238 to 338.

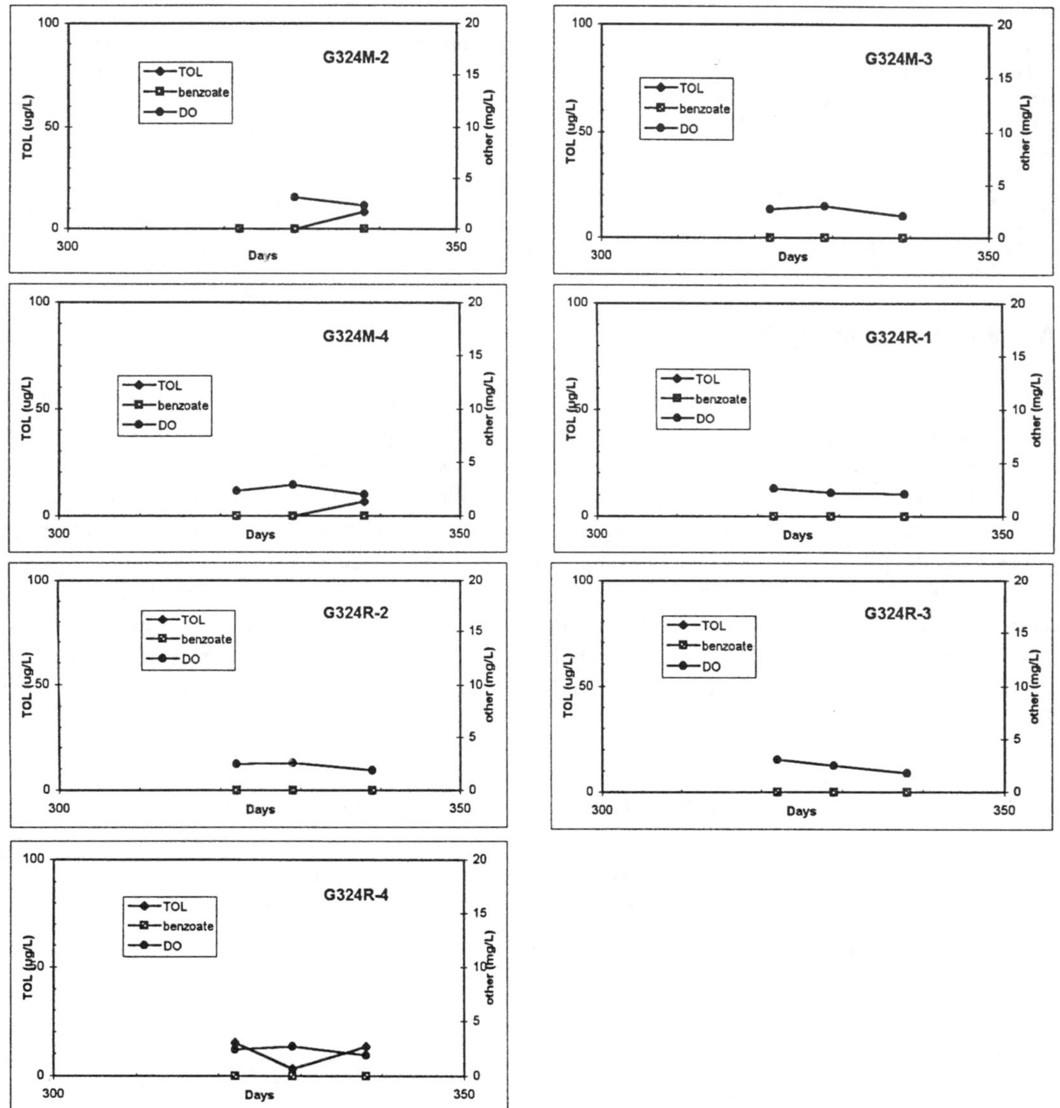

Figure A20.4 (continued) Toluene, sodium benzoate, and dissolved oxygen concentrations in groundwater from fences G316 to G324 from day 238 to 338.

upgradient anaerobic segment of Gate 3. No VC or TCE was detected beyond fence G315 from day 272 to 338.

*c*DCE was not found in the biosparge zone on day 273 (July 14, 1997). By day 315 concentrations up to almost 40 μg/L were found in the biosparge zone. On day 338, *c*DCE concentrations were distributed throughout the aerobic section of Gate 3, as illustrated by Table A20.2.

It was expected that *c*DCE could biodegrade in the aerobic zone, but with the potential to lose some or most of the *c*DCE via volatilization, it is not possible to evaluate the aerobic biodegradation rate. In any event, *c*DCE was found in only one point in fence G324 on the last sampling day, suggesting little *c*DCE was exiting Gate 3.

Table A20.2 *c*DCE Concentrations (μg/L) in Aerobic Zone Monitoring Points, Day 338 (October 23, 1997)

	Left Monitor (L-)				Middle Monitor (M-)				Right Monitor (R-)			
Fence	**-1**	**-2**	**-3**	**-4**	**-1**	**-2**	**-3**	**-4**	**-1**	**-2**	**-3**	**-4**
316	NS	67	29	ND					NS	68	18	ND
317	NS	<L	39	ND					NS	35	18	ND
318	NS	<L	17	ND	18	<L	31	ND	ND	ND	17	NS
320	ND	ND	21	22	ND	<L	17	15	ND	ND	24	<L
324	ND	ND	ND	23	ND	<L	ND	ND	ND	ND	ND	ND

Notes: NS - no sample, likely because the point was above the water table. ND - not detected. <L - detected, but less than LOQ of about 13 μg/L.

A20.3.2 Unsaturated Zone

Field calibrations of the PhotoVac GC were completed using TOL and PCE, the only VOCs known to be in the groundwater entering the biosparge gate at the start of the sampling. Neither TOL nor PCE were detected in the unsaturated zone headspace gas on any occasion. No other VOCs (i.e., TCE, *c*DCE, *t*DCE, and 1,1-DCE) were indicated in the chromatograms. The soil gas probes were sampled directly using gas-tight Hamilton syringes, in addition to using glass syringes in case the sampling protocol was faulty, but no VOCs were detected.

Limits of quantification (LOQs) for TOL and PCE were determined on each of the sampling dates using the approach of Devlin (1996). Method detection limits (MDLs) were then estimated as one half of the LOQs. The MDL for *c*DCE was estimated to be about the same as for PCE, although the GC was not calibrated for *c*DCE. LOQs and estimated MDLs are summarized in Table A20.3.

Table A20.3 TOL and PCE Limits of Quantification and Estimated Method Detection Limits for TOL, PCE, and *c*DCE

	TOL (μg/L)		PCE (μg/L)		*c*DCE (μg/L)
Date	**LOQ**	**MDL**	**LOQ**	**MDL**	**MDL**
September 30, 1997	159	80	54	30	30
October 7, 1997	123	60	45	20	20
October 14, 1997	132	70	66	30	30

The potential loss of volatiles into the headspace and potentially out of the zone as off gas was evaluated. The flux of TOL and PCE that could be represented by off gas with TOL and PCE concentrations at the MDLs (Table A20.3) was estimated as follows. The daily flux of gas out of the headspace was estimated as 40 L, based on the approximate volume of O_2 gas sparged into the biosparge zone daily. This estimate was based on a measurement of 17.1 L of O_2 gas released from the gas cylinder for the November 6, 1997 sparging. This volume was measured at a line pressure of 20 psig (2.36 atm, absolute) and, with conversion to atmospheric pressure, resulted in an estimate of 40 L being sparged each time. The daily flux of chemical potentially lost in the off gas is the air flux times concentration (MDL, in this case). This daily flux is compared to the estimated flux of TOL and PCE entering the biosparge zone, calculated as the flux of groundwater (0.46 m^3/day) times the average concentration at fence G315, immediately upgradient of the biosparge zone. The results are summarized in Table A20.4.

Using this calculation, the proportion of contaminant potentially volatilized was calculated to be <1% for TOL, <5% for PCE, and, except for day 315 when influent *c*DCE was still low, <4% for *c*DCE. These are likely maximum proportions, given that no VOCs were actually detected in the headspace gas measurements.

Table A20.4 Calculation of the Potential Loss of TOL, PCE, and *c*DCE in Off Gas Generated by Sparging

	Flux In via Groundwater (mg/day)			Maximum Possible Flux Out via Off Gas (mg/day)			Maximum Possible Loss via Off Gas (% of influx)		
Day	**TOL**	**PCE**	**DCE**	**TOL**	**PCE**	**DCE**	**TOL**	**PCE**	**DCE**
315	320	26	12	3.2	1.2	1.2	1.0	4.6	10
322	340	28	23	2.4	0.8	0.8	0.9	2.9	3.5
329	340	26	39	2.8	1.2	1.2	0.8	4.6	3.1

A20.4 SUMMARY

The DO distribution within the biosparge zone was not uniform, and likely <60% of the groundwater in the biosparge zone was aerated. The bottom 0.1 to 0.5 m of the biosparge zone was poorly aerated, since the O_2 bubbles delivered from the vent pipes did not spread significantly near the pipes. Benzoate and TOL were more persistent in this deep, less well-aerated segment in and downgradient of the biosparge zone.

Up to day 277, insufficient oxygen was provided by the twice-weekly sparge schedule. This finding was confirmed by the oxygen flow test in November 1997, which revealed that only about 41 L (at STP) of oxygen was being provided by each sparge event instead of the designed volume of almost 400 L.

CT and CF did not persist into the biosparge zone during this time of detailed assessment (days 273 to 338). PCE apparently persisted into the aerobic segment of Gate 3 only sporadically, and had essentially gone below the LOQ before fence G320. A dechlorination product, *c*DCE, was more persistent and occurred at low concentrations at a number of points in the aerobic zone. A maximum concentration of only 23 µg/L was found in G324L-4 on day 338. No TCE or VC was detected in the aerobic zone from day 273 to 338.

Overall, TOL and benzoate showed fairly extensive declines from G316 to G318, while DO values remained relatively high (i.e., above 5 mg/L). The few monitoring points with residual TOL at the end of the aerobic zone (G324L-4 and traces at G324M-2 and -4) on the final day of sampling contained sufficient DO to support aerobic biodegradation of the remaining TOL. Thus, the treatment of remaining PCE, its dechlorination products, TOL, and residual sodium benzoate was essentially complete by the end of Gate 3.

A20.5 REFERENCE

Devlin, J.F. 1996. A method to access analytical uncertainties over large concentration ranges with reference to volatile organics in water. *Ground Water Monit. Rem.* 16: 179-185.

Appendix 21

Benzoate–Conductivity Comparison

Initially, electrical conductivity of groundwater samples was used to track the progress of nutrient pulses (Devlin, 1994), containing sodium benzoate salt as an added solute. To validate this procedure, a laboratory study was undertaken in February 1997 to assess the electrical conductivity–benzoate concentration relationship over the range of benzoate concentrations anticipated in Gate 3 groundwater.

A concentration series of benzoate solutions was prepared using uncontaminated Borden groundwater. The equipment used to analyze the samples was identical to that used in field (see Appendix 4). Results are summarized in Table A21.1 and are shown in Figure A21.1 (indicated trendline is a linear regression). A strong, linear relationship was found, supporting the use of groundwater electrical conductivity measurements as a surrogate for benzoate analyses.

Table A21.1 Summary of Benzoate–Conductivity Comparison Results

Sodium benzoate (mg/L)	Conductivity (μs)	Temp. (°C)
0	386	16.2
50	462	15.1
100	491	15.7
150	508	16.4
200	529	16.7
250	576	17.1
300	602	17.2

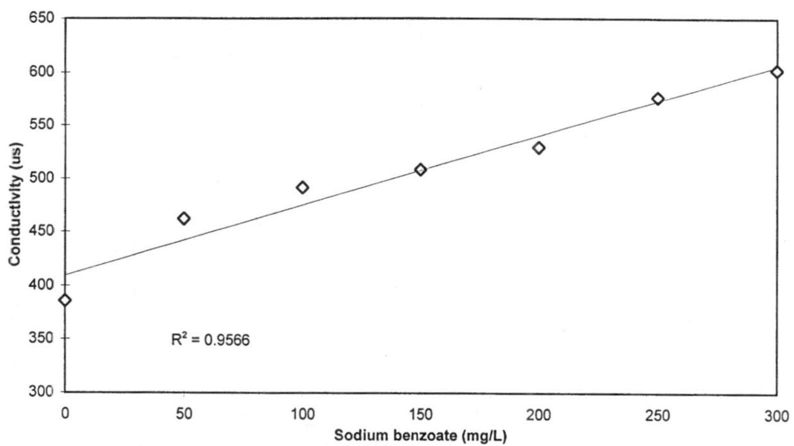

Figure A21.1 Benzoate–conductivity plot illustrating linearity of response in lab analysis.

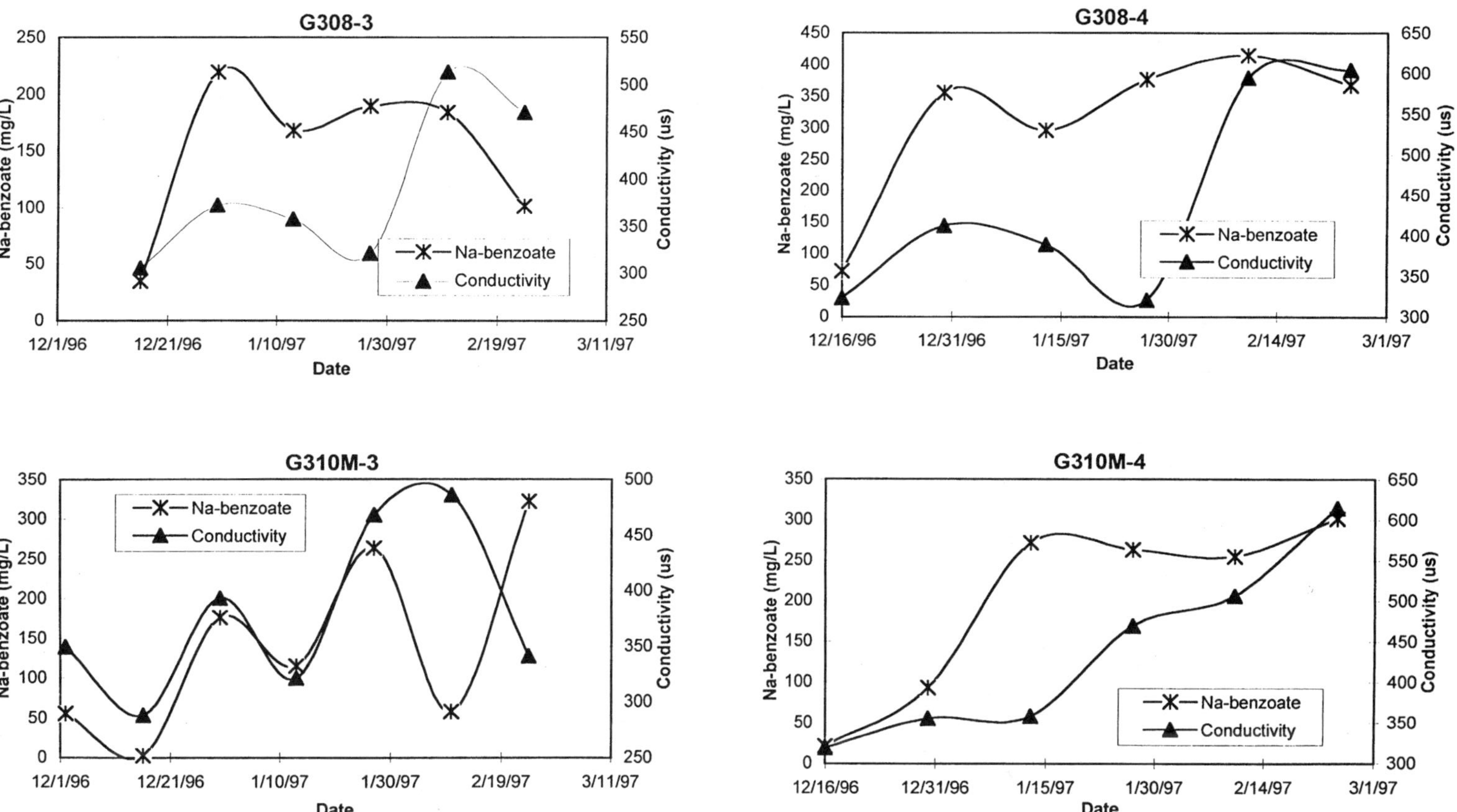

Figure A21.2 Sodium benzoate–conductivity comparisons for selected monitoring points over a series of different dates.

Subsequent analyses of benzoate samples from monitoring points for which electrical conductivity was measured in the field (using flow-through cell apparatus) exhibited a poorer correlation of these two parameters (Figure A21.2). A cross-plot of the benzoate–conductivity data is shown in Figure A21.3 (indicated trendline is a linear regression).

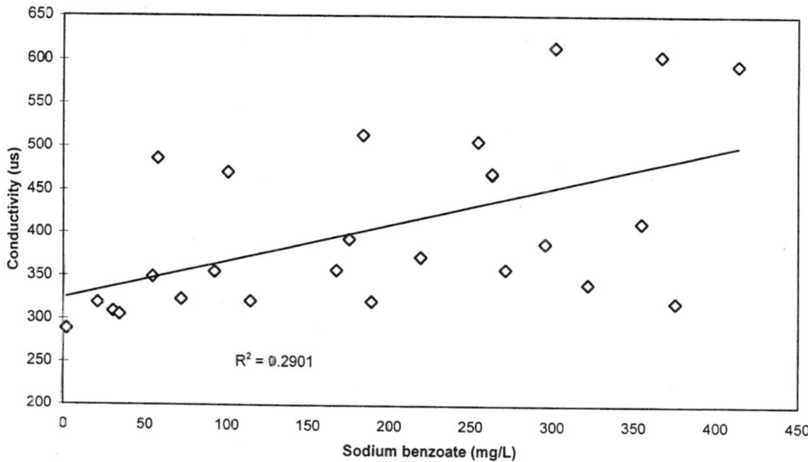

Figure A21.3 Correlation of benzoate and conductivity values from Figure A21.2.

Poor correlation may be partially attributed to variable dissolved carbon dioxide contents within the groundwater, which can contribute to conductivity in the form of ionized carbonates. Other factors which may have produced interferences include variations in chloride and sulfate contents.

Given the poor correlation between electrical conductivity benzoate concentration for field samples, it was decided that addition of inorganic tracers to the benzoate concentrate and subsequent tracer monitoring would be a more reliable method of characterizing pulse progress. Results of this Pulse Tracer Test are given in Appendix 17.

REFERENCES

Devlin, J.F. 1994. Enhanced *In Situ* Biodegradation of Carbon Tetrachloride and Trichloroethene Using a Permeable Wall Injection System. Ph.D. dissertation, Department of Earth Sciences, University of Waterloo. Waterloo, Ontario, Canada.

Appendix 22

Laboratory Sparging Experiments

A22.1 INTRODUCTION

In order to quantify the operational parameters for the aerobic biosparge system, a laboratory test was completed to estimate the residual gas phase content ($K_{r(g)}$) of pea gravel. This same gravel was used to fill the saturated zone of the biosparge treatment system at the CFB Borden AATDF research site.

The study was done to estimate the volume of gas the medium can hold in an entrapped state. With this knowledge, the frequency of sparging required to satisfy the total oxygen demand of the gate could be determined. Introduction of oxygen greater than residual quantities in any given sparge would be detrimental, since the objective is to deliver sufficient oxygen to stimulate biotransformation of all aerobically degradable compounds, while keeping volatilization losses to a minimum. Excess gas could cause partitioning of VOCs into the vapor phase and migration of VOCs out of the saturated zone during sparging.

A22.2 EXPERIMENTAL SETUP AND METHODOLOGY

A schematic diagram illustrating the column apparatus is given in Figure A22.1. Air was used in the lab and oxygen was used in the field. The metering valve was opened gradually until the desired pressure of 2 psig (pounds per square inch gauge) was reached on the column pressure gauge, then sparging continued for 2 to 4 minutes. During sparging, the maximum rise in the water table relative to the initial, presparge level was measured. The reservoir was adjusted both during and after the sparge so that its water table continuously matched that in the column, the objective being for water to flow freely in and out of the column as a result of water displacement, and not be driven by differences in head between the two resevoirs. The water table level was measured immediately following the end of the sparge, as was the final water level in the reservoir.

To determine if higher delivery pressures increased $K_{r(g)}$, the initial sparge was followed immediately by two additional sparges at increasing delivery pressures into the now unsaturated column. Increases in residual water table levels were indicative of greater $K_{r(g)}$.

$K_{r(g)}$ as a fraction of the pea gravel void volume for any given delivery pressure was determined from

$$K_{r(g)} = \frac{\delta Vc + \delta Vr}{Vc_{(g)} \times \eta} \qquad (1)$$

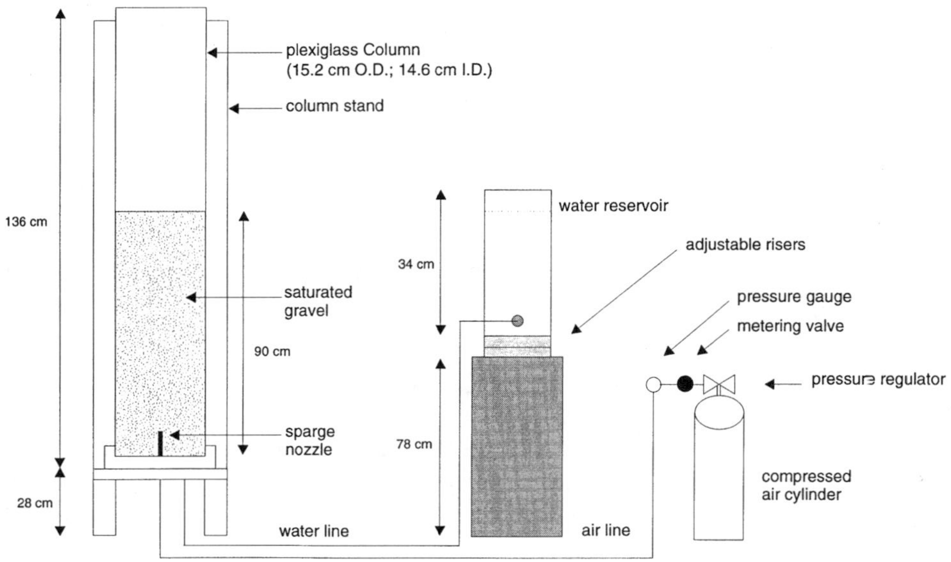

Figure A22.1 Lab column sparging test apparatus.

where $K_{r(g)}$ = gas-occupied volume, as a percentage of the total void volume
δVc = apparent change in volume of water in the column, based on elevated water levels in the column
δVr = change in volume of water in the reservoir
$Vc_{(g)}$ = total volume of saturated gravel in the column, approximately 15, 150 cm^3
η = pea gravel porosity, 0.463; determined gravimetrically

The porosity was determined by drying a fully saturated sample of pea gravel of known mass at 105°C for 24 hours, using the difference in pre- and post-drying weight to determine the volume of water lost assuming its density is 1.0 kg/L. Results are given below.

A22.3 RESULTS

Porosity measurements were made for two separate samples of pea gravel and were determined to be 46.5 and 46.1%, with an average of 46.3%. $K_{r(g)}$ values are summarized in Table A22.1. Measurements were not duplicated.

Table A22.1 Summary of $K_{r(g)}$ Results from Column Sparging Experiment

Sparge pressure (psig)	$K_{r(g)}$,%
2	7.9
4	8.7
6	9.5

The gauge pressures used were approximately equal to twice and three times the estimated pressure required to initiate gas flow (i.e., 2, 4, and 6 psig) for the type of media and the depth of water in the column (calculation below). Measurements were done sequentially; that is, after the initial sparge at 2 psi into a water-saturated gravel column, sparging at 4 psig and then sparging at 6 psig were done without resaturating the column with water.

These gas pressures were selected following consideration of the theoretical breakthrough pressure required for gas to enter the column. The gas pressure must exceed the hydrostatic pressure of the water overlying the sparge point at the bottom of the column. This is calculated using

$$H = \rho_w g h \tag{2}$$

where ρ_w = density of water, taken as 1000 kg/m^3
g = the gravitational constant, 9.8 m/s^2
h = the height of the water column above the sparge point, m

For each 1 m of water column, 1.4 psig (9800 kg m^{-1} s^{-2}, or 0.097 atm) pressure of gas is required to overcome the hydrostatic pressure. The gravel was saturated to about 90 cm and so the hydrostatic pressure was about 1.3 psig.

Nyer and Suthersan (1993) suggest that 1 psig of pressure is required for every 150 cm of head for coarse gravels. Therefore, the initial 90 cm of saturated gravel in the column, 0.6 psig gas pressure would be required in excess of the hydrostatic pressure of 1.3 psig. So the inital pressure used in these tests was 2 psig, marginally above the total pressure required of 1.9 psig as calculated above.

A22.4 FIELD OPERATIONAL PARAMETERS FOR SPARGING

Using the calculations above, the gas pressure required to induce flow was determined to be 6.3 psig, given an average saturated aquifer thickness above the sparge points of 3.0 m. Field tests in early August 1996 showed gas breakthrough at the water table surface was obtained at 4 to 5 psig, and so this lower pressure was adopted, primarily to prevent formation of macropores or channels (C. Austrins, personal communication). Visual observations of bubble distribution at the water table surface were obtained in late December 1996 by excavating the gravel to the water table and then delivering bursts of gas at increasing pressures. About 60% of the water table surface within the sparge gate area was observed to show some bubbles escaping. Higher delivery pressures did not appear to significantly increase lateral distribution at the surface of the water table.

For a given sparge at 4 psig, 5.2% of the porosity was assumed to be occupied by oxygen. This is based on the laboratory-determined $K_{r(g)}$ of 8.7% for 4 psig pressure (Table A22.1) and the observed influence of only 60% of the gate cross-sectional area (above). The volume of gas that can be entrapped beneath the water table depends on the water level within the gate. For a water table 3.0 m above the bottom of the gate (and an internal length and width of 2.9 m and 1.9 m, respectively), 16.5 m^3 of the gravel is below the water table, and so about 5.2% of the saturated porosity, or 0.86 m^3 of oxygen, may be entrapped after a sparge event.

Overall, sparge durations were timed to deliver approximately the residual volume of oxygen gas that the gravel within the biosparge zone could hold. This was based on the observations above that $K_{r(g)}$ was a fixed percentage of the saturated porosity (for a given delivery pressure).

A22.5 REFERENCES

Nyer, E.K. and Suthersan, S.S. 1993. Air sparging: savior of groundwater remediations or just blowing bubbles in the bath tub? *Groundwater Monitoring and Remediation*, 13(4): 87-91.

APPENDIX 23

Gate 3 Nutrient Concentrate

Nutrient concentrates prepared for injection into the permeable wall of Gate 3 from September 6, 1996 to March 21, 1997 contained only sodium benzoate; after this date, a mineral salts medium was also added. The types and ratio of mineral salts used were derived from, although not identical to, that of Mueller et al. (1991). A summary of solutes contained in each concentrate is given in Table A23.1.

Concentrates were prepared to achieve a particular injection concentration in the aquifer according to

$$\begin{array}{c}\text{Solute in concentration}\\\text{(mg solute/kg solution)}\end{array} = \begin{array}{c}\text{Injection concentration}\\\text{(mg/L)}\end{array} \times \frac{\text{Pump circulation flow rate (L/min)}}{\text{APF amendment rate (L/min)}} \quad (1)$$

$$\begin{array}{c}\text{Solute in concentration}\\\text{(mg solute/kg solution)}\end{array} = \begin{array}{c}\text{Injection concentration}\\\text{(mg/L)}\end{array} \times \text{Dilution factor} \quad (2)$$

where APF amendment rate is the flowrate at which the automatic proportional feeder (APF) added nutrient concentrate to the recirculating groundwater. The density of the amended groundwater within the wall was assumed to be 1.0 kg/L following dilution of the concentrate during injection.

After the quantity of each solute to be used in the concentrate was determined, a correction factor was applied to compensate for the total mass of dissolved solutes in the concentrate relative to the mass of water used in preparation of the concentrate.

Example. If 100 g of sodium benzoate were required per kilogram of concentrate, then 900 g of water would be needed. Thus for every kilogram of water used in the concentrate:

$$\frac{\text{100 g sodium benzoate}}{X} = \frac{\text{900 g water}}{\text{1000 g water}} \quad (3)$$

X = 111 g sodium benzoate

approximately 111 g of sodium benzoate would be mixed. All other solutes added to the concentrate would have their proportionate mass (per kilogram of concentrate) included with sodium benzoate, the mass of water (per kilogram of concentrate) determined, and the subsequent correction made as in (3).

The total mass of concentrate used for each flush (beginning April 18, 1997) was determined by weighing the concentrate prior to and after use in the field. The proportion of each solute's mass

Table A23.1 Summary of Constitution of Nutrient Concentrates Used for Nutrient Flushes

		Solute Concentration (g/L)							
Date	Flush #	Sodium Benzoate	KBr	KCl	K_2HPO_4	KH_2PO_4	NH_4NO_3	$MgSO_4 \cdot 7H_2O$	Approximate Volume of Water Used to Prepare Concentrate (L)
Sept. 26/96	1	72	—	—	—	—	—	—	25
Oct. 2/96	1	72	—	—	—	—	—	—	25
Oct. 7/96	1	72	—	—	—	—	—	—	25
Oct. 26/96	2	189	—	—	—	—	—	—	9.5
Oct. 27/96	2	189	—	—	—	—	—	—	9.5
Dec. 2/96	3	n/d	—	—	—	—	—	—	n/d
Dec. 13/96	4	n/d	—	—	—	—	—	—	n/d
Jan. 2/97	5	n/d	—	—	—	—	—	—	n/d
Jan. 24/97	6	168	—	—	—	—	—	—	10.2
Feb. 21/97	7	n/d	—	—	—	—	—	—	n/d
Mar. 21/97	8	167	—	—	—	—	—	—	10
Apr. 18/97	9	99	114	—	—	—	—	—	10
May 16/97	10	132	—	184	—	—	—	—	10
Jun. 13/97	11	115	64	—	6.4	6.4	6.4	1.3	11
Jul. 11/97	12	113	63	—	6.3	6.3	6.3	1.2	11
Aug. 8/97	13	104	—	—	5.8	5.8	5.8	1.2	10
Aug. 29/97	14	103	—	—	6	6.5	6	1.5	2
Sept. 25/97	15	109	—	—	6.1	6.1	6.1	1.2	9

Notes: — not added. n/d: not determined.

in the concentrate allowed calculation of the concentration of each solute in the wall by rearrangement of Eq. (2), and the mass introduced via

$$\text{Solute mass in wall} = \begin{matrix}\text{Weight fraction of solute} \\ \text{in concentrate}\end{matrix} \times \begin{matrix}\text{Mass concentration used} \\ \text{during injection}\end{matrix} \tag{4}$$

Objective concentrations for sodium benzoate injected into the aquifer are outlined in Chapter 4, with actual concentrations (and masses) of each solute introduced into the aquifer summarized in Table 4.1.

REFERENCE

Mueller, J.G., S.E. Lantz, B.O. Blattman, and P.J. Chapman. 1991. Bench-scale evaluation of alternative biological treatment processes for the remediation of pentachlorophenol- and creosote-contaminated materials: solid phase remediaton. *Environ. Sci. Technol.* 25(6): 1045-1055.

APPENDIX 24

Breakthrough Curve Analysis, Velocity Estimation, and Concentration Averaging

A24.1 INTRODUCTION

The majority of the concentration data collected over the course of this project was gathered from multilevel monitors. Detailed data such as these are useful for assessing the effects of geologic heterogeneities on groundwater flow and solute transport, as well as providing insight on how the treatment system performance is affected by these complexities. However, the assessment of overall system performance requires data in a reduced form suitable for calculating such bulk parameters as average groundwater velocity within a gate, or fence-averaged concentrations. The integration of concentration data in space and in time was performed using the algorithms incorporated in the FENCE program, described below. In some cases, the calculations were performed using spreadsheets instead of FENCE, but regardless of the platform, the basis for the calculations was the same.

A24.2 BREAKTHROUGH CURVE ANALYSIS: THE FENCE PROGRAM

The program FENCE integrates breakthrough curves (BTC) (e.g., Figure A24.1) from sampling points along a monitoring transect and calculates the total solute mass passing the transect, as well as reporting the mass transported past each sampling point. A transect is considered to be a collection of multilevel monitors arranged across a contaminant flow path in such a way that all the contaminant mass being transported through the aquifer must pass through or between the monitoring points (Figure A24.2).

The integration at each sampling point is represented in the following equation:

$$M = \int_{t_1}^{t_2} C \, \phi \, A \, v \, dt \tag{1}$$

where M = sampling point mass (M), C = concentration (M/L^3), ϕ = porosity (-), A = cross-sectional area (L^2), v = average linear velocity (L/T), t = time (T). Each monitoring point is considered to represent a finite cross-sectional area of the aquifer. The total contaminant mass transported past the transect between times t_1 and t_2 is the total of the masses determined for all the sampling points on the fence, over the same time period:

$$Mass_{tot} = \Sigma M_i \tag{2}$$

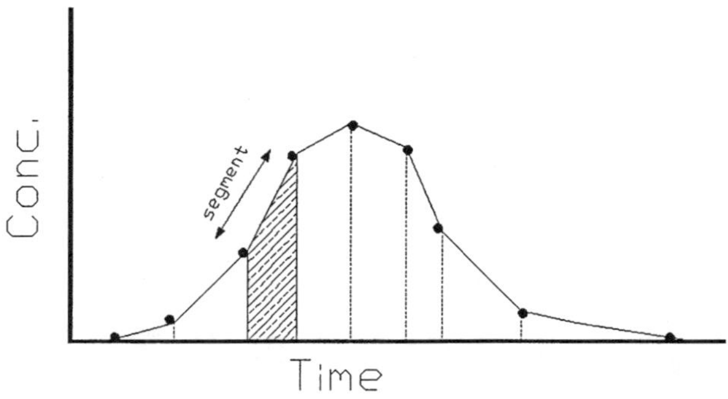

Figure A24.1 A typical breakthrough curve showing linear line segments and the areas beneath them.

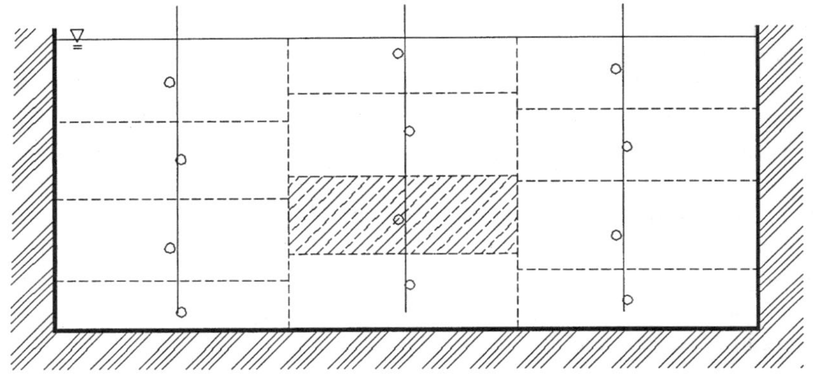

Figure A24.2 A monitoring transect through a section of aquifer confined on three sides. Each monitoring point (open circles) is assumed to represent a finite cross-sectional area. In this example, the area boundaries are denoted by broken lines.

A24.2.1 The Integration Function for Calculating Breakthrough Mass

The integrations are carried out assuming linear interpolation functions between data points (Figure A24.2). The interpolation functions defining each segment of the curve are determined individually, as illustrated in Figure A24.3, using the following relationships:

$$C = st + b \tag{3}$$

$$s = \frac{C_2 - C_1}{t_2 - t_1} \tag{4}$$

$$b = C_1 - st_1 \tag{5}$$

where C = concentration and subscripts denote data points, s = the slope of the line, t = time, and b = the line intercept. The area beneath each segment can be determined by integrating the interpolation functions. Assuming porosity, cross-sectional areas, and velocities are constant in time, for a given segment the integration leads to the following:

$$M = \phi A v \int_{t_1}^{t_2} C dt \tag{6}$$

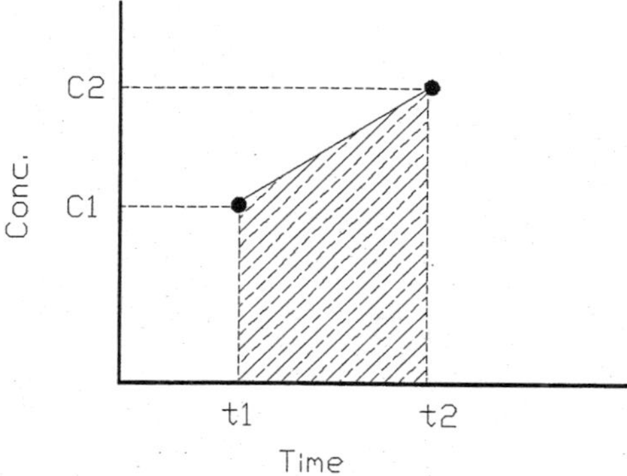

Figure A24.3 A single segment defined by a concentration range and a time range. The slope and intercept of this line are calculated as discussed in the text, and the area beneath it (shaded) is determined by integrating the line from t_1 to t_2.

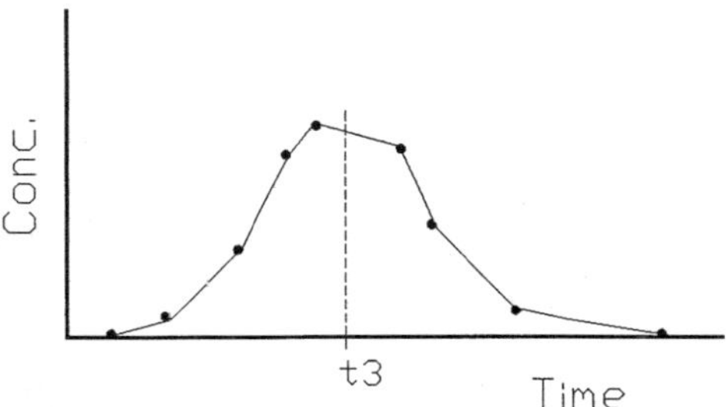

Figure A24.4 The time of arrival, t_3, of the center of mass.

$$M = \phi A v \int_{t_1}^{t_2} st + b\ dt \tag{7}$$

$$M = \phi A v \left[\frac{1}{2} s(t_2^2 - t_1^2) + b(t_2 - t_1) \right] \tag{8}$$

The sum of all the segment masses is the mass of solute having broken through at the sampling point. In cases where cross-sectional areas and/or velocities varied in time, due to changing water tables, these terms are simply treated as being segment specific during the summation of the segment masses.

A24.2.2 Estimating Velocities (the Center of Mass Method)

In cases where complete BTCs have been monitored, and where the contaminant or tracer behaves conservatively, it is possible to estimate flow velocities from the integrated mass. The average linear velocity is defined as the distance of the monitor from the source (d) divided by the arrival time of the center of mass of the solute pulse (t_3),

$$v = \frac{d}{t_3} \tag{9}$$

If the entire BTC is integrated to obtain an estimate of the total mass, it can be integrated again up to the time where the cumulative mass equals one half of the total. By definition, that time is

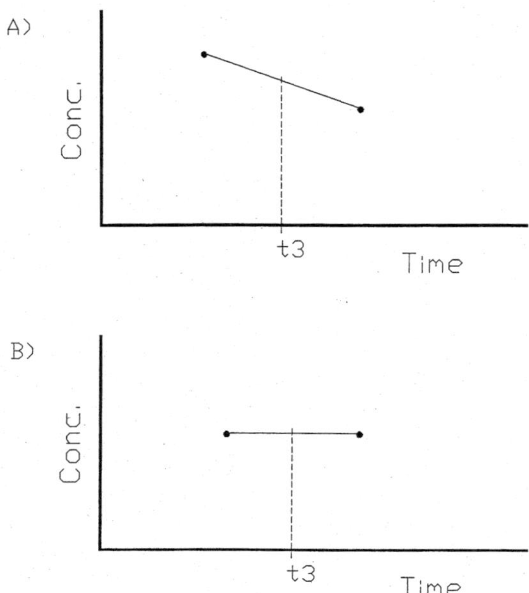

Figure A24.5 The method of estimating t_3 depends on the slope of the segment line. Non-zero (A) and zero (B) slopes must be distinguished in the estimation of t_3.

t_3 (Figure 24.4). In some cases, t_3 will fall on a data point and the velocity can be calculated immediately. More often t_3 will fall between data points and will have to be calculated by interpolation. The (linear) interpolation can occur one of two ways, depending on the slope of the line connecting the two points on either side of t_3 (Figure 24.5). If the slope is non-zero, then the area (HA) between t_3 and t_2 is given by

$$HA = \int_{t_3}^{t_2} st + b\ dt \tag{10}$$

where s and b are the slope and intercept of the segment line connecting the data points on either side of t_3. Since the cumulative area under the BTC to t_2, A_{t2}, is known and equals HA plus half the total area beneath the entire BTC (also known), the quantity HA can be calculated (Figure A24.6). Therefore, in the integral above, the only unknown quantity is t_3. Solving the integral, the following relation is obtained:

$$-\frac{1}{2}st_3^2 - bt_3 + \left(\frac{1}{2}st_2^2 + bt_2 - HA\right) = 0 \tag{11}$$

This is seen to be a quadratic equation that can be solved for t_3 as follows:

$$t_3 = \frac{-B \pm \sqrt{B^2 + 4AC}}{2A} \tag{12}$$

where $A = -1/2s$, $B = -b$, and $C = 1/2st_2^2 + bt_2 - HA$. If the slope of the segment is zero, the above expression for t_3 is undefined and another equation is needed. Fortunately, this is easily obtained from the proportionality of the time axis to the area under the segment. The segment can be divided

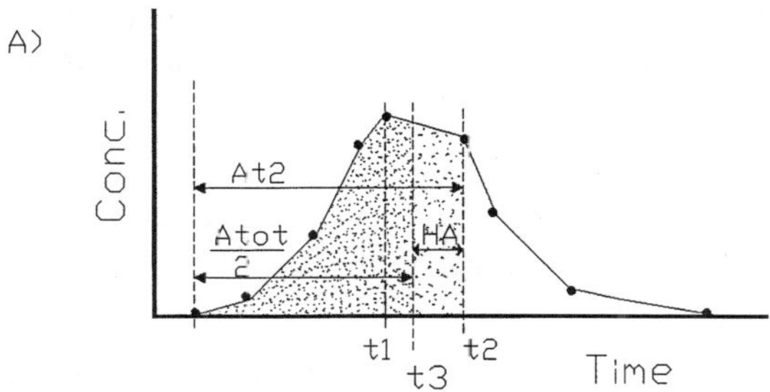

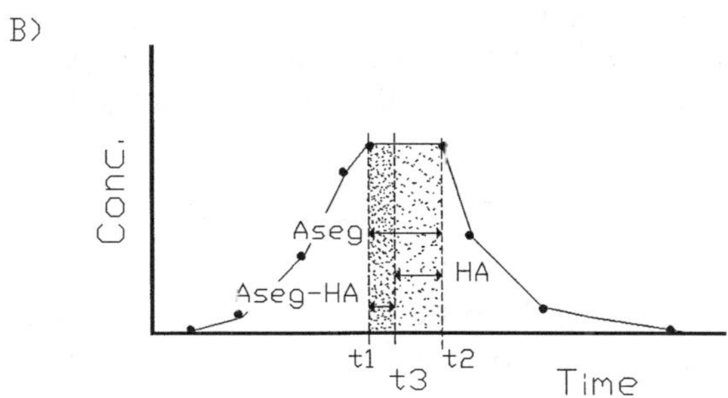

Figure A24.6 Relationships between integrated areas and times t_1, t_2, and t_3.

into two parts: t_1 to t_3 and t_3 to t_2. The area under $t_3 - t_2$ is HA and can be calculated as previously described. The total area under the segment is A_{seg} and is calculated in the course of obtaining A_{tot}, so that the area beneath $t_1 - t_3$ is obtainable as $A_{seg} - HA$ (Figure A24.6). The following proportionality equation can now be written:

$$\frac{t}{t'} = \frac{A_{seg}}{A_{seg} - HA} \tag{13}$$

where $t = t_2 - t_1$ and $t' = t_3 - t_1$. Making the appropriate substitutions, t_3 is seen to be given by

$$t_3 = \frac{(t_2 - t_1)(A_{seg} - HA)}{A_{seg}} + t_1 \tag{14}$$

A24.3 FENCE-AVERAGED CONCENTRATIONS: COMPARISON OF METHODS

The fence-averaged concentrations reported here were based on the following equation for calculating area-weighted averages:

$$C_{avg} = \sum_j \sum_i w_{ij} C_{ij} \tag{15}$$

$$w_{ij} = \frac{x_{ij} y_{ij}}{\sum_j \sum_i x_{ij} y_{ij}} \tag{16}$$

where x and y are length dimensions in the vertical and horizontal directions defining the areas represented by each sampling point, and the subscripts i and j denote monitoring point locations in the two-dimensional plane of the sampling fence. The weighting term, w_{ij}, is the fraction of the total cross-sectional area of the fence represented by a sampling point. If all the sampling points represent equal cross-sectional areas, the weighted average reduces to a simple arithmetic average of concentrations across the fence,

$$C_{avg} = \sum_j \sum_i \frac{x_{ij}y_{ij}}{\sum_j \sum_i x_{ij}y_{ij}} C_{ij} = \frac{x_{ij}y_{ij}}{nx_{ij}y_{ij}} \sum_j \sum_i C_{ij} = \frac{\sum_j \sum_i C_{ij}}{n} \tag{17}$$

where $n = i \times j$, the number of sampling points in the fence. This approximation was used throughout the report, where noted, on the basis of close agreement (within 10%) between the arithmetically averaged and the area-weighted average concentrations.

As discussed previously, the FENCE program calculates total mass (rather than average concentrations) crossing a fence over a period of time. The total mass is given by

$$M_{tot} = \sum_j \sum_i \sum_t M_{jit} \tag{18}$$

where the subscript t denotes time. If FENCE is supplied with concentrations at a fence corresponding to a single sampling event (i.e., only one time), then the mass returned, M_{tot}, can be converted to the average concentration on that fence,

$$C_{avg} = \frac{M_{tot}}{\phi \sum_j \sum_i x_{ij}y_{ij}} = \frac{\sum_j \sum_i M_{ij}}{\phi \sum_j \sum_i x_{ij}y_{ij}} \tag{19}$$

where ϕ = porosity and we assume a unit dimension in the direction of flow (z_{ij}). Expanding the mass term,

$$C_{avg} = \frac{\phi \sum_j \sum_i C_{ij}x_{ij}y_{ij}}{\phi \sum_j \sum_i x_{ij}y_{ij}} \tag{20}$$

The porosity terms cancel, and since the denominator is a constant, equal to the total cross-sectional area of the fence, it can be lumped with the area term in the numerator ($x_{ij}y_{ij}$) to form the weighting term w_{ij}, introduced above.

$$C_{avg} = \sum_j \sum_i w_{ij}C_{ij} \tag{21}$$

Thus, the the method of integrating data for the calculation of mass, used by the FENCE program, and the concentration averaging methods used in this report are seen to be equivalent.

APPENDIX 25

Alameda Schedule

Date	Activity
22-27 Jul-96	Initial site characterization
1-30 Sep-96	Developed and wrote the first draft of the Work Plan
1-31 Oct-96	External review of the Work Plan
15-30 Oct-96	Wrote the Request for Proposal (RFP)
21-23 Oct-96	Second site characterization
25-Oct-96	Send out RPF for Funnel-and-Gate construction
1-30 Nov-96	Corrections to Work Plan, submitted Revised Work Plan
8-Nov-96	Contractors meeting at Alameda, NAS, Site 1
27-Nov-96	Award construction contract to Remedial Solutions, Inc. (RSI)
9-27 Dec-96	Construction of Funnel-and-Gate
14-Jan-97	Calibrated ISPFS
	- lowered sparged units 4 feet below the water table to document bubble distribution
15-Jan-97	Assembled control gate multilevels
	- RSI graded site prior to installing the control gate wells
16-Jan-97	Developed multilevels in the remedial gate
	- Installed control gate and row 1 multilevels and fully screened wells
	- Extracted two continuous cores, one in row 1, the other in the control gate
17-Jan-97	Wastewater baker tank # 2 arrived
	- Tested sparged units again at a shallow depth
18-Jan-97	Lowered sparged units to the bottom of the biosparge zone
	-Took background conductivity samples from R1PA and R1PB
19-Jan-97	Developed R1PA
	-Installed temporary power pole for electrical hook-up
	-Poured bioballs into biosparge unit
20-Jan-97	Developed the rest of the fully screened wells in both gates
22-Jan-97	Background sampling from the fully screened wells
	-Took conductivity samples from all the multilevels in rows 2 and 3
23-Jan-97	Tried to redevelop the plugged multilevels
	-Placed sand around shed due to heavy rains, regraded site
	-Carbon drum was delivered
28-Jan-97	Recorded DO in biosparge unit
	-Discussed plans to construct bioball cages
30-Jan-97	Sparged biosparge zone with oxygen then measured DO
	-Replaced tubing in extraction pump head
31-Jan-97	Recorded water levels
	-Hooked up carbon drum to biosparge unit (treat off gas)
3-Feb-97	Turn extraction pumps on, flow rate = 35 ft^3/day due to smal-diameter tubes
	-Recorded water levels
4-Feb-97	Extraction pumps leaked, tube diameter was too small
	-Sparged biosparge zone with oxygen, measured DO
	-Flow rate = 36 ft^3/day (highest pump setting)

Date	Activity
10-Feb-97	Flow rate = 45 ft^3/day
	-Recorded water levels
	-NAS requested air sampling plan to be implemented
	-Began biosparge lid construction; completely finish April 17
	-Site clean-up
	-Mark control gate with stakes
12-Feb-97	Flow rate = 45 ft^3/day
	-Recorded water levels
	-Added two more boxes of bioballs to biosparge unit
	-Collected soil samples from piles for organic and metal analyses
	-Oxygen tank empty, suspect leak
	-Received OVA meter from NAS to measured off gas from biosparge unit
14-Feb-97	Sampled multilevels for organics and Fe
15-Feb-97	Replaced tubing in pump
18-Feb-97	Adjusted flow rate to 45 ft^3/day
24-Feb-97	Removed all row 1 and control gate wells (T and P)
	-Installed R1TA and R1TB multilevels
	-Developed new multilevels
26-Feb-97	Sampled multilevels for organics, and BOD (rows 6, 7, and 8)
27-Feb-97	Measured DO and pH in all multilevels in the remedial gate
	-Recorded water levels
	-Adjusted flow to 45 ft^3/day
28-Feb-97	Repair perimeter fence around soil piles
	-Recorded water levels
4-Mar-97	Sent microcosm samples to UW (from row 6)
	-Sampled all fully screened wells in both gates for organics and inorganics and BOD (rows 6, 7, and 8)
5-Mar-97	Photovac and Microtip arrived from Hazco — used for air sampling; measured air without biosparge unit sparging, = 0 ppm
	-PID not working - detector light intensity is low - talked to Hazco
6-Mar-97	Measured pH and DO from fully screened wells in both gates
	-Replaced tubing in pump head and extraction tubing from row 8
	-Flow rate = 45 ft^3/day
10-Mar-97	NAS makes a request for both soil piles to be removed by April 1, 1997
	-Adjusted flow rate to be 45 ft^3/day
	-Measured volume of wastewater baker tank #2
11-Mar-97	Installed control gate multilevels and R1TC multilevel
	-Developed new multilevels
	-Recorded water levels
	-Recorded DO and pH in all multilevels and rows 7 and 8
12-Mar-97	Began sampling from multilevels in the remedial gate for organics, dissolved hydrocarbon gases and iron
	-Small cation filter would plug, therefore, samples were collected and filtered at the inorganic lab
13-Mar-97	Finished sampling multilevels
	-Retarped soil piles
17-Mar-97	NAS said EBMUD required a water permit to discharge water
	-Replaced tubing in pump head
	-Flow rate = 45 ft^3/day
	-Bag excess bioballs from the biosparge zone
19-Mar-97	Sampled from all fully screened wells for organics, DHGs, and inorganics
	-RSI on site to finish building bioball cages
20-Mar-97	Measured DO and pH in all fully screened wells
	-Recorded water levels
	-NAS on site to inspect soil piles
	-NAS requests copies of wastewater and soil analyses
21-Mar-97	Replaced tubing in pump head
	-Flow rate = 45 ft^3/day
	-Removed more bioballs from biosparge zone

Date	Activity
24-Mar-97	RSI completed construction on the bioball cages
	-Most of the bioballs have been removed from the biosparge zone
25-Mar-97	Sampling from the multilevels in the remedial gate for organics and dissolved hydrocarbon gases
26-Mar-97	Recorded water levels
	-Measured pH and DO in all multilevels in the remedial gate and in rows 7 and 8
	-Replaced tubing in pump head
	-Flow rate = 45 ft^3/day
31-Mar-97	Repair perimeter fence around soil piles
	-BOD and TOC samples taken from row 5 and 6 (not all wells)
	-Adjusted pumping rate to 45 ft^3/day
	-NAS on site with Bay Area Air Quality Management District to inspect soil piles
1-Apr-97	Measured DO and pH from all multilevels in the remedial gate and from rows 7 and 8
	-Recorded water levels
	-Adjusted flow to 45 ft^3/day
2-Apr-97	Re-tarped soil piles and repaired perimeter fence
4-Apr-97	PRC on site to inspect which soil pile they are to remove
	-Extraction pumps are turned off at 12:30 PM
	-Turn power off on ISPFS as per request of Robert Langdon
	-Recorded water levels prior to turning pumps off, then monitor rebound of water table due to no pumping
6-Apr-97	Recorded water levels
7-Apr-97	Recorded water levels
8-Apr-97	Experiment with different entry pressure for oxygen in the biosparge zone
9-Apr-97	Recorded water levels
14-Apr-97	Turn extraction pumps back on to 45 ft^3/day
	-Sparged biosparge unit with oxygen, measured DO
15-Apr-97	NAS informs UW that Beale Air Force Base wants extra bags of granular iron
	-Sparged biosparge unit with oxygen, measured DO
16-Apr-97	Measured DO in row 6
	-Recorded water levels
	-Adjusted flow rate to 45 ft^3/day
	-ETI on site with EFW to calibrate geoflow meters
	-FID measurements - outside (4 ppm), inside with no sparging (5 ppm), inside with oxygen sparging at 30 psi (4-5 ppm), interior of biosparge unit while oxygen is being sparged at 30 psi (12 ppm)
	-Surveyed all fully screened wells in both gates with EFW
	-Carbon dioxide cylinder arrived on site
	-Conductivity sampling from row 1 to row 5 (all multilevels)
	-Geoflow meter did not work, wells were never developed
17-Apr-97	Local landfill accepts disposal of last soil pile
	-NAS is requested by UW to complete soil manifest forms
18-Apr-97	Sampled all fully screened wells for inorganics
	-Measured pH in all fully screened wells
21-Apr-97	Sparged biosparge unit with oxygen, measured DO
22-Apr-97	Reduced pumping rate to 12 ft^3/day
	-Sampled all multilevels from the remedial gate and rows 7 and 8 for organics and dissolved hydrocarbon gases
23-Apr-97	RSI on site to remove last soil pile
	-Iron bags are removed
	-Perimeter fence around Funnel-and-Gate and soil piles were taken down
	-Measured DO and pH in all multilevels in the remedial gate as well as rows 7 and 8
	-Recorded water levels
24-Apr-97	Sampled biosparge headspace with a summa canister
25-Apr-97	New wastewater baker tank arrives (#3)
	-Rob Langdon on site to decide how to inject tracer for tracer test
	-Flow rate = 12 ft^3/day
28-Apr-97	Flow rate = 12 ft^3/day
	-Sparged biosparge unit with oxygen and carbon dioxide, measured DO and pH

Date	Activity
30-Apr-97	Sparged biosparge unit with oxygen, measured DO -Review method of treating wastewater with a carbon drum
2-May-97	Sparged biosparge unit with oxygen, measured DO
5-May-97	Began treating wastewater from baker tank #2 into baker tank #3 using granular activated carbon -Sparged biosparge unit with oxygen, measured DO and pH
6-May-97	Sparged biosparge unit with oxygen and carbon dioxide, measured DO and pH -Adjusted rate of wastewater treatment system -Flow rate = 12 ft^3/day
7-May-97	Injected potassium bromide in R1PB (200 gallons) and extracted or pumped from R1PC to pull tracer slug toward the north
8-May-97	Assembled off gas carbon treatment system - Sparged biosparge unit with oxygen, measured DO and pH -Flow rate = 24 ft^3/day (was accidentally bumped up for 24 hours) then adjusted down to 12 ft^3/day
9-May-97	Sparged biosparge unit with oxygen, measured DO -Conductivity samples were taken through to June 16, 1997 (approximately 2–5 times per week)
12-May-97	Sparged biosparge unit with oxygen, measured DO
14-May-97	Flow rate = 12 ft^3/day
15-May-97	Sparged biosparge unit with oxygen, measured DO -Collected and sent microcosm water to UW, from row 6 -Flow rate = 12 ft^3/day
19-May-97	Sparged biosparge unit with oxygen and carbon dioxide, measured DO and pH -Investigated parts required to assemble automated timers for oxygen and carbon dioxide cylinders
20-May-97	Sparged biosparge unit with oxygen and carbon dioxide, measured DO and pH
21-May-97	Measured DO and pH from all the multilevels in the remedial gate as well as rows 7 and 8
22-May-97	Sampled water in wastewater baker tank #3 for organic and metal analyses -Sampled multilevels in the remedial gate for organics and dissolved hydrocarbon gases
12-Jun-97	Set up automated timers for carbon dioxide and oxygen cylinders -Set up initial sparging pattern -Recorded DO and pH in the multilevels from rows 4-6 -Inquired whether carbon drums will treat dissolved chromium
13-Jun-97	Removed garbage from site
14-Jun-97	Measured DO and pH in row 6 — adjusted accordingly -Adjusted flow to 12 ft^3/day -Sampled Bay water for conductivity, pH, and DO -Continued to adjust sparging rates for both gases
15-Jun-97	Tube in pump head collapsed, therefore not pumping -Replaced tubing in pump head and adjusted flow rate to 12 ft^3/day -Recorded pH and DO in row 6 -Changed sparging pattern
16-Jun-97	Adjusted flow to 12 ft^3/day -Recorded water levels -Began sampling fully screened wells and multilevels for organics and DHGs
17-Jun-97	Flow rate = 260 mL/min, adjusted it to 240 mL/min -Carbon dioxide and oxygen cylinders empty — incorrect setting on timers -Continued to sample fully screened wells and multilevels
18-Jun-97	Flow rate = 250 mL/min -New carbon dioxide and oxygen cylinders were delivered -Carbon dioxide sparging pattern was changed
19-Jun-97	Flow rate = 240 mL/min, each line was 120 mL/min -Recorded DO and pH in row 6
22-Jun-97	Flow rate = 230 mL/min, adjusted to 240 mL/min -Measured pH in row 6 -Recorded water levels

Date	Activity
23-Jun-97	Flow rate = 225 mL/min -Recorded pH in row 6
24-Jun-97	Flow rate = 250 mL/min -Recorded water levels -Measured pH in row 6
25-Jun-97	Flow rate = 240 mL/min -Clean up site -Sampled baker tank #1 for metals and baker tank #3 for organics and metal analyses -Recorded pH in row 6
26-Jun-97	Flow rate = 250 mL/min
27-Jun-97	Flow rate = 250 mL/min -Recorded pH in row 6
1-Jul-97	Flow rate = 230 mL/min
2-Jul-97	Flow rate = 250 mL/min -Measured pH and E_h in rows 6 and 7 -Recorded water levels
3-Jul-97	Flow rate = 260 mL/min, adjusted down to 240 mL/min
4-Jul-97	Recorded DO, pH, and EC and temperature from all multilevels in the remedial gate -Flow rate = 260 mL/min, adjusted it to 240 mL/min -Recorded water levels
6-Jul-97	Flow rate = 240 mL/min -Collected organics, BOD, and dissolved hydrocarbon gases (DHG) from rows 5, 6, 7, and 8 -Measured pH, DO, and EC from rows 5, 6, 7, and 8
9-Jul-97	Measured pH, DO, and EC from rows 5, 6, 7, and 8 -Emptied carbon dioxide cylinder due to incorrect sparging pattern -Oxygen cylinder emptied
11-Jul-97	Flow rate = 240 mL/min -Recorded pH and DO in row 6 -New oxygen cylinder arrived -Due to low pH readings in row 6, increased pumping rate to 930 mL/min to quickly rid the biosparge zone of low pH water. This rate continued for 2 hours, then rate was decreased to 610 mL/min -Precision on site to install R1PD and R1PE
12-Jul-97	Flow rate = 650 mL/min -After new oxygen tank was set up, measured DO in row 6
13-Jul-97	Flow rate = 550 mL/min; decreased it to 240 mL/min -pH in row 6 still low, excessive pumping did not seem to rid the system of the low pH -Recorded DO from row 6
14-Jul-97	Flow rate 260 mL/min -Sampled headspace in biosparge zone with summa canister
15-Jul-97	Flow rate = 260 mL/min -Recorded pH and DO in row 6
16-Jul-97	Flow rate = 260 mL/min
17-Jul-97	Measured pH and DO in rows 6, 7, and 8
20-Jul-97	Flow rate = 230 mL/min -Recorded pH and DO in rows 6, 7 and 8
24-Jul-97	Flow rate = 240 mL/min Measured pH and DO in row 6
28-Jul-97	Flow rate = 240 mL/min -Measured DO in row 6
31-Jul-97	Measured pH, DO, and E_h in all multilevels in the remedial gate and in rows 7 and 8 -Developed R1PD and R1PE
1-Aug-97	Flow rate = 245 mL/min -Finished measuring pH, DO, and E_h in the multilevels -Sampled for organics and DHG in the control gate
2-Aug-97	Sampled the multilevels in the remedial gate for organics and DHGs -Flow rate = 245 mL/min
4-Aug-97	Sampled fully screened wells in both gates for organics and DHGs -Recorded water levels

Date	Activity
5-Aug-97	8:30 AM - 12:30 PM power was off -Flow rate prior to no power was 240 mL/min -Pumps on 1:30 PM, flow rate = 220 mL/min
6-Aug-97	Flow rate = 240 mL/min -Recorded DO in row 6
8-Aug-97	Flow rate = 265 mL/min -Measured DO in row 6
9-Aug-97	Flow rate = 230 mL/min -Measured DO in row 6
10-Aug-97	Flow rate = 260 mL/min -Sampled row 5 multilevels for organics and DHGs Flow rate = 235 mL/min -Measured DO in row 6
16-Aug-97	Flow rate = 230 mL/min -Recorded water levels -Measured pH and DO in row 6
21-Aug-97	Flow rate = 245 mL/min Measured pH and DO in row 6
23-Aug-97	Flow rate = 220 mL/min -Set up wastewater treatment system for baker tank #1, using two granular-activated carbon drums in series
24-Aug-97	Flow rate = 270 mL/min, adjusted to 240 mL/min -Measured DO in row 6
25-Aug-97	Flow rate = 225 mL/min -Sampled row 1 multilevels and fully screened wells for organics and DHGs -Recorded water levels -Measured DO in row 6
27-Aug-97	Flow rate = 240 mL/min -New oxygen tank arrived and was set up -Sampled rows 5, 6, 7, and 8 for organics and DHGs -Measured DO in row 6
28-Aug-97	Flow rate = 240 mL/min -Measured DO, pH, and E_h in all multilevels in the remedial gate
29-Aug-97	Flow rate = 248 mL/min -Continued to measure DO, pH, and E_h in all the multilevels
30-Aug-97	Measured DO in row 6 -Installed Dwyer flow meter on oxygen line; reading was 0.0–0.2 SCFH -Replaced tubing in pumps -Flow rate = 245 mL/min
31-Aug-97	Flow rate = 220 mL/min -Measured DO and pH in row 6
10-Sep-97	Flow rate = 240 mL/min -Flow rate of oxygen = <0.2 SCFH -Recorded water levels -Measured DO and pH in rows 5, 6, 7, and 8 -Sampled organics and DHGs in rows 5, 6, 7, and 8
15-Sep-97	Flow rate = 240 mL/min -Flow rate of oxygen = 0.35 SCFH -Recorded water levels -Measured DO and pH in rows 6 and 7 -Sampled for organics and DHGs in rows 6 and 7
22-Sep-97	Flow rate = 240 mL/min -Flow rate of oxygen = 10 SCFH -Recorded water levels -Measured DO and pH in rows 6 and 8 -Sampled for organics and DHGs in rows 6 and 8
28-Sep-97	Flow rate = 255 mL/min, adjusted it to 240 mL/min -Measured DO in row 6

Date	Activity
30-Sep-97	Flow rate of oxygen = 1 SCFH -Measured DO in row 6 -Flow rate = 235 mL/min -Sampled fully screened wells in both gates for organics and inorganics -Recorded water levels
1-Oct-97	Flow rate = 245 mL/min -Measured DO in row 6 -Began sampling the multilevels in both gates for organics and DHGs -Collected 10 bioballs from the biosparge zone for aerobic microcosm -Collected 10 liters of row 6 water for aerobic microcosm
2-Oct-97	Sampled headspace gas from the biosparge zone using a summa canister -Flow rate of oxygen = 10 SCFH -Continued sampling the multilevels
3-Oct-97	Finished sampling the multilevels -Collected water for the second iron treatability study (column test) -Flow rate = 240 mL/min -Measured DO and pH in row 6
4-Oct-97	Set up new oxygen tank -Measured DO, pH, and E_h in all multilevels in both gates and in rows 7 and 8 -Measured depth of all fully screened wells
6-Oct-97	Flow rate = 240 mL/min -Flow rate of oxygen = 12 SCFH -Recorded water levels -Measured DO and pH in rows 6 and 8 -Sampled for organics and DHGs in rows 6 and 8
10-Oct-97	Flow rate = 240 mL/min -Recorded water levels -Measured DO and pH in rows 6 and 7 -Sampled for organics and DHGs in rows 6 and 7
17-Oct-97	Flow rate = 240 mL/min -Flow rate of oxygen = 25 SCFH -Recorded water levels -Measured DO and pH in rows 6 and 8 -Sampled for organics and DHGs in rows 6 and 8
21-Oct-97	Flow rate = 240 mL/min -Flow rate of oxygen = 15 SCFH -Recorded water levels -Measured DO and pH in rows 5, 6, 7, and 8 -Sampled for organics and DHGs in rows 5, 6, 7, and 8

Appendix 26

Sampling Methods — Alameda

A26.1 GROUNDWATER MONITORING PROGRAMS

Groundwater monitoring was divided into the following categories:

1. Background sampling
2. Performance sampling
3. Treatment sampling
4. Biosparge sampling
5. Tracer test sampling

The following paragraphs detail the rationale for each of the groundwater monitoring categories, the parameters analyzed, and the approximate timeline for sampling events.

Background sampling — Prior to removing the upgradient sheet piling, the remedial gate was filled with water collected from a local water hydrant. Once full, the upgradient sheet piling was removed allowing contact between the potable water and the aquifer water. Six weeks later, all fully screened wells in both the remedial and control gates were sampled for organic contaminants to determine the initial background contamination before the extraction pumps were turned on.

Performance sampling — Performance data were used to assess the overall performance of the system with respect to the remedial objectives, emphasizing the fate of target organics, and possible breakdown products. Performance data were also used to study groundwater chemical changes due to the two remedial technologies. Water samples were collected only from fully screened wells during a performance sampling event and the data represent vertically averaged concentrations. Analyses included organics, inorganics, occasionally biochemical oxygen demand (BOD), and field parameters.

Treatment sampling — Treatment data were used to assess the specific processes operating within the granular iron wall and the biosparge zone. These data were collected from three-point multilevel wells that provided depth-discrete groundwater data. During a treatment sampling event, two rows of fully screened wells, rows 7 and 8, were also sampled. These wells were included as they represented water downgradient of the biosparge zone (row 7) and the final compliance point (row 8). There were no multilevels present in rows 7 or 8. Analyses included organics, dissolved hydrocarbon gases, occasionally BOD, and field parameters.

Biosparge sampling — An intensive sampling schedule was initiated from July 1997 to December 1997 to study the aerobic biodegradation occurring within the biosparge zone. Approximately every

1–2 weeks, the multilevels in rows 5 and 6 were sampled as well as the fully screened wells in rows 7 and 8. Samples were analyzed for organics, dissolved hydrocarbon gases, and field parameters.

Tracer test sampling — A tracer test was conducted to provide insight into groundwater flow velocities and the existence of preferential pathways within the remedial gate. Bromide (added as potassium bromide) was used as a conservative tracer and was injected May 7, 1997. Samples were collected from selected multilevel points in rows 1 through 3 until June 17, 1997.

Table A26.1 provides a summary of the types of analyses completed for each monitoring category. Table A26.2 lists the specific sampling events for each of the groundwater monitoring categories. Table A26.3 lists air/headspace sampling events.

Table A26.1 Summary of Groundwater Monitoring Categories

Program Number	Monitoring Program Description	Type of Analyses Conducted	Type of Well P or T
1	Background Sampling	VOCs	P
2	Performance Sampling	VOCs, Inorganics, Field Parameters	P
3	Treatment Sampling	VOCs, DHGs, Field Parameters	P & T
4	Biosparge Sampling	VOCs, DHGs, Field Parameters	P & T
5	Tracer Test	Bromide	T

Notes:

P	Fully screened wells
T	Multilevel piezometers
VOCs	Volatile organic compounds:
	Tetrachloroethene (PCE), Trichloroethene (TCE), 1,1-Dicloroethene (1,1-DCE), *cis*-1,2-Dichloroethene (*c*DCE), *trans*-1,2-Dichloroethene (*t*DCE), Vinyl chloride (VC), Benzene, Toluene, Ethybenzene, and Xylenes (BTEX)
DHGs	Dissolved hydrocarbon gases:
	Ethane, Ethene, Methane, Propane, Propene
BOD	Biochemical oxygen demand (BOD)
Inorganics	Major Ions: Cations (Ca, Fe, Na, K, Mg, Mn)
	Anions (Cl, Br, NO_3, NO_2, PO_4, SO_4) and Alkalinity
	Field Parameters DO and pH

Types of Analyses — A variety of analytes were monitored over the duration of the field demonstration:

1. Volatile organic compounds (VOCs)
2. Dissolved hydrocarbon gases (DHGs)
3. Cations
4. Anions and alkalinity
5. Biochemical oxygen demand (BOD)
6. Field parameters
7. Water level measurements

Table A26.4 details the specific sample collection and handling requirements for the above analytes.

A26.1.1 VOCS

Volatile organic compound (VOC) sampling was completed during the background, performance, treatment, and biosparge sampling events. The organic contaminants of interest included PCE, TCE, 1,1-DCE, *c*DCE, *t*DCE, VC, and BTEX.

Table A26.2 Groundwater Sampling Events

Date	Monitoring Category	Types of Analyses Conducted
23-Jan-97	Background Sampling	VOCs
14-Feb-97	Treatment Sampling	VOCs, Inorganic (Fe only), Field Parameters
26-Feb-97	Treatment Sampling	VOCs, BOD, Field Parameters
6-Mar-97	Performance Sampling	VOCs, Inorganics, BOD, Field Parameters
12-Mar-97	Treatment Sampling	VOCs, DHGs, Inorganics (Fe only), Field Parameters
19-Mar-97	Performance Sampling	VOCs, DHGs, Inorganics, Field Parameters
25-Mar-97	Treatment Sampling	VOCs, DHGs, Field Parameters
18-Apr-97	Performance Sampling	Inorganics, Field Parameters (pH only)
22-Apr-97	Treatment Sampling	VOCs, DHGs, Field Parameters
22-May-97	Treatment Sampling	VOCs, DHGs, Field Parameters
16-Jun-97	Treatment Sampling and	VOCs, DHGs, Field Parameters
	Performance Sampling	VOCs, Inorganics, Field Parameters
	Biosparge Sampling	VOCs, DHGs, Field Parameters
6-July-97	Biosparge Sampling	VOCs, DHGs, BOD, Field Parameters
4-Aug-97	Treatment Sampling and	VOCs, DHGs, BOD, Field Parameters
	Performance Sampling	VOCs, Inorganics, BOD, Field Parameters
	Biosparge Sampling	VOCs, DHGs, Field Parameters
25-Aug-97	Biosparge Sampling	VOCs, DHGs, Field Parameters
10-Sept-97	Biosparge Sampling	VOCs, DHGs, Field Parameters
30-Sept-97	Treatment Sampling and	VOCs, DHGs, Field Parameters
	Performance Sampling	VOCs, Inorganics, Field Parameters
	Biosparge Sampling	VOCs, DHGs, Field Parameters
21-Oct-97	Biosparge Sampling	VOCs, DHGs, Field Parameters

Table A26.3 Air/Headspace Sampling Events

Date	Types of Analyses Conducted
24-Apr-97	*c*DCE, VC, BTEX
14-Jul-97	*c*DCE, VC, BTEX
2-Oct-97	*c*DCE, VC, BTEX

Table A26.4 Summary of Sampling Specifications

Type of Analysis	Sample Volume (mL)	Type of Sample Bottle	Preservation	Holding Time (days)	Analyst: Field or Lab
VOCs	40	a	d	14	lab
DHGs	40	a	4°C	7	lab
BOD	500	b	4°C	4	lab
Cations	60	plastic	c	28	lab
Anions and	125	plastic	4°C	Anions — 28	lab
Alkalinity				Alk — 4	lab
Bromide	20	plastic	none	28	field
Field Parameters	100	N/A	N/A	none	field

Notes: a: Glass hypovial, capped with a Teflon-faced septum. b: Glass hypovial, capped with a Teflon-faced septum. c: Filtered (0.45 μm) and acidified to pH <2 with concentrated HNO_3. d: External lab: HCl (bottle is provided containing the preservative); UW lab: 0.44 cc of 10% sodium azide stock solution. NA: Not applicable.

During a performance sampling event, VOCs were collected using a dedicated Waterra Inertial Lift Pump and were sent to an external laboratory, Intertek Testing Services, for analysis by GC/MS according to EPA Method 82 60. Intertek Testing Services, formally Inchcape Testing Services, supplied the 40-mL glass hypovials with preservative already present, and so no additional preservative was added. For the treatment sampling events, VOCs were collected from both the multilevel wells and fully screened wells (rows 7 and 8 only). Sampling from the multilevels was done using the UW sampling apparatus (Appendix 4, Section A4.3.1) and the dedicated Waterra Inertial Lift Pumps were used to sample from the fully screened wells. Any VOC samples sent to the Organic Geochemistry Lab at UW were preserved with 0.4 cc of 10% sodium azide stock solution and shipped in coolers full of blue ice. For the biosparge sampling event, VOCs were collected from both the fully screened wells, using the dedicated Waterras, and the multilevels, using UW's sampling apparatus, and all samples were analyzed at the University of Waterloo's Organic Geochemistry lab.

A26.1.2 DHGs

Dissolved hydrocarbon gas (DHG) sampling was completed during the performance sampling, treatment sampling, and biosparge sampling events. Groundwater was analyzed for ethane, ethene, methane, propane, and propene.

During a performance sampling event, DHG samples were collected using the dedicated Waterra, but during a treatment sampling event, the samples were collected using the UW sampling apparatus (except for rows 7 and 8 where the dedicated Waterras were used). Depending on the type of well being sampled for the biosparge sampling events, either the Waterra or the UW sampling apparatus method was used. DHG samples were organic and were kept at 4°C. All DHG samples were analyzed at the University of Waterloo's Organic Geochemistry lab.

A26.1.3 Cations

Cation samples were only collected during performance sampling events. Cation analyses reported concentrations of calcium, sodium, magnesium, manganese, iron, and potassium.

Since the samples were collected from a fully screened well, the dedicated Waterra Inertial Lift Pump was used. Once collected, samples were filtered to remove any suspended particles and then preserved using concentrated nitric acid to lower the solution pH to <2 to prevent the formation of precipitates. Therefore, the analytical results report a dissolved concentration, not a total concentration. After collection, samples were stored in coolers with blue ice.

A26.1.4 Anions and Alkalinity

Anion and alkalinity samples were only collected during performance sampling events. Anion analyses reported concentrations of chloride, bromide, nitrate, nitrite, sulfate, and phosphate. Alkalinity concentrations were expressed in terms of milligrams per liter calcium carbonate.

A dedicated Waterra Inertial Lift Pump was used to collect the samples. Anion and alkalinity samples were not filtered or preserved, but were kept cool using frozen blue ice.

A26.1.5 BOD

BOD samples were periodically collected during the performance, treatment, and biosparge sampling events. Depending on which well was being sampled, the appropriate sampling method was used (either dedicated Waterra or UW apparatus). BOD samples were not preserved, but remained chilled on frozen blue ice until analyzed.

A26.1.6 Field Parameters

pH and dissolved oxygen (DO) were measured in the field for every performance, treatment, and biosparge sampling event. Field parameters in the biosparge zone were extensively monitored in row 6 with the onset of regular oxygen and carbon dioxide sparging. During the tracer test, samples collected from multilevels were analyzed for bromide. Probe measurements for pH, DO, and bromide were completed in the field using a flow-through cell as described in Appendix 4, Section A4.4.1. Although the flow-through cell used at Alameda was slightly different then the one used at Borden, the principles are the same; hence, the reader is referenced to Appendix 4 for details of the method.

A26.1.7 Water Level Measurements

Water level measurements were taken regularly to document fluctuations in the water table and to demonstrate that the required gradient had been achieved.

A26.2 SAMPLE COLLECTION METHODOLOGY

A detailed description of the various sample collection methodologies and handling is presented below. Any wastewater resulting from sampling activities was disposed in the on-site wastewater baker tanks located south of the shed. This water was subsequently treated in an aboveground system (Appendix 31).

A26.2.1 UW Sampling Apparatus Methodology

Groundwater samples collected using the UW sampling apparatus methodology include VOCs, DHGs, and BOD. For a complete description of the method, refer to Appendix 4, Section A4.3.1. Refer to Appendix 4, Section A4.5.1 for details on calculating purge volumes from multilevels. Note the depths for the multilevel wells installed at Alameda were 2.8 m, 3.7 m, and 4.9 m bgs.

Once the samples were collected, vial labels were checked and then added to the appropriate chain-of-custody forms. Samples were shipped on ice back to the University of Waterloo Organic Geochemistry Lab.

Detailed description of laboratory analytical methods are provided in Appendix 3.

A26.2.2 Dedicated Waterra Inertial Lift Pump

Each fully screened well was equipped with a dedicated Waterra Inertial Lift Pump made of 3/8 in. ID Teflon tubing with a hard plastic bottom check valve. Dedicated Waterra pumps were used to collect water for VOC, inorganic, and BOD analyses. Below is a brief description detailing the specific method of collection. Refer to Appendix 4, Section A4.5.1 for details on calculating the required purge volume from fully screened wells. Note that the diameter of the fully screened wells in Alameda is 5 cm and the total depth is approximately 6.1 m bgs.

A26.2.2.1 VOCs

After the required purge volume had been pumped, samples were collected by positioning the end of the 3/8-in. Teflon tube on the bottom of the vial and slowly withdrawing as the vial filled. This method minimized water agitation, thereby preventing unnecessary losses due to volatilization. Sample vials were completely filled until a positive meniscus formed. As the 40-mL hypovials already contained a preservative, the bottles were capped, inverted, then tapped to verify that they did not contain any air bubbles. All samples were labeled and then added to the appropriate chain-of-custody form. Samples were kept cool with blue ice until received by the external laboratory, Intertek Testing Services.

A26.2.2.2 Cations

Once the well was purged, a Waterra Hydropure Filter, Model FHT-700, 0.45 μm pore space, was attached to the end of the 3/8-in. Teflon tube and groundwater was forced through the filter before being collected in the sample bottle. Once the sample bottle was full (a positive meniscus was not necessary), the sample was acidified to a pH <2 by adding 2 drops of concentrated nitric acid. The sample label was verified and was then recorded on the appropriate chain-of-custody form. All samples were shipped on blue ice to the Water Technology Institute.

A26.2.2.3 Anions and Alkalinity

Samples for anions and alkalinity analyses were collected using the same method as described for cation analyses. Anion and alkalinity samples were not preserved or filtered, hence the bottles were filled and then capped (filling to a positive meniscus was not necessary, but often was done). Once the label was verified and the sample added to the chain-of-custody form, samples were shipped on blue ice to the Water Technology Institute.

A26.2.3 Field Analysis Methods

A26.2.3.1 pH and DO

Field parameters pH, dissolved oxygen (DO), and bromide measurements were all completed in the field using a peristaltic pump and a flow-through cell. Although the flow-through cells for Alameda and Borden were not identical, the methodology was similar. Refer to Appendix 4, Section A4.4 for a complete description.

With respect to DO and pH, DO measurements were made first using the flow-through cell. Once the reading stabilized and was recorded, the pH probe was inserted into the flow-through cell. Prior to moving to the next sampling point, the flow-through cell was emptied of its contents. Bromide measurements were only taken during the tracer test. The following probes were used:

pH Orion model 250 meter, with Orion model 9107 triode
DO Orion model 835 meter, with Orion model 083010 probe

A26.2.3.2 Field Analyses for Bromide Tracer

Samples for bromide tracer were collected in 20-mL plastic scintillation vials. Measurements were conducted in the field by placing the ion-selective bromide probe into the sample container. Refer to Appendix 4, Section A4.3.3 for a complete description of sample collection methodology.

A26.2.3.3 Water Level Measurements

Water level measurements were completed in both gates on a regular basis throughout the duration of this field demonstration. An electric water level sounder was used to measure the depth to water from the top of casing in all fully screened wells. Top of casing elevations for each fully screened well were surveyed with respect to mean sea level on April 16, 1997. "Stick-up" measurements for each well were also measured to permit the correction of surveyed data to ground surface.

A26.3 AIR MONITORING

Head space monitoring was done after selected sparging events to try and quantify the concentration of contaminants within the headspace. Grab air samples were taken on three separate occasions using a Summa Canister. Each Summa Canister was analyzed for VC, *c*DCE, and BTEX (refer to Table A26.3).

A26.3.1 Sampling Methodology

The following steps were followed to take a grab sample using a Summa Canister. Refer to Figure A26.1.

1. Using a 9/16-in. wrench, the brass cap above the valve on top of the Summa Canister was removed.
2. The canister was securely attached to the source using 1/4-in. tubing and a Swagelok fitting.
3. The knob was turned 1 1/4 turns counterclockwise to open the valve. As the canister was evacuated and pressure checked at Performance Analytical (PA), a hissing noise as heard as the air flowed in.
4. Once the hissing noise stopped, the valve was closed by turning the knob 1 1/4 turns clockwise and then the brass cap was replaced.
5. The sample was labeled, then the attached chain-of-custody record was completed.

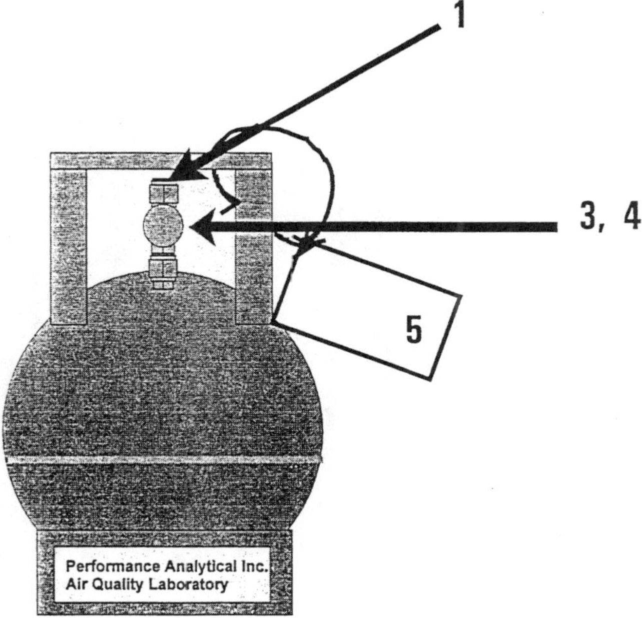

Figure A26.1 Summa canister.

APPENDIX 27

Tracer Test and *In Situ* Permeable Flow Sensors — Alameda

A27.1 TRACER TEST INTRODUCTION

In order to estimate groundwater velocity through the treatment gate, a tracer test was conducted from May 7 through June 16, 1997. In general, during a tracer test, a volume of water spiked with a known constituent is added to groundwater upgradient of a groundwater monitoring network. Plots of tracer concentration vs. time elapsed since injection (breakthrough curves) for each downgradient monitoring point are prepared. Average groundwater velocity is calculated by dividing the distance the tracer has traveled (i.e., the distance from injection well to sample point) by the time it took the tracer to arrive at the sample point (velocity = distance/time). Travel time is estimated from each sample point's breakthrough curve.

In general, a tracer is selected that stands out from site groundwater chemistry, moves with the groundwater, does not sorb to soil particles, does not biodegrade, and can be detected over a wide range of concentrations using field instrumentation or laboratory analyses. Ionic compounds, such as salts, have been used extensively as groundwater tracers. Anions such as bromide (Br^-), the tracer used in this study, are generally not affected by aquifer medium, do not decompose over time, and are relatively inexpensive to use as groundwater tracers.

A27.2 METHODS

A27.2.1 Tracer Injection

On May 7, 1997, a 200-gallon potassium bromide solution (470 mg/L as bromide) was injected into fully screened well R1PB at 2.5 L/min. The solution was pumped from 55-gal drums into well R1PB using polyethylene tubing attached to a peristaltic pump. The tubing within the well was perforated with 1/8-in.-diameter holes to help spread the tracer evenly over the saturated interval of the well. Injection flow rates were monitored using a float-type flowmeter installed inline between the peristaltic pump and well R1PB (refer to Figure A27.1 for the location of R1PB). To spread the tracer laterally, groundwater was simultaneously extracted from fully screened well R1PC at 1.7 L/min using a second peristaltic pump, tubing, and flowmeter. Extracted water was sampled for Br^-. Extraction was discontinued after significant bromide concentrations were detected in extraction well R1PC. It was hoped that by spreading the tracer across the entrance of the gate there would be a good probability that the tracer would intercept several of the downgradient monitoring points as it progressed toward the rear of the gate.

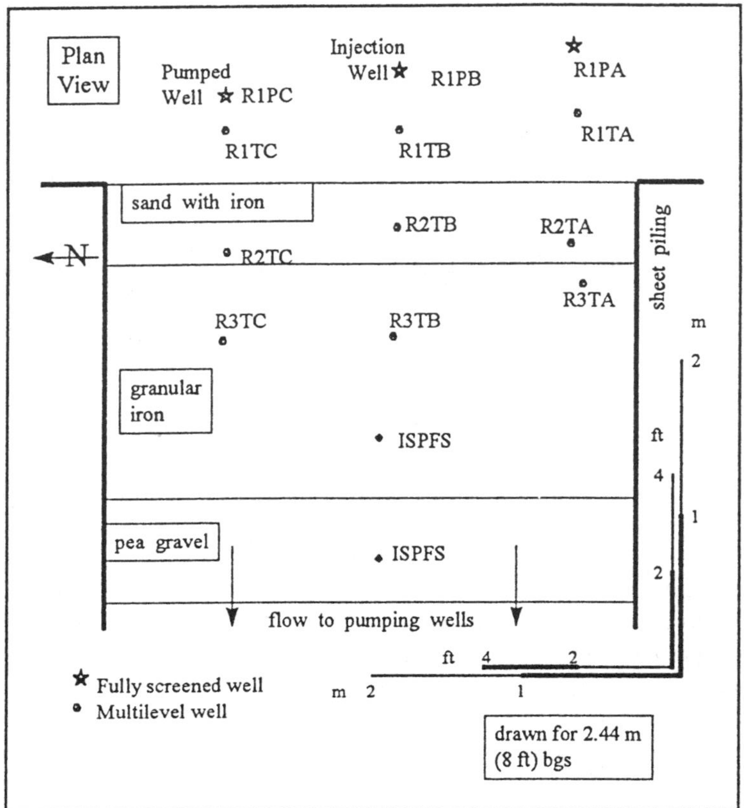

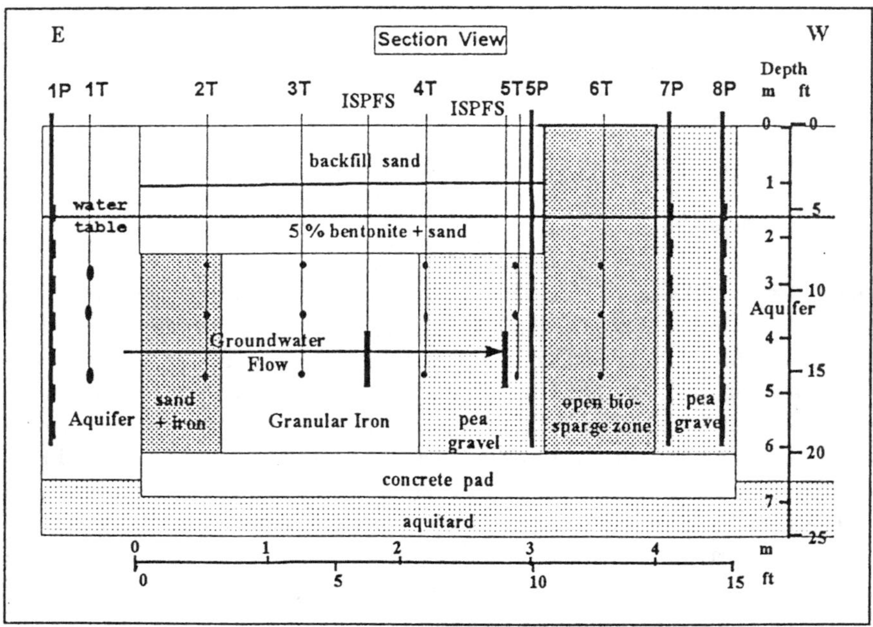

Figure A27.1 Plan view and section view of the remedial gate showing the well locations.

A27.3 MONITORING

To monitor for the presence of tracer (Br^-), a hand-held Orion multimeter, equipped with an ion-selective electrode for bromide (bromide electrode), was used. Refer to Appendix 26 for a detailed description of the sampling procedure. The bromide electrode was placed in a flow-through cell attached to the outflow side of a peristaltic pump using 1/8-in.-diameter flexible tubing. The inflow side of the pump was connected to each sample point.

To sample, groundwater was pumped from each sample point through the flow-through cell, past the bromide electrode, until a steady millivolt (mV) reading was registered on the hand-held meter. Prior to each sampling event, the bromide electrode was "calibrated" by measuring and recording the potential in several calibration solutions of known bromide concentration (i.e., 5, 50, 100, 500, and 1000 mg/L bromide solutions). A calibration plot of bromide concentration vs. potential (mV) was constructed for each sampling event. At the completion of each sampling event, millivolt readings from each sample point were converted to bromide concentration using the calibration plot for that event. An example of a typical calibration curve is presented in Figure A27.2. Table A27.1 summarizes calibration results for each sampling event. Each calibration produced a distinct slope and intercept used to convert electrode potential to bromide concentration. Table A27.2 summarizes measured electrode potential and resulting bromide concentration for each sample point with respect to elapsed time.

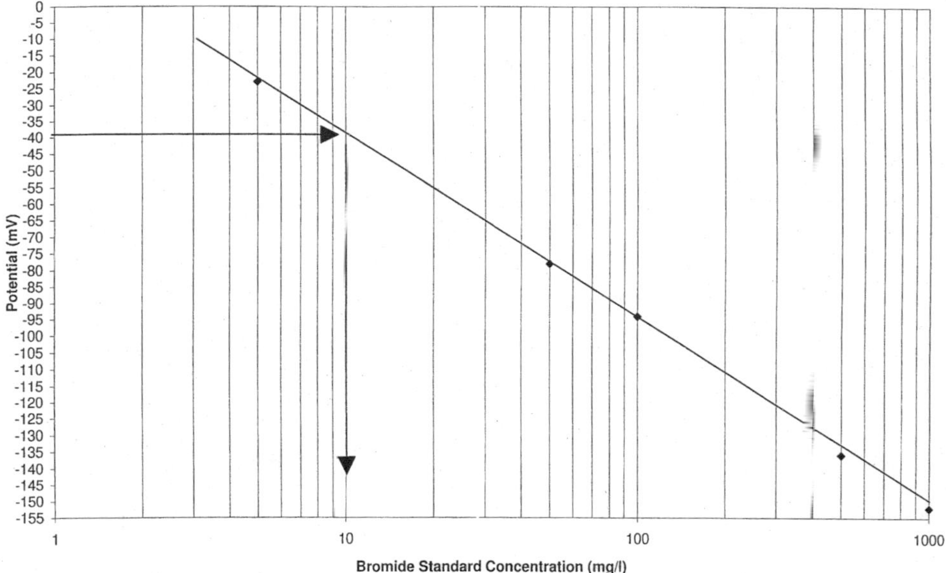

Figure A27.2 Standard concentrations vs. potential.

Prior to injection of the bromide tracer, samples were collected from several monitoring points to measure background Br^- in the groundwater. Calculated levels of Br^- measured with the probe ranged from 20 to 100 mg/L. Note that this may not reflect actual background concentrations of Br^- in groundwater, since dissolved chloride can also cause an electrical response in the probe. The apparent background variation in Br^- measured with the probe prior to injecting the tracer was considered when reviewing the data.

After injection, sampling points closest to the injection point (R1PB) were sampled most frequently for the bromide tracer (every 1 to 3 days). Sampling frequency was adjusted to follow the tracer as it moved through the gate. For example, multilevel sampling points in rows 1 and 2

Table A27.1 Ion Selective Probe Standards

	Standard #	**1**	**3**	**5**	**7**	**9**	**11**	**13**	**15**
	Date	**8 May 97**	**9 May 97**	**10 May 97**	**13 May 97**	**14 May 97**	**15 May 97**	**16 May 97**	**17 May 97**
(Br)[a] mg/L	**Log(Br)[b]**	**(rnV)[c]**	**(mV)**	**(mV)**	**(mV)**	**(mV)**	**(mV)**	**(mV)**	**(mV)**
1000	3.0	–2.9	-17.4	7.0	-12.20	-152.20	-154.2	-152.70	-49.0
500	2.7	16.3	1.8	30.0	5.00	-136.00	-136.6	-135.10	-31.6
100	2.0	67.1	55.3	85.0	50.10	-94.00	-97.0	-90.70	-86.5
50	1.7	89.8	82.8	110.0	66.50	-78.00	-80.0	-73.00	-67.4
5	0.7	177.4	166.0	176.0	119.20	-23.00	-19.50	-8.00	-8.50
	Slope	-78.31	-80.32	-74.02	-57.57	-56.36	-58.34	-62.95	-61.48
	Intercept	227.68	219.89	231.07	161.97	17.21	20.36	35.22	35.56
	Standard #	**17**	**19**	**21**	**23**	**25**	**27**	**28**	**29**
	Date	**19 May 97**	**20 May 97**	**22 May 97**	**24 May 97**	**28 May 97**	**31 May 97**	**3 Jun 97**	**16 Jun 97**
(Br) mg/L	**Log(Br)**	**(rnV)**	**(mV)**	**(mV)**	**(mV)**	**(mV)**	**(mV)**	**(mV)**	**(mV)**
1000	3.0	-147.00	-144.80	-145.30	-145.40	-133.00	-139.50	-142.1	-134
500	2.7	-128.50	-125.50	-125.10	-126.00	-111.00	-119.50	-123	-109
100	2.0	-82.00	-79.30	-82.20	-75.30	-60.00	-67.00	-73.3	-45
50	1.7	-62.00	-59.20	-60.50	-52.00	-32.00	-51.00	-56.2	-25
5	0.7	5.50	8.00	5.10	13.00	21.00	11.00	3.4	30
	Slope	-66.45	-66.47	-65.22	-69.50	-67.94	-65.72	-63.61	-72.24
	Intercept	51.38	54.06	50.10	63.20	74.20	59.51	50.21	89.27

[a] Concentration of bromide calibration sample.
[b] Log_{10} of bromide calibration sample.
[c] Millivolt (mV) reading for bromide calibration sample.

Table A27.2 Sample Point Bromide Concentrations

Elapsed Time (hours)	Standard Number	Sample Point 1a9		1a12		1a16		1b9		1b12		1b16		1c9		1c12		1c16		2b9	
		mV	mg/L	mV	mg/L	mV	mg/L	mV	mg/L	mV	mg/L	mV	mg/L	mV	mg/L	mV	mg/L	mV	mg/L	mV	mg/L
0	1	91	56	97	47	100	43	107	35	87	63	100	43	97	47	119	24	105	37	123	22
20	3	98	33	97	34	93	38	98	33	53	120	93	38	96	35	104	28	95	35	124	16
44	5	95	70	102	56	96	68	95	70	39	400	89	84	92	76	106	50	78	117	124	28
68	7	77	29	83	23	80	26	83	23	20	289	65	49	80	27	84	23	71	39	107	9
124	9	-40	10	-49	15	-51	16	-92	87	-137	545	-90	80	-53	18	-34	8	-76	45	-15	4
160	11	-53	18	-56	20	-48	15	-100	116	-136	477	-94	92	-59	22	-57	21	-70	35	-29	7
185	13	-71	48	-61	33	-58	125	-96	121	-131	443	-102	42	-52	34	-54	7	-78	63	-26	40
209	15	-82	82	-75	63	-63	40	-92	119	-137	643	-111	241	-54	28	-49	24	-72	56	-33	13
231	17	-71	NA	-56	41	-60	47	-87	89	-130	NA	-109	NA	-57	43	-50	34	-71	69	-19	11
283	19	-70	73	-54	43	-43	29	-76	91	-118	392	-112	311	-60	52	-50	36	-53	40	-27	17
304	21	-69	68	-56	42	-42	26	-79	96	-115	334	-111	290	-46	29	-48	32	-60	48	-21	12
355	24	-71	44	-71	44	-62	31	-74	94	-115	367	-120	435	-69	80	-63	65	-73	91	-31	23
404	25	-52	71	-56	83	-43	54	-59	91	-88	240	-106	454	-49	65	-46	59	-58	87	-17	22
494	26	-66	87	-59	67	-45	42	-53	54	-72	106	-107	323	-50	50	-50	49	-60	69	-30	24
569	27	NA	NA	NA	NA	NA	NA	NA	NA	NA	NA	NA	NA	NA	NA	NA	NA	NA	NA	-25	19
641	28	-64	63	-56	46	-49	36	-52	41	-64	62	-101	237	-45	32	-44	30	-58	50	-38	24
952	29	-30	45	-8	22	-23	36	0	17	-10	24	-57	106	-19	32	17	10	10	13	-40	62

continued

Table A27.2 (continued) Sample Point Bromide Concentrations

Elapsed Time (hours)	Standard Number	Sample Point 2b12		2b16		2c12		2c16		3a9		3a12		3a16		3b9		3b12		3b16		3c12		3c16	
		mV	mg/L	mV	mg/L	mV	mg/L	mV	mg/L	mV	mg/L	mV	mg/L	mV	mg/L	mV	mg/L	mV	mg/L	mV	mg/L	mV	mg/L	mV	mg/L
0	1	127	19	128	19	97	47	108	34	125	20	103	39	119	24	137	14	115	27	117	26	134	16	118	25
20	3	120	18	129	14	88	44	99	32	119	18	90	42	115	20	135	12	103	29	112	22	129	14	120	18
44	5	118	34	125	27	89	83	93	73	120	32	83	101	111	42	131	23	100	59	108	46	125	27	119	32
68	7	102	11	105	10	70	39	76	32	107	9	80	27	97	13	114	7	88	19	95	14	96	14	100	12
124	9	-20	5	-40	10	-61	24	-53	18	-25	6	-46	13	-30	7	-12	3	-37	9	-33	8	-24	5	-26	6
160	11	-20	5	-67	31	-69	34	-62	25	-23	6	-46	2	-47	14	-28	7	-40	11	-39	10	-34	9	-33	8
185	13	-27	10	-78	17	-73	53	-59	31	-33	3	-58	8	-79	18	-25	2	-32	12	-39	4	-44	18	-36	14
209	15	-29	11	-90	110	-87	98	-60	35	-29	11	-61	37	-83	86	-18	7	-24	9	-37	15	-64	42	-33	13
231	17	-21	12	-90	134	-90	134	-73	74	-25	14	-57	43	NA	na	-18	11	-97	171	-37	21	-61	49	-28	16
283	19	-21	13	-95	177	-88	136	-60	53	-18	12	-51	38	-25	16	-18	12	-116	363	-53	40	-65	63	-34	21
304	21	-22	13	-94	164	-82	108	-56	42	-18	11	-45	29	-21	12	-19	11	-110	285	-43	26	-57	44	-31	17
355	24	-49	41	-104	250	-76	101	-65	69	-37	27	-61	60	-35	26	-28	21	-91	166	-51	43	-58	55	-37	28
404	25	-71	136	-95	304	-65	113	-42	51	-32	37	-32	36	-23	27	-14	20	-89	248	-51	70	-26	30	-24	-28
494	26	-91	207	-84	164	-67	88	-36	30	-18	16	-31	25	na	na	-24	20	-99	279	-69	97	-15	14	-28	22
569	27	-96	232	-78	122	-79	129	-16	14	-14	13	-11	12	na	na	-53	51	-94	217	-80	133	-9	11	-24	18
641	28	-91	164	-69	75	-67	69	-44	2	-14	10	-52	40	-22	14	-24	15	-76	96	-80	112	-11	9	-25	15
952	29	-48	79	-51	87	0	17	-51	87	NA	NA	NA	NA	NA	NA	-40	62	-41	64	-73	176	0	17	-17	29

were sampled at a high frequency until a significant decreasing trend in tracer concentration was noted. At this time, row 3 was sampled more frequently to define peak tracer concentrations as the bromide tracer progressed past row 3 sampling points. The tracer was monitored as it traveled from row 1 to row 3 of the gate from May 7, 1997 to June 16, 1997.

A27.3.1 Tracer Velocity

Figures A27.3 through A27.5 depict breakthrough plots for rows 1 through 3, respectively. Each plot shows the concentration of bromide tracer (mg/L) versus the elapsed time since injection (hr). Plots for each row are arranged by depth and column within each row. Sample points that were plugged (i.e., row 2, column A multilevel piezometer) could not be sampled and therefore are not included in these plots. In general, higher bromide concentrations were detected directly downgradient of injection well R1PB (the tracer flow path). Sample points along column A and C, furthest from injection well R1PB, fluctuated near background bromide concentrations, as measured prior to adding the bromide tracer. Within column B, higher bromide concentrations were detected in the deepest monitoring points (depths 3.7 and 4.9 m bgs). Peak bromide concentration times varied considerably from point to point along column B. Groundwater velocities calculated for points along column B are presented in Table A27.3. Travel times were estimated from the breakthrough plots assuming that the time it took the tracer to reach a sample point was approximately equivalent to its peak concentration time at each point. Groundwater velocities ranged from 2.0 to 14.2 cm/day along column B.

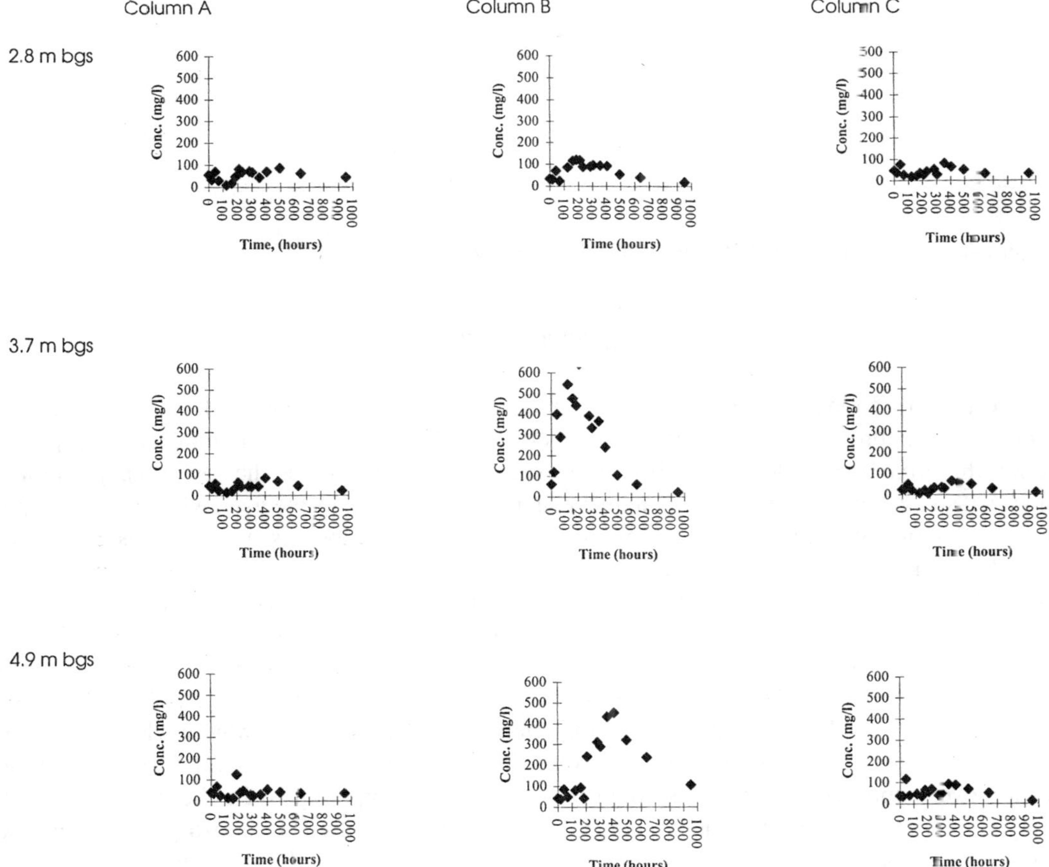

Figure A27.3 Bromide concentration vs. elapsed time in row 1.

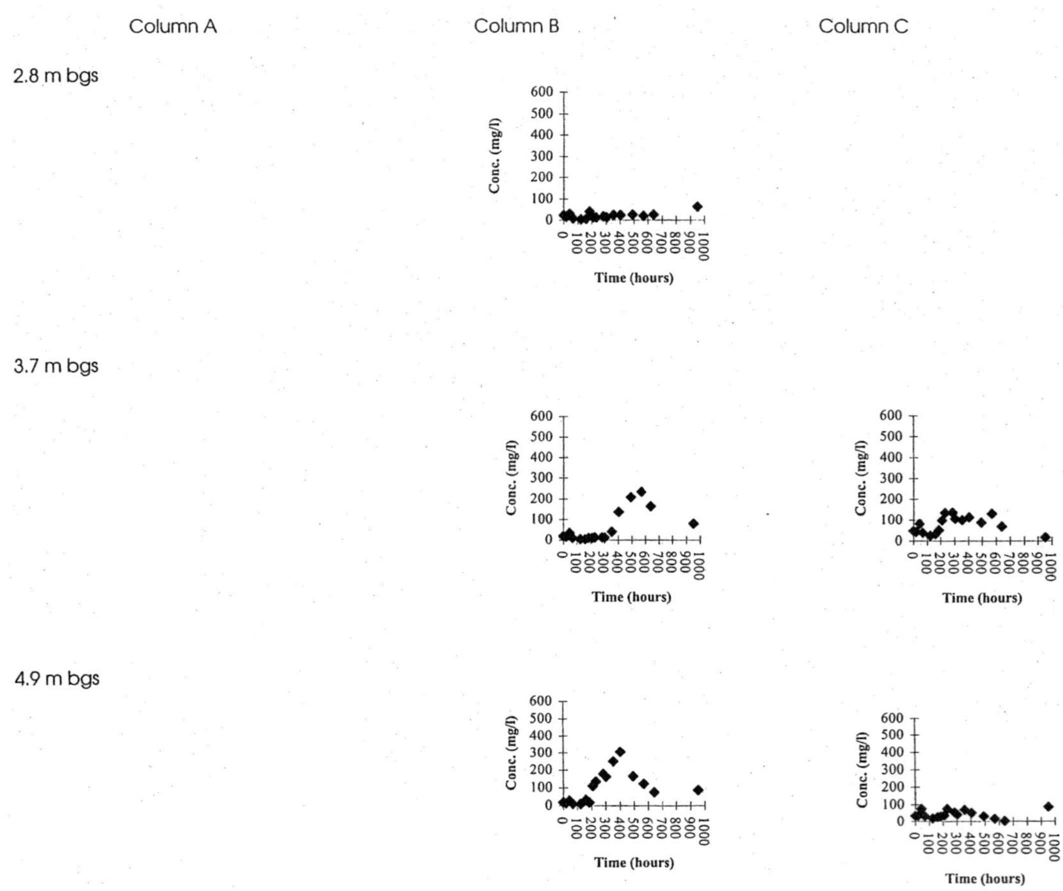

Figure A27.4 Bromide concentration vs. elapsed time in row 2.

A27.4 DISCUSSION

Porosity differences in the various materials (i.e., aquifer sand, sand and iron, and granular iron) may explain the differences in average velocity calculated for the different sampling points along the groundwater flow path. Therefore, velocities calculated for the row 1 sample points may reflect average velocity through native materials just outside of the gate, while row 2 sample point velocities would be an average of native and gate sand velocities. Velocities estimated for row 3 would be an average of velocities including native sand, gate sand, and iron velocity. The bromide tracer velocity also varied with depth. For example, peak concentrations in row 1, column B arrive at approximately 200 and 400 hours at 3.7 and 4.9 m bgs, respectively (Figure A27.3). As the tracer progresses toward row 2 (Figure A27.4), peak concentrations in column B occur at approximately 575 and 400 hours at 3.7 and 4.9 m bgs, respectively. This could be the result of preferential flow within the gate materials or an effect from the way in which the tracer slug was loaded at the gate entrance, or both. The 200 gallons of potassium bromide solution added over approximately 5 hours could have induced a temporary hydraulic gradient, which may have forced the tracer to move at a higher rate than predicted. It is also possible that the high concentrations of potassium bromide injected at the front of the gate resulted in density driven flow of the tracer. This may explain why higher concentrations were detected at the deeper sampling points along the tracer flow path.

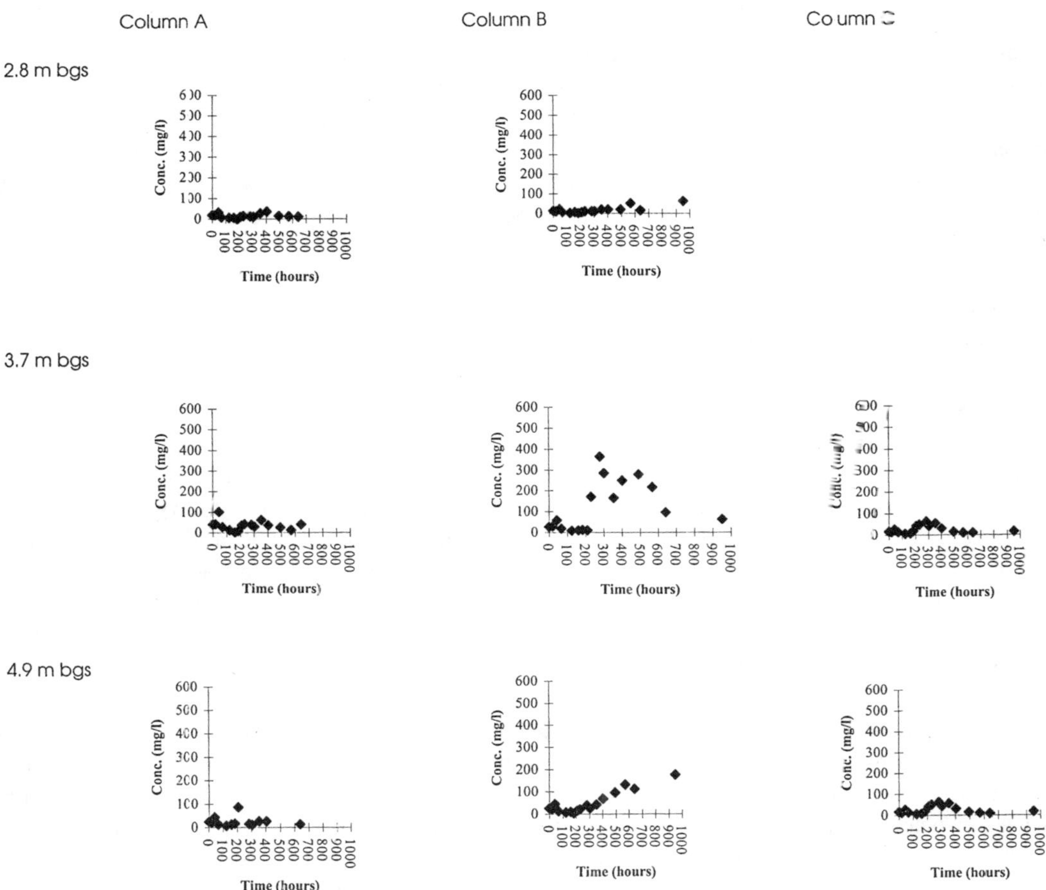

Figure A27.5 Bromide concentration vs. elapsed time in row 3.

Table A27.3 Average Groundwater Velocity for Column B Multilevel Sampling Points

Sample Point	Distance From Injection Well (cm)	Travel Time (days	Average Groundwater Velocity (cm/day)
R1TB-2.8	35.6	3	14.2
R1TB-3.7	35.6	8	4.3
R1TB-4.9	35.6	18	2.0
R2TB-2.8	97.0	—	
R2TB-3.7	97.0	24	4.0
R2TB-4.9	97.0	18	5.5
R3TB-2.8	163.0	—	
R3TB-3.7	163.0	21	7.8
R3TB-4.9	163.0	25	6.5

A27.5 *IN SITU* PERMEABLE FLOW SENSORS — INTRODUCTION

To measure groundwater velocity within the granular iron and first pea gravel section of the remedial gate, *In Situ* Permeable Flow Sensors (ISPFSs), manufactured by Hydrotechnics, Inc., were deployed. The ISPFS is unique in that it can directly measure groundwater flow in three dimensions.

Commonly, groundwater velocity is predicted using estimates of aquifer hydraulic conductivity and hydraulic potential. This approach may lead to errors in calculating groundwater velocity because it relies on a detailed knowledge of site hydrogeology and appropriate placement of wells to estimate the hydraulic potential. Furthermore, it becomes increasingly difficult to estimate groundwater velocity within smaller zones of an aquifer using this method because hydraulic potential (based on water levels) becomes too small to measure. The ISPFS measures groundwater flow within a zone of approximately 1 m^3 in size, and does not rely on knowledge of hydraulic conductivity or hydraulic potential to calculate groundwater velocity. The probe is installed in direct contact with aquifer material at the desired depth for groundwater velocity measurement.

The ISPFS relies on a thermal perturbation technique to measure groundwater velocity. The probe body of the ISPFS consists of a rod of polyurethane foam, 75 cm long by 5 cm in diameter. The polyurethane foam is surrounded by a thin film, flex circuit heater and an array of 30 thermistors (temperature sensors) covered with a waterproof jacket. In general, the heater heats the surface of the probe and the thermistors measure the temperature variation about the probe surface as it is heated. The temperature on the surface of the probe varies as a function of the direction and magnitude of groundwater flow velocity near the probe. For example, temperatures on the upgradient (facing flow) portion of the probe exhibit lower temperatures than the donwgradient (facing away from flow) portion of the probe. This results from heat being advected around the instrument by groundwater flow. Higher flow velocities result in greater temperature differences between the up and downgradient portions of the probe. Rings of thermistors surrounding the length of the probe body allow for measurement of temperature variation about the probe in three dimensions. Groundwater flow velocity and magnitude are resolved from these variations in temperature.

The thermistors and heater are connected to a datalogging system to continuously monitor each ISPFS. Wires connecting the data logger (located at ground surface) to each probe are threaded through a 2-in.-diameter polyvinylchloride (PVC) conduit A separate heater control unit is used to power the probe heater. The datalogger is programmed to collect thermistor readings (millivolts, mV) and heater output (watts) at a desired time interval (i.e., every 15 min). Periodically, the readings collected by the datalogger are retrieved and processed using the FLOW© program by Hydrotechnics, Inc. This software calculates groundwater velocity from millivolt and watt readings collected from the probe thermistors and heater.

A27.6 INSTALLATION

Two ISPFSs were deployed (one within the iron zone and one within the first pea gravel zone of the gate) on December 23, 1996. Prior to installation, the probes were assembled by attaching approximately 4 m of 2-in.-diameter PVC extension conduit to each probe, after threading the thermistor and heater wires through the conduit. Open-ended 3-in.-diameter PVC protective casings were suspended within the iron and pea gravel zones prior to backfilling. Once backfilling was complete, the ISPFSs were then suspended within the protective casings at the desired depth (i.e., in the middle of the iron and pea gravel units, centered at approximately 4.3 m bgs). The protective casings surrounding each probe were then withdrawn, allowing the granular iron and pea gravel to collapse against each of the probes. Probes were installed using the protective casing to reduce the potential of damaging the probe tips during backfilling of the gate.

A27.7 MONITORING

On February 7, 1997 both ISPFSs were attached to a Cambel Scientific CR10X datalogging system and activated by switching on the probe heaters. Thermistor and heater readings from both probes were periodically collected and processed until May 1997. During this time it was discovered that the heater output power for both probes was too high for the iron and pea gravel media. Heat produced at the probe tips was causing convection of water upward past the probes. High vertical flow velocities past the probes "washed out" the distribution of surface temperatures needed to calculated the horizontal flow velocity at each probe. The inversion process, which converts the measured probe surface temperatures into flow velocity, was unable to converge on a velocity. This resulted in erratic velocity estimates. To reduce the effects of convection on the probes, power supplies for the probe heaters were replaced with lower output units to reduce the heat produced at the probe tips. Subsequently, the iron ISPFS was able to converge on a smaller range of flow velocities. However, natural convection could not be overcome completely at this probe, and the resulting vertical velocities for the iron ISPFS are likely erroneously high (Ballard, personal communication).

The pea gravel ISPFS power could not be lowered enough to reduce the extreme convection near this probe. Therefore, groundwater flow velocity magnitude and direction in the pea gravel zone of the gate could not be calculated.

Plots of iron ISPFS velocity magnitude (vertical and horizontal), and flow direction are presented in Figures A27.6, A27.7, and A27.8, respectively. Note that vertical velocity measured at the iron ISPFS is upward. Vertical and horizontal velocities, as shown in Figures A27.6 and A27.7, are extremely erratic, jumping more than 61 cm/day between vertical flow readings and almost 30 cm/day between horizontal flow readings. This effect is shown in Figures A27.6 and A27.7 as two distinct trends over time for both the horizontal and vertical velocities measured with the iron zone ISPFS. This occurred until approximately May 9, 1997, after which the heater power for the probe was lowered enough to significantly reduce convection at the probe. This resulted in less erratic groundwater flow velocity measurements. Vertical groundwater flow velocities in the iron treatment zone of the gate measured after May 9, 1997 ranged from approximately 45.7 to 61 cm/day. Horizontal groundwater flow velocities in the iron during this time ranged from approximately 9 to 13.7 cm/day.

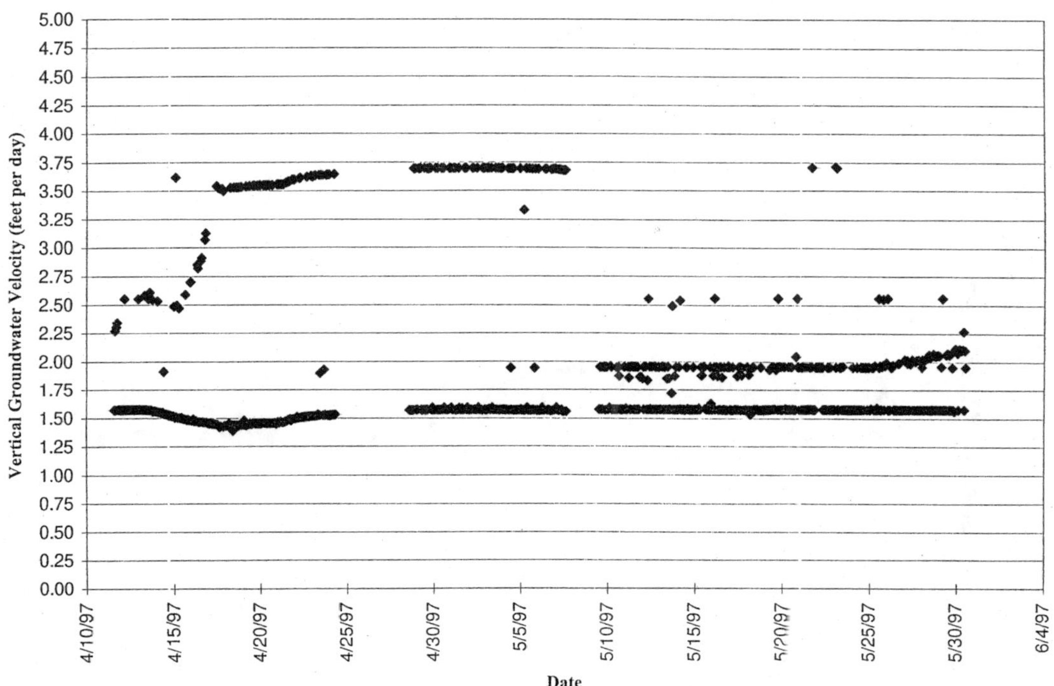

Figure A27.6 Vertical groundwater velocity — iron treatment zone.

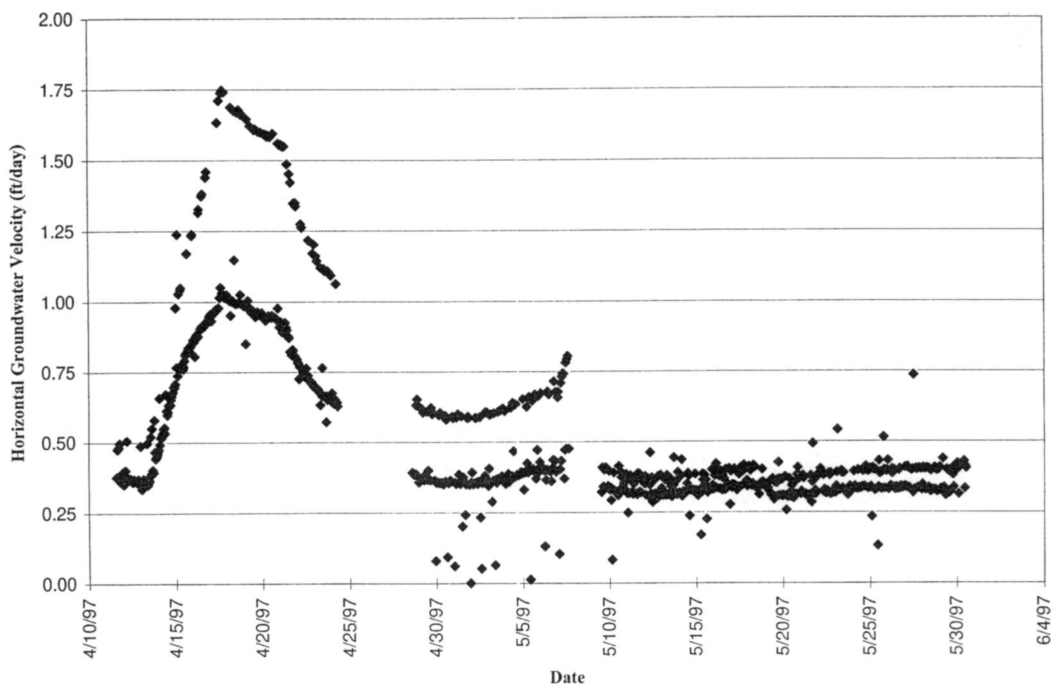

Figure A27.7 Horizontal groundwater velocity — iron treatment zone.

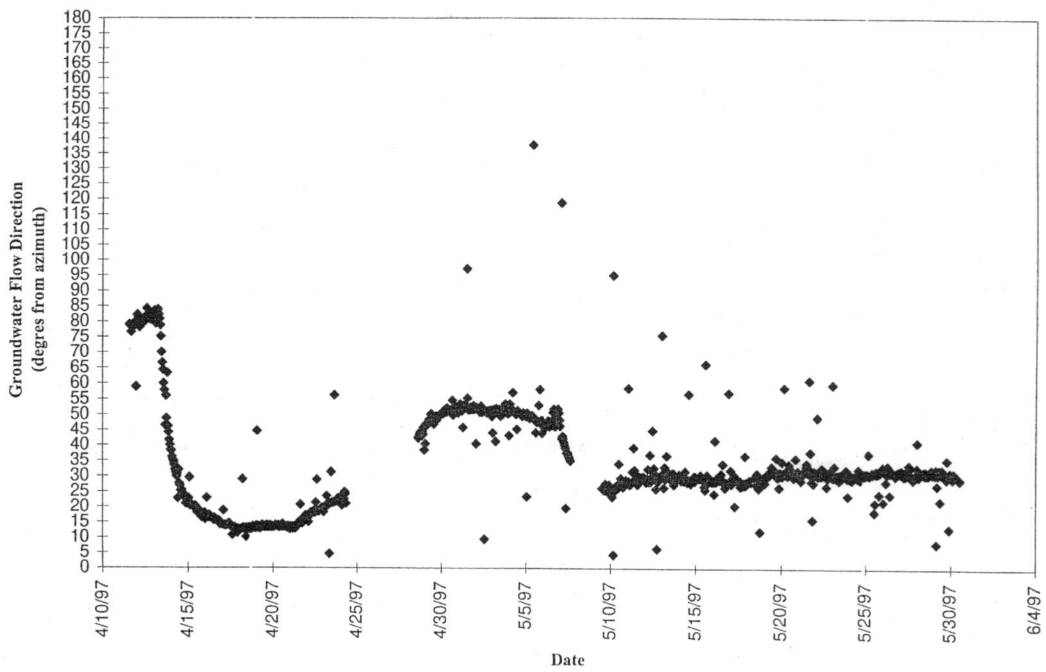

Figure A27.8 Groundwater flow direction — iron treatment zone.

Flow direction measured using the ISPFS is calculated as an angle, clockwise, relative to the reference direction of the probe. The probes were installed with the reference directions facing the back of the gate. Flow direction, as measured by the iron ISPFS shown in Figure A27.8, indicates flow toward the back of the gate at an angle of approximately 25° to the north of the longitudinal gate axis. That is, flow measurements indicate that groundwater (at the point of measurement) is moving at a slight angle, relative to the gate, as it travels through the gate.

A27.8 DISCUSSION

Of the two ISPFSs deployed within the treatment zones of the Funnel-and-Gate, only the probe placed in the iron zone could measure groundwater flow velocity magnitude and direction. The horizontal groundwater velocity measured within the iron zone of the gate after May 9, 1997, using the ISPFS (9 to 13.7 cm/day), is similar to that calculated using gate dimensions, estimated iron porosity, and pumping rates during the same time period (6 cm/day).

Vertical groundwater flow velocity, as measured by the iron zone ISPFS (45.7 to 61.0 cm/day upward), was higher than that anticipated for groundwater moving through the iron portion of the gate. Although vertical flow within this zone cannot be ruled out, it is likely that the rates measured by the iron zone ISPFS are erroneously high as a result of convection produced by the probe heater.

Groundwater flow direction within the iron zone of the gate was shown to be moving at a slight northward angle relative to the gate (approximately 25° north, relative to the longitudinal axis of the gate). Although it was anticipated that groundwater would flow straight through the gate parallel to the longitudinal gate axis, it is possible that variations in materials near the iron zone ISPFS could have had an effect. It is also possible the alignment of the probes was not exactly in the direction of groundwater flow. The probe has a fixed zero azimuth and installation must place this mark in the exact direction of groundwater flow to obtain a groundwater flow with zero degrees of variation. Subsequent checking of flow rates from each of the two extraction wells indicates they were operating at very similar rates, suggesting that perhaps the probes were not aligned exactly in the direction of groundwater flow.

The ISPFS technology is unique in that it can measure, *in situ*, three-dimensional groundwater flow velocity within unconsolidated media. However, care should be taken when applying this technology within various saturated materials. The thermal perturbation technique employed using this technology may not be suitable for high permeability materials, such as the iron and pea gravels used within the Funnel-and-Gate.

APPENDIX 28

Oxygen and Carbon Dioxide Sparging Protocol for the Biosparge Zone

Effluent from the granular iron zone was anaerobic and had a high pH (8–11). To correct for these conditions, both oxygen (grade 2.8) and carbon dioxide (grade 2.8) were sparged into the biosparge zone to increase the DO levels and decrease the pH. This appendix summarizes oxygen and carbon dioxide additions to the biosparge zone for the duration of the field demonstration. Figure A28.1 illustrates a cross section and plan view of the biosparge zone.

During the first operational phase (OP1, flux of 1.27 m^3/day), oxygen and carbon dioxide were occasionally sparged to determine whether conditions suitable for aerobic biodegradation could be achieved (target conditions were DO = ~20 mg/L, pH ~7). The frequency and length of each sparge event were originally going to be based on the flux of BOD entering the biosparge zone. The early BOD data showed a very low oxygen demand in the biosparge zone; hence, a calculated amount of oxygen that should be sparged was not determined. Instead, a variety of sparge patterns were explored. At this time, the objective was to determine the feasibility of achieving aerobic conditions in the biosparge zone, not yet to maintain them. This approach is apparent the first three pages of Table A28.1 (data from January 30 to April 28, 1997); sparge intervals for both the O_2 and CO_2 were not consistent. It was concluded that the required conditions for aerobic biodegradation could be attained and this was clearly demonstrated (see the second page of Table A28.1). Table A28.1 summarizes the random sparging events. (Refer to Figure A28.2 for an explanation of this table).

Later, after June 1997, it was decided that consistent levels of DO and pH were required to maintain conditions suitable for aerobic biodegradation. To do this, automated timers were installed on the oxygen and carbon dioxide cylinders. A sparging pattern consisted of setting the timers to sparge x number of times in 24 hours, for y amount of minutes at z delivery pressure. Table A28.2 summarizes the details of these sparge events. (Refer to Figure A28.3 for an explanation of this table).

It was determined that allowing CO_2 to sparge for 5 minutes, once a day at 10 psi, was sufficient to control the pH for several weeks at a time (Table A28.1).

Determining a suitable sparging frequency for oxygen was not as straightforward. To overcome the hydrostatic pressure, an entry pressure of 5 psi was required. The low delivery pressure had the benefit of limiting the amount of oxygen introduced in one event, thereby decreasing the chance of volatilizing contaminants from the water column. However, it was found that over time as the biosparge zone was routinely sparged with oxygen, the pressure required to deliver sufficient oxygen to maintain the target DO level increased. It was thought the increase in delivery pressure was the result of biofouling of the ports (1–2 mm in size) on the sparge hose.

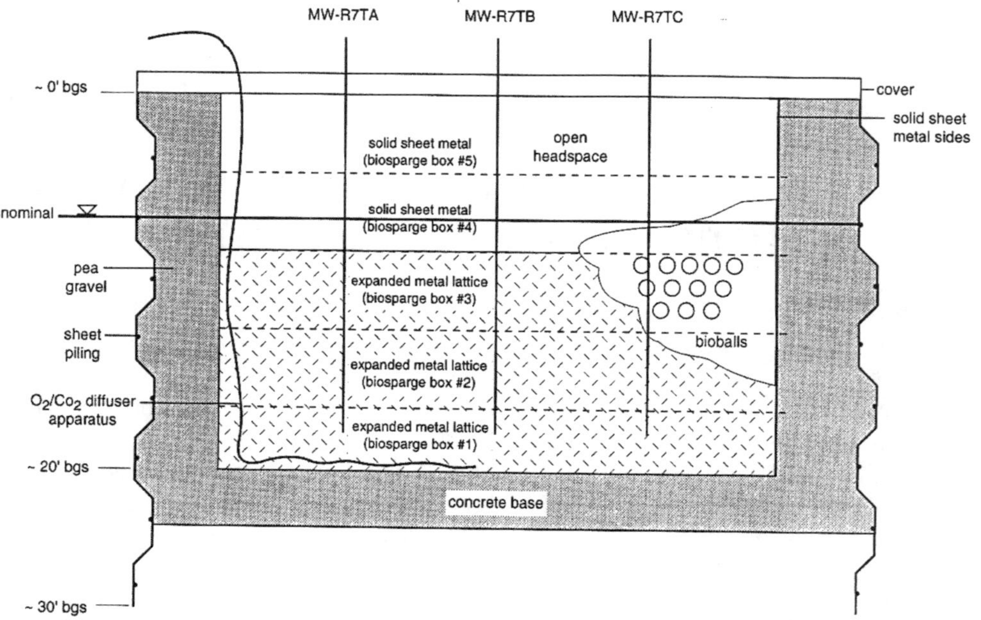

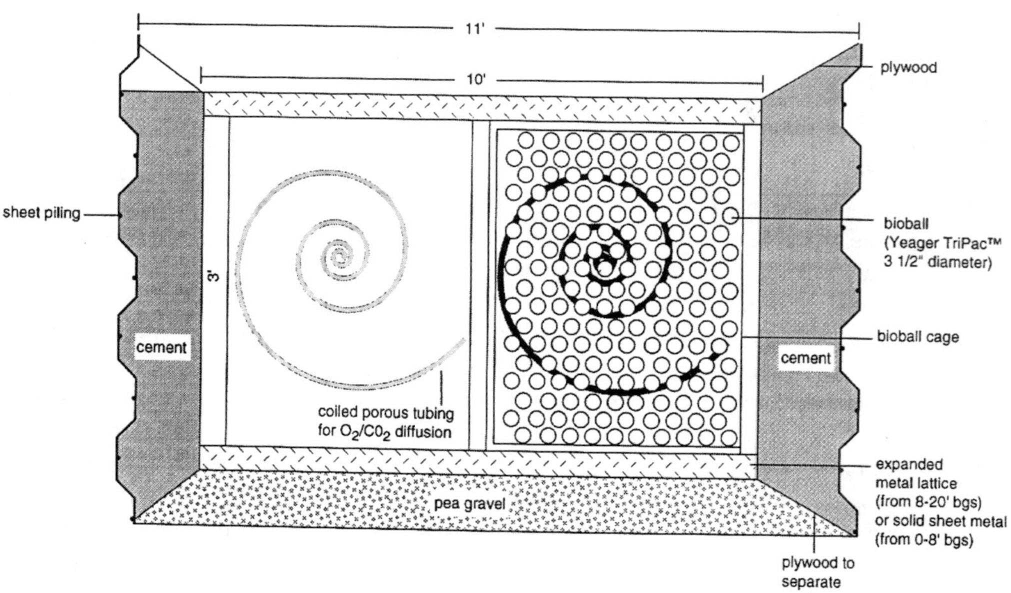

Figure A28.1 Cross section and plan view of the biosparge zone.

Table A28.1 Sparging Protocols for OP-1, Alameda, NAS

Date	Activity	Time	Well	Reading (mg/L)
30-Jan-97	Initial readings: DO	NR	R6TB-2.8	0.2
			R6TB-3.7	0.2
			R6TB-4.9	0.3
	1-sparge: 2, 5 second bursts @30 psi - OXYGEN	NR	R6TA-2.8	1.2
	spaced 1 minute apart		R6TA-3.7	0.8
	measured 5 minutes after last sparge		R6TA-4.9	1.1
			R6TB-2.8	0.9
			R6TB-3.7	4.8
			R6TB-4.9	1.3
	2-sparge: 10, 5 second bursts @ 30 psi - OXYGEN	NR	R6TA-2.8	3.9
	spaced 1 minute apart		R6TB-3.7	5.4
	measured 5 minutes after last sparge		R6TC-2.8	5.4
			R6TC-3.7	5.4
			R6TC-4.9	6.1
	3-sparge: 10, 10 second bursts @ 30 psi - OXYGEN	NR	R6TA-2.8	14
	spaced 1 minute apart		R6TA-4.9	13
	measured 5 minutes after last sparge		R6TB-2.8	14
			R6TB-4.9	13
			R6TC-3.7	13
	4-sparge: 10, 5 second bursts @ 40 psi - OXYGEN	NR	R6TA-2.8	17
	spaced 1 minute apart		R6TB-3.7	17
	measured 15 minutes after last sparge		R6TC-2.8	21
			R6TC-3.7	18
			R6TC-4.9	17
4-Feb-97	Initial readings: DO	NR	R6TB-2.8	12
			R6TB-3.7	12
			R6TB-4.9	14
	1-sparge: 5, 15 second bursts @ 30 psi - OXYGEN	NR	R6TB-2.8	12
	spaced 1 minute apart		R6TB-3.7	11
	measured 1 hour after last sparge		R6TB-4.9	11
	2-sparge: 5, 15 second bursts @ 30 psi - OXYGEN	NR	R6TB-2.8	15
	spaced 1 minute apart		R6TB-3.7	13
	measured 1 hour after last sparge		R6TB-4.9	14
	3-sparge: 10, 15 second bursts @ 30 psi - OXYGEN	NR	R6TB-2.8	24
	spaced 1 minute apart		R6TB-3.7	21
	measured 1 hour after last sparge		R6TB-4.9	21
14-Apr-97	Initial readings: DO	NR	R6TA-2.8	0.1
			R6TA-3.7	0.0
			R6TA-4.9	0.2
			R6TB-2.8	0.5
			R6TB-3.7	0.1
			R6TB-4.9	0.0
			R6TC-2.8	0.0
			R6TC-3.7	0.2
			R6TC-4.9	0.0
	1-sparge: 1, 90 second burst @ 10 psi - OXYGEN	10:15 AM	R6TA-all	0.0
	measured 1 hour after sparge		R6TB-all	0.0
			R6TC-all	0.0
14-Apr-97	2-sparge: 1, 10 minutes @ 10psi - OXYGEN	11:30 AM	R6TA-2.8	0.8
	measured 1 hour after sparge		R6TA-4.9	0.6
			R6TB-2.8	6.5
			R6TB-4.9	1.9
			R6TC-3.7	0.3

continued

Table A28.1 (continued) Sparging Protocols for OP-1, Alameda, NAS

Date	Activity	Time	Well	Reading (mg/L)
	3-sparge: 1, 20 minutes @ 10 psi - OXYGEN	1:00 PM	R6TA-2.8	12
	measured 1 hour after sparge		R6TA-4.9	15
			R6TB-3.7	15
			R6TB-4.9	16
			R6TC-2.8	12
			R6TC-4.9	14
	4-sparge: 1, 20 minutes @ 10 psi - OXYGEN	3:15 PM	R6TA-2.8	5.2
	measured 19 hours after sparge (on April 15, 1997)		R6TA-4.9	8.4
			R6TB-3.7	5.6
			R6TB-4.9	8.1
			R6TC-3.7	4.7
15-Apr-97	Initial readings: - DO	10:30 AM		
	The 4th sparge completed on April 14, 1997 was measured			
	April 15 which were considered the initial readings.			
	1-sparge: 1, 20 minutes @ 15 psi - OXYGEN	10:50 AM	R6TA-3.7	16
	measured 50 minutes after sparge		R6TA-4.9	18
			R6TB-2.8	15
			R6TB-4.9	15
			R6TC-3.7	15
			R6TC-4.9	15
	2-sparge: 1, 20 minutes @ 15 psi - OXYGEN	12:40 PM	R6TA-3.7	22
	measured 1 hour after sparge		R6TA-4.9	23
			R6TB-2.8	24
			R6TB-4.9	22
			R6TC-3.7	23
	3-sparge: 1, 20 minutes @ 12 psi - OXYGEN	2:00 PM		
	measured 19 hours later on April 16, 1997			
16-Apr-97	Initial readings: DO	9:30 AM	R6TA-2.8	7.4
	19 hours after last sparge on April 16, 1997		R6TA-4.9	7.5
			R6TB-3.7	6.1
			R6TC-2.8	6.1
			R6TC-4.9	4.5
21-Apr-97	1-sparge: 1, 30 minutes @ 10 psi - OXYGEN	11:30 AM	R6TA-2.8	4.4
	measure 1 hour after sparge		R6TA-4.9	7.8
			R6TB-3.7	4.3
			R6TC-2.8	8.2
			R6TC-4.9	4.4
	2-sparge: 1, 30 minutes 2 15 psi - OXYGEN	NR	R6TA-2.8	9.6
	measure 5 minutes after sparge		R6TA-4.9	9.8
			R6TB-2.8	10
			R6TB-3.7	10
			R6TB-4.9	10
			R6TC-3.7	9.5
			R6TC-4.9	10
21-Apr-97	3-sparge: 1, 30 minutes @ 15 psi - OXYGEN	NR	R6TA-3.7	15.95
	measure 5 minutes after sparge		R6TB-2.8	16
			R6TB-4.9	14
			R6TC-3.7	15
	4-sparge: 1, 25 minutes @ 10 psi - OXYGEN		R6TA-3.7	~20
	measure 5 minutes after sparge		R6TB-2.8	~20
			R6TB-4.9	~20
			R6TC-3.7	~20
	5-sparge: 1, 20 minutes @ 8 psi - OXYGEN	5:00 PM		
	Results were not recorded in field book.			

Table A28.1 (continued) Sparging Protocols for OP-1, Alameda, NAS

Date	Activity	Time	Well	Reading (mg/L)
28-Apr-97	Initial readings:	9:30 AM	R6TA-2.8	9.92
	pH READINGS		R6TA-3.7	9.92
			R6TA-4.9	9.45
			R6TB-2.8	10.23
			R6TB-3.7	10.21
			R6TB-4.9	10.91
			R6TC-2.8	9.96
			R6TC-3.7	10.2
			R6TC-4.9	10.86
	1-sparge: 1, 15 minutes @ 20psi - CARBON DIOXIDE	10:00 AM	R6TA-2.8	9.85
	hose froze		R6TA-3.7	9.67
	measured 30 minutes after sparge		R6TA-4.9	6.35
			R6TB-2.8	9.85
			R6TB-3.7	8.6
			R6TB-4.9	6.28
			R6TC-2.8	9.94
			R6TC-3.7	10.09
			R6TC-4.9	6.21
	2-sparge: 1, 10 minutes @ 15 psi - CARBON DIOXIDE	11:00 AM	R6TA-2.8	9.73
	measured 1.5 hours after sparge		R6TA-3.7	9.4
			R6TA-4.9	5.85
			R6TB-2.8	9.66
			R6TB-3.7	8.11
			R6TB-4.9	5.92
			R6TC-2.8	9.83
			R6TC-3.7	10.12
			R6TC-4.9	6.15
	3-sparge: 1, 15 minutes @ 15 psi - CARBON DIOXIDE	1:00 PM	R6TA-2.8	6.12
	measure 1 hour after sparge		R6TA-3.7	6.13
	**these readings were taken while oxygen was being sparged in, therefore the bubbling oxygen might have stirred up the water column, hence more homogeneous readings of pH.		R6TA-4.9	6.17
			R6TB-2.8	6.06
			R6TB-3.7	6.16
			R6TB-4.9	6.01
			R6TC-2.8	6.04
			R6TC-3.7	6
			R6TC-4.9	6
29-Apr-97	Initial readings: DO	11:00 AM	R6TA-2.8	17
			R6TA-3.7	14
			R6TA-4.9	34
			R6TB-2.8	12
29-Apr-97	Initial readings: DO	11:00 AM	R6TB-3.7	31
			R6TB-4.9	30
			R6TC-2.8	8
			R6TC-3.7	25
			R6TC-4.9	14
	Initial readings: pH	11:00 AM	R6TA-2.8	6.71
			R6TA-3.7	6.47
			R6TA-4.9	6.36
			R6TB-2.8	6.63
			R6TB-3.7	6.44
			R6TB-4.9	6.49

continued

Table A28.1 (continued) Sparging Protocols for OP-1, Alameda, NAS

Date	Activity	Time	Well	Reading (mg/L)
			R6TC-2.8	6.9
			R6TC-3.7	6.49
			R6TC-4.9	6.47
	1-sparge: 1, 10 minutes @ 10 psi - OXYGEN measure 1.25 hours after sparge	12:00 PM	R6TA-2.8	24
			R6TA-3.7	29
			R6TA-4.9	30
			R6TB-2.8	18
			R6TB-3.7	17
			R6TB-4.9	28
			R6TC-2.8	15
			R6TC-3.7	29
			R6TC-4.9	30
	pH READINGS ** no recent sparge, measuring residual effects of last sparge.	1:15 PM	R6TA-2.8	6.56
			R6TA-3.7	6.58
			R6TA-4.9	6.47
			R6TB-2.8	6.55
			R6TB-3.7	6.48
			R6TB-4.9	6.52
			R6TC-2.8	6.34
			R6TC-3.7	6.54
			R6TC-4.9	6.49
	2-sparge: 1, 15 minutes @ 15 psi - OXYGEN measure 19.5 hours later on April 30, 1997	2:30 PM		
30-Apr-97	Initial readings: DO 19.5 hours since last sparge	10:00 AM	R6TA-2.8	17
			R6TA-3.7	28
			R6TA-4.9	17
			R6TB-2.8	6.1
			R6TB-3.7	20
			R6TC-2.8	6.7
			R6TC-4.9	12
	pH READINGS ** no recent sparge, measuring residual effects of last sparge.	10:00 AM	R6TA-2.8	6.96
			R6TA-3.7	6.65
			R6TA-4.9	6.57
			R6TB-2.8	6.91
			R6TB-3.7	6.71
			R6TC-2.8	7.09
			R6TC-4.9	6.61
30-Apr-97	1-sparge: 1, 20 minutes @ 15 psi - OXYGEN measure 1 hour after sparge	12:20 PM	R6TA-2.8	12
			R6TA-3.7	15
			R6TA-4.9	14
			R6TB-2.8	9
			R6TB-4.9	30
			R6TC-2.8	16
			R6TC-4.9	11
	pH READINGS ** no recent sparge, measuring residual effects of last sparge.	12:20 PM	R6TA-2.8	6.7
			R6TA-3.7	6.71
			R6TA-4.9	6.72
			R6TB-2.8	6.71
			R6TB-4.9	6.72
			R6TC-2.8	6.7
			R6TC-4.9	6.72
	2-sparge: 1, 20 minutes @ 15 psi - OXYGEN measure 19 hours after sparge, on May 1, 1997	2:30 PM		
1-May-97	Initial readings: DO	10:00 AM	R6TA-2.8	6.8
			R6TA-4.9	22
			R6TB-3.7	18

Table A28.1 (continued) Sparging Protocols for OP-1, Alameda, NAS

Date	Activity	Time	Well	Reading (mg/L)
			R6TC-2.8	4.75
			R6TC-4.9	15
	pH READINGS	10:00 AM	R6TA-2.8	7.11
	** no recent sparge, measuring residual effects of		R6TA-4.9	6.82
	last sparge.		R6TB-3.7	6.71
			R6TC-2.8	7.1
			R6TC-4.9	6.81
	1-sparge: 1, 25 minutes @ 15 psi - OXYGEN	11:00 AM		
	measure on May 2, 1997.			
2-May-97	Initial Readings: DO	NR	R6TA-2.8	0.3
			R6TA-3.7	6.7
			R6TB-2.8	0.5
			R6TC-3.7	6.9
			R6TC-4.9	6.0
	pH READINGS	NR	R6TA-2.8	7.17
	** no recent sparge, measuring residual effects of		R6TA-3.7	6.61
	last sparge.		R6TB-2.8	7.21
			R6TC-3.7	6.82
			R6TC-4.9	6.76
	1-sparge: 1, 25 minutes @ 20 psi			
	measurements not recorded			
5-May-97	1-sparge: 1, 25 minutes @ 15 psi - OXYGEN	7:00 AM	R6TA-2.8	0.4
	measure 4 hours after sparge		R6TA-4.9	1.0
			R6TB-3.7	0.9
			R6TC-2.8	2.0
			R6TC-4.9	1.0
	pH READINGS	11:15 AM	R6TA-2.8	7.54
	** no recent sparge, measuring residual effects of		R6TA-4.9	7.13
	last sparge.		R6TB-3.7	7.15
			R6TC-2.8	7.7
			R6TC-4.9	7.61
5-May-97	1-sparge: 1 hour @ 18 psi - OXYGEN	NR		
	measurements of DO were not taken			
6-May-97	1-sparge: 1 hour @ 20 psi - OXYGEN	11:00 AM	R6TA-2.8	2.0
	measure 2 hours later		R6TA-3.7	0.4
			R6TB-3.7	0.8
			R6TC-2.8	2.0
			R6TC-4.9	6.1
	pH READINGS	1:00 PM	R6TA-2.8	7.26
	** no recent sparge, measuring residual effects of		R6TA-3.7	7.22
	last sparge.		R6TB-3.7	7.29
			R6TC-2.8	7.38
			R6TC-4.9	7.15
	2-sparge: 1 hour @ 15 psi - OXYGEN	1:30 PM	R6TA-2.8	0.4
	measure 1 hour later		R6TA-3.7	0.4
			R6TB-3.7	0.9
			R6TC-2.8	2.2
			R6TC-4.9	6.2
	pH READINGS	2:35 PM	R6TA-2.8	7.29
	** no recent sparge, measuring residual effects of		R6TA-3.7	7.21
	last sparge.		R6TB-3.7	7.25
			R6TC-2.8	7.36
			R6TC-4.9	7.1
	3-sparge: 20 minutes @ 20 psi - OXYGEN			
	measurements not recorded			

continued

Table A28.1 (continued) Sparging Protocols for OP-1, Alameda, NAS

Date	Activity	Time	Well	Reading (mg/L)
8-May-97	Initial readings: DO	12:00 PM	R6TA-2.8	0.3
			R6TA-4.9	7.1
			R6TB-3.7	5.9
			R6TB-4.9	8.3
			R6TC-2.8	0.2
			R6TC-3.7	3.0
			R6TC-4.9	6.4
	pH READINGS	12:00 PM	R6TA-2.8	7.58
	** no recent sparge, measuring residual effects of		R6TA-4.9	6.97
	last sparge.		R6TB-3.7	7.12
			R6TB-4.9	6.98
			R6TC-2.8	7.46
			R6TC-3.7	7.08
			R6TC-4.9	6.81
	1-sparge: 15 minutes @ 15 psi - OXYGEN	12:30 PM	R6TA-2.8	0.5
	measure 15 minutes after sparge		R6TA-4.9	4.1
			R6TB-2.8	7.0
			R6TB-4.9	7.0
			R6TC-2.8	0.9
			R6TC-4.9	6.0
	pH READINGS	12:45 PM	R6TA-2.8	7.26
	** no recent sparge, measuring residual effects of		R6TA-4.9	7.11
	last sparge.		R6TB-2.8	7.16
			R6TB-4.9	7.1
			R6TC-2.8	7.3
			R6TC-4.9	7.07
8-May-97	2-sparge: 60 minutes @ 20 psi : OXYGEN	2:30 PM	R6TA-2.8	13
	measure 1 hour after sparge		R6TA-4.9	10
	DO readings		R6TB-2.8	9
			R6TB-4.9	12
			R6TC-2.8	14
			R6TC-4.9	19
9-May-97	Initial readings: DO	12:00 PM	R6TA-2.8	8.5
			R6TA-4.9	13
			R6TB-3.7	10
			R6TC-2.8	9.1
			R6TC-4.9	7.2
	1- sparge: 30 minutes @ 20 psi - OXYGEN	12:40 PM	R6TA-2.8	16
	measure 1 hour 10 minutes after sparge		R6TA-4.9	18
	DO readings		R6TB-3.7	13
			R6TC-2.8	10
			R6TC-4.9	23
	2- sparge: 30 minutes @ 18 psi - OXYGEN	1:50 PM	R6TA-2.8	15
	measure 40 minutes after sparge		R6TA-4.9	23
	DO readings		R6TB-3.7	25
			R6TC-2.8	14
			R6TC-4.9	25
	3- sparge: 30 minutes @ 20 psi - OXYGEN			
	measurements were not recorded in field book.			
12-May-97	1-sparge: 30 minutes @ 20 psi - OXYGEN			
	measurements were not recorded in field book.			
15-May-97	1-sparge: 30 minutes @ 20 psi			
	measurements were not recorded in field book.			
19-May-97	1-sparge: 30 minutes @ 20 psi - OXYGEN	9:30 AM	R6TA-2.8	0.2
	measured 45 minutes after sparge		R6TA-3.7	0.6
	DO readings		R6TA-4.9	5.0

Table A28.1 (continued) Sparging Protocols for OP-1, Alameda, NAS

Date	Activity	Time	Well	Reading (mg/L)
			R6TB-2.8	3.8
			R6TB-4.9	1.7
			R6TC-2.8	7.1
			R6TC-3.7	7.8
			R6TC-4.9	7.8
	pH READINGS	10:15 AM	R6TA-2.8	8.27
	** no recent sparge, measuring residual effects of		R6TA-4.9	7.76
	last sparge.		R6TB-2.8	8.3
			R6TB-4.9	8.3
			R6TC-2.8	8.6
			R6TC-3.7	8.45
			R6TC-4.9	8.3
	2a- sparge: 30 minutes @ 20 psi - OXYGEN	11:00 AM	R6TA-2.8	1.1
	2b- sparge: 10 minutes @ 15 psi - CARBON DIOXIDE		R6TA-3.7	1.5
	measure DO and PH 10 minutes after carbon dioxide sparge		R6TA-4.9	3.3
	DO readings		R6TB-3.7	8.7
			R6TC-2.8	4.8
			R6TC-4.9	5.5
	pH readings		R6TA-2.8	8.1
			R6TA-3.7	7.9
19-May-97	pH readings		R6TA-4.9	7.72
			R6TB-3.7	7.78
			R6TC-2.8	7.28
			R6TC-4.9	7.59
	3-sparge: 60 minutes @ 18 psi - OXYGEN	12:20 PM	R6TA-2.8	15
	measure 55 minutes after sparge		R6TA-4.9	9
	DO readings		R6TB-3.7	11
			R6TC-2.8	11
			R6TC-4.9	9.4
	4- sparge: 40 minutes @ 20 psi	3:05 PM	R6TA-2.8	19
	measure 30 minutes after sparge		R6TA-4.9	16
	DO readings		R6TB-3.7	16
			R6TC-2.8	23
			R6TC-4.9	18
20-May-97	Initial readings: DO	9:30 AM	R6TA-2.8	2.3
			R6TA-4.9	15
			R6TB-3.7	16
			R6TC-2.8	1.4
			R6TC-4.9	11
	Initial readings: pH	9:30 AM	R6TA-2.8	8.52
	** no recent sparge, measuring residual effects of		R6TA-4.9	8.31
	last sparge.		R6TB-3.7	8.19
			R6TC-2.8	8.3
			R6TC-4.9	8.17
	1a- sparge: 40 minutes @ 20 psi - OXYGEN	10:00 AM	R6TA-2.8	10
	1b- sparge: 15 minutes @ 20 psi - CARBON DIOXIDE	10:50 AM	R6TA-4.9	12
	measure at 1:00 PM		R6TB-3.7	8
	DO readings		R6TC-2.8	9.8
			R6TC-4.9	8.3
	pH readings		R6TA-2.8	8.1
			R6TA-4.9	8.1
			R6TB-3.7	8.09
			R6TC-2.8	7.95
			R6TC-4.9	7.99

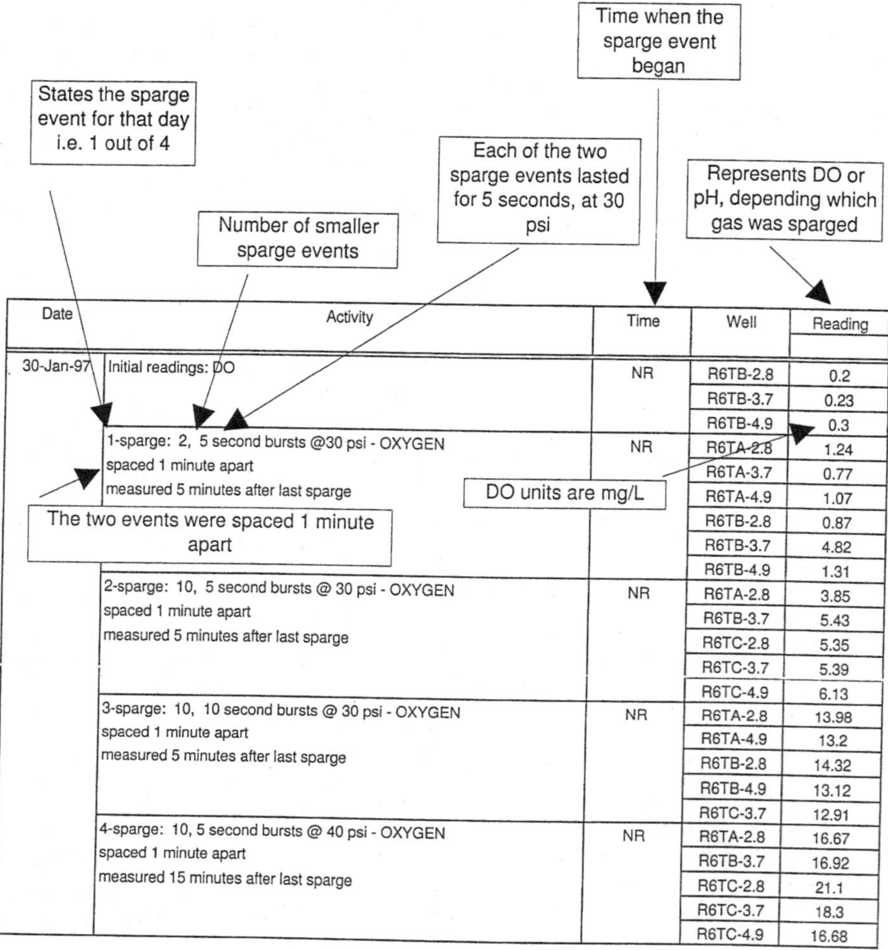

Figure A28.2 Text explanations for Table A28.1.

Due to this biofouling, the optimal sparging timetable was not finalized. Instead, the sparging frequency was continuously modified to take into account the biofouling conditions that prevailed. The sparging frequency was typically modified by one of the following methods:

1. Increasing the delivery pressure over time to maintain the required DO.
2. Cleaning the ports in the sparge hose by sparging the system for 1–2 min at an elevated pressure, e.g., 70 psi.
3. Delivering oxygen at an excess pressure, e.g., 20 psi, to ensure that oxygen would always be delivered, even though volatilization of some of the contaminants would occur.

The last option was the preferred method when the system was not monitored on a daily basis (after August 1997). On November 17, 1997, a flowmeter/controller was installed on the biosparge unit. This flowmeter regulated the flow of oxygen into the unit, keeping it at a predetermined, constant rate. The initial flow rate was 4.7 L/min, which was similar to the flow rate that had been used since September 1997. A Dwyer flow meter installed on the oxygen line monitored the oxygen flow rate during a sparge event. The delivery pressure on the regulator was set at 45 psi, which is more than enough pressure to maintain the required flow even as the system becomes biofouled. Eventually, two flowmeters will be installed, one on each side.

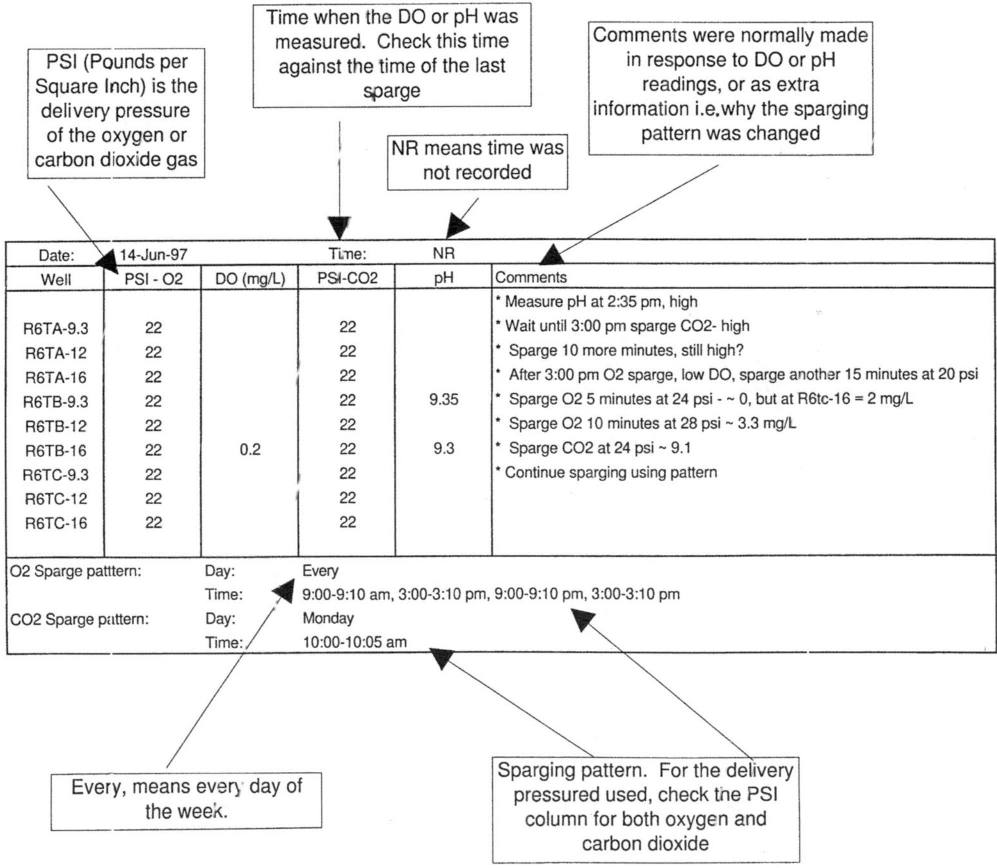

Figure A28.3 Text explanations for Table A28.2.

Table A28.2 Sparging Summary from OP-2

Date:	14-Jun-97		Time:	NR	
Well	**PSI - O_2**	**DO (mg/L)**	**PSI-CO_2**	**pH**	**Comments**
					* Measure pH at 2:35 pm, high
R6TA-9.3	22		22		* Wait until 3:00 pm sparge - high
R6TA-12	22		22		* Sparge 10 more minutes, still high?
R6TA-16	22		22		* After 3:00 pm O_2 sparge, low DO, sparge another 15 minutes at 20 psi
R6TB-9.3	22		22	9.35	* Sparge O_2 5 minutes at 24 psi ~ 0, but at R6tc-16 = 2 mg/L
R6TB-12	22		22		* Sparge O_2 10 minutes at 28 psi ~ 3.3 mg/L
R6TB-16	22	0.2	22	9.3	* Sparge CO_2 at 24 psi ~ 9.1
R6TC-9.3	22		22		* Continue sparging using pattern
R6TC-12	22		22		
R6TC-16	22		22		

O_2 Sparge patttern:	Day:	Every
	Time:	9:00-9:10 am, 3:00-3:10 pm, 9:00-9:10 pm, 3:00-3:10 pm
CO_2 Sparge pattern:	Day:	Monday
	Time:	10:00-10:05 am

continued

Table A28.2 (continued) Sparging Summary from OP-2

Date:	15-Jun-97		Time:	NR	
Well	**PSI - O_2**	**DO (mg/L)**	**PSI-CO_2**	**pH**	**Comments**
					* Readings based on previous sparge pattern
R6TA-9.3	22	o	22	8.98	* New sparge pattern:
R6TA-12	22	o	22	8.94	O_2: Every day: 9 am, 12 pm, 3 pm, 9 pm, 12 am, 3 am each for 20 minutes
R6TA-16	22	0.16	22	9	
R6TB-9.3	22	0	22	9.05	CO_2: Every day: 10 am, 4 pm, 10 pm, 4 am each for 5 minutes
R6TB-12	22	0	22	8.98	
R6TB-16	22	0.2	22	8.9	
R6TC-9.3	22	0	22		
R6TC-12	22		22		
R6TC-16	22		22		

O_2 Sparge patttern:	Day:	Every
	Time:	9:00-9:10 am, 3:00-3:10 pm, 9:00-9:10 pm, 3:00-3:10 pm
CO_2 Sparge pattern:	Day:	Monday
	Time:	10:00-10:05 am

Date:	16-Jun-97		Time:	NR	
Well	**PSI - O_2**	**DO (mg/L)**	**PSI-CO_2**	**pH**	**Comments**
					By mistake, the entire tank of O_2 and CO_2 was added, do not know how long it sparged or amount.
R6TA-9.3	22		22		
R6TA-12	22		22		
R6TA-16	22		22		
R6TB-9.3	22	16.31	22	7.61	
R6TB-12	22	17.95	22	7.66	
R6TB-16	22	18.03	22	7.7	
R6TC-9.3	22	17.3	22	7.9	
R6TC-12	22	17.8	22	7.8	
R6TC-16	22	20.4	22	8	

O_2 Sparge patttern:	Day:	Every
	Time:	9 am, 12 pm, 3 pm, 9 pm, 12 am, 3 am each for 20 minutes
CO_2 Sparge pattern:	Day:	Every
	Time:	10 am, 4 pm, 10 pm, 4 am each for 5 minutes

Appendix 29

Equations Used to Calculate Degradation Rates for Compounds at Alameda

EQUATIONS USED TO CALCULATE THE DEGRADATION RATES FOR COMPOUNDS IN THE GRANULAR IRON WALL

Degradation rates were calculated with field data for TCE, *c*DCE, *t*DCE, and 1,1-DCE using an analytical solution to the chain decay equation, assuming first-order kinetics applied. A satisfactory analytical solution for VC could not be obtained in time for this report, so a numerical solution was used to solve for the VC degradation rate. Analytical equations have the advantage of providing exact solutions to problems. Numerical solutions are based on approximations and are not exact. In order to gain confidence in the rate constants estimated from the numerical solution, in particular that for VC, the time increment (Δt) was reduced until the rate constants for TCE and the DCE isomers estimated numerically were within about 2.5% of those obtained analytically. Using the solver option in EXCEL 5.0, these degradation rates were calculated by allowing the calculated masses to be fit to the FENCE-generated masses. There were only three data points representing the three FENCE-generated masses from rows 1, 3, and 4. The first data point represented row 4. These data points were plotted and then the degradation rate constants calculated by using the analytical and numerical solutions. Originally, each of the DCE isomers were to be individually solved based on the influent masses, but a solution was not possible due to the large concentrations of *c*DCE compared to the other DCE isomers. Therefore, the ratio of the DCE isomers (*i, j,* and *m* in the equations) were set to zero, forcing all the TCE to degrade without passing through the DCE isomers or VC.

The following differential equations describe the TCE degradation sequence as far as VC.

$$\frac{dA}{dt} = -K_a A \tag{1}$$

$$\frac{dB}{dt} = -K_b B + iK_a A \tag{2}$$

$$\frac{dC}{dt} = -K_c C + jK_a A \tag{3}$$

$$\frac{dD}{dt} = -K_d D + mK_a A \tag{4}$$

$$\frac{dE}{dt} = -K_e E + K_b B + K_c C + K_d D \tag{5}$$

where A = concentration of TCE
B = concentration of *c*DCE
C = concentration of *t*DCE
D = concentration of 1,1-DCE
E = concentration of VC
K_a = degradation rate of TCE
K_b = degradation rate of *c*DCE
K_c = degradation rate of *t*DCE
K_d = degradation rate of 1,1-DCE
K_e = degradation rate of VC
i = fraction of *c*DCE
j = fraction of *t*DCE
m = fraction of 1,1-DCE

Equation (1) was solved analytically to calculate the degradation of TCE.

$$A = A_o e^{-K_a t} \tag{6}$$

where A = concentration of TCE that has degraded
A_o = initial concentration of TCE
t = time
K_a = degradation rate for TCE

Equations (7) through (9) were also solved analytically to calculate the pseudo-first-order degradation rate constants of *c*DCE, *t*DCE, and 1,1-DCE, respectively.

$$B = \frac{iK_a A_o}{K_b - K_a} \times (e^{-K_a t} - e^{-K_b t}) + B_o e^{-K_b t} \tag{7}$$

$$C = \frac{jK_a A_o}{K_c - K_a} \times (e^{-K_a t} - e^{-K_c t}) + C_o e^{-K_c t} \tag{8}$$

$$D = \frac{mK_a A_o}{K_d - K_a} \times (e^{-K_a t} - e^{-K_d t}) + D_o e^{-K_d t} \tag{9}$$

where B = degraded concentration of *c*DCE
C = degraded concentration of *t*DCE
D = degraded concentration of 1,1-DCE
B_o = initial concentration of *c*DCE
C_o = initial concentration of *t*DCE
D_o = initial concentration of 1,1-DCE
K_b = degradation rate of *c*DCE
K_c = degradation rate of *t*DCE
K_d = degradation rate of 1,1-DCE
i = fraction of TCE to *c*DCE
j = fraction of TCE to *t*DCE
m = fraction of TCE to 1,1-DCE

The assumptions made with the above equations were as follows: *i, j,* and *m* are nonzero, and $i + j + m = 1$, i.e., TCE degrades entirely through the DCE intermediates.

Once the pseudo-first-order degradation rate constants were solved (K_a, K_b, K_c, and K_d), half-lives could be calculated using Equation (10):

$$t_{1/2} = \frac{\ln 2}{K} \tag{10}$$

Equations (1) through (5) were then solved numerically by finite differences to estimate the VC degradation rate constant.

The data points being fit did not generally lie on an ideal first-order curve. So, a range of rate constants was obtained by solving the analytical and numerical solutions in two ways. The first method involved fitting the solution to the first and last data points, ignoring the middle data points. Since the calculated masses at row 3 were lower than the masses at row 4, this method calculated rate constants that were the minimum of the range, and hence the most conservative (spreadsheets 1 and 2 represent the analytical and numerical solutions for this fitting method). The second method involved fitting all three data points, a procedure that was thought to provide a more realistic estimate of the rate constants since all the data were considered (spreadsheets 3 and 4 represent the analytical and numerical solution for this fitting method).

EQUATIONS USED TO CALCULATE THE DEGRADATION RATES FOR COMPOUNDS IN THE BIOSPARGE ZONE

Modeling degradation throughout the biosparge zone was undertaken by presenting it as a large, completely mixed bioreactor that accounted for instantaneous partitioning of VOCs into the sparge bubbles, as well as storage in the unsaturated headspace. The equation was developed from the mass balance as follows: the mass in the bioreactor at any time, Mass_t, equals the mass that was present at the previous time step, Mass_{t-1}, plus whatever mass entered the reactor over the intervening time increment, Δt, due to advection from the granular iron zone, Mass_{in}, minus the mass that was removed from the reactor by the process of advection, Mass_{out}, volatilization, Mass_{vol}, and biodegradation, Mass_{bio}.

$$Mass_t = Mass_{t-1} + Mass_{in} - Mass_{out} - Mass_{bio} - Mass_{vol} \tag{11}$$

The mass in the bioreactor at time *t* can be subdivided into mass in the headspace (M_{at}) and mass in the water (M_{wt}):

$$Mass_{wt} + Mass_{at} = Mass_{wt-1} + Mass_{at-1} + Mass_{in} - Mass_{out} - Mass_{bio} - Mass_{vol} \tag{12}$$

Mass entering the system can be calculated from the influent concentration and the pumping rate times the time over which the pumping occurred:

$$Mass_{in} = C_{in} \times Q \times \Delta t \tag{13}$$

Mass leaving the system can be calculated from the pumping rate times the concentration in the bioreactor at the earlier time times the time interval:

$$Mass_{out} = C_{t-1} \times Q \times \Delta t \tag{14}$$

Mass in the system at any time is calculated from the concentrations of contaminants in the water and gas phase, and the volumes of those phases:

$$Mass_w = C_{wt} \times V_w \tag{15}$$

$$Mass_a = C_{at} \times V_a \tag{16}$$

Henry's law provides a means of reducing the number of concentration variables, since

$$C_a = H \times C_w \tag{17}$$

Employing Henry's law, we find

$$Mass_a = H \times C_w \times V_a \tag{18}$$

When sparging occurs, the mass stripped is the mass that partitions into the sparge bubbles. If this is assumed to occur instantaneously, then, using Henry's law to express all concentrations as water concentrations,

$$Mass_{vol} = Q_a \times H \times C_{wt-1} \times \Delta t \tag{19}$$

Biodegradation is assumed to occur by first-order kinetics; therefore, the rate of decline in concentration is equal to the rate constant times the concentration, or

$$\frac{\Delta C}{\Delta t} = -K_1 \times C \tag{20}$$

Rewriting this in terms of mass,

$$\Delta Mass = Mass_{bio} = -K_1 \times C_{t-1} \times \Delta t \times V_w \tag{21}$$

Now, assembling the parts of the mass balance equation, we obtain

$$\begin{aligned} C_{wt} \times V_w + H \times C_{wt} \times V_a = C_{wt-1} \times V_w + H \times C_{wt-1} \times V_a + C_{in} \times Q \times \Delta t - C_{wt-1} \times \\ Q \times \Delta t - Q_a \times H \times C_{wt-1} \times \Delta t - K_1 \times C_{wt-1} \times \Delta t \times V_w \end{aligned} \tag{22}$$

From this equation, it can be shown that

$$\frac{dC}{dt} = \frac{Q \times C_{in}}{(V_w + V_a \times H)} - \left(\frac{Q + K_1 \times V_w + Q_a \times H}{(V_w + V_a \times H)}\right) \times C \tag{23}$$

The differential equation, when solved, yields

$$\begin{aligned} C = \frac{Q_w \times C_{in}}{Q_w + K_1 \times V_w + Q_a \times H} + \left(C_o - \frac{Q_w \times C_{in}}{Q_w + K_1 V_w + Q_a \times H}\right) \\ \exp\left(-\left(\frac{Q_w + K_1 \times V_w + Q_a \times H}{V_w + V_a \times H}\right)t\right) \end{aligned} \tag{24}$$

Pseudo-first-order rate constants or apparent biodegradation rates were calculated for TCE, *cis*-1,2-DCE, and VC using both the analytical solution (Equation (24)) and the numerical solution (Equation (22)). The assumptions made with Equations (22) and (24) were as follows:

- First-order kinetics apply to the biodegradation.
- There was instantaneous equilibrium partitioning to gas phase, described by Henry's law.
- Biosparge zone was completely mixed resembling a mixed bioreactor.
- Input concentrations were constant (analytical solution only).

These rate constants were calculated by using the solver option in EXCEL 5.0 as described in the previous section. The pseudo-first-order rate constants were calculated using an oxygen flow rate of 4.7 L/min (0.564 m^3/day). Spreadsheets 5 and 6 present the bioreactor calculations for *c*DCE and VC.

Spreadsheet #1
Analytical Solution for the Reductive Dechlorination Reaction Rates - Fitting first and last data point

Initial Conc. TCE	ao	TCEo	107500	mg					
Initial conc. cDCE	bo	cDCEo	1347000	mg					
Initial Conc. tDCE	co	tDCEo	2997	mg					
Initial Conc. 11DCE	do	11DCEo	3595	mg					
						distance	time	time	
			1/m	1/day		m	days	hours	
Degradation of TCE		ka	3.857254	0.231435	Half LIfe	0.179661	2.994358	71.8646	
Degradation of cDCE		kb	1.621022	0.097261	Half LIfe	0.427508	7.125133	171.0032	
Degradation of tDCE		kc	2.53119	0.151871	Half Life	0.273784	4.563072	109.5137	
Degradation of 11DCE		kd	2.164271	0.129856	Half LIfe	0.3202	5.336671	128.0801	
Ratio of TCE degrading to cDCE		0	0.00E+00						
Ratio of TCE degrading to tDCE		0	0.00E+00						
Ratio of TCE degrading to 11DCE		0	0.00E+00						
		dt	0.2						
Residual Sum of Squares		RSS	0.019826						
Distance from front of gate		dist							
Calculated concs (see formulas)		Calc							
FENCE generated concs		Data							

	Calc.				Data			
dist	TCE	cDCE	tDCE	11DCE	TCE -d	cDCE -d	tDCE -d	11DCE -d
0	107500	1347000	2997	3595	107500	1347000	2997	3595
0.2	49701.75	974018.9	1806.469	2331.915				
0.4	22979.2	704315.4	1088.865	1512.609	1979	230300	317.4	425.5
0.6	10624.24	509292.1	656.3232	981.1614				
0.8	4912.032	368270.3	395.6047	636.4353				
1	2271.038	266297.1	238.4543	412.827				
1.2	1049.996	192560.1	143.7304	267.7824				
1.31	686.9376	161111.4	108.7997	211.0521				
1.4	485.457	139240.7	86.63481	173.6984				
1.51	317.6	116500	65.58	136.9	278.4	116500	65.58	136.9

Note: Data points used in this calculation were generated using FENCE.

Spreadsheet #2

Numerical Solution for the Reductive Dechlorination Reaction Rate - Fitting first and last point

Initial conc. TCE	ao	107500	mg		i	0		
Initial conc. cDCE	bo	1347000	mg		j	0		
Initial conc. tDCE	co	2997	mg		m	0		
Initial conc. 11DCE	do	3595	mg					
Initial conc. VC	eo	571900	mg					
						Distance	Time	Time
		1/m	1/day			m	days	hours
Degradation of TCE	ka	3.983649	0.239019		Half Life	0.173961	2.899352	69.58444
Degradation of cDCE	kb	1.62761	0.097657			0.425778	7.096297	170.3111
Degradation of tDCE	kc	2.547275	0.152836			0.272055	4.534258	108.8222
Degradation of 11DCE	kd	2.176023	0.130561			0.318471	5.307848	127.3884
Degradation of VC	ke	6.742883	0.404573			0.102775	1.712917	41.11001
Residual Sum of Squares	RSS	3.71E-15						
Distance	d	0.005						
Calculated concs	Calc							
FENCE generated concs	Data							

Note: Only a portion of the solution is shown

	Calc					Data				
Distance	TCE	cDCE	tDCE	11DCE	VC	TCE-d	cDCE - d	tDCE - d	11DCE - d	VC - d
0	107500	1347000	2997	3595	571900	107500	1347000	2997	3595	571900
0.005	105400.6	1336127	2959.309	3556.307	563840.3					
0.01	103342.2	1325341	2922.092	3518.03	555957.7					
0.015	101324	1314642	2885.343	3480.166	548247.1					
0.02	99345.23	1304030	2849.057	3442.709	540703.6					
0.025	97405.09	1293503	2813.227	3405.655	533322.4					
0.03	95502.81	1283062	2777.847	3369	526099					
0.035	93637.74	1272704	2742.912	3332.739	519028.7					
0.04	91809.07	1262431	2708.417	3296.869	512107.4					
0.045	90016.1	1252240	2674.355	3261.385	505330.8					
0.05	88258.16	1242131	2640.722	3226.282	498694.9					
0.055	86534.54	1232104	2607.512	3191.558	492195.6					
0.06	84844.58	1222158	2574.719	3157.207	485829.3					
0.065	83187.63	1212293	2542.339	3123.226	479592.2					

Note: Data points used in this calculation were generated by FENCE.

Spreadsheet #3
Analytical Solution for the Reductive Dechlorination Reaction Rates - Fitting 3 data points

Initial Conc. TCE	ao	TCEo	107500	mg
Initial conc. cDCE	bo	cDCEo	1347000	mg
Initial Conc. tDCE	co	tDCEo	2997	mg
Initial Conc. 11DCE	do	11DCEo	3595	mg

		1/m	1/day		distance m	time days	time hours
Degradation of TCE	ka	9.98622	0.599173	Half LIfe	0.069396	1.156594	27.75825
Degradation of cDCE	kb	4.258592	0.255515	Half LIfe	0.16273	2.712164	65.09194
Degradation of tDCE	kc	5.515882	0.330953	Half Life	0.125637	2.093954	50.25489
Degradation of 11DCE	kd	5.25104	0.315062	Half LIfe	0.131974	2.199564	52.78954

Ratio of TCE degrading to cDCE	0	0.00E+00
Ratio of TCE degrading to tDCE	0	0.00E+00
Ratio of TCE degrading to 11DCE	0	0.00E+00
	dt	0.2

Residual Sum of Squares	RSS	3.929033
Distance from front of gate	dist	
Calculated concs (see formulas)	Calc	
FENCE generated concs	Data	

	Calc.				Data			
dist	TCE	cDCE	tDCE	11DCE	TCE -d	cDCE -d	tDCE -d	11DCE -d
0	107500	1347000	2997	3595	107500	1347000	2997	3595
0.2	14588.7	574739.5	994.4509	1257.764				
0.4	1979.814	245230.5	329.9742	440.0477	1979	230300	317.4	425.5
0.6	268.6782	104635.2	109.4905	153.9573				
0.8	36.46199	44645.88	36.33065	53.86425				
1	4.948213	19049.55	12.05507	18.84521				
1.2	0.671516	8128.085	4.000059	6.59328				
1.31	0.223867	5087.977	2.180518	3.700384				
1.4	0.091131	3468.1	1.327282	2.306758				
1.51	0.030381	2170.944	0.723529	1.294635	278.4	116500	65.58	136.9

Note: Data points used in this calculation were generated using FENCE.

Spreadsheet #4

Numerical Solution for the Reductive Dechlorination Reaction Rate - Fitting all 3 points

				i	0		
Initial conc. TCE	ao	107500	mg	j	0		
Initial conc. cDCE	bo	1347000	mg	m	0		
Initial conc. tDCE	co	2997	mg				
Initial conc. 11DCE	do	3595	mg				
Initial conc. VC	eo	571900	mg				
					Distance	Time	Time
		1/m	1/day		m	days	hours
Degradation of TCE	ka	10.23973	0.614384	Half Life	0.067678	1.127959	27.07102
Degradation of cDCE	kb	3.952885	0.237173		0.175315	2.921917	70.126
Degradation of tDCE	kc	5.592543	0.335553		0.123915	2.06525	49.566
Degradation of 11DCE	kd	5.320385	0.319223		0.130254	2.170895	52.10149
Degradation of VC	ke	10.1536	0.609216		0.068252	1.137528	27.30067
Residual Sum of Squares	RSS	4.832052					
Distance	d	0.005					
Calculated concs	Calc						
FENCE generated concs	Data						

Note: Only a portion of the solution is shown.

	Calc					Data				
Distance	TCE	cDCE	tDCE	11DCE	VC	TCE -d	cDCE - d	tDCE - d	11DCE - d	VC - d
0	107500	1347000	2997	3595	571900	107500	1347000	2997	3595	571900
0.005	102264.2	1320893	2915.475	3501.844	569280.2					
0.01	97283.43	1295293	2836.168	3411.102	566301					
0.015	92545.24	1270188	2759.019	3322.712	562989.2					
0.02	88037.82	1245570	2683.968	3236.612	559370.3					
0.025	83749.93	1221429	2610.958	3152.743	555468					
0.03	79670.89	1197756	2539.935	3071.047	551305					
0.035	75790.52	1174542	2470.843	2991.468	540002.7					
0.04	72099.14	1151778	2403.631	2913.951	542281.2					
0.045	68587.55	1129455	2338.247	2838.443	537459.3					
0.05	65246.99	1107565	2274.642	2764.892	532455.2					
0.055	62069.13	1086098	2212.767	2693.246	527285.6					
0.06	59046.05	1065048	2152.575	2623.457	521966.5					
0.065	56170.21	1044406	2094.021	2555.477	516512.8					

Note: Data points used in this calculation were generated by FENCE.

Spreadsheet #5

Bioreactor Calculations for cDCE

Remedial Sum of Squares	RSS	241321.7	
Time Step	t	1.25	days
Extraction Rate (pumps)	Q	0.34	m3/day
Flow rate of oxygen (measured)		**4.7**	**L/min**
Flux of air (Qa*t)	Qa	0.564	m3/day
Volume of Water	V	10.83	m3
Volume of Headspace	Va	6.17	m3
Pseudo 1st Order Rate Constant	**K**	**0.045397**	**1/day**
Henry's Law Constant	H	0.312	
(1 + H*Va/V)	b	1.177751	
Half-Life	**15.26529**		
Influent Concentrations	Ci		
Measured Concentrations	Data		
Fitted line	Fit		

Ci	T-anl	t	Data	Num	Anal
4737	0	0	3827.12	3827.12	3827.12
4737	1.25	1.25		3607.041	3617.558
4737	2.5	2.5		3408.693	3427.7
4737	3.75	3.75		3229.932	3255.694
4737	5	5		3068.823	3099.862
4737	6.25	6.25		2923.624	2958.683
4737	7.5	7.5		2792.762	2830.778
4737	8.75	8.75		2674.822	2714.9
4737	10	10		2568.529	2609.917
4737	11.25	11.25		2472.732	2514.806
4737	12.5	12.5		2386.394	2428.639
4737	13.75	13.75		2308.582	2350.573
4737	15	15		2238.454	2279.848
4737	16.25	16.25		2175.251	2215.773
4737	17.5	17.5		2118.289	2157.722
4737	18.75	18.75		2066.951	2105.13
4737	20	20	1955.06	2020.683	2057.484
4737	21.25	21.25		1978.984	2014.317
998	1.25	22.5		1816.819	1856.579
998	2.5	23.75		1670.667	1713.672
998	3.75	25		1538.947	1584.203
998	5	26.25		1420.234	1466.908
998	6.25	27.5		1313.243	1360.642
998	7.5	28.75		1216.818	1264.367
998	8.75	30		1129.914	1177.146
998	10	31.25		1051.592	1098.125
998	11.25	32.5		981.0037	1026.535

Note: Only a portion of the solution is shown.

Spreadsheet #6

Bioreactor Calculations for VC

Remedial Sum of Squares	RSS	21269.44	
Time Step	t	1.25	days
Extraction Rate (pumps)	Q	0.34	m3/day
Flow rate of oxygen (measured)		**4.7**	**L/min**
Flux of air (Qa*t)	Qa	0.564	m3/day
Volume of Water	V	10.83	m3
Volume of Headspace	Va	6.17	m3
Pseudo 1st Order Rate Constant	**K**	**0.009732**	**1/day**
Henry's Law Constant	H	0.964	
(1 + H*Va/V)	b	1.549204	
Half-Life	**71.20601**	**day**	
Influent Concentrations	Ci		
Measured Concentrations	Data		
Fitted line	Fit		

Ci	T-anl	t	Data	Num	Anal
1311.84	0	0	788.3	788.3	788.3
1311.84	1.25	1.25		763.4399	764.3338
1311.84	2.5	2.5		740.4118	742.0702
1311.84	3.75	3.75		719.0806	721.3882
1311.84	5	5		699.3213	702.1755
1311.84	6.25	6.25		681.0181	684.3276
1311.84	7.5	7.5		664.0637	667.7477
1311.84	8.75	8.75		648.3587	652.3457
1311.84	10	10		633.8109	638.0378
1311.84	11.25	11.25		620.3352	624.7463
1311.84	12.5	12.5		607.8526	612.3991
1311.84	13.75	13.75		596.2898	600.9291
1311.84	15	15		585.579	590.2739
1311.84	16.25	16.25		575.6576	580.3756
1311.84	17.5	17.5		566.4672	571.1806
1311.84	18.75	18.75		557.9541	562.6387
1311.84	20	20	622.33	550.0683	554.7037
1311.84	21.25	21.25		542.7637	547.3324
719.41	1.25	22.5		520.9905	526.0175
719.41	2.5	23.75		500.8217	506.2169
719.41	3.75	25		482.1393	487.8229
719.41	5	26.25		464.8335	470.7356
719.41	6.25	27.5		448.803	454.8622
719.41	7.5	28.75		433.9538	440.1165
719.41	8.75	30		420.1989	426.4184
719.41	10	31.25		407.4576	413.6933
719.41	11.25	32.5		395.6551	401.8723
719.41	12.5	33.75		384.7224	390.891

Note: Only a portion of the solution is shown

16-Jun-97

Day 0	VC	wt-VC	CIS	wt-CIS
R6TA-2.8	692.3	92.0759	3732	496.356
R6TA-3.7	852.1	97.1394	3985	454.29
R6TA-4.9	599.8	79.7734	3682	489.706
R6TB-2.8	897.7	64.6344	4003	288.216
R6TB-3.7	784.4	48.6328	3621	224.502
R6TB-4.9	727.5	52.38	3796	273.312
R6TC-2.8	850.6	123.337	3816	553.32
R6TC-3.7	836.6	103.7384	3780	468.72
R6TC-4.9	873	126.585	3991	578.695
average	**790.4444**	**788.2963**	**3822.889**	**3827.117**

Units: ppb

6-Jul-97

Day 20	VC	wt-VC	CIS	wt-CIS
R6TA-2.8	681.7	90.6661	2113	281.029
R6TA-3.7	477.6	54.4464	1777	202.578
R6TA-4.9	637.8	84.8274	2090	277.97
R6TB-2.8	622.1	44.7912	1907	137.304
R6TB-3.7	640.7	39.7234	1917	118.854
R6TB-4.9	638.8	45.9936	1930	138.96
R6TC-2.8	634.9	92.0605	1935	280.575
R6TC-3.7	642.3	79.6452	1961	243.164
R6TC-4.9	621.9	90.1755	1894	274.63
average	**621.9778**	**622.3293**	**1947.111**	**1955.064**

Units: ppb

31-Jul-97

Day 45	VC	wt-VC	CIS	wt-CIS
R6TA-2.8	1016	135.128	1619	215.327
R6TA-3.7	238.1	27.1434	975.2	111.1728
R6TA-4.9	95.8	12.7414	826.1	109.8713
R6TB-2.8	520.1	37.4472	1046	75.312
R6TB-3.7	136.9	8.4878	799.6	49.5752
R6TB-4.9	95.06	6.84432	808	58.176
R6TC-2.8	218.4	31.668	747.7	108.4165
R6TC-3.7	177.7	22.0348	829.3	102.8332
R6TC-4.9	64.05	9.28725	741.2	107.474
average	**284.6789**	**290.7822**	**932.4556**	**938.158**

Units: ppb

27-Au-97

Day 72	VC	wt-VC	CIS	wt-CIS
R6TA-2.8	0	0	23.45	3.11885
R6TA-3.7	0	0	29.95	3.4143
R6TA-4.9	0	0	18.66	2.48178
R6TB-2.8	0	0	46.16	3.32352
R6TB-3.7	0	0	26.22	1.62564
R6TB-4.9	0	0	46.69	3.36168
R6TC-2.8	33.8	4.901	54.72	7.9344
R6TC-3.7	25.3	3.1372	27.49	3.40876
R6TC-4.9	18.9	2.7405	49.2	7.134
average	**8.666667**	**10.7787**	**35.83778**	**35.80293**

Units: ppb

Note: The first average is an arithmetic average and the second average is an area weighted average. The area weighted average is the data used in the degradation rate calculations.

10-Sep-97

Day 86	VC	wt-VC	CIS	wt-CIS
R6TA-2.8	140.6	18.6998	91.65	12.18945
R6TA-3.7	156.2	17.8068	367.4	41.8836
R6TA-4.9	62.84	8.35772	99.23	13.19759
R6TB-2.8	151.2	10.8864	126	9.072
R6TB-3.7	156.2	9.6844	123.7	7.6694
R6TB-4.9	62.84	4.52448	73.55	5.2956
R6TC-2.8	226.7	32.8715	139.6	20.242
R6TC-3.7	267.8	33.2072	200.4	24.8496
R6TC-4.9	18.89	2.73905	65.61	9.51345
average	**138.1411**	**138.7774**	**143.0156**	**143.9127**

Units: ppb

22-Se-97

Day 98	VC	wt-VC	CIS	wt-CIS
R6TA-2.8	368.1	48.9573	230.8	30.6964
R6TA-3.7	68.29	7.78506	95.19	10.85166
R6TA-4.9	37.9	5.0407	68.94	9.16902
R6TB-2.8	355.1	25.5672	217.8	15.6816
R6TB-3.7	89.22	5.53164	87.35	5.4157
R6TB-4.9	35.44	2.55168	78.48	5.65056
R6TC-2.8	348.8	50.576	218.7	31.7115
R6TC-3.7	74.36	9.22064	79.37	9.84188
R6TC-4.9	33.74	4.8923	49	7.105
average	**156.7722**	**160.1225**	**125.07**	**126.1233**

Units: ppb

20-Se-97

Day 106	VC	wt-VC	CIS	wt-CIS
R6TA-2.8	32.34	4.30122	51.25	6.81625
R6TA-3.7	21.25	2.4225	49.65	5.6601
R6TA-4.9	20.51	2.72783	48.32	6.42656
R6TB-2.8	33.68	2.42496	48.84	3.51648
R6TB-3.7	26.14	1.62068	48.32	2.99584
R6TB-4.9	0	0	47.01	3.38472
R6TC-2.8	29.61	4.29345	47.26	6.8527
R6TC-3.7	25.46	3.15704	49.62	6.15200
R6TC-4.9	26.26	3.8077	51.21	7.42545
average	**23.91667**	**24.75538**	**49.05333**	**49.23098**

Units: ppb

10-Oct-97

Day 116	VC	wt-VC	CIS	wt-CIS
R6TA-2.8	0	0	37.91	5.04203
R6TA-3.7	20.55	2.3427	37.24	4.24536
R6TA-4.9	28.66	3.81178	22.01	2.92733
R6TB-2.8	18.67	1.34424	37.27	2.68344
R6TB-3.7	0	0	31.49	1.95238
R6TB-4.9	0	0	33.54	2.41488
R6TC-2.8	21.39	3.10155	36.29	5.26205
R6TC-3.7	0	0	28.47	3.53028
R6TC-4.9	0	0	35.02	5.0779
average	**9.918889**	**10.60027**	**33.24889**	**33.13565**

Units: ppb

17-Oct-97

Day 123	VC	wt-VC	CIS	wt-CIS
R6TA-2.8	367	48.811	123.6	16.4388
R6TA-3.7	135.6	15.4584	66.19	7.54566
R6TA-4.9	128.9	17.1437	64.53	8.58249
R6TB-2.8	152.5	10.98	68.92	4.96224
R6TB-3.7	92.11	5.71082	30.19	1.87178
R6TB-4.9	99.65	7.1748	55.76	4.01472
R6TC-2.8	199.8	28.971	68.92	9.9934
R6TC-3.7	125.5	15.562	61.57	7.63468
R6TC-4.9	101.1	14.6595	55.89	8.10405
average	**155.7956**	**164.4712**	**66.17444**	**69.14782**

Units: ppb

Day 127	VC	wt-VC	CIS	wt-CIS
R6TA-2.8	78.22	10.40326	45.17	6.00761
R6TA-3.7	81.32	9.27048	44.89	5.11746
R6TA-4.9	68.72	9.13976	41.85	5.56605
R6TB-2.8	70.72	5.09184	42.75	3.078
R6TB-3.7	64.49	3.99838	41.05	2.5451
R6TB-4.9	76.84	5.53248	42.05	3.0276
R6TC-2.8	70.94	10.2863	44.27	6.41915
R6TC-3.7	70.21	8.70604	43.34	5.37416
R6TC-4.9	71.52	10.3704	43.56	6.3162
average	**72.55333**	**72.79894**	**43.21444**	**43.45133**

Units: ppb

16-Jun	VC	wt-VC	CIS	wt-CIS
R4TA-2.8	1793	111.166	108.2	6.7084
R4TA-3.7	3591	262.143	9879	721.167
R4YA-4.9	6526	476.398	26920	1965.16
R4TB-2.8	636.8	56.0384	2818	247.984
R4TB-3.7	598.6	61.0572	2622	267.444
R4TB-4.9	787.8	80.3556	3507	357.714
R4TC-2.8	299.9	44.985	1270	190.5
R4TC-3.7	689.6	120.68	3114	544.95
R4TC-4.9	565.8	99.015	2491	435.925
average	**1720.944**	**1311.838**	**5858.8**	**4737.552**

Units: ppb

31-Jul	VC	wt-VC	CIS	wt-CIS
R4TA-2.8	2299	142.538	2547	157.914
R4TA-3.7	864.1	63.0793	1142	83.366
R4TA-4.9	2694	196.662	200.6	14.6438
R4TB-2.8	374.2	32.9296	626.5	55.132
R4TB-3.7	423.2	43.1664	1061	108.222
R4TB-4.9	438.3	44.7066	1311	133.722
R4TC-2.8	302.7	45.405	407.6	61.14
R4TC-3.7	295.1	51.6425	944.1	165.2175
R4TC-4.9	567.3	99.2775	1251	218.925
average	**917.5444**	**719.4069**	**1054.533**	**998.2823**

Units: ppb

30-Sep	VC	wt-VC	CIS	wt-CIS
R4TA-2.8	2179	135.098	789.2	48.9304
R4TA-3.7	138.2	10.0886	100.7	7.3511
R4YA-4.9	105	7.665	13.15	0.95995
R4TB-2.8	687.9	60.5352	229.8	20.2224
R4TB-3.7	26.6	2.7132	63.49	6.47598
R4TB-4.9	61.98	6.32196	64.42	6.57084
R4TC-2.8	317.3	47.595	194.6	29.19
R4TC-3.7	37.69	6.59575	67.71	11.84925
R4TC-4.9	38.91	6.80925	60.03	10.50525
average	**399.1756**	**283.422**	**175.9**	**142.0552**

Units: ppb

APPENDIX 30

Plots of Contaminant Distribution in the Remedial Gate For OP2, Alameda

Figure 30.1 shows the plan view of the remedial gate at Alameda NAS. The columns A, B, and C represent three sections through the remedial gate that are in the line of fully screened wells. In this appendix, data along each column are plotted for different contaminants. Figures A30.2 to A30.11 illustrate the contaminant concentrations, sampled from fully screened wells only on June 16, 1997, July 31, 1997, and September 30, 1997 (during OP2 only). Performance data plots were organized to represent the trends along the three columns of the remedial gate, A, B, and C. The contaminants that were plotted include

- TCE (Figure A30.2),
- 1,1-DCE (Figure A30.3),
- *c*DCE (Figure A30.4),
- *t*DCE (Figure A30.5),
- VC (Figure A30.6),
- Benzene (Figure A30.7),
- Toluene (Figure A30.8),
- Ethylbenzene (Figure A30.9),
- Xylenes (*p*/*m*) (Figure A30.10), and
- Xylene (*o*) (Figure A30.11).

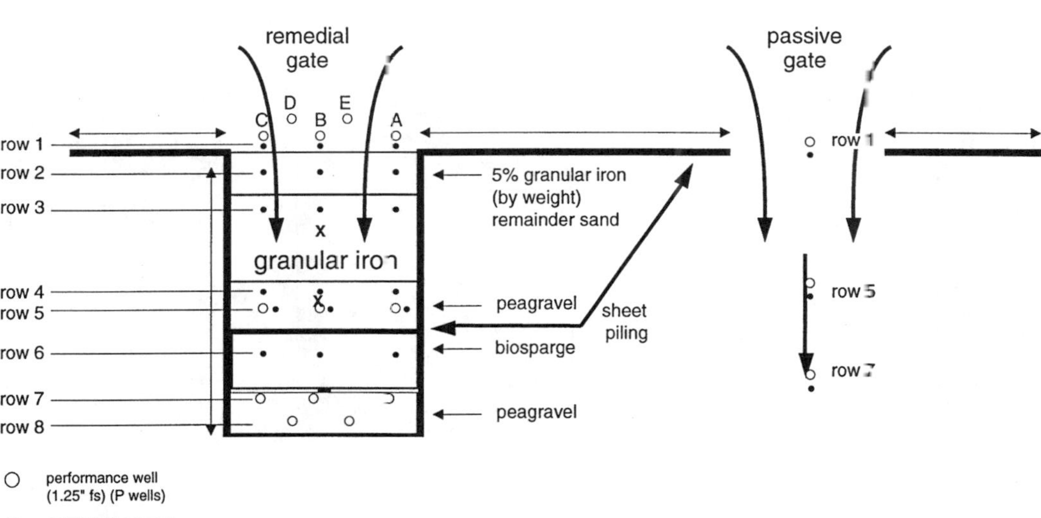

Figure A30.1 Remedial gate, NAS Alameda.

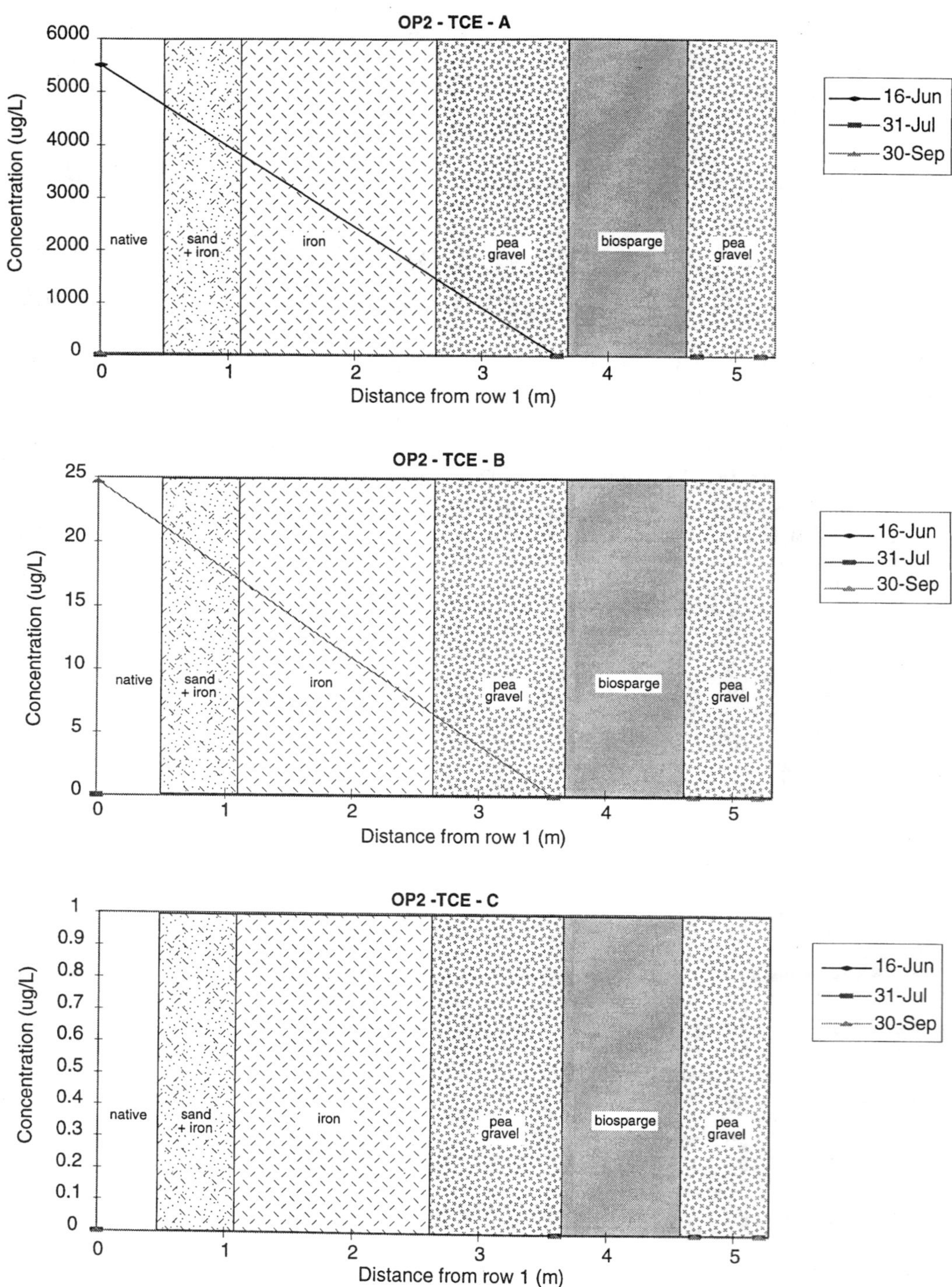

Figure A30.2 TCE distribution in remedial gate during OP2, NAS Alameda.

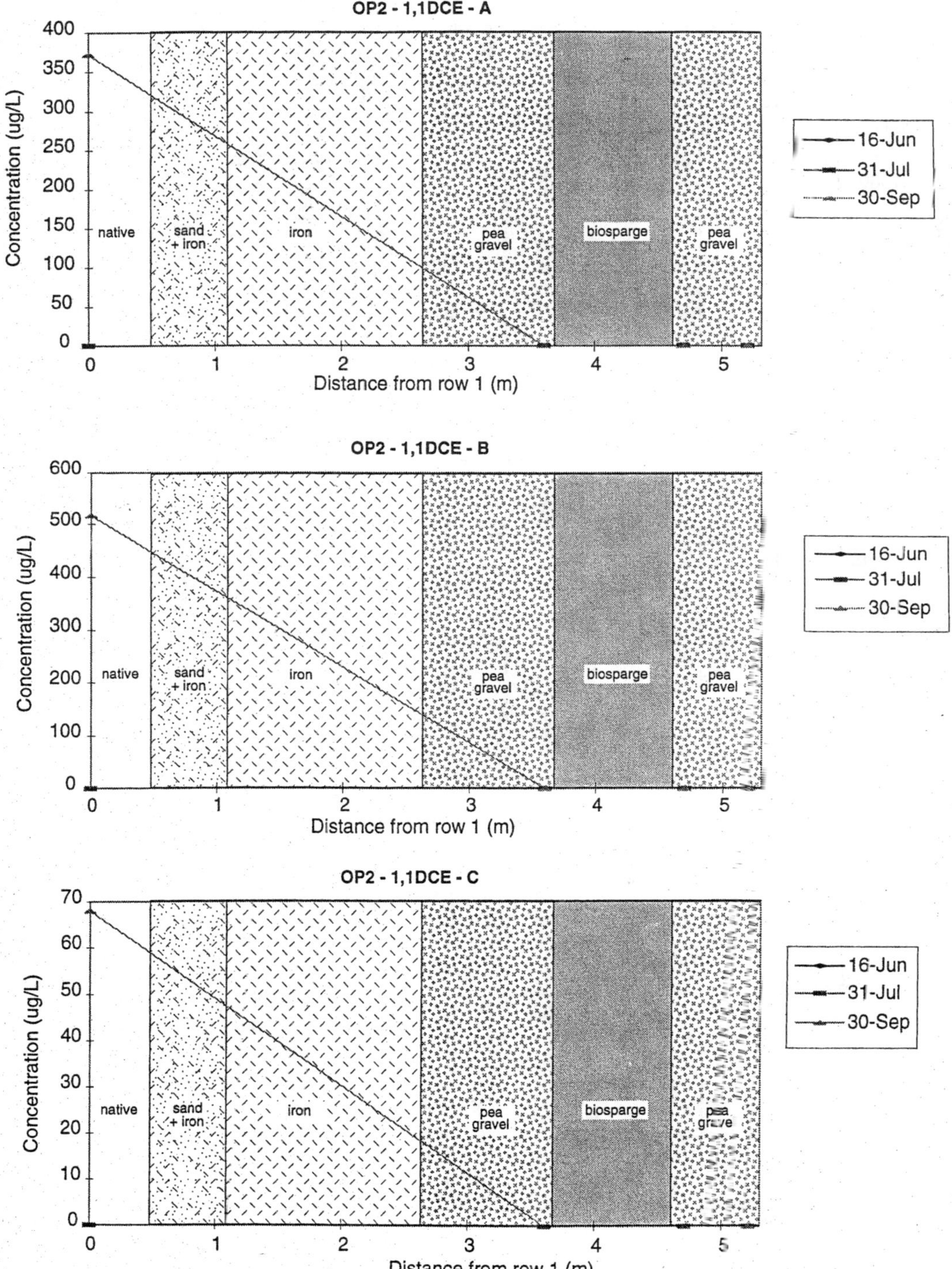

Figure A30.3 1,1-DCE distribution in remedial gate during OP2, NAS Alameda.

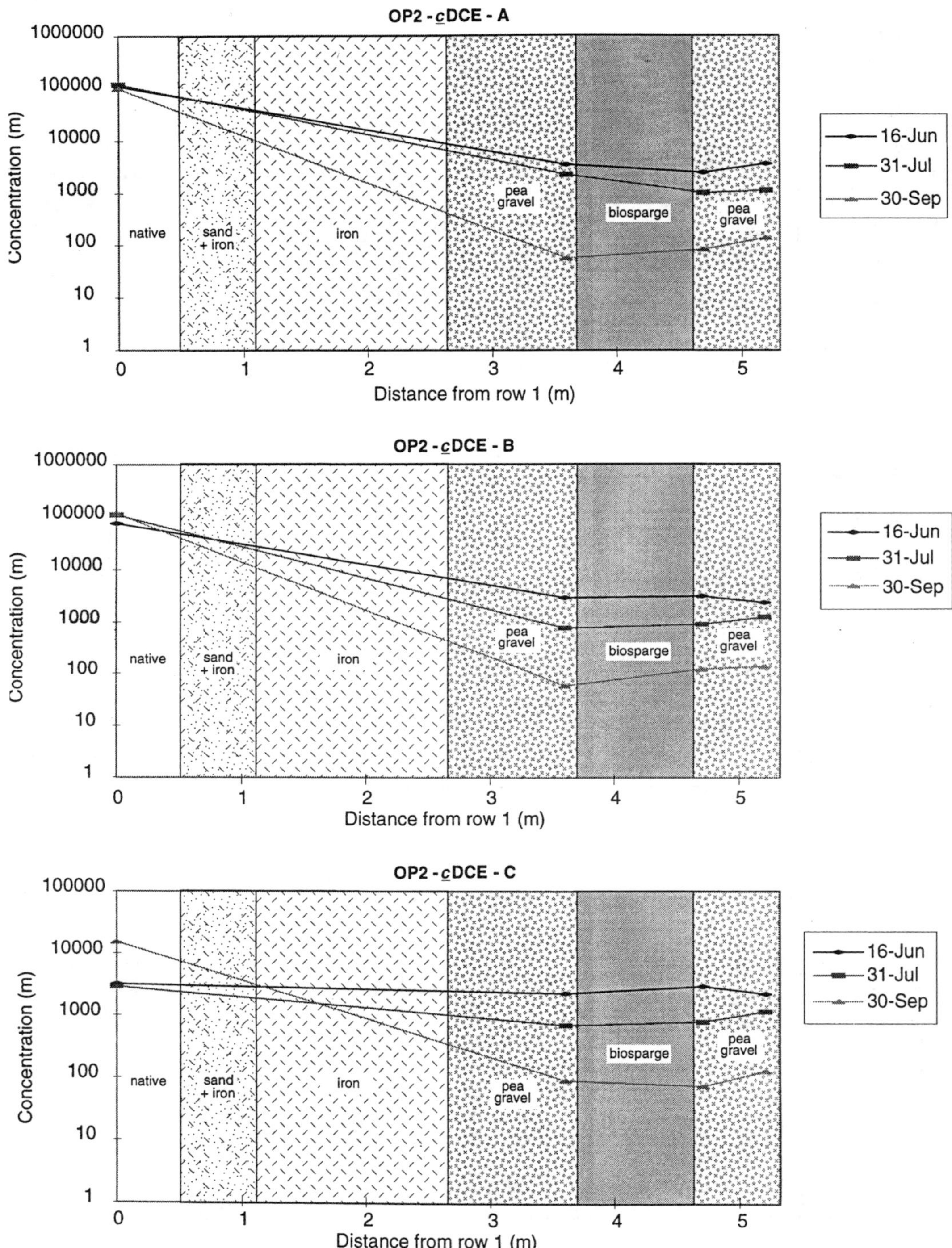

Figure A30.4 *c*DCE distribution in remedial gate during OP2, NAS Alameda.

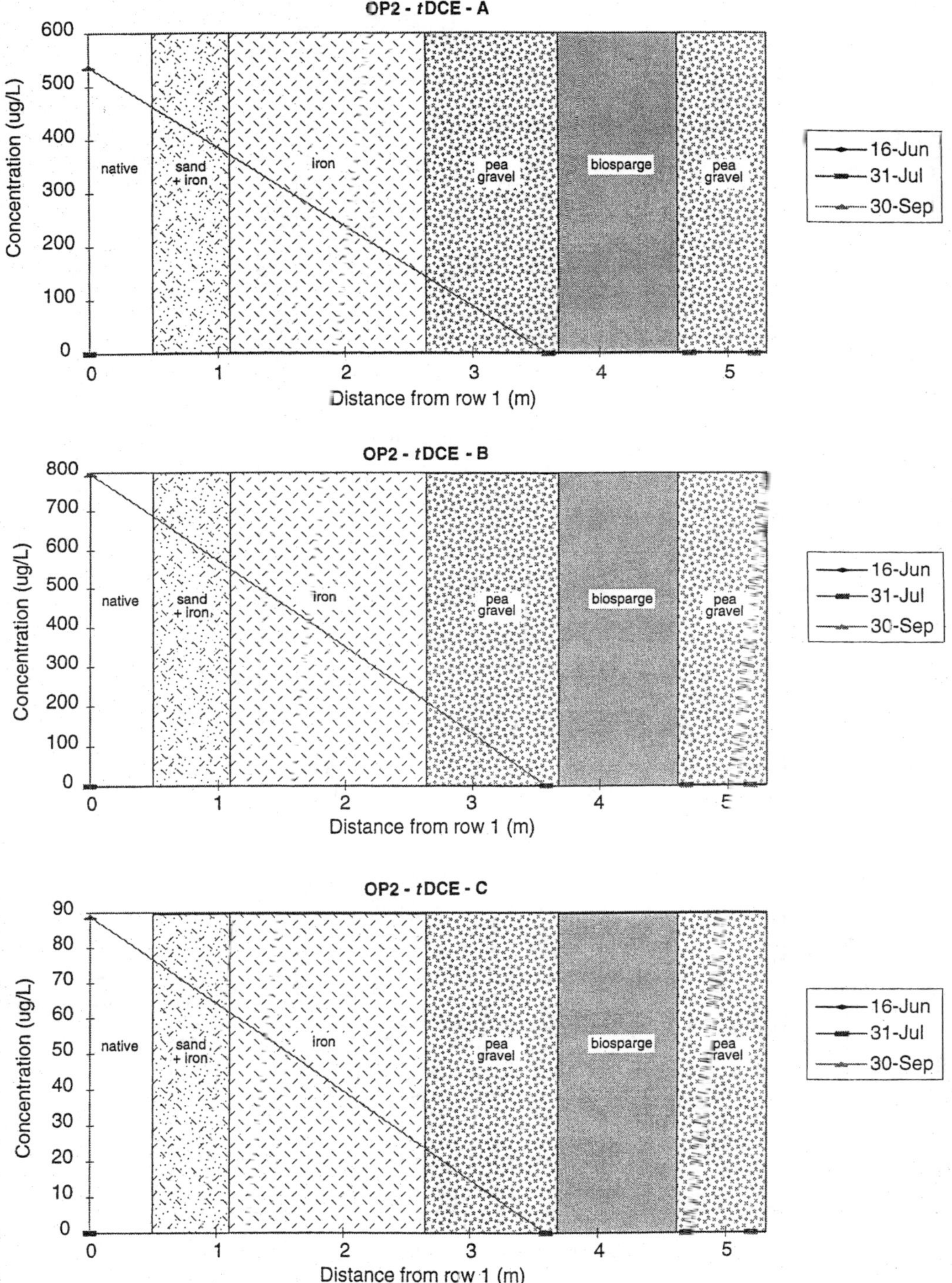

Figure A30.5 *t*DCE distribution in remedial gate during OP2, NAS Alameda.

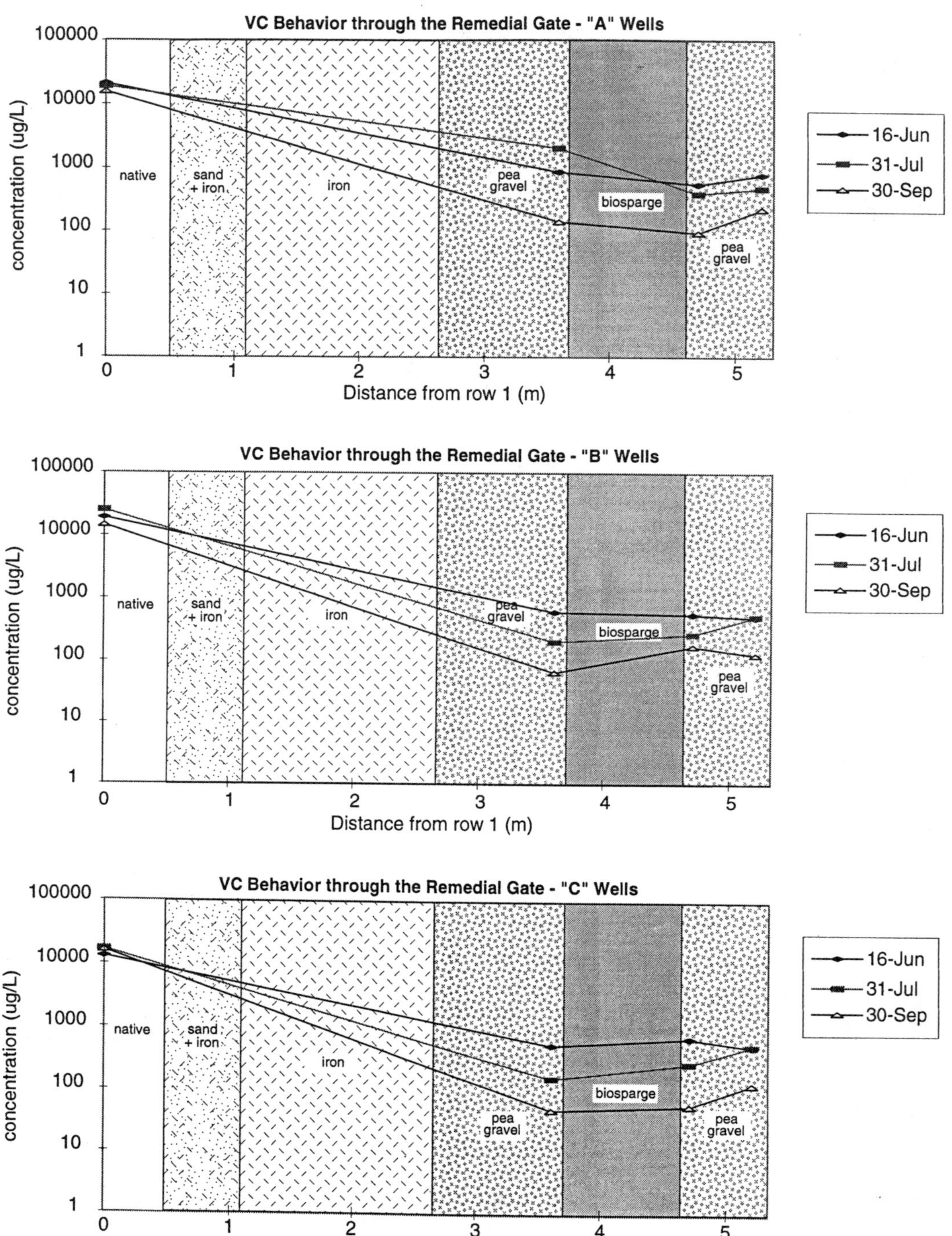

Figure A30.6 VC distribution in remedial gate during OP2, NAS Alameda.

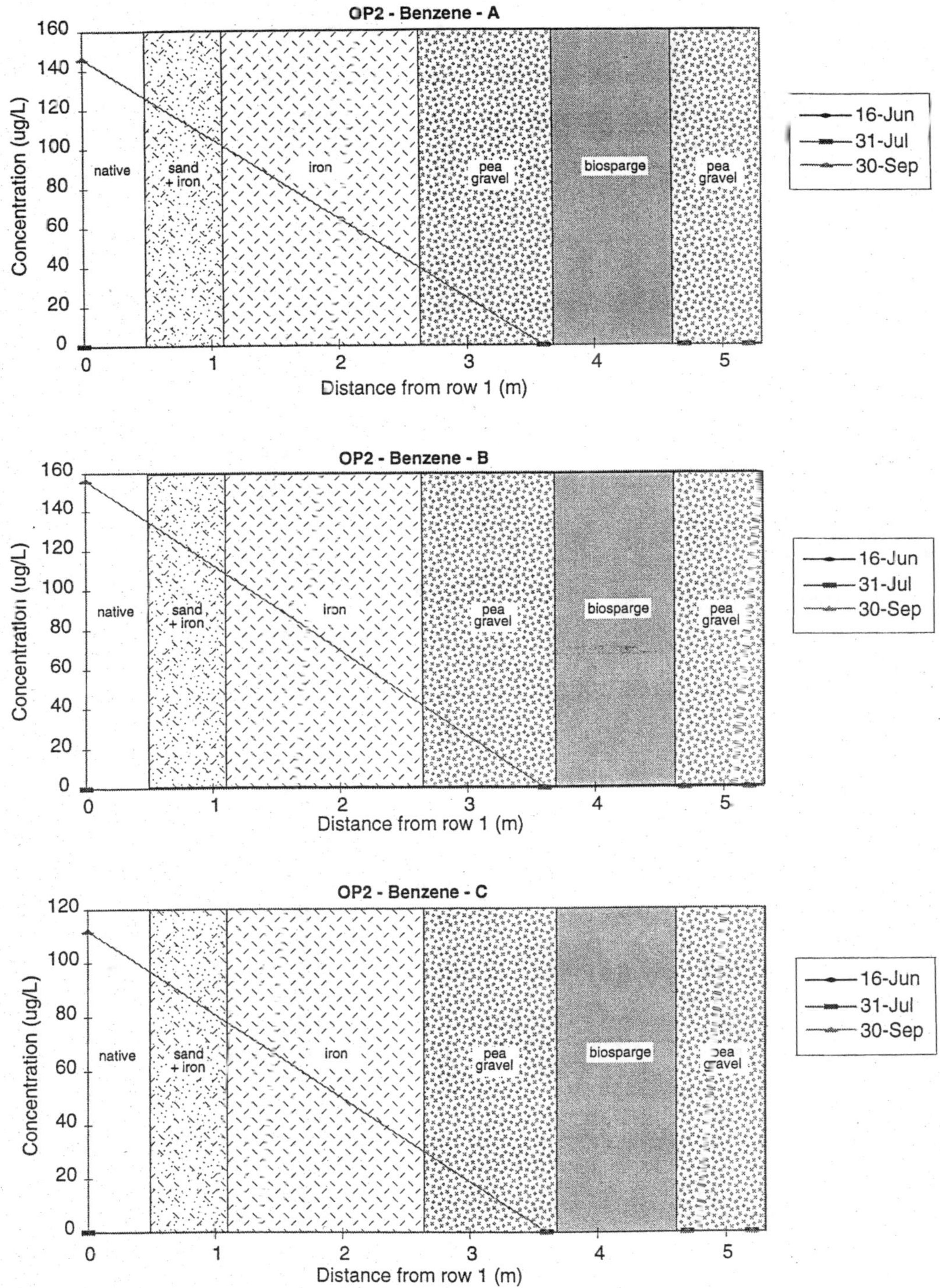

Figure A30.7 Benzene distribution in remedial gate during OP2, NAS Alameda

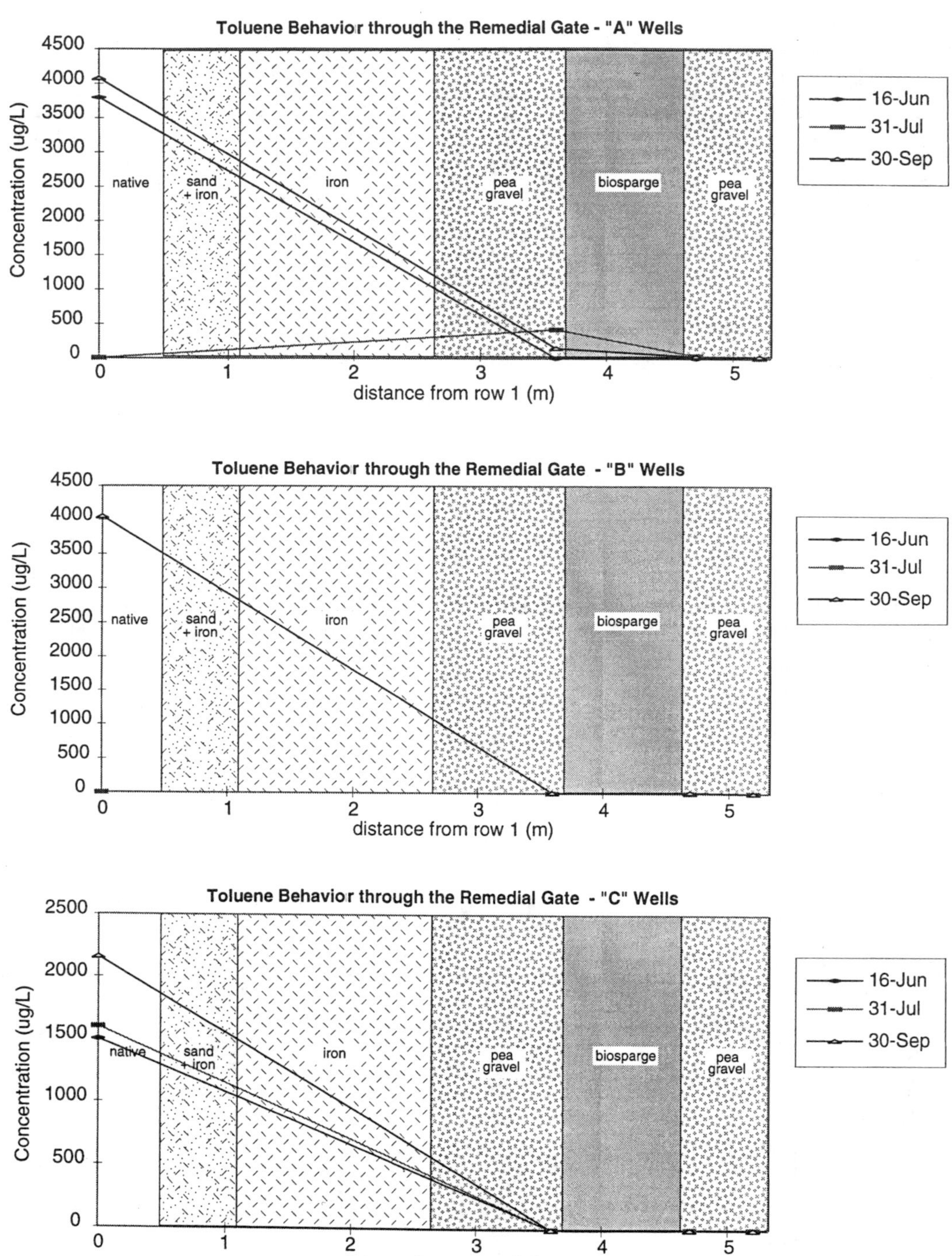

Figure A30.8 Toluene distribution in remedial gate during OP2, NAS Alameda.

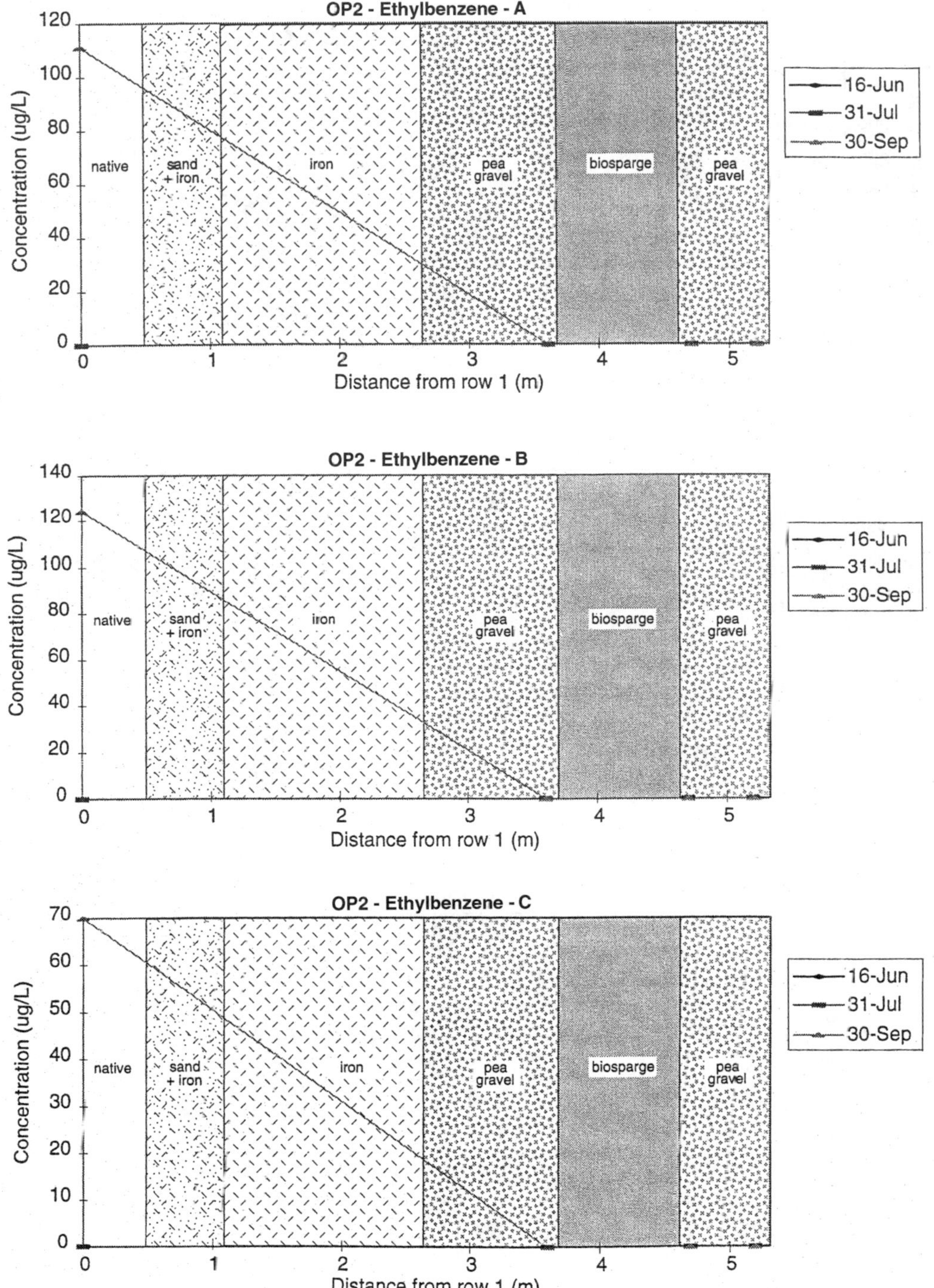

Figure A30.9 Ethylbenzene distribution in remedial gate during OP2, NAS Alameda.

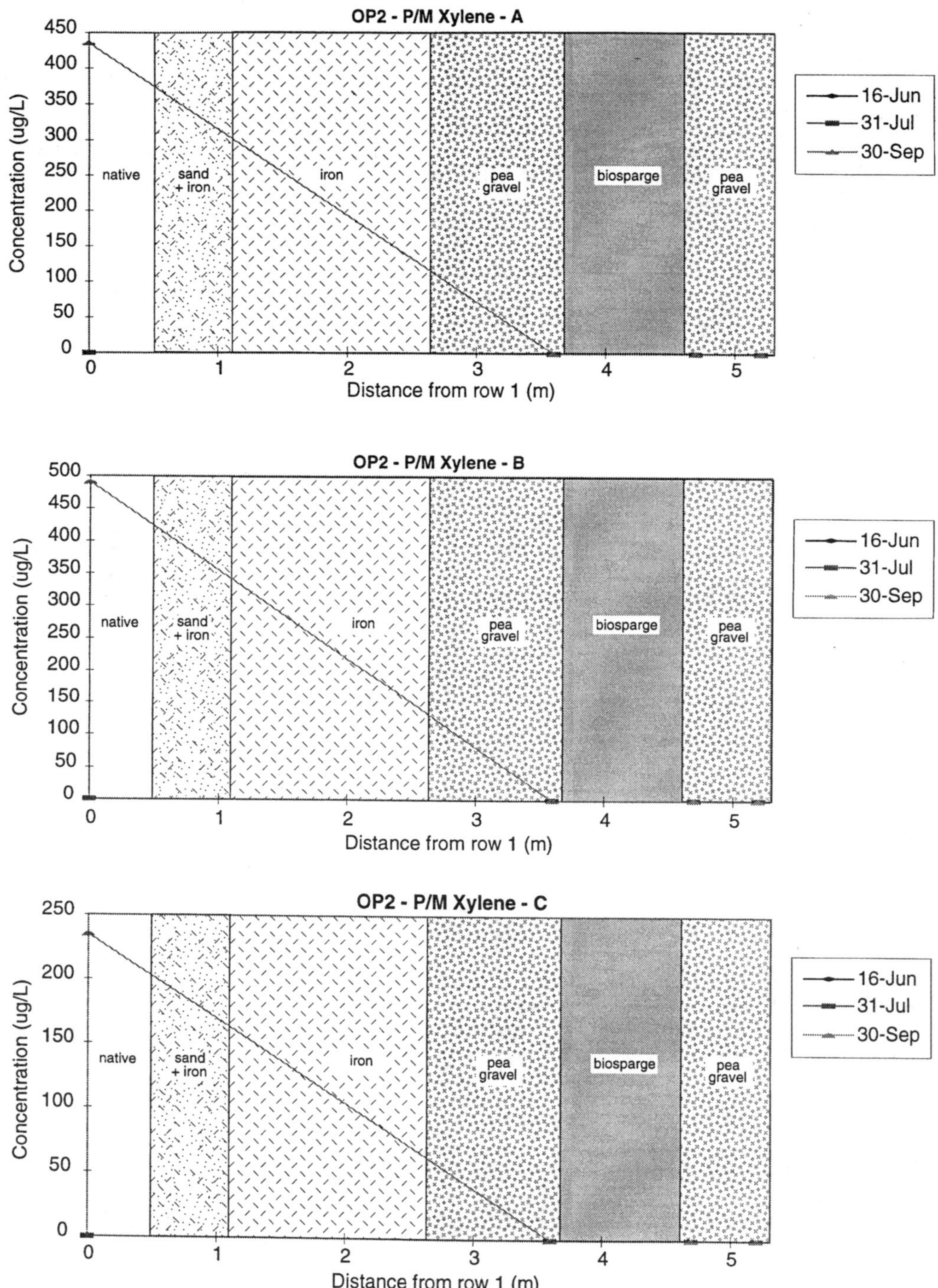

Figure A30.10 Xylene (para and meta) distribution in remedial gate during OP2, NAS Alameda.

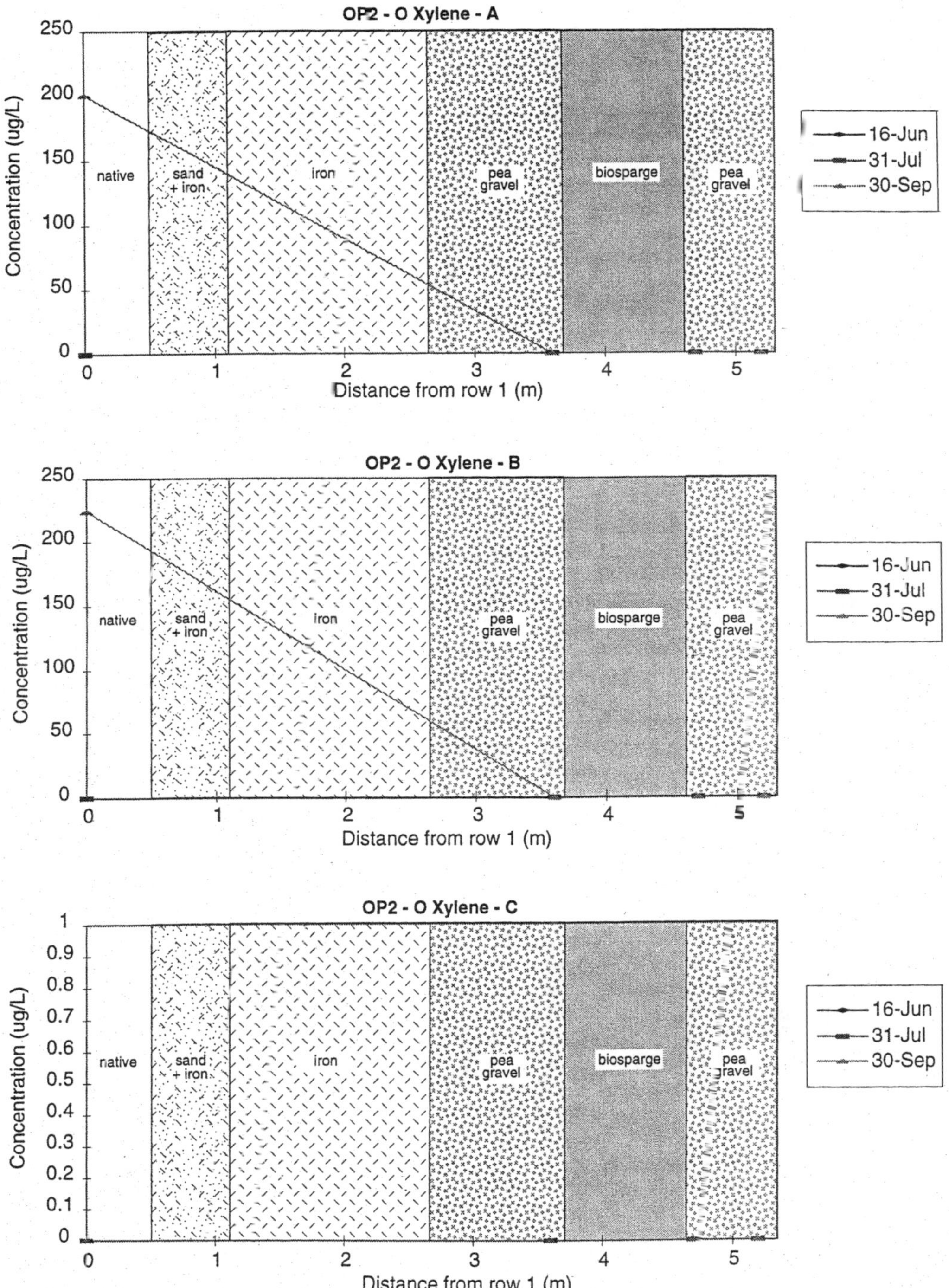

Figure A30.11 O-Xylene (C) distribution in remedial gate during OP2, NAS Alameda.

APPENDIX 31

Aboveground Treatment of Groundwater Extracted from Remedial Gate, Alameda

Groundwater extracted from the remedial gate, as well as any wastewater resulting from sampling or purging activities, was treated prior to disposal. The treatment system consisted of a storage tank, submersible pump, peristaltic pump, treatment drum (containing granular activated carbon GAC), and a treated water storage tank. A schematic of the treatment system is provided in Figure A31.1. Storage tanks were 20,000 USG Baker tanks.

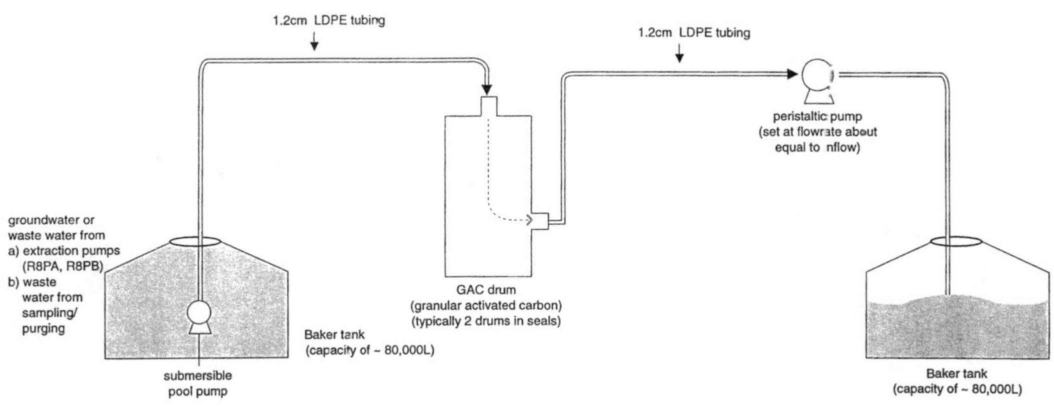

Figure A31.1 Aboveground treatment system, Alameda.

Water extracted from the remedial gate was stored in the storage tank. The treatment system was only operated when the storage tank was full. Once full, the submersible pump was placed into the storage tank and pumping started. The water was pumped through the GAC drum and then via peristatilc pumping was placed into a clean Baker tank. The peristaltic pump controlled the level of water within the GAC drum and was operated to prevent saturation of the GAC drum.

The East Bay Municipal Utility District (EBMUD) granted UW a permit to discharge groundwater to a local storm drain. Wastewater samples were collected prior to disposal to ensure that the GAC treatment was complete and that no detectable VOCs remained. All wastewater was disposed in accordance with the EBMUD permit.

Appendix 32

Numerical Simulations Alameda

A32.1 INTRODUCTION

The purpose of the numerical simulations was to provide estimates of groundwater discharge rates through the gates(s) and to provide additional insight into the expected flow system and capture zones. The software package Visual MODFLOW (version 2.5, 1997) was used to simulate the three-dimensional flow of groundwater with particle tracking.

A32.2 SITE CHARACTERIZATION

Description of the site hydrogeology can be found in Section 2 of Chapter 5, and only a brief summary will be presented here. The unconfined silty sand aquifer is relatively homogeneous, with an average hydraulic conductivity of 1.73 m/day, based on slug tests in the vicinity of MO28-A (UW, 1996b). The underlying bay mud at a depth of about 6.7 m bgs represents a lower confining layer.

The hydraulic gradient varies in magnitude and direction primarily due to seasonal and tidal influences. A report by PRC (1993) characterized the tidal influence in April 1992. The maximum groundwater level fluctuations observed within 244 m of the Funnel-and-Gate site was 1.1 m in well M-026A (PRC, 1993). The well closest to the site (M028-A) experienced only a 5-cm fluctuation. From the tidally filtered data the mean direction of groundwater flow in the vicinity of the Funnel-and-Gate was determined to be perpendicular to the shore, flowing from east to west toward the bay. This direction is consistent with the *c*DCE and BTEX plume shape delineated by EFW (UW, 1996b). A hydraulic gradient of 0.0017 was calculated from the PRC (April 1993), filtered data set. Updated filtered data from May 2, 1997 and June 27, 1997 (PRC, unpublished data) provided additional estimates of the hydraulic gradient of 0.0034 and 0.0018, respectively (based on the head difference between M028-A and M034-A). Therefore, a gradient of 0.002 was chosen to represent the mean hydraulic gradient. A maximum gradient of 0.0044 was observed in September of 1997 (M. Morkin, unpublished data), based on single-time measurements (at M028-A and M034-A). Therefore, to give a conservative estimate of the gate residence time a hydraulic gradient of 0.005 was chosen. Information regarding the influence of seasonal fluctuations on the groundwater flow velocity and direction was not available, and remains a source of uncertainty with regards to the applicability of the results.

A32.3 CONCEPTUAL MODEL

The modeled domain consisted of a 6.7-m-thick unit, representing an unconfined aquifer with the underlying bay mud providing an impermeable base. The area of consideration was 91 m by 91 m centered around the Funnel-and-Gate structure (Figure A32.1). This large area was chosen to minimize uncharacterized boundary effects because the San Francisco Bay represents the only physically defined vertical boundary in the model domain. The eastern domain boundary as well as the western boundary were represented as constant head boundary conditions while the remaining north and south boundaries were considered as no-flow. Both the top and bottom of the domain were also considered no-flow (i.e., no recharge and no leakage) boundaries.

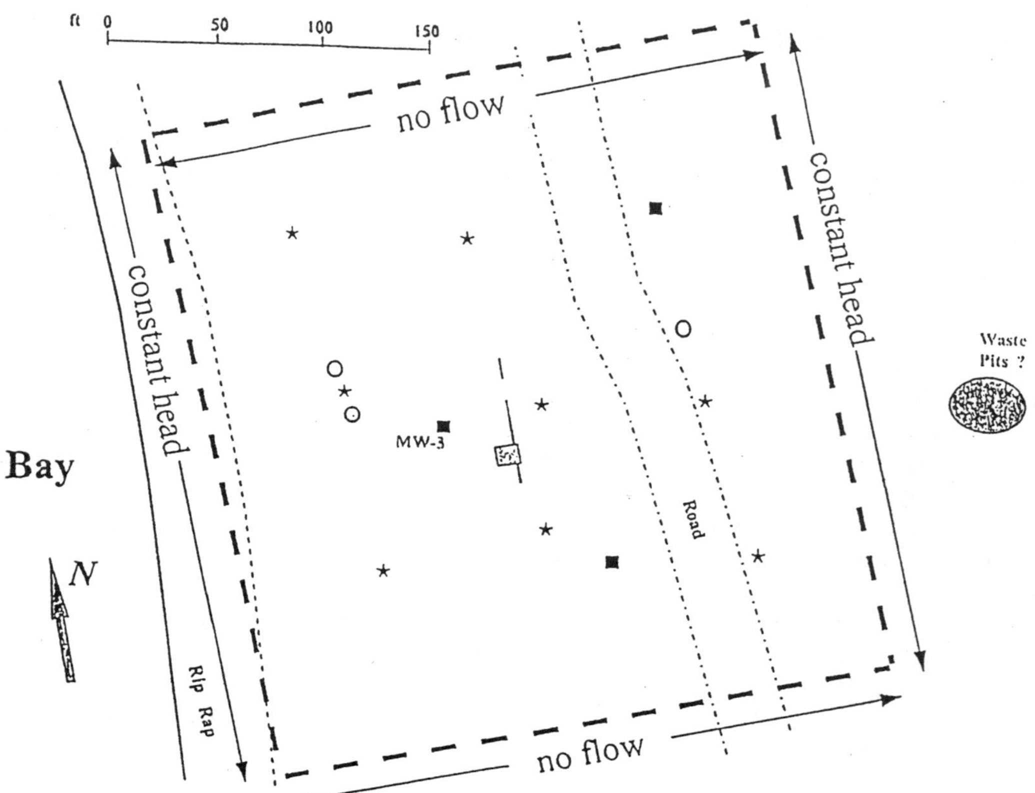

Figure A32.1 Model domain and boundary conditions.

A32.4 MODEL DESIGN

The software package Visual MODFLOW (version 2.5, 1997) was used to simulate the flow of groundwater with particle tracking. The 91 m by 91 m model domain was discretized with 1.5 m spacing in both the *x* and *y* directions, with refinement to 0.3 m spacing beginning at a distance of 6 m adjacent to the Funnel-and-Gate system, and 0.15 m at zones of intense convergence and divergent flow. The vertical dimension was represented as 5 layers. The total number of cells was approximately 75,000.

The hydraulic conductivity, *K*, of the aquifer was assumed to be homogeneous and isotropic with a value of 2×10^{-5} m/s. Assumed values for porosity of 0.30 (Freeze and Cherry, 1979), specific storage of 1×10^{-3} ft^{-1} (Anderson and Woessner, 1992), and 0.32 for specific yield (Anderson and Woessner, 1992) were chosen based on typical sandy aquifer values.

The hydraulic gradient was imposed by setting the eastern and the western boundary to constant head conditions. The constant head values were calculated so as to place the water table about 2.2 m bgs at the gravel-iron zone interface within the remedial gate. For example, a hydraulic gradient of 0.002 was imposed by applying a constant head on the western and eastern boundaries of 4.5 m and 4.7 m. This induced flow within the aquifer from east to west at a velocity of 1.15 cm/day. No recharge was allowed on the top boundary. This simple hydrogeologic rendering was not subject to model validation.

The Funnel-and-Gate design was approximated using the hydraulic properties seen in Table A32.1. The remedial Funnel-and-Gate consists of funnel walls (both parallel and orthogonal to the ambient groundwater flow direction) to direct contaminated water first through a zone of sand and iron mixture, followed by granular iron, then through a pea gravel zone, followed by the biosparge zone and then a second pea gravel zone (Figure A32.2). The control gate was simulated by simply leaving an opening between two orthogonal funnel walls, with aquifer material remaining in the gate and no side funnel walls parallel to the ambient groundwater flow direction (Figure A32.2).

Table A32.1 Hydraulic Properties Used to Represent the Funnel-and-Gate

Property	*K* (ft/day)	Source	Porosity	Source
Aquifer	5.7	Revised Work Plan	0.30	Freeze & Cherry
Sand and iron mixture	145.6	Austrins, permeameter	0.30	Freeze & Cherry
Granular iron	141.7	ETI, permeameter	0.50	ETI, personal communication
Pea gravel	2830	ETI, personal communication	0.40	Katic, personal communication
Biosparge zone	2.83×10^{8}	assumed	0.90	assumed
Funnel wall	2.83×10^{-8}	ETI, personal communication	0.10	assumed
Bentonite cap	same as above		same as above	
Concrete pad	same as above		same as above	

Large contrasts in *K*, which become an issue when simulating the funnel walls and the biosparge zone, may have caused numerical difficulties. These numerical problems may manifest themselves by inducing a poor water balance or anomalous trends in hydraulic head data. Representation of the funnel walls using the Horizontal-Flow-Barrier (HFB) package available in MODFLOW instead of the hydraulic properties listed in Table A32.1 did not improve the already low water balance error and allowed more leakage through the wall itself, thus it was not invoked. The biosparge zone was represented by assigning a large *K* value. There was no evidence of physically unrealistic results or poor mass balances using the hydraulic properties listed in Table A32.1. It should be noted that porosity values are only utilized in the numerical code when velocities are calculated and therefore do not affect the equipotential head distribution.

A32.5 MODEL APPLICATION

To determine the gate residence time, the Zone Budget package in MODFLOW was utilized. It calculates the volumetric discharge through a specified zone, such as the gate entrance, giving Q_{gate}. The groundwater velocity, v, from which the travel time could be calculated was obtained by evaluating $v = Q_{gate} / n \times A$, where n is the porosity of the gate zone and A is the cross-sectional area for flow.

The first three zones in the remedial gate have a consistent A based on the depth between the bottom of the overlying bentonite cap and the top of the underlying cement pad (3.6 m) multiplied by the gate width (3 m). In other cases (i.e., the biosparge zone and the control gate), the saturated thickness for flow varied for each simulation. The depth of flow was determined

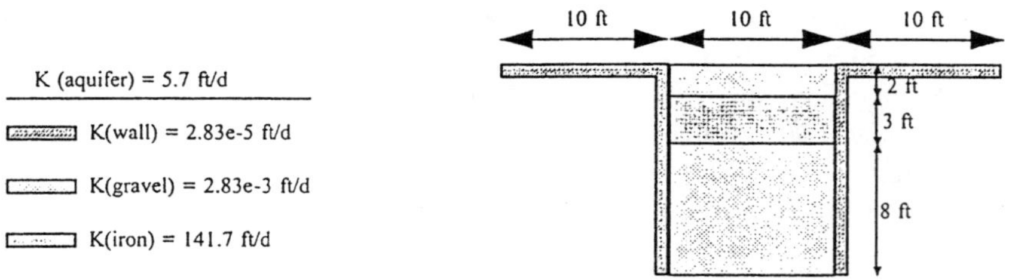

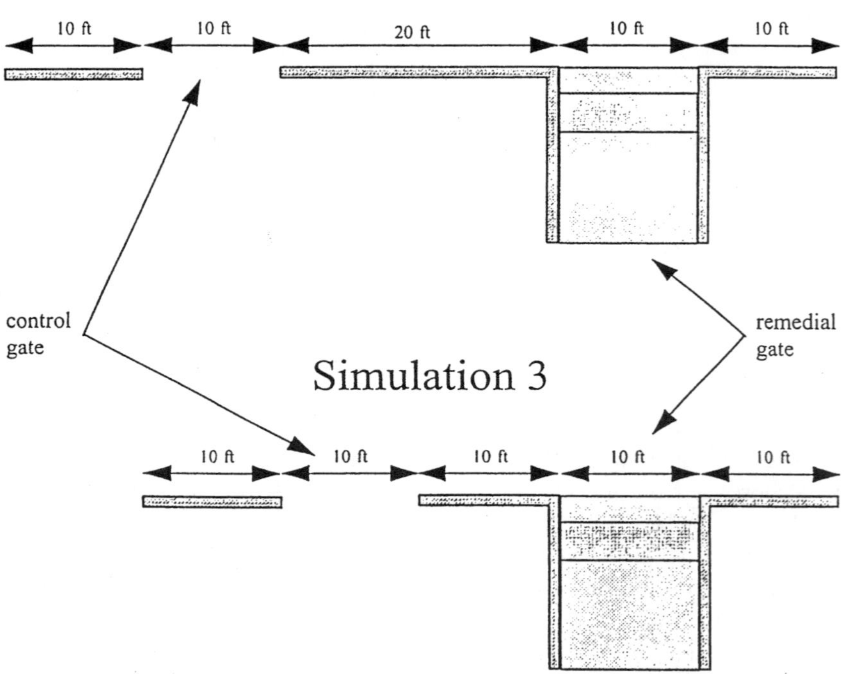

Figure A32.2 Funnel-and-Gate configuration.

by plotting equipotentials (to determine the water table elevation) specific to the zone in question. Using this value, the saturated thickness between the water table and the top of the clay aquitard or the cement pad could be determined. Multiplying this value by the gate width gave the cross-sectional area for flow.

Numerous simulations were run with different hydraulic gradients and pumping rates, as well as simulations involving elimination of the downgradient sheet piles and pumping wells in the remedial gate. Only two simulations are specifically referred to in the main body of the report. Simulation 1 (base case) involves a natural gradient of 0.002, with the total pumping well extraction rate at 0.34 m^3/day (12 ft^3/day), Figure A32.3a. Simulation 2 is similar, except the total pumping well extraction rate was 1.27 m^3/day (45 ft^3/day), Figure A32.3b. For all simulations, the model was run to steady state with the Waterloo Hydrogeologic Software solver.

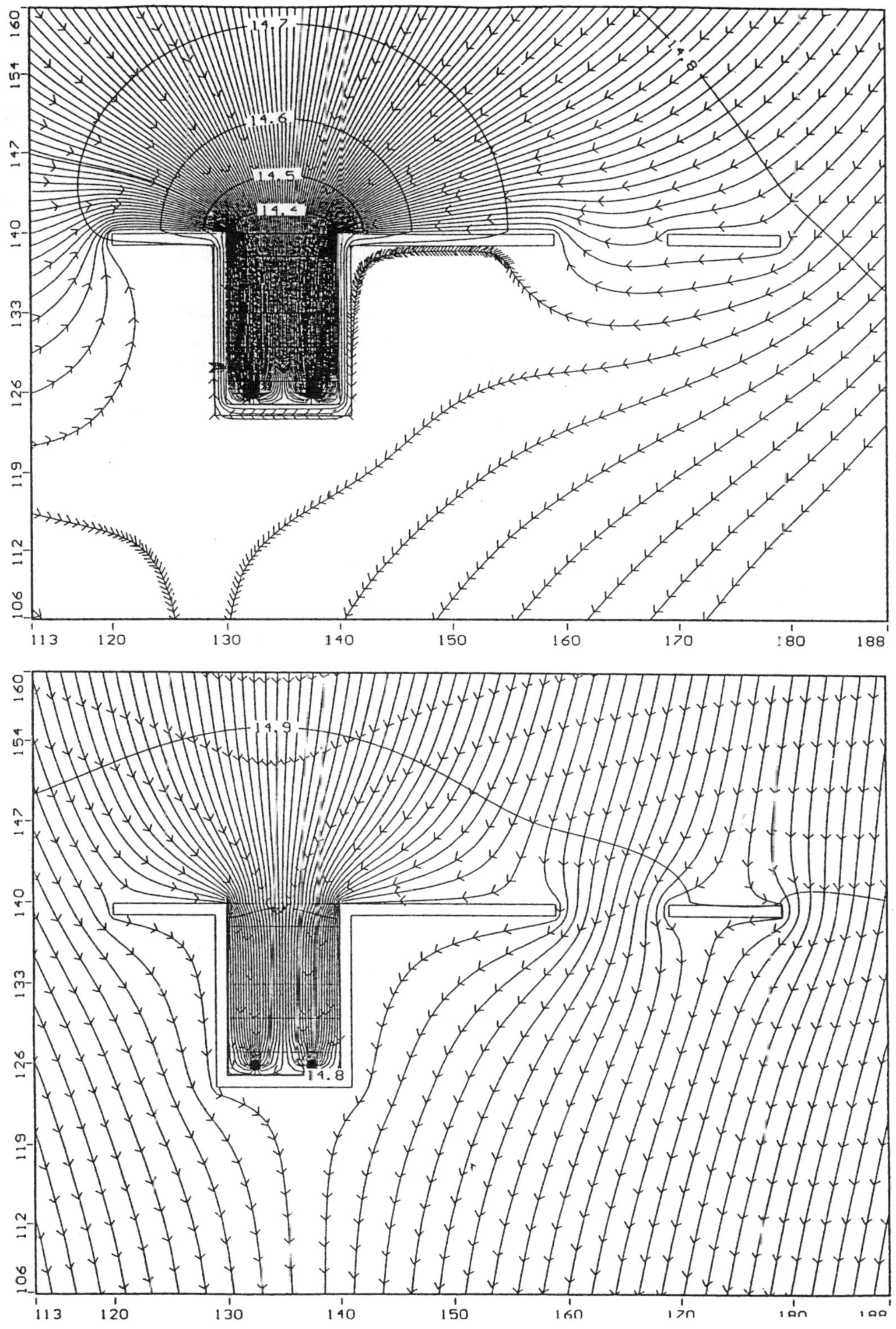

Figure A32.3 Close-up of view of flow lines under different pumping conditions.

A32.6 RESULTS

In general, the groundwater flux through the remedial gate is controlled by the pumping rate, whereas the control gate flux is controlled mainly by the imposed natural gradient. However, the flux through the control gate may also be greatly affected by the pumping rate in the remedial gate. For example, with an imposed natural gradient of 0.002 and an extraction rate of 1.27 m^3/day, groundwater flowed perpendicular and even in the opposite direction to the natural gradient in the vicinity of the control gate (Figure A32.3b). It should be noted again that these are steady-state flow lines when, in fact, the high extraction rate of 1.27 m^3/day only took place for 67 days. Additionally, with an increased natural gradient of 0.003, water flowed through the control gate as anticipated, although still influenced by the extraction wells.

Other aspects of interest for simulation 1 included no apparent mounding in front of the gates as a result of the Funnel-and-Gate structure. A close-up view of the remedial gate also showed a symmetric distribution across the gate width. In addition, the Q_{gate} value was not very sensitive to the magnitude of the high K zones within the gate (i.e., biosparge zone). This could be expected because the flow through the gate will be governed by the bulk gate harmonic K, or, in other words, the lowest K zone in the system, namely, the aquifer material beyond the gate exit.

A32.7 REFERENCES

Anderson, M.P. and W.W. Woessner. 1992. *Applied Groundwater Modeling: Simulations of Flow and Advective Transport*. Academic Press, New York.

Freeze, R.A. and J.A. Cherry. 1979. *Groundwater*. Prentice Hall, Englewood Cliffs, NJ.

PRC. 1993. Solid Waste Assessment Test (SWAT) Report, NAS Alameda, Alameda, California. Private consultants report.

Waterloo Hydrogeologic Software. 1997. Visual MODFLOW. Version 2.5. 180 Columbia Street, Unit 1104. Waterloo, Ontario.

University of Waterloo. 1996b. Revised Workplan for Semipassive Groundwater Remediation Demonstration Project at Site 1, Alameda Naval Air Station, California.

Appendix 33

Technology Evaluation — Modeling for Hypothetical Site

33.1 INTRODUCTION

The purpose of the numerical simulations was to provide estimates of groundwater discharge rates through the gate(s) and provide additional insight into the expected flow system and capture zones to aid in the design of potential Funnel-and-Gate (F&G) configurations for scale-up purposes.

The simulations were developed to produce conservative estimates of gate flux to ensure adequate residence time within the Fe^0 treatment zone. Two approaches for scale-up were taken. The first involved extending the existing F&G structure to capture the core of the plume. For this, the model domain from the existing pilot-scale system was utilized (UW, 1997). The second approach involved an expansive plume capture zone using multiple gates and reactive curtain trench schemes. For this enlarged area of interest, another model domain was created.

A33.2 SITE CHARACTERIZATION

Description of the site hydrogeology can be found in Part I of this book, and only a brief summary will be presented here. The unconfined silty sand aquifer is relatively homogeneous, with an average hydraulic conductivity of $K = 5.7$ ft/d, based on slug tests in the vicinity of M028-A (UW, 1996). The underlying bay mud at a depth of about 22 ft bgs represents a lower confining layer.

The hydraulic gradient varies in magnitude and direction primarily due to seasonal and tidal influences. A report by PRC (1993) characterized the tidal influences in April of 1992. The maximum groundwater level fluctuation observed within 800 ft of the F&G site was 3.5 ft in well M-026A (PRC, 1993). The well closest to the site (M-028A) experienced only a 2.0-in. fluctuation. From the tidally filtered data, the mean direction of groundwater flow in the vicinity of the F&G site was determined to be perpendicular to the shore, flowing from east to west toward the bay. This direction is consistent with the *c*DCE, and BTEX plume shape delineated by EFW. A hydraulic gradient of 0.0017 was calculated from the PRC (April 1992) filtered data set. Updated filtered data from May 2, 1997 and June 27, 1997 (PRC, unpublished data) provided additional estimates of the hydraulic gradient of 0.0034 and 0.0018, respectively (based on the head difference between M-028A and M-034A). Therefore, a gradient of 0.002 was chosen to represent the mean hydraulic gradient. A maximum gradient of 0.0044 was observed in January of 1995 (M. Morkin, unpublished data), based on single measurements (at M-028A and M-034A). Therefore, to give a conservative estimate of the gate residence time a hydraulic gradient of 0.005 was chosen. Information regarding the influence of seasonal fluctuations on the groundwater flow velocity and direction was not available and remains a source of uncertainty with regards to the applicability of the results.

A33.3 CONCEPTUAL MODEL

The scale-up scenario involving the existing structure was conducted with a detailed grid. The modeled domain consisted of a 22-ft-thick unit, representing an unconfined aquifer with the underlying bay mud providing an impermeable base. The area of consideration was 300 ft by 300 ft, centered around the F&G structure (Figure A33.1). This large area was chosen to minimize uncharacterized boundary effects, because the San Francisco Bay represents the only physically defined vertical boundary in the model domain. This eastern domain boundary as well as the western boundary are represented as constant head boundary conditions, while the remaining north and south boundaries are considered as no-flow. Both the top and bottom of the domain were also considered no-flow (i.e., no recharge and no leakage) boundaries.

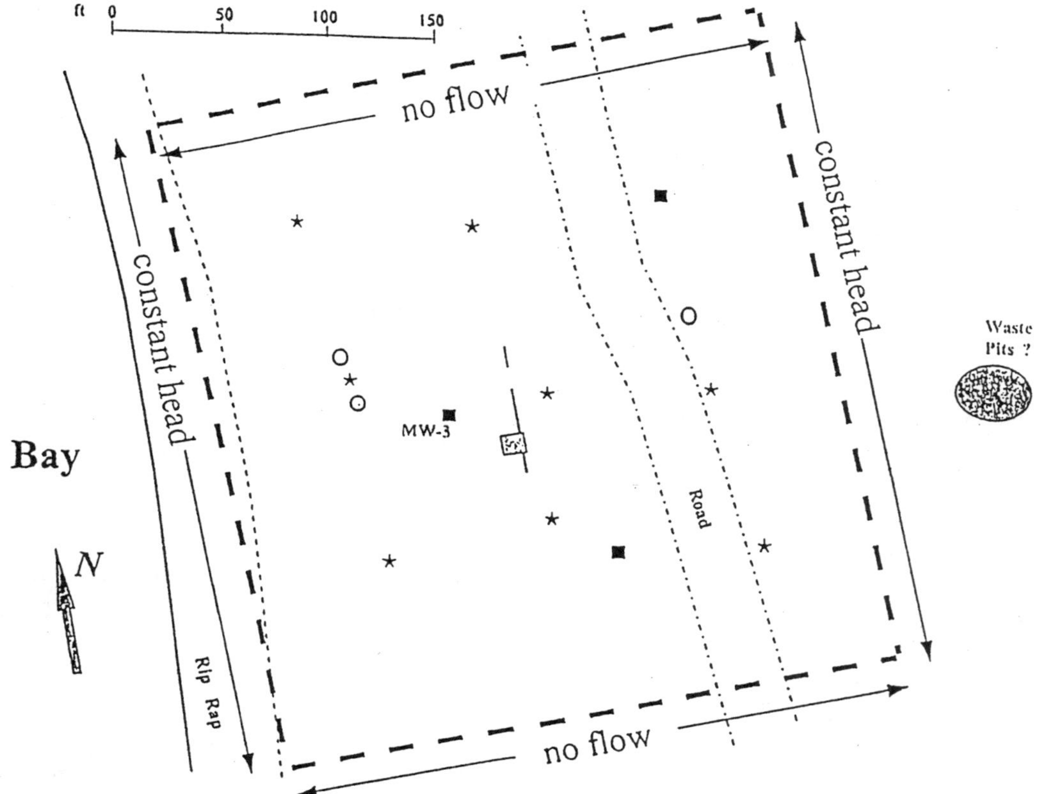

Figure A33.1 Model domain and boundary conditions.

The scenarios involving multiple gates and reactive curtain trenches necessitated an enlarged domain with coarser grid spacing. As such, only a bulk representation of the gate material could be rendered. To obtain a first approximation of expected capture zone widths and velocities within the gates, the existing conceptual model was simply enlarged to 2000 ft in the N–S direction, and 1000 ft in the E–W direction. Effects of the northwestern shore intersecting the domain and interfering with the assigned boundary conditions were disregarded. The modeled Funnel-and-Gate structures will therefore capture water that would, in reality, be diverted to the NW shore. These results will therefore overestimate both the volumetric flux through the gates and the gas capture zone widths.

A33.3.1 Model Design

The 300 ft by 300 ft model domain was discretized with 5 ft spacing in both the x and y directions, with refinement to 1 ft spacing beginning at distance of 20 ft adjacent to the F&G system, and 0.5 ft at zones of intense convergent and divergent flow. The vertical dimension was represented as five layers. The total number of cells was approximately 75,000. The Funnel-and-Gate design simulated using this domain consisted of converting the existing control gate to an identical remedial gate, with 50-ft-long flank walls (Figure A33.2, Simulation 1).

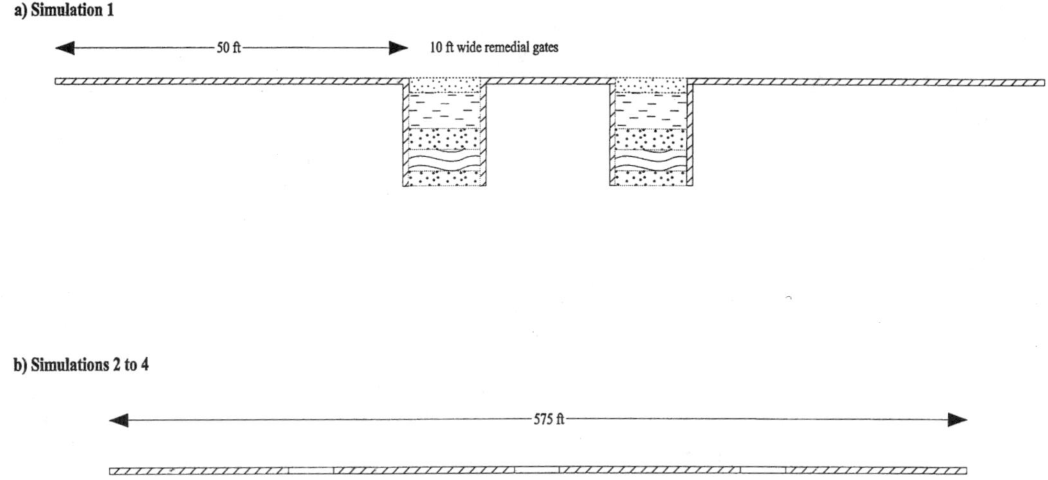

Figure A33.2 Funnel-and-Gate configurations for scale-up scenarios.

The 1000 ft by 2000 ft model domain was discretized with 25 ft spacing, with refinement to 2.5 ft spacing adjacent to the F&G structure. The vertical dimension was represented as a single layer. The total number of cells was approximately 7,000. The Funnel-and-Gate designs simulated using this domain consisted of a multiple-gate scheme and a reactive curtain scheme (Figure A33.2, Simulations 2 to 5). Three different K zones were simulated using the multiple-gate design to represent different construction techniques. For example, Simulation 2 represents Fe^0 emplacement using a vibrating beam technique, with passive wells to supply O_2. Simulation 3 represents emplacement by excavation and backfill for the Fe^0 zone and passive wells to supply O_2. Simulation 4 represents emplacement by excavation and backfill for the Fe^0 and creation of an open biosparge zone for O_2 supply via excavation. Simulation 5 represents a reactive curtain emplacement using the same techniques as in Simulation 4.

The hydraulic conductivity, K, of the aquifer was assumed to be homogeneous and isotropic with a value of 5.7 ft/d (2×10^{-3} cm/s). Assumed values for porosity of 0.30 (Freeze and Cherry, 1979), specific storage of 1×10^{-3} ft^{-1} (Anderson and Woessner, 1992), and 0.32 for specific yield (Anderson and Woessner, 1992) were chosen based on typical sandy aquifer values.

The hydraulic gradient was imposed by setting the eastern and western boundary to constant head conditions. The constant head values were calculated so as to place the water table about 7.2

ft bgs at the gravel-iron zone interface within the remedial gate. For example, a hydraulic gradient of 0.002 was imposed by applying a constant head on the western and eastern boundaries of 14.649 ft and 15.249 ft, respectively (i.e., 7.35 ft and 6.75 ft bgs, respectively). This gradient induced flow within the aquifer from east to west at a velocity of 0.038 ft/day (1.15 cm/day). No recharge was allowed on the top boundary. This simple hydrogeologic rendering was not subjected to model validation. For the expanded domain, the boundary conditions were set up in a similar manner to achieve a water level within the gate of around 7.1 ft bgs.

The F&G scale-up designs were approximated using the hydraulic properties seen in Table A33.1. The remedial gates consisted of funnel walls (both parallel and orthogonal to the ambient groundwater flow direction) to direct contaminated water first through a zone of gravel, followed by iron, and then through another zone of gravel (Figure A33.2).

Table A33.1 Hydraulic Properties Used to Represent the F&G

Property	*K* (ft/d)	Source	Porosity	Source
Aquifer	5.7	Revised Work Plan	0.30	Freeze & Cherry
Gate gravel	2830	ETI, personal communication	0.40	Katic, personal communication
Fe	141.7	ETI, permeameter	0.50	ETI, personal communication
Sand and Fe	145.6	Austrins, permeameter	0.30	Freeze & Cherry
Biosparge	2.83×10^{8}	assumed	0.90	assumed
Funnel wall	2.83×10^{-5}	ETI, personal communication	0.10	assumed
Bentonite cap	same as above		same as above	
Concrete pad	same as above		same as above	

Large contrasts in *K*, which become an issue when simulating the funnel walls and the O_2 biosparge zone, may cause numerical difficulties. These numerical problems may manifest themselves by inducing a poor water balance or anomalous trends in hydraulic head data. Representation of the funnel walls using the Horizontal-Flow-Barrier (HFB) Package available in MODFLOW instead of the hydraulic properties listed in Table A33.1 did not improve the already low water balance error and allowed more leakage through the wall itself, thus it was not invoked. The O_2 biosparge zone was represented by assigning a large *K* value. There was no evidence of physically unrealistic results or poor mass balances using the hydraulic properties listed in Table A33.1. It should be noted that porosity values are only utilized in the numerical code when velocities are calculated and therefore do not affect the equipotential head distribution.

The hydraulic parameters for the gate material in the multiple gate scenarios were approximated by calculating the harmonic *K* average using equal thicknesses of each property, giving a total thickness (d) of 2.5 ft:

$$K_{harm} = \frac{d}{\sum \frac{d_i}{K_i}} \tag{1}$$

From this equation the resultant K_{harm} values (Table A33.2) were calculated from the *K* values listed in Table A33.1.

Table A33.2 Hydraulic Properties of the Gate Material Used in the Multiple Gate Simulations

	K_{harm} (ft/d)	Porosity
Sim 2: aquifer	5.7	0.40
Sim 3: Fe and aquifer	10.9	0.40
Sim 4,5: Fe and gravel	270	0.40

A33.3.2 Model Application

To determine the gate residence time, the Zone Budget Package in MODFLOW was utilized. It calculates the volumetric discharge through a specified zone, such as the gate entrance, giving Q_{gate}. The groundwater velocity, $\bar{v}$, from which the travel time could be calculated was obtained by evaluating

$$\bar{v} = \frac{Q_{gate}}{n \times A} \tag{2}$$

where n is the porosity and A is the cross-sectional area for flow.

For Simulation 1, the first three zones in the remedial gates have a consistent A based on the depth between the bottom of the overlying bentonite cap and the top of the underlying cement foundation (12 ft) multiplied by the gate width (10 ft). For the biosparge zone (Simulation 1) and all the gates in the multiple gate simulations (i.e., Simulations 2 to 5), the saturated thickness for flow varied for each simulation. The depth of flow was determined by plotting equipotentials (to determine the water table elevation) specific to the zone in question. Using this value, the saturated thickness between the water table and the top of the clay aquitard or the cement foundation could be determined. Multiplying this value by the gate width gave the cross-sectional area for flow. In all cases, the model was run to steady state with the Waterloo Hydrogeologic Software solver.

A33.4 RESULTS

Simulation 1 has an estimated capture zone of about 83 ft (Figure A33.3). Results from the existing Funnel-and-Gate design along with Simulation 1 values are seen in Table A33.3.

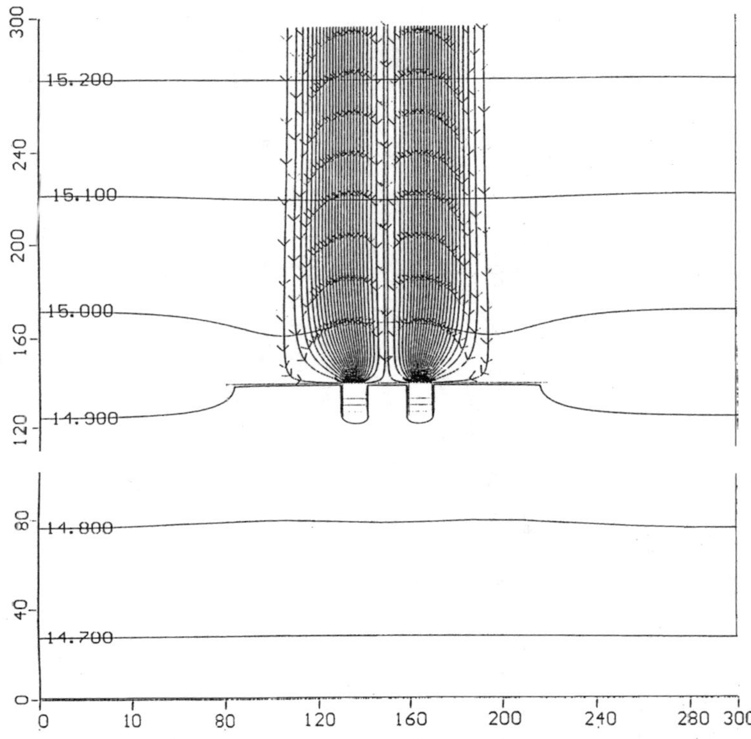

Figure A33.3 Simulation 3, 50-ft flank walls added to existing Funnel-and-Gate structure.

Table A33.3 Simulation 1 Results

Simulation	Description	Natural Gradient	Q_{pump} (ft^3/day)	Remedial g (ft^3/day)	Control g (ft^3/day)	Capture Zone (ft)	Velocity (ft/day) Fe	Velocity (ft/day) Control Gate	(Pore Vol.)/Day Fe
	Existing F&G structure	0.002	-6.0	12.00	3.18	77 (54)	0.20	0.07	0.04
	and pumping conditions	0.005	-6.0	12.00	10.19	49 (26)	0.20	0.23	0.04
1a	Control gate converted	0.002	0.0	6.88	6.88	83	0.11	0.11	0.02
1b	to a remedial gate with extended flank walls (50 ft long in total)	0.005	0.0	17.27	17.27	84	0.29	0.29	0.06

Notes: Q_{pump} = pumping rate for each of the two extraction wells.

Q_{pump} = 0 ft^3/d refers to natural flow through the gate, i.e., back end of the remedial gate removed.

Capture zone = total width of the capture zone (value in brackets is the remedial gate capture zone).

Simulation 1 = control gate modified to be identical to the remedial gate.

All simulations steady state.

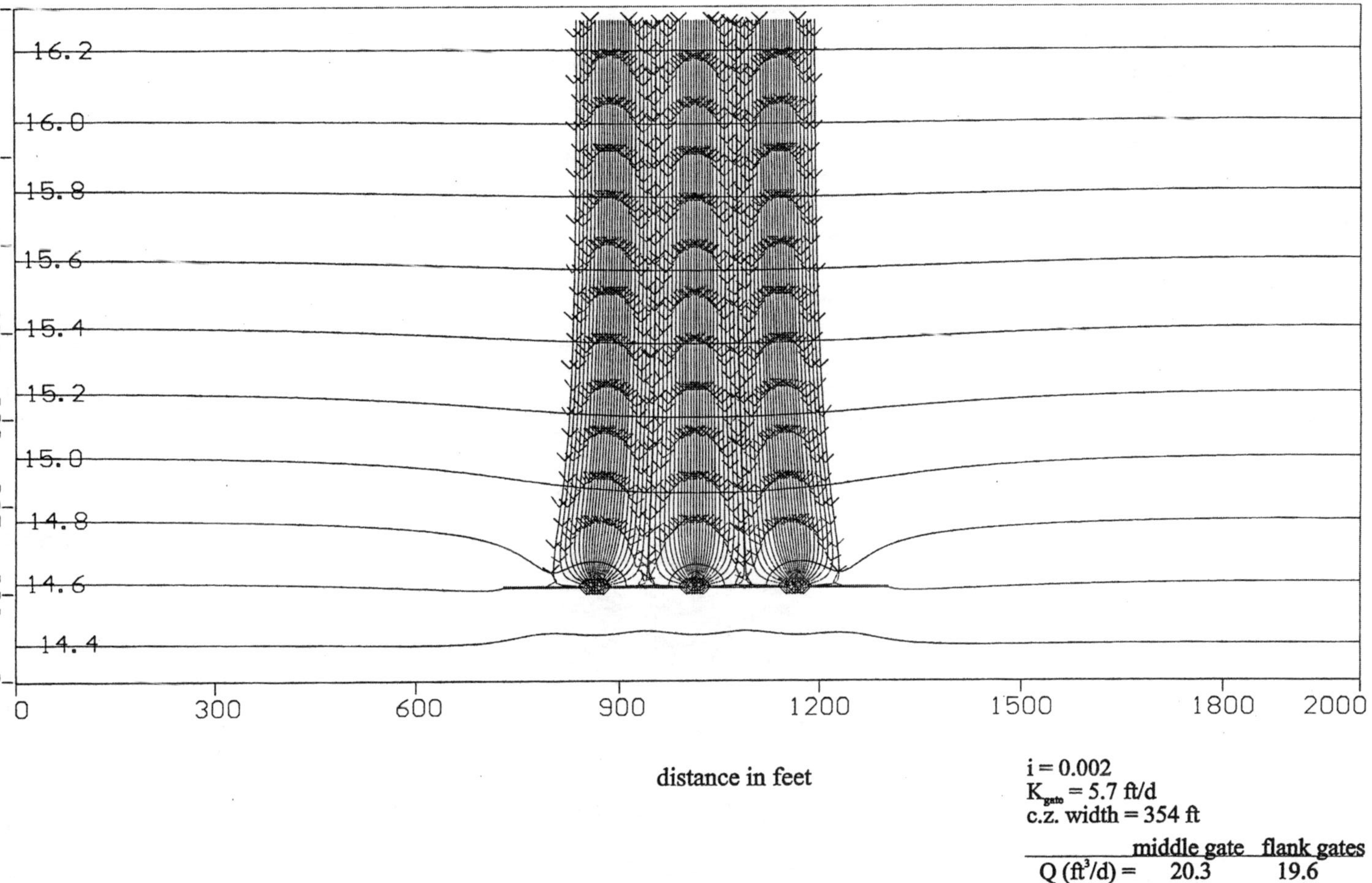

Figure A33.4 Simulation 2a, gate capture zones.

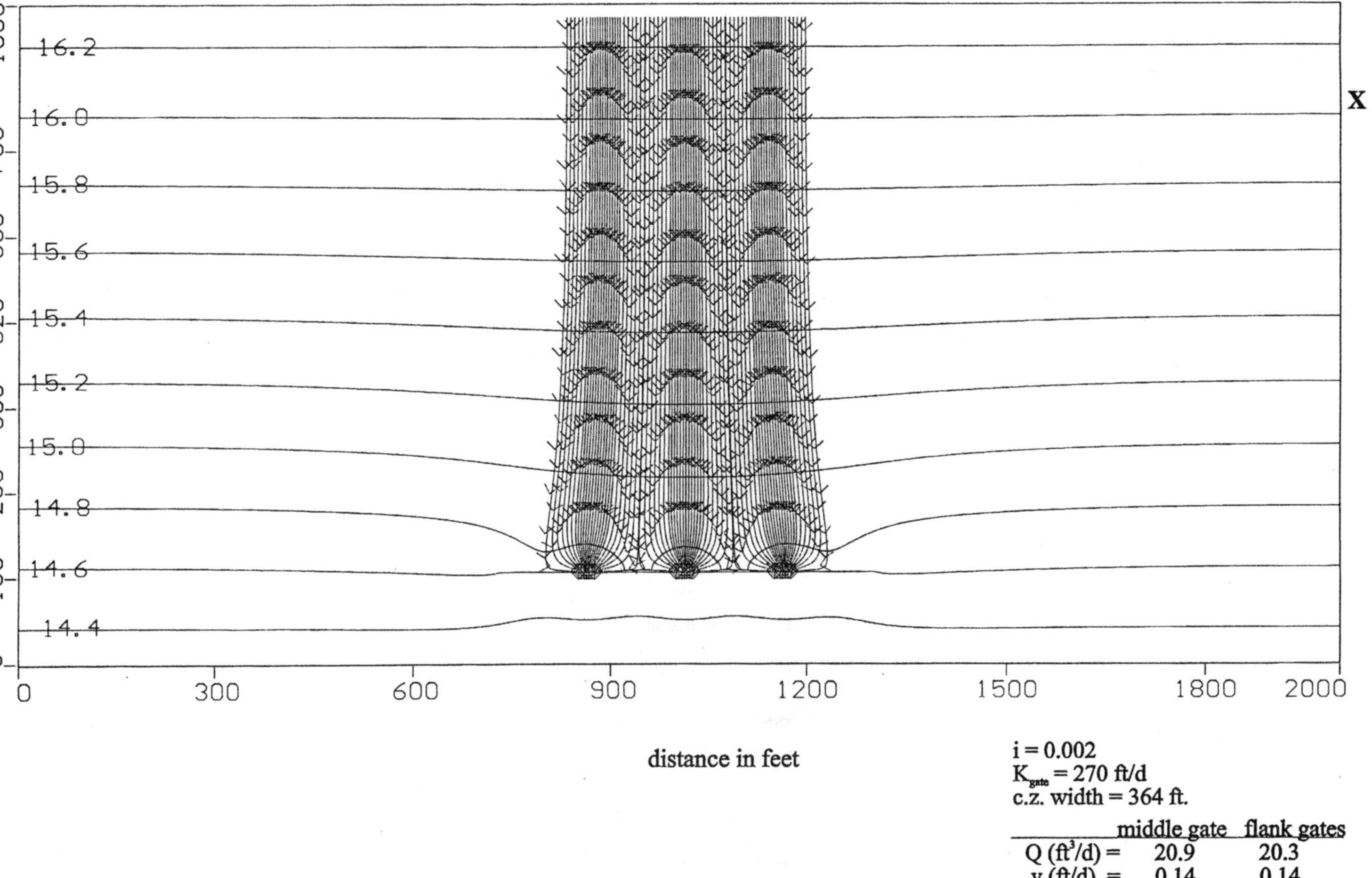

Figure A33.5 Simulation 4a, gate capture zones.

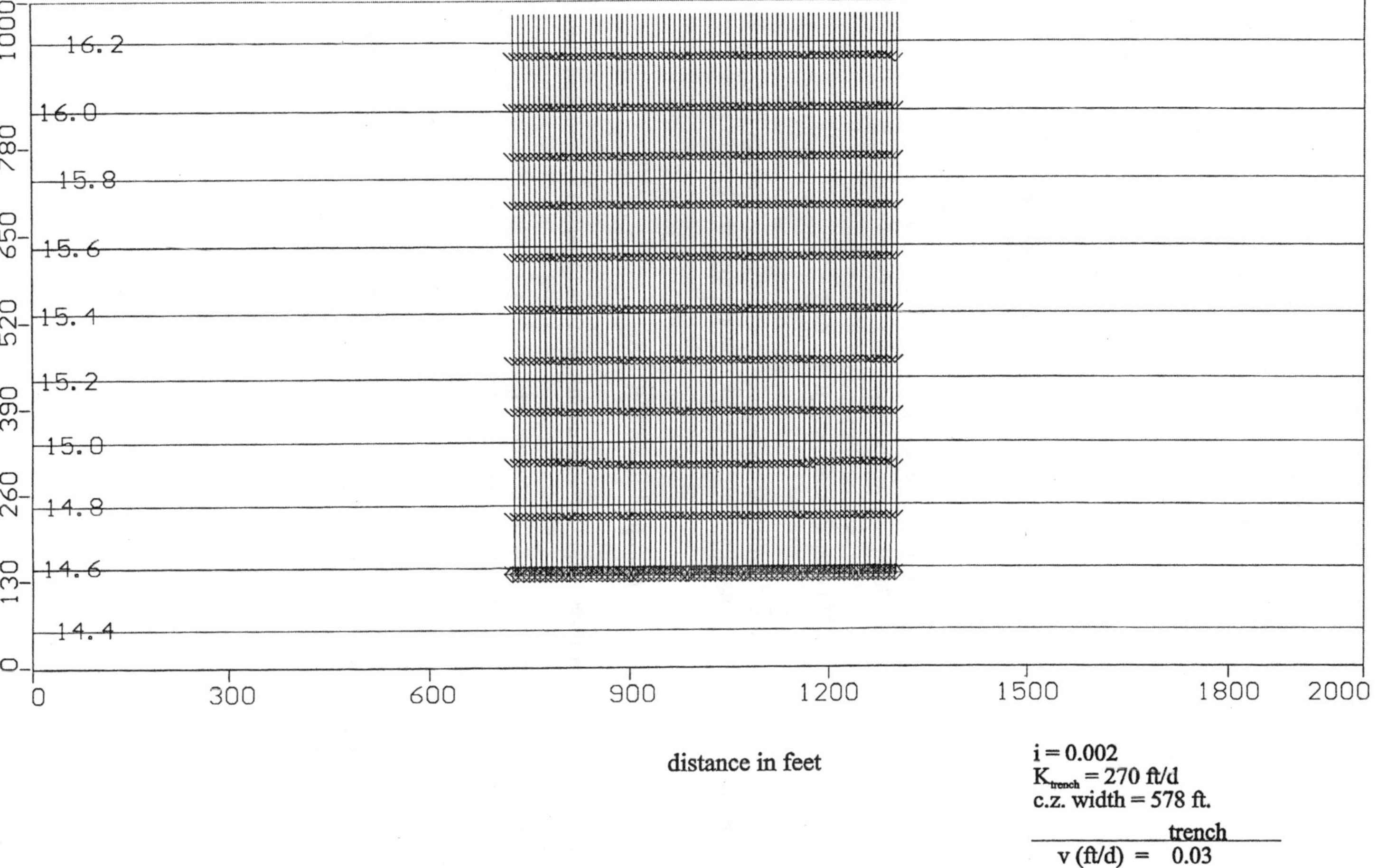

Figure A33.6 Simulation 5a, trench capture zone.

The multiple gate and reactive trench designs are found in Figures A33.4 to A33.6. Results from these simulations are seen in Table A33.4.

Table A33.4 Multiple Gate Simulation Results

Simulation	K_{gate}	Natural Gradient	Total C.Z. Width (ft)	Flank Gate		Middle Gate	
				Q (ft³/day)	v (ft/d)	Q (ft³/day)	v (ft/d)
2a	5.7	0.002	354	19.6	0.13	20.3	0.14
2b	5.7	0.005	356	53.9	0.35	55.8	0.37
3a	10.9	0.002	359	19.9	0.14	20.6	0.14
3b	10.9	0.005	360	54.8	0.36	56.5	0.37
4a	270	0.002	364	20.3	0.14	20.9	0.14
4b	270	0.005	364	55.6	0.37	57.3	0.38
5a	270	0.002	578	---	---	---	0.03
5b	270	0.005	578	---	---	---	0.08

Note: Simulations 2 to 4 consist of three, 25'-wide gates, bounded by four, 125' for a total length of 575' (F:G = 6.7:1).

K_{gate} values represent the harmonic K average of various combinations of gate material.

1: aquifer only, i.e., beam emplacement of Fe, and passive wells for O_2 delivery.

2: iron + aquifer, i.e., backfilled Fe zone, and passive wells for O_2 delivery.

3: iron + gravel, i.e., backfilled Fe zone, and biosparging for O_2 delivery.

F&G operation under natural flow conditions (i.e., no extraction wells).

The total capture zone (C.Z.) width includes both gate C.Z.s.

Overall, the flux through the gates appears relatively insensitive to changes in K_{gate}. This is likely because the highest K value modeled still restricts flow through the gate to a large extent.

A33.5 REFERENCES

Anderson, M.P. and W.W. Woessner. 1992. *Applied Groundwater Modeling: Simulation of Flow and Advective Transport*. Academic Press, New York.

Freeze, R.A. and J.A. Cherry. 1979. *Groundwater*. Prentice Hall, Englewood Cliffs, NJ.

PRC. 1993. Solid Waste Assessment Test (SWAT) Report, NAS Alameda, Alameda, California. Private consultant's report.

Waterloo Hydrogeologic Software. 1997. Visual MODFLOW. Version 2.5. 180 Columbia Street, Unit 1104. Waterloo, Ontario.

University of Waterloo. 1996. Revised Workplan for Semi-Passive Groundwater Remediation Demonstration Project at Site 1, Alameda Naval Air Station, California.

University of Waterloo. 1997. Passive and Semi-Passive Techniques for Groundwater Remediation. Phase 3 Installation Report.

Appendix 34

Technology Evaluation — Calculations for BTEX Degradation

The calculations to determine how much molecular oxygen was needed to degrade BTEX were based on the example shown in Cookson (1995), with some modifications.

Assumptions:

- Equal mole ratio of each compound (e.g., 2 to 5 mg/L of each B, T, E, and X)
- $f_e = f_s = 0.5$
- Ammonia is nitrogen source

The cell synthesis equation

$$f_e + f_s = 1 \tag{1}$$

where f_e = fraction of organic matter oxidized for energy and f_s = fraction of organic matter associated with conversion to microbial cell matter, is based on McCarty (1987) for factors for energy distribution in an aerobic environment.

Since most sites are a mixture of organics it is easier to write one redox reaction for the mixture's mean chemical structure rather than a reaction for each compound. Table A34.1 shows the average chemical structure of the mixed contaminants.

The average number of atoms in the mixture's chemical structure is the total mass divided by the atomic weight. Therefore, the structure of this mixture is C_7H_9.

Assuming that ammonia is the nitrogen source, then solving the reaction (2)

$$H_D + f_e H_A + C_s \tag{2}$$

where H_D = half reaction for electron donor, H_A = half reaction for electron acceptor, and C_s = cell synthesis will determine the mass ratios of oxygen needed to degrade the mixed compound (C_7H_9).

According to Cookson (1995), the half reaction for H_D is

$$1/Z(C_aH_bO_cN_a) + 12a - c/Z(H_2O) = a/Z(CO_2) + d/Z(NH_3) + H^+ + e^- \tag{3}$$

where $Z = 4a + b - 2c - 3d$, and a, b, c, d represent the average number of atoms for C, H, O and N, respectively, in the organic contaminant mixture. In this case $c = 0$ and $d = 0$, so the equation simplifies to

$$1/Z(C_aH_b) + 2a/Z(H_2O) = a/Z(CO_2) + H^+ + e^- \tag{4}$$

Table A34.1 Average Chemical Structure of Mixed Contaminants

Chemical	Mass Proportion of Atomic Weight[a]		Fraction of Total Mass Based on Concentration[b]	Total Molecular Atomic Weight Contribution[c]	
	C	H		C	H
Benzene (C_6H_6)	72	6	2.5/10 = 25%	18	1.5
Toluene (C_7Hg)	84	8	25%	21	2.0
Ethylbenzene (C_8H_{10})	96	10	25%	24	2.5
Xylenes (C_8H_{10})	96	10	25%	24	2.5
Total Mass Contributors				87	8.5

[a] Atomic weight times # atoms of each.

[b] Average concentration divided by total concentration.

[c] Mass proportion (e.g., for carbon) multiplied by fraction of total mass.

The half reaction for H_A under aerobic conditions, with oxygen as electron acceptor, is

$$1/4\ O_2 + H^+ + e^- = 1/2\ H_2O \tag{5}$$

The cell synthesis equation (C_5) when ammonia is the nitrogen source is

$$1/4\ CO_2 + 1/20\ NH_3 + H^+ + e^- = 1/20\ C_5H_7O_2N + 2/5\ H_2O \tag{6}$$

Writing the overall reaction (3) for C_7H_9 (our mixed chemical contaminant) yields H_D

$$1/37\ C_7H_9 + 2/37\ H_2O = 7/37\ CO_2 + H^+ + e^- \tag{7}$$

Replacing the reaction pairs listed in (2) with 7, 5, and 6 generates

$$\begin{aligned} &1/37\ C_7H_9 + 14/37\ H_2O + 1/8\ O_2 + 1/8\ CO_2 + 1/40\ NH_3 \\ &= 1/40\ C_5H_7O_2N + 1/20\ H_2O + 1/4\ H_2O + 7/37\ CO_2 \end{aligned} \tag{8}$$

Reduction of common terms generates

$$C_7H_9 + 19/10\ H_2O + 45/8\ O_2 + 37/40\ NH_3 = 37/40\ C_5H_7O_2N + 23/8\ CO_2 \tag{9}$$

To establish the mass ratios one needs to calculate molecular weights.

$$C_7H_9 = 93 \qquad O_2 = 32$$

Thus,

$$C_7H_9{:}O_2$$

$$1{:}45/8\ (32/93)$$

or 1:1.59, so for every milligram of BTEX, 1.59 mg of O_2 is needed. Given that there are other exertions on oxygen demand, this value was increased to a 1:6 ratio for the calculations in Chapter 8.

REFERENCES

Cookson, J.T. 1995. *Bioremediation Engineering*. McGraw-Hill, New York.

McCarty, P.L. 1987. Bioengineering issues related to in-situ remediation of contaminated soils and groundwater. In *Environmental Biotechnology,* G.S. Omenn (Ed.). Plenum Press, New York, pp. 143-162.

Appendix 35

Supplemental Information on Pump-and-Treat System Costs

A35.1 INTRODUCTION

Cost estimates for two pump-and-treat systems were developed for comparison to cost estimates for the two sequenced permeable reactive barriers (SPRBs) at NAS Alameda, CA. These pump-and-treat systems were designed to capture and treat the same plume as the SPRBs. The smaller of the two systems was sized to capture the core of the plume (80 ft) (Case A), and the larger system to capture a larger 350-ft plume (Case B).

A35.2 PUMP-AND-TREAT SYSTEM DESIGN

The site-specific numerical groundwater model developed by the University of Waterloo (UW) (Appendix 32) was modified to estimate the groundwater extraction rates required to capture the required plumes. Based on the modified model results, a combined groundwater extraction rate of 0.4 gpm (1.5 L/min) from two wells would capture the 80-ft plume. A three-well system with a combined extraction rate of 2.6 gpm (9.8 L/min) would capture the 350-ft plume (Appendix 33). In the event that higher extraction rates would be required, both groundwater treatment systems (GTSs) were designed with a maximum capacity of 5 gpm (19 L/min).

For the 80-ft plume, influent *c*DCE and VC concentrations requiring treatment were assumed to be 100 and 11 mg/L, respectively. Although higher flow rates would be required to capture the 350-ft plume, influent *c*DCE and VC concentrations from the additional capture zone would be lower due to lower concentration groundwater from the plume margins. Since VOC concentrations outside of the 80-ft core of the plume are relatively low, both treatment systems were assumed to treat the same total VOC mass. Treatment was assumed to be via air stripping, with off gas treatment using activated carbon.

GTS performance monitoring includes collection and analysis of influent and effluent samples on a monthly basis to ensure the GTS is providing adequate removal of the contaminants prior to discharge.

Treated groundwater was discharged to the local publicly owned treatment works via the municipal sanitary sewer system. Consequently, groundwater discharge fees were incorporated into the operating, maintenance, and monitoring (O&M) costs of the system.

The cost estimates for the pump-and-treat system included five monitoring wells for the 80-ft plume system and eight monitoring wells for the 350-ft system. These monitoring wells would provide the data necessary to ascertain regulatory compliance.

A35.3 COST ESTIMATES

Two construction and O&M cost estimates were developed for both the 80-ft plume and the 350-ft plume. The first construction and O&M cost estimate assumed *c*DCE was the primary contaminant in the extracted groundwater as discussed above. The second construction and O&M cost estimate assumed that TCE was the primary contaminant.

The capital cost estimates for the described treatment scenarios above included extraction and monitoring wells, pumps, piping, a feed tank, an air stripper, a blower, activated carbon vessels, and process instrumentation. The O&M costs included in the cost estimates included groundwater monitoring, treatment system performance monitoring, operator labor, activated carbon replacement, energy consumption, groundwater discharge fees, and reporting. Groundwater monitoring was assumed to occur with the same frequency as for the SPRBs.

Cost estimates are tabulated in the following worksheets. For the 80-ft plume with *c*DCE as the primary contaminant, the capital cost, O&M cost for the first 2 years, and O&M cost for the remaining 28 years were $120,250, $116,040, and $97,780, respectively. For the 80-ft plume with TCE as the primary contaminant, the capital cost, O&M cost for the first 2 years, and O&M cost for the remaining 28 years were $120,250, $108,550, and $84,290, respectively.

For the 350-ft plume with *c*DCE as the primary contaminant, the capital cost, O&M cost for the first 2 years, and O&M cost for the remaining 28 years were $157,950, $120,206 and $94,526, respectively. For the 350-ft plume with TCE as the primary contaminant, the capital cost, O&M cost for the first 2 years, and O&M cost for the remaining 28 years were $157,950, $112,716, and $87,036, respectively.

ATTACHMENT A35.1: COST ESTIMATE WORKSHEETS FOR A PUMP-AND-TREAT SYSTEM TO CAPTURE AND TREAT THE 80-FT-WIDE PLUME

BEAK INTERNATIONAL INC.

CONSTRUCTION COST ESTIMATE SHEET

Client: 7304.10

Description: Case A- 80 ft Plume Containment

Alternative: GW Extraction / Air Stripping

Prepared by: JC **Revision No.** A

Checked by: JC **Date Updated:** 15-May-98

Sec. No.	Sub. No.	Description/Activity	Material Required		Unit Costs		Total Cost		Sub Totals ($)	Total Cost ($)
			Quantity	Unit	Material ($)	Labour ($)	Material ($)	Labour ($)		
1.0		**Groundwater Extraction System**								
	1.1	Two (2) 6 " Extraction Wells	2		10,000	incl.	20,000	incl.	20,000	
	1.2	Two (2) GW pumps	2		1,400	600	2,800	1,200	4,000	
	1.3	Excavation, Pipe Bedding & Backfill to Tmt. Bld.	100	lin. ft.	incl.	9.00	incl.	900	900	
	1.4	Piping Placed in Trench	100	ft.	1.00	incl.	100	incl.	100	
	1.5	Well Testing	2		incl.	4,500	incl.	9,000	9,000	
										34,000
2.0		**Groundwater Treatment System**	1		35,000		35,000	0	35,000	
	2.1	Skid Mounted Building (7'x10')								
		Building Heater and Light								
		Feed Tank								
		Low Profile Air Stripper								
		Transfer Pump								
		80 CFM Blower								
		Flow Control Valve								
		Totalizing Flowmeter								
		Control Panel (PLC)								
		Piping For Gas Phase Carbon Vessels								
	2.2	Installation	1			10,000		10,000	10,000	
										45,000
3.0		**Piping to Sanitary Sewer**	1	L.S.	5,000	incl	5,000	incl	5,000	5,000
4.0		**Monitoring Wells**	7	L.S.	incl	500	incl	3,500	3,500	3,500
5.0		**Mobilization, Site Preparation, Demobilization**	1	L.S.	5,000	incl.	5,000	incl.	5,000	5,000
CONSTRUCTION SUBTOTAL									92,500	92,500

BEAK INTERNATIONAL INC.			Client:		7304.10	
			Description:	Case A- 80 ft Plume Containment		
OPERATING COST ESTIMATE (Year 1 and 2)			Alternative:	GW Extraction / Air Stripping		
			Prepared by:	JC	Revision No. A	
			Checked by:	JC	Date Updated: 15-May-98	
Sec. No.	**Sub. No.**	**Description/Activity**	**Unit Cost**	**Basis**	**Cost ($/year)**	**Total Cost ($/year)**
1.0		**Groundwater Monitoring**				
	1.1	Sample Collection	$65/hr	12hr x 4 times per year x $65/hr	3,120	
	1.2	Sample Analysis	$150/sample	7 wells (+2 QA/QC) x 4 times per year x $150/sample	5,400	
						8,520
2.0		**Groundwater Extraction**				
	2.1	EW Pump Power (2)	$ 0.05 / KWH	2x 0.5 HP x 0.7457 x 24hr x 365 d	327	
	2.2	EW Pump Maintenance	25% of cost	25% x4000	1,000	
	2.3	Operating Labor	$50/hr	8hr x 12 times per year x $50/hr	4,800	
						6,127
3.0		**Groundwater Treatment System**				
	3.1	Feed Pump Power (1 HP)	$ 0.05 / KWH	1 HP x 0.7457 x 24hr x 365 d	327	
	3.2	Blower Power (1.0 HP)	$ 0.05 / KWH	1.0 HP x 0.7457 x 24hr x 365 d	327	
	3.3	Operating Labor	$50/hr	14 hr x 52 times per year x $50/hr	36,400	
	3.4	Equipment Maintenance	7.5% of cost	7.5% x 35,000	2,625	
	3.5	Performance Monitoring Analytical	$200/sample	12 per year x $200 x 2 samples x 1.2 (QA/QC)	5,760	
	3.6	Performance Monitoring Smpl Collection & Rpt.	$75/hr	2 hr/sampling event x 12 sampling events x $75/hr	1,800	
	3.7	Carbon Consumption	$535/180 lbs drum	20 drums per year	10,700	
						57,938
4.0		**Groundwater Discharge**				
	4.1	Annual Permit Fee	$2,500 per year	$2,500 per year x 1 year	2,500	
	4.2	Monitoring and Testing Charges	$70 per month	$70 per month x 12 months	840	
	4.3	Disposal Charges	$0.41 per 100 ft^3	$0.41 x (28,107/100 ft^3)	115	
						3,455
5.0		**Reporting**	$80/hr	125 hr x 4 times per year x $80/hr	40,000	
						40,000
OPERATING COSTS (Year 1 and 2)					116,040	116,040

BEAK INTERNATIONAL INC.

OPERATING COST ESTIMATE (Year 3 through 30)

Client: | **7304.10**

Description: Case A- 80 ft Plume Containment

Alternative: GW Extraction / Air Stripping

Prepared by: JC | **Revision No.** A

Checked by: JC | **Date Updated:** 15-May-98

Sec. No.	Sub. No.	Description/Activity	Unit Cost	Basis	Cost ($/year)	Total Cost ($/year)
1.0		**Groundwater Monitoring**				
	1.1	Sample Collection	$65/hr	12hr x 2 times per year x $65/hr	1,560	
	1.2	Sample Analysis	$150/sample	7 wells (+2 QA/QC) x 2 times per year x $150/sample	2,700	
						4,260
2.0		**Groundwater Extraction**				
	2.1	EW Pump Power (2)	$ 0.05 / KWH	2x 0.5 HP x 0.7457 x 24hr x 365 d	327	
	2.2	EW Pump Maintenance	25% of cost	25% x4000	1,000	
	2.3	Operating Labor	$50/hr	8hr x 12 times per year x $50/hr	4,800	
						6,127
3.0		**Groundwater Treatment System**				
	3.1	Feed Pump Power (1 HP)	$ 0.05 / KWH	1 HP x 0.7457 x 24hr x 365 d	327	
	3.2	Blower Power (1.0 HP)	$ 0.05 / KWH	1.0 HP x 0.7457 x 24hr x 365 d	327	
	3.3	Operating Labor	$50/hr	14 hr x 52 times per year x $50/hr	36,400	
	3.4	Equipment Maintenance	7.5% of cost	7.5% x 35,000	2,625	
	3.5	Performance Monitoring Analytical	$200/sample	12 per year x $200 x 2 samples x1.2 (QA/QC)	5,760	
	3.6	Performance Monitoring Smpl Collection & Rpt.	$75/hr	2 hr/sampling event x 12 sampling events x $75/hr	1,800	
	3.7	Carbon Consumption	$535/180 lbs drum	20 drums per year	10,700	
						57,938
4.0		**Groundwater Discharge**				
	4.1	Annual Permit Fee	$2,500 per year	$2,500 per year x 1 year	2,500	
	4.2	Monitoring and Testing Charges	$70 per month	$70 per month x 12 months	840	
	4.3	Disposal Charges	$0.41 per 100 ft^3	$0.41 x (28,107/100 ft^3)	115	
						3,455
5.0		**Reporting**	$80/hr	125 hr x 2 times per year x $80/hr	20,000	
						20,000
OPERATING COSTS (Year 3 through 30)					91,780	91,780

BEAK INTERNATIONAL INC.	Client:			7304.10
	Description:	Case A- 80 ft Plume Containment		
TOTAL CAPITAL & NPV COSTS ESTIMATE	Alternative:	GW Extraction / Air Stripping		
	Prepared by:	JC	Revision No.	A
	Checked by:	JC	Date Updated:	15-May-98

TOTAL CAPITAL COST ESTIMATE

A	Construction Subtoal	from above	92,500
B	Permitting & Legal	5% of D	4,625
C	Services During Construction	10% of D	9,250
D	Other Costs	15% of D	13,875
E	TOTAL IMPLEMENTATION COSTS	B + C + D	27,750
F	**TOTAL CAPITAL COSTS**	**A + E**	**120,250**

NET PRESENT VALUE CALCULATIONS

TOTAL CAPITAL COSTS	F above	120,250
First Year Operating Costs	from above	116,040
Second Year Operating Costs	from above	116,040
Operating Costs for Each Year (3 to 30)	from above	91,780
Interest Rate		8%
NET PRESENT VALUE		1,196,752

ATTACHMENT A35.2: COST ESTIMATE WORKSHEETS FOR A PUMP-AND-TREAT SYSTEM TO CAPTURE AND TREAT THE 80-FT-WIDE PLUME WITH TCE AS THE MAJOR CONTAMINANT

BEAK INTERNATIONAL INC.

CONSTRUCTION COST ESTIMATE SHEET

Client: | **7304.10**

Description: Case A- 80 ft Plume Containment (TCE Primarily)

Alternative: GW Extraction / Air Stripping

Prepared by: JC | **Revision No.** A

Checked by: JC | **Date Updated:** 15-May-98

Sec. No.	Sub. No.	Description/Activity	Material Required		Unit Costs		Total Cost		Sub Totals ($)	Total Cost ($)
			Quantity	Unit	Material ($)	Labour ($)	Material ($)	Labour ($)		
1.0		**Groundwater Extraction System**								
	1.1	Two (2) 6 " Extraction Wells	2		10,000	incl.	20,000	incl.	20,000	
	1.2	Two (2) GW pumps	2		1,400	600	2,800	1,200	4,000	
	1.3	Excavation, Pipe Bedding & Backfill to Tmt. Bld.	100	lin. ft.	incl.	9.00	incl.	900	900	
	1.4	Piping Placed in Trench	100	ft.	1.00	incl.	100	incl.	100	
	1.5	Well Testing	2		incl.	4,500	incl.	9,000	9,000	
										34,000
2.0		**Groundwater Treatment System**	1		35,000		35,000	0	35,000	
	2.1	Skid Mounted Building (7'x10')								
		Building Heater and Light								
		Feed Tank								
		Low Profile Air Stripper								
		Transfer Pump								
		80 CFM Blower								
		Flow Control Valve								
		Totalizing Flowmeter								
		Control Panel (PLC)								
		Piping For Gas Phase Carbon Vessels								
	2.2	Installation	1			10,000		10,000	10,000	.
										45,000
3.0		**Piping to Sanitary Sewer**	1	L.S.	5,000	incl	5,000	incl	5,000	5,000
4.0		**Monitoring Wells**	7	L.S.	incl	500	incl	3,500	3,500	3,500
5.0		**Mobilization, Site Preparation, Demobilization**	1	L.S.	5,000	incl.	5,000	incl.	5,000	5,000
CONSTRUCTION SUBTOTAL									92,500	92,500

BEAK INTERNATIONAL INC.

OPERATING COST ESTIMATE (Year 1 and 2)

Client: **7304.10**

Description: **Case A- 80 ft Plume Containment (TCE Primarily)**

Alternative: **GW Extraction / Air Stripping**

Prepared by:	JC	**Revision No.**	A
Checked by:	JC	**Date Updated:**	15-May-98

Sec. No.	Sub. No.	Description/Activity	Unit Cost	Basis	Cost ($/year)	Total Cost ($/year)
1.0		**Groundwater Monitoring**				
	1.1	Sample Collection	$65/hr	12hr x 4 times per year x $65/hr	3,120	
	1.2	Sample Analysis	$150/sample	7 wells (+2 QA/QC) x 4 times per year x $150/sample	5,400	
						8,520
2.0		**Groundwater Extraction**				
	2.1	EW Pump Power (2)	$ 0.05 / KWH	2x 0.5 HP x 0.7457 x 24hr x 365 d	327	
	2.2	EW Pump Maintenance	25% of cost	25% x4000	1,000	
	2.3	Operating Labor	$50/hr	8hr x 12 times per year x $50/hr	4,800	
						6,127
3.0		**Groundwater Treatment System**				
	3.1	Feed Pump Power (1 HP)	$ 0.05 / KWH	1 HP x 0.7457 x 24hr x 365 d	327	
	3.2	Blower Power (1.0 HP)	$ 0.05 / KWH	1.0 HP x 0.7457 x 24hr x 365 d	327	
	3.3	Operating Labor	$50/hr	14 hr x 52 times per year x $50/hr	36,400	
	3.4	Equipment Maintenance	7.5% of cost	7.5% x 35,000	2,625	
	3.5	Performance Monitoring Analytical	$200/sample	12 per year x $200 x 2 samples x 1.2 (QA/QC)	5,760	
	3.6	Performance Monitoring Smpl Collection & Rpt.	$75/hr	2 hr/sampling event x 12 sampling events x $75/hr	1,800	
	3.7	Carbon Consumption	$535/180 lbs drum	6 drums per year	3,210	
						50,448
4.0		**Groundwater Discharge**				
	4.1	Annual Permit Fee	$2,500 per year	$2,500 per year x 1 year	2,500	
	4.2	Monitoring and Testing Charges	$70 per month	$70 per month x 12 months	840	
	4.3	Disposal Charges	$0.41 per 100 ft^3	$0.41 x (28,107/100 ft^3)	115	
						3,455
5.0		**Reporting**	$80/hr	125 hr x 4 times per year x $80/hr	40,000	
						40,000
OPERATING COSTS (Year 1 and 2)					108,550	108,550

BEAK INTERNATIONAL INC.

OPERATING COST ESTIMATE (Year 3 through 30)

Client: 7304.10

Description: Case A- 80 ft Plume Containment (TCE Primarily)

Alternative: GW Extraction / Air Stripping

Prepared by: JC **Revision No.** A

Checked by: JC **Date Updated:** 15-May-98

Sec. No.	Sub. No.	Description/Activity	Unit Cost	Basis	Cost ($/year)	Total Cost ($/year)
1.0		**Groundwater Monitoring**				
	1.1	Sample Collection	$65/hr	12hr x 2 times per year x $65/hr	1,560	
	1.2	Sample Analysis	$150/sample	7 wells (+2 QA/QC) x 2 times per year x $150/sample	2,700	
						4,260
2.0		**Groundwater Extraction**				
	2.1	EW Pump Power (2)	$ 0.05 / KWH	2x 0.5 HP x 0.7457 x 24hr x 365 d	327	
	2.2	EW Pump Maintenance	25% of cost	25% x4000	1,000	
	2.3	Operating Labor	$50/hr	8hr x 12 times per year x $50/hr	4,800	
						6,127
3.0		**Groundwater Treatment System**				
	3.1	Feed Pump Power (1 HP)	$ 0.05 / KWH	1 HP x 0.7457 x 24hr x 365 d	327	
	3.2	Blower Power (1.0 HP)	$ 0.05 / KWH	1.0 HP x 0.7457 x 24hr x 365 d	327	
	3.3	Operating Labor	$50/hr	14 hr x 52 times per year x $50/hr	36,400	
	3.4	Equipment Maintenance	7.5% of cost	7.5% x 35,000	2,625	
	3.5	Performance Monitoring Analytical	$200/sample	12 per year x $200 x 2 samples x1.2 (QA/QC)	5,760	
	3.6	Performance Monitoring Smpl Collection & Rpt.	$75/hr	2 hr/sampling event x 12 sampling events x $75/hr	1,800	
	3.7	Carbon Consumption	$535/180 lbs drum	6 drums per year	3,210	
						50,448
4.0		**Groundwater Discharge**				
	4.1	Annual Permit Fee	$2,500 per year	$2,500 per year x 1 year	2,500	
	4.2	Monitoring and Testing Charges	$70 per month	$70 per month x 12 months	840	
	4.3	Disposal Charges	$0.41 per 100 ft^3	$0.41 x (28,107/100 ft^3)	115	
						3,455
5.0		**Reporting**	$80/hr	125 hr x 2 times per year x $80/hr	20,000	
						20,000
OPERATING COSTS (Year 3 through 30)					84,290	84,290

BEAK INTERNATIONAL INC.	**Client:**				**7304.10**
	Description:	**Case A- 80 ft Plume Containment (TCE Primarily)**			
TOTAL CAPITAL & NPV COSTS ESTIMATE	**Alternative:**	**GW Extraction / Air Stripping**			
	Prepared by:	JC		**Revision No.**	A
	Checked by:	JC		**Date Updated:**	15-May-98

TOTAL CAPITAL COST ESTIMATE

A	Construction Subtoal	from above	92,500
B	Permitting & Legal	5% of D	4,625
C	Services During Construction	10% of D	9,250
D	Other Costs	15% of D	13,875
E	TOTAL IMPLEMENTATION COSTS	B + C + D	27,750
F	**TOTAL CAPITAL COSTS**	**A + E**	**120,250**

NET PRESENT VALUE CALCULATIONS

TOTAL CAPITAL COSTS	F above	120,250
First Year Operating Costs	from above	108,550
Second Year Operating Costs	from above	108,550
Operating Costs for Each Year (3 to 30)	from above	84,290
Interest Rate		8%
NET PRESENT VALUE		1,112,432

ATTACHMENT A35.3: COST ESTIMATE WORKSHEETS FOR A PUMP-AND-TREAT SYSTEM TO CAPTURE AND TREAT THE 350-FT-WIDE PLUME

BEAK INTERNATIONAL INC.

CONSTRUCTION COST ESTIMATE SHEET

Client: | **7304.10**

Description: Case B- 350 ft Plume Containment

Alternative: GW Extraction / Air Stripping

Prepared by: JC | **Revision No.** A

Checked by: JC | **Date Updated:** 15-May-98

Sec. No.	Sub. No.	Description/Activity	Material Required: Quantity	Material Required: Unit	Unit Costs: Material ($)	Unit Costs: Labour ($)	Total Cost: Material ($)	Total Cost: Labour ($)	Sub Totals ($)	Total Cost ($)
1.0		**Groundwater Extraction System**								
	1.1	Three (3) 6 " Extraction Wells	3		10,000	incl.	30,000	incl.	30,000	
	1.2	Three (3) GW pumps	3		1,400	600	4,200	1,800	6,000	
	1.3	Excavation, Pipe Bedding & Backfill to Tmt. Bld.	350	lin. ft.	incl.	9.00	incl.	3,150	3,150	
	1.4	Piping Placed in Trench	350	ft.	1.00	incl.	350	incl.	350	
	1.5	Well Testing	3		incl.	4,500	incl.	13,500	13,500	
										53,000
2.0		**Groundwater Treatment System**	1		35,000		35,000	0	35,000	
	2.1	Skid Mounted Building (7'x10')								
		Building Heater and Light								
		Feed Tank								
		Low Profile Air Stripper								
		Transfer Pump								
		80 CFM Blower								
		Flow Control Valve								
		Totalizing Flowmeter								
		Control Panel (PLC)								
		Piping For Gas Phase Carbon Vessels								
	2.2	Installation	1			10,000		10,000	10,000	
										45,000
3.0		**Piping to Sanitary Sewer**	1	L.S.	8,000	incl	8,000	incl	8,000	8,000
4.0		**Monitoring Wells**	10	L.S.	incl	500	incl	5,000	5,000	5,000
5.0		**Mobilization, Site Preparation, Demobilization**	1	L.S	incl.	10,500	incl.	10,500	10,500	10,500
CONSTRUCTION SUBTOTAL									121,500	121,500

BEAK INTERNATIONAL INC.

OPERATING COST ESTIMATE (Year 1 and 2)

Client: 7304.10

Description: Case B- 350 ft Plume Containment

Alternative: GW Extraction / Air Stripping

Prepared by:	JC	**Revision No.**	A
Checked by:	JC	**Date Updated:**	15-May-98

Sec. No.	Sub. No.	Description/Activity	Unit Cost	Basis	Cost ($/year)	Total Cost ($/year)
1.0		**Groundwater Monitoring**				
	1.1	Sample Collection	$65/hr	16hr x 4 times per year x $65/hr	4,160	
	1.2	Sample Analysis	$150/sample	10 wells (+2 QA/QC) x 4 times per year x $150/sample	7,200	
						11,360
2.0		**Groundwater Extraction**				
	2.1	EW Pump Power (3)	$ 0.05 / KWH	3x 0.5 HP x 0.7457 x 24hr x 365 d	490	
	2.2	EW Pump Maintenance	25% of cost	25% x6000	1,500	
	2.3	Operating Labor	$50/hr	8hr x 12 times per year x $50/hr	4,800	
						6,790
3.0		**Groundwater Treatment System**				
	3.1	Feed Pump Power (1 HP)	$ 0.05 / KWH	1 HP x 0.7457 x 24hr x 365 d	327	
	3.2	Blower Power (1.0 HP)	$ 0.05 / KWH	1.0 HP x 0.7457 x 24hr x 365 d	327	
	3.3	Operating Labor	$50/hr	14 hr x 52 times per year x $50/hr	36,400	
	3.4	Equipment Maintenance	7.5% of cost	7.5% x 35,000	2,625	
	3.5	Performance Monitoring Analytical	$200/sample	12 per year x $200 x 2 samples x1.2 (QA/QC)	5,760	
	3.6	Performance Monitoring Smpl Collection & Rpt.	$75/hr	2 hr/sampling event x 12 sampling events x $75/hr	1,800	
	3.7	Carbon Consumption	$535/180 lbs drum	20 drums per year	10,700	
						57,938
4.0		**Groundwater Discharge**				
	4.1	Annual Permit Fee	$2,500 per year	$2,500 per year x 1 year	2,500	
	4.2	Monitoring and Testing Charges	$70 per month	$70 per month x 12 months	840	
	4.3	Disposal Charges	$0.41 per 100 ft^3	$0.41 x (189,722/100 ft^3)	778	
						4,118
5.0		**Reporting**	$80/hr	125hr x 4 times per year x $80/hr	40,000	
						40,000
OPERATING COSTS (Year 1 and 2)					120,206	120,206

BEAK INTERNATIONAL INC.

OPERATING COST ESTIMATE (Year 3 through 30)

Client: 7304.10

Description: Case B- 350 ft Plume Containment

Alternative: GW Extraction / Air Stripping

Prepared by: JC **Revision No.** A

Checked by: JC **Date Updated:** 15-May-98

Sec. No.	Sub. No.	Description/Activity	Unit Cost	Basis	Cost ($/year)	Total Cost ($/year)
1.0		**Groundwater Monitoring**				
	1.1	Sample Collection	$65/hr	16hr x 2times per year x $65/hr	2,080	
	1.2	Sample Analysis	$150/sample	10 wells (+2 QA/QC) x 2 times per year x $150/sample	3,600	
						5,680
2.0		**Groundwater Extraction**				
	2.1	EW Pump Power (3)	$ 0.05 / KWH	3x 0.5 HP x 0.7457 x 24hr x 365 d	490	
	2.2	EW Pump Maintenance	25% of cost	25% x6000	1,500	
	2.3	Operating Labor	$50/hr	8 hr x 12 times per year x $50/hr	4,800	
						6,790
3.0		**Groundwater Treatment System**				
	3.1	Feed Pump Power (1 HP)	$ 0.05 / KWH	1 HP x 0.7457 x 24hr x 365 d	327	
	3.2	Blower Power (1.0 HP)	$ 0.05 / KWH	1.0 HP x 0.7457 x 24hr x 365 d	327	
	3.3	Operating Labor	$50/hr	14 hr x 52 times per year x $50/hr	36,400	
	3.4	Equipment Maintenance	7.5% of cost	7.5% x 35,000	2,625	
	3.5	Performance Monitoring Analytical	$200/sample	12 per year x $200 x 2 samples x1.2 (QA/QC)	5,760	
	3.6	Performance Monitoring Smpl Collection & Rpt.	$75/hr	2 hr/sampling event x 12 sampling events x $75/hr	1,800	
	3.7	Carbon Consumption	$535/180 lbs drum	20 drums per year	10,700	
						57,938
4.0		**Groundwater Discharge**				
	4.1	Annual Permit Fee	$2,500 per year	$2,500 per year x 1 year	2,500	
	4.2	Monitoring and Testing Charges	$70 per month	$70 per month x 12 months	840	
	4.3	Disposal Charges	$0.41 per 100 ft^3	$0.41 x (189,722/100 ft^3)	778	
						4,118
5.0		**Reporting**	$80/hr	125hr x 2 times per year x $80/hr	20,000	
						20,000
OPERATING COSTS (Year 3 through 30)					94,526	94,526

BEAK INTERNATIONAL INC.	**Client:**			**7304.10**
	Description:	**Case B- 350 ft Plume Containment**		
TOTAL CAPITAL & NPV COSTS ESTIMATE	**Alternative:**	**GW Extraction / Air Stripping**		
	Prepared by:	JC	**Revision No.**	A
	Checked by:	JC	**Date Updated:**	15-May-98

TOTAL CAPITAL COST ESTIMATE

A	Construction Subtoal	from above	121,500
B	Permitting & Legal	5% of D	6,075
C	Services During Construction	10% of D	12,150
D	Other Costs	15% of D	18,225
E	TOTAL IMPLEMENTATION COSTS	B + C + D	36,450
F	**TOTAL CAPITAL COSTS**	**A + E**	**157,950**

NET PRESENT VALUE CALCULATIONS

TOTAL CAPITAL COSTS	F above	157,950
First Year Operating Costs	from above	120,206
Second Year Operating Costs	from above	120,206
Operating Costs Each Year (3 to 30)	from above	94,526
Interest Rate		8%
NET PRESENT VALUE		1,267,898

ATTACHMENT A35.4: COST ESTIMATE WORKSHEETS FOR A PUMP-AND-TREAT SYSTEM TO CAPTURE AND TREAT THE 350-FT-WIDE PLUME WITH TCE AS THE MAJOR CONTAMINANT

BEAK INTERNATIONAL INC.

CONSTRUCTION COST ESTIMATE SHEET

Client: 7304.10

Description: Case B- 350 ft Plume Containment (TCE Primarily)

Alternative: GW Extraction / Air Stripping

Prepared by: JC **Revision No.** A

Checked by: JC **Date Updated:** 15-May-98

Sec. No.	Sub. No.	Description/Activity	Material Required Quantity	Material Required Unit	Unit Costs Material ($)	Unit Costs Labour ($)	Total Cost Material ($)	Total Cost Labour ($)	Sub Totals ($)	Total Cost ($)
1.0		**Groundwater Extraction System**								
	1.1	Three (3) 6 " Extraction Wells	3		10,000	incl.	30,000	incl.	30,000	
	1.2	Three (3) GW pumps	3		1,400	600	4,200	1,800	6,000	
	1.3	Excavation, Pipe Bedding & Backfill to Tmt. Bld.	350	lin. ft.	incl.	9.00	incl.	3,150	3,150	
	1.4	Piping Placed in Trench	350	ft.	1.00	incl.	350	incl.	350	
	1.5	Well Testing	3		incl.	4,500	incl.	13,500	13,500	
										53,000
2.0		**Groundwater Treatment System**	1		35,000		35,000	0	35,000	
	2.1	Skid Mounted Building (7'x10')								
		Building Heater and Light								
		Feed Tank								
		Low Profile Air Stripper								
		Transfer Pump								
		80 CFM Blower								
		Flow Control Valve								
		Totalizing Flowmeter								
		Control Panel (PLC)								
		Piping For Gas Phase Carbon Vessels								
	2.2	Installation	1			10,000		10,000	10,000	
										45,000
3.0		**Piping to Sanitary Sewer**	1	L.S.	8,000	incl	8,000	incl	8,000	8,000
4.0		**Monitoring Wells**	10	L.S.	incl	500	incl	5,000	5,000	5,000
5.0		**Mobilization, Site Preparation, Demobilization**	1	L.S	incl.	10,500	incl.	10,500	10,500	10,500
CONSTRUCTION SUBTOTAL									121,500	121,500

BEAK INTERNATIONAL INC.			Client:		7304.10	
			Description:	Case B- 350 ft Plume Containment (TCE Primarily)		
OPERATING COST ESTIMATE (Year 1 and 2)			Alternative:	GW Extraction / Air Stripping		
			Prepared by:	JC	Revision No. A	
			Checked by:	JC	Date Updated: 15-May-98	
Sec. No.	**Sub. No.**	**Description/Activity**	**Unit Cost**	**Basis**	**Cost ($/year)**	**Total Cost ($/year)**
1.0		**Groundwater Monitoring**				
	1.1	Sample Collection	$65/hr	16hr x 4 times per year x $65/hr	4,160	
	1.2	Sample Analysis	$150/sample	10 wells (+2 QA/QC) x 4 times per year x $150/sample	7,200	
						11,360
2.0		**Groundwater Extraction**				
	2.1	EW Pump Power (3)	$ 0.05 / KWH	3x 0.5 HP x 0.7457 x 24hr x 365 d	490	
	2.2	EW Pump Maintenance	25% of cost	25% x6000	1,500	
	2.3	Operating Labor	$50/hr	8 hr x 12 times per year x $50/hr	4,800	
						6,790
3.0		**Groundwater Treatment System**				
	3.1	Feed Pump Power (1 HP)	$ 0.05 / KWH	1 HP x 0.7457 x 24hr x 365 d	327	
	3.2	Blower Power (1.0 HP)	$ 0.05 / KWH	1.0 HP x 0.7457 x 24hr x 365 d	327	
	3.3	Operating Labor	$50/hr	14 hr x 52 times per year x $50/hr	36,400	
	3.4	Equipment Maintenance	7.5% of cost	7.5% x 35,000	2,625	
	3.5	Performance Monitoring Analytical	$200/sample	12 per year x $200 x 2 samples x1.2 (QA/QC)	5,760	
	3.6	Performance Monitoring Smpl Collection & Rpt.	$75/hr	2 hr/sampling event x 12 sampling events x $75/hr	1,800	
	3.7	Carbon Consumption	$535/180 lbs drum	6 drums per year	3,210	
						50,448
4.0		**Groundwater Discharge**				
	4.1	Annual Permit Fee	$2,500 per year	$2,500 per year x 1 year	2,500	
	4.2	Monitoring and Testing Charges	$70 per month	$70 per month x 12 months	840	
	4.3	Disposal Charges	$0.41 per 100 ft^3	$0.41 x (189,722/100 ft^3)	778	
						4,118
5.0		**Reporting**	$80/hr	125hr x 4 times per year x $80/hr	40,000	
						40,000
OPERATING COSTS (Year 1 and 2)					112,716	112,716

BEAK INTERNATIONAL INC.

OPERATING COST ESTIMATE (Year 3 through 30)

Client: 7304.10

Description: Case B- 350 ft Plume Containment (TCE Primarily)

Alternative: GW Extraction / Air Stripping

Prepared by: JC **Revision No.** A

Checked by: JC **Date Updated:** 15-May-98

Sec. No.	Sub. No.	Description/Activity	Unit Cost	Basis	Cost ($/year)	Total Cost ($/year)
1.0		**Groundwater Monitoring**				
	1.1	Sample Collection	$65/hr	16hr x 2 times per year x $65/hr	2,080	
	1.2	Sample Analysis	$150/sample	10 wells (+2 QA/QC) x 2 times per year x $150/sample	3,600	
						5,680
2.0		**Groundwater Extraction**				
	2.1	EW Pump Power (3)	$ 0.05 / KWH	3x 0.5 HP x 0.7457 x 24hr x 365 d	490	
	2.2	EW Pump Maintenance	25% of cost	25% x6000	1,500	
	2.3	Operating Labor	$50/hr	8hr x 12 times per year x $50/hr	4,800	
						6,790
3.0		**Groundwater Treatment System**				
	3.1	Feed Pump Power (1 HP)	$ 0.05 / KWH	1 HP x 0.7457 x 24hr x 365 d	327	
	3.2	Blower Power (1.0 HP)	$ 0.05 / KWH	1.0 HP x 0.7457 x 24hr x 365 d	327	
	3.3	Operating Labor	$50/hr	14 hr x 52 times per year x $50/hr	36,400	
	3.4	Equipment Maintenance	7.5% of cost	7.5% x 35,000	2,625	
	3.5	Performance Monitoring Analytical	$200/sample	12 per year x $200 x 2 samples x1.2 (QA/QC)	5,760	
	3.6	Performance Monitoring Smpl Collection & Rpt.	$75/hr	2 hr/sampling event x 12 sampling events x $75/hr	1,800	
	3.7	Carbon Consumption	$535/180 lbs drum	6 drums per year	3,210	
						50,448
4.0		**Groundwater Discharge**				
	4.1	Annual Permit Fee	$2,500 per year	$2,500 per year x 1 year	2,500	
	4.2	Monitoring and Testing Charges	$70 per month	$70 per month x 12 months	840	
	4.3	Disposal Charges	$0.41 per 100 ft^3	$0.41 x (189,722/100 ft^3)	778	
						4,118
5.0		**Reporting**	$80/hr	125hr x 2 times per year x $80/hr	20,000	
						20,000
OPERATING COSTS (Year 3 through 30)					87,036	87,036

BEAK INTERNATIONAL INC.	**Client:**			**7304.10**
	Description:	**Case B- 350 ft Plume Containment (TCE Primarily)**		
TOTAL CAPITAL & NPV COSTS ESTIMATE	**Alternative:**	**GW Extraction / Air Stripping**		
	Prepared by:	JC	**Revision No.**	A
	Checked by:	JC	**Date Updated:**	15-May-98

TOTAL CAPITAL COST ESTIMATE

A	Construction Subtoal	from above	121,500
B	Permitting & Legal	5% of D	6,075
C	Services During Construction	10% of D	12,150
D	Other Costs	15% of D	18,225
E	TOTAL IMPLEMENTATION COSTS	B + C + D	36,450
F	**TOTAL CAPITAL COSTS**	**A + E**	**157,950**

NET PRESENT VALUE CALCULATIONS

TOTAL CAPITAL COSTS	F above	157,950
First Year Operating Costs	from above	112,716
Second Year Operating Costs	from above	112,716
Operating Costs Each Year (1 to 30)	from above	87,036
Interest Rate		8%
NET PRESENT VALUE		1,183,577

Index

A

D

E

F

G

N

O

P

R

S

T

V

X